# Vehicle Vibrations

Reza N. Jazar • Hormoz Marzbani

# Vehicle Vibrations

## Linear and Nonlinear Analysis, Optimization, and Design

Springer

Reza N. Jazar
Civil Engineering and Architecture
Xiamen University of Technology
Fuzhou, China

Hormoz Marzbani
School of Engineering
RMIT University
Melbourne, Australia

ISBN 978-3-031-43485-3 ISBN 978-3-031-43486-0 (eBook)
https://doi.org/10.1007/978-3-031-43486-0



This Springer imprint is published by the registered company Springer Nature Switzerland AG
The registered company address is: Gewerbestrasse 11, 6330 Cham, Switzerland

Paper in this product is recyclable.

*to:*

*Mojgan*

*Pargol*

Darius the Great King, King of Kings, King in Persia, King of countries, grand grand grand grandchild of Achaemenes. For this reason we are called Achaemenians. These are the countries which came to me; by the favor of Ahuramazda I was king of them: Persia, Elam, Babylonia, Assyria, Arabia, Egypt, who are beside the sea, Sardis, Ionia, Media, Armenia, Cappadocia, Parthia, Drangiana, Aria, Chorasmia, Bactria, Sogdiana, Gandara, Scythia, Sattagydia, Arachosia, Maka: in all, 23 provinces. I ordered to write books, many books, books to teach my people,

I ordered to make schools, many schools, to educate my people. Ahuramazda, the lord of the gods, said burning books is the greatest sin. I, Darius, and my people, and my army will protect books and schools. They will fight whoever burns books and burns schools, the great sin.

# Preface

Vehicle vibrations is an active research topic in industry and academia since the beginning of the twentieth century. There is a suspension design department in every Research and Development section of vehicle design and manufacturing company, and there are several academics in every engineering school of universities who actively study vehicle vibrations. For several years, our colleagues and students requested us to use our teaching and research experiences to write a book dedicated to Vehicle Vibrations, because there is no unique book to cover vehicle vibrations from analytic, design, application, research, and educational viewpoints. It seems to be a gap in the arsenal of research material on such a popular and important subject. This book is prepared to address those requests.

This book is specially prepared for designers, practicing engineers, researchers, and graduate students of engineering. It introduces the fundamental and advanced knowledge used in *vehicle vibrations*. This knowledge can be utilized to develop computer programs for analyzing, designing, and optimizing vibration problems in vehicle systems. The main problem in the study of vibrations is to determine the response of a system subjected to a given excitation. The excitation and the response can be expressed by either kinematic quantities such as displacements, velocities, accelerations, or by forces. However, displacement road excitation is the main source of excitation in vehicles.

We deeply appreciate the extensive helps from my colleague, Nguyen Dang Quy, for his valuable contribution in chapter "Tire-Road Separation." This chapter would have not been prepared without his contributions.

## Level of the Book

This book has evolved from nearly two decades of research and working on engineering vehicle vibrating systems, and teaching courses in fundamental and advanced topics in vibrations, vehicle dynamics, and vehicle vibrations. It is primarily designed to help vehicle vibration researchers; however, it can also be used to teach from the last year of undergraduate study up to the final year of graduate study in engineering. It provides the reader with both fundamental and advanced topics. If the book is being used for teaching, it would be possible to jump over some sections and cover the book in one course. The readers are assumed to know the fundamentals of kinematics and dynamics, as well as basic knowledge of linear algebra, differential equations, and numerical methods.

The contents of the book have been kept at a theoretical-practical level. Many concepts are deeply explained and their application emphasized, and most of the related theories and formal proofs have been explained. The book places a strong emphasis on the physical meaning and applications of the concepts. Topics that have been selected are of high interest in the field. An attempt has been made to expose readers to a broad range of topics and approaches.

The subjects marked by an asterisk ★ can be dropped in the first reading.

## Organization of the Book

The text is organized such that it can be used for teaching or for self-study. It is organized in to four parts:

### Part I, Vehicle Vibration Fundamentals

This part is elementary to review the physical elements in vehicle vibration such as "Spring," "Damper," and "Suspension Systems." They are combined mathematically in the last chapter of this part to remind "Mechanical Vibrations Modeling."

### Part II, Mechanical Vibrations Analysis

Vibrations is a science under Mechanical Engineering. The analytic methods of Mechanical Vibrations are covered in this part to make the base of analysis for the next part.

### Part III, Vehicle Vibrations Analysis

This is the main part of the book that covers time and frequency responses of all vibrating models of vehicle. The models are base-excited one-degree-of-freedom system, quarter car, bicycle car, half car, and full car.

### Part IV, Advanced Vehicle Vibration Problems

This part discusses special and rare phenomena in vehicle vibrations. Flat ride tuning and tire-road separation detail analysis are covered in this part.

## Method of Presentation

The structure of presentation is in a "*fact-reason-application*" fashion. The "fact" is the main subject that introduces the statement in each section. Then the reason is given as a "proof." Finally the application of the fact is examined in some "examples." The "examples" are a very important part of the book. They show how to implement the knowledge introduced in "facts." They also cover some other facts that are needed to expand the subject.

## Prerequisites

The book is written for researchers, engineers, and students in engineering who are working on vehicle vibrations. The assumption is that users are familiar with matrix algebra as well as basic dynamics, and differential equations. Prerequisites are the fundamentals of kinematics, dynamics, calculus, fundamentals of differential equations, numerical analysis, and matrix theory. These topics are usually taught in the first three undergraduate years.

## Unit System

The system of units adopted in this book is, unless otherwise stated, the international system of units (*SI*). The units of degree (deg) or radian ( rad) are utilized for variables representing angular quantities.

## How to Use This Book

Researchers who are familiar with fundamental of mechanical vibrations may go directly to Part III to learn vehicle vibrations modeling and analysis. Those readers who need to learn vehicle vibrations from the beginning can start from Part I to learn the fundamental mechanical elements, and then Part II to learn how to model and analysis vibrating systems. Then, they are ready to read the vehicle vibrations in Part III. Part IV is on special topic in vehicle vibrations for researchers who wish to study more advanced and uncommon phenomena in vehicle vibrations.

To use the book for teaching, instructors can teach Parts II and III in one subject under a title such as: Applied Vibrations, Advanced Vibrations, Special Topic in Vibrations, etc. Also, Part III can be used as the applied section for related subjects such as: Advanced Vibrations and Vehicle Dynamics.

Fuzhou, China
Melbourne, Australia
July 2023

Reza N. Jazar
Hormoz Marzbani

# Contents

# About the Authors

**Reza N. Jazar** is a Professor of Mechanical Engineering. Reza received his Ph.D. degree from Sharif University of Technology in Nonlinear Vibrations and Applied Mathematics; he received his MSc in Mechanical Engineering and Robotics and BSc in Mechanical Engineering and Internal Combustion Engines from Tehran Polytechnic. His areas of expertise include Nonlinear Dynamic Systems and Applied Mathematics. He obtained original results in non-smooth dynamic systems, applied nonlinear vibrating problems, time optimal control of dynamic systems, and mathematical modeling of vehicle dynamics stability. He authored several monographs in vehicle dynamics, robotics, dynamics, vibrations, and mathematics and published numerous professional articles, as well as several book chapters in research volumes. Most of his books have been adopted by many universities for teaching and research, and by many research agencies as standard model for research results.

Dr. Jazar had the pleasure to work in several Canadian, American, Asian, Middle Eastern, and Australian universities, as well as several years in Automotive Industries all around the world. Working in different engineering firms and educational systems provide him with a vast experience and knowledge to publish his researches on important topics in engineering and science. His unique revolutionized style of writing helps readers to learn the topics deeply in the easiest possible way.

**Hormoz B. Marzbani** is a distinguished mechanical engineer and academic with notable contributions to the field of Autonomous Land Vehicle Technology. He received his Master's degree in Mechanical Engineering in 2010 and later his Doctor of Philosophy degree, in 2015, both from RMIT University in Melbourne, Australia. Prior to embarking on his academic career, he honed his professional skills in the automotive industry, serving as a product engineer. His experience in the field has grounded his subsequent academic pursuits in practical industry experience.

In his academic role since 2015, he has been imparting his extensive knowledge through various engineering-related lectures. Simultaneously, he is dedicated to advancing research on Autonomous Land Vehicle Technology, with a particular emphasis on enhancing public acceptance of this emerging technology. His research interests revolve around critical aspects of autonomous vehicle technology. These include exploring ways to minimize motion sickness in autonomous vehicles, mitigating uncertainties inherent to the technology, and improving the safety of other road users in scenarios involving autonomous vehicles. Through his practical experience and academic pursuits, he is continually working to bridge the gap between technology and its acceptance in the public domain.

# Part I

# Vehicle Vibration Fundamentals

The study of vehicle vibrations is a multidisciplinary field of science that relies on various tools to gain a comprehensive understanding. Key tools for comprehending vehicle vibrations include linear algebra, trigonometry, principles of dynamics, differential equations, and principles of mechanical design. These tools are essential for analyzing and studying vibrations. In this part, we will review the fundamental topics necessary to grasp the concept of vehicle vibrations.

Starting with the basics, it is crucial to acknowledge that vibrations are inherent in every vehicle system due to the presence of mechanical components. These vibrations arise from the cyclic conversion of kinetic and potential energies. One of the critical components in any vehicle is its suspension system, which greatly affects its performance. In automobile suspension systems, coil and leaf springs are commonly utilized. An effectively designed suspension system is key to achieving a smooth and jerk-free ride, enhancing driving pleasure, and minimizing motion sickness. By analyzing the dynamics of the suspension's mechanical elements, we can gain insights into their behavior during vibrations. The utilization of mathematical modeling empowers us with the necessary tools to express and manipulate representations of vibrations, facilitating precise analysis. Through mathematical modeling, we gain a deeper understanding of the periodic nature of vibrations and the associated waveforms they generate. This enables us to study and interpret vibrations with greater accuracy.

Differential equations play a crucial role in capturing the dynamic behavior of vibrating systems. They allow us to formulate mathematical models that describe the motion and interactions of various vehicle components. Furthermore, an understanding of the principles of mechanical design is essential for designing and optimizing vehicle systems to minimize unwanted vibrations. By incorporating appropriate design considerations, such as damping mechanisms and stiffness characteristics, we can enhance the overall comfort, performance, and durability of a vehicle.

In summary, the study of vehicle vibrations necessitates the utilization of several fundamental tools, including linear algebra, trigonometry, principles of dynamics, differential equations, and principles of mechanical design. These tools enable us to comprehend and analyze the complex phenomena associated with the transformation of kinetic and potential energies that underlie mechanical vibrations in vehicles.

# Springs

# 1

A spring is any elastic body whose function is to collect potential energy by elastic deformation and recover its original shape and dimensions after the removal of the load. The primary purpose of springs in vehicle suspensions is to extend the period over which energy is absorbed and subsequently released. In vehicle applications, the most commonly used springs for suspension systems are helical springs, leaf springs, and torsion bars. Their deformation under loads are either in bending or twisting, and hence, the stored potential strain energy will be made by normal stress in bending or shear stress in torsion. Figure 1.1 illustrates a coil spring.

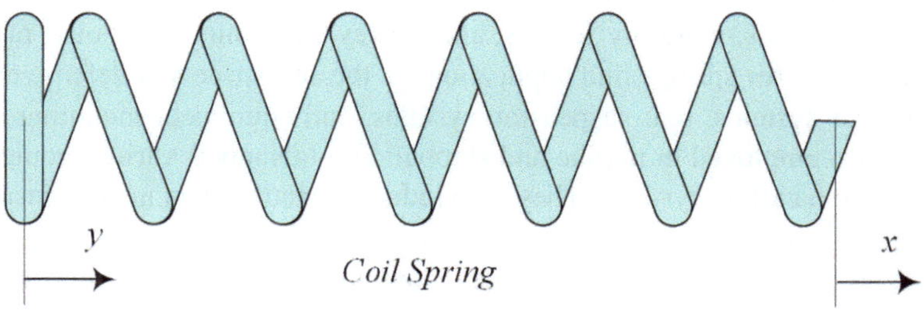


**Fig. 1.1** A coil spring

The spring is designed to store and release energy in response to external forces acting on the suspension. When a wheel encounters a bump on an uneven surface, the spring compresses or extends, absorbing the impact and minimizing the transfer of vibrations to the vehicle body. Springs in suspension systems can take various forms, including coil springs, leaf springs, torsion bars, or air springs. Each type has its own characteristics and advantages, depending on the specific application and requirements of the vehicle.

Coil springs (Fig. 1.1) consist of a helically shaped wire wound into a cylindrical shape. Coil springs are widely used in modern suspension systems to achieve optimal ride quality. Another type of spring is the leaf spring, which comprises multiple curved metal strips or leaves arranged in a stacked configuration. Leaf springs are commonly utilized in the rear suspension setups of trucks and some older vehicle models. Torsion bars represent another category of springs, leveraging the twisting properties of a solid or hollow metal bar to provide support for the suspension. Torsion bars are instrumental in offering suspension stability and resilience. Air springs, also known as airbags, utilize compressed air to bear the vehicle's weight and offer adjustable suspension characteristics. These springs are capable of enhancing ride comfort by adapting to varying load conditions.

Irrespective of their specific types, springs in suspension systems serve several essential functions. They maintain the appropriate ride height, distribute the vehicle's weight, keep the wheels in contact with the road, absorb shocks and impacts, and enhance overall ride comfort and handling. By providing controlled resistance to vertical movements, springs play a pivotal role in ensuring a smooth and stable ride for a vehicle's occupants.

The main factor to be considered in the design of springs is the strain energy of a particular geometric design and the particular material used. Specific strain energy $U$ in the material is expressed as



R. N. Jazar, H. Marzbani, *Vehicle Vibrations*, https://doi.org/10.1007/978-3-031-43486-0_1

$$U = \frac{\sigma^2}{\rho E}. \tag{1.1}$$

This indicates that a material with a lower Young's modulus $E$ or density $\rho$ will have a higher specific strain energy under the same stress condition. Springs can be practically groped into categories of helical springs, leaf springs, and air springs.

**Example 1** Air springs are adjustable springs and suitable for active spring designs.

Air springs consist of a cylindrical chamber of air positioned between the wheel and the vehicle's body and use the compressive qualities of air to absorb wheel vibrations. The technology is used in many luxury vehicles today, but the concept is more than a century old and was found on horse-drawn buggies. Air springs from this era were made from air-filled, leather diaphragms, much like a bellows; they were replaced by molded-rubber air springs in the 1930s.

Air springs, also known as airbags or pneumatic springs, are devices that use compressed air to provide support, cushioning, and vibration isolation in various applications. They consist of a flexible and airtight rubber bladder or bellows that is inflated with air. When the air pressure inside the spring changes, it affects its stiffness and ability to absorb shocks and vibrations.

Air springs have several advantages and are commonly used in various industries and applications. Air springs are widely used in vehicle suspension systems, particularly in commercial trucks, buses, and recreational vehicles. They provide adjustable ride height, improved comfort, and better load-carrying capacity compared to traditional steel springs. The air pressure can be adjusted to accommodate varying loads, ensuring a smooth and stable ride. Air springs are utilized in heavy-duty applications such as construction equipment, agricultural machinery, and off-road vehicles. They help absorb shocks and vibrations caused by uneven terrain, improving operator comfort and reducing stress on equipment.

Air springs are also employed in industrial machinery and equipment to isolate vibrations and dampen shocks. They are used in presses, stamping machines, vibratory feeders, and conveyors to minimize noise, protect delicate components, and improve machine performance. Air springs find applications in the aerospace and defense and related industries. They are used in aircraft landing gear systems, seat suspension systems, and equipment mounting to provide cushioning and vibration isolation. They are also employed in marine and shipbuilding to support various equipment and machinery, such as engines, generators, and cargo handling systems. They help reduce vibrations and noise transmission, ensuring smoother and quieter operation. Air springs are used in railway systems, including locomotives, passenger trains, and metro cars, to provide vibration isolation, improve ride comfort, and protect equipment from shock and impact.

Air springs offer several benefits, including adjustability, consistent performance over time, durability, and ease of installation and maintenance. They can be customized to specific requirements and offer a cost-effective solution in many applications. However, it's important to consider factors such as load capacity, pressure range, temperature sensitivity, and compatibility with other components when selecting and using air springs. Proper maintenance, regular inspection, and monitoring of air pressure are essential for optimal performance and longevity.

**Example 2** Materials for Springs.

The use of conventional steel as spring material increases the overall weight of the automobile suspension system and, hence, the weight of the entire assembly or vehicle. Helical springs, which are widely used for suspension systems, are generally manufactured with high carbon steel springs such as $EN47$. This material has all the properties that are suitable for springs of suspension systems and also possesses some properties that meet the principal requirements for suspension systems. However, the material has some major drawbacks, among which are its heavy weight, lack of resistance to corrosion, the fact that it conducts electricity. Hence, the idea emerged that new materials should be designed to address the aforementioned drawbacks. The new materials include composites or plastics or new efficient design of springs, such as hollow tubes.

Conventionally used spring material; Springs are generally manufactured from plain carbon steel or simply carbon steel. Carbon steel has properties mainly due to its carbon content and does not contain more than 0.5% of silicon and 1.5% of manganese.

Plain carbon steels varying from 0.06% carbon to 1.5% carbon are divided into the following types depending on the carbon content:

1. Dead mild steel,
2. Low carbon or mild steel,
3. Medium carbon steel,
4. High carbon steel.

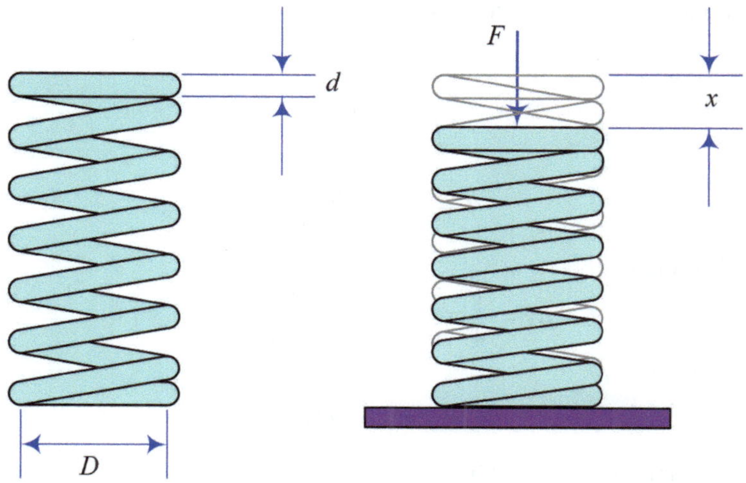


**Fig. 1.2** A coil spring with closed ends

Carbon steels, which are used for spring materials, have the following properties useful for spring materials:

1. Allowable shear stress: 420 MPa (average service)
2. Modulus of rigidity G: 80 kN/mm Superscript 2
3. Modulus of elasticity E: 210 kN/mm Superscript 2

Following an examination of the properties of various materials that make them suitable for manufacturing springs, E-glass was selected for the design and analysis of new springs.

## 1.1 Coil Springs

Coil or helical springs are formed by wrapping wire in which the coil of the wire is in helical form, and the idea is to use it in compressive or tensile loads. Although it may have different cross sections such as circular, square, or rectangular, the circular is the most commonly use form. Coil springs store elastic energy by means of torsion and bending of wire. Figure 1.1 illustrates a coil spring with a closed–open end configuration.

Figure 1.2 illustrates a coil spring with closed ends. The approximate equivalent stiffness of the coil spring is

$$k = k_{coil} = \frac{Gpd^4}{8nlD^3}, \tag{1.2}$$

where $G$ is the shear modulus of the spring material, $D$ is the mean diameter of the coil spring measured from the centers of the wire cross sections, $d$ is the diameter of the wire, $n$ is the number of active coils, and $p$ is the pitch of the helix:

$$p = \frac{l}{n}. \tag{1.3}$$

***Proof*** The deflection $x$ of a coil spring under a force $F$ and with $n$ active coils is

$$x = \frac{8FD^3n}{Gd^4}, \tag{1.4}$$

which causes the equivalent stiffness $k$ to be

$$k = \frac{F}{x} = \frac{Gd^4}{8nD^3}. \tag{1.5}$$

If $l$ is the free length of the spring and $p$ is its pitch, then we have

$$p = \frac{l}{n} \tag{1.6}$$

and then

$$k = \frac{F}{x} = \frac{Gpd^4}{8nlD^3}. \tag{1.7}$$

The energy stored in a coil spring is

$$E_s = \frac{1}{2}kx^2 = \frac{1}{2}Fx = \frac{4F^2D^3n}{Gd^4}. \tag{1.8}$$

The volume of material in a spring is

$$V = \frac{1}{4}\pi d^2 Dn, \tag{1.9}$$

and hence, the energy stored per unit of length will be

$$\frac{E_s}{V} = \frac{16}{\pi}\frac{F^2D^2}{Gd^6}. \tag{1.10}$$

■

**Example 3** What would be the stiffness of a coil spring that was cut in half?

When we cut a coil spring in half, its stiffness doubles:

$$k_{l/2} = \frac{Gpd^4}{8n\,(l/2)\,D^3} = 2\frac{Gpd^4}{8nlD^3} = 2k_{coil}. \tag{1.11}$$

Let us assume that a coil spring of length $l$ is made of a series of two half-length similar coil springs. The stiffness of the spring would be half that of a short spring:

$$k_{coil} = \frac{1}{\dfrac{1}{k_{l/2}} + \dfrac{1}{k_{l/2}}} = \frac{1}{2}k_{l/2}. \tag{1.12}$$

**Example 4** Coil spring features, advantages, and problems for suspension systems.

The coil spring is the preferred choice for both front and rear suspension systems and has widespread application. Coil springs are crafted from a coiled steel bar. When a vehicle encounters uneven terrain, these coil springs compress, absorbing the impact of the road, and subsequently rebound to their original static height.

While coil springs are engineered to bear substantial loads, they must also remain lightweight. Despite their effectiveness, coil springs are susceptible to certain failure conditions, which include consistent overloading, prolonged jounce and rebound motion, metal fatigue, and surface layer and coating cracks. Coil springs also do not possess significant resistance against lateral movement. Hence, when coil springs are utilized on drive wheels, specialized bars are integrated into the suspension system to prevent the lateral motion of springs.

Although all coil springs act as nonlinear-rate springs, depending on their structures, we may separate them as near-linear-rate coil springs and nonlinear-rate coil springs. Linear coil springs have exactly helical shapes. When a linear spring is subjected to an increased load, it undergoes compression, causing the coils to twist and deflect. Conversely, as the load is removed, the coils unwind and flex, returning to their static position. The spring rate refers to the amount of load needed to deflect the spring by one unit of length, such as a meter or an inch. Linear coil springs maintain a consistent spring rate regardless of the applied load.

Nonlinear-rate or variable-rate coil springs exhibit a diverse range of wire sizes and shapes. Among the most prevalent types are those with a consistent wire diameter, cylindrical shape, and unequally spaced coils. Figure 1.3 illustrates a few nonlinear coil spring geometric designs. The variations of variable-rate springs include truncated cone, double cone, and barrel shapes.

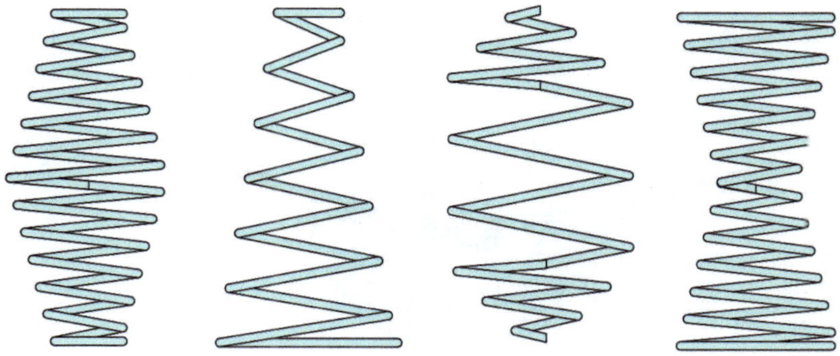

**Fig. 1.3** A few nonlinear coil spring designs

Because coil springs are mainly subjected to torsion, the failure of the spring occurs due to shear. The shear stress varies linearly from a maximum at the surface to zero at the center of the circular cross section. The average stress is equal to two-thirds of the maximum.

**Example 5** Helical springs. Helical springs or coil springs possess several characteristic features.

1. Shape: Helical springs are formed in a helix or spiral shape, with a wire wound around a cylindrical, convex, or concave barrel form. This helical structure allows for efficient energy storage and distribution.
2. Elasticity: Helical springs exhibit elastic properties, meaning they can be deformed under a load and return to their original shape when the load is removed. This elasticity enables the spring to absorb and dissipate potential energy.
3. Extension and compression: Two of the most common types of helical springs are compression and extension springs. These helical mechanisms are most often made of metal but occasionally are made of other materials as well. Extension springs are coiled more tightly than compression springs, while both may have hooks or loops on either end to attach to other objects. The compression and extension names refer to the state in which the springs contain the most potential energy; compression springs have the most potential energy when they are compressed, and extension springs have the most potential energy when they are extended.
4. Helical springs can operate in both compression and extension modes. In compression, the spring is compressed or squashed, while in extension it is stretched or elongated. This versatility makes helical springs suitable for a wide range of suspension and mechanical systems.
5. Load-bearing capacity: Helical springs are designed to bear and support various loads, depending on their size, material, and design. They can be manufactured to accommodate light- to heavy-duty applications.
6. Stiffness: The stiffness or spring rate of a helical spring determines its resistance to deformation under a given load. It defines how much the spring will compress or extend in response to an applied force. The stiffness of a helical spring can be adjusted by altering factors such as the wire diameter, number of coils, and material used.
7. Material selection: Helical springs can be manufactured from various materials, including steel, stainless steel, alloy steel, or nonferrous metals such as bronze. The material selection depends on factors such as the required strength, corrosion resistance, temperature resistance, and cost considerations.
8. Damping characteristics: Helical springs can possess damping characteristics, which refers to their ability to absorb and dissipate vibrations or oscillations. Damping can be achieved by incorporating additional components such as rubber inserts or by modifying the design of the spring.
9. Durability: Helical springs are designed to withstand repeated loading and unloading cycles without significant loss of performance. They are engineered to maintain their shape and mechanical properties over an extended period.
10. Torsion: Torsion springs are wound tightly like an extension spring, although the ends of the spring typically extend away from the spring in a nonhelical shape. Instead of being compressed or extended, a torsion spring is twisted to store potential energy. Common applications of torsion springs are those found in clothespins and in traditional mouse traps. Torsion springs obey Hooke's law, in angular form:

$$M = k\varphi. \tag{1.13}$$

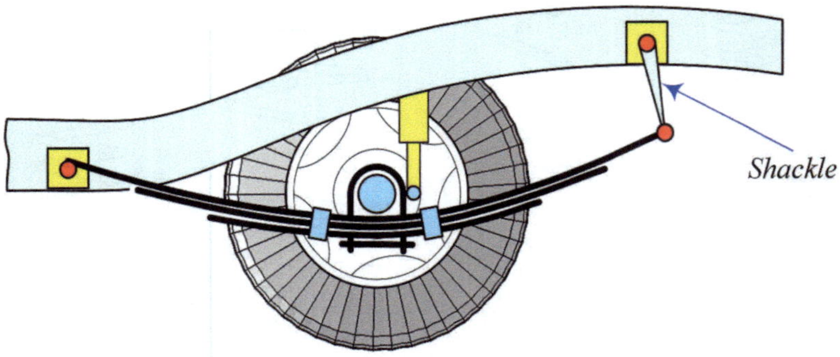


**Fig. 1.4** A leaf spring

For torsion springs, torque replaces force, and angular distance in radians replaces linear distance. In the context of automobiles, the same concept is used in the form of torsion bars. Torsion bars use the twisting properties of a steel bar to provide coil-spring-like performance. This is how they work: One end of a bar is anchored to the vehicle frame. The other end is attached to a wishbone, which acts like a lever that moves perpendicular to the torsion bar. When the wheel hits a bump, vertical motion is transferred to the wishbone and then, through the levering action, to the torsion bar. The torsion bar then twists along its axis to provide the spring force. European carmakers used this system extensively, as did Packard and Chrysler in the United States, through the 1950s and 1960s.

11. Constant force: Constant-force springs exert a near-constant force and they are typically made of a thin sheet metal that is tightly wrapped around a drum. One end is attached to the drum, and the free end is attached to a loading force. As the spool is unwound, it exerts a near-constant force because the geometry of the coil is nearly maintained by the drum. Different applications for constant-force springs include systems that require retraction, such as seat belts, power cords, or tape measures, and power for clockwork devices.

Overall, helical springs offer reliable and versatile suspension and mechanical support, providing controlled resistance, shock absorption, and a more comfortable ride and more stability in various applications.

## 1.2 Leaf Springs

The leaf spring is a commonly employed component in the suspension systems of automobiles and railroad cars. It offers the advantage of serving as both a structural member and a spring within the suspension mechanism. When a leaf spring is used for the suspension of vehicles, it is installed on the vehicle by a shackle, as illustrated in Fig. 1.4.

The value of stiffness $k$ is a combined property of the material and geometric characteristics of a spring. The linear approximated force–displacement equation of a leaf spring, $y = Fl^3/(3EI)$, indicates that the equivalent stiffness $k$ of the spring is,

$$k = k_{leaf} = \frac{3EI}{l^3}, \tag{1.14}$$

where $E$ is the Young's modulus of the material, $I$ is the geometrical moment of the cross-sectional area of the beam about the lateral neutral axis, and $l$ is the length of the beam. When we cut a leaf spring in half, its stiffness increases eightfold:

$$k_{l/2} = \frac{3EI}{(l/2)^3} = 8\frac{3EI}{l^3} = 8k_{leaf}. \tag{1.15}$$

**Example 6** Leaf springs were the first invented type of spring and are grouped into different types and possess several applied characteristics.

Leaf springs, commonly used in the rear suspensions of trucks, pickups, and some older vehicle models, possess distinctive features. Here are the key characteristics of leaf springs:

1. Construction: Leaf springs consist of multiple curved metal strips or leaves, typically made of spring steel. The leaves are stacked on top of each other, with the longest leaf at the bottom, known as the main leaf, and progressively shorter leaves above it. The leaves are held together by clamps or bolts at the center and ends.

2. Flexibility: Leaf springs offer a high degree of flexibility. When a load is applied, the leaves bend and flatten out, absorbing the energy. This flexibility helps provide a smoother ride by allowing the springs to absorb shocks and bumps.
3. Load-bearing capacity: Leaf springs are known for their ability to handle heavy loads. The stacking of multiple leaves allows them to distribute loads across the entire length of the springs effectively. The number and thickness of leaves can be adjusted to meet the specific load requirements.
4. Progressive spring rate: Leaf springs have a progressive spring rate, meaning that as the load increases, more leaves come into contact and provide additional support. This progressive nature allows the springs to adjust their stiffness based on the load, offering improved load-carrying capacity and stability.
5. Longevity: Leaf springs are known for their durability and longevity. The materials used, such as high-quality spring steel, provide excellent resistance to fatigue and deformation, ensuring the spring can withstand repeated loading and unloading cycles.
6. Simple design: Leaf springs have a relatively simple design, consisting of stacked leaves and clamps or bolts to hold them together. This simplicity makes them durable and less prone to mechanical failure compared to more complex suspension systems.
7. Lateral stability: Leaf springs offer inherent lateral stability due to their wide design. They help maintain proper alignment and resist sideways movement, contributing to the stability of the vehicle during cornering or when carrying off-center loads.
8. Cost-effectiveness: Leaf springs are generally cost-effective compared to other suspension systems. Their simple design and ease of manufacture contribute to their affordability and widespread use in certain vehicle applications.

While leaf springs may have some limitations, such as a potentially rougher ride compared to more advanced suspension systems, their robustness, load-carrying capacity, and suitability for specific applications make them a preferred choice in certain contexts, especially for heavy-duty vehicles.

**Example 7** Leaf springs can be divided into several classes based on shape and functionality.

1. Semi-elliptical leaf springs. These are the most popular leaf spring in automobiles. They are made from steel leaves of different lengths but the same width and thickness. The uppermost/longest leaf at the two ends is the master leaf. The arrangement of the steel leaves resembles a semi-elliptical shape. Semi-elliptical leaf springs have one end rigidly fixed to the vehicle frame and the other to the shackle. This helps in varying the lengths and absorbing shock when traveling on rough terrain. Semi-elliptical leaf springs require little maintenance, are easy to repair, and have a long life.

   Using the following equation, we calculate the spring stiffness for a single-leaf semi-elliptical leaf spring:

$$k = \frac{2F}{x} = \frac{Ebt^3}{4L^3}, \tag{1.16}$$

   where $L$ [m] is half the overall length of the longest leaf spring, $F$ [N] is the force applied at each mounting point to the chassis, $b$ [m] is the leaf spring width at the center point, $t$ [m] is the vertical depth of the leaf spring at the center point where it mounts to the axle, $E$ [Pa] is Young's modulus for the material, and $x$ [m] is the spring vertical displacement. The equation varies slightly in the case of a laminated or multiple-leaf spring:

$$k = \frac{2F}{x} = \frac{(2 + \frac{n'}{n})Enbt^3}{6L^3}, \tag{1.17}$$

   where $n$ is the number of leaves stacked and $n\prime$ is the number of leaves directly at the spring ends.
2. Elliptical leaf springs. An elliptical leaf spring is constructed by connecting two semi-elliptical springs in opposite directions. This forms an elliptical shape. The axle and frame attach the elliptical leaf springs. There is no need for spring shackles as the two semi-elliptical springs are elongated by the same amount during compression. These springs are not very common in modern cars.
3. Quarter elliptical leaf springs. Also known as cantilever-type leaf springs, quarter elliptical leaf springs are also considered an outdated leaf spring. They have one end fixed on the side member of the frame using a U-clamp or I-bolt. The other is freely connected to the front axle. When the front axle beam is subjected to a shock load, the leaves straighten to absorb the shock. The spring rate for the quarter elliptical leaf spring is equal to $\frac{1}{16}$ that of a semi-elliptical spring.

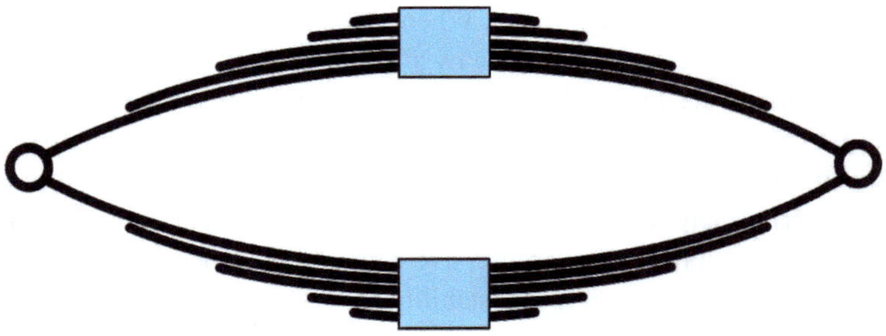

**Fig. 1.5** Elliptic spring

4. Three-quarter elliptical leaf spring. A simple example of this spring's application is a door hinge. Here, when you open the door, the spring will store its rotational energy; when you release the door, the spring uses the stored energy to bring the door back to its original position. The rotation force depends on the rotation of the spring. This type of leaf spring is a combination of the quarter elliptical spring and semi-elliptical spring. One end of the semi-elliptical part is attached to the vehicle frame, while the other is attached to the quarter elliptical spring. The other end of the quarter elliptical spring is attached to the frames and head by an I-bolt.
5. Transverse leaf spring. The transverse leaf spring results from transversely mounting a semi-elliptical leaf spring along the vehicle width. The arrangement is such that the spring's longest leaf is located at the bottom, and the midportion is fixed to the frame using a U-bolt. Transverse leaf springs utilize two shackles. However, they can cause rolling, making them unsuitable for automobiles.

**Example 8** Elliptic spring. The elliptical shape of leaf springs were used in carriages as well as early vehicle models.

Figure 1.5 illustrates an elliptic spring, of a type used on carriages, automobiles, and railroad cars. The available space for a spring may determine its shape and size. Often, a long and wide elliptic spring cannot be used. The main disadvantage of elliptic springs that caused their disuse today is their weakness in bearing lateral loads.

There is a lack of evidence to suggest that scientific calculations were ever utilized in the design of the original forms of elliptic springs as commonly used in ordinary road carriages. While these springs were found to be satisfactory, their design was not based on engineering principles. Rather, their manufacture was passed down through generations or acquired through extensive experience, where craftsmen accumulated the knowledge and techniques of the trade. Manufacturers in this field did not rely on mathematical calculations to arrive at design outcomes. Instead, they gained an understanding within their trade that specific types of carriages necessitated certain styles of springs.

## 1.3 ★ Torsion Bars

In a few suspension systems, usually front suspension, torsion bars serve as alternatives to coil springs. These torsion bars undergo twisting during wheel jounce and subsequently unwind during wheel rebound, returning to their static position. One end of the torsion bar is connected to the vehicle frame, and the opposite end is linked to the suspension control arm. Some light-duty trucks and sport utility vehicles (SUVs) utilize torsion bars in their front suspensions.

Figure 1.6 illustrates a control arm attached to a torsion bar. As the wheel encounters road irregularities, the suspension control arm experiences vertical movement, resulting in the torsion bar twisting. The inherent resistance of the bar to twisting causes it to revert back to its original position. Compared to loaded coil or leaf springs, torsion bars have the capacity to store higher maximum energy. Shorter and thicker torsion bars exhibit greater load-carrying capacity compared to longer and thinner bars.

When a straight bar of uniform cross section is subjected to a torque $M$, constant along the bar, its torsional stiffness $k_\varphi$ will be

$$M = \frac{GJ}{l}\varphi, \tag{1.18}$$

$$k_\varphi = \frac{GJ}{l}, \tag{1.19}$$

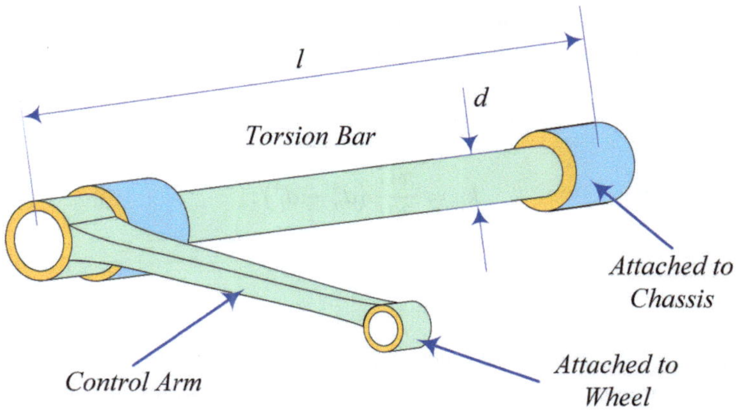


**Fig. 1.6** A torsion bar and control arm

where $G$ is the shear module of elasticity, $J$ is the geometric polar moment of area of the cross section, and $l$ is the length of the torsion bar.

***Proof*** The small twist angle $d\varphi$ when a torque $M$ is applied on a torsion bar of length $dx$, cross-sectional area moment $J$, and shear module of elasticity $G$ is

$$d\varphi = \frac{M\,dx}{GJ}; \tag{1.20}$$

therefore, integrating with respect to $dx$, for the length of the bar, we obtain

$$\varphi = \int_0^l \frac{M\,dx}{GJ} = \frac{M\,l}{GJ}. \tag{1.21}$$

Assuming a circular cross section of radius $r = d/2$,

$$J = \frac{\pi}{2}r^4 = \frac{\pi}{32}d^4, \tag{1.22}$$

the torsional stiffness of the bar will be

$$k_\varphi = \frac{\pi G d^4}{32l}. \tag{1.23}$$

The energy stored in a torsion bar is

$$E_\varphi = \frac{1}{2}k\varphi^2 = \frac{1}{2}M\varphi = \frac{1}{2}\frac{M^2\,l}{GJ}. \tag{1.24}$$

The volume of material in a spring is

$$V = \frac{1}{4}\pi d^2 l, \tag{1.25}$$

and, hence, the energy stored per unit of length will be

$$\frac{E_s}{V} = \frac{2M^2}{\pi G J d^2} = \frac{64}{\pi^2 G}\frac{M^2}{d^6}. \tag{1.26}$$

■

**Example 9** Tube as torsion bar. The shear stress in shafts under torque linearly increases from center to the surface radially. Hence, the center of shafts does not contribute in strain energy that much. To be mass effective, we may use a tube instead of solid shafts.

If a tube is used instead of a solid shaft, a torsion bar will be more mass efficient as the parts of materials that provide less strain and stress are removed. To derive the stiffness formula for a tube, we only need to calculate the polar area moment of the cross section:

$$J = \frac{\pi}{2}\left(r_o^4 - r_i^4\right) = \frac{\pi}{32}\left(d_o^4 - d_i^4\right), \tag{1.27}$$

$$k_\varphi = \frac{\pi G}{32l}\left(d_o^4 - d_i^4\right). \tag{1.28}$$

## 1.4 Summary

A spring is an elastic body, usually metallic, whose function is to collect potential energy by elastic deformation and recover its original shape following the removal of load. The primary purpose of springs in vehicle suspensions is to extend the duration over which energy is absorbed and, subsequently, released. In vehicle applications, the most commonly used springs for suspension systems are helical springs, leaf springs, and torsion bars. Their deformation under loads are either in bending or twisting, and hence, the stored potential strain energy will be made by normal stress in bending or shear stress in torsion.

## 1.5 Key Symbols

| | |
|---|---|
| $b$ | leaf spring width at center point |
| $c$ | damping |
| $c_e$ | equivalent damping |
| $d$ | wire diameter |
| $d_1$ | wave length |
| $d_i$ | inner diameter |
| $d_o$ | outer diameter |
| $D$ | mean diameter of coil, dissipation function |
| $E$ | mechanical energy |
| $E$ | Young's modulus of elasticity |
| $E_s$ | energy stored in a coil spring |
| $f = 1/T$ | cyclic frequency [ Hz] |
| $f, F, \mathbf{F}$ | force |
| $f_k$ | spring force |
| $g$ | gravitational acceleration, function |
| $G$ | shear modulus of elasticity, modulus of rigidity |
| $I$ | moment of cross-sectional area |
| $J$ | polar moment of cross-sectional area |
| $k$ | stiffness |
| $k_e$ | equivalent stiffness |
| $k_{coil}$ | equivalent stiffness of coil spring |
| $k_{leaf}$ | equivalent stiffness of leaf spring |
| $k_\varphi$ | torsional stiffness |
| $k_R$ | antiroll bar torsional stiffness |
| $K$ | kinetic energy |
| $l, L$ | length |
| $m$ | mass |
| $M$ | torque, moment |
| $n$ | number of coils, number of decibels, number of note |
| $n'$ | number of leaves directly at spring ends |
| $N$ | normal force |

| | |
|---|---|
| $p$ | coil pitch |
| $P$ | potential energy |
| $P$ | power |
| $r$ | radius, frequency ratio |
| $r_i$ | inner radius |
| $r_o$ | outer radius |
| $\mathbf{r}$ | position vector |
| $s$ | exponent, eigenvalue |
| $t$ | time |
| $t$ | vertical depth of leaf spring at center |
| $T$ | period |
| $U$ | special strain energy |
| $v \equiv \dot{x}$ | velocity |
| $V$ | volume |
| $W$ | work |
| $W_D$ | dissipated energy in one cycle |
| $x, y, z, \mathbf{x}$ | displacement |
| $x, y$ | displacement |
| $x_i$ | initial displacement |
| $x_f$ | final position |
| $\dot{x}_0$ | initial velocity |
| $\dot{x}, \dot{y}, \dot{z}$ | velocity, time derivative of $x, y, z$ |
| $\ddot{x}$ | acceleration |

Greek

| | |
|---|---|
| $\alpha, \beta, \gamma$ | angle, angle of spring with respect to displacement |
| $\delta$ | deflection |
| $\delta_s$ | static deflection |
| $\varepsilon$ | mass ratio |
| $\varepsilon$ | small coefficient |
| $\rho$ | mass density |
| $\sigma$ | stress |
| $\varphi$ | torsion angle, twist angle |
| $\theta$ | angular motion, phase angle |
| $\pi$ | 3.141592653589793... |
| $\omega, \Omega$ | angular frequency |
| $\omega_n$ | natural frequency |
| $\tau$ | torque |

Symbols

| | |
|---|---|
| $\forall$ | for all |
| [ ] | matrix |
| $\int$ | integral |
| \| \| | absolute value |
| Hz | Hertz |
| $d$ | differential |
| $\partial$ | partial derivative |
| ! | factorial |
| $\Delta$ | small amount |

## References


1. J. Agyris, H.P. Mlejnek, *Computational Mechanics* (Elsevier, New York, 1991)
2. S. Axler, P. Bourdon, W. Ramey, *Harmonic Function Theory* (Springer, New York, 2001)
3. B. Balachandran, E.B. Magrab, *Vibrations* (Brooks/Cole, Pacific Grove, 2003)
4. G.M. Ballou, *Handbook for Sound Engineers*, 4th edn. (Focal Press, Elsevier, Oxford, 2008)
5. H. Benaroya, *Mechanical Vibration: Analysis, Uncertainities, and Control* (Dekker, New York, 2004)
6. D.J. Benson, *Music: A Mathematical Offering* (Cambridge University Press, London, 2007)
7. G. Buzdugan, E. Mihailescu, M. Rades, *Vibration Measurement* (Springer, Dordrecht, 1986)
8. H.S. Carslaw, *An Interoduction to the Theory of Fourier's Series and Integrals* (Macmillan, London, 1921)
9. E.A. Coddington, *Ordinary Differential Equations* (Prentice Hall, New Jersey, 1961)
10. T.S. Chihara, *An Introduction to Orthogonal Polynomials* (Gordon and Breach, Science Publishers, New York, 1978)
11. H.K. Dass, E.R. Verma, *Higher Mathematical Physics* (S CHAND & Company, New Delhi, 2014)
12. R. Dean-Averns, *Automobile Chassis Design* (Iliffe & Sons, London, 1948)
13. M. Del Pedro, P. Pahud, *Vibration Mechanics* (Kluwer Academic Publishers, The Netherland, 1991)
14. J.P. Den Hartog, *Mechanical Vibrations* (McGraw-Hill, New York, 1934)
15. A. Dimarogonas, *Vibration for Engineers* (Prentice Hall, New Jersey, 1996)
16. E. Esmailzadeh, R.N. Jazar, Periodic solution of a mathieu-duffing type equation. Int. J. Nonlinear Mech. **32**(5), 905–912 (1997)
17. E. Esmailzadeh, B. Mehri, R.N. Jazar, Periodic solution of a second order, autonomous, nonlinear system. J. Nonlinear Dyn. **10**(4), 307–316 (1996)
18. J. Fauvel, R. Flood, R. Wilson, *Music and Mathematics: From Pythagoras to Fractals* (Oxford University Press, New York, 2003)
19. J.B.J. Fourier, *The Analytical Theory of Heat* (Translated by: A. Freeman, Digitally Printed Version 2009) (Cambridge University Press, New York, 1878)
20. E. Grigorieva, *Methods of Solving Nonstandard Problems* (Springer, New York, 2015)
21. G. Gunther, *The Physics of Music and Color* (Springer, New York, 2012)
22. C.M. Harris, A.G. Piersol, *Harris' Shock and Vibration Handbook* (McGraw-Hill, New York, 2002)
23. M.H. Holmes, *Introduction to the Foundations of Applied Mathematics* (Springer, New York, 2009)
24. D. Inman, *Engineering Vibrations* (Prentice Hall, New York, 2007)
25. R.N. Jazar, Stability chart of parametric vibrating systems using energy-rate method. Int. J. Non-Linear Mech. **39**(8), 1319–1331 (2004)
26. R.N. Jazar, *Advanced Dynamics: Rigid Body, Multibody, and Aero-space Applications* (Wiley, New York, 2011)
27. R.N. Jazar, *Advanced Vibrations: A Modern Approach* (Springer, New York, 2013)
28. R.N. Jazar, *Vehicle Dynamics: Theory and Application*, 3rd edn. (Springer, New York, 2017)
29. R.N. Jazar, *Perturbation Methods in Science and Engineering* (Springer, New York, 2021)
30. R.N. Jazar, *Advanced Vibrations: Theory and Application* (Springer, New York, 2023)
31. R.N. Jazar, M. Kazemi, S. Borhani, *Mechanical Vibrations, (in Persian)* (Ettehad Publications, Tehran, 1992)
32. R.N. Jazar, M. Mahinfalah, N. Mahmoudian, M.R. Aagaah, B. Shiari, Behavior of mathieu equation in stable regions. Int. J. Mech. Solids **1**(1), 1–18 (2006)
33. R.N. Jazar, M. Mahinfalah, N. Mahmoudian, M.R. Aagaah, Energy-rate method and stability chart of parametric vibrating systems. J. Brazilian Soc. Mech. Sci. Eng. **30**(3), 182–188 (2008)
34. R.N. Jazar, M. Mahinfalah, N. Mahmoudian, M.R. Aagaah, Effects of nonlinearities on the steady state dynamic behavior of electric actuated microcantilever-based resonators. J. Vibrat. Control **15**(9), 1283–1306 (2009)
35. F. Jedrzejewski, *Mathematical Theory of Music* (Ircam Centre Pompidou, Delatour, 2006)
36. V. Kobelev, *Durability of Springs* (Springer, Nature Switzerland, 2021)
37. D. Knowles, *Classroom Manual for Automotive Suspension & Steering Systems* (Cengage Delmar Learning, Clifton Park, 2007)
38. L. Meirovitch, *Analytical Methods in Vibrations* (Macmillan, New York, 1967)
39. L. Meirovitch, *Principles and Techniques of Vibrations* (Prentice Hall, Ney Jersey, 1997)
40. L. Meirovitch, *Fundamentals of Vibrations* (McGraw-Hill, New York, 2002)
41. R.E. Mickens, *Mathematical Methods for the Natural and Engineering Sciences* (World Scientific Publishing, Singapore, 2004)
42. R.E. Mickens, *Generalized Trigonometric and Hyperbolic Functions* (CRC Press, Boca Raton, 2019)
43. V.P. Minofsky, *Problems in Higher Mathematics* (Mir Publishers, Moscow, 1975)
44. K. Ogata, *System Dynamics* (Prentice Hall, Ney Jersey, 2004)
45. Z. Osinski, *Damping of Vibrations* (A. A. Balkema Publishers, Rotterdam, 1998)
46. W.J. Palm, *Mechanical Vibration* (Wiley, New York, 2006)
47. E. Prestini, *The Evolution of Applied Harmonic Analysis: Models of the Real World* (Springer, New York, 2004)
48. S.S. Rao, *Mechanical Vibrations* (Pearson Prentice Hall, Harlow, 2018)
49. J.W.S. Rayleigh, *The Theory of Sound* (Dover Publication, New York, 1945)
50. M. Roseau, *Vibrations in Mechanical Systems* (Springer, Berlin, 1987)
51. G. Sansone, *Orthogonal Functions* (Translated From the Italian by A.H. Diamond) (Dover Publications, New York, 1991)
52. A.A. Shabana, *Vibration of Discrete and Continuous Systems* (Springer, New York, 1997)
53. H. Shima, T. Nakayama, *Higher Mathematics for Physics and Engineering* (Springer, Berlin, 2010)
54. M. Shimoseki, T. Hamano, T. Imaizumi (eds.), *FEM for Springs* (Springer, Berlin, 2003)
55. G.F. Simmons, *Differential Equations with Applications and Historical Notes* (CRC Press, Boca Raton, 2017)
56. J.C. Snowdon, *Vibration and Shock in Damped Mechanical Systems* (Wiley, New York, 1968)
57. E.C. Titchmarsh, *Introduction to the Theory of Fourier Integrals* (Oxford University Press, London, 1948)
58. W.T. Thomson, M.D. Dahleh, *Theory of Vibration with Applications* (Prentice Hall, Ney Jersey, 1997)
59. B.H. Tongue, *Principles of Vibration* (Oxford University Press, New York, 2001)
60. F.S. Tse, I.E. Morse, R.T. Hinkle, *Mechanical Vibrations Theory and Applications* (Allyn and Bacon, Boston, 1978)

# 2 Dampers

A liquid shock absorber is typically an oil pump placed between the frame of the vehicle and the wheels. The upper mount of the shock connects to the frame, i.e., the sprung mass, while the lower mount connects to the axle near the wheel, i.e., the unsprung mass as is shown in Fig. 2.1.

The role of damping is crucial in removal energy form a vibrating vehicle. In vehicle suspension system, unless a dampening structure is present, a car spring will extend and release the energy it absorbs from a bump at an uncontrolled rate. The spring will continue to bounce until all of the energy originally put into it is used up in a very low rate by internal hysteresis and external frictions. Hence, every suspension system requires a device that controls unwanted spring motion through a process known as dampening. The most common designs of industrial dampers are twin-tube and monotube shock absorbers.

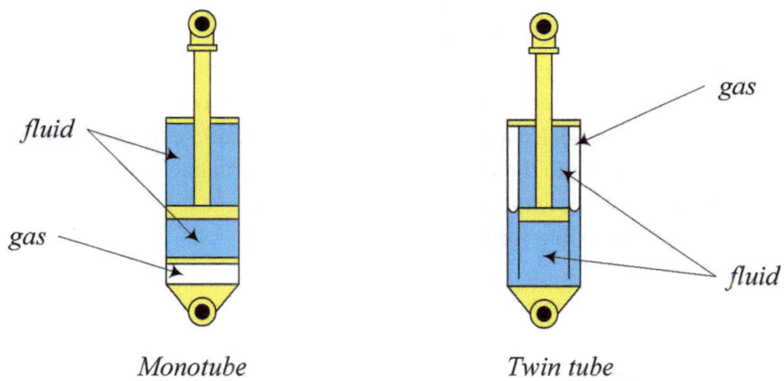


**Fig. 2.1** A twin-tube and a monotube design shock absorbers

In a twin-tube design, the most common types of shock absorbers, the upper mount is connected to a piston rod, which in turn is connected to a piston, which in turn sits in a tube filled with hydraulic fluid. The inner tube is known as the pressure tube, and the outer tube is known as the reservoir tube. The reservoir tube stores excess hydraulic fluid. When the wheel encounters a bump in the road and causes the spring to coil and uncoil, the energy of the spring is transferred to the shock absorber through the upper mount, down through the piston rod and into the piston. Holes perforate the piston and allow fluid to leak through as the piston moves up and down in the pressure tube. Because the holes are relatively tiny, only a small amount of fluid, under great pressure, passes through. This slows down the piston, which in turn slows down the spring. Shock absorbers work in two cycles, the compression cycle and the extension cycle. The compression cycle occurs as the piston moves downward, compressing the hydraulic fluid in the chamber below the piston. The extension cycle occurs as the piston moves toward the top of the pressure tube, compressing the fluid in the chamber above the piston. A typical street car will have more resistance during its extension cycle than its compression cycle. Shock absorbers are velocity-sensitive and the faster the suspension moves, the more resistance the shock absorber provides. This enables shocks to adjust to road conditions and to control all of the unwanted motions that can occur in a moving vehicle, including bounce, sway, brake dive, etc.



R. N. Jazar, H. Marzbani, *Vehicle Vibrations*, https://doi.org/10.1007/978-3-031-43486-0_2

In a monotube design, the oscillation of the suspension system will be dampen in a similar way that twin-tube shock absorbers work. The difference between them is that monotube shock absorbers consist of a single tube, as the name suggests. This tube contains both the hydraulic fluid and the piston assembly. Inside the monotube shock absorber, there is a piston that moves up and down in response to the suspension movement. The piston is attached to a rod that extends out from the top of the shock absorber. The hydraulic fluid fills the monotube shock absorber and is responsible for absorbing and dissipating the energy generated by the suspension movement. The fluid provides resistance to the piston's movement. When the suspension compresses due to a bump on the road, the piston is pushed into the tube, compressing the hydraulic fluid. This compression causes resistance, which helps control the speed at which the suspension compresses. As the suspension extends after compression, the piston moves in the opposite direction. The hydraulic fluid flows through specific valves in the piston assembly, allowing controlled fluid flow back into the main chamber. This helps regulate the speed of extension and prevents the suspension from rebounding too quickly. The difference of the volumes of fluid movement between the upper and lower chambers will be compensated by moving a round plate at the bottom of the tube, separating fluid and a chamber filled with a compressible gas.

Monotube shock absorbers have better heat dissipation capabilities compared to twin-tube shock absorbers. The single-tube design allows for more efficient heat transfer, reducing the risk of overheating during extended use. Overall, monotube shock absorbers provide better performance and more precise control over suspension movement compared to twin-tube shock absorbers. They are commonly used in high-performance vehicles and applications where superior handling and stability are desired.

## 2.1 ★ Shock Absorbers

Shock absorbers slow down and reduce the magnitude of vibratory motions by turning the kinetic energy of suspension movement into heat energy that can be dissipated through hydraulic fluid. Hydrodynamic and aerodynamic damping are popular types of damping used in shock absorbers in automotive industry in which case the motion of an element in a liquid or in a gas is damped by these media. The energy dissipation can also be done using various other techniques considering the type of shock absorber in use. The vibration energy of the system is converted to heat and dissipates into the environment. In the viscous damper, energy is converted to heat via a viscous fluid. In hydraulic cylinders, the hydraulic fluid is heated up, while in air cylinders, the hot air is emitted into the atmosphere. The electromagnetic damper is different; here the vibration energy is converted into electricity via an electric motor, such as an induction machine, or DC motor, or synchronous machine, and stored in a condenser or battery for possible further use.

### 2.1.1 ★ Chronicle and Evolution

The current world-wide production of vehicle dampers, or so-called shock absorbers, is difficult to estimate with accuracy but is around 200 million units per annum with a retail value well in excess of a couple of billion dollars per annum. On the average, a European country has a demand for over 20 million units per year on new cars and over 2 million replacement units. The US market is several times of that. If all is well, these suspension dampers do their work quietly and without fuss. Dampers are at their best when they are not noticed while driving, as drivers and passengers just want the dampers to be trouble-free.

The need for dampers arises because of the roll and pitch associated with vehicle maneuvering and from the roughness of roads. The fitting of damping devices to vehicle suspensions followed rapidly on the heels of the arrival of the motor car itself. Since those early days the damper has passed through a century of evolution, the basic stages of which can be considered as follows:

1. Dry friction
2. Blow-off hydraulics
3. Progressive hydraulics
4. Adjustable (manual alteration)
5. Slow adaptive (automatic alteration)

6. Fast adaptive (semi-active)
7. Electrofluidic

Some of the major historic advances made over the life of automotive dampers are listed below:

- Up to 1910 dampers were hardly used at all.
- From 1910 to 1925 mostly dry snubbers were used.
- From 1925 to 1980 there was a long period of dominance by hydraulics, initially simply constant-force blow-off, then through progressive development to a more proportional characteristic, and then adjustable, leading to a mature modern product.
- From 1980 to 1985 there was excitement about the possibilities for active suspension, which could effectively eliminate the ordinary damper, but little has come of this commercially in practice because of the cost.
- From 1985 it became increasingly apparent that a good deal of the benefit of active suspension could be obtained much more cheaply by fast auto-adjusting dampers, and the damper suddenly became an interesting, developing, component again.
- From about 2000 the introduction of controllable magnetorheological dampers.

Development of the adaptive damper has occurred rapidly to be effective. Fully active suspension offers some performance advantages with some extra cost. Continuous developments can be expected for detail refinement of design, control strategies, and production costs. Fast acting control, requiring extra sensors and controls, will continue to be more expensive, so simple fixed dampers, adjustable and slow adaptive types, will continue to dominate the market numerically for the foreseeable future.

The basic suspension using the simple spring and damper is not ideal, but it is good enough for most of the purposes. For low-cost vehicles, it is the most cost-effective system. Therefore much emphasis remains on improvement of operating life, reliability, and low-cost production rather than on refinement of performance by technical development.

The variable damper, in several forms, has now found quite wide application on mid-range and expensive vehicles. On the most expensive passenger and sports cars, magnetorheologically controlled dampers are now a popular fitment, at significant expense. The purpose of dampers is to dissipate any energy in the vertical motion of body or wheels, such motion having arisen from rough roads. Here "vertical" motion includes body heave, pitch and roll, and wheel hop.

The mathematical theory of vibrating systems largely uses the concept of a linear damper, with force proportional to extension speed, mainly because it gives equations for which the solutions are well understood and documented and usually realistic. There is no obligation on a damper to exhibit such a characteristic; nevertheless the typical modern hydraulic damper does so approximately. This is because the vehicle and damper manufacturers consider this to be desirable for good physical behavior. The desired characteristics are achieved only by some effort from the manufacturer in the detail design of the valves.

### 2.1.2 ★ Mechanical Design and Performance

The full specification of a damper can be immensely complex, covering all the dimensional data, plus solid material specifications, manufacturing methods, liquid specifications, gas pressurization, and performance specifications with tolerances. However, for a normal damper many of these are fairly standard and may be taken for granted. Essentially, the damper must be connected to the vehicle and exert the desired forces. Hence the primary or functional specification features may be considered to be:

- End fitting design
- Length range
- $f(v)$ curve

From a performance point of view specifications may include:

- Configuration
- Diameter
- Oil properties

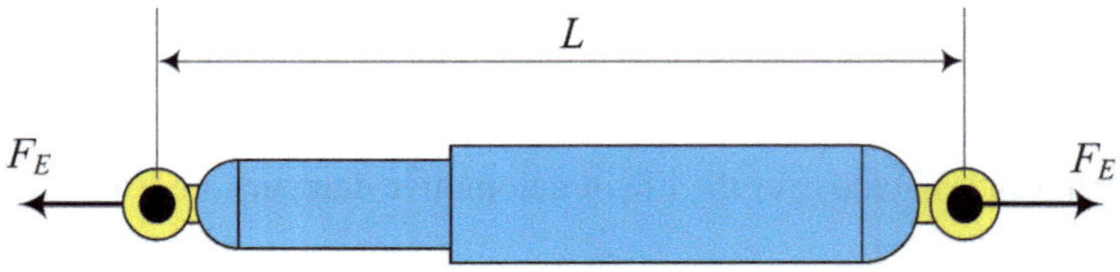


**Fig. 2.2** Positive velocity and force in extension. The damper is in tension, with its length increasing

To help to achieve durability, the specification may include information on:

- Seals
- Surface finish
- Corrosion resistance

Dimensional data include the stroke, the minimum and maximum length between mountings, diameters, mounting method, etc. Force characteristics indicate how the force varies with compression and extension velocities, production tolerances on these forces, and any effect of position. Other factors include limitations on operating temperature, power dissipation, cooling requirements, etc.

In considering the required characteristics of dampers, it is desirable to express their complex behaviors in a few simple parameters that can be correlated with subjective ride and handling quality. Thus the complexity of the speed characteristics might be reduced to the following parameters:

1. Overall mean damping coefficient
   The most fundamental parameter is the total average damping coefficient. The damping ratio for vehicles varies considerably according to the type of vehicle and the philosophy of the particular vehicle manufacturer. Typical overall damping ratios are 0.2–0.4 for a passenger car and 0.4–0.8 for a performance-oriented passenger car or competition car. Considering the variation in vehicle mass and spring stiffness, the required damping coefficient per wheel varies from 1 to 5 kN s/ m for passenger cars and higher for commercial vehicles.
2. Asymmetry, the transfer factor
   The second fundamental parameter is the asymmetry, the relative amounts of bound and rebound damping, which on passenger cars tends to be around 30/70, although not narrowly constrained, varying between 20/80 and 50/50. On motorcycles it is even more asymmetric, from 20/80 to 5/95.
3. $f(v)$ shape, the progressive factor
   The third parameter, the shape of the force against speed curve, can be represented by the progressivity factor. With intelligent choice of valve parameters it is possible to achieve a wide range of force-speed graph shapes. In general the preference is for a force that decreases less than proportionally with speed, so that the damping ratio is higher at low damper speed. This is to provide good control of handling motions while avoiding unacceptable harshness on sudden bumps.

Considering the cyclic characteristic, it is important for the $f(x)$ loop to be smooth in shape. Basically this means having smooth valve characteristics and avoiding cavitation in the usual range of operation. The force exerted by a damper depends on its velocity, as well as its time history of operation which influences the temperature and fluid properties.

The basic characteristic of a complete damper is represented by a graph of force against velocity. The damper extension velocity is

$$v_{DE} = \frac{dL}{dt} \tag{2.1}$$

where $L$, as is shown in Fig. 2.2, is the length between mounts. The compression velocity is then defined as

$$v_{DC} = -v_{DE} = -\frac{dL}{dt} \tag{2.2}$$

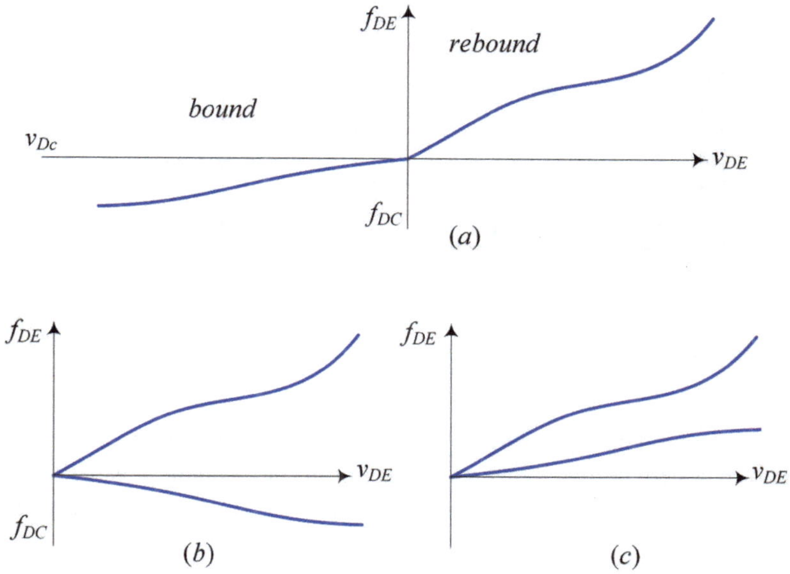


**Fig. 2.3** Damper characteristics: (**a**) force vs velocity, (**b**) force vs absolute velocity, and (**c**) absolute force vs absolute velocity

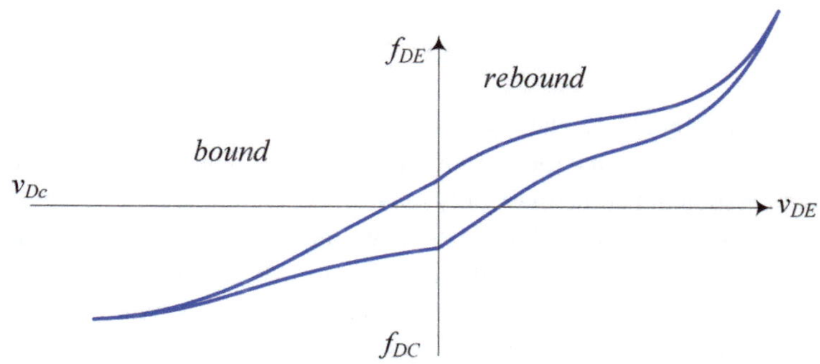


**Fig. 2.4** Example damper $f(v)$ loop resulting from stiffness

For drawing the force function $f(v)$, it is often convenient to plot the forces against absolute velocity. The normal installation of a suspension damper is such that suspension bump causes damper compression, so bump velocity is a common alternative term for compression velocity and rebound velocity for extension velocity.

Figure 2.3 shows the form of a typical $f(v)$ characteristic, with the extension (rebound) force upward and the compression (bump) force downward. The abscissa in Fig. 2.3b is the magnitude of velocity. This is sometimes also done for the forces if the compression and extension forces are sufficiently distinct to avoid confusion, as in Fig. 2.3c.

The graph of Fig. 2.3 assumes that the position of the damper along its stroke is not important, which means that the force is to be dependent on velocity rather than position. This is approximately true for a conventional damper but ceases to be true for a combined spring-damper unit. Any positional dependence of force turns the $f(v)$ graph's line into a loop such as the one shown in Fig. 2.4 because the speed is zero at the two extreme positions.

## 2.2 Mathematical Modeling

A vibrating mass–spring–damper model is illustrated in Fig. 2.5, in which $c$ is the damping constant in Newton-second per meter and $v = \dot{x}$ is the velocity of the mass $m$ in meters per second. The damping force is proportional to the mass velocity which also opposes the mass velocity.

$$f_{damping} = -cv \tag{2.3}$$

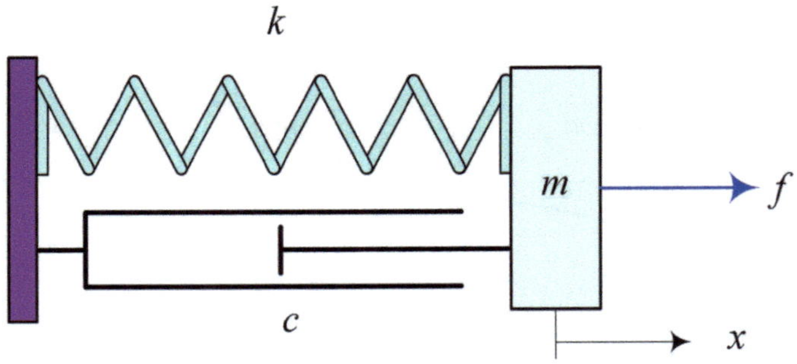


**Fig. 2.5** Mass, spring, and damper system

A liner (viscous) damping description can be incorporated to examine damping phenomena in many mechanical systems. Because an analysis of a given system becomes easy in the case of linear damping, it is very often used as an approximate and local description of more complex systems. So the total force acting on the mass–spring–damper system is

$$f = f_{spring} + f_{damping} = -kx - cv \tag{2.4}$$

Applying Newton's law of motion, $f = ma$, the equation of motion has the following form:

$$m\ddot{x} + c\dot{x} = kx = 0 \tag{2.5}$$

where $m$ denotes the mass of vibrating body, $c$ is the viscous damping coefficient, and $k$ stands for the spring coefficient. The simple mass–spring–damper model is the foundation of vibration analysis. This is defined as the single degree of freedom (*DOF*) model, since it has been assumed that the mass only moves in one direction.

Let us consider the one *DOF* system in Fig. 2.5. To evaluate the process of vibration disappearance, the concept of logarithmic decrement is introduced. It is defined as the logarithm of the ratio of two successive maximum deflections (amplitudes). There exist two different definitions of the decrement. The first is based on the ratio of the absolute values of two successive extreme deflections, while the second employs two maximum deflections appearing after the vibration period. According to the first definition, we have

$$\delta = \ln \frac{X_n}{X_{(n+1)}} = 2\pi\xi = 2\pi \frac{c}{c_{cr}} = 2\pi Q \tag{2.6}$$

The second can be written as

$$\delta_T = \ln \frac{X_n}{X_{(n+2)}} \tag{2.7}$$

The relationship between them is

$$\delta_T = \ln \left( \frac{X_n}{X_{n+1}} . \frac{X_{n+1}}{X_{n+2}} \right) = \delta_1 + \delta_2 \tag{2.8}$$

Since successive deflections usually do not reveal considerable differences, a distance of $m$ half-periods (or full periods) is taken into account.

$$\delta = \frac{1}{m} \ln \frac{X_n}{X_{n+m}} \tag{2.9}$$

For small values of the decrement, it is convenient to expand the logarithmic function into a power series and truncate it at a term of higher order. If the difference between two successive amplitudes is

$$\Delta X_n = X_n - X_{n+1} \tag{2.10}$$

then

$$\delta = \ln \frac{X_n}{X_{n+1}} = \ln \frac{X_n}{X_n - \Delta X_n} \tag{2.11}$$

The level of vibration damping can also be determined by the dissipation coefficient. This coefficient is defined as the ratio of dissipated energy within one period to maximum energy observed in that period:

$$\psi = \frac{\Delta W}{W} \tag{2.12}$$

In the case of systems characterized by linear elasticity, a simple relationship exists between the logarithmic decrement and dissipation coefficient. The envelope of the function $X = X(t)$ corresponding to damped vibration determines the energy of a given system. This energy can be calculated with the help of the following formula:

$$W_t = k\frac{X^2(t)}{2} \tag{2.13}$$

where $k$ denotes the stiffness of the system. Hence

$$\psi = -\int_{W_{t+T}}^{W_t} \frac{dW}{W} = -2\int_{a_{n+2}}^{a_n} \frac{dX}{X} = 2\delta_T \tag{2.14}$$

and consistently

$$\psi = 2(\delta_1 + \delta_2) \tag{2.15}$$

and in the case of constant $\delta$ we have

$$\psi = 4\delta \tag{2.16}$$

Vibration damping can be expressed in terms of quality factors being the ratio of the damping coefficient (viscous, linear) to the critical value of this coefficient:

$$Q = \frac{c}{c_{cr}} \tag{2.17}$$

The reciprocal of the quality factor constitutes the loss factor:

$$\eta = Q^{-1} \tag{2.18}$$

Both factors can be determined through examination of the resonant curves. For this purpose we have to find the frequencies $\omega_1$ and $\omega_2$ for which the vibration amplitude is $\sqrt{2}$ less than the resonant one. Now the loss factor can be calculated as

$$\eta = \frac{\omega_{02} - \omega_{01}}{\omega_{0res}} \tag{2.19}$$

## 2.3 ★ Hysteresis Loops of Vibrating Systems

In the study of vibration, hysteresis loops are often applied as a measure of energy dissipation. A hysteresis loop can be obtained by recording the magnitude of a force versus the cyclic displacement. Hysteresis exists in the presence of any damping in a mechanical system which could be seen in the force–displacement graph. The area enclosed by the loop expresses the amount of waste energy in a cycle.

Let us introduce the following three types of hysteresis loops:

1. External loop: reflecting the relation between externally applied force and displacements measured in the system.
2. Internal loop: if displacement is related to internal forces appearing in the systems.
3. Damping hysteresis loop: when damping forces are taken into consideration.

Let us regard the following equation of motion of a vibrating system:

$$\ddot{x} = F_i(x, \dot{x}) + F_e(t) \tag{2.20}$$

where $F_i(x, \dot{x})$ and $F_e(t)$ correspond to internal and external forces, respectively. The external and the internal loops can then be described by

$$F_e = F_e(x), F_i = F_i(x) \tag{2.21}$$

We can write down the following relationship between them:

$$F_i(x) = \ddot{x}(x) - F_e(x) \tag{2.22}$$

Owing to this relationship we can convert the external loop into an internal one by subtracting the values corresponding to the external loop from acceleration dependent on displacement. Multiplication of Eq. (2.22) by $dx = \dot{x}dt$ enables us to obtain the infinitesimal work done by existing forces. Integration over the entire period leads to

$$\int_0^T F_i[x(t)]\dot{x}(t)dt = \int_0^T \ddot{x}(t)\dot{x}(t)dt - \int_0^T F_e[x(t)]\dot{x}(t)dt \tag{2.23}$$

The first integral on the right-hand side of Eq. (2.23) equals zero for periodic motion. Hence, we can derive the well-known formula:

$$L_i = L_e \tag{2.24}$$

which states that dissipated energy is equal to the work done by external force. If it is possible to split the internal force into a restitution (elastic or gravitational) and a damping force such as

$$F_i = F_s(x) + F_d(x) \tag{2.25}$$

then for a given restitution force $F_s(x)$ we can determine the damping hysteresis loop:

$$F_d(x) = F_i(x) - F_s(x) = \ddot{x}(x) - F_e(x) - F_s(x) \tag{2.26}$$

For a very slow motion, we obtain a static hysteresis loop. Consider a linear system with viscous damping and harmonic excitation:

$$\ddot{x} + 2h\dot{x} + \omega_0^2 = q \sin vt \tag{2.27}$$

In this case we have

$$\begin{aligned} F_e &= q \sin vt \qquad F_s = -\omega_0^2 x \qquad F_d = -2h\dot{x} \\ F_i &= -\omega_0^2 x - ah\dot{x} \qquad x = qX \sin(vt - \delta) \end{aligned} \tag{2.28}$$

where

$$X = \frac{1}{\sqrt{(\omega_0^2 - v^2)^2 + (2hv)^2}} \tag{2.29}$$

$$\delta = \arctan \frac{-2hv}{\omega_0^2 - v^2} \tag{2.30}$$

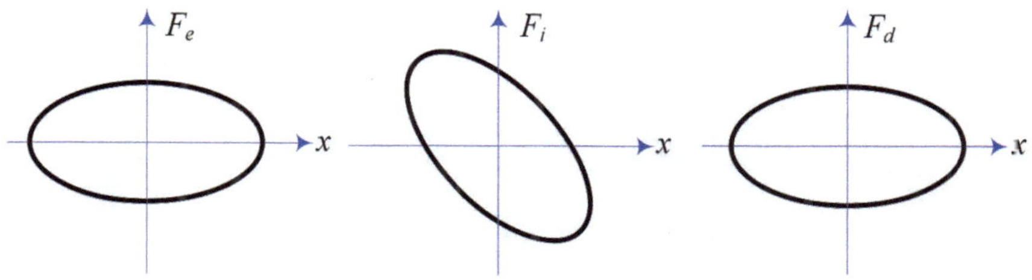


**Fig. 2.6** Hysteresis loops

$$\dot{x} = qXv\cos(vt - \delta) = \mp v\sqrt{q^2X^2 - x^2} \tag{2.31}$$

$$\ddot{x} = -qXv^2\sin(vt - \delta) = -v^2x \tag{2.32}$$

The external loop can be expressed by

$$F_e = (\omega_0^2 - v^2)x \mp 2hv\sqrt{q^2X^2 - x^2} \tag{2.33}$$

For the internal one, we write

$$F_i = -\omega_0^2 x \mp 2hv\sqrt{q^2X^2 - x^2} \tag{2.34}$$

while the damping loop is given by

$$F_d = \mp 2hv\sqrt{q^2X^2 - x^2} \tag{2.35}$$

The damping hysteresis loop has the form of an ellipse:

$$\left(\frac{f_d}{2hvqX}\right)^2 + \left(\frac{x}{qX}\right)^2 = 1 \tag{2.36}$$

The internal loop can be obtained by adding $-\omega^2 x$ to ordinates of the ellipse. The external loop can be found by adding $(-\omega^2 + v^2)x$ to the ellipse ordinates. All the hysteresis loops are presented in Fig. 2.6.

## 2.4 ★ Advanced Dampers

Suspensions may be classified by type and speed of response as follows:

- Passive—slow (manual adjustment)
- Adaptive—slow (roughness and speed) and fast (individual bumps)
- Active—very slow (load leveling), slow (roughness and speed), and fast (individual bumps)

An active suspension requires a large power input from a pump. An adaptive suspension requires power for the valves only. The basic purposes of adjustment are as follows:

- To optimize damper characteristics for varying conditions of road roughness and driving style
- To compensate for wear

### 2.4.1 Fast Adaptive Systems

Fast adaptive damping requires timely information from sensors on the body and from suspension vertical motions. Several operating strategies have been proposed. The underlying concept of fast adaptive damping is that the damper is rapidly altered according to whether the force that it can produce is deemed desirable or not. Considering a heave-only (quarter-car) model,

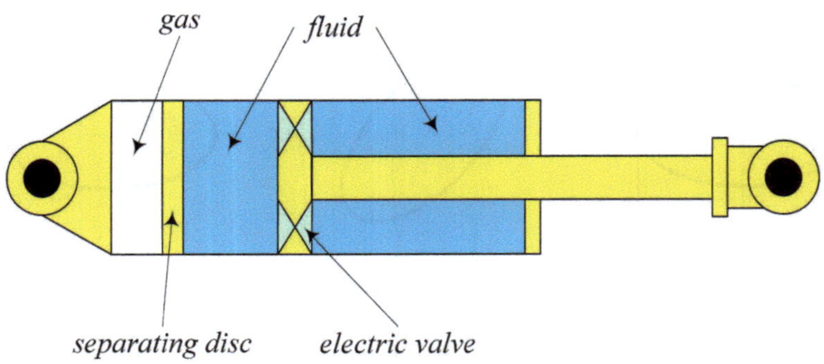


**Fig. 2.7** Schematic representation of an electrohydraulic shock absorber

the body (unsprung mass) vertical displacement is $x_s$ with velocity $\dot{x}_s$ and acceleration $\ddot{x}_s$, all positive upwards. The wheel has corresponding $x_u$. The suspension bump deflection is

$$z = x_u - x_s \tag{2.37}$$

with corresponding derivative equations for the suspension velocity and acceleration. The suspension deflection $z$ is positive for bump, which corresponds to damper compression, i.e., negative extension velocity, for a normal installation. For an ideal damper with force dependent on velocity only, the direction of damper force (i.e., damper tension or compression) is governed exclusively by the direction of the suspension bump velocity. For positive $\dot{z}$, the damper can produce a force upward on the sprung mass and downward on the unsprung mass. We may choose to exert such a force or not, according to some sensor information. Because the passenger discomfort is based on vehicle body accelerations, one strategy is to choose to exercise a damper force if it will reduce the body acceleration. Consider a bilinear damper with high damping when $\dot{z} > 0$ and low damping when $\dot{z} < 0$. Theoretical and practical studies show that such a strategy can indeed improve vehicle ride. However there may be a deleterious effect on road holding. If vehicle handling quality is chosen as the criterion, rather than ride, then the damper state selection strategy would be based on making the tire vertical force as steady as possible.

### 2.4.2 Electrohydraulic Dampers (EH Dampers)

The concept of electrohydraulic damper is depicted in Fig. 2.7. Compared to the classical passive element, the electrohydraulic device comprises electronic valves instead of passive valves. The ideal characteristic of electrohydraulic dampers is represented by the following relation:

$$F_d(\dot{x}, I) = c(I)\dot{x} + F_0 sign(\dot{x}) \tag{2.38}$$

where

$$c(I) = \gamma I + c_0 \qquad 0 \geq I \geq I_{max} \tag{2.39}$$

where $x$ is the suspension deflection, and $I$ is the electrical command (e.g. a current for solenoid valves) limited between $I = 0$ and $I = I_{max}$. $c_0$ is the minimum damping achieved when the electronic command is off, $\gamma$ is the characteristic gain that turns the electronic command into a damping, and $F_0$ is the internal friction of the shock absorber, herein assumed to be constant.

### 2.4.3 ER and MR Dampers

Electrorheological (ER) dampers and related devices have been under development for many years. However, since around year 2000 magnetorheological (MR) dampers have come into commercial use on some passenger vehicles. Rheology is the science of the deformation of solids and the flow of fluids under stress. ER (electrorheological) and MR (magnetorheological) liquids have properties dependent on the electric field or magnetic field, respectively. Because of these properties, ER and MR fluids, also known as smart or intelligent materials. By adjusting the electric or magnetic field as appropriate, the liquid properties are changed, controlling the damping force, which is no longer governed solely by the extension or compression speed. The damper itself is considerably different from conventional design, lacking the usual variable-area valves. Therefore, the damper force must be continuously controlled. A conventional damper oil is considered to be a Newtonian liquid, in that it has a simple viscosity, albeit temperature dependent. ER and MR liquids have a yield stress and a post-yield marginal viscosity, both dependent on the applied field. Hence, they are Bingham plastics, characterized by two parameters, the yield shear stress and the subsequent marginal viscosity. In practical use, it is the controlled variation of the yield stress that is the main operational parameter.

Small electrorheological effects have long been known, but large-scale effects with possible practical applications were first studied by W. H. Winslow in 1947. The first important work on MR appeared in 1951 by the Ukrainian-American engineer, Jacob Rabinow (1910–1999). For 40 years significant efforts were made on ER, largely neglecting MR, but since 1990, when work on MR increased, it has become apparent that MR may be much more practical because of the lower operating voltage, lower power requirement, higher shear yield stresses achievable, broader operating temperature range, and greater tolerance of the liquid to contamination, particularly water. Against this must be weighed greater expense and some significant hazards in manufacture. Winslow's 1947 patent is on various configurations of clutches and relays using electro-fluids. Rabinow's 1951 patent considers various configurations of clutch and emphasizes the low operating voltage, large force generation, and rapid response. Winslow took out a further patent in 1959, in which Winslow suggests that the coupling force for ER fluids is proportional to the field strength squared.

### 2.4.4 ER Dampers

Electrorheological dampers can vary the damping exploiting the physical property of the ER fluid flowing inside the shock absorber. ER fluid is a mixture of oil and micron-sized particles which are sensitive to electric field. When no electric field is applied, the fluid is almost free to flow in the damper orifices; when an electric field is applied, the particles react as dipoles and form chains, so that the flowing of the fluid turns from free to visco-plastic.

The ideal and simplified characteristics of ER dampers are represented by the following relation:

$$F_d(\dot{x}, v) = c_0\dot{x} + F_{ER}(v, \dot{x}) \qquad 0 \geq v \geq v_{max} \tag{2.40}$$

Electrorheological dampers are of distinct internal design. At zero field, the ER liquid has negligible yield stress and behaves as a Newtonian fluid with constant viscosity. To add the ER effect, it is necessary to produce a strong electric field over a significant area. The practical approach is to use an annular flow design. In this design, it is assumed that the force will be controlled entirely by the electrical field, i.e., there are no conventional valves in the piston.

#### Mechanism of ER Viscosity Change

Considering a polar molecule or particle with a dipole moment, with an electric field, in a static or very low-speed fluid flow the particle will align with the field. In a slowly shearing flow, this particle will then add to the effective viscosity by drag effect on the fluid due to the different fluid velocities relative to the particle at the two ends, the particle drifting along at a mean speed. As the shear rate increases, at constant voltage, the particle will become partially aligned with the flow, having somewhat less effect on the effective viscosity, i.e., there would be a nonlinear resistance to flow. Finally, at a high shear rate, of value depending on the field strength, the particle will become aligned and there will be little extra viscosity effect. Considering a constant velocity shear rate with varying field strength, as the voltage increases the particle will be pulled more out of alignment with the flow and will give a greater effective viscosity. ER materials may also exhibit a reduction of marginal viscosity with field strength.

#### Mechanism of ER Shear Strength Change

Dipole molecules can cluster together in chain form. They become entangled, giving the former liquid some of the properties of a solid, in particular in this case some shear yield strength. The yield stress of this material is very low compared to structural steel or solid polymers. The yield stress has been observed to be subject to a stick-slip phenomenon, with a force overshoot as the flow changes direction. The response time for ER fluid is short, being the time required for two responses to occur: (1) the individual particles may align with the field by rotation and (2) the particles must form any fibrils or bunches. These responses are slowed by the viscosity of the host oil, which should therefore be kept low. The response time is temperature sensitive, becoming rather slow at low temperature, due to the high oil viscosity. Leakage current due to electrical conductivity limits the upper temperature, requiring high current and power from the power supply to create sufficient working voltage. ER materials have a pre-yield region at low stress with some strain and appear to be conventional viscoelastic materials in this range.

### 2.4.5 MR Dampers

Magnetorheological dampers (MR dampers) exploit the physical properties of magnetorheological fluids (MR fluids). MR fluids change their viscosity when subject to a magnetic field. An MR fluid consists of a mixture of oil (usually a silicon oil) and micro-particles sensitive to the magnetic field (e.g., iron particles). The MR fluid behaves as a liquid when no field is applied. In the case of a magnetic field applied to the fluid, the particles form chains and the fluid becomes very viscous.

The ideal and simplified characteristics of MR dampers are represented by the following relation:

$$F_d(\dot{x}, I) = c_0\dot{x} + F_{MR}(I, \dot{x}) \qquad 0 \geq I \geq I_{max} \tag{2.41}$$

where $c_0$ is the minimum damping given by the free flowing of the fluid through the piston orifices, $x$ is the suspension deflection, and $I$ is the electronic command (usually current) given to the coils, which must be between $I = 0$ and $I = I_{max}$. $F_{MR}$ is the friction force between the MR fluid and the piston orifices, regarded as a nonlinear function of the current command $I$ and of the deflection speed $\dot{x}$. MR characteristics may be affected by harmful phenomena such as large hysteresis, and variable friction.

In an MR damper the piston includes coils capable of delivering a magnetic field in the orifices. In these terms, the piston may be viewed as a magnetorheological valve and the damping is the result of the friction between the fluid and the orifices.

An MR (magnetorheological) material is fundamentally different in practice from an ER material, because the realistically achievable shear stress is much higher for MR, more than one order of magnitude better than ER. As a result, the design of an MR damper can be more conventional, with the valve in the piston. The design of a magnetic circuit is more complex than for an electric current circuit because of the nonlinear behavior of the materials and nowadays is likely to be done using a suitable software package.

For the MR and ER dampers, the force-velocity maps are far from linear.

## 2.5 Summary

The role of damping is crucial in reducing energy form a vibrating vehicle. In vehicle suspension system, unless a dampening structure is present, a car spring will extend and release the energy it absorbs from a bump at an uncontrolled rate. A liquid damper for automotive application is an oil pump placed between the frame of the vehicle and the wheels. The upper mount of the damper connects to the frame of the vehicle, while the lower mount connects to the axle near the wheel. The most common designs of industrial dampers are twin-tube and monotube shock absorbers.

In a twin-tube design, the upper mount is connected to a piston rod, which in turn is connected to a piston, which in turn sits in a tube filled with hydraulic fluid. The inner tube is known as the pressure tube, and the outer tube is known as the reservoir tube. The reservoir tube stores excess hydraulic fluid. Holes perforate the piston and allow fluid to leak through as the piston moves up and down in the pressure tube and slows down the piston.

The monotube shock absorber contains both the hydraulic fluid and the piston assembly. Inside the monotube shock absorber, there is a piston that moves up and down in response to the suspension movement. The piston is attached to a rod that extends out from the top of the shock absorber. The difference of the volumes of fluid movement between the upper and lower chambers will be compensated by moving a round plate at the bottom of the tube, separating fluid and a chamber filled by a compressible gas. Monotube shock absorbers have better heat dissipation capabilities compared to twin-tube shock absorbers.

Approximating the damping of a shock absorber, $c$, to be constant, the damping force would be proportional to the relative velocity of the two ends of the shock absorber, $v$.

$$f_{damping} = -cv \tag{2.42}$$

A liner damping description can be incorporated to examine damping phenomena in many mechanical systems.

## 2.6 Key Symbols

| | |
|---|---|
| $a \equiv \ddot{x}$ | acceleration |
| $A$ | amplitude |
| $c$ | damping |
| $c_e$ | equivalent damping |
| $C$ | coefficient |
| $d$ | diameter of wire |
| $E$ | mechanical energy, Young modulus of elasticity |
| $E_c$ | consumed energy of a damper per cycle |
| $f = 1/T$ | cyclic frequency [ Hz] |
| $f, F, \mathbf{F}$ | force |
| $f, g$ | function, periodic function |
| $f_H$ | hydraulic damping force |
| $f_c$ | damper force |
| $f_e$ | equivalent force |
| $f_k$ | spring force |
| $f_m$ | required force to move a mass $m$ |
| $f_x$ | $x$-component of a force $f$ |
| $F$ | amplitude of a harmonic force $f = F \sin \omega t$, total force |
| $g$ | gravitational acceleration, function |
| $G$ | shear modulus of elasticity |
| $i$ | $i^2 = -1$, imaginary unit number |
| $I$ | area moment, mass moment |
| $J$ | polar moment of cross sectional area |
| $k$ | stiffness |
| $k_e$ | equivalent stiffness |
| $K$ | kinetic energy |
| $l, L$ | length |
| $m$ | mass |
| $m_d$ | damper mass |
| $m_e$ | eccentric mass, equivalent mass |
| $m_s$ | spring mass |
| $M$ | mass dimension symbol |
| $n$ | number of coils, number of decibels, number of note |
| $O$ | fixed point, origin |
| $P$ | potential energy, power |
| $P_D$ | dissipated power in one cycle |
| $Q$ | quality factor |
| $r$ | frequency ratio |
| $t$ | time |
| $t_0$ | initial time |
| $T$ | period |
| $T_n$ | natural period |
| $v \equiv \dot{x}$ | velocity |
| $W$ | work |
| $W_D$ | dissipated energy in one cycle |
| $x, y, z, \mathbf{x}$ | displacement |
| $X$ | amplitude |
| $x_0$ | initial displacement |
| $x_i$ | initial position |
| $x_f$ | final position |

| | |
|---|---|
| $\dot{x}_0$ | initial velocity |
| $\dot{x}, \dot{y}, \dot{z}$ | velocity, time derivative of $x$, $y$, $z$ |
| $\ddot{x}$ | acceleration |
| $X$ | amplitude of $x$ |
| $z$ | relative displacement |
| $Z$ | amplitude of $z$ |

Greek

| | |
|---|---|
| $\delta$ | deflection |
| $\delta$ | logarithmic decrement |
| $\delta_s$ | static deflection |
| $\varepsilon$ | mass ratio |
| $\varepsilon$ | small coefficient |
| $\rho$ | length mass density |
| $\theta$ | angular motion, phase angle |
| $\pi$ | 3.141592653589793... |
| $\omega$, $\Omega$ | angular frequency |
| $\omega_n$ | natural frequency |
| $\varphi$, $\Phi$ | phase angle |

Symbols

| | |
|---|---|
| [ ] | dimension |
| [ ] | matrix |
| $\int$ | integral |
| \| \| | absolute value |
| Hz | Hertz |
| $d$ | differential |
| $\partial$ | partial derivative |
| $\Delta$ | difference, small amount |

## References


1. J. Agyris, H.P. Mlejnek, *Computational Mechanics* (Elsevier, New York, 1991)
2. S. Axler, P. Bourdon, W. Ramey, *Harmonic Function Theory* (Springer, New York, 2001)
3. F. Jedrzejewski, *Mathematical Theory of Music* (IRCAM Centre Pompidou, Delatour, France, 2006)
4. B. Balachandran, E.B. Magrab, *Vibrations* (Brooks/Cole, Pacific Grove, CA, 2003)
5. G.M. Ballou, *Handbook for Sound Engineers*, 4th edn. (Focal Press, Elsevier, Oxford, UK, 2008)
6. D.J. Benson, *Music: A Mathematical Offering* (Cambridge University Press, London, UK, 2007)
7. H. Benaroya, *Mechanical Vibration: Analysis, Uncertainties, and Control* (Marcel Dekker, New York, 2004)
8. G. Buzdugan, E. Mihailescu, M. Rades, *Vibration Measurement* (Springer, Dordrecht, 1986)
9. E.A. Coddington, *Ordinary Differential Equations* (Prentice Hall, New Jersey, 1961)
10. H.S.Carslaw *An Introduction to the Theory of Fourier's Series and Integrals* (Macmillan, London UK, 1921)
11. T.S. Chihara, *An Introduction to Orthogonal Polynomials* (Gordon and Breach, Science Publishers, New York, NY, 1978)
12. H.K. Dass, E.R. Verma, *Higher Mathematical Physics* (S CHAND & Company, New Delhi, India, 2014)
13. M. Del Pedro, P. Pahud, *Vibration Mechanics* (Kluwer Academic Publishers, The Netherland, 1991)
14. J.P. Den Hartog, *Mechanical Vibrations* (McGraw-Hill, New York, 1934)
15. A. Dimarogonas, *Vibration for Engineers* (Prentice Hall, New Jersey, 1996)
16. J.C. Dixon, *The Shock Absorber Handbook*, 2nd edn. (Wiley, West Sussex, England, 2007)
17. E. Esmailzadeh, R.N. Jazar, Periodic solution of a Mathieu-Duffing type equation. Int. J. Nonlinear Mech. **32**(5), 905–912 (1997)
18. E. Esmailzadeh, B. Mehri, R.N. Jazar, Periodic solution of a second order, autonomous, nonlinear system. J. Nonlinear Dyn. **10**(4), 307–316 (1996)
19. J. Fauvel, R. Flood, R. Wilson, *Music and Mathematics: From Pythagoras to Fractals* (Oxford University Press, New York, NY, 2003)
20. J.B.J. Fourier, *The Analytical Theory of Heat* Translated by: Freeman A., digitally printed version 2009 (Cambridge University Press, New York, NY, 1878)
21. E. Grigorieva, *Methods of Solving Nonstandard Problems* (Springer, New York, NY, 2015)
22. G. Gunther, *The Physics of Music and Color* (Springer, New York, 2012)

23. C.M. Harris, A.G. Piersol, *Harris' Shock and Vibration Handbook* (McGraw-Hill, New York, NY, 2002)
24. M.H. Holmes, *Introduction to the Foundations of Applied Mathematics* (Springer, New York, NY, 2009)
25. D. Inman, *Engineering Vibrations* (Prentice Hall, New York, NY, 2007)
26. R.N. Jazar, *Analysis of Nonlinear Parametric Vibrating Systems*. Ph.D. Thesis, Mechanical Engineering Department, Tehran: Sharif University of Technology (1997)
27. R.N. Jazar, Stability chart of parametric vibrating systems using energy-rate method. Int. J. Non-linear Mech. **39**(8), 1319–1331 (2004)
28. R.N. Jazar, *Advanced Dynamics: Rigid Body, Multibody, and Aero-space Applications* (Wiley, New York, NY, 2011)
29. R.N. Jazar, *Advanced Vibrations: A Modern Approach* (Springer, New York, NY, 2013)
30. R.N. Jazar, *Advanced Vibrations: Theory and Application* (Springer, New York, NY, 2023)
31. R.N. Jazar, *Vehicle Dynamics: Theory and Application*, 3rd edn. (Springer, New York, NY, 2017)
32. R.N. Jazar, *Approximation Methods in Science and Engineering* (Springer, New York, NY, 2020)
33. R.N. Jazar, *Perturbation Methods in Science and Engineering* (Springer, New York, 2021)
34. R.N. Jazar, M. Kazemi, S. Borhani, *Mechanical Vibrations* (in Persian) (Ettehad Publications, Tehran, 1992)
35. R.N. Jazar, M. Mahinfalah, N. Mahmoudian, M.R. Aagaah, Energy-rate method and stability chart of parametric vibrating systems. J. Braz. Soc. Mech. Sci. Eng. **30**(3), 182–188 (2008)
36. R.N. Jazar, M. Mahinfalah, N. Mahmoudian, M.R. Aagaah, Effects of nonlinearities on the steady state dynamic behavior of electric actuated microcantilever-based resonators. J. Vib. Control **15**(9), 1283–1306 (2009)
37. R.N. Jazar, M. Mahinfalah, N. Mahmoudian, M.R. Aagaah, B. Shiari, Behavior of Mathieu equation in stable regions. Int. J. Mech. Solids **1**(1), 1–18 (2006)
38. L. Meirovitch, *Analytical Methods in Vibrations* (Macmillan, New York, 1967)
39. L. Meirovitch, *Principles and Techniques of Vibrations* (Prentice Hall, Ney Jersey, 1997)
40. L. Meirovitch, *Fundamentals of Vibrations* (McGraw-Hill, New York, 2002)
41. R.E. Mickens, *Generalized Trigonometric and Hyperbolic Functions* (CRC Press, Boca Raton, FL, 2019)
42. R.E. Mickens, *Mathematical Methods for the Natural and Engineering Sciences* (World Scientific Publishing, Singapore, 2004)
43. V.P. Minofsky, *Problems in Higher Mathematics* (Mir Publishers, Moscow, 1975)
44. K. Ogata, *System Dynamics* (Prentice Hall, Ney Jersey, 2004)
45. Z. Osinski, *Damping of Vibrations* (A. A. Balkema Publishers, Rotterdam, Netherlands, 1998)
46. W.J. Palm, *Mechanical Vibration* (Wiley, New York, 2006)
47. P. Polak, R.T. Burton, Improving suspension damping. J. Automotive Eng. **2**(2), 13–17 (1971)
48. E. Prestini, *The Evolution of Applied Harmonic Analysis: Models of the Real World* (Springer, New York, NY, 2004)
49. S.S. Rao, *Mechanical Vibrations* (Pearson Prentice Hall, Harlow, UK, 2018)
50. J.W.S. Rayleigh, *The Theory of Sound* (Dover Publication, New York, 1945)
51. M. Roseau, *Vibrations in Mechanical Systems* (Springer, Berlin, 1987)
52. A.A. Shabana, *Vibration of Discrete and Continuous Systems* (Springer, New York, 1997)
53. G. Sansone, *Orthogonal Functions*, Translated From the Italian by Diamond A. H. (Dover Publications, New York, 1991)
54. H. Shima, T. Nakayama, *Higher Mathematics for Physics and Engineering* (Springer, Berlin, Heidelberg, 2010)
55. G.F. Simmons, *Differential Equations with Applications and Historical Notes* (CRC Press, Boca Raton, FL, 2017)
56. J.C. Snowdon, *Vibration and Shock in Damped Mechanical Systems* (Wiley, New York, 1968)
57. E.C. Titchmarsh, *Introduction to the Theory of Fourier Integrals* (Oxford University Press, London UK, 1948)
58. W.T. Thomson, M.D. Dahleh, *Theory of Vibration with Applications* (Prentice Hall, Ney Jersey, 1997)
59. B.H. Tongue, *Principles of Vibration* (Oxford University Press, New York, 2001)
60. F.S. Tse, I.E. Morse, R.T. Hinkle, *Mechanical Vibrations Theory and Applications* (Allyn and Bacon Inc., Boston, MA, 1978)

# 3 Suspension Systems

Every multi-wheel vehicle requires a suspension system to provide road contact to all wheels and provide ride comfort. Due to physical and engineering imperfections, it is impossible that all wheels of a multi-wheel rigid vehicle touch the ground at the same time. Flexible suspension is required to make all wheels to be in contact with the ground on road irregularities, wheels' unequal loads, and under forward and lateral accelerating conditions. Figure 3.1 illustrates side view of a solid axle suspension with leaf spring. The mechanical requirements of suspensions and their design will be studied in this chapter.

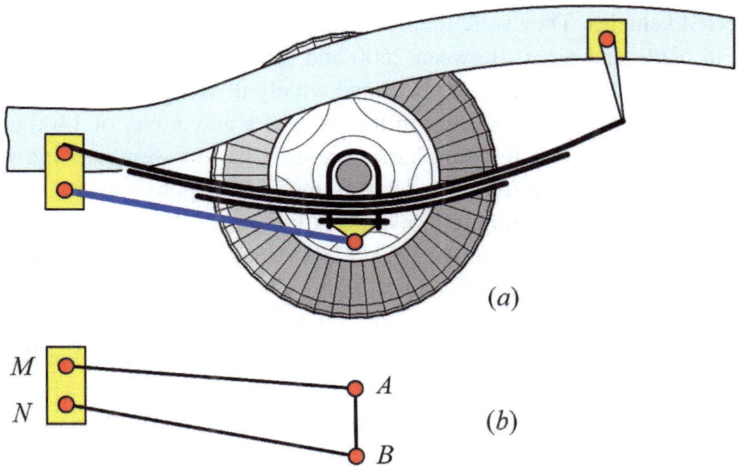


**Fig. 3.1** Side view of a solid axle suspension with leaf spring

The job of vehicle suspensions is to maximize the contact and, hence, the friction between the tires and the road surface. This phenomenon provides steering stability with good handling to ensure controllability of the vehicle. The suspension system also provides flexibility to act as a vibration filter to give ride comfort to the passengers. A bump in the road causes the wheel to move up and down with respect to the road surface and with respect to the vehicle body. The wheel experiences a vertical acceleration as it passes over an imperfection. Without an intervening structure, all of the vertical movement of the wheels are transferred to the frame of the vehicle. In such a situation, the tires can lose contact with the road, known as "tire–road separation." All of these phenomena highlight the need for a system that would absorb the energy of the wheel movement and in turn allows the frame and the body of the vehicle to ride undisturbed, while the tires follow the bumps on the road. Generally speaking, the suspension system in a vehicle needs to have two major characteristics:

1. **Ride**: A car's ability to smooth out the transferred vibrations from the road.
   Ride comfort refers to the vehicle's ability to absorb or isolate road bumps from the passengers. Here the role of the suspension system is to allow the vehicle body to ride undisturbed while traveling over rough roads. To do so the vehicle suspensions system should be able to absorb energy from road bumps and dissipate it without causing undue oscillation in the vehicle.



R. N. Jazar, H. Marzbani, *Vehicle Vibrations*, https://doi.org/10.1007/978-3-031-43486-0_3

2. **Handling**: A car's ability to safely maneuver on the road.
   Road handling refers to a vehicle maintaining contact with the road surface in various types of directional changes as well as in a straight line. Here the suspension system must keep the tires in contact with the ground at all times ensuring no compromise to the car's ability to steer, brake, or accelerate. Suspension system also has the responsibility of minimizing body roll in the case of cornering over curved paths.

The suspension of a car is part of the chassis, which comprises all of the important systems located beneath the vehicle's body. These systems include:

- **The frame**: structural, load-carrying component that supports the vehicle's body, engine, and components, which are in turn supported by the suspension
- **The suspension system**: setup that supports weight, absorbs and dampens road shocks, and helps maintain tire contact
- **The steering system**: mechanism that enables the driver to guide and direct the vehicle
- **The tires and wheels**: components that make vehicle motion possible by way of grip and friction with the road

## 3.1 ★ Origin of Suspension System

Modern horses have descended from the Hyracotherium, a small creature that was around 30 cm high and lived around 50 million years ago. Among the first horse-drawn vehicles was the chariot, invented by the Mesopotamians in Early Bronze Age about 3000 BC in the third millennium. Bronze is produced when 90% copper is mixed with 10% tin. The chariot was light horse-drawn, two-wheeled vehicles. They were the first vehicles designed to carry cargo, but they were also good for speed and maneuverability in battle and races. Between 2500 and 700 BC, the use of chariots spread from Mesopotamia to Persia, Egypt, Greece, India, and China. They were built and widely driven by Indo-Iranian people during the first half of the second millennium. The earliest evidence comes in the form of a clay tablet of 1400 BC, found in eastern lands of Persian empire that is about training program for war chariot horses. Chariots were prestige vehicles of the nobles. From the beginning of the fifth century BC, there are reliefs in detail from the Persepolis showing contemporary Persian chariots. Persepolis was the capital of the Achaemenid Empire (550–330 BC). Figure 3.2 illustrates Darius' war chariot, engraved in Darius cylinder seal, held in British Museum. Darius the Great (550–486 BC) was the third Kings of the Persian Achaemenid Empire. Darius is the king who conquered the eastern Greek lands.

Chariots are not suitable to move off road, and that was a reason Achaemenid Empire developed a network of Royal Road all over the ever largest empire, spanning a total of 5.5 million square kilometers from the Balkans and Egypt in the west to the Indus River Valley in the east. The cargo carriages and wagons have four wheels. The wheels were solid or made of tripartite discs cut from planks. Such four-wheeled wagons weighed few hundreds kilograms and could not be pulled by a team of oxen with a speed more than 3 km/ h. The wheels were solid at first and then came crossbar wheels and finally spoked wheels. Use of spoked wheel instead of solid wheel lightened the chariots and increased their speed. The earliest evidence of crossbar wheel comes from North-Western Persia, and this artefact could be dated to third millennium BC.

Following the development of the Royal Road and the mass production of horse-drawn chariots and carriages, Achaemenid invented courier-house to use the chariots and carriages practically. Chapar Khaneh (courier-house) is the Persian-language term that refers to the postal service system used throughout the Achaemenid Empire. It was created by Cyrus the Great (600–530 BC) and later developed by Darius the Great (550–486 BC) as the royal method of communication throughout the empire. Each Chapar Khaneh was a station mainly located along the Royal Road, an ancient highway that was rebuilt by Darius the Great. The royal road network stretched from Ephesus and Sardis, in modern-day Türkiye, to Sindomana, in modern Pakistan, and from Cyrene in modern Lybia to Cyropolis in modern Tajikistan, connecting most of the major cities of the Empire to the administrative center of the Empire at Susa, in modern-day Iran. Cyropolis (the city of Cyrus) was an ancient city founded by Cyrus the Great to mark the northeastern border of his Empire.

The main path of the royal road was connecting nine greatest cities of the Persian empire: Pasargadae, the religious and royal center of Cyrus' Achaemenid Empire; Persepolis, the jewel in the Achaemenid crown; Susa, administrative center of the Empire; Ecbatana, the first conquest of the Empire and preferred summer residence of several Persian kings; Sardis, mint of the Achaemenid empire; Babylon, the oldest and most important city in Mesopotamia; Memphis, Persian capital of Egypt; Tyre, naval base of Persian Phoenicia; Miletus, where the first Greek philosopher, Thales (624/623–548/545 BC), was born.

**Fig. 3.2** A redrawing of the chariot on Darius Seal

Manufacturing and maintenance of chariots required a lot of innovations, improvements, and resources that could only be generated by the rich agricultural riverine civilizations. Large-scale deployment of chariots required skilled craftsmen for metallurgy, leather work, wood-working, and tanning and taming a large number of horses. To ride chariots at higher speed and for longer distance, besides flat and smooth pavements, they needed some sort of suspension system. By suspension system, we mean separating the wheels or axles from the chassis by a flexible material. Leather and ropes were the first tools to suspend axles before metallic spring appeared. The cockpit floor of the light chariots, carrying two persons, was manufactured with interwoven leather strips caused less vibrations compared to solid wood floors. These chariots had crossbar wheel that enabled the vehicles to move at 40 km/ h and to negotiate curves at speed of 20 km/ h.

An early form of suspension on ox-drawn carts had the platform swing on iron chains attached to the wheeled frame of the carriage. Leaf springs have also been used for suspension of chariot by Egyptians and Persians. This system remained the basis for all suspension systems until the turn of the nineteenth century. Automobiles were initially developed as self-propelled versions of horse-drawn vehicles. However, horse-drawn vehicles had been designed for relatively slow speeds, and their suspension was not well suited to the higher speeds permitted by the internal combustion engines. This fact made a proper suspension system much needed, which was made possible using a combination of springs and dampers.

Ancient military engineers used leaf springs in the form of bows to power their siege engines. The use of leaf springs in catapults was later refined and made to work years later. Springs were not only made of metal, and a sturdy tree branch could also be used as a spring, such as with a bow. Wooden leaf springs in one form or another have been used for vehicles since the Persians suspended two-wheeled vehicles on elastic wooden poles.

The first steel spring that successfully put on a vehicle was a single flat plate installed on carriages by the French in the eighteenth century. Horse -drawn carriages and the Ford Model *T* used this system. It is still being used today in larger vehicles, especially in the rear suspension. The first workable spring suspension required advanced metallurgical knowledge and skills and became possible only with the advent of industrialization. Obadiah Elliott (1763–1838), a British inventor, registered the first patent for a spring-suspension vehicle in 1804. Each wheel of the suspension had two durable steel leaf springs on each side, and the body of the carriage was fixed directly to the springs attached to the axles.

Within a decade, most British horse carriages were equipped with springs: wooden springs in the case of light one-horse vehicles to avoid taxation and steel springs in larger vehicles. These were often made of low-carbon steel and usually took the form of multiple layer leaf springs. The elliptic spring consisted of steel plates piled on top of one another and pinned together; it is the same method still used. Elliott's invention was a major breakthrough in carriage suspension design and it inspired a boom in the construction and sale of lightweight private carriages. This was the first modern suspension system and, along with advances in the construction of paved roads, heralded the single greatest improvement in road transport until the advent of the automobile. The first steel springs were not well suited for use on America's rough roads of the time, so the Abbot Downing Company re-introduced modified leather strap suspensions, which gave a swinging motion instead of the jolting up and down of a spring suspension. Abbot-Downing Company was a coach and carriage builder in Concord, New Hampshire, which became known throughout the United States for its products, in particular the first Concord stagecoach.

In 1906, Brush brothers, William and Alanson, two American designers and inventors, released a revolutionary suspension system with front coil springs and shock absorbers mounted on a flexible axle. This design was seen unusual, as most car manufacturers were still using leaf springs, which were cheaper and easier to be reshaped for different vehicles. The innovation was the combination of a coil spring and a shock absorber where both invented before. The first patent for coil springs was issued to Richard Tredwell in 1763 in Great Britain, although coil springs have been used for several centuries. Interestingly, the invention of leaf springs has been credited in 1804 to Obadiah Elliott (1763–1838), a British engineer. The first rubber mount shock absorber introduced by the British engineer, A. Gimmig in 1897, and the first hydraulic shock absorber has been credited to M. Houdaille (1880–1953) of France in 1909. An important step in suspension development was the invention of MacPherson strut in 1940 by Earle S. MacPherson (1891–1960), an American engineer of General Motors company.

In the first decades of the twentieth century and beginning of automotive industries, engineers found the lack of damping characteristic in suspensions. The amount of damping provided by leaves friction of leaf spring was limited and variable according to the conditions of the springs. The first application of frictional shock absorbers was fitting on a racing bicycle in 1898 by a French engineer, Jules Michel Marie Truffault. The Truffault-Hartford unit was not only the first automotive shock absorber, but also the first adjustable shock. Motorcycle front suspension also adopted coil sprung from 1906, and similar designs later added rotary friction dampers. Andre Hartford brought the car to America, where he opened his own plant, the Hartford Suspension Co., in New Jersey. Brush brothers installed Hartford shock absorbers along with front coil springs on the 1906 Brush Runabout. One of the earliest hydraulic dampers to go into production was the Telesco Shock Absorber, exhibited at the 1912 Olympia Motor Show. This design contained a spring inside the telescopic unit with oil and an internal valve so that the oil damped in the rebound direction. The Telesco unit was fitted at the rear end of the leaf spring, in place of the rear spring to chassis mount.

After the introduction of coil springs in suspension systems in 1910 by Brush brothers, some European car makers tried coil springs too, with Gottlieb Daimler (1834–1900) in Germany being the leading exponent. However, most manufacturers stood fast with leaf springs. They were less costly, and by simply adding leaves or changing the shape from full elliptic to three-quarter or half elliptic, the spring could be made to support varying weights. Ford's 1908 Model of T featured old-fashioned leaf springs with a novel twist. Henry Ford (1863–1947) used only one spring at each axle, mounted transversely, instead of one at each wheel. Ford's adaptation of high-strength vanadium steel from a French racing car allowed him to save weight and cut costs in many areas of the Model T without compromising its durability.

With the exception of a car here and there, independent coil spring front suspension remained in Limbo for 25 years after the introduction of the Brush Runabout. Then suddenly in 1934, General Motors, Chrysler, Hudson, and others re-introduced coil spring front suspension, this time with each wheel sprung independently. In that year, most cars started using hydraulic shock absorbers and balloon (low-pressure) tires. Not all cars used coil springs at first. Some had independently suspended leaf springs. But soon after World War II, most manufacturers switched to coil springs for the front wheels. Buick became the first U.S. car manufacturer to use back-end coil springs in 1938. Manufacturers have switched back and forth from model to model between leaf and coil springs since then. Generally, large, heavy cars are equipped with leaf springs, while small light cars have coil springs. Independent rear suspension became popular on the rough, twisty roads of Europe because it can offer improved ride and handling. The cheapest method is the swing axle, for which early Volkswagens were infamous. In 1920, Leyland Motors used torsion bars in a suspension system. In 1922, independent front suspension was pioneered on the Lancia Lambda and became more common in mass market cars from 1932. Today most cars have independent suspension on all four wheels.

## 3.2 Components of Suspension Systems

Suspension systems in vehicles are composed of various components that work together to provide a comfortable ride, ensure stability, and enhance vehicle handling. The key components of a suspension system include: Springs, Shock Absorbers/Dampers, Control Arms, Struts, Bushings, Sway Bars/Stabilizer Bars, Ball Joints, Strut Mounts and Bearings, Steering Linkages, Wheel Hubs and Bearings, Control-Arm Bushings, and Subframe. Considering vehicle vibrations, we use the concept of sprung mass suspended by spring and damper on unsprung mass.

### 3.2.1 Sprung and Unsprung Mass

The sprung mass, $m_s$, is the mass of the vehicle supported on the springs, and the unsprung mass, $m_u$, is loosely defined as the mass between the road and the suspension springs. The stiffness of the springs affects how the sprung mass responds, while the car is being driven. The mass ratio of sprung mass to unsprung mass, $m_s/m_u$, is an indicator of how ride can be comfort. Generally speaking, the higher $m_s/m_u$, the higher ride comfort can be achieved. The sprung cars with higher $m_s/m_u$ can absorb bumps better and provide a smoother ride; however, such a car is prone to dive and squat during braking and acceleration and tend to experience body sway or roll during cornering. Tightly sprung cars with lower $m_s/m_u$, such as sports cars, are less forgiving on bumpy roads, but they minimize body motion, which means they can be driven aggressively around corners. This trade-off was addressed by the British-American engineer, Maurice Olley (1889–1972), who called this the flat ride tuning of a suspension. Flat ride tuning will be discussed in later chapters. So, while springs by themselves are simple devices, designing and implementing them on a vehicle to balance passenger comfort with handling is a complex task. However, springs alone cannot provide a perfectly smooth ride. This is because although springs are great at absorbing energy, they are not so good at dissipating energy. Other structures, known as dampers, are required to do the energy dissipation.

### 3.2.2 Springs

At their most basic definition, springs are devices that store potential energy. Springs are mechanisms that have the capacity to absorb, store, and release potential energy through a change in shape. The most common use for a spring is to return a mechanism to its starting position and/or to add cushioning. While helical coiled springs are typically the most common form of mechanical springs, they may come in many shapes and materials, such as a wooden bow, as they used with arrows, or serpentine springs that are made of wire and used in furniture.

Robert Hooke (1635–1703) was a British physicist who stated that there is a proportional relationship between the force required to extend or compress a spring, and the distance that the spring is extended or compressed. This relationship is expressed by the following equation:

$$F = k\,x \tag{3.1}$$

where $F$ is the force generated by the spring after application of displacement $x$, and $k$ is the spring's stiffness. Hooke's law holds for most solid bodies, as long as the forces are small enough and large deformations do not occur. If something is described as Hookean, it means the proportional relationship (3.1) exists for that elastic material. Elastic materials are those that are able to spontaneously regain their normal or relaxed shape after force is applied.

### 3.2.3 Dampers

A vehicle spring will extend and release the potential energy it absorbs from a bump at an uncontrolled rate, unless a dampening structure is present. Dissipation of energy is done by converting into heat, and transferring to ambient. The practical way of converting energy into heat is by friction. The hydraulic shock absorbers make a fluid to move in long narrow tubes or passing through small orifices to make use of hydraulic friction to convert energy into heat. In other words, shock absorbers slow down and reduce the magnitude of vibratory motions by turning the kinetic energy of suspension movement into heat energy through hydraulic fluid.

A shock absorber is basically an oil pump placed between the frame of the vehicle and the wheels. The upper mount of the shock connects to the frame (i.e., the sprung mass), while the lower mount connects to the axle near the wheel (i.e., the unsprung mass). When the wheel encounters a bump in the road and causes the wheel to move up and down, the shock absorber will bound and rebound. A typical shock absorber will have more resistance during its extension cycle than its compression cycle. All hydraulic shock absorbers are velocity-sensitive that means, the faster the suspension moves, the more resistance the shock absorber provides. This positive behavior is because of nature of fluid resistance.

The shock absorber provides a resistance force, proportional to the relative velocity of its ends,

$$F = -cv = -c\dot{x} \tag{3.2}$$

where $c$ is the damping constant and $v$ is its ends' relative velocity. In metric system of units, $c$ is expressed in Newton-second per meter and $v$ is expressed in meters per second.

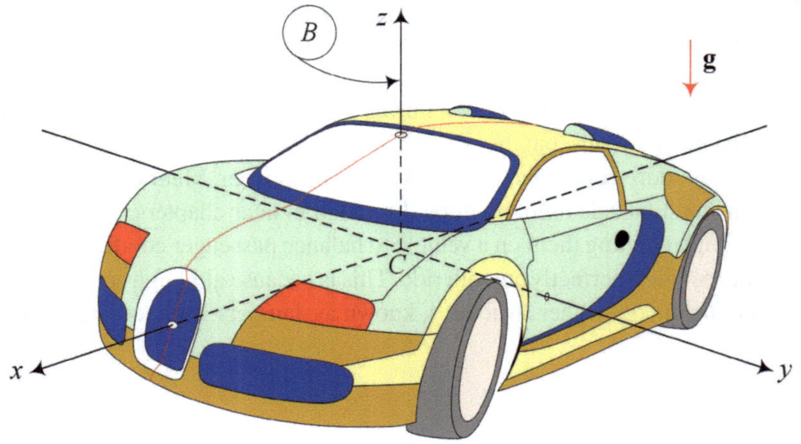


**Fig. 3.3** The vehicle body coordinate frame is a principal frame attached to the vehicle at mass center $C$

## 3.3 Suspension Systems

Considering a wheel and the body of a vehicle as rigid bodies, the wheel can theoretically have 6 degrees of freedom with respect to the body. The degrees of freedom indicate the type of possible relative motions. The wheel can have three displacements: up and down along the $z$-axis, front and back along the $x$-axis, and left and right along the $y$-axis. The wheel can also have three rotations: spin about the $y$-axis, camber about the $x$-axis, and steer about the $z$-axis. The axes $x$, $y$, $z$, are indicated by the coordinate frame $B(x, y, z)$, which is the vehicle coordinate frame attached to the vehicle at its mass center $C$, as suggested in Fig. 3.3.

Suspension mechanism is a combination of links and joints, connecting the wheels to the body such that there is some relative motion between the wheel and the body of the vehicle. If the wheel is steerable, then rotation about the $y$- and $z$-axes plus displacement along the $z$-axis must be allowed. If the wheel is not steerable, then only the spin about the $y$-axis and displacement along the $z$-axis must be allowed. The suspension mechanism is supposed to prevent all other possible motions.

### 3.3.1 Mechanisms

The four wheels of a passenger vehicle work together in two independent systems: the two wheels connected by the front axle and the two wheels connected by the rear axle. That means a car can and usually does have a different type of suspension on the front and back. Much is determined by whether a rigid axle binds the wheels or if the wheels are permitted to move independently. The former arrangement is known as a dependent suspension system, while the latter arrangement is known as an independent suspension system. In the following sections, we will look at some of the common types of front and back suspensions typically used on mainstream cars.

#### Dependent Suspension

Dependent suspensions have a rigid axle that connects the wheels and could be used in the front or rear of the vehicle. Basically, this looks like a solid bar under the car, kept in place by springs and shock absorbers. Front dependent axles are common on trucks but have not been used in mainstream cars for years. If a solid axle connects the rear wheels of a car, then the suspension is usually quite simple based on either leaf or coil springs. In the former design, the leaf springs clamp directly to the drive axle. The ends of the leaf springs attach directly to the frame, and the shock absorber is attached at the clamp that holds the spring to the axle. For many years, car manufacturers preferred this design because of its simplicity. The same basic design can be achieved with coil springs replacing the leaves. In this case, the spring and shock absorber can be mounted as a single unit or as separate components.

#### Solid Axle Suspension

The simplest way to attach a pair of wheels to a chassis is to mount them at two ends of a solid axle, such as the one shown in Fig. 3.4.

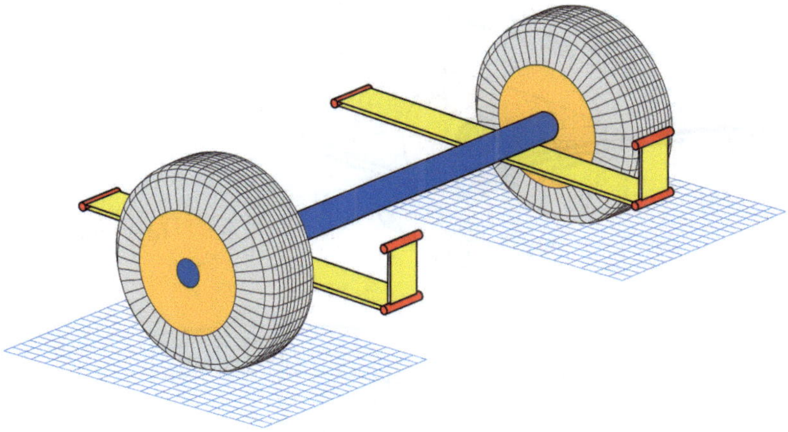

**Fig. 3.4** A solid axle with leaf springs

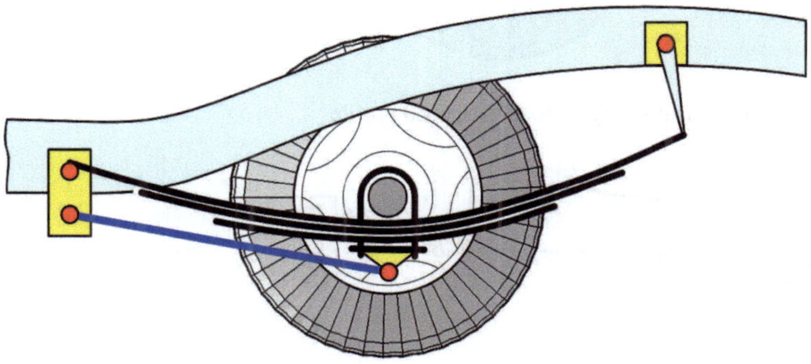

**Fig. 3.5** A side view of a multi-leaf spring and solid axle suspension

The solid axle must be attached to the body such that an up and down motion in the $z$-direction, as well as a roll rotation about the x-axis, is possible. Therefore, no forward and lateral translation, and also no rotation about the axle and the $z$-axis, is allowed. There are many combinations of links and springs that can provide the kinematic and dynamic requirements. The simplest design is to clamp the axle to the middle of two leaf springs with their ends tied and shackled to the chassis as shown schematically in Fig. 3.4. A side view of a multi-leaf spring and solid axle is shown in Fig. 3.5. A suspension with a solid axle between the left and right wheels is among the dependent suspension type.

The performance of a solid axle with leaf spring suspension can be improved by adding a linkage to guide the axle kinematically and provide dynamic support to carry the non-$z$-direction forces. The solid axle with leaf spring combination came to vehicle industry from horse-drawn vehicles. When a live solid axle is connected to the chassis with nothing but two leaf springs, it is called the Hotchkiss drive, which is the name of the car that used it first. Figure 3.5 illustrates a Hotchkiss suspension. The main problems of Hotchkiss drive are locating the axle under lateral and longitudinal forces and having a low mass ratio $\varepsilon = m_s/m_u$, where $m_s$ is the sprung mass and $m_u$ is the unsprung mass.

The solid axle suspension systems with longitudinal leaf springs have many drawbacks. The main problem lies in the fact that springs themselves should act as locating members. Springs are supposed to flex under load, but their flexibility is needed in only one direction. However, it is the nature of leaf springs to twist and bend laterally and, hence, flex and twist in planes other than the tireplane. Leaf springs are not suited for taking up the driving and braking traction forces. These forces tend to push the springs into an $S$-shaped profile, as shown in Fig. 3.6.

The driving and braking flexibility of leaf springs may generate a negative caster and increase instability. Long springs provide a better ride. However, long springs exaggerate their bending and twisting under different load conditions. To reduce the effect of a horizontal force and $S$-shaped profile appearance in a solid axle with leaf springs, the axle may be attached to the chassis by a longitudinal bar as Fig. 3.5 shows. Such a bar is called an *anti-tramp* bar, and the suspension is the simplest cure for longitudinal problems of a Hotchkiss drive. A solid axle with an anti-tramp bar may kinematically be approximated by a four-bar linkage, as shown in Fig. 3.1. Although an anti-tramp bar may control the shape of the leaf spring, it introduces

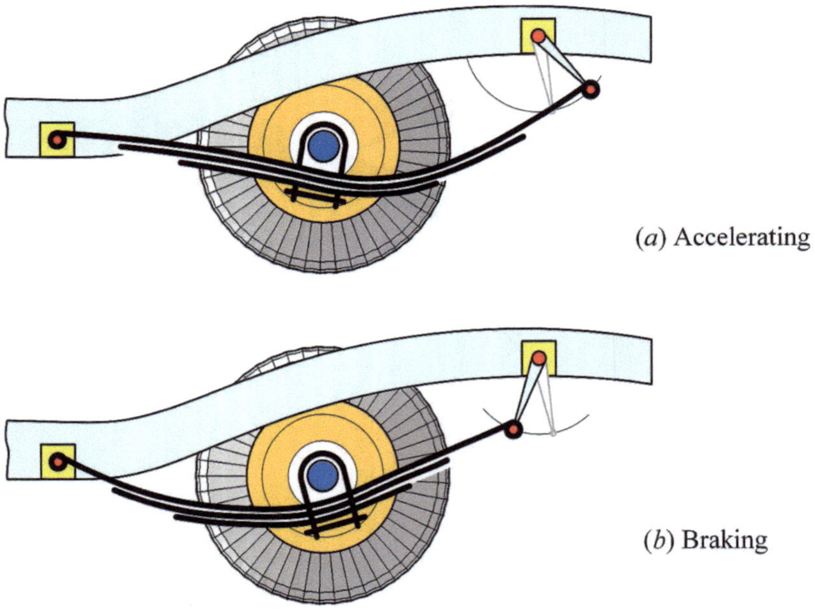


**Fig. 3.6** A driving and braking trust, force leaf springs into an $S$-shaped profile

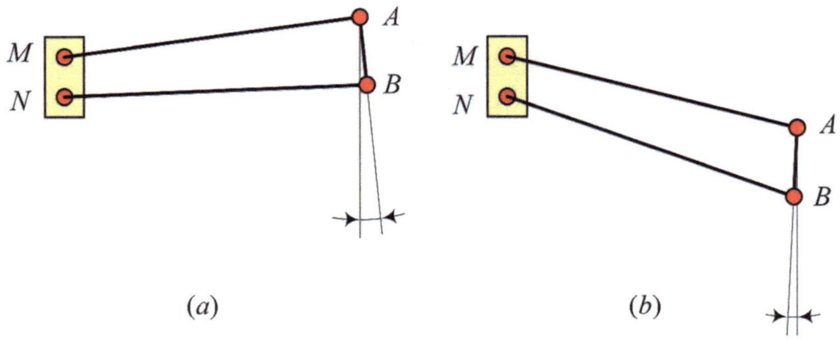


**Fig. 3.7** An anti-tramp bar introduces a twisting angle problem. (**a**) The wheel moves up and (**b**) the wheel moves down

a twisting angle problem when the axle is moving up and down, as shown in Fig. 3.7. Twisting the axle and the wheel about the axle is called caster. The solid axle is frequently used to help keeping the wheels perpendicular to the road.

A triangulated linkage, as shown in Fig. 3.8, may be attached to a solid axle to provide lateral resistance in turning maneuvers and twist resistance during acceleration and braking.

High spring rate is a problem of leaf springs. Reducing their stiffness by narrowing them and using fewer leaves reduces the lateral stiffness and affects the directional stability of the suspension. A Panhard arm is a bar that attaches a solid axle suspension to the chassis laterally. Figure 3.9 illustrates a solid axle and a Panhard arm to limit the lateral displacement of the axle. Figure 3.10 shows a triangular linkage and a Panhard arm combination for guiding a solid axle. A double triangle mechanism, as shown in Fig. 3.11, is an alternative design to guide the axle and support it laterally.

There are many mechanisms that provide a straight line motion to be used in suspension design. The simplest mechanisms are four-bar linkages with a coupler point moving straight. Some of the most applied and famous linkages are shown in Fig. 3.12. By having proper lengths, the Watt, Robert, Chebyshev, and Evance linkages can make the coupler point $C$ move on a straight line vertically. Such a mechanism and straight motion can be used to guide a solid axle.

Solid axles are usually heavier than independent suspensions for the same size vehicles. A solid axle is a part of unsprung members, and hence, the unsprung mass is increased where using solid axle suspensions. A heavy unsprung mass ruins the ride and handling quality of a vehicle. Lightening the solid axle makes it weaker and increases the most dangerous problem in vehicles: axle breakage. The solid axle must be strong enough to make sure it will not break under any loading conditions at any age. As a rough estimate, 90% of the leaf spring mass may also be counted as unsprung mass, which makes the problem

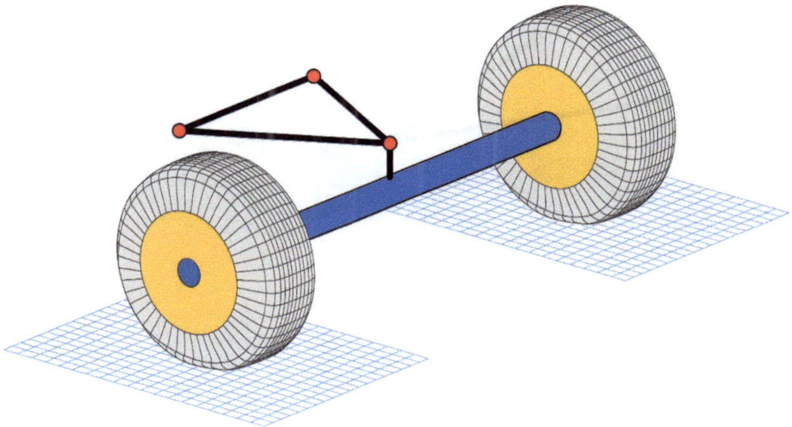

**Fig. 3.8** A solid axle suspension with a triangulated linkage

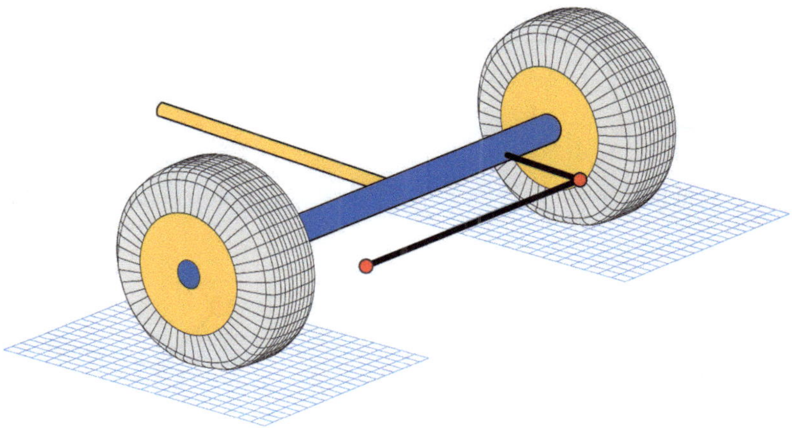

**Fig. 3.9** A solid axle and a Panhard arm to guide the axle

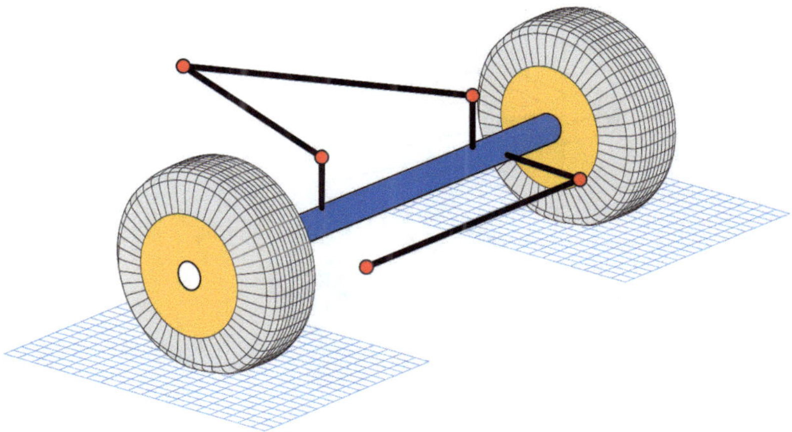

**Fig. 3.10** A triangle mechanism and a Panhard arm to guide a solid axle

worse. The unsprung mass problem is worse in front, and it is the main reason that solid axles are no longer being used in street cars. However, front solid axles are still common on trucks and buses. These are heavy vehicles, and solid axle suspension does not reduce the mass ratio much. When a vehicle is rear-wheel drive and a solid axle suspension is used in the back, the suspension is called live axle. A live axle is a casing that contains a differential, and two drive shafts. The drive shafts are connected to the wheel hubs. A live axle can be three to four times heavier than a dead $I$-beam axle. It is called live axle because of rotating gears and shafts inside the axle.

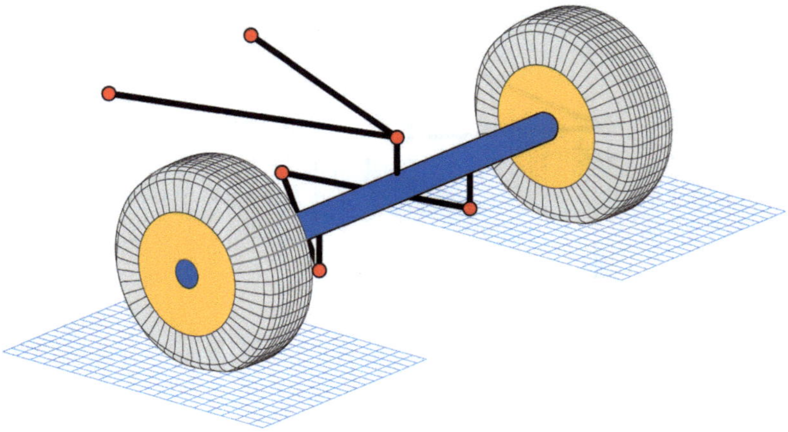

**Fig. 3.11** Double triangle suspension mechanism

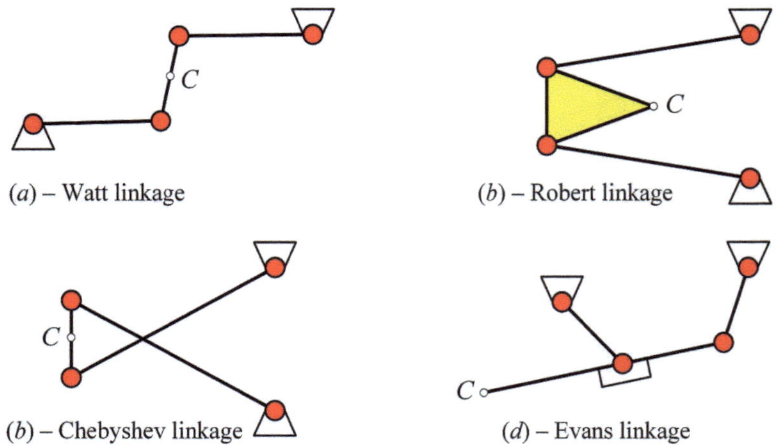


**Fig. 3.12** Some linkages with straight line motion

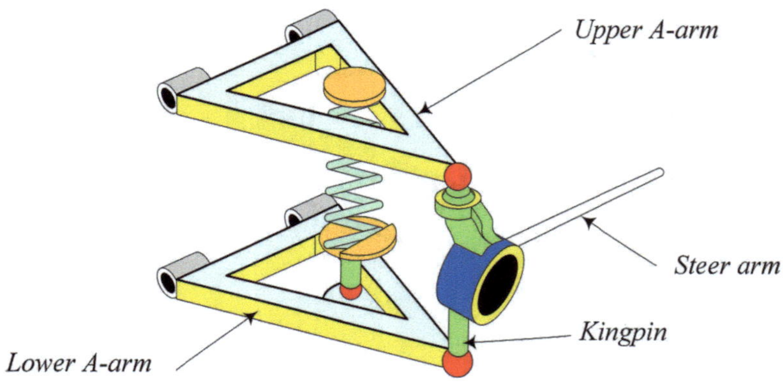


**Fig. 3.13** A double $A$-arm suspension

### Independent Suspension

In this setup, the front wheels are allowed to move independently. Independent suspensions are introduced to allow a wheel to move up and down without affecting the opposite wheel. There are many forms and designs of independent suspensions. However, double $A$-arm and MacPherson strut suspensions are the simplest and the most common designs.

Figure 3.13 illustrates a sample of a double $A$-arm, and Fig. 3.14 shows a MacPherson suspension. The double-wishbone suspension, also known as an $A$-arm suspension or control-arm suspension, is another common type of front independent suspension. While there are several different possible configurations, this design typically uses two wishbone-shaped arms

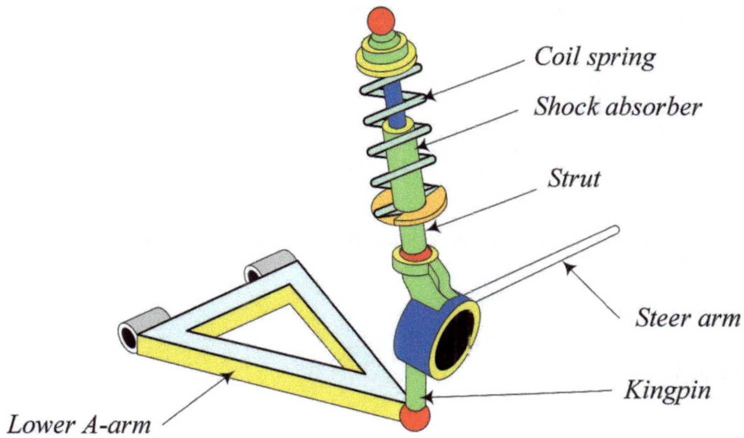


**Fig. 3.14** A McPherson suspension

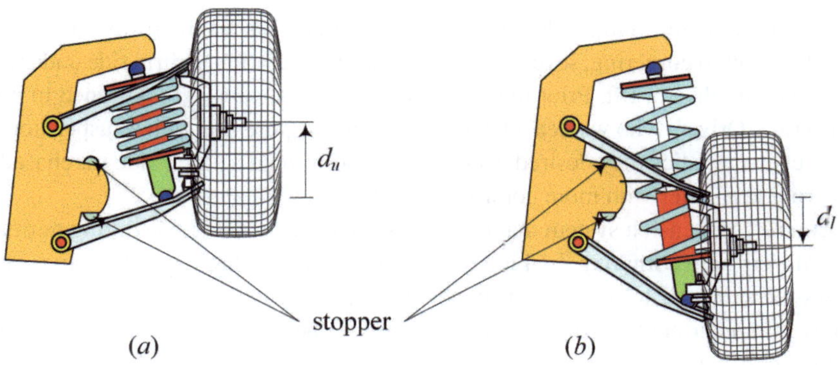


**Fig. 3.15** The stoppers for a double $A$-arm suspension

to locate the wheel. Each wishbone, which has two mounting positions to the frame and one at the wheel, bears a shock absorber and a coil spring to absorb vibrations. Double-wishbone suspensions allow for more control over the camber angle of the wheel, which describes the degree to which the wheels tilt in and out. They also help minimize roll or sway and provide for a more consistent steering feel. The MacPherson strut, which is developed by American automotive engineer Earle S. MacPherson (1891–1960) of General Motors in 1947, is the most widely used front suspension system. The MacPherson strut combines a shock absorber and a coil spring into a single unit. This provides a more compact and lighter suspension system that can be used for front-wheel drive vehicles. A MacPherson suspension is an inverted slider mechanism that has the chassis as the ground link and the coupler as the wheel carrying link. Kinematically, a double $A$-arm suspension mechanism is a four-bar linkage with the chassis as the ground link and coupler as the wheel carrying link.

**Independent Rear Suspension**

If both the front and back suspensions are independent, then all of the wheels are mounted and sprung individually. Any suspension that can be used on the front of the car can also be used on the rear. In the rear of the car, the steering rack, the assembly that includes the pinion gear wheel and enables the wheels to turn from side to side, is absent. This means that rear independent suspensions can be simplified versions of front ones, although the basic principles remain the same.

### 3.3.2 System Nonlinearities

Suspension travel is limited to certain maximum values of expansion and compression. Ideal suspension system would have infinite travel in both directions; however, practical concerns such as ground can chassis clearances provide a trade-off in the suspension system designs. At the limits of suspension travel, there are mechanisms to stop the suspension from traveling beyond limits. These could be metallic or hard rubber stoppers. They may be modeled as springs with very high stiffness $k_r$. Figure 3.15 illustrates the stoppers for a double $A$-arm suspension.

To include these stopper rates, the following nonlinear model of suspension spring force could be considered,

$$F_{k_s} = \begin{cases} k_s(x_s - x_u) & d_u < x_s - x_u < d_l \\ k_s d_l + k_{rl}(x_s - x_u - d_l) & x_s - x_u > d_l \\ k_s d_u + k_{ru}(x_s - x_u - d_u) & x_s - x_u < d_u \end{cases} \tag{3.3}$$

where $d_l$ and $d_u$ are expansion and compression limits, respectively. It is customary to use a symmetric saturation ($d_l = d_u$ and $k_{rl} = k_{ru} = k_r$).

### 3.3.3 Controllability Level

Suspension systems are categorized as passive, active, and semi-active considering their level of controllability. Although all the types of the suspension systems have different advantages and disadvantages, all of them utilize the spring and damper units.

Many drivers enjoy a soft and comfortable ride while driving normally on the highway. However, many of these same drivers prefer a firm ride during hard cornering, severe braking, or fast acceleration. A firm ride under these driving conditions reduces body sway and front-end dive or lift. Prior to the age of electronics, cars were designed to provide either a soft and comfortable ride or a firm ride. Drivers who wanted a firm ride selected a sports car with a suspension designed to supply the type of ride and handling characteristics they desired. Car buyers who wanted a softer ride purchased a family sedan with a suspension designed to provide a softer with more comfortable ride.

Thanks to computer control, suspension system manufacturers can provide a soft ride during normal highway driving and then switch to a firm ride during hard cornering, braking, fast acceleration, and high-speed driving. The computer-controlled suspension system allows the same car to meet the demands of both the driver who desires a soft ride and the driver who wants a firm ride. Because computer-controlled suspension systems reduce body sway during hard cornering, these systems provide improved steering control as well.

Some computer-controlled suspension systems also supply a constant vehicle riding height regardless of the vehicle passenger or cargo load. This action maintains the vehicle's cosmetic appearance as the passenger and/or cargo load is changed. Maintaining a constant riding height also supplies more constant suspension alignment angles, which may provide improved steering control. Ideally, a control action may be introduced at different levels of the suspension system: at the level of the dissipative unit, by a modulation of the damping force; at the level of the elastic unit, by a modulation of the spring force; at the full level of the suspension, by replacing both the elastic and the damping devices with a force actuator.

The classification of controllable suspension may be carried out according to the energy input and the bandwidth of the actuator. More specifically, three features may be observed: the controllability range, in other words the range of forces that the actuators deliver; the control bandwidth that is a measure of how fast the actuator acts; the power request that is mainly due to the mix of controllability range and control bandwidth. Below are the five families of ideal controllable suspensions:

- Adaptive Suspension: The control action is represented by a relatively slow modulation of damping, so the control range is limited by the passivity constraint. The adaptive shock absorber is characterized by a bandwidth of a few Hertz. Since no energy is introduced in the system and the bandwidth is relatively small, the power request is usually limited to a few Watts.
- Semi-active suspension: This system features an electronic shock absorber that may vary the damping with a relatively large bandwidth (usually around 30–40 Hz). The deliverable forces follow the passivity constraint of the damper, and thus no energy may be introduced into the system. Due to these features, the requested power is relatively low, around tens of Watts.
- Load-leveling suspension: This class of suspensions may be considered the first attempt at active suspensions, since they are capable of introducing energy in the system to change the steady-state condition as a response of a variation of the static load. The control acts on the parameter of the springs (usually an air spring). The bandwidth is usually within 0.1–1 Hz, but the power request is usually some hundreds of Watts.
- Slow-active suspension: In active suspensions, the passivity constraint is fully overcome and energy may be injected into the system. The control input is the suspension force $F$ delivered by an actuator that replaces the passive devices of the

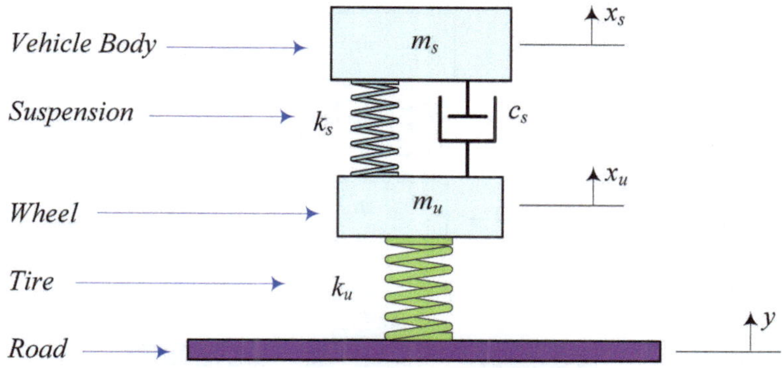


**Fig. 3.16** Passive suspension system

suspension. The bandwidth is limited to a few Hertz. Note that the vast controllability range is paid in terms of power request (around some kilo-Watts).

- Fully active suspension: The difference between slow- and fully active suspensions is in terms of bandwidth. The fully active actuator is able to react in milliseconds (bandwidth of 20–30 Hz). As in slow-active suspension systems, the control variable is the suspension force $F$. Interestingly enough, the available bandwidth is the same as semi-active suspensions. However, since the controllability range is beyond the passivity constraint, the overall power request is relatively high-demanding, around tens of kilo-Watts.

Considering the practicality and availability of suspensions systems, they could be roughly categorized as passive, active, and semi-active considering their level of controllability. Although all the types of the suspension systems have different advantages and disadvantages, all of them utilize the spring and damper units.

### Passive Suspension Systems

Passive suspension systems are composed of conventional springs and oil dampers with constant damping properties. Figure 3.16 illustrates passive suspension model. In this model, $m_u$ and $m_s$ represent the unsprung mass and sprung mass, respectively, $k_u$ is the tire stiffness coefficient or tire spring constant, and $k_s$ is the suspension stiffness or suspension spring constant. $c_s$ is the suspension damping constant. Damping of tires is much smaller than $c_s$ and can be ignored in calculations. $y$, $x_u$, and $x_s$ represent road profile input, displacement of unsprung mass, and displacement of sprung mass, respectively.

In most instances, passive suspension systems are less complex, more Reliable, and less costly compared to active or semi-active suspension systems. The constant damping characteristic is the main disadvantage of passive suspension systems. For a passive suspension, the use of soft springs and moderate to low damping rates is needed, but the use of stiff springs and high damping rates is needed to reduce the effects of dynamic forces. Designers utilize soft springs and a damper with low damping rate for applications that need a smooth and comfortable ride such as in a luxury automobile. On the other hand, sports cars incorporate stiff springs and a damper with high damping rates to gain greater stability and control at the expense of comfort. Therefore, the performance in each area is limited to the two opposing goals. There is always a compensation need to be made between ride comfort and ride handling in the passive suspension system as spring and damper characteristics cannot be changed according to the road profile.

### Semi-active Suspension Systems

Semi-active automotive suspension systems are systems that can adjust the damping of the suspension in real-time to provide improved handling and ride comfort. Unlike passive suspension systems, which have fixed damping characteristics, semi-active suspension systems can be adjusted in real-time based on driving conditions and passenger comfort preferences.

The components of a semi-active automotive suspension system include:

- Sensors: Sensors are used to measure the vehicle's body acceleration, road roughness, and suspension deflection.
- Actuators: Actuators are used to adjust the damping of the suspension based on the control input.
- Control algorithm: A control algorithm is used to determine the appropriate damping setting for the suspension in real-time based on the sensor data and the desired suspension behavior.

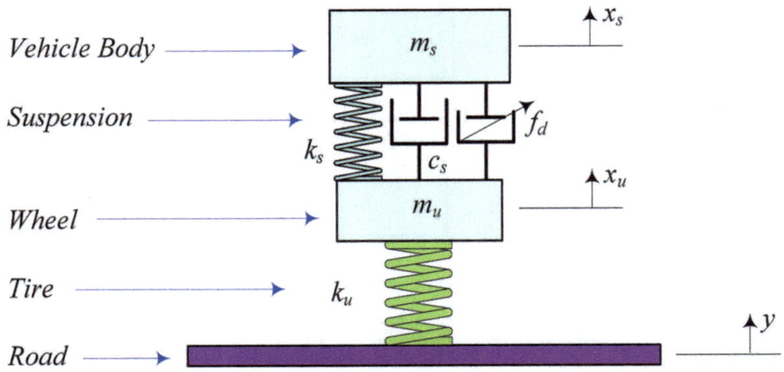


**Fig. 3.17** A semi-active suspension model

Figure 3.17 shows a semi-active suspension model. Here the component $f_d$ can generate an active actuating force by an intelligent controller. Since then, semi-active suspension systems have continued to acquire popularity in vehicular suspension system applications, due to their better performance and advantageous characteristics over passive suspension systems. In semi-active suspension systems, the damping properties of the damper can be changed and adjusted. The adjustable damping characteristics are achieved through a variety of technologies, such as electrorheological ($ER$) and magnetorheological ($MR$) fluids, solenoid valves, and piezoelectric actuators. As it is safe, economical, and does not need a large power supply, the semi-active suspension has been adopted for use in high-performance automobiles. $MR$ degradation with time, sealing problems, and temperature sensitivity are some crucial issues of the $MR$ dampers that need development.

From a vibration engineering viewpoint, the suspension is a mechanical low-pass filter that attenuates the effects of a disturbance from irregular road profile on an output variable. The output variable is the body acceleration when comfort is the main objective; the tire deflection when the design goal is road holding. These two objectives are somehow conflicting: the tuning and the design of a mechanical suspension try to find the best compromise between these two goals.

This critical trade-off is worsened by the fact that a suspension has a limited wheel travel; when the end stop of a suspension is reached, both the comfort and road-holding performances are dramatically deteriorated, and the occurrence of this situation must be avoided. The main job of the design problem of a classical mechanical suspension consists in the definition of a spring stiffness and a damping ratio in order to deliver a good compromise between comfort and handling with an additional bound on the suspension travel. Given such a tricky set of trade-offs, it is not surprising that when the early age of car manufacturing passed, suspension designers began to look for possible ways to reduce the problem of compromising between opposite goals.

The best compromise of cost (component cost, weight, electronics and sensors, power consumption, etc.) and performance (comfort, handling, safety) lay in another technology of electronically controllable suspensions, namely, the variable-damping suspensions or, in brief, the semi-active suspensions.

Electronically controlled suspensions can be classified according to two main features:

- Energy input: When energy is added into the suspension system, the suspension is classified as active; however, when the suspension is electronically modified without energy insertion, apart from a small amount of energy used to drive the electronically controlled element, the suspension is semi-active. Roughly speaking, a suspension is active when it can lift the vehicle, semi-active otherwise.
- Bandwidth: The electronically controlled element of the suspension can be modified with a specific reaction time; this feature strongly characterizes the suspension system, since it inherently defines the maximum achievable bandwidth of the corresponding closed-loop control system.

According to the above features, there are two main classes of semi-active suspension systems:

- Adaptive suspensions: Suspension with slowly modified damping ratio; typically, this modification is made with an open-loop architecture.
- Semi-active suspensions: Suspensions with a damping ratio modified in a closed-loop configuration over a large bandwidth.

Semi-active suspensions are mix of appealing features; among others, the most interesting are the following: *Negligible power demand*, since they are based on the regulation of the damping ratio only, the power absorption is limited to a few Watts required to modify the hydraulic orifices or the fluid viscosity. *Safety* in a semi-active suspension is always guaranteed by the fact that the whole system remains dissipative, whatever the damping ratio is. *Low cost, low weight*, the main damping–modulation technologies, such as electrohydraulic, magnetorheological, electrorheological, air damping, can be produced for large volumes at low cost and with compact packaging. *Significant impact on the vehicle performance* by changing the damping ratio of a suspension, the overall comfort and road-holding performance can be significantly modified.

Changing the damping ratio represents an interesting opportunity for a suspension designer; however, the selection of the best damping ratio is not an easy task, even in the simple case when the damping ratio is not subject to fast switching. The task becomes extremely challenging when the suspension designer has the opportunity to change, possibly with a feedback control scheme, using vehicle dynamics sensors such as accelerometers and potentiometers, the damping ratio every few milliseconds. In this case, the real key problem becomes the control algorithm design problem.

The control algorithms used in semi-active suspension systems can be broadly categorized as passive, semi-active, and active. Some common control algorithms used in semi-active suspension systems include:

- Skyhook control algorithm: This control algorithm adjusts the damping of the suspension based on the vehicle's body acceleration, providing a more stable and comfortable ride.
- Groundhook control algorithm: This control algorithm adjusts the damping of the suspension based on the road roughness, providing improved handling and ride comfort on rough roads.
- Fuzzy logic control algorithm: This control algorithm uses a fuzzy logic system to determine the appropriate damping setting for the suspension based on the sensor data and desired suspension behavior.

In semi-active suspension systems, the control algorithm adjusts the damping of the suspension based on driving conditions and passenger comfort preferences. This provides improved handling and ride comfort compared to passive suspension systems, while still allowing the suspension to be adjusted in real-time.

An example of a semi-active automotive suspension system using the Skyhook control algorithm is as follows:

- System Model: A mathematical model of the vehicle's suspension system is developed, which takes into account the relationship between the suspension, wheels, and road, as well as the effect of road roughness on the suspension.
- Sensors: Sensors are used to measure the vehicle's body acceleration, road roughness, and suspension deflection.
- Actuators: Actuators are used to adjust the damping of the suspension in real-time based on the control input.
- Control Algorithm: The Skyhook control algorithm is used to determine the appropriate damping setting for the suspension based on the sensor data and the desired suspension behavior. The algorithm adjusts the damping of the suspension based on the vehicle's body acceleration, providing a more stable and comfortable ride.
- Implementation: The control algorithm is implemented in real-time using a microcontroller, which receives sensor data and sends control signals to the actuators.

In this example, the Skyhook control algorithm adjusts the damping of the suspension based on the vehicle's body acceleration, providing improved handling and ride comfort. The algorithm takes into account factors such as the vehicle's speed, body acceleration, and road roughness to provide an optimal suspension setting.

When the vehicle encounters a bump or rough road, the sensors measure the body acceleration and send these data to the microcontroller. The microcontroller then calculates the appropriate damping setting based on the Skyhook control algorithm and sends a signal to the actuators to adjust the damping of the suspension.

An example of a semi-active automotive suspension system using the Groundhook control algorithm is as follows:

- System Model: A mathematical model of the vehicle's suspension system is developed, which takes into account the relationship between the suspension, wheels, and road, as well as the effect of road roughness on the suspension.
- Sensors: Sensors are used to measure the vehicle's body acceleration, road roughness, and suspension deflection.
- Actuators: Actuators are used to adjust the damping of the suspension in real-time based on the control input.
- Control Algorithm: The Groundhook control algorithm is used to determine the appropriate damping setting for the suspension in real-time based on the sensor data and the desired suspension behavior. The algorithm adjusts the damping of the suspension based on the road roughness, providing improved handling and ride comfort on rough roads.

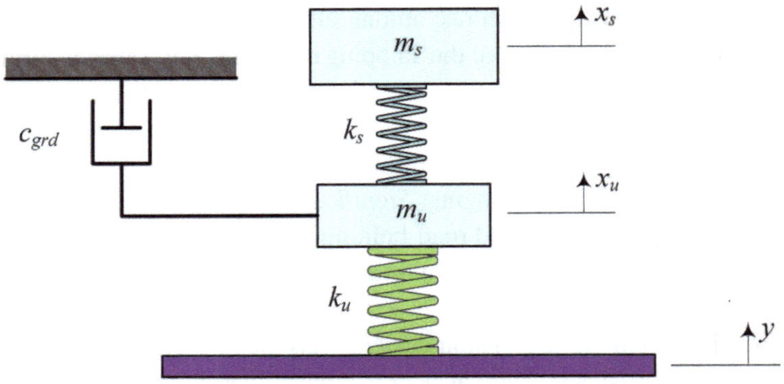


**Fig. 3.18** Ideal Groundhook control

- Implementation: The control algorithm is implemented using a microcontroller, which receives sensor data and sends control signals to the actuators.

The mechanical model for the use of the Groundhook model is shown in Fig. 3.18 on a quarter car model. It adopts a hypothetical damper, $C_{grd}$, connected between the unsprung mass and the fixed fictitious frame on the ground. Theoretically, Groundhook control will improve the responses of the unsprung system.

The Groundhook control is given by

$$if \quad -\dot{x}_u\,\dot{z} \geq 0 \quad then \quad f = c_{gnd}\,\dot{x}_u \tag{3.4}$$

$$if \quad -\dot{x}_u\,\dot{z} < 0 \quad then \quad f = 0 \tag{3.5}$$

$$z = x_s - x_u \tag{3.6}$$

where $\dot{x}_u$ is the unsprung mass velocity, $\dot{z}$ is the relative velocity between sprung mass and unsprung mass, $f$ is the semi-active suspension force, and $c_{gnd}$ is the Groundhook damping coefficient. The forces generated by the semi-active controller are added to the passive vehicle model.

An example of a semi-active automotive suspension system using the fuzzy logic control algorithm is as follows:

- System Model: A mathematical model of the vehicle's suspension system is developed, which takes into account the relationship between the suspension, wheels, and road, as well as the effect of road roughness on the suspension.
- Sensors: Sensors are used to measure the vehicle's body acceleration, road roughness, and suspension deflection.
- Actuators: Actuators are used to adjust the damping of the suspension in real-time based on the control input.
- Control Algorithm: The fuzzy logic control algorithm is used to determine the appropriate damping setting for the suspension in real-time based on the sensor data and the desired suspension behavior. The algorithm adjusts the damping of the suspension based on the vehicle's body acceleration and road roughness, using a set of rules to provide an optimal suspension setting.
- Implementation: The control algorithm is implemented in real-time using a microcontroller, which receives sensor data and sends control signals to the actuators.

The fuzzy controller operates similar to a feedforward control system by determining the ideal operating point for the shock absorbers based on the actions of the vehicle driver. Its knowledge base is determined by the relationship between the level of damping and various driving conditions, as gathered by a set of sensors placed in the vehicle. The inputs for the control system include the position of the steering wheel, rotation speed of the steering wheel, vertical acceleration of the vehicle, degree of longitudinal acceleration (such as accelerator pedal position or throttle valve position), speed of the vehicle, and brake contact.

The damping level is calculated on a continuous scale to achieve optimal performance from the semi-active damping system. This system is relatively robust, even if the precision and repeatability of fixing the shock absorber's damping level are not always high. The primary control objective is comfort, but safety is given priority over comfort if it is deemed to be

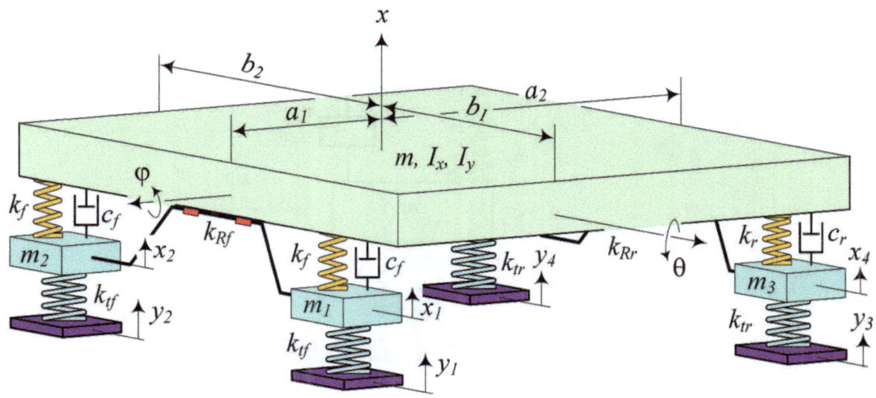


**Fig. 3.19** 7 degrees of freedom vibration model

severely impacted. In such situations, the shock absorbers are set to operate at their hardest setting, prioritizing safety over comfort during sudden maneuvers.

**Example 10** ★ How the fuzzy logic contributes in vehicle vibrations.

For an example of a system using fuzzy logic controller, a 7 degree of freedom model is used, which takes into account the mutual relationships between the vibrations transmitted through the various sections of the vehicle. Additionally, it provides information about the roll and pitch angles, which are crucial in determining driving comfort. The model is depicted in Fig. 3.19. To facilitate a comprehensive understanding of the model, the figure includes a thorough explanation of the unsprung masses ($m_1$, $m_2$, $m_3$, and $m_4$) and their corresponding displacements. The dynamic equation for the suspension model can be written as follows:

$$m\,\ddot{x} = -[F_{1S} + F_{2S} + F_{3S} + F_{4S} + F_{1D} + F_{2D} + F_{3D} + F_{4D}] \tag{3.7}$$

where $m$ is the vehicle body mass, $\ddot{x}$ acceleration of the center of vehicle mass, and $F_{1..4S}$, $F_{1..4D}$ are the forces generated by springs and dampers, respectively. The forces expressed in Eq. (3.7) can be defined as follows:

$$F_{iS} = K_{iS}(x_{is} - x_{iu}) \tag{3.8}$$

$$F_{iD} = c_{iD}(\dot{x}_{is} - \dot{x}_{iu}) \tag{3.9}$$

where $i$ is the wheel index, $k_{iS}$, $c_{iD}$ spring coefficient and suspension damper coefficient, respectively, $x_{is}$, $x_{iu}$ are the sprung and unsprung mass displacements, respectively, and $\dot{x}_{is}$, $\dot{x}_{iu}$ are the sprung and unsprung masses velocities, respectively. The dynamic equation for the unsprung mass acceleration is given as

$$m_i\ddot{x}_{iu} = F_{iS} + F_{iD} - k_{iT}(x_{iu} - y_i) \tag{3.10}$$

where $m_i$ the unsprung mass, $k_{iT}$ tire stiffness coefficient, and $y_i$ is the road profile excitation. Torque equations for the roll $\varphi$ and pitch $\theta$ are defined as

$$\begin{aligned} I_{FR}\ddot{\phi} &= r[F_{2S} + F_{3S} + F_{2D} + F_{3D}]\cos\varphi \\ &\quad -l[F_{1S} + F_{4S} + F_{1D} + F_{4D}]\cos\varphi \end{aligned} \tag{3.11}$$

$$\begin{aligned} I_{LR}\ddot{\theta} &= f[F_{1S} + F_{2S} + F_{1D} + F_{2D}]\cos\theta \\ &\quad -b[F_{3S} + F_{4S} + F_{3D} + F_{4D}]\cos\theta \end{aligned} \tag{3.12}$$

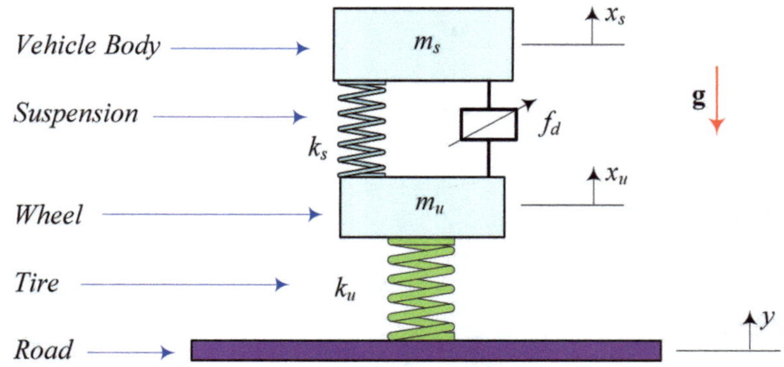


**Fig. 3.20** Active suspension systems

where $J_{FR}$, $J_{LR}$ are the mass moment of the front–rear axle and the left right axis, respectively, and $f, b, l, r$ is the length of chassis from mass center to the front, rear, left, and right corners, respectively.

Fuzzy controller is a linguistic base-oriented system, which allows for synthesis of the control signal on the basis of an expert knowledge. The expert knowledge is stored in the form of IF-THEN rules. In our example model here, the Takagi–Sugeno inference system is used, with triangle membership functions defined as $N$, $Z$, $P$ on the input set $[-1, 1]$ corresponding to negative, zero, and positive linguistic fuzzy inputs description.

## Active Suspension Systems

Active automotive suspension control refers to a system in vehicles that monitors and adjusts the suspension in real-time to provide improved handling and ride comfort. The active suspension system uses sensors, actuators, and control algorithms to continuously adjust the suspension in response to driving conditions and passenger comfort preferences.

An active suspension system typically consists of several components:

- Sensors: These are used to measure various parameters such as vehicle body acceleration, road roughness, and suspension deflection.
- Actuators: These are used to control the suspension, such as hydraulic or electromechanical actuators.
- Control algorithms: These are used to process the sensor data and determine the appropriate suspension settings.

One of the key benefits of active suspension control is improved handling. By continuously adjusting the suspension, the vehicle's body roll, pitch, and yaw can be controlled, providing a more stable and controlled ride. Additionally, the system can provide improved ride comfort by reducing road-induced vibrations and minimizing the impact of rough road surfaces.

The active suspension system, as illustrated in Fig. 3.20, actuates the suspension system by extending or contracting the sprung and unsprung masses through an active power source as required. Automotive suspension designs have been a compromise between the three contradictory criteria of road handling, suspension travel, and passengers' comfort. The use of active suspension systems has allowed car manufacturers to achieve all three desired criteria independently. A similar approach has also been used in train bogies to improve the curving behavior of the trains and decrease the acceleration perceived by passengers. The essential function of the vehicle suspension is to connect the vehicle body with the wheels. Thereby it is possible to carry the body along the drive way and to transmit forces in the horizontal plane. The suspension gives the wheel a primary vertically aligned movement possibility. As a result, the wheel follows a route with uneven road surfaces. By using spring and damping elements, the resulting body movements are reduced, and driving safety and comfort are ensured.

Furthermore, the vehicle suspension influences the position of the wheel relative to the road by its geometry and the spring and damping rate. This allows a systematic influence on the dynamic driving characteristics of the vehicle. The adjustment of these characteristics takes up a compromise, because the requirements of a good driving behavior and a high comfort are most time inconsistent with one another. Therefore, in designing the control law for a suspension system, usually we need to take the following aspects into consideration:

- Ride comfort: It is an important performance for vehicle design, which is usually evaluated by the body acceleration in the vertical, longitudinal, and lateral directions.

- Road-holding ability: In order to ensure a firm uninterrupted contact of wheels to road, the dynamic tire load should not exceed the static ones.
- Maximum suspension deflection: Because of the constraint of mechanical structure, the maximum allowable suspension strokes have to be taken into consideration to prevent excessive suspension bottoming, which result in deterioration of ride comfort and even structural damage.
- Saturation effect of the actuator: In view of the limited power of the actuator, the control force for the suspension system should be confined to a certain range.
- Reliability of closed-loop systems: The closed-loop systems should be reliable when meeting with non-ideal situations caused by actuators, such as the problems of actuator input delay, sampled data, and fault accommodation for unknown actuator failures.

A typical active suspension system includes the following sensors and actuators:

- Steering sensor: The steering sensor is mounted on the steering column. This sensor contains a pair of light emitting diodes (LEDs) and a matching pair of photo diodes. A slotted disc attached to the steering shaft rotates between the LEDs and photo diodes when the steering wheel is turned. A signal is sent from the steering sensor to the control module in relation to the amount and speed of steering wheel rotation.
- Brake Sensor: The brake sensor is a normally open (NO) switch mounted in the brake control valve assembly. When the brake fluid pressure reaches 400 pounds per square inch ($psi$) or 2758 kilopascals ( kPa), the brake pressure switch closes and sends a signal to the control module.
- Speed sensor: The vehicle speed sensor is usually mounted in the speedometer cable outlet of the transaxle or transmission. This sensor sends a vehicle speed signal to the control module.
- Struts and shocks with electric actuators: An actuator is positioned in the top of each strut and shock. Each actuator contains a single pole armature, a pair of permanent magnets, and a position switch. When current is applied through the plush relay to the armature, the magnetic fields of the armature and the permanent magnets repel each other. This action causes clockwise armature rotation until the armature hits the internal stop. Under this condition, the leaf spring switch is open in the position sensor circuit, and no signal is returned to the control module. If current is applied through the firm relay to the armature, there is an attraction between the magnetic fields of the armature and permanent magnets. This attraction causes counterclockwise armature rotation until the armature contacts the internal stop. The armature rotates an internal strut or shock valve to restrict oil movement and provide increased suspension damping. In the firm position, the leaf spring switch closes and sends a feedback signal to the control module. The armature movement is 60 Å, and armature response time is 30 milliseconds ( ms).

There are several different types of active suspension control systems, including hydraulic, electromechanical, and pneumatic systems. Hydraulic systems use a hydraulic actuator to control the suspension, while electromechanical systems use an electric motor and gear mechanism. Pneumatic systems use air pressure to control the suspension.

One of the key factors in the design of active suspension control systems is the control algorithm. The control algorithm determines how the suspension will be adjusted in response to various driving conditions. There are several different control algorithms that can be used, including proportional–integral–derivative ($PID$) control, linear quadratic regulator ($LQR$) control, and model predictive control ($MPC$).

An example of an active suspension control algorithm using linear quadratic regulator control is one that adjusts the suspension of a vehicle in real-time to provide improved handling and ride comfort. The $LQR$ algorithm solves a quadratic optimization problem to determine the optimal control input for the suspension system.

The $LQR$ algorithm for active suspension control typically includes the following steps:

- Measurement of the vehicle's body acceleration, road roughness, and suspension deflection using sensors.
- System model: The $LQR$ algorithm uses a mathematical model of the vehicle's suspension system to predict its behavior based on the current state and control inputs. The model takes into account the relationship between the suspension, wheels, and road, as well as the effect of road roughness on the suspension.
- Cost function: The $LQR$ algorithm minimizes a cost function that represents a trade-off between the deviation of the suspension from the desired state and the control effort required to achieve this deviation. The cost function takes into account factors such as the deviation of the suspension from the desired state, the control effort required to achieve this deviation, and the smoothness of the suspension motion.

- Optimization: The $LQR$ algorithm solves the quadratic optimization problem to determine the control input that minimizes the cost function and determines the appropriate suspension settings.
- Control action: The $LQR$ algorithm sends the control input to the actuators, which adjust the suspension in real-time to provide improved handling and ride comfort.

The $LQR$ algorithm uses a combination of sensor data and a predictive model to continuously adjust the suspension in real-time, providing a more stable and controlled ride. The algorithm takes into account factors such as the vehicle's body acceleration, road roughness, and suspension deflection to provide an optimal suspension setting.

Overall, the $LQR$ algorithm for active suspension control provides improved handling and ride comfort by continuously adjusting the suspension in response to driving conditions and passenger comfort preferences. This type of algorithm is widely used in modern vehicles, as it provides a sophisticated and effective solution for active suspension control.

The $MPC$ algorithm predicts the future behavior of the suspension based on the current state and control inputs and calculates the control inputs that minimize a cost function representing the deviation from the desired state and the control effort required to achieve this deviation. The $MPC$ algorithm calculates the control inputs that minimize a cost function representing the deviation from the desired state and the control effort required to achieve this deviation.

The $MPC$ algorithm for active suspension control typically includes the following steps:

- Measurement of the vehicle's body acceleration, road roughness, and suspension deflection using sensors.
- Predictive model: The $MPC$ algorithm uses a mathematical model of the vehicle's suspension system to predict its behavior based on the current state and control inputs. The model takes into account the relationship between the suspension, wheels, and road, as well as the effect of road roughness on the suspension.
- Cost function: The $MPC$ algorithm minimizes a cost function that represents a trade-off between the deviation of the suspension from the desired state and the control effort required to achieve this deviation. The cost function takes into account factors such as the deviation of the suspension from the desired state, the control effort required to achieve this deviation, and the smoothness of the suspension motion.
- Optimization: The $MPC$ algorithm calculates the control inputs that minimize the cost function and determine the appropriate suspension settings.
- Control action: The $MPC$ algorithm sends the control inputs to the actuators, which adjust the suspension to provide improved handling and ride comfort.

In this method, the $MPC$ algorithm uses a combination of sensor data and a predictive model to continuously adjust the suspension, providing a more stable and controlled ride. The algorithm takes into account factors such as the vehicle's body acceleration, road roughness, and suspension deflection to provide an optimal suspension setting.

Overall, the $MPC$ algorithm for active suspension control provides improved handling and ride comfort by continuously adjusting the suspension in response to driving conditions and passenger comfort preferences. This type of algorithm is becoming increasingly popular in modern vehicles, as it provides a more sophisticated and effective solution for active suspension control compared to traditional control algorithms.

$MPC$ finds an optimal control sequence by minimizing a performance index over the feasible domain. Once the optimal control sequence is computed, the first element of the sequence is applied to the actual plant and the process is repeated. One of the important features of $MPC$ is its ability to incorporate any constraints on the process variables explicitly in the optimization problem. Various formulations of $MPC$ differ in the form of the internal model used as well as the optimization scheme. Assume that the process model can be represented by

$$x(t+1) = Ax(t) + Bu(t) \quad y(t) = Cx(t) \tag{3.13}$$

At each sample time, a set of $M$ control moves $u(t)$, $u(t+1)$, ..., $u(t+M-1)$ are computed by solving the following optimization problem:

$$min_{u(t),u(t+1),\ldots,u(t+m-1)} \sum_{i=1}^{T} Q_i \left[y_p(t+i) - y_r(t+i)\right]^2 + \sum_{i=1}^{M} R_i \left[\Delta u(t+i)\right]^2 \tag{3.14}$$

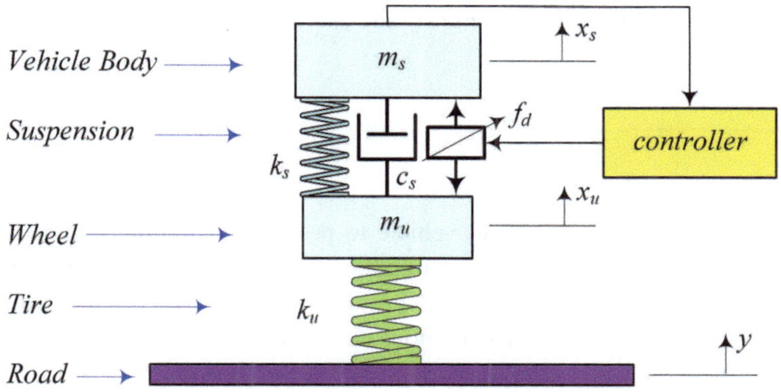


**Fig. 3.21** Active suspension system of quarter car model

where $y_p(t+i)$ are the predicted values of the controlled variable at the $i^{th}$ future sample time and $u$ is the vector of control moves to be computed. The inequalities in the above equation represent constraints such as rate and/or absolute bounds on control/output variables.

An example of an active suspension control algorithm using proportional–integral–derivative ($PID$) control is one that adjusts the suspension of a vehicle to provide improved handling and ride comfort. The $PID$ algorithm uses three parameters to control the suspension: proportional gain, integral gain, and derivative gain. A $PID$ algorithm for active suspension control typically includes the following steps:

- Measurement of the vehicle's body acceleration, road roughness, and suspension deflection using sensors.
- Error calculation: The $PID$ algorithm calculates the error between the current suspension state and the desired state. This error represents the deviation of the suspension from the desired state.
- Proportional control: The proportional gain determines the response of the suspension to changes in the input signal (the error). The control input is proportional to the error, meaning that the larger the error, the larger the control input.
- Integral control: The integral gain determines the response of the suspension to long-term changes in the input signal (the error). The control input accumulates over time, providing a sustained correction to the error.
- Derivative control: The derivative gain determines the response of the suspension to changes in the rate of the input signal (the error). The control input is proportional to the rate of change of the error, providing a correction to the error that depends on how quickly the error is changing.
- Total control input: The total control input is the sum of the proportional, integral, and derivative control inputs.
- Control action: The total control input is sent to the actuators, which adjust the suspension in real-time to provide improved handling and ride comfort.

In this method, the $PID$ algorithm uses sensor data to continuously adjust the suspension, providing a more stable and controlled ride. The algorithm takes into account factors such as the vehicle's body acceleration, road roughness, and suspension deflection to provide an optimal suspension setting.

Overall, the $PID$ algorithm for active suspension control provides improved handling and ride comfort by continuously adjusting the suspension in response to driving conditions and passenger comfort preferences. This type of algorithm is widely used in modern vehicles, as it provides a simple and effective solution for active suspension control. Figure 3.21 illustrates an example of an active suspensions system that will be used below to discuss a $PID$ active suspension controller.

Active suspension systems add hydraulic actuators to the passive components of suspension system. The advantage of such a system is that even if the active hydraulic actuator or the control system fails, the passive components come into action. The equations of motion are written as

$$m_s\ddot{x}_s + c_s(\dot{x}_s - \dot{x}_u) + k_s(x_s - x_u) - u_a = 0 \tag{3.15}$$

$$m_u\ddot{x}_u - c_s(\dot{x}_s - \dot{x}_u) - k_s(x_s - x_u) + k_u(x_u - y) + u_a = 0 \tag{3.16}$$

where $u_a$ is the control force from the hydraulic actuator. In the case of $u_a = 0$, the equations will become the equation of passive suspension system.

## 3.4 Summary

Suspension system is required for every multi-wheel vehicle to provide road contact to all wheels. Flexible suspension makes all wheels to be in contact with the ground on road irregularities, wheels' unequal loads, and under forward and lateral accelerating conditions. The job of vehicle suspensions is to provide flexibility to act as a vibration filter to give ride comfort to the passengers. A bump in the road causes the wheel to move up and down with respect to the road surface and with respect to the vehicle body.

Suspension system absorbs the energy of the wheel movement and in turn allows the frame and the body of the vehicle to ride undisturbed, while the tires follow the bumps on the road. Suspension systems in vehicles are composed of various components that work together to provide a comfortable ride, ensure stability, and enhance vehicle handling. The key components of a suspension system include: Springs, Shock Absorbers/Dampers, Control Arms, Struts, Bushings, Sway Bars/Stabilizer Bars, Ball Joints, Strut Mounts and Bearings, Steering Linkages, Wheel Hubs and Bearings, Control-Arm Bushings, and Subframe. Considering vehicle vibrations, we use the concept of sprung mass suspended by spring and damper on unsprung mass. The sprung mass, $m_s$, is the mass of the vehicle supported on the springs, and the unsprung mass, $m_u$, is defined as the mass between the road and the suspension springs.

Suspension mechanism is a combination of links and joints, connecting the wheels to the body such that there is some relative motion between the wheel and the body of the vehicle. If the wheel is steerable, then rotation about the $y$- and $z$-axes plus displacement along the $z$-axis must be allowed. If the wheel is not steerable, then only the spin about the $y$-axis and displacement along the $z$-axis must be allowed. The suspension mechanism is supposed to prevent all other possible motions.

A car can and usually have a different type of suspension on the front and back. Much is determined by whether a rigid axle binds the wheels or if the wheels are permitted to move independently. The former arrangement is known as a dependent suspension system, while the latter arrangement is known as an independent suspension system.

Suspension systems are categorized as passive, active, and semi-active considering their level of controllability. Although all the types of the suspension systems have different advantages and disadvantages, all of them utilize the spring and damper units.

## 3.5 Key Symbols

| | |
|---|---|
| $a \equiv \ddot{x}$ | acceleration |
| $A, B$ | weight factor |
| $c$ | damping coefficient |
| $C$ | coefficient |
| $d_l$ | lower wheel travel |
| $d_u$ | upper wheel travel |
| $E$ | mechanical energy |
| $E_c$ | consumed energy of a damper |
| $f = 1/T$ | cyclic frequency [Hz] |
| $f, F, \mathbf{F}$ | force |
| $f, g$ | function, periodic function |
| $f_c$ | damper force |
| $f_d$ | semi-active actuator force |
| $f_e$ | equivalent force |
| $f_k$ | spring force |
| $f_m$ | required force to move a mass $m$ |
| $F$ | amplitude of the harmonic force $f = F \sin \omega t$ |
| $g$ | gravitational acceleration, function |
| $k$ | stiffness coefficient |
| $k_e$ | equivalent stiffness |

| | |
|---|---|
| $k_r$ | stopper stiffness |
| $k_s$ | spring stiffness |
| $k_u$ | unsprung spring stiffness, tire stiffness |
| $K$ | kinetic energy |
| $l, L$ | length |
| $m$ | mass |
| $m_e$ | eccentric mass, equivalent mass |
| $P$ | potential energy |
| $Q$ | quality factor, general amplitude |
| $r$ | frequency ratio |
| $t$ | time |
| $T$ | period, time dimension |
| $v \equiv \dot{x}$ | velocity |
| $x, y, z, \mathbf{x}$ | displacement, coordinate axes |
| $x_s$ | sprung mass displacement |
| $x_u$ | unsprung mass displacement |
| $X$ | steady-state amplitude of $x$ |
| $X_s$ | steady-state amplitude of $x_s$ |
| $X_u$ | steady-state amplitude of $x_u$ |
| $\dot{x}, \dot{y}, \dot{z}$ | velocity, time derivative of $x, y, z$ |
| $X, Y, Z$ | amplitude |
| $Y$ | amplitude of base displacement |
| $\dot{y}$ | velocity |
| $z$ | relative displacement |
| $Z$ | amplitude of relative displacement |
| $Z_1, Z_2$ | short notation |

Greek

| | |
|---|---|
| $\delta$ | difference |
| $\delta_s$ | static deflection |
| $\Delta$ | difference |
| $\lambda$ | eigenvalue |
| $\varepsilon$ | mass ratio |
| $\varepsilon$ | small coefficient |
| $\theta$ | angular motion |
| $\xi$ | damping ratio |
| $\omega$ | angular frequency |
| $\omega_n$ | natural frequency |
| $\omega = 2\pi f$ | angular frequency [ rad/ s] |
| $\omega_s = \sqrt{k_s/m_s}$ | sprung mass frequency |
| $\omega_u = \sqrt{k_u/m_u}$ | unsprung mass frequency |

Subscript

| | |
|---|---|
| $n$ | natural |
| $s$ | sprung |
| $u$ | unsprung |

Symbols

| | |
|---|---|
| $DOF$ | degree of freedom |
| $FBD$ | free body diagram |
| $ODF$ | one degree of freedom |
| Re | real part |
| $RMS$ | root mean square |

## References


1. B. Jacobs, R. Rollinger (Eds.), *A Companion to the Achaemenid Persian Empire* (Wiley, New York, 2021)
2. A. Cotterell, *Chariot: The Astounding Rise and Fall of the World's First War Machine* (The Overlook Press, Peter Mayer Publishers, Woodstock, New York, 2005)
3. J.H. Crouwel, *Chariots and Other Wheeled Vehicles in Italy Before the Roman Empire* (Oxbow Books, Oxford, UK, 2012)
4. V.K. Doulatani, S.V. Chaitanya, A review on technologies of passive suspension system with variable stiffness. Int. J. Eng. Res. Technol. **3**(6), 917–921 (2014)
5. E. Yarshater (Ed.), *Encyclopaedia Iranica* (Encyclopedia Iranica Foundation, New York, NY, 2022)
6. G. Genta, L. Morello, F. Cavallino, L. Filtri, *The Motor Car: Past, Present and Future* (Springer, Dordrecht, 2014)
7. A.A. Harms, B.W. Baetx, R.R. Volti, *Engineering in Time: The Systematics of Engineering History and Its Contemporary Context* (Imperial College Press, London, 2004)
8. D. Hill, *A History of Engineering in Classical and Medieval Times* (Croom Helm, Great Britain, 1984)
9. J. Poolos, *Darius the Great* (Chelsea House, New York, NY, 2008)
10. R.N. Jazar, *Advanced Vibrations: Theory and Applications* (Springer, New York, NY, 2023)
11. K. Roy, *A Global History of Warfare and Technology from Slings to Robots* (Springer, Singapore, 2022)
12. A. Kammenhuber, *Hippologia Hethitica* (Otto Harrassowitz, Wiesbaden, 1961)
13. M.A. Littauer, J.H. Crouwel, *Wheeled Vehicles and Ridden Animals in the Ancient Near East* (Brill, Leiden, 1979)
14. B. Marsden, C. Smith, *Engineering Empires: A Cultural History of Technology in Nineteenth-Century Britain* (Palgrave Macmillan, New York, NY, 2005)
15. P.R.S. Moorey, The emergence of the light horse-drawn chariot in the near east c. 2000–1500 BC. World Archaeol. **18**(2), 196–215 (1986)
16. N. Fields, *The Bronze Age War Chariot* (Osprey Publisher, Oxford, UK, 2006)
17. R. Straus, *Carriages & Coaches, Their History & Their Evolution* (Martin Secker, London, 1912)
18. E. Schmidt, *Persepolis I-II* (The University of Chicago, Oriental Institute Publications, Chicago, 1953, 1957)
19. J.M. Silverman, *Persian Royal–Judaean Elite Engagements in the Early Teispid and Achaemenid Empire* (T&T CLARK Bloomsbury Publishing, New York, NY, 2020)
20. R. Straus, *Carriage & Coaches: Their History and Their Evolution* (Martin Secker, London, 1912)
21. A.J. Veldmeijer, *Chariot in Ancient Egypt* (Sidestone Press, Leiden, 2018)
22. V.S. Curtis, S. Stewart, *Birth of the Persian Empire* (I.B.Tauris & Co Ltd., New York NY, 2005)
23. W.H. Ward, *The Seal Cylinders of Western Asia* (Carnegie Institution of Washington, Philadelphia, 1910)

# 4 Mechanical Vibrations Modeling

In this chapter, we study 1—mechanical elements of vibrating systems, 2—physical causes of mechanical vibrations, 3—kinematics of vibrations, 4—simplification methods of complex vibrating systems, and 5—mathematical review of mechanical vibrations. We will solve and review a general mass–spring–damper system and then will make other vibrating systems to be reduced to an equivalent system. Such equivalence yields a generalization in mathematical treatment of vibrating systems. The first section defines all fundamental terms we need to study vehicle vibrations.

## 4.1 Mechanical Vibration Elements

The repeated transformation of kinetic energy $K$ to potential energy $P$ and potential energy to kinetic energy is called vibrations. When the potential energy is at its maximum, the kinetic energy is zero, and when the kinetic energy is at its maximum, the potential energy is minimum. When such fluctuation in kinetic energy appears as a periodic motion of a massive body, we call such energy transformations *mechanical vibrations*.

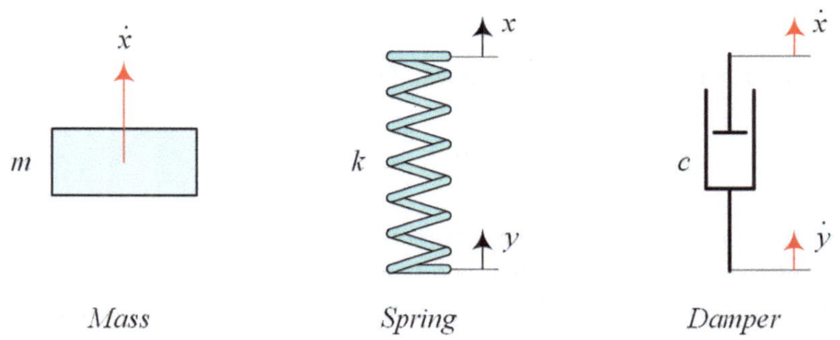


**Fig. 4.1** A mass $m$, spring $k$, and damper $c$

The kinetic and potential energies are stored in physical elements. Any element that stores kinetic energy is called the *mass* and *inertia*, and any element that stores potential energy is called the *spring* and *restoring element*. If the total value of *mechanical energy* $E = K + P$ decreases during vibrations, there is also a phenomenon or an element that dissipates energy. The element that causes energy to dissipate is called *damper*. The symbolic illustration of mass, spring, and damper is shown in Fig. 4.1.

### 4.1.1 Mass

In mechanical vibrations, the mass $m$ is an element that acts as a container to store kinetic energy. The amount of stored kinetic energy $K$ in a mass $m$ is proportional to the square of its velocity, $v^2$, $v \equiv \dot{x} = dx/dt$.


R. N. Jazar, H. Marzbani, *Vehicle Vibrations*, https://doi.org/10.1007/978-3-031-43486-0_4

$$K = \frac{1}{2}mv^2 \tag{4.1}$$

In Newtonian mechanics, a mass $m$ is the proportionality coefficient between an applied force on an object and the acceleration of the object. The mass of an object is the most important characteristic of the object and that is why we can refer to an object only by its mass $m$. The required force $f$ to move a mass $m$ is proportional to its acceleration $a \equiv \ddot{x}$. The mass $m$ is the proportionality coefficient.

$$f = ma \tag{4.2}$$

The mass $m$ is the resistance of an object to acceleration. The higher the mass, the lower the acceleration of the object under a constant force. The linearity and proportionality properties of the Newton equation of motion make the mass to be an invariant characteristic of the object.

$$f_1 = ma_1 \qquad f_2 = ma_2 \tag{4.3}$$

$$f_1 + f_2 = m\,(a_1 + a_2) \tag{4.4}$$

We assume an object to have a constant mass regardless of motion or rest or the applied force. This fact suggested to accept mass as a fundamental unit and hence define dimension of force in terms of mass and acceleration.

$$[f] = [ma] = [m]\,[a] = M\ LT^{-2} \tag{4.5}$$

### 4.1.2 Spring

In mechanical vibrations, a spring is any element that acts as a container to store potential energy. The amount of stored potential energy $P$ in a spring is equal to the work done by the spring force $f$ during the deflection of the spring, proportional to the relative displacement of its ends $z = x - y$.

$$P = -\int f\,dz = -\int -kz\,dz \tag{4.6}$$

The spring potential energy is then a function of displacement. For linear Springs, we have

$$P = \frac{1}{2}kz^2 \tag{4.7}$$

A spring with stiffness $k$ is an element to make a force $f$ proportional and in opposite direction of the relative displacement of its ends $z = x - y$. The stiffness $k$ is the constant of proportionality. The curve of the force $f$ versus relative displacement $z$ is called the *spring characteristic curve*. If the stiffness of a spring, $k$, is constant, it is called a *linear spring*.

$$f = -kz = -k(x - y) \tag{4.8}$$

$$z = x - y \tag{4.9}$$

$$[k] = [f/x] = [f]\,/\,[x] = MT^{-2} \tag{4.10}$$

### 4.1.3 Damper

In mechanical vibrations, a damper is any element, device, or mechanism that makes the system to lose mechanical energy $E = P + K$. The *viscous damping model* is the simplest and the most adopted model of damping in mechanical vibrations. A viscose damper with damping $c$ is an element to make a resistance force $f$ proportional to the relative velocity of its ends $\dot{z} = \dot{x} - \dot{y}$. The damping $c$ is the constant of proportionality. The curve of the force $f$ versus relative velocity $\dot{z} = \dot{x} - \dot{y}$ is called the *damper characteristic curve*. If the damping of a damper, $c$, is constant, it is called a *linear damper*.

$$f = -c\,\dot{z} = -c(\dot{x} - \dot{y}) \tag{4.11}$$

Dampers are also called *shock absorber*. Loss of energy can be measured by the average of dissipation power in one cycle. Assuming a harmonic relative displacement,

$$z = A\cos\omega t \tag{4.12}$$

the dissipated energy $W_D$ in one cycle equals the work done by the damping force.

$$\begin{aligned} W_D &= \int_0^T f\,dz = \int_0^T c\dot{z}\,dz = \int_0^{2\pi/\omega} c\dot{z}^2\,dt \\ &= cA^2\omega^2 \int_0^{2\pi/\omega} \sin^2\omega t\,dt = \pi c\omega A^2 \end{aligned} \tag{4.13}$$

$$c = \frac{W_D}{\pi\omega A^2} \tag{4.14}$$

The dimension of damping $c$ is

$$[c] = [f/\dot{x}] = [f]\,/\,[\dot{x}] = MT^{-1} \tag{4.15}$$

The absolute value of damping force $f_d$ for a harmonic oscillator is

$$\begin{aligned} f_d &= c\dot{x} = c\omega A\sin\omega t \\ &= c\omega A\sqrt{1-\cos^2\omega t} = c\omega\sqrt{A^2 - z^2} \end{aligned} \tag{4.16}$$

This equation can be rearranged in the form of an ellipse in the coordinate $(x, f_d)$.

$$\left(\frac{x}{A}\right)^2 + \left(\frac{f_d}{c\omega A}\right)^2 = 1 \tag{4.17}$$

The ellipse is called the hysteresis loop. The area enclosed by the ellipse equals the dissipated energy during one cycle, $\pi c\omega A^2$. The average of dissipation power $P_D$ of viscus damping in one cycle will be

$$\begin{aligned} P_D &= \frac{1}{T}\int_0^T f\,\dot{z}\,dt = \frac{\omega}{2\pi}\int_0^{2\pi/\omega} c\dot{z}\,\dot{z}\,dt \\ &= \frac{c\omega^3}{2\pi}A^2 \int_0^{2\pi/\omega} \sin^2\omega t\,dt = \frac{1}{2}cA^2\omega^2 \end{aligned} \tag{4.18}$$

A more general model of damping is governed by the following mathematical equation:

$$f = -c\,\dot{z}\;|\dot{z}|^{s-1} \tag{4.19}$$

For $s = 1$, we have the linear viscous damping model (4.11). For $s = 0$, we have the Coulomb damping to represent contact friction losses.

$$f = -\mu N\,\frac{\dot{z}}{|\dot{z}|} = -\mu N\,\operatorname{sgn}(\dot{z}) \tag{4.20}$$

For $s = 2$, we have quadratic damping to represent hydrodynamic damping.

$$f = -c\,\dot{z}\;|\dot{z}| \tag{4.21}$$

These models are all functions of velocity only. It may define other damping models to be functions of velocity as well as displacement.

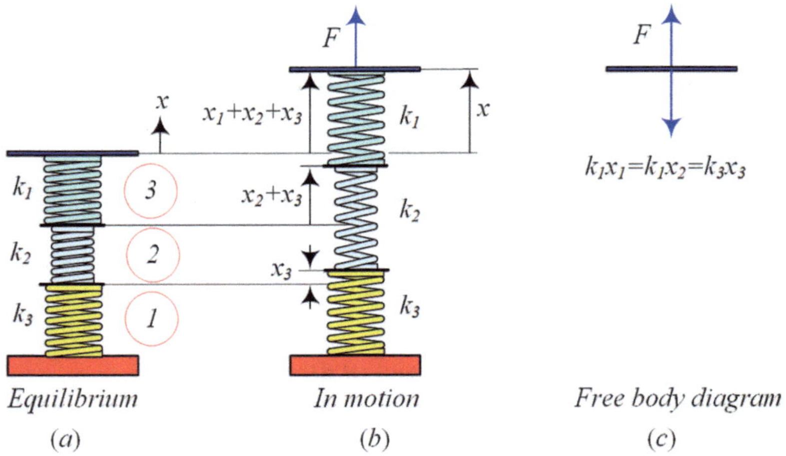


**Fig. 4.2** Serial springs

### 4.1.4 Vibration Modes

When there is no applied external *excitation* force on a vibrating system, any possible oscillation of the system is called a *free vibration*. A free vibrating system happens only if any one of the kinematic states of position, or speed, $x$, $\dot{x}$, is not zero. When we apply an excitation, then any possible motion of the system is called a *forced vibration*. There are four types of applied excitation: *harmonic*, *periodic*, *transient*, and *random*. The harmonic and transient excitations are more applied than the others and more predictable than the periodic and random types. When the excitation is a sinusoidal function of time, it is called a *harmonic excitation*, and when the excitation force disappears after a while or stays steady, it is a *transient excitation*. Any periodic excitation can theoretically be decomposed into a series of harmonic excitations with different coefficients and multiple frequencies. Therefore, a *periodic excitation* is a combination of infinite number of harmonic excitations with decreasing weight factor coefficients. The Fourier analysis will reveal the harmonic functions and their coefficients of the periodic excitation. A *random excitation* has no short-term pattern and is not predictable; however, we may define some long-term averages to characterize a random excitation.

Any periodic motion is characterized by a *period* $T$, which is the required time for one complete cycle of vibration, starting from and ending at the same conditions such as ($\dot{x} = 0, 0 < \ddot{x}$). The *frequency* $f$ is the number of cycles in one $T$.

$$f = \frac{1}{T} \tag{4.22}$$

$$x(t) = x(t + T) \tag{4.23}$$

$$\dot{x}(t) = \dot{x}(t + T) \tag{4.24}$$

In theoretical vibrations, we usually work with *angular frequency* $\omega\,[\mathrm{rad/s}]$, and in applied vibrations we use the *cyclic frequency* $f\,[\mathrm{Hz}]$:

$$\omega = 2\pi f \tag{4.25}$$

If a variable $x$ is periodic in $T$, it is also periodic in $2T, 3T, \cdots$. Therefore, the period of a periodic variable, $x$, is the least value of required time to make a repetition.

We use $f$ to indicate a force. If $f$ is a harmonically variable force, we show its amplitude by $F$, to be consistent with a harmonic motion $x$ with amplitude $X$. We also use $f$ for cyclic frequency; however, $f$ is a force unless it is indicated that it is a frequency.

**Example 11** Serial springs and dampers. A series of springs can mathematically be replaced by a single equivalent spring.

**Example 12** Serial springs are attached head to tail to each other as is shown in Fig. 4.2. Serial springs have the same force and a resultant displacement equal to the sum of individual displacements. Figure 4.2 illustrates three serial springs attached to a massless plate and the ground.

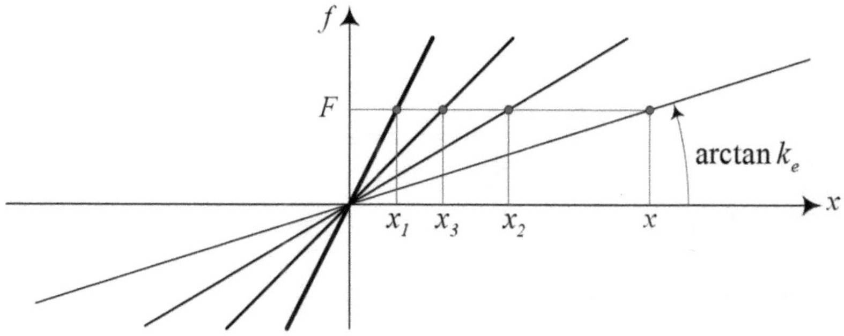


**Fig. 4.3** The characteristic curves of 3 serial springs with stiffnesses $k_1 > k_2 > k_3$ and their equivalent stiffness $k_{eq}$

The equilibrium position of the springs is their unstretched configuration in Fig. 4.2a. Because springs are assumed to be massless, serial attachments of springs will not generate any displacement in them. Applying a displacement $x$, as is shown in Fig. 4.2b, generates the free body diagram of the upper plate as shown in Fig. 4.2c. Each spring makes a force $f_i = -k_i x_i$, where $x_i$ is the length change in the spring number $i$. The total displacement of the springs, $x$, is the sum of their individual displacements $\sum x_i$.

$$x = \sum x_i = x_1 + x_2 + x_3 \tag{4.26}$$

We may substitute a set of serial springs with only one equivalent spring of stiffness $k_e$ that produces the same displacement $x$ under the same force $F = f_k$.

$$F = f_k = -k_1 x_1 = -k_2 x_2 = -k_3 x_3 = -k_e x \tag{4.27}$$

Substituting (4.27) into (4.26) yields

$$\frac{F}{k_e} = \frac{F}{k_1} + \frac{F}{k_2} + \frac{F}{k_3} = \left(\frac{1}{k_1} + \frac{1}{k_2} + \frac{1}{k_3}\right) F \tag{4.28}$$

It shows that the inverse of the equivalent stiffness, $1/k_e$, of the serial springs is the sum of their inverse stiffness, as $\sum 1/k_i$:

$$\frac{1}{k_e} = \frac{1}{k_1} + \frac{1}{k_2} + \frac{1}{k_3} \tag{4.29}$$

We assume that the force of a linear spring is not affected by kinematic variables $x$, $\dot{x}$, $\ddot{x}$, and time $t$. The characteristic curves of springs $k_1$, $k_2$, $k_3$, and $k_e$ and their force–displacement behaviors are illustrated in Fig. 4.3. The equivalent stiffness of a serial spring is always less than the stiffness of every individual springs.

Serial dampers have similar equivalent. They have the same force, $F = f_c$, and a resultant velocity $\dot{x}$ that is equal to the sum of individual velocities, $\sum \dot{x}_i$. We can substitute a set of serial dampers with only one equivalent damping $c_e$ that with the same velocity $\dot{x}$ produces the same force $f_c$. For three parallel dampers, the velocity and force balance equations are

$$\dot{x} = \dot{x}_1 + \dot{x}_2 + \dot{x}_3 \tag{4.30}$$

$$F = f_c = -c_1 \dot{x} = -c_2 \dot{x} = -c_3 \dot{x} = -c_e \dot{x} \tag{4.31}$$

It shows that the inverse of the equivalent damping of the serial dampers, $1/c_e$, is the sum of their inverse dampings, $\sum 1/c_i$:

$$\frac{1}{c_e} = \frac{1}{c_1} + \frac{1}{c_2} + \frac{1}{c_3} \tag{4.32}$$

We assume that the force of a linear damper is not affected by kinematics $x$, $\dot{x}$, $\ddot{x}$, and time $t$. The equivalent damping of a serial dampers is always less than the damping of every individual dampers.

**Example 13** Parallel springs and dampers. Parallel springs can be replaced by an equivalent spring.

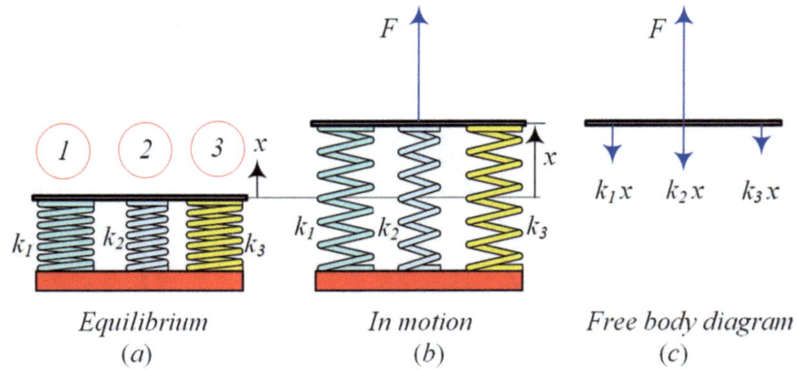


**Fig. 4.4** Parallel springs

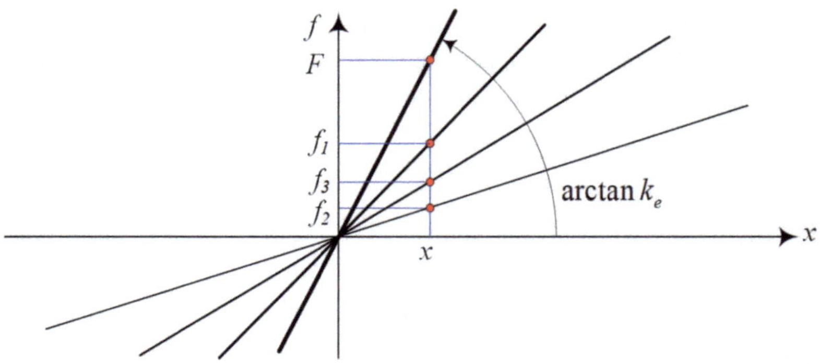


**Fig. 4.5** The characteristic curves of 3 parallel springs with stiffnesses $k_1 < k_2 < k_3$ and their equivalent $k_{eq}$

Parallel springs refer to their geometrical attachments as is shown in Fig. 4.4. Parallel springs have the same displacement $x$, and a resultant force $F$, that is equal to the sum of the individual forces, $\sum f_i$. Figure 4.4 illustrates three parallel springs between a massless plate and the ground. The equilibrium position of the springs is the unstretched configuration, shown in Fig. 4.4a. Applying a displacement $x$ to all the springs in Fig. 4.4b generates the free body diagram of Fig. 4.4c. Each spring makes a force $-kx$. The resultant force of the springs $F = f_k$ is

$$F = f_k = -k_1 x - k_2 x - k_3 x = -(k_1 + k_2 + k_3)\, x \tag{4.33}$$

We can substitute parallel springs with only one equivalent spring of stiffness $k_e$ that produces the same force $F$ under the same displacement $x$.

$$f_k = -k_e x \qquad k_e = k_1 + k_2 + k_3 \tag{4.34}$$

Therefore, the equivalent stiffness, $k_e$, of parallel springs is the sum of their stuffiness $\sum k_i$. The characteristic curves of springs $k_1$, $k_2$, $k_3$, and $k_e$ and their force–displacement behaviors are illustrated in Fig. 4.5.

Parallel dampers will similarly be analyzed. They have the same speed $\dot{x}$, and a resultant force $F = f_c$ is equal to the sum of the individual forces $\sum f_i$. We can substitute parallel dampers with only one equivalent damping $c_e$ that produces the same force $F$ under the same velocity $\dot{x}$. Consider three parallel dampers for which their force balance and equivalent damper are

$$f_c = -c_e \dot{x} = -c_1 \dot{x} - c_2 \dot{x} - c_3 \dot{x} = -(c_1 + c_2 + c_3)\, \dot{x} \tag{4.35}$$

$$c_e = c_1 + c_2 + c_3 \tag{4.36}$$

**Example 14** Different length parallel springs. Assembling springs with different length happens frequently in engineering application. How to determine the equivalent spring of two parallel springs is important in application.

In practical applications, it frequently happens that we should adjust and fit a shorter or longer spring in a physical space. Figure 4.6 illustrates a mass $m$ that is supposed to sit on two springs with unequal lengths. To examine the system, let us

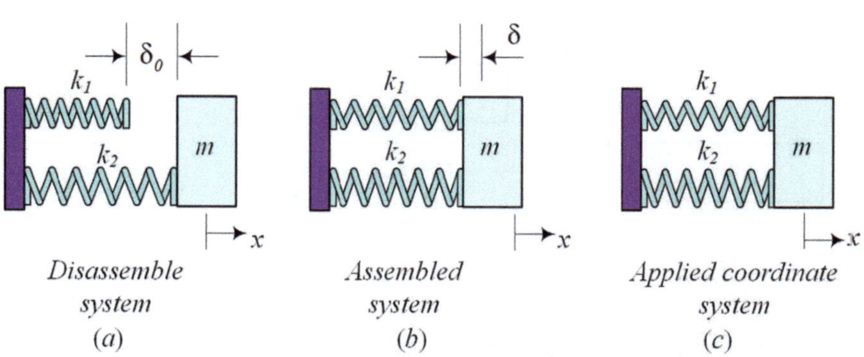


**Fig. 4.6** Parallel assembling two springs with different lengths

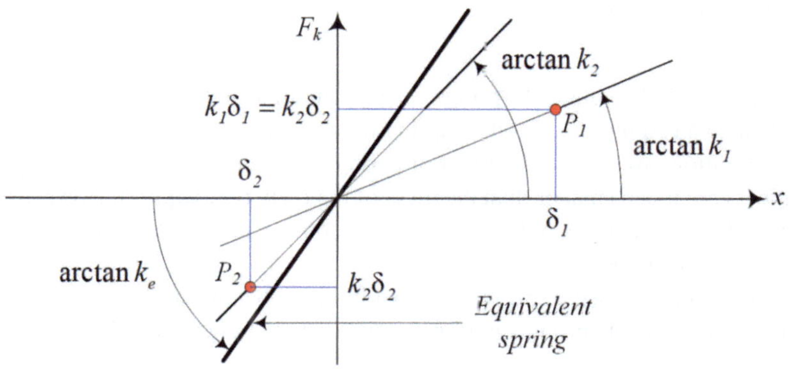


**Fig. 4.7** Characteristic curves of springs $k_1$ and $k_2$ and their equivalent replacement

assume that $m$ can only move in the $x$ direction and the length of spring $k_1$ is shorter than $k_2$ by $\delta_0$. When we attach $k_1$ to $m$, the equilibrium position of $m$ will move to have a distance of $\delta$ from its original position. At this equilibrium position, the spring $k_1$ is elongated by $\delta_1$ and $k_2$ is shortened by $\delta_2$. There are two equations to determine $\delta_1$ and $\delta_2$. Force balance on $m$ indicates that

$$-k_1\delta_1 + k_2\delta_2 = 0 \tag{4.37}$$

and geometric compatibility provides us with

$$\delta_1 = \delta_0 - \delta \qquad \delta_2 = \delta \tag{4.38}$$

Therefore, $\delta_1$ and $\delta_2$ can be calculated.

$$\delta_1 = \frac{\delta_0}{1 + k_1/k_2} \qquad \delta_2 = \frac{\delta_0}{k_2/k_1 + 1} \tag{4.39}$$

The assembled system indicates a mass $m$ that is attached to two springs which are not at their neutral lengths. The working points of $k_1$ and $k_2$ are shown by $P_1$ and $P_2$ in characteristic curves of $k_1$ and $k_2$ of Fig. 4.7. To examine the possible equivalent spring $k_e = k_1 + k_2$, let us apply a displacement $x$ to $m$ and determine the force–displacement equation:

$$F_1 = k_1\delta_1 + k_1 x \tag{4.40}$$

$$F_2 = -k_2\delta_2 + k_2 x = -k_1\delta_1 + k_2 x \tag{4.41}$$

$$F_k = F_1 + F_2 = (k_1 + k_2)\,x = k_e x \tag{4.42}$$

Therefore, as long as the displacements of the assembled springs are equal, there is an equivalent spring $k_e = k_1 + k_2$, to be substituted for the two springs with the equilibrium point at $x = 0$ of the assembled system.

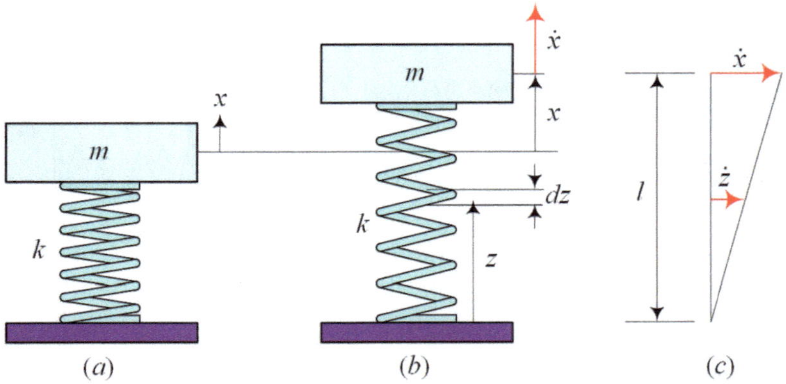


**Fig. 4.8** A vibrating system with a massive spring

**Example 15** ★ Massive spring. If the mass of springs is high enough, they collect some kinetic energy. Here is an approximate method to calculate an equivalent massless spring for a massive spring.

In modeling of vehicle vibrations we ignore the mass of springs and dampers. This assumption is valid as long as the masses of springs and dampers are much smaller than the mass of the body they support. However, when the mass of spring $m_s$ or damper $m_d$ is comparable with the mass of body $m$, we may define a new model with an equivalent mass $m_e$ which is supported by a massless spring and damper.

Consider a vibrating system with a massive spring as shown in Fig. 4.8a. When the system is at equilibrium, the spring has a mass $m_s$ and a length $l$. The mass of spring is uniformly distributed along its length, so we may define a mass density $\rho$.

$$\rho = \frac{m}{l} \tag{4.43}$$

The equivalent mass $m_e$ of an equivalent massless spring system is

$$m_e = m + \frac{1}{3} m_s \tag{4.44}$$

To verify this equation, we seek a system which keeps the same amount of kinetic energy as the original system. Figure 4.8b illustrates the original system when the mass $m$ is at position $x$ and has a velocity $\dot{x}$. The spring is extended between the mass $m$ and the ground. So, the base of the spring has no velocity, while the other end has the same velocity as $m$. Let us define a coordinate position $z$ that goes from the grounded base of the spring to the end point at $m$. An element of spring at $z$ has a length $dz$ and a mass $dm$.

$$dm = \rho \, dz \tag{4.45}$$

Assuming $x << l$ and a linear velocity distribution of the elements of spring, as shown in Fig. 4.8c, we find the velocity $\dot{z}$ of $dm$.

$$\dot{z} = \frac{z}{l} \dot{x} \tag{4.46}$$

The kinetic energy of the system is a summation of kinetic energy of the mass $m$ plus the kinetic energy of the spring.

$$\begin{aligned} K &= \frac{1}{2} m \dot{x}^2 + \frac{1}{2} \int_0^l \left( dm \, \dot{z}^2 \right) = \frac{1}{2} m \dot{x}^2 + \frac{1}{2} \int_0^l \rho \left( \frac{z}{l} \dot{x} \right)^2 dz \\ &= \frac{1}{2} m \dot{x}^2 + \frac{1}{2} \frac{\rho}{l^2} \dot{x}^2 \int_0^l z^2 \, dz = \frac{1}{2} m \dot{x}^2 + \frac{1}{2} \frac{\rho}{l^2} \dot{x}^2 \left( \frac{1}{3} l^3 \right) \\ &= \frac{1}{2} \left( m + \frac{1}{3} m_s \right) \dot{x}^2 = \frac{1}{2} m_e \dot{x}^2 \end{aligned} \tag{4.47}$$

Therefore, an equivalent system should have a massless spring attached to a mass $m_e = m + \frac{1}{3} m_s$ to keep the same amount of kinetic energy. Because the potential energy of a spring is not a function of its mass, the equivalent spring will collect

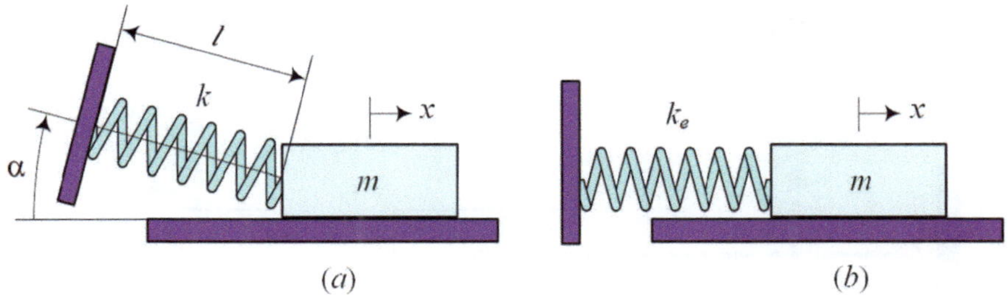


**Fig. 4.9** A tilted spring and its equivalent stiffness

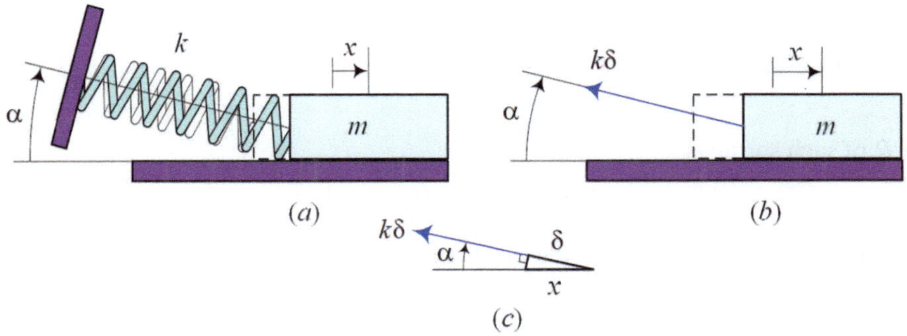


**Fig. 4.10** A mass–spring system such that the spring makes an angle $\alpha$ with direction of mass translation

the same amount of potential energy as the original spring and the equivalent system will vibrate in similar to the original system.

**Example 16** Tilted spring. This is to show how we can replace a tilted spring with an equivalent colinear spring.

Consider a mass–spring system such that the spring makes an angle $\alpha$ with the axis of mass translation, as shown in Fig. 4.9a. We may substitute such a tilted spring with an equivalent spring $k_e$ that is on the same axis of mass translation, as shown in Fig. 4.9b. We have

$$k_e \approx k \cos^2 \alpha \tag{4.48}$$

When the mass $m$ is in motion, such as is in Fig. 4.10a, its free body diagram is as shown in Fig. 4.10b. If the motion of mass $m$ is much smaller than the length of the spring, $x << l$, we ignore any changes in $\alpha$ and calculate the spring elongation $\delta$ as shown in Fig. 4.10c.

$$\delta \approx x \cos \alpha \tag{4.49}$$

Therefore, the spring force $f_k$ is

$$f_k = k\delta \approx kx \cos \alpha \tag{4.50}$$

The spring force should be projected on the $x$-axis to find the $x$ component, $f_x$, that causes the mass $m$ to move.

$$f_x = f_k \cos \alpha \approx \left(k \cos^2 \alpha\right) x \tag{4.51}$$

Therefore, the tilted spring can be substituted by an equivalent spring $k_e$ on the $x$-axis that needs the same force $f_x$ to elongate the same amount $x$ as the mass moves.

$$f_x = k_e x \qquad k_e \approx k \cos^2 \alpha \tag{4.52}$$

Equivalent masses can be found by equating kinetic energies $K$, and equivalent springs can be found by equating potential energies $P$. Here we can also find equivalent spring for a tilted spring by equating potential energies. Consider a spring that makes an angle $\alpha$ with the direction of motion as shown in Fig. 4.10a. When the mass $m$ translates $x$, the elongation of the spring would be $\delta$.

$$\delta \approx x \cos \alpha \tag{4.53}$$

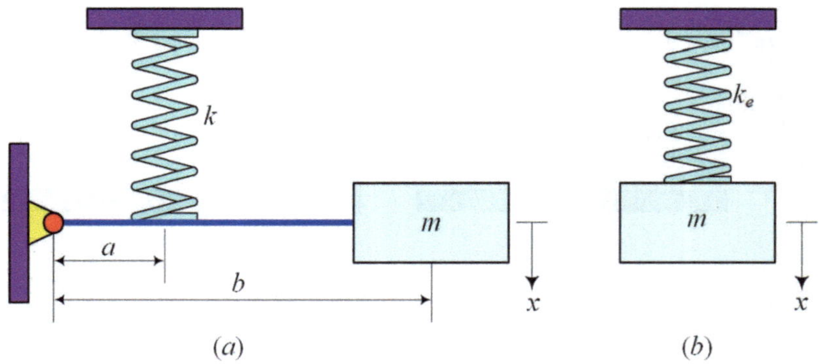


**Fig. 4.11** A mass $m$ attached to the tip of a massless bar with length $b$

The potential energy $P$ of such spring is

$$P = \frac{1}{2}k\delta^2 = \frac{1}{2}\left(k\cos^2\alpha\right)x^2 \tag{4.54}$$

An equivalent spring with stiffness $k_e$ must collect the same amount of potential energy for the same displacement $x$.

$$P = \frac{1}{2}k_e x^2 \tag{4.55}$$

Therefore, the equivalent stiffness $k_e$ is

$$k_e = k\cos^2\alpha \tag{4.56}$$

**Example 17** Displaced spring. Springs may be installed offset with respect to the measuring displacement point. Here is an example to determine an equivalent model.

Figure 4.11a illustrates a mass $m$ attached to the tip of a massless bar with length $b$. The bar is pivoted to a wall and a spring $k$ is attached to the bar at a distance $a$ from the pivot. When the mass oscillates with displacement $x << b$, the elongation $\delta$ of the spring is

$$\delta \approx \frac{a}{b}x \tag{4.57}$$

We may substitute the system with a translational mass–spring system such as shown in Fig. 4.11b. The new system has the same mass $m$ and an equivalent spring $k_e$.

$$k_e = \left(\frac{a}{b}\right)^2 k \tag{4.58}$$

The equivalent spring provides us with the same potential energy as the original spring.

$$P = \frac{1}{2}k_e x^2 = \frac{1}{2}k\delta^2 = \frac{1}{2}k\left(\frac{a}{b}x\right)^2 = \frac{1}{2}k\left(\frac{a}{b}\right)^2 x^2 \tag{4.59}$$

**Example 18** Equivalent spring and damper of a MacPherson suspension. Here is an example to model a vehicle suspension with equivalent mass–spring–damper.

Figure 4.12 illustrates a MacPherson strut mechanism and its equivalent model of vibrating system. We assume the tire is stiff, and therefore, the wheel center gets the same motion $y$ as the tireprint of the wheel. Furthermore, we assume the wheel and the body of the vehicle to move only vertically.

To find the equivalent parameters for the vibrating model, we use $m$ equal to $1/4$ of the vehicle body mass. The spring $k$ and damper $c$ make an angle $\alpha$ with the direction of wheel motion. They are also displaced $b - a$ from the wheel center, laterally. So, the equivalent spring $k_e$ and damper $c_e$ are

$$k_e = k\left(\frac{a}{b}\cos\alpha\right)^2 \qquad c_e = c\left(\frac{a}{b}\cos\alpha\right)^2 \tag{4.60}$$

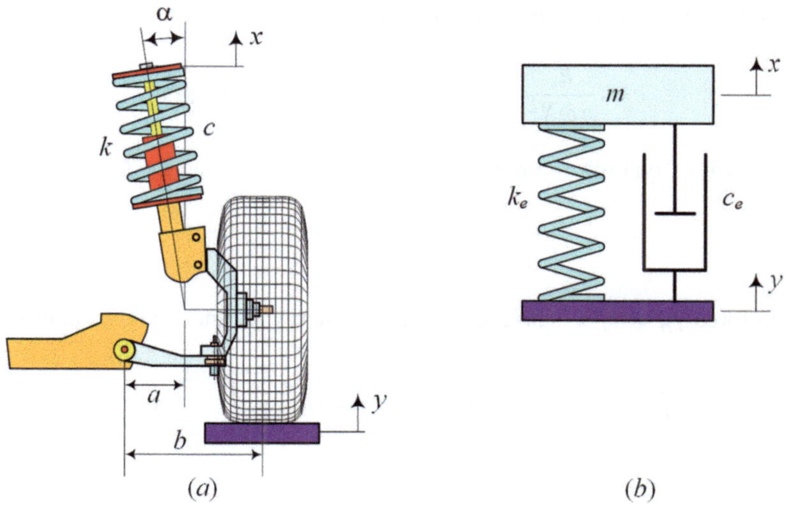


**Fig. 4.12** A MacPherson suspension and its equivalent vibrating system

As an application assume we have determined the following stiffness and damping as a result of an optimization algorithm.

$$k_e = 9869.6\,\text{N/m} \qquad c_e = 87.965\,\text{N s/m} \tag{4.61}$$

The actual $k$ and $c$ for a MacPherson suspension with

$$a = 19\,\text{cm} \qquad b = 32\,\text{cm} \qquad \alpha = 27\,\text{deg} \tag{4.62}$$

will be

$$k = 35264\,\text{N/m} \qquad c = 314.3\,\text{N s/m} \tag{4.63}$$

**Example 19** Equivalent viscous damping and energy loss. It is to show how we approximate damping of real systems.

Consider the system in Fig. 4.10b when $y = 0$ and $x = X \sin \omega t$. The required force to produce this motion is

$$f = kx + c\dot{x} = k\,(X \sin \omega t) + c\,(X\omega \cos \omega t) \tag{4.64}$$

In one period of vibration, the consumed energy per cycle $E_c$ is proportional to the viscous damping of the system.

$$\begin{aligned} E_c &= \int_0^T f x dt = \int_0^{2\pi/\omega} (k\,(X \sin \omega t) + c\,(X\omega \cos \omega t))\,(X\omega \cos \omega t)\,dt \\ &= c\pi\omega X^2 \end{aligned} \tag{4.65}$$

The definition of linear viscous damping is then equal to the dissipated energy in one cycle divided by $\pi\omega X^2$:

$$c = \frac{E_c}{\pi\omega X^2} \tag{4.66}$$

This equation is being used to calculate equivalent damping of other types of damping.

$$c_e = \frac{E_c}{\pi\omega X^2} \tag{4.67}$$

As an example consider a system with

$$m = 2\,\text{kg} \qquad k = 10000\,\text{N/m} \qquad x = 0.01 \sin 100t \;\; \text{m} \tag{4.68}$$

which consumes $E_c = 3.1416\,\text{J}/cycle$. The equivalent linear damping of the system is

$$c = \frac{E_c}{\pi \omega X^2} = \frac{3.1416}{\pi \times 100 \times 0.01^2} = 100\,\text{N s/m} \tag{4.69}$$

For Coulomb friction model with Coulomb force $f_C = \mu N$ that dissipates energy $E_c = 4 f_C X$, the equivalent viscous damper is

$$c_e = \frac{4\mu N}{\pi \omega X} \tag{4.70}$$

For hydraulic damping with force $f_H = c_L\, \dot{x}^2 \text{sgn} \dot{x}$, the dissipated energy $E_c$ and the equivalent viscous damper are

$$\begin{aligned} E_c &= \int_0^T c_L \dot{x}^2 dx = c_L \int_0^T \dot{x}^3 dt = 4 c_L \omega^3 X^3 \int_0^{\pi/(2\omega)} \cos^3 \omega t \, dt \\ &= \frac{8}{3} c_L \omega^3 X^3 \end{aligned} \tag{4.71}$$

$$c_e = \frac{8}{3} \frac{c_L \omega^2 X}{\pi} \tag{4.72}$$

## 4.2 Kinematics of Vibrations

The *fundamental equation of vibration* is the free and undamped equation of motion of a mass $m$ attached to a linear spring with stiffness $k$.

$$m\ddot{x} + kx = 0 \qquad x(0) = x_0 \qquad \dot{x}(0) = \dot{x}_0 \tag{4.73}$$

$$\ddot{x} + \omega_n^2 x = 0 \qquad \omega_n = \sqrt{\frac{k}{m}} \tag{4.74}$$

The solution of the fundamental equation of vibration is

$$x = A \sin \omega_n t + B \cos \omega_n t \tag{4.75}$$

$$= x_0 \cos \omega_n + \frac{1}{\omega_n} \dot{x}_0 \sin \omega_n \tag{4.76}$$

$$= X \sin(\omega_n t + \varphi) \tag{4.77}$$

$$= X \cos(\omega t + \theta) \tag{4.78}$$

The solution (4.77) or (4.78) is called the *harmonic* or *wave* solution, and the solution (4.75) is called the *weighted harmonic* or *complete harmonic* solution. The $X$ is the *amplitude*, $\omega$ is the *angular frequency*, $\varphi$ is the *phase* of the harmonic solution, and the coefficients $A$ and $B$ are the *weight factors* of the weighted harmonic solution. The parameters $X$, $\varphi$, $\theta$, $A$, and $B$ are functions of the initial conditions $x_0$ and $\dot{x}_0$.

$$A = x_0 \qquad B = \frac{1}{\omega} \dot{x}_0 \tag{4.79}$$

$$X = \sqrt{A^2 + B^2} = \sqrt{x_0^2 + \frac{\dot{x}_0^2}{\omega^2}} \tag{4.80}$$

The parameters $X$, $\varphi$, $A$, and $B$ are related by the following equations:

$$A = X \cos \varphi \qquad B = X \sin \varphi \qquad \tan \varphi = \frac{B}{A} \tag{4.81}$$

$$A = X \sin\theta \qquad B = -X\cos\theta \qquad \tan\theta = \frac{A}{-B} \tag{4.82}$$

***Proof*** The fundamental equation of vibration (4.73) is a linear second-order ordinary deferential equation and hence its solution will be exponential function. To find the solution, we substitute a general exponential function with unknown coefficients and unknown power and determine the unknowns.

$$x = X\, e^{st} \tag{4.83}$$

It is done in two steps:

1– We determine the exponent $s$. The value depends on the parameters of the system, and the number of $s$ is equal to the order of the differential equation. In this case, we will have two values for $s$ and they will be functions of the parameters $k$ and $m$.

$$m\left(Xs^2e^{st}\right) + k\left(Xe^{st}\right) = \left(ms^2 + k\right)Xe^{st} = 0 \tag{4.84}$$

$$ms^2 + k = 0 \tag{4.85}$$

$$s_1 = \sqrt{-\frac{k}{m}} = i\sqrt{\frac{k}{m}} = i\omega_n \tag{4.86}$$

$$s_2 = -\sqrt{-\frac{k}{m}} = -i\sqrt{\frac{k}{m}} = -i\omega_n \tag{4.87}$$

$$\omega_n = \sqrt{\frac{k}{m}} \tag{4.88}$$

2– We determine the coefficient $X$. Every value of $s$ provides us with an exponential function. All these exponential functions are linearly independent, and hence the general solution of the equation would be a linear combination of the exponential functions, each with an unknown coefficient. The coefficients depend on the initial conditions of the system. In this case, we have two exponential functions with coefficients $X_1$ and $X_2$.

$$x = X_1e^{s_1t} + X_2e^{s_2t} \tag{4.89}$$

$$\dot{x} = X_1s_1e^{s_1t} + X_2s_2e^{s_2t} \tag{4.90}$$

The initial conditions (4.73) provide us with two equations to find $X_1$ and $X_2$.

$$x_0 = X_1 + X_2 \tag{4.91}$$

$$\dot{x}_0 = X_1s_1 + X_2s_2 \tag{4.92}$$

$$\begin{bmatrix} x_0 \\ \dot{x}_0 \end{bmatrix} = \begin{bmatrix} 1 & 1 \\ s_1 & s_2 \end{bmatrix}\begin{bmatrix} X_1 \\ X_2 \end{bmatrix} \tag{4.93}$$

$$\begin{aligned}\begin{bmatrix} X_1 \\ X_2 \end{bmatrix} &= \begin{bmatrix} 1 & 1 \\ s_1 & s_2 \end{bmatrix}^{-1}\begin{bmatrix} x_0 \\ \dot{x}_0 \end{bmatrix} \\ &= \begin{bmatrix} \frac{1}{s_1-s_2}\left(\dot{x}_0 - s_2x_0\right) \\ -\frac{1}{s_1-s_2}\left(\dot{x}_0 - s_1x_0\right) \end{bmatrix}\end{aligned} \tag{4.94}$$

The complete solution is achieved.

$$\begin{aligned} x &= X_1e^{s_1t} + X_2e^{s_2t} \\ &= \frac{\dot{x}_0 - s_2x_0}{s_1 - s_2}e^{s_1t} - \frac{\dot{x}_0 - s_1x_0}{s_1 - s_2}e^{s_2t}\end{aligned} \tag{4.95}$$

Although it is the complete solution, it is not in the shape we want to work with. To rewrite it with harmonic functions, we employ the Euler identity.

$$e^{i\theta} = \cos\theta + i\sin\theta \tag{4.96}$$

$$e^{i\omega_n t} = \cos\omega_n t + i\sin\omega_n t \tag{4.97}$$

$$e^{-i\omega_n t} = \cos\omega_n t - i\sin\omega_n t \tag{4.98}$$

$$\begin{aligned} x &= X_1 e^{i\omega_n t} + X_2 e^{-i\omega_n t} \\ &= X_1(\cos\omega_n t + i\sin\omega_n t) + X_2(\cos\omega_n t - i\sin\omega_n t) \\ &= (X_1 + X_2)\cos\omega_n t + (X_1 - X_2)\, i\sin\omega_n t \end{aligned} \tag{4.99}$$

The coefficients $X_1$ and $X_2$ from (4.94) are complex conjugates.

$$\begin{bmatrix} X_1 \\ X_2 \end{bmatrix} = \begin{bmatrix} \frac{1}{s_1 - s_2}(\dot{x}_0 - s_2 x_0) \\ -\frac{1}{s_1 - s_2}(\dot{x}_0 - s_1 x_0) \end{bmatrix} = \frac{1}{2}\begin{bmatrix} x_0 - i\frac{1}{\omega_n}\dot{x}_0 \\ x_0 + i\frac{1}{\omega_n}\dot{x}_0 \end{bmatrix} \tag{4.100}$$

Therefore,

$$X_1 + X_2 = x_0 \tag{4.101}$$

$$X_1 - X_2 = -i\frac{1}{\omega_n}\dot{x}_0 \tag{4.102}$$

and it makes the solution $x$ to be a full harmonic function with real coefficients.

$$x = x_0\cos\omega_n t + \frac{1}{\omega_n}\dot{x}_0\sin\omega_n t \tag{4.103}$$

The parameter $\omega_n$ is the frequency of both functions $\cos\omega_n t$ and $\sin\omega_n t$ and hence the frequency of $x$. This is the frequency of response with no external excitation and no damping. That is the reason for calling this the natural frequency.

This solution suggests to begin with a solution made up of a full harmonic function with unknown coefficients and determine the coefficients. This solution is a full harmonic because it has both $\cos\omega_n t$ and $\sin\omega_n t$. These two functions are orthogonal, and hence, they are linearly independent. Each of them satisfies the differential equation, and therefore, a complete solution would be a linear combination of them.

$$x = A\sin\omega_n t + B\cos\omega_n t \tag{4.104}$$

Employing the initial conditions determines the coefficients and retrieves the same solution (4.104).

$$A = x_0 \qquad B = \frac{1}{\omega_n}\dot{x}_0 \tag{4.105}$$

Any full harmonic function such as the solution (4.104) may be combined into a single cos or sin function with a phase lag.

$$x = X\sin(\omega_n t + \varphi) \tag{4.106}$$

$$x = X\cos(\omega_n t + \theta) \tag{4.107}$$

The relationship between the amplitude $X$, phase $\varphi$, and weight factors $A$ and $B$ can be found by equating Eqs. (4.75) and (4.77):

$$\begin{aligned} A\sin\omega_n t + B\cos\omega_n t &= X\sin(\omega_n t + \varphi) \\ &= X\cos\varphi\sin\omega_n t + X\sin\varphi\cos\omega_n t \end{aligned} \tag{4.108}$$

$$A = X\cos\varphi \qquad B = X\sin\varphi \tag{4.109}$$

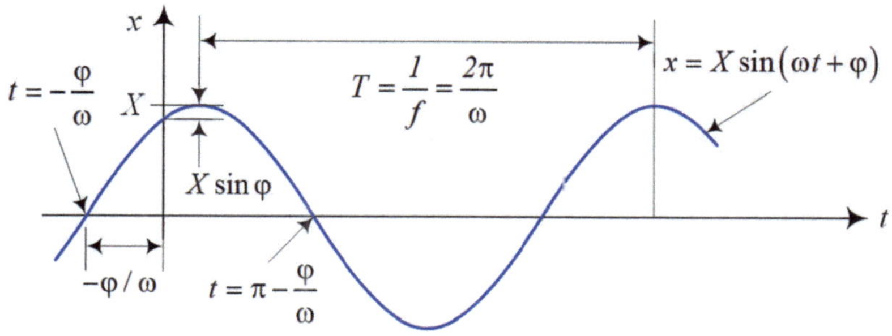


**Fig. 4.13** A harmonic wave $x = X \sin(\omega_n t + \varphi)$ and its characteristics

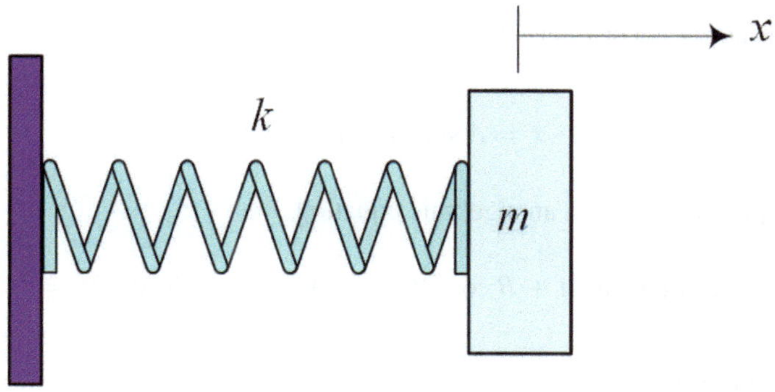


**Fig. 4.14** A mass–spring, single degree-of-freedom vibrating system

$$X = \sqrt{A^2 + B^2} \qquad \tan\varphi = \frac{B}{A} \tag{4.110}$$

Similarly, the relationship between the amplitude $X$, phase $\theta$, and weight factors $A$ and $B$ will be found by equating Eqs. (4.75) and (4.78):

$$\begin{aligned} A\sin\omega_n t + B\cos\omega_n t &= X\cos(\omega_n t + \theta) \\ &= X\sin\theta\sin\omega_n t - X\cos\theta\cos\omega_n t \end{aligned} \tag{4.111}$$

$$A = X\sin\theta \qquad B = -X\cos\theta \tag{4.112}$$

$$X = \sqrt{A^2 + B^2} \qquad \tan\theta = \frac{A}{-B} \tag{4.113}$$

A harmonic wave $x = X \sin(\omega_n t + \varphi)$ and its characteristics are plotted in Fig. 4.13.

There are three characteristics for a harmonic vibration:

1. *Amplitude* $X$ is the maximum deviation of $x$ from the $x = 0$. Amplitude is the size of vibration and is equivalent to *loudness* and *intensity* in music.
2. *Frequency* $\omega_n$ is the number of periodic motion in unit of time. Frequency is equivalent to *pitch* in music.
3. *Period* $T$, *length*, or *duration* is the elapsed time for a complete wave. Period is equivalent to *rhythm* in music.

■

**Example 20** Natural frequency. The natural frequency is a unique parameter at which a free vibrating system will oscillate, regardless of the initial conditions.

Consider the mass–spring system of Fig. 4.14. The system is undamped and force-free, with a harmonic equation of motion.

$$m\ddot{x} + kx = 0 \tag{4.114}$$

To find the solution, let us try a harmonic solution with an unknown frequency $\Omega$.

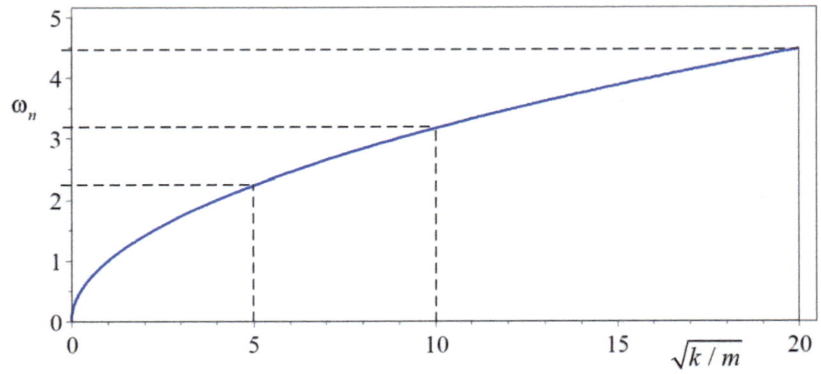


**Fig. 4.15** Variation of the natural frequency as a function of the ratio of stiffness to mass

$$x = A\sin\Omega t + B\cos\Omega t \tag{4.115}$$

Substituting (4.115) in (4.114) provides us with an algebraic equation

$$-\Omega^2 m\,(A\sin\Omega t + B\cos\Omega t) + k\,(A\sin\Omega t + B\cos\Omega t) = 0 \tag{4.116}$$

which can be collected as a full harmonic equation.

$$\left(Bk - Bm\Omega^2\right)\cos\Omega t + \left(Ak - Am\Omega^2\right)\sin\Omega t = 0 \tag{4.117}$$

The functions $\sin\Omega t$ and $\cos\Omega t$ are orthogonal, and hence, this equation can be correct only if their coefficients be zero. These conditions provide us with $\Omega$ and the possible solution $x$.

$$\Omega = \sqrt{\frac{k}{m}} \qquad x = A\sin\sqrt{\frac{k}{m}}t + B\cos\sqrt{\frac{k}{m}}t \tag{4.118}$$

The parameter $\Omega = \sqrt{k/m}$ is the frequency of vibration of a free and undamped mass–spring system. It is the natural frequency and is shown by a special character $\omega_n$.

$$\omega_n = \sqrt{\frac{k}{m}} \tag{4.119}$$

Figure 4.15 illustrates how the natural frequency varies as a function of the ratio of stiffness to mass. Natural frequency is very sensitive to the ratio of $k/m$ at low values and loses its sensitivity at higher values of $k/m$. Doubling the ratio of $k/m$ makes $\omega_n$ to increase by a factor of 1.414. As an example, a system with $k/m = 5$ has the natural frequency of $\omega_n = 2.236\,\mathrm{rad/s}$. Doubling $k/m$ to $k/m = 10$ will increase the natural frequency to $\omega_n = 3.162\,\mathrm{rad/s}$, and another doubling to $k/m = 20$ will make $\omega_n = 4.472$. An engineering point of this analysis is that it is harder to change the natural frequency of a heavy mechanical system than a light system. A system has as many natural frequencies as it has degrees of freedom.

From another viewpoint, we see that a system with no external excitation is a *free system* and with no damping is an *undamped system*. A single $DOF$ undamped-free system is governed by Eq. (4.114). As long as the mass and stiffness are constant, any nonzero response of the free system is harmonic.

$$x = A\sin\omega_n t + B\cos\omega_n t \tag{4.120}$$

Substituting the solution in the equation provides us with two equations:

$$-\omega_n^2 m\,(A\sin\omega_n t + B\cos\omega_n t) + k\,(A\sin\omega_n t + B\cos\omega_n t) = 0 \tag{4.121}$$

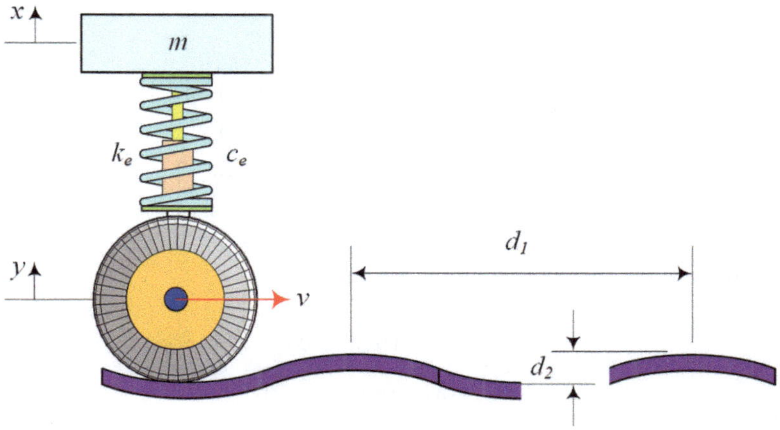


**Fig. 4.16** A $1/8$ car model moving with speed $v$ on a wavy road

$$Ak - Am\omega_n^2 = 0 \tag{4.122}$$

$$Bk - Bm\omega_n^2 = 0 \tag{4.123}$$

which shows that $A$ and $B$ can be nonzero if

$$\omega_n = \sqrt{\frac{k}{m}} \tag{4.124}$$

The frequency of the harmonic solution (4.120) is called the natural frequency of the system, just because any disturbance of the system puts it in a vibration with this frequency.

**Example 21** Wavy road and excitation frequency. The motion of a wheel on a wavy road is a practical example of harmonic motion. Here is to show the parameters of a wavy road to be compared with a harmonic function.

Figure 4.16 illustrates a $1/8$ car model moving with speed $v$ on a wavy road with length $d_1$ and peak-to-peak height $d_2$. Assuming a stiff tire with a small radius compared to the road waves, we may consider $y$ as the fluctuation of the road. The required time $T$ to pass one wave length $d_1$ is the period of the excitation.

$$T = \frac{d_1}{v} \tag{4.125}$$

Therefore the frequency of excitation is

$$\omega = \frac{2\pi}{T} = \frac{2\pi v}{d_1} \tag{4.126}$$

and the excitation $y = Y \sin \omega t$ is

$$y = \frac{d_2}{2} \sin \frac{2\pi v}{d_1} t \tag{4.127}$$

**Example 22** Boundary conditions. To determine the coefficients of the solution (4.75), we need the value of position $x$ or velocity $\dot{x}$ at two different times or a combination of them. Having the states $x$ and $\dot{x}$ at any time $t_0$ other time than $t_0 = 0$ is called boundary conditions.

If the state of the harmonic solution (4.75) is given at a time $t_0$,

$$x(t_0) = x_0 \qquad \dot{x}(t_0) = \dot{x}_0 \tag{4.128}$$

then we must have

$$x_0 = A \sin \omega t_0 + B \cos \omega t_0 \tag{4.129}$$

$$\dot{x}_0 = A\omega \cos \omega t_0 - B\omega \sin \omega t_0 \tag{4.130}$$

which yield

$$\begin{aligned}\begin{bmatrix} A \\ B \end{bmatrix} &= \begin{bmatrix} \sin \omega t_0 & \cos \omega t_0 \\ \omega \cos \omega t_0 & -\omega \sin \omega t_0 \end{bmatrix}^{-1} \begin{bmatrix} x_0 \\ \dot{x}_0 \end{bmatrix} \\ &= \frac{1}{\omega} \begin{bmatrix} \dot{x}_0 \cos \omega t_0 + \omega x_0 \sin \omega t_0 \\ -\dot{x}_0 \sin \omega t_0 + \omega x_0 \cos \omega t_0 \end{bmatrix}\end{aligned} \tag{4.131}$$

Substituting the initial conditions in the harmonic solution, we have

$$x_0 = X \sin(\omega t_0 + \varphi) \tag{4.132}$$

$$\dot{x}_0 = X\omega \cos(\omega t_0 + \varphi) \tag{4.133}$$

which provide us with

$$X = \sqrt{x_0^2 + \frac{\dot{x}_0^2}{\omega^2}} \qquad \tan(\omega t_0 + \varphi) = \frac{x_0}{\dot{x}_0/\omega} \tag{4.134}$$

## 4.3 Summary

In this chapter, we delve into the following topics:

1. The mechanical components of vibrating systems
2. The underlying physical causes of mechanical vibrations
3. The kinematics of vibrations
4. Simplification methods for handling complex vibrating systems

Mechanical vibration occurs due to the periodic conversion of kinetic energy ($K$) into potential energy ($P$). At the peak of potential energy, the kinetic energy is at its lowest and vice versa. The oscillation of a massive object resulting from these alternating energy transformations is what we refer to as mechanical vibrations. The kinetic and potential energies are stored within physical elements of the system. The components responsible for storing kinetic energy are referred to as masses or inertias, while the elements storing potential energy are known as springs or restoring elements. Additionally, if there is a decrease in the total mechanical energy ($E = K + P$) during vibrations, there exists an element or a phenomenon that dissipates energy. This energy-dissipating element is called a damper.

A periodic motion $x(t) = x(t + T)$ is characterized by a *period* $T$, which is the required time for one complete cycle of vibration, starting from and ending at the same conditions such as ($\dot{x} = 0, 0 < \ddot{x}$). The *frequency* $f$ is the number of cycles in one $T$.

$$f = \frac{1}{T}$$

In theoretical vibrations, we usually work with *angular frequency* $\omega\,[\mathrm{rad/s}]$, and in vehicle vibrations we use the *cyclic frequency* $f\,[\mathrm{Hz}]$.

$$\omega = 2\pi f \tag{4.135}$$

## 4.4 Key Symbols

| | |
|---|---|
| $a \equiv \ddot{x}$ | acceleration |
| $a, b$ | distance, length |
| $A$ | amplitude |
| $A, B$ | weight factor, coefficients for frequency responses |
| $c$ | damping |
| $c_e$ | equivalent damping |
| $C$ | coefficient, amplitude |

| | |
|---|---|
| $d_1$ | wave length |
| $d_2$ | peak-to-peak height of a wave |
| $E$ | mechanical energy |
| $E$ | Young modulus of elasticity |
| $E_c$ | consumed energy of a damper per cycle |
| $f = 1/T$ | cyclic frequency [ Hz] |
| $f, F, \mathbf{F}$ | force |
| $f, g$ | function, periodic function |
| $f_c$ | damper force |
| $f_d$ | damping force |
| $f_e$ | equivalent force |
| $f_k$ | spring force |
| $f_m$ | required force to move a mass $m$ |
| $F$ | amplitude of a harmonic force $f = F \sin \omega t$, total force |
| $F_0$ | constant force |
| $g$ | gravitational acceleration, function |
| $I$ | area moment, mass moment |
| $J$ | polar moment of cross sectional area |
| $k$ | stiffness |
| $k_e$ | equivalent stiffness |
| $k_R$ | antiroll bar torsional stiffness |
| $K$ | kinetic energy |
| $l$ | length |
| $L$ | length dimension symbol |
| $m$ | mass |
| $m_d$ | damper mass |
| $m_e$ | eccentric mass, equivalent mass |
| $m_s$ | spring mass |
| $m_k$ | spring mass |
| $M$ | mass dimension symbol |
| $N$ | normal force |
| $P$ | potential energy |
| $P_D$ | dissipated power in one cycle |
| $r$ | frequency ratio |
| $s$ | exponent, eigenvalue |
| $t$ | time |
| $t_0$ | initial time |
| $T$ | period |
| $T$ | Time dimension symbol |
| $T_n$ | natural period |
| $v \equiv \dot{x}$ | velocity |
| $W$ | work |
| $W_D$ | dissipated energy in one cycle |
| $x, y, z, \mathbf{x}$ | displacement |
| $x_0$ | initial displacement |
| $\dot{x}_0$ | initial velocity |
| $\dot{x}, \dot{y}, \dot{z}$ | velocity, time derivative of $x, y, z$ |
| $\ddot{x}$ | acceleration |
| $X$ | amplitude of $x$ |
| $z$ | relative displacement |
| $Z$ | amplitude of $z$ |

Greek

| | |
|---|---|
| $\alpha, \beta, \gamma$ | angle, angle of spring with respect to displacement |
| $\delta$ | deflection |
| $\delta_s$ | static deflection |
| $\varepsilon$ | mass ratio |
| $\varepsilon$ | small coefficient |
| $\rho$ | length mass density |
| $\theta$ | angular motion, phase angle |
| $\varphi$ | angular motion, phase angle |
| $\pi$ | $3.141592653589793\ldots$ |
| $\omega$, $\Omega$ | angular frequency |
| $\omega_n$ | natural frequency |
| $\mu$ | friction coefficient |

Symbols

| | |
|---|---|
| $\forall$ | for all |
| [ ] | dimension, matrix |
| $\int$ | integral |
| $\vert\ \vert$ | absolute value |
| Hz | Hertz |
| $d$ | differential |
| d | differential |

## References


1. J. Agyris, H.P. Mlejnek, *Computational Mechanics* (Elsevier, New York, 1991)
2. S. Axler, P. Bourdon, W. Ramey, *Harmonic Function Theory* (Springer, New York, 2001)
3. F. Jedrzejewski, *Mathematical Theory of Music* (IRCAM Centre Pompidou, Delatour, France, 2006)
4. B. Balachandran, E.B. Magrab, *Vibrations* (Brooks/Cole, Pacific Grove, CA, 2003)
5. G.M. Ballou, *Handbook for Sound Engineers*, 4th edn. (Focal Press, Elsevier, Oxford, UK, 2008)
6. D.J. Benson, *Music: A Mathematical Offering* (Cambridge University Press, London, UK, 2007)
7. H. Benaroya, *Mechanical Vibration: Analysis, Uncertainties, and Control* (Marcel Dekker, New York, 2004)
8. G. Buzdugan, E. Mihailescu, M. Rades, *Vibration measurement* (Springer, Dordrecht, 1986)
9. E.A. Coddington, *Ordinary Differential Equations* (Prentice Hall, New Jersey, 1961)
10. H.S. Carslaw, *An Introduction to the Theory of Fourier's Series and Integrals* (Macmillan, London UK, 1921)
11. T.S. Chihara, *An Introduction to Orthogonal Polynomials* (Gordon and Breach, Science Publishers, New York, NY, 1978)
12. H.K. Dass, E.R. Verma, *Higher Mathematical Physics* (S CHAND & Company, New Delhi, India, 2014)
13. M. Del Pedro, P. Pahud, *Vibration Mechanics* (Kluwer Academic Publishers, The Netherland, 1991)
14. J.P. Den Hartog, *Mechanical Vibrations* (McGraw-Hill, New York, 1934)
15. A. Dimarogonas, *Vibration for Engineers* (Prentice Hall, New Jersey, 1996)
16. E. Esmailzadeh, R.N. Jazar, Periodic solution of a Mathieu-Duffing type equation. Int. J. Nonlinear Mech. **32**(5), 905–912 (1997)
17. E. Esmailzadeh, B. Mehri, R.N. Jazar, Periodic solution of a second order, autonomous, nonlinear system. J. Nonlinear Dyn. **10**(4), 307–316 (1996)
18. J. Fauvel, R. Flood, R. Wilson, *Music and Mathematics: From Pythagoras to Fractals* (Oxford University Press Inc., New York, NY, 2003)
19. J.B.J. Fourier, *The Analytical Theory of Heat* Translated by: Freeman A., digitally printed version 2009 (Cambridge University Press, New York, NY, 1878)
20. E. Grigorieva, *Methods of Solving Nonstandard Problems* (Springer, New York, NY, 2015)
21. G. Gunther, *The Physics of Music and Color* (Springer, New York, 2012)
22. C.M. Harris, A.G. Piersol, *Harris' Shock and Vibration Handbook* (McGraw-Hill, New York, NY, 2002)
23. M.H. Holmes, *Introduction to the Foundations of Applied Mathematics* (Springer, New York, NY, 2009)
24. D. Inman, *Engineering Vibrations* (Prentice Hall, New York, NY, 2007)
25. R.N. Jazar, *Analysis of Nonlinear Parametric Vibrating Systems*. Ph.D. Thesis, Mechanical Engineering Department, Tehran: Sharif University of Technology (1997)
26. R.N. Jazar, Stability chart of parametric vibrating systems using energy-rate method. Int. J. Non-linear Mech. **39**(8), 1319–1331 (2004)
27. R.N. Jazar, *Advanced Dynamics: Rigid Body, Multibody, and Aero-space Applications* (Wiley, New York, NY, 2011)
28. R.N. Jazar, *Advanced Vibrations: A Modern Approach* (Springer, New York, NY, 2013)
29. R.N. Jazar, *Vehicle Dynamics: Theory and Application*, 3rd edn. (Springer, New York, NY, 2017)
30. R.N. Jazar, *Approximation Methods in Science and Engineering* (Springer, New York, NY, 2020)
31. R.N. Jazar, *Perturbation Methods in Science and Engineering* (Springer, New York, 2021)
32. R.N. Jazar, *Advanced Vibrations: Theory and Application* (Springer, New York, NY, 2023)

33. R.N. Jazar, M. Kazemi, S. Borhani, *Mechanical Vibrations* (in Persian) (Ettehad Publications, Tehran, 1992)
34. R.N. Jazar, M. Mahinfalah, N. Mahmoudian, M.R. Aagaah, Energy-rate method and stability chart of parametric vibrating systems. J. Braz. Soc. Mech. Sci. Eng. **30**(3), 182–188 (2008)
35. R.N. Jazar, M. Mahinfalah, N. Mahmoudian, M.R. Aagaah, Effects of nonlinearities on the steady state dynamic behavior of electric actuated microcantilever-based resonators. J. Vib. Control **15**(9), 1283–1306 (2009)
36. R.N. Jazar, M. Mahinfalah, N. Mahmoudian, M.R. Aagaah, B. Shiari, Behavior of Mathieu equation in stable regions. Int. J. Mech. Solids **1**(1), 1–18 (2006)
37. L. Meirovitch, *Analytical Methods in Vibrations* (Macmillan, New York, 1967)
38. L. Meirovitch, *Principles and Techniques of Vibrations* (Prentice Hall, Ney Jersey, 1997)
39. L. Meirovitch, *Fundamentals of Vibrations* (McGraw-Hill, New York, 2002)
40. R.E. Mickens, *Generalized Trigonometric and Hyperbolic Functions* (CRC Press, Boca Raton, FL, 2019)
41. R.E. Mickens, *Mathematical Methods for the Natural and Engineering Sciences* (World Scientific Publishing, Singapore, 2004)
42. V.P. Minofsky, *Problems in Higher Mathematics* (Mir Publishers, Moscow, 1975)
43. D.E. Newland, *Mechanical Vibration Analysis and Computation* (Wiley, New York, NY, 1989)
44. K. Ogata, *System Dynamics* (Prentice Hall, Ney Jersey, 2004)
45. Z. Osinski, *Damping of Vibrations* (A. A. Balkema Publishers, Rotterdam, Netherlands, 1998)
46. W.J. Palm, *Mechanical Vibration* (Wiley, New York, 2006)
47. E. Prestini, *The Evolution of Applied Harmonic Analysis: Models of the Real World* (Springer, New York, NY, 2004)
48. S.S. Rao, *Mechanical Vibrations* (Pearson Prentice Hall, Harlow, UK, 2018)
49. J.W.S. Rayleigh, *The Theory of Sound* (Dover Publication, New York, 1945)
50. M. Roseau, *Vibrations in Mechanical Systems* (Springer, Berlin, 1987)
51. A.A. Shabana, *Vibration of Discrete and Continuous Systems* (Springer, New York, 1997)
52. G. Sansone, *Orthogonal Functions*. Translated From the Italian by Diamond A. H. (Dover Publications, New York, 1991)
53. H. Shima, T. Nakayama, *Higher Mathematics for Physics and Engineering* (Springer, Berlin, Heidelberg, 2010)
54. G.F. Simmons, *Differential Equations with Applications and Historical Notes* (CRC Press, Boca Raton, FL, 2017)
55. J.C. Snowdon, *Vibration and Shock in Damped Mechanical Systems* (Wiley, New York, 1968)
56. E.C. Titchmarsh, *Introduction to the Theory of Fourier Integrals* (Oxford University Press, London UK, 1948)
57. W.T. Thomson, M.D. Dahleh, *Theory of Vibration with Applications* (Prentice Hall, Ney Jersey, 1997)
58. B.H. Tongue, *Principles of Vibration* (Oxford University Press, New York, 2001)
59. F.S. Tse, I.E. Morse, R.T. Hinkle, *Mechanical Vibrations Theory and Applications* (Allyn and Bacon Inc., Boston, MA, 1978)

# Part II

# Mechanical Vibrations Analysis

In this part, we will explore the various analytical tools used to study vehicle vibrations mathematically. Vehicle vibrations is a complex field that requires a range of tools to fully understand. Among these tools, linear algebra, trigonometry, principles of dynamics, mechanical vibrations, differential equations, and principles of mechanical design play crucial roles in the study of vehicle vibrations. Of these tools, mechanical vibrations stands out as the most essential, and we will focus on this topic in this part.

The first chapter of this part introduces the dynamics of vibration, providing an overview of the methods used to derive the equations of motion for mechanical systems, including vehicles. Mechanical vibrations are explained as a phenomenon that arises from the repeated conversion of kinetic and potential energies. Understanding this dynamic process is fundamental to comprehending vehicle vibrations.

Moving on, the following chapter delves into the time response of single and multiple degrees of freedom ($DOF$) systems. The analysis of these systems differs significantly, with multiple $DOF$ systems relying on the concept of mode shapes, which is absent in single $DOF$ systems. Mode shapes play a vital role in characterizing the behavior and response of multiple $DOF$ systems, and their understanding is crucial for an accurate analysis.

Finally, the last chapter of this part focuses on the analysis method of frequency response for both single and multiple $DOF$ systems. The frequency response analysis allows us to examine how a mechanical system, such as a vehicle, responds to different frequencies of excitation. This analysis technique is essential for evaluating the dynamic behavior and performance of the system under varying conditions.

By comprehensively reviewing these analytical tools and their applications to vehicle vibrations, we gain a deeper understanding of the subject and enhance our ability to analyze and design vehicle systems that effectively manage vibrations.

# 5 Vibration Dynamics

In this chapter, we review the dynamics of vibrations and the methods of deriving the equations of motion of vibrating systems. The Newton–Euler and Lagrange methods are the most applied methods of deriving the equations of motion. Having symmetric coefficient matrices for multi-degrees-of-freedom linear vibrating systems is the main advantage of using the Lagrange method in mechanical vibrations. For example, Fig. 5.1 illustrates a one degree of freedom vibrating system.

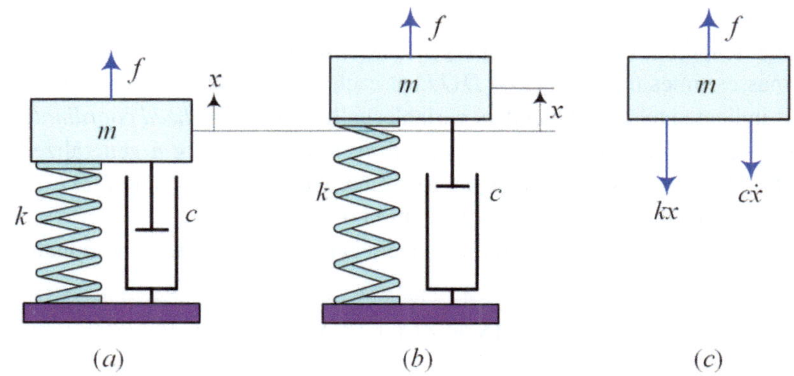


**Fig. 5.1** A one $DOF$ vibrating system and its $FBD$. (**a**) Equilibrium. (**b**) In motion. (**c**) Free body diagram

## 5.1 Newton–Euler Method

When a vibrating system is modeled as a combination of masses $m_i$, dampers $c_i$, and springs $k_i$, it is called a *discrete* or *lumped* model of the system. To derive the equations of motion of a low degree of freedom ($DOF$) discrete model of a vibrating system, the Newton–Euler method works very well. We move all the masses $m_i$ out of their equilibria at positions $x_i$ with velocities $\dot{x}_i$. Then, a free body diagram ($FBD$) of the lumped masses indicates the total forces $\mathbf{F}_i$ on mass $m_i$. Employing the momentum $\mathbf{p}_i = m_i\,\mathbf{v}_i$ of the mass $m_i$, the Newton equation provides us with the equation of motion of the system:

$$\mathbf{F}_i = \frac{d}{dt}\,\mathbf{p}_i = \frac{d}{dt}\,(m_i\,\mathbf{v}_i) \tag{5.1}$$

When the motion of a massive body with mass moment $I_i$ is rotational, then its equation of motion will be determined by the Euler equation, in which we employ the moment of momentum $\mathbf{L}_i = I_i\,\boldsymbol{\omega}$ of the mass $m_i$:

$$\mathbf{M}_i = \frac{d}{dt}\,\mathbf{L}_i = \frac{d}{dt}\,(I_i\,\boldsymbol{\omega}) \tag{5.2}$$


R. N. Jazar, H. Marzbani, *Vehicle Vibrations*, https://doi.org/10.1007/978-3-031-43486-0_5

For example, Fig. 5.1 illustrates a one degree of freedom vibrating system. Figure 5.1b depicts the system when $m$ is out of the equilibrium position at $x$ and moving with velocity $\dot{x}$, both in positive direction. The free body diagram of the system is as shown in Fig. 5.1c. The Newton equation generates the equations of motion:

$$f(x, v, t) - c\dot{x} - kx = m\ddot{x} \tag{5.3}$$

We measure displacement of a vibrating system from its equilibrium configuration. The *equilibrium position* of a vibrating system is where the potential energy of the system, $P$, is extremum:

$$\frac{\partial P}{\partial x} = 0 \tag{5.4}$$

We set $P = 0$ at the equilibrium position. Linear systems have only one equilibrium configuration or infinity equilibria, while nonlinear systems may have multiple equilibria. An equilibrium is *stable* if the second spatial derivative of the potential energy $P$ is positive,

$$\frac{\partial^2 P}{\partial x^2} > 0 \tag{5.5}$$

and is *unstable* if $P$ is negative.

$$\frac{\partial^2 P}{\partial x^2} < 0 \tag{5.6}$$

The geometric arrangement and the number of employed mechanical elements can be used to classify discrete vibrating systems. The number of masses times the number of $DOF$ of each mass makes the total $DOF$ of the vibrating system $n$. Each independent $DOF$ is indicated by an independent variable, called the *generalized coordinate*. The final set of equations of motion would be $n$ coupled second-order differential equations to be solved for $n$ generalized coordinates. When every moving mass has one $DOF$, the system's $DOF$ is equal to the number of masses. The $DOF$ of a system is the minimum number of independent coordinates that defines the configuration of a system.

The equation of motion of an $n$ $DOF$ linear vibrating system can always be arranged in matrix form of a set of second-order differential equations

$$[m]\,\ddot{\mathbf{x}} + [c]\,\dot{\mathbf{x}} + [k]\,\mathbf{x} = \mathbf{F} \tag{5.7}$$

where $\mathbf{x}$ is a column array of generalized coordinates of the system, and $\mathbf{F}$ is a column array of the associated applied forces. The *square matrices* of $[m]$, $[c]$, $[k]$ are the mass, damping, and stiffness matrices, respectively.

**Example 23** The one, two, and three $DOF$ vibration model of vehicles. Vertical vibration of vehicles may be modeled by base vibrating system with one or two or three $DOF$ systems by increasing complexity and accuracy.

The one, two, and three $DOF$ models for analysis of vertical vibrations of a vehicle are shown in Fig. 5.2a–c. The system in Fig. 5.2a is called the quarter car model, in which $m_s$ represents a quarter mass of the body, and $m_u$ represents a wheel. The parameters $k_u$ and $c_u$ are models for tire stiffness and damping. Similarly, $k_s$ and $c_s$ are models for the stiffness and damping of the main suspension vehicle. Figure 5.2c is called the 1/8 car model, which does not show the wheel of the car. Figure 5.2b is a quarter car with a driver $m_d$. The stiffness and damping of the driver's seat are modeled by $k_d$ and $c_d$.

The 1/8 car model is the simplest vibrating model of vehicles to study and optimize the main characteristics of vehicle suspensions. To derive the equation of motion of the model, Fig. 5.3a illustrates the mathematical model in which the mass $m_s$ represents one quarter of the car's body. The mass $m_s$ is mounted on a suspension made of a spring $k_s$ and a damper $c_s$. When $m_s$ is at an off-equilibrium position such as shown in Fig. 5.3b, its free body diagram is as shown in Fig. 5.3c.

Applying Newton's method on the free body diagram, the equation of motion would be

$$m_s\ddot{x} = -k_s\left(x_s - y\right) - c_s\left(\dot{x}_s - \dot{y}\right) \tag{5.8}$$

which can be simplified to

$$m_s\ddot{x} + c_s\dot{x}_s + k_s x_s = k_s y + c_s\dot{y} \tag{5.9}$$

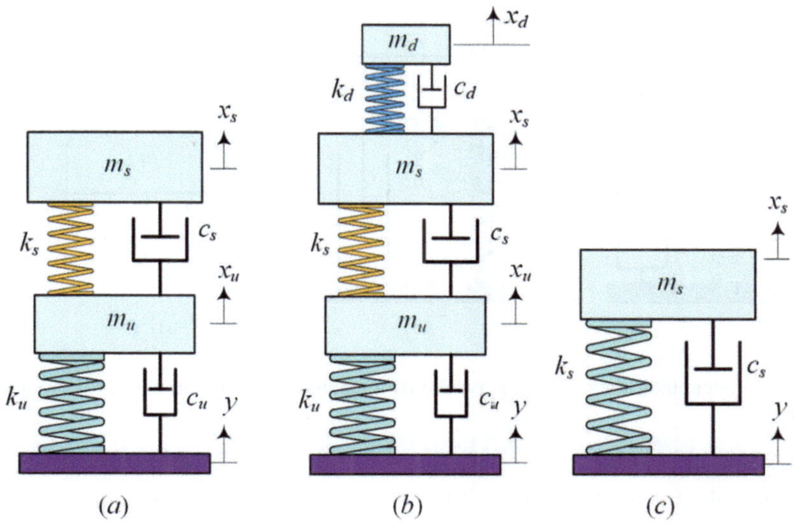


**Fig. 5.2** Two, three, and one $DOF$ models for vertical vibrations of vehicles

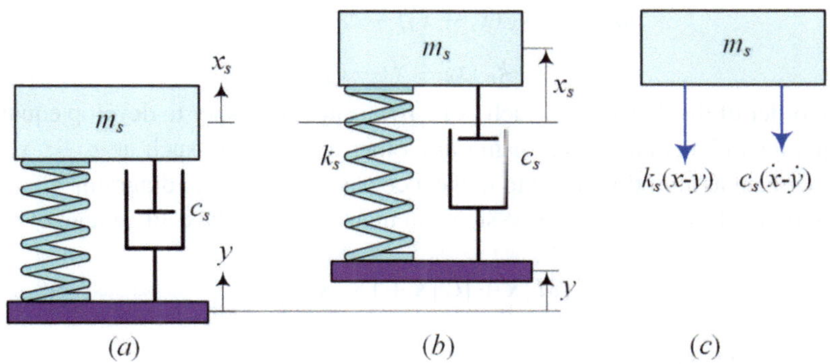


**Fig. 5.3** A 1/8 car model and its free body diagram

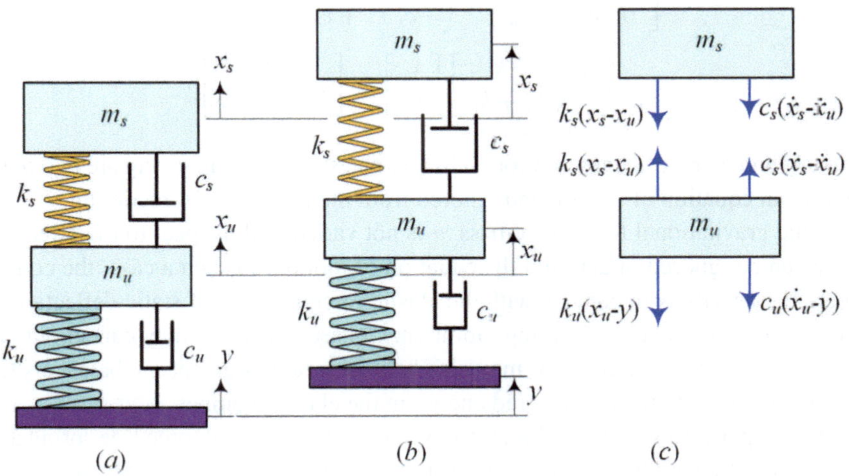


**Fig. 5.4** A 1/4 car model and its free body diagram

The coordinate $y$ indicates the input from the road, and $x$ indicates the absolute displacement of the body. Absolute displacement refers to displacement with respect to the motionless background.

Figure 5.4a–c illustrates the equilibrium, motion, and $FBD$ of a two $DOF$ system for quarter car model. The $FBD$ is plotted based on the following assumption:

$$x_s > x_u > y \tag{5.10}$$

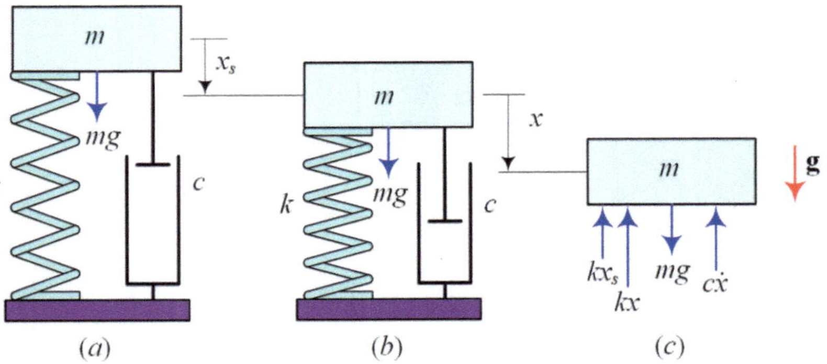


**Fig. 5.5** A mass–spring–damper system indicating that the gravitational force in rectilinear vibrations provides a static deflection. (**a**) In touch. (**b**) Equilibrium. (**c**) In motion

Applying Newton's method on the free body diagram of Fig. 5.4c provides us with two equations of motion:

$$m_s \ddot{x}_s = -k_s (x_s - x_u) - c_s (\dot{x}_s - \dot{x}_u) \tag{5.11}$$

$$m_u \ddot{x}_u = k_s (x_s - x_u) + c_s (\dot{x}_s - \dot{x}_u) - k_u (x_u - y) - c_u (\dot{x}_u - \dot{y}) \tag{5.12}$$

An assumption for relative order of displacements, such as (5.10), is not necessary to develop equations of motion correctly. We can find the same Equations (5.11) and (5.12) using any other assumption, such as $x_s < x_u > y$, $x_s > x_u < y$, or $x_s < x_u < y$. However, having an assumption helps to make a consistent free body diagram.

We usually arrange equations of motion of linear systems in matrix form to take advantage of matrix calculus.

$$[M]\,\dot{\mathbf{x}} + [C]\,\dot{\mathbf{x}} + [K]\,\mathbf{x} = \mathbf{F} \tag{5.13}$$

Rearrangement of Equations (5.11) and (5.12) yields

$$\begin{bmatrix} m_s & 0 \\ 0 & m_u \end{bmatrix} \begin{bmatrix} \ddot{x}_s \\ \ddot{x}_u \end{bmatrix} + \begin{bmatrix} c_s & -c_s \\ -c_s & c_s + c_u \end{bmatrix} \begin{bmatrix} \dot{x}_s \\ \dot{x}_u \end{bmatrix} + \begin{bmatrix} k_s & -k_s \\ -k_s & k_s + k_u \end{bmatrix} \begin{bmatrix} x_s \\ x_u \end{bmatrix} = \begin{bmatrix} 0 \\ k_u y + c_u \dot{y} \end{bmatrix} \tag{5.14}$$

**Example 24** Gravitational force in rectilinear vibrations. Gravity and static force in spring are two opposite and equal forces that may be canceled out from equation of motion and ignored from the beginning of modeling.

When the direction of the gravitational force on a mass $m$ is not varied with respect to the direction of motion of $m$, the effect of the weight force can be ignored in deriving the equation of motion. In such a case, the equilibrium position of the system is at a point where the gravity is in balance with the elastic force because of static deflection in the elastic member. This force balance equation will not be altered during vibration. Consequently, we may cancel out and ignore both forces: the gravitational force and the static elastic force. It may also be interpreted as an energy balance situation where the work of gravitational force is always equal to the extra stored energy in the elastic member.

Consider a spring $k$ and damper $c$ as is shown in Fig. 5.5a. A mass $m$ is put on the force free spring and damper. The weight of $m$ compresses the spring a static length $x_s$ to bring the system at equilibrium in Fig. 5.5b. When $m$ is at equilibrium, it is under the balance of two forces, $mg$ and $-kx_s$:

$$mg - kx_s = 0 \tag{5.15}$$

While the mass is in motion, its $FBD$ is as shown in Fig. 5.5c and its equation of motion is

$$m\ddot{x} = -kx - c\dot{x} + mg - kx_s = -kx - c\dot{x} \tag{5.16}$$

It shows that if we examine the motion of the system from equilibrium, we can ignore both the gravitational force and the initial compression of the elastic member of the system.

The equation of motion for a vibrating system is a balance between four different forces: a force proportional to displacement, $-kx$, a force proportional to speed, $-c\dot{x}$, a force proportional to acceleration, $ma$, and an applied external force $f(x, \dot{x}, t)$. The external force may be a function of displacement, velocity, and time. Based on Newton method, the force proportional to acceleration, $ma$, is always equal to the sum of all the other forces.

$$ma = -c\dot{x} - kx + f(x, \dot{x}, t) \tag{5.17}$$

## 5.2 Energy Method

In Newtonian mechanics, the acting forces on a system of bodies can be divided into *internal* and *external forces*. Internal forces are acting between bodies of the system, and external forces are acting from outside of the system. External forces and moments are called *load*. Principle of work and energy is a strong method to derive equations of motion of energy conserved dynamic system. The equations we find for such systems are first-order differential equations, much simpler to solve.

### 5.2.1 Force System

The resultant of acting forces and moments on a rigid body is called a *force system*. The *resultant* or *total force* $\mathbf{F}$ is the vectorial sum of all the external forces acting on the body, and the *resultant* or *total moment* $\mathbf{M}$ is the vectorial sum of all the moments of the external forces about a point, such as the origin of a coordinate frame.

$$\mathbf{F} = \sum_i \mathbf{F}_i \qquad \mathbf{M} = \sum_i \mathbf{M}_i \tag{5.18}$$

The moment of a force $\mathbf{F}$ an acting on a point $P$ at $\mathbf{r}_P$ is

$$\mathbf{M} = \mathbf{r}_P \times \mathbf{F} \tag{5.19}$$

The moment of the force $\mathbf{F}$ about a point $Q$ at $\mathbf{r}_Q$ is

$$\mathbf{M}_Q = \left(\mathbf{r}_P - \mathbf{r}_Q\right) \times \mathbf{F} \tag{5.20}$$

The *moment of the force* about a directional line $l$ with unit vector $\hat{u}$ passing through the origin is

$$\mathbf{M}_l = \hat{u} \cdot (\mathbf{r}_P \times \mathbf{F}) \tag{5.21}$$

The moment of a force may also be called *torque* or *moment* although they are conceptually different.

The effect of a force system is equivalent to the effect of the resultant force and resultant moment of the force system. Any two force systems are equivalent if their resultant forces and resultant moments are equal. If the resultant force of a force system is zero, the resultant moment of the force system will be independent of the origin of the coordinate frame. Such a resultant moment is called a *couple*.

When a force system is reduced to a resultant $\mathbf{F}_P$ and $\mathbf{M}_P$ with respect to a reference point $P$, we may change the reference point to another point $Q$ and find the new resultants as

$$\mathbf{F}_Q = \mathbf{F}_P \tag{5.22}$$

$$\mathbf{M}_Q = \mathbf{M}_P + \left(\mathbf{r}_P - \mathbf{r}_Q\right) \times \mathbf{F}_P = \mathbf{M}_P + {}_Q\mathbf{r}_P \times \mathbf{F}_P \tag{5.23}$$

The application of a force system is by *Newton's second* and *third laws of motion*. The second law of motion, also called the *Newton's equation of motion*, states that the global rate of change of *linear momentum* is proportional to the global *applied force*.

$${}^G\mathbf{F} = \frac{{}^Gd}{dt}\,{}^G\mathbf{p} = \frac{{}^Gd}{dt}\left(m\,{}^G\mathbf{v}\right) \tag{5.24}$$

The third Newton's law of motion states that the action and reaction forces acting between two bodies are equal and opposite.

The second law of motion can be expanded to include rotational motions. Hence, the second law of motion also states that the global rate of change of *angular momentum* is proportional to the global *applied moment.*

$$ {}^{G}\mathbf{M} = \frac{{}^{G}d}{dt}\,{}^{G}\mathbf{L} \tag{5.25} $$

***Proof*** Differentiating the angular momentum (5.29) shows that

$$ \begin{aligned} \frac{{}^{G}d}{dt}\,{}^{G}\mathbf{L} &= \frac{{}^{G}d}{dt}\left(\mathbf{r}_C \times \mathbf{p}\right) = \left(\frac{{}^{G}d\mathbf{r}_C}{dt} \times \mathbf{p} + \mathbf{r}_C \times \frac{{}^{G}d\mathbf{p}}{dt}\right) \\ &= {}^{G}\mathbf{r}_C \times \frac{{}^{G}d\mathbf{p}}{dt} = {}^{G}\mathbf{r}_C \times {}^{G}\mathbf{F} = {}^{G}\mathbf{M} \end{aligned} \tag{5.26} $$

■

### 5.2.2 Momentum

The *momentum* **p** of a moving body is a vector quantity equal to the total mass of the body times the translational velocity of the mass center of the body.

$$ \mathbf{p} = m\mathbf{v} \tag{5.27} $$

The momentum **p** may also be called the *translational momentum* or *linear momentum.*

Consider a rigid body with momentum **p**. The *moment of momentum,* **L**, about a directional line $l$ passing through the origin is

$$ \mathbf{L}_l = \hat{u} \cdot \left(\mathbf{r}_C \times \mathbf{p}\right) \tag{5.28} $$

where $\hat{u}$ is a unit vector indicating the direction of the line, and $\mathbf{r}_C$ is the position vector of the mass center $C$. The moment of momentum about the origin is

$$ \mathbf{L} = \mathbf{r}_C \times \mathbf{p} \tag{5.29} $$

The moment of momentum **L** is also called *angular momentum.*

### 5.2.3 Mechanical Energy

*Kinetic energy* $K$ of a moving body point with mass $m$ at a position ${}^{G}\mathbf{r}$, and having a velocity ${}^{G}\mathbf{v}$, in the global coordinate frame $G$ is

$$ K = \frac{1}{2} m\, {}^{G}\mathbf{v} \cdot {}^{G}\mathbf{v} = \frac{1}{2} m\, {}^{G}\mathbf{v}^2 \tag{5.30} $$

where $G$ indicates the global coordinate frame in which the velocity vector **v** is expressed.

The work ${}_1W_2$ done by the applied force ${}^{G}\mathbf{F}$ on $m$ in moving from spatial point 1 to point 2 on a path, indicated by a vector ${}^{G}\mathbf{r}$, is

$$ {}_1W_2 = \int_1^2 {}^{G}\mathbf{F} \cdot d\,{}^{G}\mathbf{r} \tag{5.31} $$

However,

$$ \begin{aligned} \int_1^2 {}^{G}\mathbf{F} \cdot d\,{}^{G}\mathbf{r} &= m \int_1^2 \frac{{}^{G}d}{dt}\,{}^{G}\mathbf{v} \cdot {}^{G}\mathbf{v}dt = \frac{1}{2} m \int_1^2 \frac{d}{dt} v^2 dt \\ &= \frac{1}{2} m \left(v_2^2 - v_1^2\right) = K_2 - K_1 \end{aligned} \tag{5.32} $$

which shows that ${}_1W_2$ is equal to the difference of the kinetic energy of terminal and initial points.

$$ {}_1W_2 = K_2 - K_1 \tag{5.33} $$

Equation (5.33) is called the *principle of work and energy*. If there is a scalar *potential field function* $P = P(x, y, z)$ such that

$$ \mathbf{F} = -\nabla P = -\frac{dP}{d\mathbf{r}} = -\left(\frac{\partial P}{\partial x}\hat{\imath} + \frac{\partial P}{\partial y}\hat{\jmath} + \frac{\partial P}{\partial z}\hat{k}\right) \tag{5.34} $$

then the principle of work and energy (5.33) simplifies to the principle of *conservation of energy*,

$$ K_1 + P_1 = K_2 + P_2 \tag{5.35} $$

The value of the potential field function $P = P(x, y, z)$ is the *potential energy* of the system.

The sum of kinetic and potential energies, $E = K + P$, is called the mechanical energy. Mechanical energy of an energy conserved system is a constant of motion, and hence, its time derivative is zero, $\dot{E} = 0$.

***Proof*** The spatial integral of Newton equation of motion is

$$ \int_1^2 \mathbf{F} \cdot d\mathbf{r} = m \int_1^2 \mathbf{a} \cdot d\mathbf{r} \tag{5.36} $$

We can simplify the right-hand side of the integral (5.36) by the change of variable

$$ \int_{\mathbf{r}_1}^{\mathbf{r}_2} \mathbf{F} \cdot d\mathbf{r} = m \int_{\mathbf{r}_1}^{\mathbf{r}_2} \mathbf{a} \cdot d\mathbf{r} = m \int_{t_1}^{t_2} \frac{d\mathbf{v}}{dt} \cdot \mathbf{v}dt $$
$$ = m \int_{\mathbf{v}_1}^{\mathbf{v}_2} \mathbf{v} \cdot d\mathbf{v} = \frac{1}{2}m\left(\mathbf{v}_2^2 - \mathbf{v}_1^2\right) = K_2 - K_1 \tag{5.37} $$

The kinetic energy of a point mass $m$ at position ${}^G\mathbf{r}$ and having a velocity ${}^G\mathbf{v}$ is defined by (5.30). Whenever the global coordinate frame $G$ is the only involved frame, we may drop the superscript $G$ for simplicity. The work done by the applied force ${}^G\mathbf{F}$ on $m$ in going from point $\mathbf{r}_1$ to $\mathbf{r}_2$ is defined by (5.31). Hence, the spatial integral of equation of motion (5.36) reduces to the principle of work and energy (5.33), which says that the work ${}_1W_2$ done by the applied force ${}^G\mathbf{F}$ on $m$ during the displacement $\mathbf{r}_2 - \mathbf{r}_1$ is equal to the difference of the kinetic energy of $m$.

$$ {}_1W_2 = K_2 - K_1 \tag{5.38} $$

If the force $\mathbf{F}$ is the gradient of a potential function $P$,

$$ \mathbf{F} = -\nabla P \tag{5.39} $$

then $\mathbf{F} \cdot d\mathbf{r}$ in Equation (5.36) is an exact differential.

$$ \int_1^2 \mathbf{F} \cdot d\mathbf{r} = \int_1^2 dP = -(P_2 - P_1) \tag{5.40} $$
$$ E = K_1 + P_1 = K_2 + P_2 \tag{5.41} $$

Therefore, the work done by a potential force $\mathbf{F} = -\nabla P$ is independent of the path of motion between $\mathbf{r}_1$ and $\mathbf{r}_2$ and depends only on the value of the potential $P$ at start and end points of the path. The function $P$ is the potential energy, Equation (5.41) is called the principle of conservation of energy, and the force $\mathbf{F} = -\nabla P$ is called a potential or a conservative force. The kinetic plus potential energy of a dynamic system is called the mechanical energy of the system and is denoted by $E = K + P$. The mechanical energy $E$ is a constant of motion if all the applied forces are conservative. Hence, $\dot{E} = 0$ will make the equations of motion.

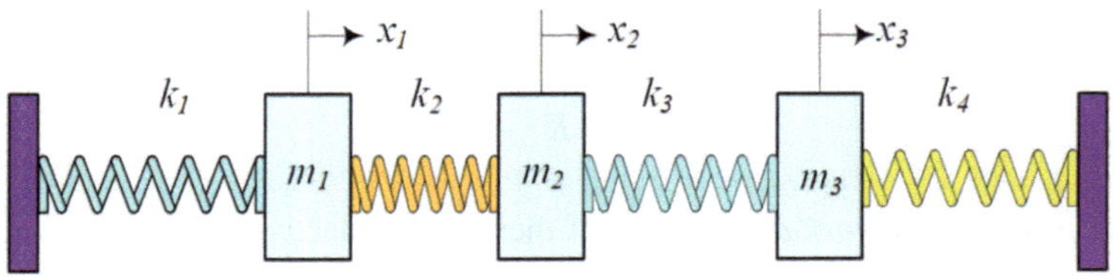


**Fig. 5.6** A multi-$DOF$ conservative vibrating system

A force **F** is conservative only if it is the gradient of a stationary scalar function. The components of a conservative force will only be functions of space coordinates.

$$\mathbf{F} = F_x(x, y, z)\hat{\imath} + F_y(x, y, z)\hat{\jmath} + F_z(x, y, z)\hat{k} \tag{5.42}$$

■

**Example 25** Energy and equation of motion. If a vibrating system is energy conserved, derivative of mechanical energy will provide us with the equation of motion.

Whenever there is no loss of energy in a mechanical vibrating system, the sum of kinetic and potential energies is a constant of motion.

$$E = K + P = const. \tag{5.43}$$

A system with constant energy is called a conservative system. The time derivative of a constant of motion must be zero at all times.

Consider a mass–spring system that is a conservative system with the total mechanical energy $E$.

$$E = \frac{1}{2}m\dot{x}^2 + \frac{1}{2}kx^2 \tag{5.44}$$

Having a zero rate of energy,

$$\dot{E} = m\dot{x}\ddot{x} + kx\dot{x} = \dot{x}(m\ddot{x} + kx) = 0 \tag{5.45}$$

and knowing that $\dot{x}$ cannot be zero at all times provide us with the equation of motion as

$$m\ddot{x} + kx = 0 \tag{5.46}$$

**Example 26** Energy and multi-$DOF$ systems. Here is a conservative system whose equation of motion can be found by principle of conservation of energy.

We may use energy method and determine the equations of motion of multi-$DOF$ conservative systems. Consider the 3 $DOF$ system in Fig. 5.6 whose mechanical energy is

$$\begin{aligned} E = K + P = {} & \frac{1}{2}m_1\dot{x}_1^2 + \frac{1}{2}m_2\dot{x}_2^2 + \frac{1}{2}m_3\dot{x}_3^2 \\ & + \frac{1}{2}k_1x_1^2 + \frac{1}{2}k_2(x_1 - x_2)^2 + \frac{1}{2}k_3(x_2 - x_3)^2 + \frac{1}{2}k_4x_3^2 \end{aligned} \tag{5.47}$$

Because the system has 3 degrees of freedom, we must find three equations of motions. To find the first equation of motion associated to $x_1$, we assume $x_2$ and $x_3$ and their time rates are constant and take a time directive from $E$.

$$\dot{E} = m_1\dot{x}_1\ddot{x}_1 + k_1x_1\dot{x}_1 + k_2(x_1 - x_2)\dot{x}_1 = 0 \tag{5.48}$$

Because $\dot{x}_1$ cannot be zero at all times, we may factor $\dot{x}_1$ out and find the first equation of motion.

$$m_1\ddot{x}_1 + k_1x_1 + k_2(x_1 - x_2) = 0 \tag{5.49}$$

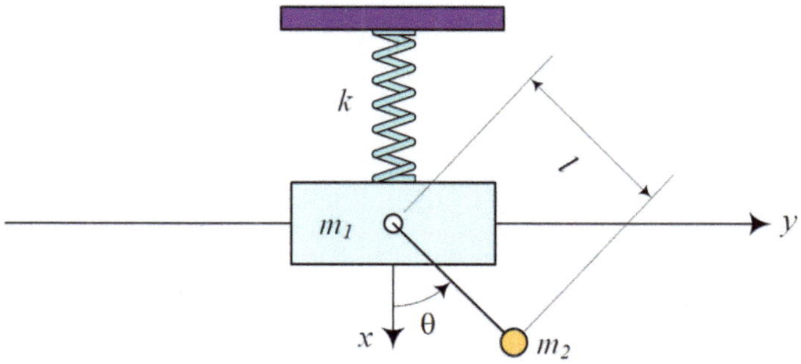


**Fig. 5.7** A two $DOF$ conservative nonlinear vibrating system

To find the second equation of motion associated to $x_2$, we assume that $x_1$ and $x_3$ and their time rates are constant and take a time directive of $E$,

$$\dot{E} = m_2\dot{x}_2\ddot{x}_2 - k_2(x_1 - x_2)\dot{x}_2 + k_3(x_2 - x_3)\dot{x}_2 = 0 \tag{5.50}$$

which provides us with the second equation of motion.

$$m_2\ddot{x}_2 - k_2(x_1 - x_2) + k_3(x_2 - x_3) = 0 \tag{5.51}$$

To find the third equation of motion associated to $x_2$, we assume that $x_1$ and $x_3$ and their time rates are constant and we take the time directive of $E$,

$$\dot{E} = m_3\dot{x}_3\ddot{x}_3 - k_3(x_2 - x_3)\dot{x}_3 + k_4x_3\dot{x}_3 = 0 \tag{5.52}$$

to find the third equation of motion.

$$m_3\ddot{x}_3 - k_3(x_2 - x_3) + k_4x_3 = 0 \tag{5.53}$$

We may set up the equations in a matrix form.

$$\begin{bmatrix} m_1 & 0 & 0 \\ 0 & m_2 & 0 \\ 0 & 0 & m_3 \end{bmatrix} \begin{bmatrix} \ddot{x}_1 \\ \ddot{x}_2 \\ \ddot{x}_3 \end{bmatrix} + \begin{bmatrix} k_1 + k_2 & -k_2 & 0 \\ -k_2 & k_2 + k_3 & -k_3 \\ 0 & -k_3 & k_3 + k_4 \end{bmatrix} \begin{bmatrix} x_1 \\ x_2 \\ x_3 \end{bmatrix} = 0 \tag{5.54}$$

Alternatively, we may take a time derivative of the total mechanical energy $E$.

$$\begin{aligned} \dot{E} &= m_1\dot{x}_1\ddot{x}_1 + m_2\dot{x}_2\ddot{x}_2 + m_3\dot{x}_3\ddot{x}_3 + k_1x_1\dot{x}_1 + k_2(x_1 - x_2)\dot{x}_1 \\ &\quad -k_2(x_1 - x_2)\dot{x}_2 + k_3(x_2 - x_3)\dot{x}_2 - k_3(x_2 - x_3)\dot{x}_3 + k_4x_3\dot{x}_3 \\ &= (m_1\dot{x}_1 + k_1x_1 + k_2(x_1 - x_2))\dot{x}_1 \\ &\quad + (m_2\ddot{x}_2 - k_2(x_1 - x_2) + k_3(x_2 - x_3))\dot{x}_2 \\ &\quad + (m_3\dot{x}_3 - k_3(x_2 - x_3) + k_4x_3)\dot{x}_3 \end{aligned} \tag{5.55}$$

The variables $\dot{x}_1$, $\dot{x}_2$, $\dot{x}_3$ are independent, and because $\dot{x}_1$, $\dot{x}_2$, $\dot{x}_3$ cannot always be zero, their coefficients must be zero independently. Those coefficients will provide us with the same equations of motion.

$$m_1\ddot{x}_1 + k_1x_1 + k_2(x_1 - x_2) = 0 \tag{5.56}$$

$$m_2\ddot{x}_2 - k_2(x_1 - x_2) + k_3(x_2 - x_3) = 0 \tag{5.57}$$

$$m_3\ddot{x}_3 - k_3(x_2 - x_3) + k_4x_3 = 0 \tag{5.58}$$

**Example 27** Energy and nonlinear multi-$DOF$ systems. The energy method to derive the equation of motion works on all conservative systems regardless of linearity or nonlinearity.

Figure 5.7 illustrates a two $DOF$ nonlinear conservative system. The coordinates of $m_1$ and $m_2$ are

$$x_1 = x \qquad y_1 = 0$$
$$x_2 = x + l\cos\theta \qquad y_2 = l\sin\theta \tag{5.59}$$

Their time rates are

$$\dot{x}_1 = \dot{x} \qquad \dot{y}_1 = 0$$
$$\dot{x}_2 = \dot{x} - l\dot{\theta}\sin\theta \qquad y_2 = l\dot{\theta}\cos\theta \tag{5.60}$$

Hence, their velocity squares, which will contribute in kinetic energy of the system, are

$$v_1^2 = \dot{x}_1^2 + \dot{y}_1^2 = \dot{x}^2$$
$$v_2^2 = \dot{x}_2^2 + \dot{y}_2^2 = \dot{x}^2 + l^2\dot{\theta}^2 - 2l\dot{x}\dot{\theta}\sin\theta$$

The potential energies of the system are due to length change in spring and height change of $m_2$. Therefore, the kinetic and potential energies of the system are:

$$K = \frac{1}{2}m_1 v_1^2 + \frac{1}{2}m_2 v_2^2$$
$$= \frac{1}{2}m_1\dot{x}^2 + \frac{1}{2}m_2\left(\dot{x}^2 + l^2\dot{\theta}^2 - 2l\dot{x}\dot{\theta}\sin\theta\right) \tag{5.61}$$
$$P = \frac{1}{2}kx^2 - m_2 g x_2 = \frac{1}{2}kx^2 - m_2 g\left(x + l\cos\theta\right) \tag{5.62}$$

$$E = K + P = \frac{1}{2}m_1\dot{x}^2 + \frac{1}{2}m_2\left(\dot{x}^2 + l^2\dot{\theta}^2 - 2l\dot{x}\dot{\theta}\sin\theta\right)$$
$$+\frac{1}{2}kx^2 - m_2 g\left(x + l\cos\theta\right) \tag{5.63}$$

We assumed that the motionless hanging down position is the equilibrium configuration of the system and that the gravitational energy is zero at the level of $m_1$ at the equilibrium.

The system has two $DOF$ with generalized coordinates $x$ and $\theta$. Let us take a time derivative of mechanical energy $E = K + P$.

$$\frac{dE}{dt} = (m_1 + m_2)\dot{x}\ddot{x} + m_2 l^2\dot{\theta}\ddot{\theta} - m_2 l\ddot{x}\dot{\theta}\sin\theta - m_2 l\dot{x}\ddot{\theta}\sin\theta$$
$$-m_2 l\dot{x}\dot{\theta}^2\cos\theta + kx\dot{x} - m_2 g\dot{x} - m_2 g l\dot{\theta}\sin\theta = 0 \tag{5.64}$$

Factoring out $\dot{x}$ and $\dot{\theta}$ makes $\dot{E}$ to be of two terms.

$$\left((m_1 + m_2)\ddot{x} - m_2 l\ddot{\theta}\sin\theta + kx - m_2 g - m_2 l\dot{\theta}^2\cos\theta\right)\dot{x}$$
$$+\left(m_2 l^2\ddot{\theta} - m_2 l\ddot{x}\sin\theta - m_2 g l\sin\theta\right)\dot{\theta} = 0 \tag{5.65}$$

The variables $\dot{x}$ and $\dot{\theta}$ are independent and $\dot{x}$ and $\dot{\theta}$ cannot always be zero, and their coefficients must be zero independently. The coefficients are the equations of motion of the system.

$$(m_1 + m_2)\ddot{x} - m_2 l\ddot{\theta}\sin\theta + kx - m_2 g - m_2 l\dot{\theta}^2\cos\theta = 0 \tag{5.66}$$
$$m_2 l^2\ddot{\theta} - m_2 l\ddot{x}\sin\theta - m_2 g l\sin\theta = 0 \tag{5.67}$$

## 5.3 ★ Rotational Dynamics

The rigid body rotational equations of motion come from the *Euler equation*

$$\begin{aligned} {}^{B}\mathbf{M} &= \frac{{}^{G}d}{dt}\,{}^{B}\mathbf{L} = {}^{B}\dot{\mathbf{L}} + {}_{G}^{B}\boldsymbol{\omega}_{B} \times {}^{B}\boldsymbol{L} \\ &= {}^{B}I\,{}_{G}^{B}\dot{\boldsymbol{\omega}}_{B} + {}_{G}^{B}\boldsymbol{\omega}_{B} \times \left({}^{B}I\,{}_{G}^{B}\boldsymbol{\omega}_{B}\right) \end{aligned} \tag{5.68}$$

where $\boldsymbol{L}$ is the *angular momentum* and $I$ is the *mass moment* of the rigid body.

$${}^{B}\mathbf{L} = {}^{B}I\,{}_{G}^{B}\boldsymbol{\omega}_{B} \tag{5.69}$$

$$I = \begin{bmatrix} I_{xx} & I_{xy} & I_{xz} \\ I_{yx} & I_{yy} & I_{yz} \\ I_{zx} & I_{zy} & I_{zz} \end{bmatrix} \tag{5.70}$$

The expanded form of the Euler equation (5.68) is

$$\begin{aligned} M_x &= I_{xx}\dot{\omega}_x + I_{xy}\dot{\omega}_y + I_{xz}\dot{\omega}_z - \left(I_{yy} - I_{zz}\right)\omega_y\omega_z \\ &\quad - I_{yz}\left(\omega_z^2 - \omega_y^2\right) - \omega_x\left(\omega_z I_{xy} - \omega_y I_{xz}\right) \end{aligned} \tag{5.71}$$

$$\begin{aligned} M_y &= I_{yx}\dot{\omega}_x + I_{yy}\dot{\omega}_y + I_{yz}\dot{\omega}_z - \left(I_{zz} - I_{xx}\right)\omega_z\omega_x \\ &\quad - I_{xz}\left(\omega_x^2 - \omega_z^2\right) - \omega_y\left(\omega_x I_{yz} - \omega_z I_{xy}\right) \end{aligned} \tag{5.72}$$

$$\begin{aligned} M_z &= I_{zx}\dot{\omega}_x + I_{zy}\dot{\omega}_y + I_{zz}\dot{\omega}_z - \left(I_{xx} - I_{yy}\right)\omega_x\omega_y \\ &\quad - I_{xy}\left(\omega_y^2 - \omega_x^2\right) - \omega_z\left(\omega_y I_{xz} - \omega_x I_{yz}\right) \end{aligned} \tag{5.73}$$

When the body coordinate is the *principal coordinate frame*, Euler equations reduce to

$$\begin{aligned} M_1 &= I_1\dot{\omega}_1 - \left(I_2 - I_2\right)\omega_2\omega_3 \\ M_2 &= I_2\dot{\omega}_2 - \left(I_3 - I_1\right)\omega_3\omega_1 \\ M_3 &= I_3\dot{\omega}_3 - \left(I_1 - I_2\right)\omega_1\omega_2 \end{aligned} \tag{5.74}$$

The principal coordinate frame is denoted by numbers 123, instead of $xyz$, to indicate the first, second, and third *principal axes*. The parameters $I_{ij},\ i \neq j$ are zero in the principal frame. The principal body coordinate frame sits at the mass center $C$.

The kinetic energy $K$ of a rotating rigid body is

$$\begin{aligned} K &= \frac{1}{2}\left(I_{xx}\omega_x^2 + I_{yy}\omega_y^2 + I_{zz}\omega_z^2\right) \\ &\quad - I_{xy}\omega_x\omega_y - I_{yz}\omega_y\omega_z - I_{zx}\omega_z\omega_x \end{aligned} \tag{5.75}$$

$$= \frac{1}{2}\boldsymbol{\omega}\cdot\mathbf{L} = \frac{1}{2}\boldsymbol{\omega}^T I\,\boldsymbol{\omega} \tag{5.76}$$

which in the principal coordinate frame reduces to

$$K = \frac{1}{2}\left(I_1\omega_1^2 + I_2\omega_2^2 + I_3\omega_3^2\right) \tag{5.77}$$

***Proof*** Let $m_i$ be the mass of the $i$th particle of a rigid body $B$, which is made of $n$ particles. Let $\mathbf{r}_i$ be the Cartesian position vector of $m_i$ in a body coordinate frame $B\,(Oxyz)$.

$$\mathbf{r}_i = {}^B\mathbf{r}_i = \begin{bmatrix} x_i & y_i & z_i \end{bmatrix}^T \tag{5.78}$$

Assume that $\boldsymbol{\omega}$ is the angular velocity of the rigid body with respect to the global coordinate frame $G\,(OXYZ)$, expressed in the body coordinate frame.

$$\boldsymbol{\omega} = {}_G^B\boldsymbol{\omega}_B = \begin{bmatrix} \omega_x & \omega_y & \omega_z \end{bmatrix}^T \tag{5.79}$$

The angular momentum of $m_i$ is $\mathbf{L}_i$.

$$\begin{aligned}\mathbf{L}_i &= \mathbf{r}_i \times m_i\dot{\mathbf{r}}_i = m_i\left[\mathbf{r}_i \times (\boldsymbol{\omega} \times \mathbf{r}_i)\right] \\ &= m_i\left[(\mathbf{r}_i \cdot \mathbf{r}_i)\,\boldsymbol{\omega} - (\mathbf{r}_i \cdot \boldsymbol{\omega})\,\mathbf{r}_i\right] \\ &= m_i r_i^2 \boldsymbol{\omega} - m_i\,(\mathbf{r}_i \cdot \boldsymbol{\omega})\,\mathbf{r}_i\end{aligned} \tag{5.80}$$

Hence, the angular momentum of the whole rigid body would be

$$\mathbf{L} = \boldsymbol{\omega}\sum_{i=1}^{n} m_i r_i^2 - \sum_{i=1}^{n} m_i\,(\mathbf{r}_i \cdot \boldsymbol{\omega})\,\mathbf{r}_i \tag{5.81}$$

Substitution for $\mathbf{r}_i$ and $\boldsymbol{\omega}$ in $\mathbf{L}$ yields

$$\begin{aligned}\mathbf{L} = &\left(\omega_x\hat{\imath} + \omega_y\hat{\jmath} + \omega_z\hat{k}\right)\sum_{i=1}^{n} m_i\left(x_i^2 + y_i^2 + z_i^2\right) \\ &-\sum_{i=1}^{n} m_i\left(x_i\omega_x + y_i\omega_y + z_i\omega_z\right)\cdot\left(x_i\hat{\imath} + y_i\hat{\jmath} + z_i\hat{k}\right)\end{aligned} \tag{5.82}$$

which can be rearranged as

$$\begin{aligned}\mathbf{L} = &\sum_{i=1}^{n} m_i\left(y_i^2 + z_i^2\right)\omega_x\hat{\imath} + \sum_{i=1}^{n} m_i\left(z_i^2 + x_i^2\right)\omega_y\hat{\jmath} + \sum_{i=1}^{n} m_i\left(x_i^2 + y_i^2\right)\omega_z\hat{k} \\ &-\left(\sum_{i=1}^{n}(m_i x_i y_i)\,\omega_y + \sum_{i=1}^{n}(m_i x_i z_i)\,\omega_z\right)\hat{\imath} \\ &-\left(\sum_{i=1}^{n}(m_i y_i z_i)\,\omega_z + \sum_{i=1}^{n}(m_i y_i x_i)\,\omega_x\right)\hat{\jmath} \\ &-\left(\sum_{i=1}^{n}(m_i z_i x_i)\,\omega_x + \sum_{i=1}^{n}(m_i z_i y_i)\,\omega_y\right)\hat{k}\end{aligned} \tag{5.83}$$

By introducing the mass moment matrix $I$ with the following elements,

$$I_{xx} = \sum_{i=1}^{n}\left[m_i\left(y_i^2 + z_i^2\right)\right] \tag{5.84}$$

$$I_{yy} = \sum_{i=1}^{n}\left[m_i\left(z_i^2 + x_i^2\right)\right] \tag{5.85}$$

$$I_{zz} = \sum_{i=1}^{n} \left[ m_i \left( x_i^2 + y_i^2 \right) \right] \tag{5.86}$$

$$I_{xy} = I_{yx} = -\sum_{i=1}^{n} (m_i x_i y_i) \tag{5.87}$$

$$I_{yz} = I_{zy} = -\sum_{i=1}^{n} (m_i y_i z_i) \tag{5.88}$$

$$I_{zx} = I_{xz} = -\sum_{i=1}^{n} (m_i z_i x_i) \, , \tag{5.89}$$

we may rewrite the angular momentum **L** in concise form.

$$L_x = I_{xx}\omega_x + I_{xy}\omega_y + I_{xz}\omega_z \tag{5.90}$$

$$L_y = I_{yx}\omega_x + I_{yy}\omega_y + I_{yz}\omega_z \tag{5.91}$$

$$L_z = I_{zx}\omega_x + I_{zy}\omega_y + I_{zz}\omega_z \tag{5.92}$$

$$\mathbf{L} = I \cdot \boldsymbol{\omega} \tag{5.93}$$

$$\begin{bmatrix} L_x \\ L_y \\ L_z \end{bmatrix} = \begin{bmatrix} I_{xx} & I_{xy} & I_{xz} \\ I_{yx} & I_{yy} & I_{yz} \\ I_{zx} & I_{zy} & I_{zz} \end{bmatrix} \begin{bmatrix} \omega_x \\ \omega_y \\ \omega_z \end{bmatrix} \tag{5.94}$$

For a rigid body that is a continuous solid, the summations must be replaced by integrations over the volume of the body.

If ${}^B\mathbf{M}$ denotes the resultant of the external moments applied on the rigid body, the Euler equation of motion for a rigid body will be

$${}^B\mathbf{M} = \frac{{}^G d}{dt} \, {}^B\mathbf{L} \tag{5.95}$$

The angular momentum ${}^B\mathbf{L}$ is a vector quantity defined in the body coordinate frame. Hence, its time derivative in the global coordinate frame is

$$\frac{{}^G d \, {}^B\mathbf{L}}{dt} = {}^B\dot{\mathbf{L}} + {}^B_G\boldsymbol{\omega}_B \times {}^B \mathbf{L}. \tag{5.96}$$

Therefore, the vectorial form of the Euler equation of motion is

$${}^B\mathbf{M} = \frac{d\mathbf{L}}{dt} = \dot{\mathbf{L}} + \boldsymbol{\omega} \times \mathbf{L} = I\dot{\boldsymbol{\omega}} + \boldsymbol{\omega} \times (I\boldsymbol{\omega}) \tag{5.97}$$

and in expanded form is

$$\begin{aligned} {}^B\mathbf{M} = & \left( I_{xx}\dot{\omega}_x + I_{xy}\dot{\omega}_y + I_{xz}\dot{\omega}_z \right) \hat{\imath} + \omega_y \left( I_{xz}\omega_x + I_{yz}\omega_y + I_{zz}\omega_z \right) \hat{\imath} \\ & -\omega_z \left( I_{xy}\omega_x + I_{yy}\omega_y + I_{yz}\omega_z \right) \hat{\imath} \\ & + \left( I_{yx}\dot{\omega}_x + I_{yy}\dot{\omega}_y + I_{yz}\dot{\omega}_z \right) \hat{\jmath} + \omega_z \left( I_{xx}\omega_x + I_{xy}\omega_y + I_{xz}\omega_z \right) \hat{\jmath} \\ & -\omega_x \left( I_{xz}\omega_x + I_{yz}\omega_y + I_{zz}\omega_z \right) \hat{\jmath} \\ & + \left( I_{zx}\dot{\omega}_x + I_{zy}\dot{\omega}_y + I_{zz}\dot{\omega}_z \right) \hat{k} + \omega_x \left( I_{xy}\omega_x + I_{yy}\omega_y + I_{yz}\omega_z \right) \hat{k} \\ & -\omega_y \left( I_{xx}\omega_x + I_{xy}\omega_y + I_{xz}\omega_z \right) \hat{k} \end{aligned} \tag{5.98}$$

Therefore, the most general form of the Euler equations of motion for a rigid body in a body frame attached to $C$ is

$$M_x = I_{xx}\dot{\omega}_x + I_{xy}\dot{\omega}_y + I_{xz}\dot{\omega}_z - \left(I_{yy} - I_{zz}\right)\omega_y\omega_z \\ -I_{yz}\left(\omega_z^2 - \omega_y^2\right) - \omega_x\left(\omega_z I_{xy} - \omega_y I_{xz}\right) \tag{5.99}$$

$$M_y = I_{yx}\dot{\omega}_x + I_{yy}\dot{\omega}_y + I_{yz}\dot{\omega}_z - \left(I_{zz} - I_{xx}\right)\omega_z\omega_x \\ -I_{xz}\left(\omega_x^2 - \omega_z^2\right) - \omega_y\left(\omega_x I_{yz} - \omega_z I_{xy}\right) \tag{5.100}$$

$$M_z = I_{zx}\dot{\omega}_x + I_{zy}\dot{\omega}_y + I_{zz}\dot{\omega}_z - \left(I_{xx} - I_{yy}\right)\omega_x\omega_y \\ -I_{xy}\left(\omega_y^2 - \omega_x^2\right) - \omega_z\left(\omega_y I_{xz} - \omega_x I_{yz}\right) \tag{5.101}$$

Assume that we are able to rotate the body frame about its origin to find an orientation that makes. In a *principal coordinate frame*, we have $I_{ij} = 0$, for $i \neq j$, and hence, the equations simplify to

$$M_1 = I_1\dot{\omega}_1 - \left(I_2 - I_2\right)\omega_2\omega_3 \tag{5.102}$$

$$M_2 = I_2\dot{\omega}_2 - \left(I_3 - I_1\right)\omega_3\omega_1 \tag{5.103}$$

$$M_3 = I_3\dot{\omega}_3 - \left(I_1 - I_2\right)\omega_1\omega_2 \tag{5.104}$$

The kinetic energy of a rigid body may be found by the integral of the kinetic energy of the mass element $dm$, over the whole body:

$$K = \frac{1}{2}\int_B \dot{\mathbf{v}}^2 dm = \frac{1}{2}\int_B (\boldsymbol{\omega}\times\mathbf{r})\cdot(\boldsymbol{\omega}\times\mathbf{r})\,dm \\ = \frac{\omega_x^2}{2}\int_B \left(y^2+z^2\right)dm + \frac{\omega_y^2}{2}\int_B \left(z^2+x^2\right)dm + \frac{\omega_z^2}{2}\int_B \left(x^2+y^2\right)dm \\ -\omega_x\omega_y\int_B xy\,dm - \omega_y\omega_z\int_B yz\,dm - \omega_z\omega_x\int_B zx\,dm \\ = \frac{1}{2}\left(I_{xx}\omega_x^2 + I_{yy}\omega_y^2 + I_{zz}\omega_z^2\right) \\ -I_{xy}\omega_x\omega_y - I_{yz}\omega_y\omega_z - I_{zx}\omega_z\omega_x \tag{5.105}$$

The kinetic energy can be rearranged to a matrix multiplication form.

$$K = \frac{1}{2}\boldsymbol{\omega}^T I\,\boldsymbol{\omega} = \frac{1}{2}\boldsymbol{\omega}\cdot\mathbf{L} \tag{5.106}$$

When the body frame is principal, the kinetic energy will be simplified.

$$K = \frac{1}{2}\left(I_1\omega_1^2 + I_2\omega_2^2 + I_3\omega_3^2\right) \tag{5.107}$$

Kinetic energy is what we need to apply energy method, as well as developing Lagrange method. ■

**Example 28** ★ A tilted disc on a massless shaft. Rotation of a rigid body about a fixed axis is the simplest example of a rotating rigid body and is the most common case in industry. Here is to show how to develop equations of motion and determine supporting forces.

Figure 5.8 illustrates a disc with mass $m$ and radius $R$, mounted on a massless shaft. The shaft is turning with a constant angular speed $\omega$. The disc is attached to the shaft at an angle $\theta$. Because of $\theta$, the bearings at $A$ and $B$ must support a rotating force.

To model the system, we attach a principal body coordinate frame at the disc center as shown in the figure. The $G$-expression of the angular velocity is a constant vector

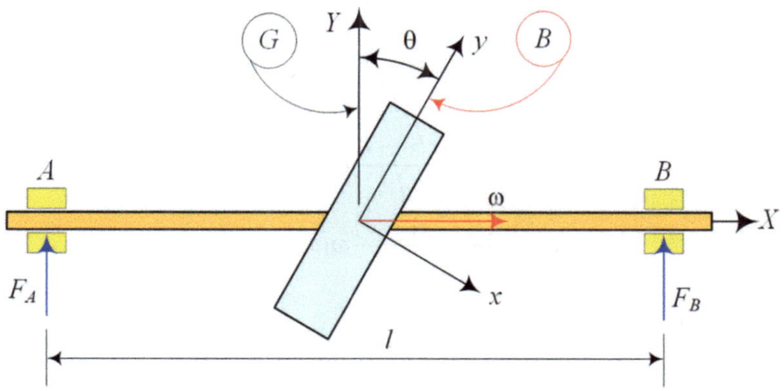


**Fig. 5.8** A disc with mass $m$ and radius $r$, mounted on a massless turning shaft

$$^G\boldsymbol{\omega}_B = \omega \hat{I} \tag{5.108}$$

and the expression of the angular velocity vector in the body frame is

$$^B_G\boldsymbol{\omega}_B = \omega \cos\theta \, \hat{\imath} + \omega \sin\theta \, \hat{\jmath} \tag{5.109}$$

The mass moment of inertia matrix of the disc in its body frame is

$$^BI = \begin{bmatrix} mR^2/2 & 0 & 0 \\ 0 & mR^2/4 & 0 \\ 0 & 0 & mR^2/4 \end{bmatrix} \tag{5.110}$$

Substituting (5.109) and (5.110) in (5.102)–(5.104), with $1 \equiv x, 2 \equiv y, 3 \equiv z$, yields

$$M_x = 0 \qquad M_y = 0 \tag{5.111}$$

$$M_z = \frac{mr^2}{4} \omega \cos\theta \sin\theta \tag{5.112}$$

Therefore, the bearing reaction forces $F_A$ and $F_B$ are

$$F_A = -F_B = -\frac{M_z}{l} = -\frac{mr^2}{4l} \omega \cos\theta \sin\theta \tag{5.113}$$

**Example 29** Steady rotation of a freely rotating rigid body. Motion of a particle with no external force is rest or steady-state motion on a straight line with constant velocity. However, motion of a rigid body with no external moment is more complicated. Here is the equations of motion and the conditions to have a steady-state rotation.

Consider a situation in which the resultant applied force and moment on a rigid body are zero.

$$^G\mathbf{F} = {}^B\mathbf{F} = 0 \tag{5.114}$$

$$^G\mathbf{M} = {}^B\mathbf{M} = 0 \tag{5.115}$$

Based on the Newton's equation,

$$^G\mathbf{F} = m \, {}^G\dot{\mathbf{v}} \tag{5.116}$$

the velocity of the mass center will be constant in the global coordinate frame. However, the Euler equation

$$^B\mathbf{M} = I \, {}^B_G\dot{\boldsymbol{\omega}}_B + {}^B_G\boldsymbol{\omega}_B \times {}^B\mathbf{L} \tag{5.117}$$

reduces to

$$\dot{\omega}_1 = \frac{I_2 - I_3}{I_1}\omega_2\omega_3 \tag{5.118}$$

$$\dot{\omega}_2 = \frac{I_3 - I_1}{I_{22}}\omega_3\omega_1 \tag{5.119}$$

$$\dot{\omega}_3 = \frac{I_1 - I_2}{I_3}\omega_1\omega_2 \tag{5.120}$$

which show that the angular velocity can be constant if

$$I_1 = I_2 = I_3 \tag{5.121}$$

or if two principal moments of inertia, say $I_1$ and $I_2$, are zero and the third angular velocity, in this case $\omega_3$, is initially zero, or if the angular velocity vector is initially parallel to a principal axis.

## 5.4 Lagrange Method

The Lagrange method in deriving the equations of motion of vibrating systems has some advantages over Newton–Euler due to its simplicity and generality, specially for multi-degree-of-freedom ($DOF$) systems. Let us assume that for some forces **F** there is a *potential energy* function $P$ such that the force is derivable from $P$. Such a force is called a *potential* or *conservative force*.

$$\begin{aligned}\mathbf{F} = -\nabla P &= \frac{\partial P}{\partial x}\hat{\imath} + \frac{\partial P}{\partial y}\hat{\jmath} + \frac{\partial P}{\partial z}\hat{k} \\ &= \begin{bmatrix} F_x & F_y & F_z \end{bmatrix}^T\end{aligned} \tag{5.122}$$

The Lagrange equation of motion of a dynamic system can be written as

$$\frac{d}{dt}\left(\frac{\partial \mathcal{L}}{\partial \dot{q}_r}\right) - \frac{\partial \mathcal{L}}{\partial q_r} = Q_r \qquad r = 1, 2, \cdots n \tag{5.123}$$

where $\mathcal{L}$ is the *Lagrangean* of the system

$$\mathcal{L} = K - P \tag{5.124}$$

and $Q_r$ is the nonpotential generalized force for which there is no potential function.

$$Q_r = \sum_{i=1}^{n}\left(F_{ix}\frac{\partial f_i}{\partial q_1} + F_{iy}\frac{\partial g_i}{\partial q_2} + F_{iz}\frac{\partial h_i}{\partial q_n}\right) \tag{5.125}$$

***Proof*** Assume that the external forces $\mathbf{F} = \begin{bmatrix} F_x & F_y & F_z \end{bmatrix}^T$ acting on the system are conservative:

$$\mathbf{F} = -\nabla P \tag{5.126}$$

The work done by these forces in an arbitrary virtual displacement $\delta q_1, \delta q_2, \delta q_3, \cdots, \delta q_n$ is

$$\partial W = -\frac{\partial P}{\partial q_1}\delta q_1 - \frac{\partial P}{\partial q_2}\delta q_2 - \cdots \frac{\partial P}{\partial q_n}\delta q_n \tag{5.127}$$

and then the Lagrange equation becomes

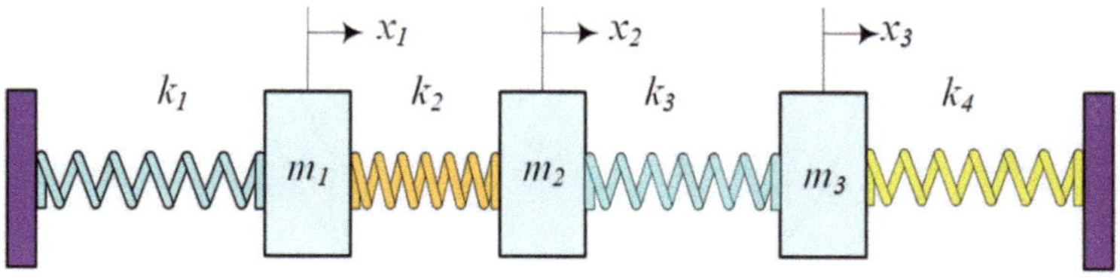


**Fig. 5.9** An undamped three $DOF$ system

$$\frac{d}{dt}\left(\frac{\partial K}{\partial \dot{q}_r}\right) - \frac{\partial K}{\partial q_r} = -\frac{\partial P}{\partial q_1} \qquad r = 1, 2, \cdots, n \tag{5.128}$$

Introducing the Lagrangean function $\mathcal{L} = K - P$ converts the Lagrange equation of a conservative system into

$$\frac{d}{dt}\left(\frac{\partial \mathcal{L}}{\partial \dot{q}_r}\right) - \frac{\partial \mathcal{L}}{\partial q_r} = 0 \qquad r = 1, 2, \cdots n \tag{5.129}$$

If there is a nonpotential force, then the virtual work done by the force is

$$\begin{aligned} \delta W &= \sum_{i=1}^{n} \left( F_{xi}\frac{\partial f_i}{\partial q_r} + F_{yi}\frac{\partial g_i}{\partial q_r} + F_{zi}\frac{\partial h_i}{\partial q_r} \right) \delta q_r \\ &= Q_r \, \delta q_r \end{aligned} \tag{5.130}$$

and the Lagrange equation of motion would be

$$\frac{d}{dt}\left(\frac{\partial \mathcal{L}}{\partial \dot{q}_r}\right) - \frac{\partial \mathcal{L}}{\partial q_r} = Q_r \qquad r = 1, 2, \cdots n \tag{5.131}$$

where $Q_r$ is the nonpotential generalized force doing work in a virtual displacement of the $r$th generalized coordinate $q_r$.

The Lagrangean $\mathcal{L}$ is also called the *kinetic potential.* ■

**Example 30** An undamped three $DOF$ system. An example of a multi-$DOF$ vibrating system and showing how to derive the equations of motion by applying Lagrange method.

Figure 5.9 illustrates an undamped three $DOF$ linear vibrating system. The kinetic and potential energies of the system are:

$$K = \frac{1}{2}m_1\dot{x}_1^2 + \frac{1}{2}m_2\dot{x}_2^2 + \frac{1}{2}m_3\dot{x}_3^2 \tag{5.132}$$

$$P = \frac{1}{2}k_1x_1^2 + \frac{1}{2}k_2\left(x_1 - x_2\right)^2 + \frac{1}{2}k_3\left(x_2 - x_3\right)^2 + \frac{1}{2}k_4x_3^2 \tag{5.133}$$

Because there is no damping in the system, we may find the Lagrangean $\mathcal{L}$ and use Equation (5.161) with $Q_r = 0$.

$$\mathcal{L} = K - P \tag{5.134}$$

$$\frac{\partial \mathcal{L}}{\partial x_1} = -k_1x_1 - k_2\left(x_1 - x_2\right) \tag{5.135}$$

$$\frac{\partial \mathcal{L}}{\partial x_2} = k_2\left(x_1 - x_2\right) - k_3\left(x_2 - x_3\right) \tag{5.136}$$

$$\frac{\partial \mathcal{L}}{\partial x_3} = k_3\left(x_2 - x_3\right) - k_4x_3 \tag{5.137}$$

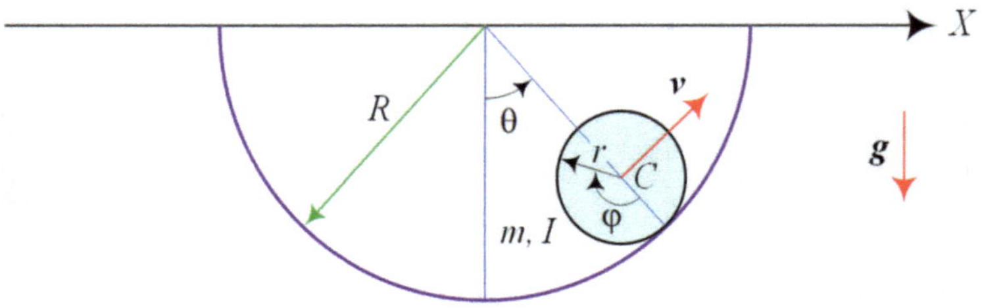


**Fig. 5.10** A uniform disc, rolling in a circular path

$$\frac{\partial \mathcal{L}}{\partial \dot{x}_1} = m_1 \dot{x}_1 \qquad \frac{\partial \mathcal{L}}{\partial \dot{x}_2} = m_2 \dot{x}_2 \qquad \frac{\partial \mathcal{L}}{\partial \dot{x}_3} = m_3 \dot{x}_3 \tag{5.138}$$

The equations of motion are

$$m_1 \ddot{x}_1 + k_1 x_1 + k_2 (x_1 - x_2) = 0 \tag{5.139}$$

$$m_2 \ddot{x}_2 - k_2 (x_1 - x_2) + k_3 (x_2 - x_3) = 0 \tag{5.140}$$

$$m_3 \ddot{x}_3 - k_3 (x_2 - x_3) + k_4 x_3 = 0 \tag{5.141}$$

These equations can be rewritten in matrix form for simpler calculation.

$$\begin{bmatrix} m_1 & 0 & 0 \\ 0 & m_2 & 0 \\ 0 & 0 & m_3 \end{bmatrix} \begin{bmatrix} \ddot{x}_1 \\ \ddot{x}_2 \\ \ddot{x}_3 \end{bmatrix} + \begin{bmatrix} k_1 + k_2 & -k_2 & 0 \\ -k_2 & k_2 + k_3 & -k_3 \\ 0 & -k_3 & k_3 + k_4 \end{bmatrix} \begin{bmatrix} x_1 \\ x_2 \\ x_3 \end{bmatrix} = 0 \tag{5.142}$$

**Example 31** A rolling disc in a circular path. It is an example of a rotating rigid body with some kinematic constraints to derive the equation of motion.

Figure 5.10 illustrates a uniform disc with mass $m$ and radius $r$. The disc is rolling without slip in a circular path of radius $R$. The disc may have free oscillations around $\theta = 0$. To find the equation of motion, we employ the Lagrange method. The energies of the system are

$$\begin{aligned} K &= \frac{1}{2} m v_C^2 + \frac{1}{2} I_c \omega^2 \\ &= \frac{1}{2} m (R - r)^2 \dot{\theta}^2 + \frac{1}{2} \left( \frac{1}{2} m r^2 \right) \left( \dot{\varphi} - \dot{\theta} \right)^2 \end{aligned} \tag{5.143}$$

$$P = -mg(R - r) \cos\theta \tag{5.144}$$

When there is no slip, there is a constraint between $\theta$ and $\varphi$,

$$R\theta = r\varphi \qquad R\dot{\theta} = r\dot{\varphi} \tag{5.145}$$

which can be used to eliminate $\varphi$ from $K$.

$$K = \frac{3}{4} m (R - r)^2 \dot{\theta}^2 \tag{5.146}$$

Based on the partial derivatives

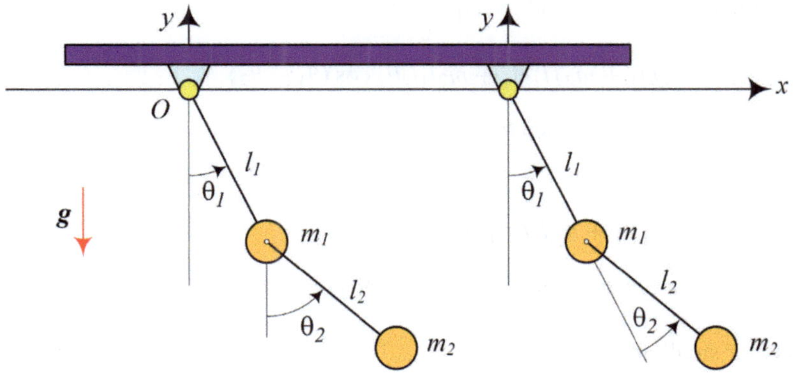


**Fig. 5.11** A double pendulum

$$\frac{d}{dt}\left(\frac{\partial \mathcal{L}}{\partial \dot{\theta}}\right) = \frac{3}{2}m\,(R-r)^2\,\ddot{\theta} \tag{5.147}$$

$$\frac{\partial \mathcal{L}}{\partial \theta} = mg(R-r)\sin\theta \tag{5.148}$$

we find the equation of motion for the oscillating disc.

$$\frac{3}{2}\,(R-r)\,\ddot{\theta} + g\sin\theta = 0 \tag{5.149}$$

When $\theta$ is very small, this equation is equivalent to a mass–spring system with $m_e = 3\,(R-r)$ and $k_e = 2g$.

$$m_e\ddot{\theta} + 2g\theta = 0 \tag{5.150}$$

**Example 32** ★ Double pendulum and dynamic coupling. Pendulums are nonlinear vibrating systems; however, assuming small oscillations, the equations of motion can be linearized. It is an example of a linearized system with dynamic coupled and statically decoupled.

The double pendulum of Fig. 5.11 is made of two massless rods with lengths $l_1$ and $l_2$ and two point masses $m_1$ and $m_2$. The absolute variables $\theta_1$ and $\theta_2$ act as the generalized coordinates to express the configuration of the system. The Lagrangean of the system is

$$\begin{aligned}\mathcal{L} = K - P &= \frac{1}{2}m_1 l_1^2\dot{\theta}_1^2 \\ &+ \frac{1}{2}m_2\left(l_1^2\dot{\theta}_1^2 + l_2^2\dot{\theta}_2^2 + 2l_1l_2\dot{\theta}_1\dot{\theta}_2\cos(\theta_1-\theta_2)\right) \\ &+ m_1 g l_1\cos\theta_1 + m_2 g\,(l_1\cos\theta_1 + l_2\cos\theta_2)\end{aligned} \tag{5.151}$$

where

$$\begin{aligned}K = \frac{1}{2}m_1 v_1^2 + \frac{1}{2}m_2 v_2^2 &= \frac{1}{2}m_1 l_1^2\dot{\theta}_1^2 \\ &+ \frac{1}{2}m_2\left(l_1^2\dot{\theta}_1^2 + l_2^2\dot{\theta}_2^2 + 2l_1l_2\dot{\theta}_1\dot{\theta}_2\cos(\theta_1-\theta_2)\right)\end{aligned} \tag{5.152}$$

and

$$\begin{aligned}P &= m_1 g y_1 + m_2 g y_2 \\ &= -m_1 g l_1\cos\theta_1 - m_2 g\,(l_1\cos\theta_1 + l_2\cos\theta_2)\end{aligned} \tag{5.153}$$

Employing the Lagrange method (5.161), we find the equations of motion.

$$\begin{aligned} &(m_1 + m_2)\, l_1^2 \ddot{\theta}_1 + m_2 l_1 l_2 \ddot{\theta}_2 \cos(\theta_1 - \theta_2) \\ &-m_2 l_1 l_2 \dot{\theta}_2^2 \sin(\theta_1 - \theta_2) + (m_1 + m_2)\, l_1 g \sin\theta_1 = 0 \end{aligned} \tag{5.154}$$

$$\begin{aligned} &m_2 l_2^2 \ddot{\theta}_2 + m_2 l_1 l_2 \ddot{\theta}_1 \cos(\theta_1 - \theta_2) \\ &+m_2 l_1 l_2 \dot{\theta}_1^2 \sin(\theta_1 - \theta_2) + m_2 l_2 g \sin\theta_2 = 0 \end{aligned} \tag{5.155}$$

The linearized equations for small angles are

$$\begin{aligned} &\begin{bmatrix} (m_1 + m_2)\, l_1 & m_2 l_2 \\ l_1 & l_2 \end{bmatrix} \begin{bmatrix} \ddot{\theta}_1 \\ \ddot{\theta}_2 \end{bmatrix} \\ &+ \begin{bmatrix} (m_1 + m_2)\, g & 0 \\ 0 & g \end{bmatrix} \begin{bmatrix} \theta_1 \\ \theta_2 \end{bmatrix} = \begin{bmatrix} 0 \\ 0 \end{bmatrix} \end{aligned} \tag{5.156}$$

which indicate the equations of the double pendulum are dynamically coupled and statically decoupled.

Employing the relative coordinates, the following equations of motion will be derived.

$$\begin{aligned} &\left((m_1 + m_2)\, l_1^2 + m_2 l_2 (l_2 + 2 l_1 \cos\theta_2)\right) \ddot{\theta}_1 \\ &+m_2 l_2 (l_2 + l_1 \cos\theta_2)\, \ddot{\theta}_2 - m_2 l_1 l_2 \dot{\theta}_2 \left(2\dot{\theta}_1 + \dot{\theta}_2\right) \sin\theta_2 \\ &+ (m_1 + m_2)\, g l_1 \sin\theta_1 + m_2 g l_2 \sin(\theta_1 + \theta_2) = 0 \end{aligned} \tag{5.157}$$

$$\begin{aligned} &(m_2 l_2 (l_2 + l_1 \cos\theta_2))\, \ddot{\theta}_1 + m_2 l_2^2 \ddot{\theta}_2 \\ &+m_2 l_1 l_2 \dot{\theta}_1^2 \sin\theta_2 + m_2 g l_2 \sin(\theta_1 + \theta_2) = 0 \end{aligned} \tag{5.158}$$

The linearized forms of these equations for small angles are both, statically and dynamically coupled.

$$\begin{aligned} &\begin{bmatrix} m_1 l_1^2 + m_2 (l_1 + l_2)^2 & m_2 l_2 (l_2 + l_1) \\ m_2 l_2 (l_2 + l_1) & m_2 l_2^2 \end{bmatrix} \begin{bmatrix} \ddot{\theta}_1 \\ \ddot{\theta}_2 \end{bmatrix} \\ &+ \begin{bmatrix} (m_1 + m_2)\, g l_1 + m_2 g l_2 & m_2 g l_2 \\ m_2 g l_2 & m_2 g l_2 \end{bmatrix} \begin{bmatrix} \theta_1 \\ \theta_2 \end{bmatrix} = \begin{bmatrix} 0 \\ 0 \end{bmatrix} \end{aligned} \tag{5.159}$$

## 5.5 Dissipation Function

The Lagrange equation based on kinetic energy $K$,

$$\frac{d}{dt}\left(\frac{\partial K}{\partial \dot{q}_r}\right) - \frac{\partial K}{\partial q_r} = F_r \qquad r = 1, 2, \cdots n \tag{5.160}$$

or based on Lagrangean $\mathcal{L}$, as introduced in Equation (5.123), can both be applied to derive the equations of motion of a vibrating system.

$$\frac{d}{dt}\left(\frac{\partial \mathcal{L}}{\partial \dot{q}_r}\right) - \frac{\partial \mathcal{L}}{\partial q_r} = Q_r \qquad r = 1, 2, \cdots n \tag{5.161}$$

However, we may use a simpler and more practical Lagrange equation, for linear vibrations:

$$\frac{d}{dt}\left(\frac{\partial K}{\partial \dot{q}_r}\right) - \frac{\partial K}{\partial q_r} + \frac{\partial D}{\partial \dot{q}_r} + \frac{\partial P}{\partial q_r} = f_r \qquad r = 1, 2, \cdots n \tag{5.162}$$

where $K$ is the kinetic energy, $P$ is the potential energy, and $D$ is the *dissipation function* of the system and $f_r$ is the applied force on the mass $m_r$.

$$K = \frac{1}{2}\dot{\mathbf{q}}^T [m] \dot{\mathbf{q}} = \frac{1}{2}\sum_{i=1}^{n}\sum_{j=1}^{n} \dot{q}_i m_{ij} \dot{q}_j \tag{5.163}$$

$$P = \frac{1}{2}\mathbf{q}^T [k] \mathbf{q} = \frac{1}{2}\sum_{i=1}^{n}\sum_{j=1}^{n} q_i k_{ij} q_j \tag{5.164}$$

$$D = \frac{1}{2}\dot{\mathbf{q}}^T [c] \dot{\mathbf{q}} = \frac{1}{2}\sum_{i=1}^{n}\sum_{j=1}^{n} \dot{q}_i c_{ij} \dot{q}_j \tag{5.165}$$

***Proof*** Consider a one $DOF$ mass–spring–damper vibrating system. When viscous damping is the only type of damping in the system, we may employ a function known as the *Rayleigh dissipation function* $D$,

$$D = \frac{1}{2} c \dot{x}^2 \tag{5.166}$$

to find the damping force $f_c$ by differentiation.

$$f_c = -\frac{\partial D}{\partial \dot{x}} \tag{5.167}$$

Remembering that the elastic force $f_k$ can be found from a potential energy $P$,

$$f_k = -\frac{\partial P}{\partial x} \tag{5.168}$$

the generalized force $F$ can be expressed as a collection of the forces.

$$F = f_c + f_k + f = -\frac{\partial D}{\partial \dot{x}} - \frac{\partial P}{\partial x} + f \tag{5.169}$$

where $f$ is the non-conservative applied force on mass $m$. Substituting (5.169) into (5.160)

$$\frac{d}{dt}\left(\frac{\partial K}{\partial \dot{x}}\right) - \frac{\partial K}{\partial x} = -\frac{\partial D}{\partial \dot{x}} - \frac{\partial P}{\partial x} + f \tag{5.170}$$

gives us the Lagrange equation for a viscous damped vibrating system.

$$\frac{d}{dt}\left(\frac{\partial K}{\partial \dot{x}}\right) - \frac{\partial K}{\partial x} + \frac{\partial D}{\partial \dot{x}} + \frac{\partial P}{\partial x} = f \tag{5.171}$$

For vibrating systems with $n$ $DOF$, the kinetic energy $K$, potential energy $P$, and dissipating function $D$ are as (5.163)–(5.165). Applying the Lagrange equation to the $n$ $DOF$ system would result in $n$ second-order differential equations (5.162). ■

**Example 33** A one $DOF$ forced mass–spring–damper system. A simple example of employing dissipation function $D$ to derive equation of motion of a damped vibrating system.

Figure 5.12 illustrates a single $DOF$ mass–spring–damper system with an external force $f$ applied on the mass $m$. The kinetic and potential energies of the system, when it is in motion, are

$$K = \frac{1}{2} m \dot{x}^2 \qquad P = \frac{1}{2} k x^2 \tag{5.172}$$

and its dissipation function $D$ is

$$D = \frac{1}{2} c \dot{x}^2 \tag{5.173}$$

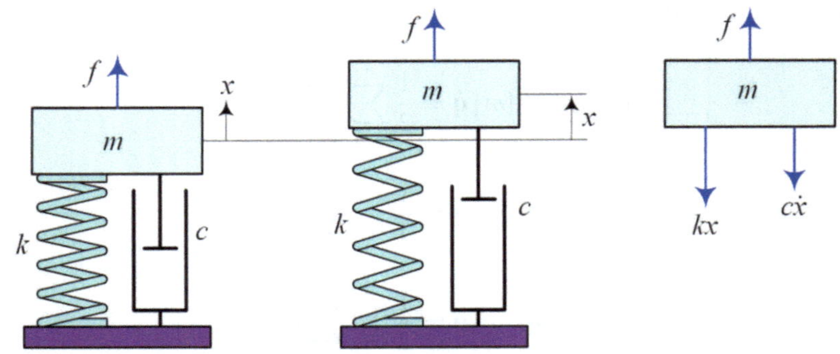


**Fig. 5.12** A one $DOF$ forced mass–spring–damper system

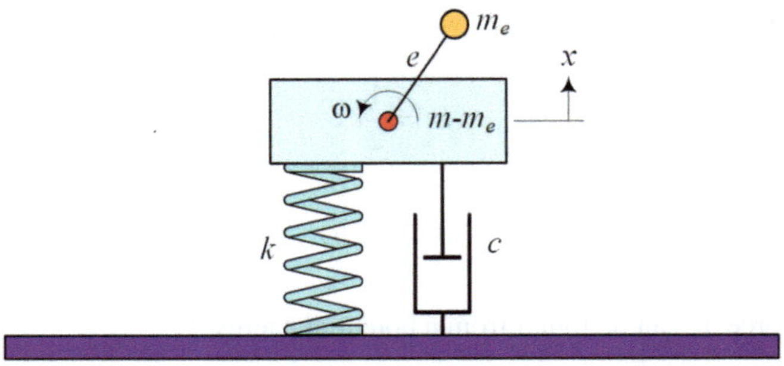


**Fig. 5.13** An eccentric excited single $DOF$ system

Substituting (5.172)–(5.173) into the Lagrange equation (5.162) provides us with the equation of motion:

$$\frac{\partial K}{\partial \dot{x}} = m\dot{x} \qquad \frac{\partial K}{\partial x} = 0 \qquad \frac{\partial D}{\partial \dot{x}} = c\dot{x} \qquad \frac{\partial P}{\partial x} = kx \tag{5.174}$$

$$\frac{d}{dt}(m\dot{x}) + c\dot{x} + kx = f \tag{5.175}$$

**Example 34** An eccentric excited one $DOF$ system. When a device with a rotating shaft is mounted on a base, there would be vibrations due to eccentric excitations. Here is to derive equation of motion.

An eccentric excited one $DOF$ system is shown in Fig. 5.13 with mass $m$ supported by a suspension of a spring $k$ and a damper $c$. There is also a mass $m_e$ at a distance $e$ that is rotating with an angular velocity $\omega$. We may find the equation of motion of the system by applying the Lagrange method. The kinetic energy $K$ of the system is

$$K=\frac{1}{2}(m-m_e)\,\dot{x}^2+\frac{1}{2}m_e(\dot{x}+e\omega\cos\omega t)^2+\frac{1}{2}m_e(-e\omega\sin\omega t)^2 \tag{5.176}$$

The velocity of the main vibrating mass $m - m_e$ is $\dot{x}$, and the velocity of the eccentric mass $m_e$ has two components $\dot{x} + e\omega\cos\omega t$ and $-e\omega\sin\omega t$. The potential energy $P$ and dissipation function of the system are

$$P = \frac{1}{2}kx^2 \qquad D = \frac{1}{2}c\dot{x}^2 \tag{5.177}$$

Applying the Lagrange equation (5.162),

$$\frac{\partial K}{\partial \dot{x}} = m\dot{x} + m_e e\omega\cos\omega t \tag{5.178}$$

$$\frac{d}{dt}\left(\frac{\partial K}{\partial \dot{x}}\right) = m\ddot{x} - m_e e\omega^2\sin\omega t \tag{5.179}$$

$$\frac{\partial D}{\partial \dot{x}} = c\dot{x} \qquad \frac{\partial P}{\partial x} = kx \tag{5.180}$$

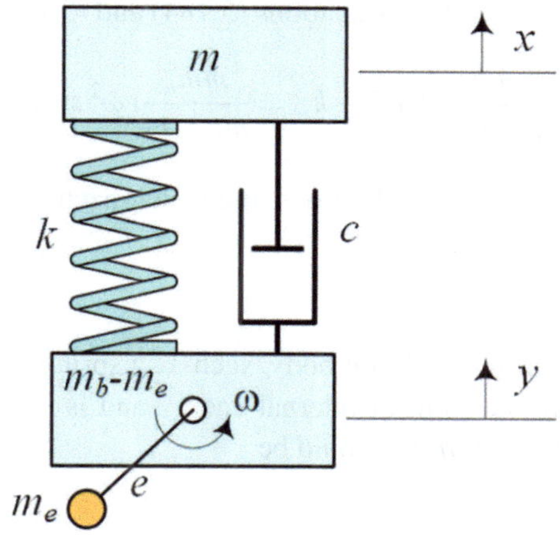


**Fig. 5.14** A one $DOF$ eccentric base excited vibrating system

provides us with the equation of motion.

$$m\ddot{x} + c\dot{x} + kx = m_e e\omega^2 \sin\omega t \tag{5.181}$$

**Example 35** An eccentric base excited vibrating system. When a device is mounted on another device with a rotating shaft, there would be vibrations due to eccentric excitations. Here is how to derive equation of motion by Lagrange method.

Figure 5.14 illustrates a one $DOF$ eccentric base excited vibrating system. A mass $m$ is mounted on an eccentric excited base by a spring $k$ and a damper $c$. The base has a mass $m_b$ with an attached unbalance mass $m_e$ at a distance $e$. The mass $m_e$ is rotating with an angular velocity $\omega$. We may derive the equation of motion of the system by applying Lagrange method. The required functions are

$$\begin{aligned} K = {} & \frac{1}{2}m\dot{x}^2 + \frac{1}{2}(m_b - m_e)\dot{y}^2 \\ & + \frac{1}{2}m_e(\dot{y} - e\omega\cos\omega t)^2 + \frac{1}{2}m_e(e\omega\sin\omega t)^2 \end{aligned} \tag{5.182}$$

$$P = \frac{1}{2}k(x-y)^2 \qquad D = \frac{1}{2}c(\dot{x}-\dot{y})^2 \tag{5.183}$$

Applying the Lagrange method (5.162) provides us with the equations

$$m\ddot{x} + c(\dot{x}-\dot{y}) + k(x-y) = 0 \tag{5.184}$$

$$m_b\ddot{y} + m_e e\omega^2\sin\omega t - c(\dot{x}-\dot{y}) - k(x-y) = 0 \tag{5.185}$$

because

$$\frac{\partial K}{\partial \dot{x}} = m\dot{x} \qquad \frac{d}{dt}\left(\frac{\partial K}{\partial \dot{x}}\right) = m\ddot{x} \tag{5.186}$$

$$\frac{\partial D}{\partial \dot{x}} = c(\dot{x}-\dot{y}) \qquad \frac{\partial P}{\partial x} = k(x-y) \tag{5.187}$$

$$\frac{\partial K}{\partial \dot{y}} = m_b\dot{y} - m_e e\omega\cos\omega t \tag{5.188}$$

$$\frac{d}{dt}\left(\frac{\partial K}{\partial \dot{y}}\right) = m_b\ddot{y} + m_e e\omega^2\sin\omega t \tag{5.189}$$

$$\frac{\partial D}{\partial \dot{y}} = -c(\dot{x}-\dot{y}) \qquad \frac{\partial P}{\partial y} = -k(x-y) \tag{5.190}$$

Using a relative variable $z = x - y$, we may combine Equations (5.184) and (5.185) to find the equation of relative motion.

$$\frac{mm_b}{m_b + m}\ddot{z} + c\,\dot{z} + kz = \frac{mm_e}{m_b + m}e\omega^2 \sin\omega t \tag{5.191}$$

**Example 36** ★ Generalized forces. In mechanical vibrations, the generalized force is usually one of two kinds of elastic and dissipative:

1. Elastic force:
   An elastic force is a recoverable force from an elastic body, such as a spring. An elastic body is the one for which any produced work is stored in the body in the form of internal energy and is recoverable. Therefore, the variation of the internal potential energy of the body, $P = P(q, t)$, would be

$$\delta P = -\delta W = \sum_{i=1}^{n} Q_i \delta q_i \tag{5.192}$$

   where $q_i$ is the generalized coordinate of the particle $i$ of the body, and $\delta W$ is the virtual work of the generalized elastic force $Q$.

$$Q_i = -\frac{\partial P}{\partial q_i} \tag{5.193}$$

2. Dissipation force:
   A dissipative force between two bodies is proportional to and in opposite direction of the relative velocity vector $\mathbf{v}$ between two bodies.

$$Q_i = -c_i\, f_i(v_i)\, \frac{\mathbf{v}_i}{v_i} \tag{5.194}$$

   The coefficient $c_i$ is assumed constant, $f_i(v_i)$ is the velocity function of the force, and $v_i$ is the magnitude of the relative velocity:

$$v_i = \sqrt{\sum_{j=1}^{3} v_{ij}^2} \tag{5.195}$$

   The virtual work of the dissipation force is

$$\delta W = \sum_{i=1}^{n_1} Q_i\, \delta q_i \tag{5.196}$$

$$Q_i = -\sum_{k=1}^{n_1} c_k\, f_k(v_k)\, \frac{\partial v_k}{\partial \dot{q}_i} \tag{5.197}$$

   where $n_1$ is the total number of dissipation forces. By introducing the dissipation function $D$ as

$$D = \sum_{i=1}^{n_1} \int_0^{v_i} c_k\, f_k(z_k)\, dz \tag{5.198}$$

   we have

$$Q_i = -\frac{\partial D}{\partial \dot{q}_i} \tag{5.199}$$

   The dissipation power $P$ of the dissipation force $Q_i$ is

$$P = \sum_{i=1}^{n} Q_i\, \dot{q}_i = \sum_{i=1}^{n} \dot{q}_i \frac{\partial D}{\partial \dot{q}_i} \tag{5.200}$$

## 5.6 ★ Quadratures

If $[m]$ is an $n \times n$ square matrix and $\mathbf{x}$ is an $n \times 1$ vector, then $S$ is a scalar function called *quadrature*.

$$S = \mathbf{x}^T [m] \mathbf{x} \tag{5.201}$$

The derivative of the quadrature $S$ with respect to the vector $\mathbf{x}$ is

$$\frac{\partial S}{\partial \mathbf{x}} = \left([m] + [m]^T\right) \mathbf{x} \tag{5.202}$$

Kinetic energy $K$, potential energy $P$, and dissipation function $D$ are quadratures.

$$K = \frac{1}{2}\dot{\mathbf{x}}^T [m] \dot{\mathbf{x}} \tag{5.203}$$

$$P = \frac{1}{2}\mathbf{x}^T [k] \mathbf{x} \tag{5.204}$$

$$D = \frac{1}{2}\dot{\mathbf{x}}^T [c] \dot{\mathbf{x}} \tag{5.205}$$

Therefore,

$$\frac{\partial K}{\partial \dot{\mathbf{x}}} = \frac{1}{2}\left([m] + [m]^T\right) \dot{\mathbf{x}} \tag{5.206}$$

$$\frac{\partial P}{\partial \mathbf{x}} = \frac{1}{2}\left([k] + [k]^T\right) \mathbf{x} \tag{5.207}$$

$$\frac{\partial D}{\partial \dot{\mathbf{x}}} = \frac{1}{2}\left([c] + [c]^T\right) \dot{\mathbf{x}} \tag{5.208}$$

Employing quadrature derivatives and the Lagrange method,

$$\frac{d}{dt}\frac{\partial K}{\partial \dot{\mathbf{x}}} + \frac{\partial K}{\partial \mathbf{x}} + \frac{\partial D}{\partial \dot{\mathbf{x}}} + \frac{\partial P}{\partial \mathbf{x}} = \mathbf{F} \tag{5.209}$$

$$\delta W = \mathbf{F}^T \partial \mathbf{x} \tag{5.210}$$

the equation of motion for a linear $n$ degree of freedom vibrating system is

$$\left[\underline{m}\right] \ddot{\mathbf{x}} + \left[\underline{c}\right] \dot{\mathbf{x}} + \left[\underline{k}\right] \mathbf{x} = \mathbf{F} \tag{5.211}$$

where $\left[\underline{m}\right]$, $\left[\underline{c}\right]$, $\left[\underline{k}\right]$ are symmetric matrices.

$$\left[\underline{m}\right] = \frac{1}{2}\left([m] + [m]^T\right) \tag{5.212}$$

$$\left[\underline{c}\right] = \frac{1}{2}\left([c] + [c]^T\right) \tag{5.213}$$

$$\left[\underline{k}\right] = \frac{1}{2}\left([k] + [k]^T\right) \tag{5.214}$$

Quadratures are also called *Hermitian forms*.

***Proof*** Let us define a general asymmetric quadrature $S$ as

$$S = \mathbf{x}^T [a] \mathbf{y} = \sum_i \sum_j x_i a_{ij} y_j \tag{5.215}$$

If the quadrature is symmetric, then $\mathbf{x} = \mathbf{y}$.

$$S = \mathbf{x}^T [a] \mathbf{x} = \sum_i \sum_j x_i a_{ij} x_j \tag{5.216}$$

The vectors $\mathbf{x}$ and $\mathbf{y}$ may be functions of $n$ generalized coordinates $q_i$ and time $t$:

$$\mathbf{x} = \mathbf{x}(q_1, q_2, \cdots, q_n, t) \tag{5.217}$$

$$\mathbf{y} = \mathbf{y}(q_1, q_2, \cdots, q_n, t) \tag{5.218}$$

$$\mathbf{q} = \begin{bmatrix} q_1 \; q_2 \cdots q_n \end{bmatrix}^T \tag{5.219}$$

The derivative of $\mathbf{x}$ with respect to $\mathbf{q}$ is a square matrix

$$\frac{\partial \mathbf{x}}{\partial \mathbf{q}} = \begin{bmatrix} \frac{\partial x_1}{\partial q_1} & \frac{\partial x_2}{\partial q_1} & \cdots & \frac{\partial x_n}{\partial q_1} \\ \frac{\partial x_1}{\partial q_2} & \frac{\partial x_2}{\partial q_2} & \cdots & \cdots \\ \cdots & \cdots & \cdots & \cdots \\ \frac{\partial x_1}{\partial q_n} & \cdots & \cdots & \frac{\partial x_n}{\partial q_n} \end{bmatrix} \tag{5.220}$$

which can also be expressed by

$$\frac{\partial \mathbf{x}}{\partial \mathbf{q}} = \begin{bmatrix} \frac{\partial \mathbf{x}}{\partial q_1} \\ \frac{\partial \mathbf{x}}{\partial q_2} \\ \cdots \\ \frac{\partial \mathbf{x}}{\partial q_n} \end{bmatrix} \tag{5.221}$$

or

$$\frac{\partial \mathbf{x}}{\partial \mathbf{q}} = \begin{bmatrix} \frac{\partial x_1}{\partial \mathbf{q}} & \frac{\partial x_2}{\partial \mathbf{q}} & \cdots & \frac{\partial x_n}{\partial \mathbf{q}} \end{bmatrix} \tag{5.222}$$

The derivative of $S$ with respect to an element of $q_k$ is

$$\begin{aligned} \frac{\partial S}{\partial q_k} &= \frac{\partial}{\partial q_k} \sum_i \sum_j x_i a_{ij} y_j \\ &= \sum_i \sum_j \frac{\partial x_i}{\partial q_k} a_{ij} y_j + \sum_i \sum_j x_i a_{ij} \frac{\partial y_j}{\partial q_k} \\ &= \sum_j \sum_i \frac{\partial x_i}{\partial q_k} a_{ij} y_j + \sum_i \sum_j \frac{\partial y_j}{\partial q_k} a_{ij} x_i \\ &= \sum_j \sum_i \frac{\partial x_i}{\partial q_k} a_{ij} y_j + \sum_j \sum_i \frac{\partial y_i}{\partial q_k} a_{ji} x_j \end{aligned} \tag{5.223}$$

and hence, the derivative of $S$ with respect to $\mathbf{q}$ is

$$\frac{\partial S}{\partial \mathbf{q}} = \frac{\partial \mathbf{x}}{\partial \mathbf{q}} [a] \mathbf{y} + \frac{\partial \mathbf{y}}{\partial \mathbf{q}} [a]^T \mathbf{x} \tag{5.224}$$

If $S$ is a symmetric quadrature, then

$$\frac{\partial S}{\partial \mathbf{q}} = \frac{\partial}{\partial \mathbf{q}} \left( \mathbf{x}^T [a] \mathbf{x} \right) = \frac{\partial \mathbf{x}}{\partial \mathbf{q}} [a] \mathbf{x} + \frac{\partial \mathbf{x}}{\partial \mathbf{q}} [a]^T \mathbf{x} \tag{5.225}$$

and if $\mathbf{q} = \mathbf{x}$, then the derivative of a symmetric $S$ with respect to $\mathbf{x}$ is

$$\begin{aligned} \frac{\partial S}{\partial \mathbf{x}} &= \frac{\partial}{\partial \mathbf{x}} \left( \mathbf{x}^T [a] \mathbf{x} \right) = \frac{\partial \mathbf{x}}{\partial \mathbf{x}} [a] \mathbf{x} + \frac{\partial \mathbf{x}}{\partial \mathbf{x}} [a]^T \mathbf{x} \\ &= [a] \mathbf{x} + [a]^T \mathbf{x} = \left( [a] + [a]^T \right) \mathbf{x} \end{aligned} \tag{5.226}$$

If $[a]$ is a symmetric matrix, then

$$[a] + [a]^T = 2 [a] \tag{5.227}$$

and if $[a]$ is not a symmetric matrix, then $\left[\underline{a}\right] = [a] + [a]^T$ is a symmetric matrix because

$$\underline{a}_{ij} = a_{ij} + a_{ji} = a_{ji} + a_{ij} = \underline{a}_{ji} \tag{5.228}$$

and therefore,

$$\left[\underline{a}\right] = \left[\underline{a}\right]^T \tag{5.229}$$

Kinetic energy $K$, potential energy $P$, and dissipation function $D$ can be expressed by quadratures.

$$K = \frac{1}{2} \dot{\mathbf{x}}^T [m] \dot{\mathbf{x}} \tag{5.230}$$

$$P = \frac{1}{2} \mathbf{x}^T [k] \mathbf{x} \tag{5.231}$$

$$D = \frac{1}{2} \dot{\mathbf{x}}^T [c] \dot{\mathbf{x}} \tag{5.232}$$

Substituting $K$, $P$, $D$ in the Lagrange equation provides us with the equations of motion:

$$\begin{aligned} \mathbf{F} &= \frac{d}{dt} \frac{\partial K}{\partial \dot{\mathbf{x}}} + \frac{\partial K}{\partial \mathbf{x}} + \frac{\partial D}{\partial \dot{\mathbf{x}}} + \frac{\partial P}{\partial \mathbf{x}} \\ &= \frac{1}{2} \frac{d}{dt} \frac{\partial}{\partial \dot{\mathbf{x}}} \left( \dot{\mathbf{x}}^T [m] \dot{\mathbf{x}} \right) + \frac{1}{2} \frac{\partial}{\partial \dot{\mathbf{x}}} \left( \dot{\mathbf{x}}^T [c] \dot{\mathbf{x}} \right) + \frac{1}{2} \frac{\partial}{\partial \mathbf{x}} \left( \mathbf{x}^T [k] \mathbf{x} \right) \\ &= \frac{1}{2} \left[ \frac{d}{dt} \left( \left( [m] + [m]^T \right) \dot{\mathbf{x}} \right) + \left( [c] + [c]^T \right) \dot{\mathbf{x}} + \left( [k] + [k]^T \right) \mathbf{x} \right] \\ &= \frac{1}{2} \left( [m] + [m]^T \right) \ddot{\mathbf{x}} + \frac{1}{2} \left( [c] + [c]^T \right) \dot{\mathbf{x}} + \frac{1}{2} \left( [k] + [k]^T \right) \mathbf{x} \\ &= \left[\underline{m}\right] \ddot{\mathbf{x}} + \left[\underline{c}\right] \dot{\mathbf{x}} + \left[\underline{k}\right] \mathbf{x} \end{aligned} \tag{5.233}$$

where

$$\left[\underline{m}\right] = \frac{1}{2} \left( [m] + [m]^T \right) \tag{5.234}$$

$$\left[\underline{c}\right] = \frac{1}{2} \left( [k] + [k]^T \right) \tag{5.235}$$

$$\left[\underline{k}\right] = \frac{1}{2} \left( [c] + [c]^T \right) \tag{5.236}$$

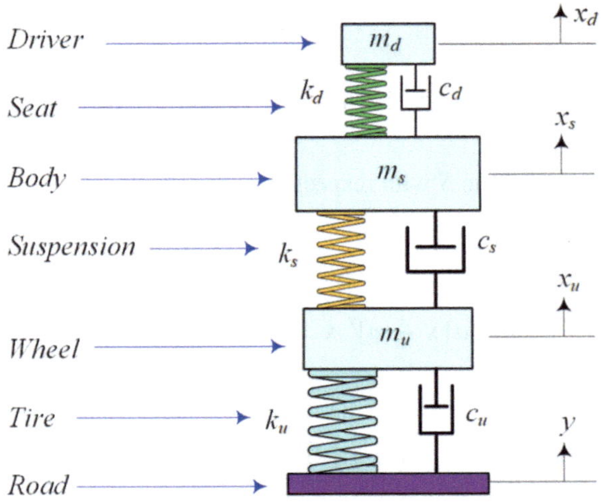


**Fig. 5.15** A quarter car model with driver

From now on, we assume that every equation of motion is found from the Lagrange method to have symmetric coefficient matrices. Hence, we show the equations of motion in the form of

$$[m]\,\ddot{\mathbf{x}} + [c]\,\dot{\mathbf{x}} + [k]\,\mathbf{x} = \mathbf{F} \tag{5.237}$$

and use $[m]$, $[c]$, $[k]$ as a substitute for $\left[\underline{m}\right]$, $\left[\underline{c}\right]$, $\left[\underline{k}\right]$:

$$[m] \equiv \left[\underline{m}\right] \tag{5.238}$$

$$[c] \equiv \left[\underline{c}\right] \tag{5.239}$$

$$[k] \equiv \left[\underline{k}\right] \tag{5.240}$$

Symmetric matrices are equal to their transpose:

$$[m] \equiv [m]^T \tag{5.241}$$

$$[c] \equiv [c]^T \tag{5.242}$$

$$[k] \equiv [k]^T \tag{5.243}$$

■

**Example 37** ★ A quarter car model with driver's chair. Quarter car is the simplest model of a vehicle to study its vertical vibrations. Adding a chair and driver makes the model to have three $DOF$s. Here is to show how to derive the equations of motion by Lagrange method.

Figure 5.15 illustrates a quarter car model plus a driver, which is modeled by a mass $m_d$ over a linear cushion above the sprung mass $m_s$. Assuming

$$y = 0 \tag{5.244}$$

we find the free vibration equations of motion. The kinetic energy $K$ of the system

$$\begin{aligned} K &= \frac{1}{2}m_u\dot{x}_u^2 + \frac{1}{2}m_s\dot{x}_s^2 + \frac{1}{2}m_d\dot{x}_d^2 \\ &= \frac{1}{2}\left[\,\dot{x}_u \;\dot{x}_s \;\dot{x}_d\,\right]\begin{bmatrix} m_u & 0 & 0 \\ 0 & m_s & 0 \\ 0 & 0 & m_d \end{bmatrix}\begin{bmatrix} \dot{x}_u \\ \dot{x}_s \\ \dot{x}_d \end{bmatrix} \\ &= \frac{1}{2}\dot{\mathbf{x}}^T\,[m]\,\dot{\mathbf{x}} \end{aligned} \tag{5.245}$$

and the potential energy $P$ are

$$
\begin{aligned}
P &= \frac{1}{2}k_u \left(x_u\right)^2 + \frac{1}{2}k_s \left(x_s - x_u\right)^2 + \frac{1}{2}k_d \left(x_d - x_s\right)^2 \\
&= \frac{1}{2}\begin{bmatrix} x_u & x_s & x_d \end{bmatrix} \begin{bmatrix} k_u + k_s & -k_s & 0 \\ -k_s & k_s + k_d & -k_d \\ 0 & -k_d & k_d \end{bmatrix} \begin{bmatrix} x_u \\ x_s \\ x_d \end{bmatrix} \\
&= \frac{1}{2}\mathbf{x}^T \left[k\right] \mathbf{x}
\end{aligned}
\tag{5.246}
$$

Similarly, the dissipation function $D$ can be expressed as

$$
\begin{aligned}
D &= \frac{1}{2}c_u \left(\dot{x}_u\right)^2 + \frac{1}{2}c_s \left(\dot{x}_s - \dot{x}_u\right)^2 + \frac{1}{2}c_d \left(\dot{x}_d - \dot{x}_s\right)^2 \\
&= \frac{1}{2}\begin{bmatrix} \dot{x}_u & \dot{x}_s & \dot{x}_d \end{bmatrix} \begin{bmatrix} c_u + c_s & -c_s & 0 \\ -c_s & c_s + c_d & -c_d \\ 0 & -c_d & c_d \end{bmatrix} \begin{bmatrix} \dot{x}_u \\ \dot{x}_s \\ \dot{x}_d \end{bmatrix} \\
&= \frac{1}{2}\dot{\mathbf{x}}^T \left[c\right] \dot{\mathbf{x}}
\end{aligned}
\tag{5.247}
$$

Employing the quadrature derivative method, we may find the derivatives of $K$, $P$, $D$ with respect to their variable vectors:

$$
\begin{aligned}
\frac{\partial K}{\partial \dot{\mathbf{x}}} &= \frac{1}{2}\left(\left[m\right] + \left[m\right]^T\right)\dot{\mathbf{x}} = \frac{1}{2}\left(\left[k\right] + \left[k\right]^T\right)\begin{bmatrix} \dot{x}_u \\ \dot{x}_s \\ \dot{x}_d \end{bmatrix} \\
&= \begin{bmatrix} m_u & 0 & 0 \\ 0 & m_s & 0 \\ 0 & 0 & m_d \end{bmatrix} \begin{bmatrix} \dot{x}_u \\ \dot{x}_s \\ \dot{x}_d \end{bmatrix}
\end{aligned}
\tag{5.248}
$$

$$
\begin{aligned}
\frac{\partial P}{\partial \mathbf{x}} &= \frac{1}{2}\left(\left[k\right] + \left[k\right]^T\right)\mathbf{x} = \frac{1}{2}\left(\left[k\right] + \left[k\right]^T\right)\begin{bmatrix} x_u \\ x_s \\ x_d \end{bmatrix} \\
&= \begin{bmatrix} k_u + k_s & -k_s & 0 \\ -k_s & k_s + k_d & -k_d \\ 0 & -k_d & k_d \end{bmatrix} \begin{bmatrix} x_u \\ x_s \\ x_d \end{bmatrix}
\end{aligned}
\tag{5.249}
$$

$$
\begin{aligned}
\frac{\partial D}{\partial \dot{\mathbf{x}}} &= \frac{1}{2}\left(\left[c\right] + \left[c\right]^T\right)\dot{\mathbf{x}} = \frac{1}{2}\left(\left[c\right] + \left[c\right]^T\right)\begin{bmatrix} \dot{x}_u \\ \dot{x}_s \\ \dot{x}_d \end{bmatrix} \\
&= \begin{bmatrix} c_u + c_s & -c_s & 0 \\ -c_s & c_s + c_d & -c_d \\ 0 & -c_d & c_d \end{bmatrix} \begin{bmatrix} \dot{x}_u \\ \dot{x}_s \\ \dot{x}_d \end{bmatrix}
\end{aligned}
\tag{5.250}
$$

Therefore, we find the system's free vibration equations of motion.

$$
\left[m\right]\ddot{\mathbf{x}} + \left[c\right]\dot{\mathbf{x}} + \left[k\right]\mathbf{x} = 0 \tag{5.251}
$$

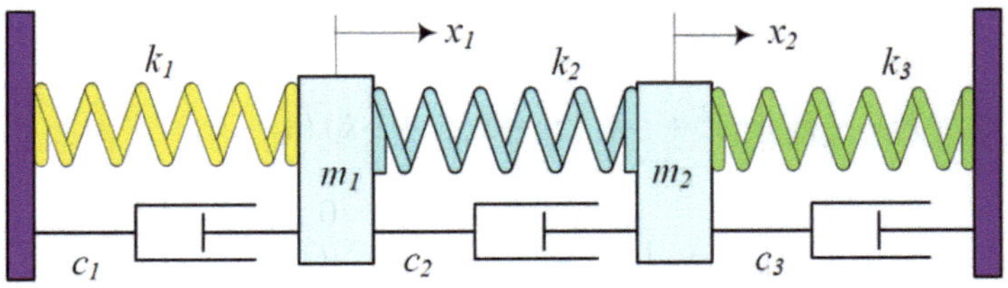


**Fig. 5.16** A two $DOF$ with absolute coordinates $x_1$ and $x_2$

$$\begin{bmatrix} m_u & 0 & 0 \\ 0 & m_s & 0 \\ 0 & 0 & m_d \end{bmatrix} \begin{bmatrix} \ddot{x}_u \\ \ddot{x}_s \\ \ddot{x}_d \end{bmatrix} + \begin{bmatrix} c_u + c_s & -c_s & 0 \\ -c_s & c_s + c_d & -c_d \\ 0 & -c_d & c_d \end{bmatrix} \begin{bmatrix} \dot{x}_u \\ \dot{x}_s \\ \dot{x}_d \end{bmatrix}$$
$$+ \begin{bmatrix} k_u + k_s & -k_s & 0 \\ -k_s & k_s + k_d & -k_d \\ 0 & -k_d & k_d \end{bmatrix} \begin{bmatrix} x_u \\ x_s \\ x_d \end{bmatrix} = 0 \tag{5.252}$$

**Example 38** ★ Discrete multiple vibrating systems. Employing dissipation function and using Lagrange method are the best methods to derive the equations of motion of multi-$DOF$ systems with linear damping.

Figure 5.16 indicates a two $DOF$ with absolute coordinates $x_1$ and $x_2$. The kinetic energy $K$, potential energy $P$, and dissipation function $D$ can be expressed as

$$K = \frac{1}{2} m_1 \dot{x}_1^2 + \frac{1}{2} m_2 \dot{x}_2^2 = \frac{1}{2} \dot{\mathbf{x}}^T [m] \dot{\mathbf{x}}$$
$$= \frac{1}{2} \begin{bmatrix} \dot{x}_1 & \dot{x}_2 \end{bmatrix} \begin{bmatrix} m_1 & 0 \\ 0 & m_2 \end{bmatrix} \begin{bmatrix} \dot{x}_1 \\ \dot{x}_2 \end{bmatrix} \tag{5.253}$$

$$P = \frac{1}{2} k_1 x_1^2 + \frac{1}{2} k_2 (x_1 - x_2)^2 + \frac{1}{2} k_3 x_2^2 = \frac{1}{2} \mathbf{x}^T [k] \mathbf{x}$$
$$= \frac{1}{2} \begin{bmatrix} x_1 & x_2 \end{bmatrix} \begin{bmatrix} k_1 + k_2 & -k_2 \\ -k_2 & k_2 + k_3 \end{bmatrix} \begin{bmatrix} x_1 \\ x_2 \end{bmatrix} \tag{5.254}$$

$$D = \frac{1}{2} c_1 \dot{x}_1^2 + \frac{1}{2} c_2 (\dot{x}_1 - \dot{x}_2)^2 + \frac{1}{2} c_3 \dot{x}_2^2 = \frac{1}{2} \dot{\mathbf{x}}^T [c] \dot{\mathbf{x}}$$
$$= \frac{1}{2} \begin{bmatrix} \dot{x}_1 & \dot{x}_2 \end{bmatrix} \begin{bmatrix} c_1 + c_2 & -c_2 \\ -c_2 & c_2 + c_3 \end{bmatrix} \begin{bmatrix} \dot{x}_1 \\ \dot{x}_2 \end{bmatrix} \tag{5.255}$$

Now consider the three $DOF$ system of Fig. 5.17 with absolute coordinates $x_1$, $x_2$, and $x_3$. The kinetic energy $K$, potential energy $P$, and dissipation function $D$ of the system are

$$K = \frac{1}{2} m_1 \dot{x}_1^2 + \frac{1}{2} m_2 \dot{x}_2^2 + \frac{1}{2} m_3 \dot{x}_3^2 = \frac{1}{2} \dot{\mathbf{x}}^T [m] \dot{\mathbf{x}}$$
$$= \frac{1}{2} \begin{bmatrix} \dot{x}_1 & \dot{x}_2 & \dot{x}_3 \end{bmatrix} \begin{bmatrix} m_1 & 0 & 0 \\ 0 & m_2 & 0 \\ 0 & 0 & m_3 \end{bmatrix} \begin{bmatrix} \dot{x}_1 \\ \dot{x}_2 \\ \dot{x}_3 \end{bmatrix} \tag{5.256}$$

$$P = \frac{1}{2} k_1 x_1^2 + \frac{1}{2} k_2 (x_1 - x_2)^2 + \frac{1}{2} k_3 (x_2 - x_3)^2 + \frac{1}{2} k_4 x_3^2$$
$$= \frac{1}{2} \mathbf{x}^T [k] \mathbf{x} \tag{5.257}$$

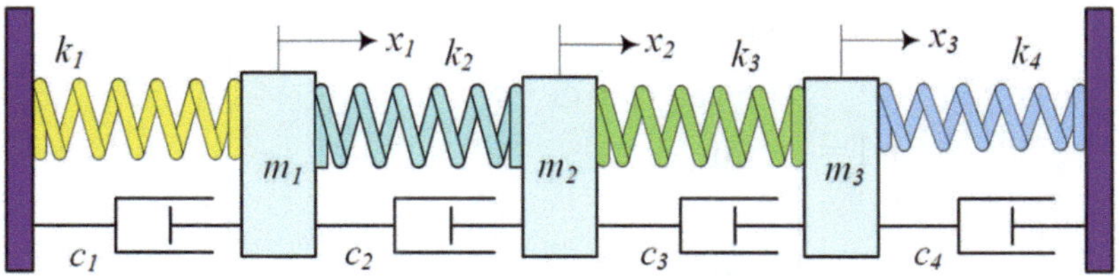


**Fig. 5.17** A three $DOF$ system with absolute coordinates $x_1$, $x_2$, and $x_3$

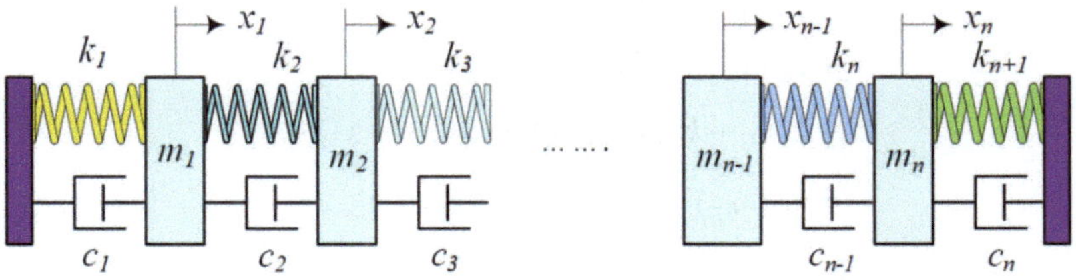


**Fig. 5.18** A discrete $n$ $DOF$ system with absolute coordinates of $x_1, x_2, \ldots, x_n$

$$[k] = \begin{bmatrix} k_1 + k_2 & -k_2 & 0 \\ -k_2 & k_2 + k_3 & -k_3 \\ 0 & -k_3 & k_3 + k_4 \end{bmatrix} \tag{5.258}$$

$$\begin{aligned} D &= \frac{1}{2}c_1\dot{x}_1^2 + \frac{1}{2}c_2\left(\dot{x}_1 - \dot{x}_2\right)^2 + \frac{1}{2}c_3\left(\dot{x}_2 - \dot{x}_3\right)^2 + \frac{1}{2}c_4\dot{x}_3^2 \\ &= \frac{1}{2}\dot{\mathbf{x}}^T\,[c]\,\dot{\mathbf{x}} \end{aligned} \tag{5.259}$$

$$[c] = \begin{bmatrix} c_1 + c_2 & -c_2 & 0 \\ -c_2 & c_2 + c_3 & -c_3 \\ 0 & -c_3 & c_3 + c_4 \end{bmatrix} \tag{5.260}$$

Similarly, if we have a discrete $n$ $DOF$ system with absolute coordinates of $x_1, x_2, \ldots, x_n$ such as shown in Fig. 5.18, then the kinetic energy $K$, potential energy $P$, and dissipation function $D$ of the system are

$$[k] = \begin{bmatrix} m_1 & 0 & 0 & \cdots & 0 \\ 0 & m_2 & 0 & \cdots & 0 \\ 0 & 0 & m_3 & \vdots & \vdots \\ \vdots & \vdots & \vdots & \ddots & \vdots \\ 0 & 0 & \cdots & \cdots & m_n \end{bmatrix} \tag{5.261}$$

$$[k] = \begin{bmatrix} k_1 + k_2 & -k_2 & 0 & \cdots & 0 \\ -k_2 & k_2 + k_3 & -k_3 & \cdots & \vdots \\ 0 & -k_3 & \ddots & \ddots & \vdots \\ \vdots & \vdots & \ddots & \ddots & -k_n \\ 0 & 0 & \cdots & -k_n & k_n + k_{n+1} \end{bmatrix} \tag{5.262}$$

$$[c] = \begin{bmatrix} c_1 + c_2 & -c_2 & 0 & \cdots & 0 \\ -c_2 & c_2 + c_3 & -c_3 & \cdots & \vdots \\ 0 & -c_3 & \ddots & \ddots & \vdots \\ \vdots & \vdots & \ddots & \ddots & -c_n \\ 0 & 0 & \cdots & -c_n & c_n + c_{n+1} \end{bmatrix} \tag{5.263}$$

Employing the Lagrange equation, we find the general equations of motion.

$$\begin{aligned} & m_i \ddot{x}_i + c_1 \dot{x}_1 + c_2 (\dot{x}_1 - \dot{x}_2) + c_3 (\dot{x}_2 - \dot{x}_3) + \cdots \\ & + c_n (\dot{x}_{n-1} - \dot{x}_n) + c_{n+1} \dot{x}_n \\ & + k_1 x_1 + k_2 (x_1 - x_2) + k_3 (x_2 - x_3) + \cdots \\ & + k_n (x_{n-1} - x_n) k_{n+1} x_n = 0 \end{aligned} \tag{5.264}$$

**Example 39** ★ Symmetric matrices. Working with symmetric matrices is critical in linear vibrations. There are many simplicities with symmetric matrices such as easy to inverse, real eigenvalues, etc.

Employing the Lagrange method guarantees that the coefficient matrices of equations of motion of linear vibrating systems are symmetric. A matrix $[A]$ is symmetric if the columns and rows of $[A]$ are interchangeable, so $[A]$ is equal to its transpose:

$$[A] = [A]^T \tag{5.265}$$

The characteristic equation of a symmetric matrix $[A]$ is a polynomial for which all the roots are real. Therefore, the eigenvalues of $[A]$ are real and distinct, and $[A]$ is diagonalizable. Any two eigenvectors that come from distinct eigenvalues of a symmetric matrix $[A]$ are orthogonal.

## 5.7 Summary

In this chapter, we examine the dynamics of vibrations and explore the techniques used to derive the equations of motion for vibrating systems. The two widely utilized methods for deriving these equations are the Newton–Euler method and the Lagrange method. These approaches have proven to be highly effective in accurately formulating the equations of motion for vibrating systems.

The translational and rotational equations of motion for a rigid body, expressed in the global coordinate frame $G$, are

$${}^G\mathbf{F} = \frac{{}^G d}{dt}\, {}^G\mathbf{p} = \frac{{}^G d}{dt} \left(m\, {}^G\mathbf{v}\right) \tag{5.266}$$

$${}^G\mathbf{M} = \frac{{}^G d}{dt}\, {}^G\mathbf{L} = \frac{{}^G d}{dt} \left(I_i\, {}_G\boldsymbol{\omega}_B\right) \tag{5.267}$$

where ${}^G\mathbf{F}$ and ${}^G\mathbf{M}$ indicate the resultant of the external forces and moments applied on the rigid body, measured at the mass center $C$.

The equation of motion of an $n$ $DOF$ linear vibrating system can always be arranged in matrix form of a set of second-order differential equations

$$[m]\, \ddot{\mathbf{x}} + [c]\, \dot{\mathbf{x}} + [k]\, \mathbf{x} = \mathbf{F} \tag{5.268}$$

where $\mathbf{x}$ is a column array of generalized coordinates of the system, and $\mathbf{F}$ is a column array of the associated applied forces. The *square matrices* of $[m]$, $[c]$, $[k]$ are the mass, damping, and stiffness matrices, respectively.

The work ${}_1W_2$ done by an applied force ${}^G\mathbf{F}$ on $m$ in moving from spatial point 1 to point 2 on a path, indicated by a vector ${}^G\mathbf{r}$, is

$${}_1W_2 = \int_1^2 {}^G\mathbf{F} \cdot d\, {}^G\mathbf{r} \tag{5.269}$$

However, ${}_1W_2$ is equal to the difference of the kinetic energy of terminal and initial points. This is called the *principle of work and energy.*

$$\int_1^2 {}^G\mathbf{F} \cdot d\,{}^G\mathbf{r} = m \int_1^2 \frac{{}^Gd}{dt}\,{}^G\mathbf{v} \cdot {}^G\mathbf{v}dt = \frac{1}{2}m \int_1^2 \frac{d}{dt}v^2 dt$$

$$= \frac{1}{2}m\left(v_2^2 - v_1^2\right) = K_2 - K_1 \tag{5.270}$$

$${}_1W_2 = K_2 - K_1 \tag{5.271}$$

If there is a scalar *potential field function* $P = P(x, y, z)$ such that

$$\mathbf{F} = -\nabla P = -\frac{dP}{d\mathbf{r}} = -\left(\frac{\partial P}{\partial x}\hat{\imath} + \frac{\partial P}{\partial y}\hat{\jmath} + \frac{\partial P}{\partial z}\hat{k}\right) \tag{5.272}$$

then the principle of work and energy simplifies to the principle of *conservation of energy.*

$$K_1 + P_1 = K_2 + P_2 \tag{5.273}$$

The sum of kinetic and potential energies, $E = K + P$, is called the mechanical energy. Mechanical energy of an energy conserved system is a constant of motion, and hence, its time derivative is zero, $\dot{E} = 0$. Derivative of energy of the energy conserved system is a simpler method that provides us with the equations of motion.

The expressions of the equations of motion in the body coordinate frame are

$${}^B\mathbf{F} = {}^G\dot{\mathbf{p}} + {}^B_G\boldsymbol{\omega}_B \times {}^B\mathbf{p} = m\,{}^B\mathbf{a}_B + m\,{}^B_G\boldsymbol{\omega}_B \times {}^B\mathbf{v}_B \tag{5.274}$$

$${}^B\mathbf{M} = {}^B\dot{\mathbf{L}} + {}^B_G\boldsymbol{\omega}_B \times {}^B\mathbf{L} = {}^BI\,{}^B_G\dot{\boldsymbol{\omega}}_B + {}^B_G\boldsymbol{\omega}_B \times \left({}^BI\,{}^B_G\boldsymbol{\omega}_B\right) \tag{5.275}$$

where ${}^BI$ is the mass moment of the rigid body, expressed in the body coordinate frame $B$.

$${}^BI = \begin{bmatrix} I_{xx} & I_{xy} & I_{xz} \\ I_{yx} & I_{yy} & I_{yz} \\ I_{zx} & I_{zy} & I_{zz} \end{bmatrix} \tag{5.276}$$

The elements of $I$ are functions of the mass distribution of the rigid body and are defined by the following equation where $\delta_{ij}$ is Kronecker's delta.

$$I_{ij} = \int_B \left(r_i^2\delta_{mn} - x_{im}x_{jn}\right) dm \qquad i, j = 1, 2, 3 \tag{5.277}$$

$$\delta_{ij} = \begin{cases} 1 & i = j \\ 0 & i \neq j \end{cases} \tag{5.278}$$

Every rigid body has a principal body coordinate frame in which the mass moment matrix is diagonal

$${}^BI = \begin{bmatrix} I_1 & 0 & 0 \\ 0 & I_2 & 0 \\ 0 & 0 & I_3 \end{bmatrix} \tag{5.279}$$

The rotational equation of motion in the principal coordinate frame simplifies to

$$\begin{aligned} M_1 &= I_1\dot{\omega}_1 - (I_2 - I_2)\,\omega_2\omega_3 \\ M_2 &= I_2\dot{\omega}_2 - (I_3 - I_1)\,\omega_3\omega_1 \\ M_3 &= I_3\dot{\omega}_3 - (I_1 - I_2)\,\omega_1\omega_2 \end{aligned} \tag{5.280}$$

The equations of motion for a mechanical system having $n$ $DOF$ can also be found by the Lagrange equation, for linear vibrations:

$$\frac{d}{dt}\left(\frac{\partial K}{\partial \dot{q}_r}\right) - \frac{\partial K}{\partial q_r} + \frac{\partial D}{\partial \dot{q}_r} + \frac{\partial P}{\partial q_r} = f_r \qquad r = 1, 2, \cdots n \tag{5.281}$$

where $K$ is the kinetic energy, $P$ is the potential energy, and $D$ is the *dissipation function* of the system and $f_r$ is the nonpotential applied force on the mass $m_r$.

$$K = \frac{1}{2}\dot{\mathbf{q}}^T [m]\, \dot{\mathbf{q}} = \frac{1}{2}\sum_{i=1}^{n}\sum_{j=1}^{n} \dot{q}_i m_{ij} \dot{q}_j \tag{5.282}$$

$$P = \frac{1}{2}\mathbf{q}^T [k]\, \mathbf{q} = \frac{1}{2}\sum_{i=1}^{n}\sum_{j=1}^{n} q_i k_{ij} q_j \tag{5.283}$$

$$D = \frac{1}{2}\dot{\mathbf{q}}^T [c]\, \dot{\mathbf{q}} = \frac{1}{2}\sum_{i=1}^{n}\sum_{j=1}^{n} \dot{q}_i c_{ij} \dot{q}_j \tag{5.284}$$

When $(x_i, y_i, z_i)$ are Cartesian coordinates in a globally fixed coordinate frame for the particle $m_i$, then its coordinates may be functions of another set of generalized coordinates $q_1, q_2, q_3, \cdots, q_n$ and possibly time $t$.

$$x_i = f_i(q_1, q_2, q_3, \cdots, q_n, t) \tag{5.285}$$

$$y_i = g_i(q_1, q_2, q_3, \cdots, q_n, t) \tag{5.286}$$

$$z_i = h_i(q_1, q_2, q_3, \cdots, q_n, t) \tag{5.287}$$

## 5.8 Key Symbols

| | |
|---|---|
| $\mathbf{1}$ | identity vector, identity matrix |
| $a \equiv \ddot{x}$ | acceleration |
| $a, b, w, h$ | distance, length |
| $a_{ij}$ | element of row $i$ and column $j$ of matrix $S$ |
| $\mathbf{a}, \ddot{\mathbf{r}}$ | acceleration |
| $A, B$ | weight factors |
| $A, B$ | coefficients for frequency responses |
| $B$ | body coordinator |
| $c$ | damping |
| $c$ | constant of integral, integral of motion |
| $[c], [C]$ | damping matrix |
| $c_e$ | equivalent damping |
| $C$ | mass center |
| $\mathbf{d}$ | position vector of the body coordinate frame |
| $D$ | dissipation function |
| $d\mathbf{f}$ | infinitesimal force |
| $dm$ | infinitesimal mass |
| $d\mathbf{m}$ | infinitesimal moment |
| $E$ | mechanical energy |
| $E$ | Young modulus of elasticity |
| $f = 1/T$ | cyclic frequency [ Hz] |
| $f$ | function, integral of motion |
| $f, F, \mathbf{f}, \mathbf{F}$ | force |
| $F_C$ | Coriolis force |
| $g, \mathbf{g}$ | gravitational acceleration |

| | |
|---|---|
| $G$ | global coordinate |
| $H$ | height |
| $I, \mathbf{I}, [I]$ | identity matrix |
| $[I]$ | mass moment matrix |
| $I_1, I_2, I_3$ | principal moment of inertia |
| $k$ | stiffness |
| $k_e$ | equivalent stiffness |
| $k_s$ | sprung spring stiffness |
| $k_u$ | unsprung spring stiffness |
| $k_{ij}$ | element of row $i$ and column $j$ of a stiffness matrix |
| $[k], [K]$ | stiffness matrix |
| $K$ | kinetic energy |
| $l$ | length, directional line |
| $L, \mathbf{L}$ | moment of momentum |
| $\mathcal{L} = K - P$ | Lagrangean |
| $m$ | mass |
| $m_e$ | eccentric mass, equivalent mass |
| $m_{ij}$ | element of row $i$ and column $j$ of a mass matrix |
| $m_k$ | spring mass |
| $m_s$ | sprung mass |
| $m_u$ | unsprung mass |
| $[m], [M]$ | mass matrix |
| $M, \mathbf{M}$ | moment |
| $n$ | number of coils, number of decibels, number of notes |
| $N$ | natural numbers |
| $O$ | order of magnitude |
| $\mathbf{p}$ | momentum, translational momentum |
| $P$ | potential energy |
| $P$ | power |
| $P, Q$ | fixed points on a rigid body |
| $q, \mathbf{q}$ | generalized coordinate |
| $Q$ | generalized force, torque |
| $r$ | frequency ratio |
| $\mathbf{r}$ | position vector |
| $r, R$ | radius |
| $R$ | rotation transformation matrix |
| $s$ | eigenvalue, characteristic value |
| $S$ | quadrature |
| $t$ | time |
| $t_0$ | initial time |
| $T$ | period |
| $T$ | tension |
| $T$ | transpose |
| $T_n$ | natural period |
| $\hat{u}$ | unit vector |
| $\mathbf{u}, \mathbf{v}$ | velocity |
| $v \equiv \dot{x},\ \mathbf{v}, \dot{\mathbf{r}}$ | velocity |
| $W$ | work |
| $x, y, z,\ \mathbf{x}$ | displacement |
| $\mathbf{x}$ | vector of variables |
| $x_0$ | initial displacement |
| $\dot{x}_0$ | initial velocity |
| $\dot{x}, \dot{y}, \dot{z}$ | velocity, time derivative of $x, y, z$ |

| | |
|---|---|
| $\ddot{x}$ | acceleration |
| $X$ | amplitude of $x$ |
| $z$ | relative displacement |

Greek

| | |
|---|---|
| $\alpha, \beta, \gamma$ | angle, angle of spring with respect to displacement |
| $\boldsymbol{\alpha}$ | angular acceleration |
| $\delta$ | deflection, angle, variation |
| $\delta$ | Kronecker delta function |
| $\delta_s$ | static deflection |
| $\lambda$ | longitude |
| $\varepsilon$ | mass ratio |
| $\theta$ | angular motion coordinate |
| $\omega$, $\boldsymbol{\omega}$, $\Omega$ | angular frequency |
| $\varphi$, $\Phi$ | phase angle |
| $\varphi$ | latitude |
| $\phi$ | angle of coordinate frame rotation |
| $\lambda$ | eigenvalue |
| $\pi$ | $3.141592653589793\ldots$ |

Symbols

| | |
|---|---|
| [ ] | matrix |
| $\int$ | integral |
| \| \| | absolute value |
| Hz | Hertz |
| $d$ | differential |
| $DOF$ | degree of freedom |
| $FBD$ | free body diagram |
| $\partial$ | partial derivative |
| $\mathcal{L}$ | Lagrangean |
| $O$ | order of magnitude |
| $\nabla$ | gradient |

## References


1. J. Angeles, *Dynamic Response of Linear Mechanical Systems: Modeling, Analysis and Simulation* (Springer, New York, 2011)
2. V.I. Arnold, *Mathematical Methods of Classical Mechanics* (Springer, New York, 1978)
3. R.H. Battin, *An Introduction to the Mathematics and Methods of Astrodynamics* (American Institute of Aeronautics and Astronautics, New York, 1987)
4. G.D. Birkoff, *Dynamical Systems* (American Mathematical Society, New York, 1927)
5. A.I. Borisenko, I.E. Tarapov, *Vector and Tensor Analysis with Applications* (translated by Silverman, R.A.) (Dover, New York, 1968)
6. H.S.M. Coxeter, *Introduction to Geometry* (Wiley, New York, 1989)
7. R. Courant, H. Robbins, *What Is Mathematics?* (Oxford University Press, London, 1941)
8. L. Dai, R.N. Jazar, *Nonlinear Approaches in Engineering Applications* (Springer, New York, 2012)
9. A. Dimarogonas, *Vibration for Engineers* (Prentice Hall, New Jersey, 1996)
10. R. Dugas, *A History of Mechanics*, Switzerland, Editions du Griffon (English translation) (Central Book, New York, 1995)
11. A.V. Efimov, Y.G. Zolotarev, V.M. Terpigoreva, *Mathematical Analysis: Advanced Topics* (Mir, Moscow, 1985)
12. E. Esmailzadeh, Design synthesis of a vehicle suspension system using multi-parameter optimization. Veh. Syst. Dyn. **7**, 83–96 (1978)
13. E. Esmailzadeh, R.N. Jazar, Periodic solution of a Mathieu-Duffing type equation. Int. J. Nonlinear Mech. **32**(5), 905–912 (1997)
14. E. Esmailzadeh, R.N. Jazar, Periodic behavior of a cantilever with end mass subjected to harmonic base excitation. Int. J. Nonlinear Mech. **33**(4), 567–577 (1998)
15. L. Euler, Du mouvement de rotation des corps solides autour d'un axe variable, in (1964) Leonhard Euler, ed. by O. Füssli. Opera Omnia: Commentationes Mechanicae ad Theoriam Corporum Rigidorum Pertinentes (1758), pp. 200–236
16. H. Goldstein, C. Poole, J. Safko, *Classical Mechanics*, 3rd edn. (Addison Wesley, New York, 2002)
17. W. Hauser, *Introduction to the Principles of Mechanics* (Addison Wesley, Reading, 1965)
18. D. Hestenes, *New Foundations for Classical Mechanics*, 2nd edn. (Kluwer Academic, New York, 1999)
19. R.N. Jazar, Stability chart of parametric vibrating systems using energy-rate method. Int. J. Non-Linear Mech. **39**(8), 1319–1331 (2004)
20. R.N. Jazar, *Advanced Dynamics: Rigid Body, Multibody, and Aero-space Applications* (Wiley, New York, 2011)

21. R.N. Jazar, Nonlinear modeling of squeeze-film phenomena, in *Nonlinear Approaches in Engineering Applications*, ed. by L. Dai, R. Jazar (Springer, New York, 2012)
22. R.N. Jazar, Nonlinear mathematical modeling of microbeam, in *Nonlinear Approaches in Engineering Applications*, ed. by L. Dai, R. Jazar (Springer, New York, 2012)
23. R.N. Jazar, Derivative and coordinate frames. J. Nonlinear Eng. **1**(1), 25–34 (2012)
24. R.N. Jazar, *Advanced Vibrations: A Modern Approach* (Springer, New York, 2013)
25. R.N. Jazar, *Vehicle Dynamics: Theory and Application*, 3rd edn. (Springer, New York, 2017)
26. R.N. Jazar, *Advanced Vehicle Dynamics* (Springer, New York, 2019)
27. R.N. Jazar, *Approximation Methods in Science and Engineering* (Springer, New York, 2020)
28. R.N. Jazar, *Perturbation Methods in Science and Engineering* (Springer, New York, 2021)
29. R.N. Jazar, *Applied Robotics: Kinematics, Dynamics, and Control*, 3rd edn. (Springer, New York, 2022)
30. R.N. Jazar, *Advanced Vibrations: Theory and Application* (Springer, New York, 2023)
31. R.N. Jazar, M. Kazemi, S. Borhani, *Mechanical Vibrations* (in Persian) (Ettehad Publications, Tehran, 1992)
32. R.N. Jazar, A. Naghshinehpour, Time optimal control algorithm for multi-body dynamical systems. IMechE Part K: J. Multi-Body Dyn. **219**(3), 225–236 (2005)
33. R.N. Jazar, M. Mahinfalah, N. Mahmoudian, M.R. Aagaah, Energy-rate method and stability chart of parametric vibrating systems. J. Braz. Soc. Mech. Sci. Eng. **30**(3), pp. 182–188 (2008)
34. R.N. Jazar, M. Mahinfalah, N. Mahmoudian, M.R. Aagaah, Effects of nonlinearities on the steady state dynamic behavior of electric actuated microcantilever-based resonators. J. Vib. Control **15**(9), 1283–1306 (2009)
35. R.N. Jazar, M. Mahinfalah, N. Mahmoudian, M.R. Aagaah, B. Shiari, Behavior of Mathieu equation in stable regions. Int. J. Mech. Solids **1**(1), 1–18 (2006)
36. E.L. Ince, Longmans, in *Ordinary Differential Equations* (Green and Co., London, 1926)
37. B.G. Korenev, L.M. Reznikov, *Dynamic Vibration Absorbers: Theory and Technical Applications* (Wiley, West Sussex, 1993)
38. W.D. MacMillan, *Dynamics of Rigid Bodies* (McGraw-Hill, New York, 1936)
39. L. Meirovitch, *Analytical Methods in Vibrations* (Macmillan, New York, 1967)
40. L. Meirovitch, *Principles and Techniques of Vibrations* (Prentice Hall, New Jersey, 1997)
41. L. Meirovitch, *Fundamentals of Vibrations* (McGraw-Hill, New York, 2002)
42. E.A. Milne, *Vectorial Mechanics* (Methuen, London, 1948)
43. D.E. Newland, *Mechanical Vibration Analysis and Computation* (Wiley, New York, 1989)
44. I. Newton, *Philosophiae Naturalis Principia Mathematica*. The Mathematical Principles of Natural Philosophy (Royal Society of London, London, 1687)
45. K. Ogata, *System Dynamics* (Prentice Hall, New Jersey, 2004)
46. B. O'Neill, *Elementary Differential Geometry* (Academic, New York, 1966)
47. Z. Osinski, *Damping of Vibrations* (Rotterdam, A. A. Balkema Publishers, 1998)
48. L.A. Pars, *A Treatise on Analytical Dynamics* (Wiley, New York, 1968)
49. W.J. Palm, *Mechanical Vibration* (Wiley, New York, 2006)
50. M. Roseau, *Vibrations in Mechanical Systems* (Springer, Berlin, 1987)
51. R.M. Rosenberg, *Analytical Dynamics of Discrete Systems* (Plenum, New York, 1977)
52. H. Schaub, J.L. Junkins, *Analytical Mechanics of Space Systems*. AIAA Educational Series (American Institute of Aeronautics and Astronautics, Reston, 2003)
53. R.A. Sharipuv, *Course of Differential Geometry* (Bashkir State University Press, Ufa (1996)
54. J.B. Shaw, *Vector Calculus: With Applications to Physics* (Van Nostrand, New York, 1922)
55. J.C. Slater, N.H. Frank, *Mechanics* (Mc-Graw-Hill, New York, 1947)
56. J.C. Snowdon, *Vibration and Shock in Damped Mechanical Systems* (Wiley, New York, 1968)
57. J.L. Synge, B.A. Griffith, *Principle of Mechanics* (McGraw-Hill, New York, 1942)
58. V. Szebehely, *Adventures in Celestial Mechanics* (Wiley, New York, (1998)
59. F.S. Tse, I.E. Morse, R.T. Hinkle, *Mechanical Vibrations Theory and Applications* (Allyn and Bacon Inc, Boston, 1978)
60. E.T. Whittaker, *A Treatise on the Analytical Dynamics of Particles and Rigid Bodies* (Cambridge University Press, London, 1965)

# 6 Time Response

The time response of a vibrating system refers to its reaction when subjected to nonzero initial conditions and/or time-varying forcing functions, known as transient excitations. This behavior is referred to as transient response because it captures the system's response during a finite period of time.

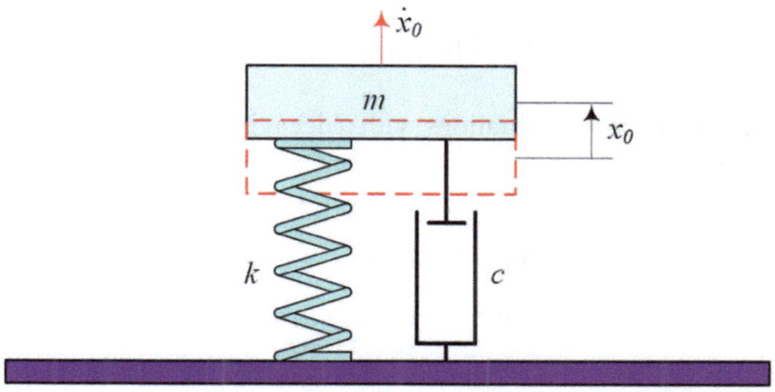


**Fig. 6.1** The model of a forced–free one degree of freedom vibrating system with initial conditions

## 6.1 Free Vibrations One $DOF$

Consider a forced–free mass–spring–damper system $(m, k, c)$ as is shown in Fig. 6.1. The equation of motion of the system and the general initial conditions are

$$m\ddot{x} + c\dot{x} + kx = 0 \tag{6.1}$$

$$x(0) = x_0 \qquad \dot{x}(0) = \dot{x}_0 \tag{6.2}$$

Using the definition of the *natural frequency* $\omega_n$ and *damping ratio* $\xi$,

$$\omega_n = \sqrt{\frac{k}{m}} \tag{6.3}$$

$$\xi = \frac{c}{2\sqrt{km}} = \frac{c}{2m\omega_n} = \frac{c\omega_n}{2k} \tag{6.4}$$

we can rewrite the equation of motion to be dependent on only two parameters $\omega_n$ and $\xi$, instead of three parameters $m, k, c$.

$$\ddot{x} + 2\xi\omega_n\dot{x} + \omega_n^2 x = 0 \tag{6.5}$$



R. N. Jazar, H. Marzbani, *Vehicle Vibrations*, https://doi.org/10.1007/978-3-031-43486-0_6

The time response of the free vibration of the system (6.1) is

$$x(t) = X_1 e^{s_1 t} + X_2 e^{s_2 t} \tag{6.6}$$

where the *characteristic parameters* $s_1$ and $s_2$ are the solutions of the *characteristic equation* of the system.

$$s^2 + 2\xi\omega_n s + \omega_n^2 = 0 \tag{6.7}$$

$$s_1 = -\xi\omega_n + \omega_n\sqrt{\xi^2 - 1} \tag{6.8}$$

$$s_2 = -\xi\omega_n - \omega_n\sqrt{\xi^2 - 1} \tag{6.9}$$

The constants of integrations $X_1$ and $X_2$ depend on the initial conditions.

$$X_1 = \frac{\dot{x}_0 - s_2 x_0}{s_1 - s_2} \tag{6.10}$$

$$X_2 = -\frac{\dot{x}_0 - s_1 x_0}{s_1 - s_2} \tag{6.11}$$

Depending on the value of the damping ratio $\xi$, we will have five types of time response for free vibrations:

1. Underdamped: $0 < \xi < 1$. The system shows vibrations with a decaying amplitude.

$$x = e^{-\xi\omega_n t}\left(A\cos\omega_d t + B\sin\omega_d t\right) \tag{6.12}$$

$$= Xe^{-\xi\omega_n t}\sin\left(\omega_d t + \varphi\right) \tag{6.13}$$

$$= Xe^{-\xi\omega_n t}\cos\left(\omega_d t + \theta\right) \tag{6.14}$$

$$A = x_0 \qquad B = \frac{\dot{x}_0 + \xi\omega_n x_0}{\omega_d} \tag{6.15}$$

$$X = \sqrt{A^2 + B^2} \tag{6.16}$$

$$\omega_d = \omega_n\sqrt{1 - \xi^2} \tag{6.17}$$

2. Critically damped: $\xi = 1$. The system approaches its equilibrium in fastest time with no vibrations.

$$x = x_0 e^{-t\omega_n}\left(1 + \omega_n t\right) + \dot{x}_0 t e^{-t\omega_n} \tag{6.18}$$

3. Overdamped: $\xi > 1$. The system approaches its equilibrium asymptotically and shows no vibrations.

$$x(t) = X_1 e^{s_1 t} + X_2 e^{s_2 t} \tag{6.19}$$

$$X_1 = \frac{\dot{x}_0 - s_2 x_0}{s_1 - s_2} \qquad X_2 = -\frac{\dot{x}_0 - s_1 x_0}{s_1 - s_2} \tag{6.20}$$

$$s_1 = \left(-\xi + \sqrt{\xi^2 - 1}\right)\omega_n < 0 \tag{6.21}$$

$$s_2 = \left(-\xi - \sqrt{\xi^2 - 1}\right)\omega_n < 0 \tag{6.22}$$

4. Undamped: $\xi = 0$. The system vibrates harmonically with a constant amplitude.

$$x = A\cos\omega_n t + B\sin\omega_n t \tag{6.23}$$

$$= X\sin(\omega_n t + \varphi) \tag{6.24}$$

$$= X\cos(\omega_n t + \theta) \tag{6.25}$$

$$A = x_0 \qquad B = \frac{\dot{x}_0}{\omega_d} \tag{6.26}$$

$$X = \sqrt{A^2 + B^2} \tag{6.27}$$

5. Negative damping: $\xi < 0$. The system is unstable and vibrates with an increasing amplitude unboundedly.

$$x = \frac{\dot{x}_0 + \left(\xi + \sqrt{\xi^2 - 1}\right)\omega_n x_0}{2\omega_n\sqrt{\xi^2 - 1}} e^{-\left(\xi - \sqrt{\xi^2-1}\right)\omega_n t} - \frac{\dot{x}_0 + \left(\xi - \sqrt{\xi^2 - 1}\right)\omega_n x_0}{2\omega_n\sqrt{\xi^2 - 1}} e^{-\left(\xi + \sqrt{\xi^2-1}\right)\omega_n t} \tag{6.28}$$

***Proof*** Dividing the free vibration equation of motion (6.1) by $m$

$$\ddot{x} + \frac{c}{m}\dot{x} + \frac{k}{m}x = 0 \tag{6.29}$$

$$x(0) = x_0 \qquad \dot{x}(0) = \dot{x}_0 \tag{6.30}$$

yields

$$\ddot{x} + 2\xi\omega_n\dot{x} + \omega_n^2 x = 0 \tag{6.31}$$

where

$$2\xi\omega_n = \frac{c}{m} \tag{6.32}$$

$$\omega_n^2 = \frac{k}{m} \tag{6.33}$$

The parameter $\omega_n$ is the natural frequency of the system, which is the frequency of vibration of the undamped system, and the dimensionless parameter $\xi$ is a measure of the damping of the system.

The free vibration of a system is a *transient response* that for a set of parameters $\omega_n$ and $\xi$ depends solely on the initial conditions $x_0 = x(0)$ and $\dot{x}_0 = \dot{x}(0)$. The natural frequency $\omega_n$ is a positive number that indicates the fluctuations of the $x$ in a unit of time. Variation of $\omega_n$ will not change the category of $x$. The value of the positive damping ratio $\xi$ indicates how quick the effects of initial conditions $x_0 = x(0)$ and $\dot{x}_0 = \dot{x}(0)$ will disappear and $x$ stops at equilibrium point. However, variation of $\xi$ will provide a different category of responses $x$.

The general solution of a second-order homogeneous linear equation

$$\ddot{x} + p(t)\,\dot{x} + q(t)\,x = 0 \tag{6.34}$$

is

$$x = X_1 x_1(t) + X_2 x_2(t) \tag{6.35}$$

where the $X_i$ are constants of integration and $x_1(t)$ and $x_2(t)$ are linearly independent functions, each satisfying Equation (6.34). There are always exactly 2 linearly independent solutions to Equation (6.34) in any region in which the coefficient functions $p(t)$ and $q(t)$ are continuous. For linear differential equation, the exponential functions $x = X\exp(st)$ are the solution. Substituting the exponential function into the differential equation will determine two values for $s$ that makes the two solutions $x_1 = X_1\exp(s_1 t)$ and $x_2 = X_2\exp(s_2 t)$. In special case when $s_1 = s_2 = s$, the two solutions would be $x_1 = X_1\exp(st)$ and $x_2 = X_2 t\exp(st)$.

The equation of vibration (6.31) is a linear second-order ordinary deferential equation, and hence, its solution will be exponential function. To find the solution, we substitute a general exponential function with unknown coefficients and unknown power and determine the unknowns.

$$x = X\,e^{st} \tag{6.36}$$

Determination of the solution is performed in two steps:

1. We determine the exponent $s$. Their value depends on the parameters of the system, and the number of $s$ is equal to the order of the differential equation. In this case, we will have two values for $s$, and they will be functions of the parameters $\omega_n$ and $\xi$.

   Substituting (6.36) in (6.31)

   $$\left(Xs^2e^{st}\right) + 2\xi\omega_n\left(Xse^{st}\right) + \omega_n^2\left(Xe^{st}\right) = \left(s^2 + 2\xi\omega_n s + \omega_n^2\right)Xe^{st} = 0 \tag{6.37}$$

   provides us with an second-degree algebraic equation, called the characteristic equation.

   $$s^2 + 2\xi\omega_n s + \omega_n^2 = 0 \tag{6.38}$$

   The solutions of the characteristic equation are the characteristic values $s_1$ and $s_2$.

   $$s_1 = -\xi\omega_n + \sqrt{\xi^2 - 1}\,\omega_n \tag{6.39}$$

   $$s_2 = -\xi\omega_n - \sqrt{\xi^2 - 1}\,\omega_n \tag{6.40}$$

2. We determine the coefficient $X$. Every value of $s$ provides us with an exponential function. All of these exponential functions are linearly independent, and hence, the general solution of the equation would be a linear combination of the exponential functions, each with unknown coefficients. The coefficients depend on the initial conditions of the system. In this case, we have two exponential functions with coefficients $X_1$ and $X_2$.

   $$x = X_1e^{s_1t} + X_2e^{s_2t} \tag{6.41}$$

   $$\dot{x} = X_1s_1e^{s_1t} + X_2s_2e^{s_2t} \tag{6.42}$$

   The initial condition (6.30) provides us with two equations to find $X_1$ and $X_2$.

   $$x_0 = X_1 + X_2 \tag{6.43}$$

   $$\dot{x}_0 = X_1s_1 + X_2s_2 \tag{6.44}$$

   $$\begin{bmatrix} x_0 \\ \dot{x}_0 \end{bmatrix} = \begin{bmatrix} 1 & 1 \\ s_1 & s_2 \end{bmatrix} \begin{bmatrix} X_1 \\ X_2 \end{bmatrix} \tag{6.45}$$

   $$\begin{aligned} \begin{bmatrix} X_1 \\ X_2 \end{bmatrix} &= \begin{bmatrix} 1 & 1 \\ s_1 & s_2 \end{bmatrix}^{-1} \begin{bmatrix} x_0 \\ \dot{x}_0 \end{bmatrix} \\ &= \begin{bmatrix} \frac{1}{s_1 - s_2}\left(\dot{x}_0 - s_2x_0\right) \\ -\frac{1}{s_1 - s_2}\left(\dot{x}_0 - s_1x_0\right) \end{bmatrix} \end{aligned} \tag{6.46}$$

Therefore, the complete solution is achieved.

$$
\begin{aligned}
x &= X_1 e^{s_1 t} + X_2 e^{s_2 t} \\
&= \frac{\dot{x}_0 - s_2 x_0}{s_1 - s_2} e^{s_1 t} - \frac{\dot{x}_0 - s_1 x_0}{s_1 - s_2} e^{s_2 t}
\end{aligned}
\tag{6.47}
$$

The complete solution will have different categories of appearance depending on the value of the damping ratio $\xi$ as classified in the following example.

The linear discrete vibrating systems are expressed by linear ordinary differential equations. If the parameters of the system are constant, the equations of motion are differential equations with constant coefficients with known general solutions. If there is transient excitations, then such systems have the following properties for their solutions:

1. Superposition of time responses is applied.

$$x(C_1 f_1 + C_2 f_2) = C_1 x(f_1) + C_2 x(f_2)$$

2. The total time response is a combination of two parts: *natural* and *forced*, or *general* and *particular*, or *homogeneous* and *inhomogenous*.

$$x(t) = x_h(t) + x_p(t)$$

3. The natural solution of linear equations is always exponential time functions.

$$x_h(t) = \sum_{i=1}^{n} C_i x(0) e^{-st}$$

4. The forced solution always has the same form as the force function with proportional magnitude.

$$x_p(t) = C f(t)$$

In this part, we review the natural, transient, and nonharmonic forced vibrations of vibrating systems in time domain. ■

**Example 40** The characteristic values. This is to show how to determine the characteristic values of a second-order equation of motion, and classifying the solutions according to the value of damping ratio $\xi$. The two characteristic values provide the two independent solutions of the equation.

The exponential function is the general solution of every linear differential equation. The equation of motion of the free vibrations of a one $DOF$ is a second-order linear equation.

$$m\ddot{x} + c\dot{x} + kx = 0 \tag{6.48}$$

Substituting an exponential function with unknown coefficient to satisfy the equation,

$$x = Ce^{st} \tag{6.49}$$

provides us with

$$\left(ms^2 + cs + k\right) Ce^{st} = 0 \tag{6.50}$$

Because $Ce^{st} \neq 0$ at any time, we must have

$$ms^2 + cs + k = 0 \tag{6.51}$$

This is a second-degree algebraic equation, called the characteristic equation of Equation (6.48). The solutions of the characteristic equation are all possible values of the exponent $s$ of the exponential solution (6.49).

$$s_1 = -\frac{1}{2m}\left(c + \sqrt{-4km + c^2}\right) \tag{6.52}$$

$$s_2 = -\frac{1}{2m}\left(c - \sqrt{-4km + c^2}\right) \tag{6.53}$$

Employing the parameters $\omega_n$ and $\xi$ from (6.3) and (6.4), we usually rewrite the equation of motion (6.48) in the following form to generalize the equation, and work with the minimum number of parameters.

$$\ddot{x} + 2\xi\omega_n\dot{x} + \omega_n^2 x = 0 \tag{6.54}$$

Changing the parameters does not change the linearity of the equation, and Therefore, the exponential function is still the solution. The characteristic equation will be changed to

$$s^2 + 2\xi\omega_n s + \omega_n^2 = 0 \tag{6.55}$$

with equivalent solutions.

$$\begin{aligned} s_1 &= -\xi\omega_n + \omega_n\sqrt{\xi^2 - 1} = \omega_n\left(-\xi + \sqrt{\xi^2 - 1}\right) \\ &= -\xi\omega_n\left(1 - \sqrt{1 - \frac{1}{\xi^2}}\right) \end{aligned} \tag{6.56}$$

$$\begin{aligned} s_2 &= -\xi\omega_n - \omega_n\sqrt{\xi^2 - 1} = \omega_n\left(-\xi - \sqrt{\xi^2 - 1}\right) \\ &= -\xi\omega_n\left(1 + \sqrt{1 - \frac{1}{\xi^2}}\right) \end{aligned} \tag{6.57}$$

The values of $s_1$ and $s_2$ depend on the natural frequency $\omega_n > 0$ and the damping ratio $\xi$ of the system. The natural frequency $\omega_n$ appears as a coefficient and hence will not change the character of $s_1$ and $s_2$. If $\xi < 1$, then $s_1$ and $s_2$ will be two complex conjugate numbers. We will have only one characteristic value $s_1 = s_2$ if $\xi = 1$. The $s_1$ and $s_2$ provide two distinct real values as long as $1 < \xi$. The behavior of the system is different for different ranges of $\xi$. The ranges and special cases for different $\xi$ are:

1. Underdamped: $0 < \xi < 1$. The system shows vibrations with a decaying amplitude.
2. Critically damped: $\xi = 1$. The system approaches its equilibrium in fastest time with no vibrations.
3. Overdamped: $1 < \xi$. The system approaches its equilibrium asymptotically and shows no vibrations.
4. Undamped: $\xi = 0$. The system vibrates harmonically with a constant amplitude.
5. Oscillatory unstable: $-1 < \xi < 0$. The system vibrates with an increasing amplitude unbounded.
6. Monotonically unstable: $\xi < -1$. The system goes to infinity and shows no vibrations.

The characteristic equation of an $n$-order real system is an $n$-degree algebraic equation with $n$ characteristic values. The solution of the equation of motion of the system would be a linear combination of the exponential time functions with the characteristic values as exponents. Therefore, there are $n$ coefficients in the linear combined solutions that should be calculated based on $n$ initial conditions of the system. The solution for a one $DOF$ with a second-order equation of motion will be

$$x = C_1 e^{s_1 t} + C_2 e^{s_2 t} \tag{6.58}$$

**Example 41** ★ State space representation of the equation of motion. Any differential equation of order $n$ can be expressed by $n$ first-order differential equations by introducing a set of new variables called state variables.

Consider a damped vibrating system.

$$\ddot{x} + 2\xi\omega_n\,\dot{x} + \omega_n^2 x = 0 \qquad x\,(0) = x_0 \qquad \dot{x}\,(0) = \dot{x}_0 \tag{6.59}$$

We can rewrite the equation of the system in state-variable form, by transforming it into a system of two first-order, linear, constant-coefficient ordinary differential equations. Let us define the following new variables:

$$x_1 = x \qquad x_2 = \dot{x} \tag{6.60}$$

Therefore,

$$\dot{x}_1 = x_2 \qquad \dot{x}_2 = -2\xi\omega_n\,x_2 - \omega_n^2 x_1 \tag{6.61}$$

$$x_1\,(0) = x_0 \qquad x_2\,(0) = \dot{x}_0 \tag{6.62}$$

Introducing a variable state vector $\mathbf{x}$, the equation of motion (6.59) will be defined by a set of first-order equations in matrix form.

$$\mathbf{x} = \begin{bmatrix} x_1 \\ x_2 \end{bmatrix} \qquad \mathbf{x}_0 = \begin{bmatrix} x_1\,(0) \\ x_2\,(0) \end{bmatrix} = \begin{bmatrix} x_0 \\ \dot{x}_0 \end{bmatrix} \tag{6.63}$$

$$\dot{\mathbf{x}} = \frac{d\mathbf{x}}{dt} = \mathbf{A}\,\mathbf{x} = \begin{bmatrix} 0 & 1 \\ -\omega_n^2 & -2\xi\omega_n \end{bmatrix} \begin{bmatrix} x_1 \\ x_2 \end{bmatrix} \tag{6.64}$$

The matrix $\mathbf{A}$ is called the coefficient matrix, and the solution of the first-order equation is

$$\mathbf{x} = \mathbf{x}_0\,e^{\mathbf{A}} \qquad 0 < t \tag{6.65}$$

The exponential functions of the coefficient matrix, $e^{\mathbf{A}}$, for underdamped, critically damped, and overdamped are as below.

**Underdamped** $\xi < 1$

$$e^{\mathbf{A}} = e^{-\xi\omega_n t} \begin{bmatrix} \cos\omega_d t + \dfrac{\xi\omega_n}{\omega_d}\sin\omega_d t & \dfrac{\sin\omega_d t}{\omega_d} \\ -\dfrac{\omega_n^2}{\omega_d}\sin\omega_d t & \cos\omega_d t - \dfrac{\xi\omega_n}{\omega_d}\sin\omega_d t \end{bmatrix} \tag{6.66}$$

$$\omega_d = \omega_n\sqrt{1-\xi^2}$$

$$x = e^{-\xi\omega_n t}\left(\cos\omega_d t + \frac{\xi\omega_n}{\omega_d}\sin\omega_d t\right)x_0 + \frac{e^{-\xi\omega_n t}\sin\omega_d t}{\omega_d}\dot{x}_0 \tag{6.67}$$

$$\dot{x} = -\frac{\omega_n^2 e^{-\xi\omega_n t}\sin\omega_d t}{\omega_d}x_0 + e^{-\xi\omega_n t}$$

$$\left(\cos\omega_d t - \frac{\xi\omega_n}{\omega_d}\sin\omega_d t\right)\dot{x}_0 \tag{6.68}$$

**Critically damped** $\xi = 1$

$$e^{\mathbf{A}} = e^{-\omega_n t} \begin{bmatrix} 1+\omega_n t & t \\ -\omega_n^2 t & 1-\omega_n t \end{bmatrix} \tag{6.69}$$

$$x = e^{-\omega_n t} \left( (1+\omega_n t) \, x_0 + t \, \dot{x}_0 \right) \tag{6.70}$$

$$\dot{x} = e^{-\omega_n t} \left( -\omega_n^2 \, t \, x_0 + (1-\omega_n t) \, \dot{x}_0 \right) \tag{6.71}$$

**Overdamped** $1 < \xi$

$$e^{\mathbf{A}} = e^{-\xi\omega_n t} \begin{bmatrix} \cosh\omega_s t + \dfrac{\xi\omega_n}{\omega_s} \sinh\omega_s t & \dfrac{\sinh\omega_s t}{\omega_s} \\ -\dfrac{\omega_n^2}{\omega_s} \sinh\omega_s t & \cosh\omega_s t - \dfrac{\xi\omega_n}{\omega_s} \sinh\omega_s t \end{bmatrix} \tag{6.72}$$

$$\omega_s = \omega_n \sqrt{\xi^2 - 1} \tag{6.73}$$

$$x = e^{-\xi\omega_n t} \left( \cosh\omega_s t + \frac{\xi\omega_n}{\omega_s} \sin\omega_s t \right) x_0 + \frac{e^{-\xi\omega_n t} \sinh\omega_s t}{\omega_s} \dot{x}_0 \tag{6.74}$$

$$\dot{x} = -\frac{\omega_n^2 e^{-\xi\omega_n t} \sinh\omega_s t}{\omega_s} x_0 + e^{-\xi\omega_n t} \left( \cosh\omega_s t - \frac{\xi\omega_n}{\omega_s} \sinh\omega_s t \right) \dot{x}_0 \tag{6.75}$$

In all three cases, the response is a linear combination of the two initial conditions $x_0$ and $\dot{x}_0$.

**Example 42** ★ One $DOF$ second-order system. Solving vibrating systems in state space provides us with position and velocity together, and it can be used to check the results.

Consider a second-order vibrating system.

$$\ddot{x} + 2\xi\omega_n \dot{x} + \omega_n^2 x = 0 \tag{6.76}$$

$$x(0) = x_0 \qquad \dot{x}(0) = \dot{x}_0 \tag{6.77}$$

This system can be rewritten in matrix form.

$$\dot{\mathbf{x}} = \mathbf{A}\mathbf{x} \qquad x_1 = x \qquad x_2 = \dot{x} \tag{6.78}$$

$$\begin{bmatrix} \dot{x}_1 \\ \dot{x}_2 \end{bmatrix} = \begin{bmatrix} 0 & 1 \\ -\omega_n^2 & -2\xi\omega_n \end{bmatrix} \begin{bmatrix} x_1 \\ x_2 \end{bmatrix} \tag{6.79}$$

The response of the system can be written as

$$\mathbf{x} = \mathbf{x}_0 \, e^{\mathbf{A}t} \tag{6.80}$$

Having the eigenvalues of **A** in (6.865) and (6.866), we have

$$\begin{bmatrix} 1 & \lambda_1 t \\ 1 & \lambda_2 t \end{bmatrix} \begin{bmatrix} \dfrac{\lambda_1 e^{\lambda_2 t} - \lambda_2 e^{\lambda_1 t}}{\lambda_1 - \lambda_2} \\ \dfrac{e^{\lambda_1 t} - e^{\lambda_2 t}}{(\lambda_1 - \lambda_2)\, t} \end{bmatrix} = \begin{bmatrix} e^{\lambda_1 t} \\ e^{\lambda_2 t} \end{bmatrix} \tag{6.81}$$

$$e^{\mathbf{A}t} = \frac{\lambda_1 e^{\lambda_2 t} - \lambda_2 e^{\lambda_1 t}}{\lambda_1 - \lambda_2} [\mathbf{I}] + \frac{e^{\lambda_1 t} - e^{\lambda_2 t}}{(\lambda_1 - \lambda_2)} \mathbf{A} \tag{6.82}$$

$$\mathbf{x} = \frac{\lambda_1 e^{\lambda_2 t} - \lambda_2 e^{\lambda_1 t}}{\lambda_1 - \lambda_2} \mathbf{x}_0 + \frac{e^{\lambda_1 t} - e^{\lambda_2 t}}{(\lambda_1 - \lambda_2)} \mathbf{A}\, \mathbf{x}_0 \tag{6.83}$$

In case of under damped system, $\xi < 1$, this solution will be simplified to the followings:

$$x_1 = x_0 e^{-\xi\omega_n t}\left(\cos\omega_d t + \frac{\xi}{\sqrt{1-\xi^2}}\sin\omega_d t\right) + \frac{\dot{x}_0 e^{-\xi\omega_n t}}{\omega_d}\sin\omega_d t \tag{6.84}$$

$$\begin{aligned} x_2 = {} & \frac{-x_0\omega_d e^{-\xi\omega_n t}}{1-\xi^2}\sin\omega_d t \\ & + \frac{\dot{x}_0 e^{-\xi\omega_n t}}{\sqrt{1-\xi^2}}\left(\sqrt{1-\xi^2}\cos\omega_d t - \xi\sin\omega_d t\right) \end{aligned} \tag{6.85}$$

### 6.1.1 Underdamped: $0 < \xi < 1$

When $0 < \xi < 1$, the two roots of the characteristic equation (6.7) are complex conjugate numbers with negative real parts. The underdamped system is the vibrating class of responses of free vibrating systems with the most important engineering applications. The response of a system with parameters of natural frequency $\omega_n$ and damping ratio $\xi$ from initial conditions $x(0) = x_0$, and $\dot{x}(0) = \dot{x}_0$ is an oscillating function with damped frequency $\omega_d = \omega_n\sqrt{1-\xi^2}$ and an exponentially decreasing amplitude $Xe^{-\xi\omega_n t}$.

$$x = e^{-\xi\omega_n t}\left(x_0\cos\omega_d t + \frac{\dot{x}_0 + \xi\omega_n x_0}{\omega_d}\sin\omega_d t\right) \tag{6.86}$$

$$= Xe^{-\xi\omega_n t}\sin(\omega_d t + \varphi) \tag{6.87}$$

$$= Xe^{-\xi\omega_n t}\cos(\omega_d t + \theta) \tag{6.88}$$

$$X = \sqrt{x_0^2 + \left(\frac{\dot{x}_0 + \xi\omega_n x_0}{\omega_d}\right)^2} \tag{6.89}$$

$$\tan\varphi = \frac{x_0\omega_d}{\dot{x}_0 + \xi\omega_n x_0} \tag{6.90}$$

$$\tan\theta = -\frac{\dot{x}_0 + \xi\omega_n x_0}{x_0\omega_d} \tag{6.91}$$

The solutions (6.86)–(6.88) are equivalent, which depending on the given problem and preference of the engineer, one of the expressions will work better than the other. This fact will be clarified by studying some examples.

***Proof*** Having $0 < \xi < 1$ makes the characteristics $s_1$ and $s_2$ to be complex conjugate.

$$\begin{aligned} s_1 &= -\xi\omega_n + \sqrt{\xi^2 - 1}\,\omega_n = -\xi\omega_n + i\sqrt{1-\xi^2}\,\omega_n \\ &= -\xi\omega_n + i\omega_d \end{aligned} \tag{6.92}$$

$$\begin{aligned} s_2 &= -\xi\omega_n - \sqrt{\xi^2 - 1}\,\omega_n = -\xi\omega_n - i\sqrt{1-\xi^2}\,\omega_n \\ &= -\xi\omega_n - i\omega_d \end{aligned} \tag{6.93}$$

$$\omega_d = \sqrt{1-\xi^2}\,\omega_n \qquad 0 < \omega_d < \omega_n \tag{6.94}$$

To rewrite it with harmonic functions, we employ the Euler identity.

$$e^{i\omega_d t} = \cos\omega_d t + i\sin\omega_d t \qquad i^2 = -1 \tag{6.95}$$

$$e^{-i\omega_d t} = \cos\omega_d t - i\sin\omega_d t \tag{6.96}$$

$$\begin{aligned} x &= X_1 e^{(-\xi\omega_n + i\omega_d)t} + X_2 e^{(-\xi\omega_n - i\omega_d)t} \\ &= e^{-\xi\omega_n t}\left(X_1\left(\cos\omega_d t + i\sin\omega_d t\right) + X_2\left(\cos\omega_d t - i\sin\omega_d t\right)\right) \\ &= \left(X_1 + X_2\right) e^{-\xi\omega_n t}\cos\omega_d t + \left(X_1 - X_2\right) i e^{-\xi\omega_n t}\sin\omega_d t \end{aligned} \tag{6.97}$$

The coefficients $X_1$ and $X_2$ from (6.46) will be complex conjugates for $0 < \xi < 1$.

$$\begin{aligned} \begin{bmatrix} X_1 \\ X_2 \end{bmatrix} &= \begin{bmatrix} \frac{1}{s_1 - s_2}\left(\dot{x}_0 - s_2 x_0\right) \\ -\frac{1}{s_1 - s_2}\left(\dot{x}_0 - s_1 x_0\right) \end{bmatrix} \\ &= \begin{bmatrix} \frac{1}{2\omega_d}\left(\omega_d x_0 - i\left(\dot{x}_0 + \xi\omega_n x_0\right)\right) \\ \frac{1}{2\omega_d}\left(\omega_d x_0 + i\left(\dot{x}_0 + \xi\omega_n x_0\right)\right) \end{bmatrix} \end{aligned} \tag{6.98}$$

Therefore,

$$X_1 + X_2 = x_0 \tag{6.99}$$

$$X_1 - X_2 = -\frac{i}{\omega_d}\left(\dot{x}_0 + \xi\omega_n x_0\right) \tag{6.100}$$

and it makes the solution $x$ to be a full harmonic function with real coefficients and exponentially decreasing amplitude.

$$x = x_0 e^{-\xi\omega_n t}\cos\omega_d t - \frac{i}{\omega_d}\left(\dot{x}_0 + \xi\omega_n x_0\right) i e^{-\xi\omega_n t}\sin\omega_d t \tag{6.101}$$

$$= e^{-\xi\omega_n t}\left(x_0\cos\omega_d t + \frac{\dot{x}_0 + \xi\omega_n x_0}{\omega_d}\sin\omega_d t\right) \tag{6.102}$$

The parameter $\omega_d$ is the frequency of both functions $\cos\omega_d t$ and $\sin\omega_d t$ and, hence, is the frequency of $x$. This is the frequency of response with damping that is called the damped frequency. The damping ratio $\xi$ reduces the angular frequency from $\omega_n$ and increases the period of the oscillations from $T_n$.

$$T_d = \frac{T_n}{\sqrt{1-\xi^2}} = \frac{2\pi}{\sqrt{1-\xi^2}\,\omega_n} = \frac{2\pi}{\omega_d} \tag{6.103}$$

The solution (6.102) includes a full harmonic function that may be combined into a single cos or sin function with a phase lag.

$$x = X e^{-\xi\omega_n t}\sin\left(\omega_d t + \varphi\right) \tag{6.104}$$

$$x = X e^{-\xi\omega_n t}\cos\left(\omega_d t + \theta\right) \tag{6.105}$$

The relationship between the amplitude $X$, phase $\varphi$, and weight factors $A$ and $B$ can be found by equating Equations (6.102) and (6.104):

$$\begin{aligned} e^{-\xi\omega_n t}\left(A\sin\omega_d t + B\cos\omega_d t\right) &= X e^{-\xi\omega_n t}\sin\left(\omega_d t + \varphi\right) \\ &= e^{-\xi\omega_n t}\left(X\cos\varphi\sin\omega_d t + X\sin\varphi\cos\omega_d t\right) \end{aligned} \tag{6.106}$$

$$A = \frac{\dot{x}_0 + \xi\omega_n x_0}{\omega_d} \qquad B = x_0 \tag{6.107}$$

$$A = X \cos\varphi \qquad B = X \sin\varphi \tag{6.108}$$

$$X = \sqrt{A^2 + B^2} = \sqrt{\left(\frac{\dot{x}_0 + \xi\omega_n x_0}{\omega_d}\right)^2 + x_0^2} \tag{6.109}$$

$$\tan\varphi = \frac{B}{A} = \frac{x_0\omega_d}{\dot{x}_0 + \xi\omega_n x_0} \tag{6.110}$$

Similarly, the relationship between the amplitude $X$, phase $\theta$, and weight factors $A$ and $B$ will be found by equating Equations (6.102) and (6.105):

$$\begin{aligned} e^{-\xi\omega_n t}\left(A\sin\omega_d t + B\cos\omega_d t\right) &= Xe^{-\xi\omega_n t}\cos\left(\omega_d t + \theta\right) \\ &= \left(X\sin\theta\sin\omega_d t - X\cos\theta\cos\omega_d t\right) \end{aligned} \tag{6.111}$$

$$A = X\sin\theta \qquad P = -X\cos\theta \tag{6.112}$$

$$X = \sqrt{A^2 + B^2} = \sqrt{x_0^2 + \left(\frac{\dot{x}_0 + \xi\omega_n x_0}{\omega_d}\right)^2} \tag{6.113}$$

$$\tan\theta = \frac{A}{-B} = -\frac{\dot{x}_0 + \xi\omega_n x_0}{x_0\omega_d} \tag{6.114}$$

$$\tan\varphi\tan\theta = -1 \tag{6.115}$$

The solution (6.102) suggests to begin with a solution made up of a full harmonic function with unknown coefficients and exponentially decreasing amplitude and then determine the coefficients by initial conditions.

$$x = e^{-\xi\omega_n t}\left(A\sin\omega_d t + B\cos\omega_d t\right) \tag{6.116}$$

When $\xi = 0$, the response is sinusoidal. Damping is a frictional force that generates heat and dissipates energy. When the damping of the system is small, it still oscillates, but with decreasing amplitude, and hence, it goes to rest at equilibrium over time. ■

**Example 43** Underdamped system $\xi < 1$. Graphical illustration of underdamped vibrating systems.

The underdamped case of a vibrating system is the most practical engineering situation. By "damped system," we usually refer to the class of underdamped vibrating systems. It may also be called "subcritically damped." An underdamped system has an oscillatory time response with a decaying amplitude as shown in Fig. 6.2 for $\xi = 0.15$, $\omega_n = 20\pi$ rad/ s, $x_0 = 1$, and $\dot{x}_0 = 0$. The exponential functions $\pm e^{-\xi\omega_n t}$ are the envelops of the response. Envelopes are curves to be tangent to a family of curves differing by family parameter.

The time response of free vibrations of an underdamped one $DOF$ system is expressed by

$$x = e^{-\xi\omega_n t}\left(\frac{\dot{x}_0 + \xi\omega_n x_0}{\omega_d}\sin\omega_d t + x_0\cos\omega_d t\right) \tag{6.117}$$

which can also be written in the following equivalent forms:

$$x = e^{-\xi\omega_n t}\left(x_0\left(\cos\omega_d t + \frac{\xi}{\omega_d}\omega_n\sin\omega_d t\right) + \frac{\dot{x}_0}{\omega_d}\sin\omega_d t\right) \tag{6.118}$$

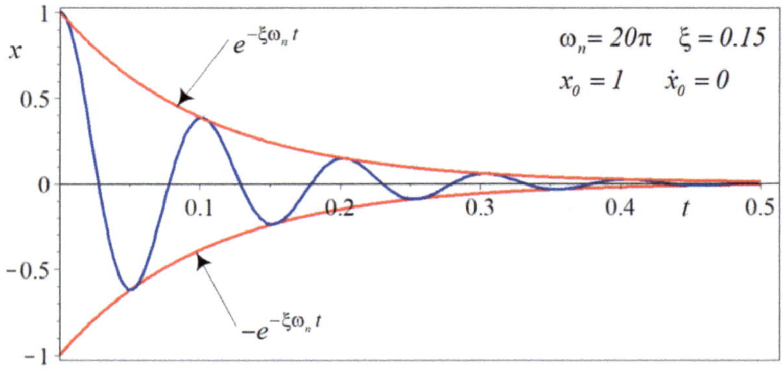


**Fig. 6.2** A sample time response for an underdamped system

$$x = \frac{x_0}{\sin\varphi} e^{-\xi\omega_n t} \sin\left(\omega_d t + \arctan\frac{\omega_d x_0}{\xi\omega_n x_0 + \dot{x}_0}\right) \tag{6.119}$$

$$\varphi = \arctan\frac{\omega_d x_0}{\xi\omega_n x_0 + \dot{x}_0} \tag{6.120}$$

$$x = \frac{e^{-\xi\omega_n t}}{\omega_d}\sqrt{(\omega_d x_0)^2 + (\dot{x}_0 + \xi\omega_n x_0)^2} \times \sin\left(\omega_d t + \arctan\frac{\omega_d x_0}{\xi\omega_n x_0 + \dot{x}_0}\right) \tag{6.121}$$

The response of the damped system is always bounded by the exponential functions $x_B$.

$$x_B = \pm e^{-\xi\omega_n t} \tag{6.122}$$

Therefore, if we are only interested in the maximum possible values of $x$, we may use $x_B$. It is possible to reach $x_B$ at any desired time $t$ by adjusting $\omega_d$.

**Example 44** The envelop of underdamped time responses. To use the envelop theory and determine the envelop curves of underdamped responses.

The time response $x$ of underdamped systems, $\xi < 1$, for different $\omega_d = \omega_n\sqrt{1-\xi^2}$ is always bounded between two envelop exponential functions $x_e$.

$$x = Xe^{-\xi\omega_n t}\sin(\omega_d t + \varphi) \tag{6.123}$$

$$x_e = \pm Xe^{-\xi\omega_n t} \tag{6.124}$$

Figure 6.3 illustrates $x$ for $\omega_d = 19.8997\pi$ and $\omega_d = 14.866\pi$, and a constant $\xi\omega_n = 2\pi$. To have the condition $\xi\omega_n = const$ and different $\omega_d$, we select two values of $\omega_d$ and calculate $\xi$ and $\omega_n$.

$$\omega_d = 19.8997\pi \;\; \xi\omega_n = 2\pi \;\; \xi = 0.1 \;\; \omega_n = 20\pi \tag{6.125}$$

$$\omega_d = 14.866\pi \;\; \xi\omega_n = 2\pi \;\; \xi = 0.1333 \;\; \omega_n = 15\pi \tag{6.126}$$

To determine the equation of envelop curves that are tangent to the solution curves $x = x(t, \omega_d)$, we need to eliminate $\omega_d$ between (6.123) and $dx/d\omega_d$.

$$\frac{dx}{d\omega_d} = Xte^{-\xi\omega_n t}\cos(\varphi + \omega_d t) = 0 \tag{6.127}$$

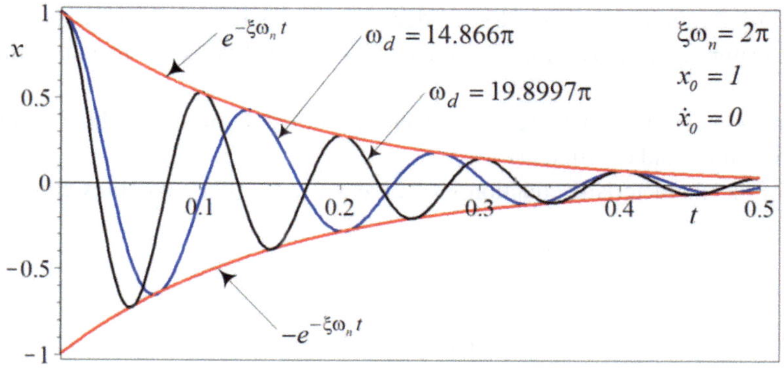


**Fig. 6.3** The graph of $x$ for $\omega_d = 19.8997\pi$ and $\omega_d = 14.866\pi$; at a constant $\xi\omega_n = 2\pi$

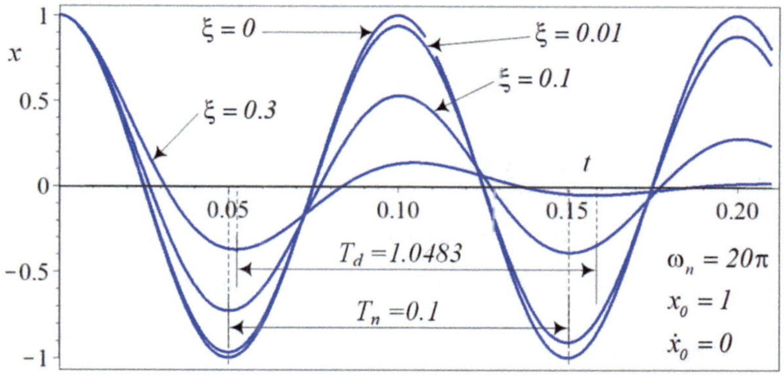


**Fig. 6.4** The response of a damped one $DOF$ vibrating system for $\omega_n = 20\pi$ and $\xi = 0, 0.01, 0.1, 0.3$

The equation $dx/d\omega_d = 0$ yields

$$\varphi + \omega_d t = \pm\frac{\pi}{2} \tag{6.128}$$

which after substituting in (6.123) provides us with the envelop curves that are tangent to the family of solution curves $x = x(t, \omega_d)$.

$$x = \pm X e^{-\xi\omega_n t} \tag{6.129}$$

**Example 45** Damped frequency $\omega_d$. To illustrate how the damping ratio $\xi$ affects the amplitude and frequency of oscillations.

When $\xi = 0$, the system is undamped, and the solution of an $(m, k, c)$-system is simplified to an undamped harmonic function.

$$x = \frac{\dot{x}_0}{\omega_d}\sin\omega_n t + x_0\cos\omega_n t \tag{6.130}$$

The frequency of the undamped solution is $\omega_n = 2\pi f_n = 2\pi/T_n$, where $T_n$ is the natural period of the response. Introducing damping to the system yields the solution of a damped harmonic function $x$.

$$x = e^{-\xi\omega_n t}\left(\frac{\dot{x}_0 + \xi\omega_n x_0}{\omega_d}\sin\omega_d t + x_0\cos\omega_d t\right) \tag{6.131}$$

The frequency of the damped solution is $\omega_d = 2\pi f_d = 2\pi/T_d$, where $\omega_d$ is the frequency of the vibration, and $T_d$ is the damping period of the response. Figure 6.4 illustrates $x$ for $\omega_n = 20\pi$ and $\xi = 0, 0.01, 0.1, 0.3$. The periods of vibrations for $\xi = 0$ and $\xi = 0.3$ are compared.

The damped frequency of vibration $\omega_d = \omega_n \sqrt{1-\xi^2}$ decreases by increasing $\xi$. So, a damped system vibrates slower than its undamped system. Damping decreases the frequency of oscillation and hence increases the period of oscillation.

**Example 46** Real solution from complex characteristic values. A detailed discussion on how a real solution will be achieved from complex characteristic values and coefficients.

When $\xi < 1$, the roots of the characteristic equation $s^2 + 2\xi\omega_n s + \omega_n^2 = 0$ are complex conjugates,

$$s_1 = -\xi\omega_n - \omega_n i\sqrt{1-\xi^2} = a - ib \tag{6.132}$$

$$s_2 = -\xi\omega_n + \omega_n i\sqrt{1-\xi^2} = a + ib \tag{6.133}$$

$$a = -\xi\omega_n \qquad b = \omega_n\sqrt{1-\xi^2} \qquad \xi < 1 \tag{6.134}$$

and the solution of the equation of motion $x$ is

$$x = C_1 e^{(a-ib)t} + C_2 e^{(a+ib)t} \tag{6.135}$$

Employing the Euler identity $e^{i\theta} = \cos\theta + i\sin\theta$ yields

$$\begin{aligned} x &= e^{at} C_1 (\cos bt - i \sin bt) + e^{at} C_2 (\cos bt + i \sin bt) \\ &= e^{at} \left( (C_1 + C_2) \cos bt + i (C_1 - C_2) \sin bt \right) \end{aligned} \tag{6.136}$$

The coefficients $C_1$ and $C_2$ must be complex conjugate to provide a real $x$.

$$C_1 = A_1 - iB_1 \qquad C_2 = A_1 + iB_1 \tag{6.137}$$

Therefore,

$$C_1 + C_2 = 2A_1 \qquad C_1 - C_2 = -2iB_1 \tag{6.138}$$

and

$$\begin{aligned} x &= e^{at} C_1 (\cos bt - i \sin bt) + e^{at} C_2 (\cos bt + i \sin bt) \\ &= 2e^{at} (A_1 \cos bt - B_1 \sin bt) = e^{at} (A \cos bt + B \sin bt) \end{aligned} \tag{6.139}$$

This is how to write the solution of the equation all with real parameters and coefficients.

$$\begin{aligned} x &= e^{at} (A \cos bt + B \sin bt) \\ &= e^{-\xi\omega_n t} (A \cos \omega_d t + B \sin \omega_d t) \end{aligned} \tag{6.140}$$

$$\omega_d = b = \omega_n\sqrt{1-\xi^2} \tag{6.141}$$

**Example 47** Damping ratio evaluation. The amplitude of free vibrations reduce because of damping in the system. The rate of reduction can be used to evaluate the damping of the system.

Damping ratio $\xi$ of an underdamped one $DOF$ system can be approximately found by the following equation that is based on a plot of $x = x(t)$ and peak amplitudes $x_i$.

$$\xi = \frac{1}{\sqrt{4(n-1)^2\pi^2 + \ln^2 \frac{x_1}{x_n}}} \ln\frac{x_1}{x_n} \approx \frac{1}{2(n-1)\pi} \ln\frac{x_1}{x_n} \tag{6.142}$$

Consider the free vibration of an underdamped one $DOF$ system.

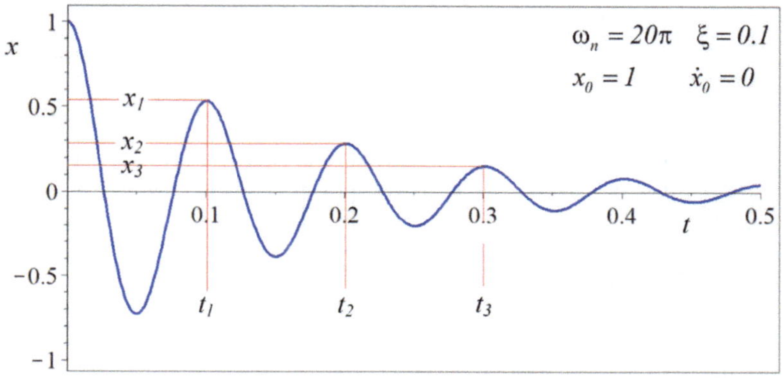


**Fig. 6.5** Free vibration of an underdamped one $DOF$ vibrating system

$$\ddot{x} + 2\xi\omega_n \dot{x} + \omega_n^2 x = 0 \tag{6.143}$$

The time response of the system is given in Equation (6.88) as

$$x = Xe^{-\xi\omega_n t}\cos(\omega_d t + \theta) \tag{6.144}$$

where the parameters $X$ and $\theta$ are dependent on initial conditions.

$$X = \sqrt{x_0^2 + \left(\frac{\dot{x}_0 + \xi\omega_n x_0}{\omega_d}\right)^2} \tag{6.145}$$

$$\tan\theta = -\frac{\dot{x}_0 + \xi\omega_n x_0}{x_0\omega_d} \tag{6.146}$$

Figure 6.5 illustrates a sample of the free vibration response of an underdamped one $DOF$ system. The peak amplitudes $x_i$ are

$$x_1 = e^{-\xi\omega_n t_1}\left(X\cos(\omega_d t_1 + \theta)\right) \tag{6.147}$$

$$x_2 = e^{-\xi\omega_n t_2}\left(X\cos(\omega_d t_2 + \theta)\right) \tag{6.148}$$

$$\vdots$$

$$x_n = e^{-\xi\omega_n t_n}\left(X\cos(\omega_d t_n + \theta)\right) \tag{6.149}$$

The ratio of the first two peaks is

$$\frac{x_1}{x_2} = e^{-\xi\omega_n(t_1 - t_2)}\frac{\cos(\omega_d t_1 + \theta)}{\cos(\omega_d t_2 + \theta)} \tag{6.150}$$

The time difference between $t_1$ and $t_2$ is the damped period of oscillations,

$$T_d = t_2 - t_1 = \frac{2\pi}{\omega_d} = \frac{2\pi}{\omega_n\sqrt{1-\xi^2}} \tag{6.151}$$

and hence, we can simplify Equation (6.150).

$$\begin{aligned}\frac{x_1}{x_2} = \frac{x_k}{x_{k+1}} &= e^{\xi\omega_n T_d}\frac{\cos(\omega_d t_k + \phi)}{\cos(\omega_d(t_k + T_d) + \phi)} \\ &= e^{\xi\omega_n T_d}\frac{\cos(\omega_d t_k + \phi)}{\cos(\omega_d t_k + 2\pi + \phi)} = e^{\xi\omega_n T_d}\end{aligned} \tag{6.152}$$

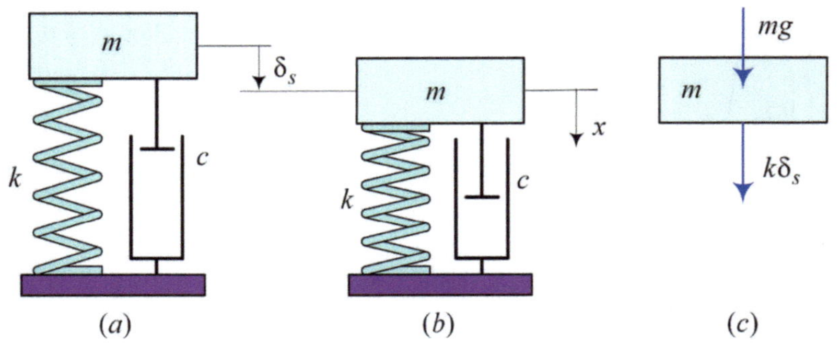


**Fig. 6.6** Measurement of natural frequency based on static deflection

This equation can be transformed to

$$\ln \frac{x_1}{x_2} = \ln \frac{x_k}{x_{k+1}} = \xi \omega_n T_d = \frac{2\pi \xi}{\sqrt{1-\xi^2}} \tag{6.153}$$

that can be used to evaluate the damping ratio $\xi$.

$$\xi = \frac{1}{\sqrt{4\pi^2 + \ln^2 \frac{x_k}{x_{k+1}}}} \ln \frac{x_k}{x_{k+1}} \tag{6.154}$$

For a better evaluation, we can measure the ratio between $x_1/x_n$ and use the following equation:

$$\xi = \frac{1}{\sqrt{4(n-1)^2 \pi^2 + \ln^2 \dfrac{x_1}{x_n}}} \ln \frac{x_1}{x_n} \tag{6.155}$$

If $\xi \ll 1$, then $\sqrt{1-\xi^2} \approx 1$, and we may evaluate $\xi$ from (6.153) with a simpler equation.

$$\xi \approx \frac{1}{2(n-1)\pi} \ln \frac{x_1}{x_n} \tag{6.156}$$

All measurable vibration parameters, such as period $T$ and amplitude $X$, may be used to identify mechanical characteristics of the vibrating system. In most vibration measurement methods, a transient or harmonically steady-state vibration will be applied to the system, and the parameter will be discovered by examination of the response. Using time and kinematic measurement devices, we usually measure amplitude and period of response and employ the analytic equations to find the required data.

**Example 48** Natural frequency determination. The natural frequency indicates the stiffness of the suspension of a vibrating system for a constant mass. Natural frequency and damping ratio are the two parameters of a one $DOF$ vibrating system. The natural frequency can be estimated by static deflection of the spring.

The natural frequency of a mass–spring–damper system can be found by measuring the static deflection of the system under gravity. Consider a one $DOF$ system shown in Fig. 6.6a that barely touches the ground. Assume that the spring is at its free length with no deflection. When the system rests on the ground as shown in Fig. 6.6b, the spring is compressed by a static deflection $\delta_s = mg/k$ because of gravity.

$$\delta_s = \frac{mg}{k} = \frac{g}{\omega_n^2} \tag{6.157}$$

We may determine the natural frequency of the system by measuring $\delta_s$.

$$\omega_n = \sqrt{\frac{g}{\delta_s}} \tag{6.158}$$

### 6.1.2 Critically Damped: $\xi = 1$

When $\xi = 1$, the damping $c$ would be equal to the critical damping $c_c$, and the equation of motion will only be dependent of one parameter, $\omega_n$.

$$\ddot{x} + 2\omega_n \dot{x} + \omega_n^2 x = 0 \tag{6.159}$$

$$x(0) = x_0 \qquad \dot{x}(0) = \dot{x}_0$$

$$\xi = \frac{c}{2m\omega_n} = \frac{c}{c_c} = 1 \tag{6.160}$$

$$c_c = 2m\omega_n = 2\sqrt{km} \tag{6.161}$$

The two characteristic values $s_1$ and $s_2$ of Equation (6.7) are equal and negative, indicating the response to be approaching equilibrium when time goes $t \to \infty$.

$$s_1 = s_2 = -\omega_n \tag{6.162}$$

The critically damped is a non-vibrating class of responses of free vibrating systems. The two linearly independent solutions of the differential equation (6.159) would be $e^{-\omega_n t}$ and $te^{-\omega_n t}$. These solutions make the general solution of the critically damped oscillator.

$$x = x_0 e^{-\omega_n t} + (\dot{x}_0 + \omega_n x_0)\, te^{-\omega_n t} \tag{6.163}$$

***Proof*** Substituting a general exponential equation $Xe^{st}$ into the linear differential equation makes the characteristic equation of the critically damped oscillator to have only one solution.

$$\left(s^2 + 2\omega_n s + \omega_n^2\right) Xe^{st} = 0 \tag{6.164}$$

$$s^2 + 2\omega_n s + \omega_n^2 = 0 \tag{6.165}$$

$$s = -\omega_n \tag{6.166}$$

Therefore, $x_1 = X_1 e^{-\omega_n t}$ is one of the solutions of Equation (6.159). The second solution of the equation would be $x_2 = Xte^{-\omega_n t}$. Let us examine $x_2$ as a solution.

$$\ddot{x} + 2\omega_n \dot{x} + \omega_n^2 x = \left((t\omega_n - 2)\,\omega_n - 2\omega_n\,(t\omega_n - 1) + \omega_n^2 t\right) Xe^{-t\omega_n} = 0 \tag{6.167}$$

Therefore, the general solution of the equation would be

$$x = X_1 e^{-\omega_n t} + X_2 te^{-\omega_n t} \tag{6.168}$$

The coefficients $X_1$ and $X_2$ will be determined by initial conditions.

$$x(0) = X_1 = x_0 \tag{6.169}$$

$$\dot{x}(0) = -\omega_n x_0 - X_2\,(t\omega_n - 1) = \dot{x}_0 \tag{6.170}$$

$$X_1 = x_0 \qquad X_2 = \dot{x}_0 + \omega_n x_0 \tag{6.171}$$

The solution of critically damped of free vibration may alternatively be proven by approaching $\xi \to 1$ from higher or lower values. Let us assume

$$\xi = 1 + \varepsilon \qquad 0 < \varepsilon \ll 1 \tag{6.172}$$

and calculate the solution of the characteristic equation (6.7).

$$s_1 \simeq -\omega_n (1+\varepsilon) \tag{6.173}$$

$$s_2 \simeq -\omega_n (1-\varepsilon) \tag{6.174}$$

Therefore, the solution of the equation of motion (6.159) will be

$$\begin{aligned} x &= X_1 e^{s_1 t} + X_2 e^{s_2 t} = \frac{\dot{x}_0 - s_2 x_0}{s_1 - s_2} e^{s_1 t} - \frac{\dot{x}_0 - s_1 x_0}{s_1 - s_2} e^{s_2 t} \\ &= \frac{s_1 e^{s_2 t} - s_2 e^{s_1 t}}{s_1 - s_2} x_0 + \frac{e^{s_1 t} - e^{s_2 t}}{s_1 - s_2} \dot{x}_0 \\ &= \frac{-\omega_n (1+\varepsilon) e^{-\omega_n (1-\varepsilon) t} + \omega_n (1-\varepsilon) e^{e^{-\omega_n (1+\varepsilon) t}}}{-2\varepsilon\omega_n} x_0 \\ &\quad + \frac{e^{e^{-\omega_n (1+\varepsilon) t}} - e^{e^{-\omega_n (1-\varepsilon) t}}}{-2\varepsilon\omega_n} \dot{x}_0 \end{aligned} \tag{6.175}$$

Expanding $e^{-\varepsilon t}$ in a power series of $\varepsilon$,

$$e^{-\varepsilon t} = 1 - t\varepsilon + \frac{1}{2} t^2 \varepsilon^2 + O\left(\varepsilon^3\right) \tag{6.176}$$

yields

$$e^{-\omega_n (1+\varepsilon) t} = e^{-\omega_n t} - \varepsilon\omega_n t e^{-\omega_n t} + \frac{1}{2}\varepsilon^2 \omega_n^2 t^2 e^{-\omega_n t} + O\left(\varepsilon^3\right) \tag{6.177}$$

$$e^{-\omega_n (1-\varepsilon) t} = e^{-\omega_n t} + \varepsilon\omega_n t e^{-\omega_n t} + \frac{1}{2}\varepsilon^2 \omega_n^2 t^2 e^{-\omega_n t} + O\left(\varepsilon^3\right) \tag{6.178}$$

and we will have

$$\begin{aligned} x &= \frac{x_0}{2\varepsilon} \left( (1+\varepsilon) \left( e^{-\omega_n t} + t\varepsilon\omega_n e^{-\omega_n t} + \frac{1}{2} t^2 \varepsilon^2 \omega_n^2 e^{-\omega_n t} + \cdots \right) \right. \\ &\quad \left. - (1-\varepsilon) \left( e^{-\omega_n t} - t\varepsilon\omega_n e^{-\omega_n t} + \frac{1}{2} t^2 \varepsilon^2 \omega_n^2 e^{-\omega_n t} + \cdots \right) \right) \\ &\quad + \frac{\dot{x}_0}{-2\varepsilon\omega_n} \left( \left( e^{-\omega_n t} - t\varepsilon\omega_n e^{-\omega_n t} + \frac{1}{2} t^2 \varepsilon^2 \omega_n^2 e^{-\omega_n t} + \cdots \right) \right. \\ &\quad \left. - \left( e^{-\omega_n t} + t\varepsilon\omega_n e^{-\omega_n t} + \frac{1}{2} t^2 \varepsilon^2 \omega_n^2 e^{-\omega_n t} + \cdots \right) \right) \end{aligned} \tag{6.179}$$

or

$$\begin{aligned} x &= x_0 \frac{e^{-\omega_n t}}{2\varepsilon} \left( 1 + \varepsilon + \varepsilon\omega_n t + \varepsilon^2 \omega_n t + \frac{1}{2}\varepsilon^2 \omega_n^2 t^2 + \frac{1}{2}\varepsilon^3 \omega_n^2 t^2 + \cdots \right. \\ &\quad \left. - \left( 1 - \varepsilon - \varepsilon\omega_n t + \varepsilon^2 \omega_n t + \frac{1}{2} t^2 \varepsilon^2 \omega_n^2 - \frac{1}{2}\varepsilon^3 \omega_n^2 t^2 + \cdots \right) \right) \\ &\quad + \frac{\dot{x}_0 e^{-\omega_n t}}{-2\varepsilon\omega_n} \left( \left( 1 - t\varepsilon\omega_n + \frac{1}{2} t^2 \varepsilon^2 \omega_n^2 + \cdots \right) \right. \\ &\quad \left. - \left( 1 + t\varepsilon\omega_n + \frac{1}{2} t^2 \varepsilon^2 \omega_n^2 + \cdots \right) \right) \end{aligned}$$

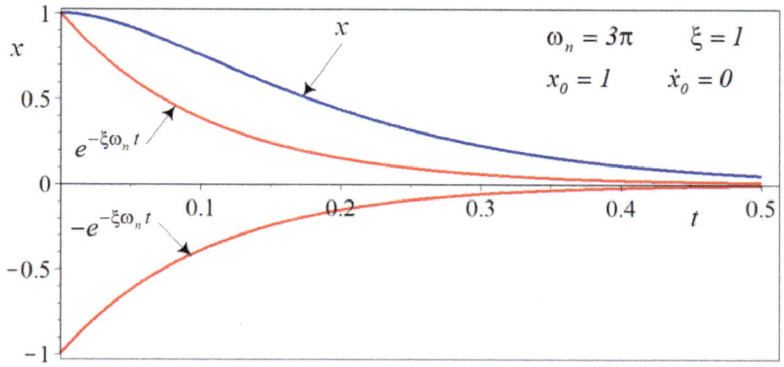


**Fig. 6.7** A sample time response of a critically damped vibrating system

$$= x_0 \frac{e^{-\omega_n t}}{2\varepsilon} \left(2\varepsilon + 2\varepsilon\omega_n t + \varepsilon^3 \omega_n^2 t^2 + \cdots\right)$$

$$+ \frac{\dot{x}_0 e^{-\omega_n t}}{-2\varepsilon\omega_n} \left(-2t\varepsilon\omega_n - \frac{1}{3} t^3 \varepsilon^3 \omega_n^3 + \cdots\right) \tag{6.180}$$

Taking the limit $\varepsilon \to 0$, the solution $x$ will be simplified to (6.163).

$$x = x_0 e^{-\omega_n t} (1 + \omega_n t) + \dot{x}_0 t e^{-\omega_n t} \tag{6.181}$$

$$x = x_0 e^{-\omega_n t} + (\dot{x}_0 + \omega_n x_0) t e^{-\omega_n t} \tag{6.182}$$

Critically damping is practically not a significant case in engineering applications. There is not practically much difference between the responses of a vibrating system for $\xi = 1$, $\xi = 1.01$, $\xi = 0.99$. It is a mathematical issue where the eigenvalues of the system become equal. The solution will then be $X_1 e^{st}$ plus $X_1 t e^{st}$. The critically damped response does not have any envelop to cover of $x$. ■

**Example 49** Critically damped system $\xi = 1$. Graphical illustration of critically damped vibrating systems.

If the system is critically damped, $\xi = 1$, then the time response to free vibrations is

$$x = e^{-\xi\omega_n t} (X_1 + X_2 t) \tag{6.183}$$

Using the initial conditions, $x(0) = x_0$, $\dot{x}(0) = \dot{x}_0$, we find the coefficients $X_1$ and $X_2$,

$$X_1 = x_0 \qquad X_2 = \dot{x}_0 + \omega_n x_0 \tag{6.184}$$

and therefore, the general response of critically damped system is calculated.

$$x = e^{-\omega_n t} (x_0 + (\dot{x}_0 + \omega_n x_0) t) \tag{6.185}$$

Figure 6.7 shows a critically damped response for $\xi = 1$, $\omega_n = 3\pi$ rad, $x_0 = 1$, and $\dot{x}_0 = 0$.

**Example 50** ★Crossing the equilibrium. The critically damped system will approach equilibrium position from any initial conditions. However, the curve of $x$ may pass the line $x = 0$ once if the initial condition is set properly.

Let us assume the following numerical values for a critically damped system.

$$x_0 = 1 \qquad \omega_n = 1 \tag{6.186}$$

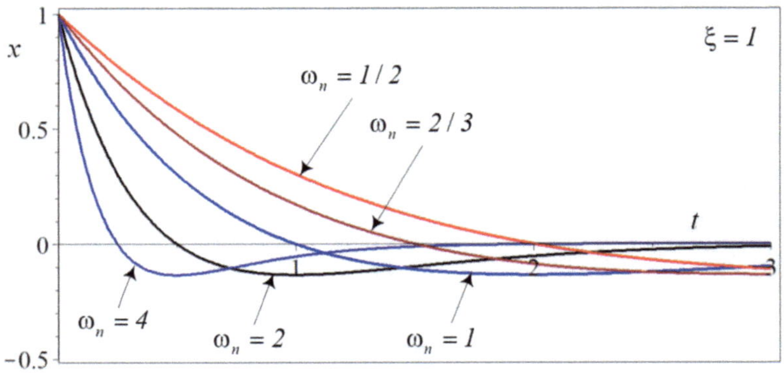


**Fig. 6.8** Time response of a critically damped system for some $\omega_n$ and their associate $\dot{x}_0$ to cross $x = 0$ at $t = 1/\omega_n$

The response $x$ of the system will be

$$\begin{aligned} x &= e^{-\omega_n t} \left( x_0 + (\dot{x}_0 + \omega_n x_0) t \right) \\ &= e^{-t} \left( 1 + (\dot{x}_0 + 1) t \right) \end{aligned} \tag{6.187}$$

Setting $x = 0$ provides us with a relationship between $\dot{x}_0$ and $t$ to cross the equilibrium $x = 0$.

$$\dot{x}_0 = -\frac{1}{t} (t + 1) \tag{6.188}$$

We are able to select the time at which we wish $x$ to cross the equilibrium position, and this chart will determine what should be the initial velocity $\dot{x}_0$. On the other hand, we are able to determine the time at which we wish $x$ to cross the equilibrium position for a given initial velocity, $\dot{x}_0$. The position variable, $x$, will cross the equilibrium position, $x = 0$, only for negative initial velocity if $x_0 > 0$.

To make general design criteria, let us begin with the solution and make it nondimensionalized.

$$x = e^{-\omega_n t} \left( x_0 + (\dot{x}_0 + \omega_n x_0) t \right) = 0 \tag{6.189}$$

Because $e^{-\omega_n t}$ cannot be zero, the equation reduces to an algebraic equation.

$$x_0 + (\dot{x}_0 + \omega_n x_0) t = 0 \tag{6.190}$$

$$\frac{x_0}{\dot{x}_0 t} = \frac{-1}{1 + \omega_n t} \tag{6.191}$$

The design criterion is a plot of $\frac{x_0}{\dot{x}_0 t}$ against $\omega_n t$. This design condition is nondimensionalized, and hence, it is valid for all critically damped free vibrating systems. Setting a value for $\omega_n t$ will provide us with the associated value for $\frac{x_0}{\dot{x}_0 t}$ to have the initial conditions to cross the equilibrium position. There would be no crossing the $x = 0$ axis for $0 < \frac{x_0}{\dot{x}_0 t} < -1$.

As an example, a point on horizontal axis at $\omega_n t = 1$ gives $\frac{x_0}{\dot{x}_0 t} = -\frac{1}{2}$ on vertical axis. A system with $\omega_n = 1$ from $x_0 = 1$ will need $\dot{x}_0 = -2x_0/t = -2$ to pass $x = 0$ at $t = 1$. Also a point on horizontal axis at $\omega_n t = 1$ gives $\frac{x_0}{\dot{x}_0 t} = -\frac{1}{2}$ on vertical axis, and a system with $\omega_n = 2$ from $x_0 = 1$ will need $\dot{x}_0 = -2x_0/t = -4$ to pass $x = 0$ at $t = 1/2$. Keeping $x_0 = 1$ and $\omega_n t = 1$, we will need $\dot{x}_0 = -2/t = -4/3$ for $t = 1.5$, and $\dot{x}_0 = -1$ for $t = 2$, and $\dot{x}_0 = -8$ for $t = 1/4$. Figure 6.8 depicts these time responses.

### 6.1.3 Overdamped: $1 < \xi$

When $1 < \xi$, the vibrating system is called overdamped. The characteristic values $s_1$ and $s_2$ of Equation (6.7) are both real and negative for $1 < \xi$.

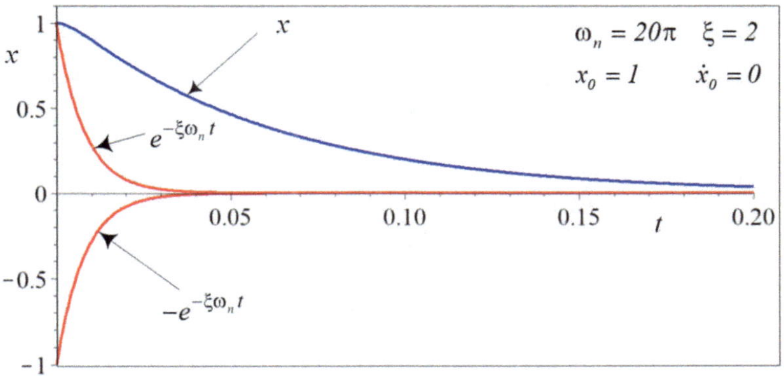


**Fig. 6.9** A sample time response for an overdamped system

$$s_1 = \left(-\xi + \sqrt{\xi^2 - 1}\right)\omega_n < 0 \tag{6.192}$$

$$s_2 = \left(-\xi - \sqrt{\xi^2 - 1}\right)\omega_n < 0 \tag{6.193}$$

$$x(t) = X_1 e^{s_1 t} + X_2 e^{s_2 t} \tag{6.194}$$

$$X_1 = \frac{\dot{x}_0 - s_2 x_0}{s_1 - s_2} \qquad X_2 = -\frac{\dot{x}_0 - s_1 x_0}{s_1 - s_2} \tag{6.195}$$

The negative characteristic values make the response of the system to have no oscillatory motion, and go to zero for large $t$.

$$\begin{aligned} x = {} & \frac{\dot{x}_0 + \left(\xi + \sqrt{\xi^2 - 1}\right)\omega_n x_0}{2\omega_n\sqrt{\xi^2 - 1}} e^{-\left(\xi - \sqrt{\xi^2 - 1}\right)\omega_n t} \\ & - \frac{\dot{x}_0 + \left(\xi - \sqrt{\xi^2 - 1}\right)\omega_n x_0}{2\omega_n\sqrt{\xi^2 - 1}} e^{-\left(\xi + \sqrt{\xi^2 - 1}\right)\omega_n t} \end{aligned} \tag{6.196}$$

The overdamped case is mathematically the simplest with least engineering applications.

**Example 51** Overdamped systems, $1 < \xi$. Graphical illustration of overdamped vibrating systems and comparison with critically and underdamped cases.

An overdamped system has damping $c$ greater than the critical dumping, $c > 2\xi\omega_2$, or $1 < \xi$. The time response of the system will be exponentially decaying to approach the equilibrium position, starting from any set of initial conditions. Figure 6.9 shows an overdamped response for $\xi = 2$, $\omega_n = 20\pi$, $x_0 = 1$, and $\dot{x}_0 = 0$.

Using the initial conditions, $x(0) = x_0$, $\dot{x}(0) = \dot{x}_0$, and the characteristic values (6.192)–(6.193), the general overdamped response is

$$\begin{aligned} x &= \frac{\dot{x}_0 - s_2 x_0}{s_1 - s_2} e^{s_1 t} + \frac{s_1 x_0 - \dot{x}_0}{s_1 - s_2} e^{s_2 t} \\ &= \left(\frac{s_1 e^{s_2 t}}{s_1 - s_2} - \frac{s_2 e^{s_1 t}}{s_1 - s_2}\right) x_0 + \left(\frac{e^{s_1 t}}{s_1 - s_2} - \frac{e^{s_2 t}}{s_1 - s_2}\right) \dot{x}_0 \\ &= \frac{e^{-\xi\omega_n t}}{2\sqrt{\xi^2 - 1}} \left( \left(\frac{\dot{x}_0}{\omega_n} + x_0\left(\xi + \sqrt{\xi^2 - 1}\right)\right) e^{\omega_n \sqrt{\xi^2 - 1} t} \right. \\ &\quad \left. + \left(-\frac{\dot{x}_0}{\omega_n} + x_0\left(-\xi - \sqrt{\xi^2 - 1}\right)\right) e^{-\omega_n \sqrt{\xi^2 - 1} t} \right) \end{aligned} \tag{6.197}$$

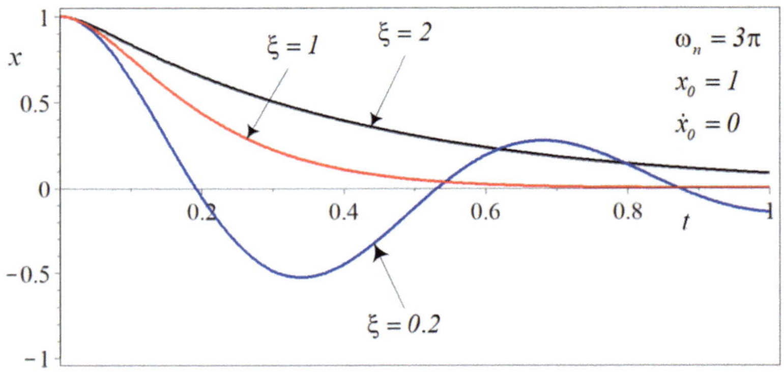


**Fig. 6.10** Comparison of underdamped, critically damped, and overdamped vibrating systems with $\omega_n = 3\pi$, and initial conditions of $x_0 = 1$, $\dot{x}_0 = 0$

Figure 6.10 compares underdamped, critically damped, and overdamped vibrating systems with $\omega_n = 3\pi$, and initial conditions of $x_0 = 1$, $\dot{x}_0 = 0$.

### 6.1.4 Negative Damping: $\xi < 0$

When $\xi < 0$, the vibrating system is called unstable. The characteristic values $s_1$ and $s_2$ of Equation (6.7) are both real and positive for negative damping $\xi < 0$.

$$s_1 = \left(\xi + \sqrt{\xi^2 - 1}\right)\omega_n > 0 \tag{6.198}$$

$$s_2 = \left(\xi - \sqrt{\xi^2 - 1}\right)\omega_n > 0 \tag{6.199}$$

$$x(t) = X_1 e^{s_1 t} + X_2 e^{s_2 t} \tag{6.200}$$

$$X_1 = \frac{\dot{x}_0 - s_2 x_0}{s_1 - s_2} \qquad X_2 = -\frac{\dot{x}_0 - s_1 x_0}{s_1 - s_2} \tag{6.201}$$

The positive characteristic values make the response of the system to have oscillatory motion with exponentially growing amplitude.

$$x = \frac{\dot{x}_0 + \left(\xi + \sqrt{\xi^2 - 1}\right)\omega_n x_0}{2\omega_n\sqrt{\xi^2 - 1}} e^{-\left(\xi - \sqrt{\xi^2 - 1}\right)\omega_n t} - \frac{\dot{x}_0 + \left(\xi - \sqrt{\xi^2 - 1}\right)\omega_n x_0}{2\omega_n\sqrt{\xi^2 - 1}} e^{-\left(\xi + \sqrt{\xi^2 - 1}\right)\omega_n t} \tag{6.202}$$

The negative damping is always due to a condition at which energy inserted into the vibrating system. An example could be wind excitation on structures. In practical vibration engineering applications, negative damping will be provided by semi-active system to control damping rate of a computer-controlled shock absorber. Such damping is not constant and will be changed according to constraints and control strategy.

**Example 52** ★ Negative damping and instability. Graphical illustration of overdamped vibrating systems and comparison with critically and underdamped cases.

When there is a mechanism that pumps energy into a vibrating system during the oscillation, the damping is negative. The equation of motion of such a system is

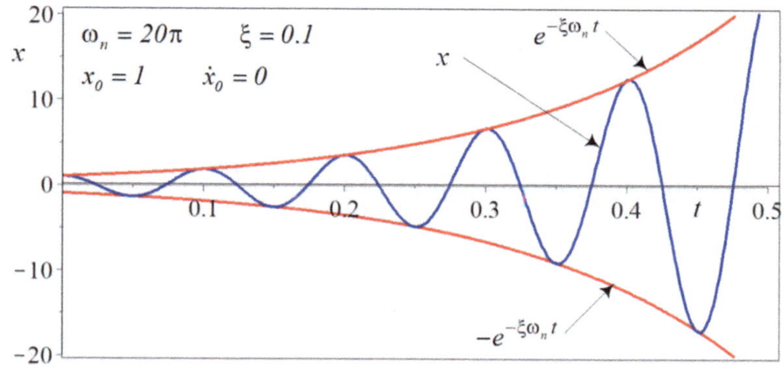


**Fig. 6.11** A sample time response of a negative damped system

$$m\ddot{x} = -kc + c\dot{x} + f(x, \dot{x}, t) \tag{6.203}$$

$$m > 0 \qquad k > 0 \qquad c > 0 \tag{6.204}$$

If we move all terms but the forcing term to the other side, the coefficient of $\dot{x}$ becomes negative, and this justifies the name of negative damping.

$$m\ddot{x} - c\dot{x} + kx = f(x, \dot{x}, t) \tag{6.205}$$

The free vibration of a system with negative damping is divergent and hence unstable. It means that the amplitude of vibration increases periodically until the system practically breaks down.

$$\ddot{x} - 2\xi\omega_n\dot{x} + \omega_n^2 x = 0 \tag{6.206}$$

$$2\xi\omega_n = \frac{c}{m} \qquad \omega_n^2 = \frac{k}{m} \tag{6.207}$$

$$x(0) = x_0 \qquad \dot{x}(0) = \dot{x}_0 \tag{6.208}$$

When $\xi > 1$, both characteristic values $s_{1,2}$ are real and positive, $s_{1,2} \in \mathbb{R}$, $s_{1,2} > 0$.

$$s^2 - 2\xi\omega_n s + \omega_n^2 = 0 \tag{6.209}$$

$$s_{1,2} = \xi\omega_n \pm \omega_n\sqrt{\xi^2 - 1} \tag{6.210}$$

Therefore, both terms of the general solution of Equation (6.206) $X_1 e^{s_1 t}$ and $X_2 e^{s_2 t}$ exponentially grow with time, with different weights and rates.

$$\begin{aligned} x &= X_1 e^{s_1 t} + X_2 e^{s_2 t} \\ &= X_1 e^{\omega_n\left(\xi+\sqrt{\xi^2-1}\right)t} + X_2 e^{\omega_n\left(\xi-\sqrt{\xi^2-1}\right)t} \end{aligned} \tag{6.211}$$

Figure 6.11 depicts the response of a negative damping system.

**Example 53** ★Temporary negative damping. Negative damping is usually a temporary effect and may disappear after certain conditions.

Let us consider a vibrating system whose damping ratio is negative for a while and change to be zero according to a safety control system. Such control system is usually shut down the damping mechanism when the displacement exceeds certain value. Here let us assume a system that is equipped with a control system that shuts down the damping of the vibrating system when it is in negative range for a period of time.

$$\ddot{x} + 2\xi\omega_n\dot{x} + \omega_n^2 x = 0 \tag{6.212}$$

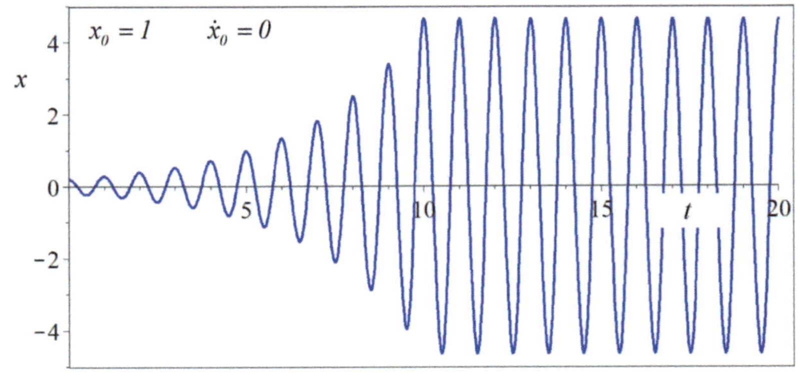


**Fig. 6.12** Vibration of a system with negative damping with a control system to shut down the damping after 10 s

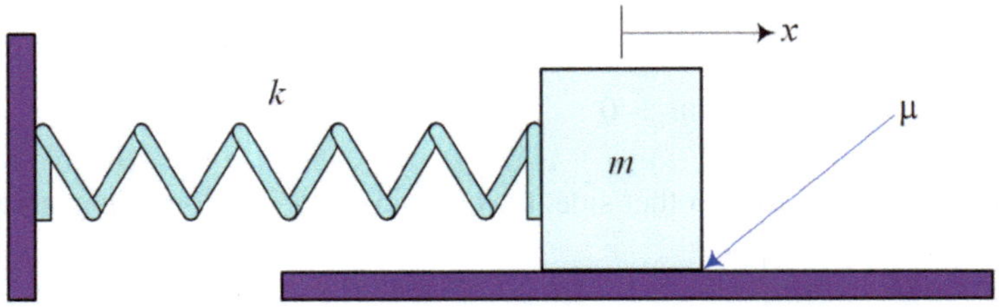


**Fig. 6.13** A mass–spring oscillator with a resisting dry friction force

$$\xi = \begin{cases} -0.5 & 0 < t \leq 10\,\text{s} \\ 0 & 10\,\text{s} < t \end{cases} \tag{6.213}$$

To define such damping ratio mathematically, we may use Heaviside, $H$, function.

$$H(t - \tau) = \begin{cases} 1 & t \geq \tau \\ 0 & t < \tau \end{cases} \tag{6.214}$$

$$\ddot{x} + 2(\xi H(10 - t))\,\omega_n \dot{x} + \omega_n^2 x = 0 \tag{6.215}$$

This will make a system to have complex dynamics. A complex dynamic system will have more than one different equation of motion depending on certain conditions. In this example, we will have

$$\ddot{x} + 2\xi\omega_n\dot{x} + \omega_n^2 x = 0 \qquad 0 < t \leq 10\,\text{s} \tag{6.216}$$

$$\ddot{x} + \omega_n^2 x = 0 \qquad 10\,\text{s} < t \tag{6.217}$$

Figure 6.12 illustrates the time history of the system starting from an initial unit displacement.

$$x(0) = 1 \qquad \dot{x}(0) = 0 \tag{6.218}$$

### 6.1.5 Coulomb Friction Damping

Consider a particle with mass $m$ moving on a rough surface. Then, there would exist a friction force $\mathbf{F}_f$ that resists the relative motion between the surfaces. The friction force $\mathbf{F}_f$ is proportional to the normal force $F_N$ that presses the two surfaces together. Consider an object with mass $m$ is moving on a rough surface as is shown in Fig. 6.13, and then,

$$\mathbf{F}_f = -\mu N \frac{\mathbf{v}}{v} \tag{6.219}$$

where $\mu$ is the coefficient of friction, $N$ is the normal force exerted by the surfaces on the particle, and $\mathbf{v}/v$ is the unit vector in the direction of the object's velocity relative to the surface. This is a simplified Coulomb theory of friction that states: The friction force is proportional to the force pressing two physical surfaces together and is in the tangent plane to the surfaces at the contact point.

If Coulomb friction is the only damping mechanism in a one $DOF$ vibrating system, then its equaticn of motion will be

$$m\ddot{x} + \mu N \text{sgn}\dot{x} + kx = \ddot{x} + \omega_n^2 x + \mu \frac{N}{m} g \frac{\dot{x}}{|\dot{x}|}$$

$$= \ddot{x} + \omega^2 x + \mu \frac{N}{m} g \text{sgn}\dot{x} = 0 \tag{6.220}$$

$$x(0) = x_0 \qquad \dot{x}(0) = 0 \qquad \omega_n^2 = \frac{k}{m} \tag{6.221}$$

The equation of motion is piecewise including two equations depending on the direction of motion.

$$\ddot{x} + \omega_n^2 x = \begin{cases} \mu \frac{N}{m} & \dot{x} < 0 \\ -\mu \frac{N}{m} & 0 < \dot{x} \end{cases} \tag{6.222}$$

The system will have an equilibrium zone $x < |\mu N/k|$ in which the spring force $kx$ is less than the maximum friction force $\mu N$.

$$-\frac{\mu N}{k} < x < \frac{\mu N}{k} \tag{6.223}$$

As long as the mass $m$ is within the equilibrium zone with zero velocity, it remains at rest. Let us count the motion starting from the stop position $x(0) = x_0 > \mu N/k$, $\dot{x}(0) = 0$ on the right side of to the equilibrium zone to the point where the mass stops again on the left side of equilibrium zone as the part 1 of motion after half a cycle of oscillation. Then, count the second half of the first cycle of motion from the left stop to the right stop of the equilibrium zone as part 2, and so on. The solutions of the equation are as below, where $j$ is the number of motions.

$$x = \left(x_0 - (2j-1)\frac{N}{k}\mu\right)\cos\omega_n t - (-1)^j \mu \frac{N}{k} \quad t_j < t < t_{j+1} \tag{6.224}$$

$$\dot{x} = -\left(x_0 - (2j-1)\frac{N}{k}\mu\right)\omega_n \sin\omega_n t \qquad t_j = \frac{j\pi}{\omega_n} \tag{6.225}$$

The extreme displacements $x_j$ on both sides of the equilibrium zone where the mass stops before returning to oscillation are

$$x_j = (-1)^j \left(x_0 - 2j\,\mu \frac{N}{k}\right) \qquad j = 1, 2, 3, \cdots, s \tag{6.226}$$

$$x_{j+1} = -x_j + 2\mu \frac{N}{k} \qquad x_0 = x(0) \tag{6.227}$$

The displacement of $m$ in every half a cycle will be shortened by $2\mu\frac{N}{k}$. The value of $2\mu\frac{N}{k}$ is equal to the total length of the equilibrium zone.

The initial energy of the system is the stored potential energy in the spring.

$$P_0 = \frac{1}{2} k x_0^2 \tag{6.228}$$

The potential energy drops continuously after every half a cycle.

$$P_i = \frac{1}{2}k\left(x_0 - 2i\,\mu\frac{N}{k}\right)^2 \tag{6.229}$$

The mass finally stops at some point in the equilibrium zone when the stop $x_i$ is within the zone.

***Proof*** The equation of motion of a mass–spring system with Coulomb friction can be written as two piecewise linear equations,

$$m\ddot{x} + kx = \begin{cases} -\mu N & 0 < \dot{x} \\ \mu N & \dot{x} < 0 \end{cases} \tag{6.230}$$

or as a single equation using signum function sgn$\dot{x}$.

$$\ddot{x} + \mu\frac{N}{m}\text{sgn}\dot{x} + \omega_n^2 x = 0 \quad x(0) = x_0 > 0 \quad \dot{x}(0) = \dot{x}_0 = 0 \tag{6.231}$$

$$\text{sgn}\dot{x} = \begin{cases} -1 & \dot{x} < 0 \\ 0 & 0 \\ 1 & 0 < \dot{x} \end{cases} \tag{6.232}$$

As the velocity changes sign, the sign of friction force that is always opposing the motion will also change. Therefore, the friction force will slow down the mass, which will eventually come to a stop. The mass will be in motion as long as the supplied force by the spring is larger than the maximum friction force, $\mu N$. Motion will eventually stop when the spring force is not larger than the friction force.

To find the time response of the system, we assume the mass starts from the initial conditions $x(0) = x_0 > 0$, $x_0 > \mu\frac{N}{m}$, $\dot{x}(0) = 0$.

The equation may equivalently be transformed to the following form.

$$\ddot{x} + \omega_n^2 x = -\mu\frac{N}{m}\text{sgn}\dot{x} = \begin{cases} \mu\frac{N}{m} & \dot{x} < 0 \\ -\mu\frac{N}{m} & 0 < \dot{x} \end{cases} \tag{6.233}$$

Assuming $x_0 > 0$ and $\dot{x}(0) = 0$, we expect the motion to start in negative direction, $\dot{x} < 0$, until the mass goes to stop. The equation for this part of motion will be

$$\ddot{x} + \omega_n^2 x = \mu\frac{N}{m} \qquad x_0 > 0 \qquad \dot{x}_0 = 0 \tag{6.234}$$

The solution of the equation will have a homogenous solution $x_h$ for $\ddot{x} + \omega_n^2 x = 0$, plus a particular solution for the external force function $f = \mu\frac{N}{m}$. The homogenous solution is the free vibrations (6.12) for $\xi = 0$.

$$x_h = A\sin\omega_n t + B\cos\omega_n t \tag{6.235}$$

Because the forcing function is constant, the particular solution $x_p$ will be a constant function that satisfies Equation (6.234).

$$x_p = \mu\frac{N}{m\omega_n^2} = \mu\frac{N}{k} \tag{6.236}$$

Therefore, the total solution of Equation (6.234) will be

$$x = x_h + x_p = A\sin\omega_n t + B\cos\omega_n t + \mu\frac{N}{k} \tag{6.237}$$

$$\dot{x} = \omega_n \left(A \cos \omega_n t - B \sin \omega_n t\right) \tag{6.238}$$

Applying the initial conditions,

$$x_0 = B + \mu \frac{N}{k} \tag{6.239}$$

$$\dot{x}_0 = \omega_n A = 0 \tag{6.240}$$

we find the solution.

$$x = \left(x_0 - \mu \frac{N}{k}\right) \cos \omega_n t + \mu \frac{N}{k} \qquad 0 < t < t_1 \tag{6.241}$$

$$\dot{x} \quad = -\left(x_0 - \mu \frac{N}{k}\right) \omega_n \sin \omega_n t \qquad t_1 = \frac{\pi}{\omega_n} \tag{6.242}$$

The mass will come to the first stop at $t = t_1 = \pi / \omega_n$, when $m$ is at $x_1$. The mass stops at a point that is $2\mu \frac{N}{k}$ closer to $x = 0$.

$$x_1 = x\left(t_1\right) = \mu \frac{N}{k} - \left(x_0 - \mu \frac{N}{k}\right) = -x_0 + 2\mu \frac{N}{k} \tag{6.243}$$

Now the problem reduces to

$$\ddot{x} + \omega_n^2 x = -\mu \frac{N}{m} \qquad x\left(t_1\right) = x_1 < 0 \qquad \dot{x}_1 = 0 \tag{6.244}$$

whose solution will be found similar to (6.241).

$$x = \left(x_0 - 3\frac{N}{k}\mu\right) \cos \omega_n t - \mu \frac{N}{k} \qquad t_1 < t < t_2 \tag{6.245}$$

$$\dot{x} = -\left(x_0 - 3\frac{N}{k}\mu\right) \omega_n \sin \omega_n t \qquad t_2 = \frac{2\pi}{\omega_n} \tag{6.246}$$

The mass will come to the second stop at $t = t_2 = 2\pi / \omega_n$, when $m$ is at $x_2$. In one cycle of oscillation, the mass will lose its amplitude by $4\mu \frac{N}{k}$.

$$x_2 = x\left(t_2\right) = x_0 - 4\mu \frac{N}{k} \tag{6.247}$$

The new problem will be similar to (6.234) with a new initial position as $x_2$. The next step will show that $x_3$ at $t_3$ is

$$x_3 = x\left(t_3\right) = -x_0 + 6\mu \frac{N}{k} \qquad t_3 = \frac{3\pi}{\omega_n} \tag{6.248}$$

Similarly, we derive the time response of $m$ and the stop position of the mass at both sides of $x = 0$ at $x_i$, $i = 1, 2, 3, \cdots, s$.

$$x = \left(x_0 - (2i - 1) \frac{N}{k}\mu\right) \cos \omega_n t - (-1)^i \mu \frac{N}{k} \; t_i < t < t_{i+1} \tag{6.249}$$

$$\dot{x} = -\left(x_0 - (2i - 1) \frac{N}{k}\mu\right) \omega_n \; \sin \omega_n t \; t_i = \frac{i\,\pi}{\omega_n} \tag{6.250}$$

$$x_i = (-1)^i \left(x_0 - 2i \; \mu \frac{N}{k}\right) \qquad i = 1, 2, 3, \cdots, s \tag{6.251}$$

$$x_{i+1} = -x_i + 2\mu \frac{N}{k} \qquad x_0 = x\,(0) \tag{6.252}$$

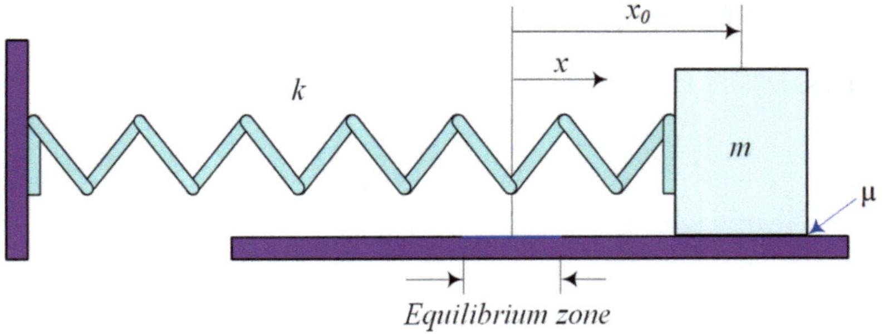


**Fig. 6.14** A mass–spring oscillator on a rough surface with Coulomb friction

The final position $x_s$ will be reached when the friction force is greater than the spring force, while the mass is at rest. The mass then stops forever. The value $s$ is the lowest integer for which we have

$$k\ |x_s| < \mu N < k\ |x_{s-1}| \tag{6.253}$$

or

$$|x_s| < \frac{\mu N}{k} < |x_{s-1}| \tag{6.254}$$

The rest or equilibrium zone of the system is $x < |\mu N/k|$ in which the spring force is less than the maximum friction force.

$$-\frac{\mu N}{k} < x < \frac{\mu N}{k} \tag{6.255}$$

Friction force is an example of forces that arise from the presence of holonomic constraints. It is a function of constraint force, but its line of action lies in the tangent plane and hence does virtual work.

Phenomena related to dry friction appear in poorly lubricated or non-lubricated contact surfaces of bodies in relative motion.

French scientist Charles Augustin de Coulomb (1736–1806) developed a series of two-term equations to model the friction force. The first term is a constant, and the second term varies with time, normal force, velocity, or other parameters. Leonardo Da Vinci (1452–1519) was one of the first scholars to systematically study friction. Friction force $\mathbf{F}_f$ is a given force that exists because of a constraint and hence is a function of constraint force $N$. ■

**Example 54** Oscillation with Coulomb friction. A step-by-step detailed example is needed to learn vibrations with friction, numerically.

Consider a mass and spring system as shown in Fig. 6.14 where the mass is sliding on a rough surface with friction coefficient $\mu$. Let us work with the following data.

$$\begin{aligned} &\mu = 0.5 \qquad N = 10\,\text{N} \qquad k = 100\,\text{N/m} \\ &\omega_n = 4\,\text{s}^{-1} \qquad x_0 = 1\,\text{m} \qquad \dot{x}_0 = 0 \end{aligned} \tag{6.256}$$

The mass is released from $x_0 = 1$ m on the right side of $x = 0$. The equation of motion and its solution will be

$$\ddot{x} + \omega_n^2 x = \mu \frac{N}{m} \qquad x_0 = 1\,\text{m} \qquad \dot{x}_0 = 0 \tag{6.257}$$

$$\ddot{x} + 16x = \mu \frac{N}{k/\omega_n^2} = 0.5 \frac{10}{100/16} = 0.8 \tag{6.258}$$

$$x = \left(x_0 - \mu \frac{N}{k}\right) \cos \omega_n t + \mu \frac{N}{k} \qquad 0 < t < t_1 \tag{6.259}$$

$$= 0.95 \cos 4t + 0.05 \qquad t_1 = \frac{\pi}{\omega_n} = \frac{\pi}{4} \tag{6.260}$$

$$\dot{x} = -3.8 \sin 4t \tag{6.261}$$

The mass will come to stop at the left-hand side of $x = 0$ at $x_1$.

$$x_1 = x(t_1) = -x_0 + 2\mu \frac{N}{k} = -0.9\,\text{m} \tag{6.262}$$

The second and other parts of oscillations are

$$x = 0.85 \cos 4t - 0.05 \qquad t_1 < t < t_2 \qquad t_2 = \frac{2\pi}{4} \; x_2 = 0.8\,\text{m} \tag{6.263}$$

$$x = 0.75 \cos 4t + 0.05 \qquad t_2 < t < t_3 \qquad t_3 = \frac{3\pi}{4} \; x_3 = -0.7\,\text{m} \tag{6.264}$$

$$x = 0.65 \cos 4t - 0.05 \qquad t_3 < t < t_4 \qquad t_4 = \frac{4\pi}{4} \; x_4 = 0.6\,\text{m} \tag{6.265}$$

$$x = 0.55 \cos 4t + 0.05 \qquad t_4 < t < t_5 \qquad t_5 = \frac{5\pi}{4} \; x_5 = -0.5\,\text{m} \tag{6.266}$$

$$x = 0.45 \cos 4t - 0.05 \qquad t_5 < t < t_6 \qquad t_6 = \frac{6\pi}{4} \; x_6 = 0.4\,\text{m} \tag{6.267}$$

$$x = 0.35 \cos 4t + 0.05 \qquad t_6 < t < t_7 \qquad t_7 = \frac{7\pi}{4} \; x_7 = -0.3\,\text{m} \tag{6.268}$$

$$x = 0.25 \cos 4t - 0.05 \qquad t_7 < t < t_8 \qquad t_8 = \frac{8\pi}{4} \; x_8 = 0.2\,\text{m} \tag{6.269}$$

$$x = 0.15 \cos 4t + 0.05 \qquad t_8 < t < t_9 \qquad t_9 = \frac{9\pi}{4} \; x_9 = -0.1\,\text{m} \tag{6.270}$$

$$x = 0.05 \cos 4t - 0.05 \qquad t_9 < t < t_{10} \qquad t_{10} = \frac{10\pi}{4} \; x_{10} = 0.0\,\text{m} \tag{6.271}$$

Figure 6.15 illustrates $m$ at the initial position $x_0$ and at the first stop at $x_1$, as well as the location of $x_i, i = 1, 2, 3, \cdots, x_{10}$. The time response of the system is shown in Fig. 6.16. Because the amplitude drops with a constant value, the response curve stays within a pair of linear envelopes. The oscillation ends up at rest position $x_{10} = 0$, in the middle of equilibrium zone.

$$-0.05 < x < 0.05 \tag{6.272}$$

The equilibrium zone is an area in which the spring force is less than friction force. In this example, the mass will stop exactly at the middle of the equilibrium point; however, other data may make $m$ to stop at other point of the zone. For example, if we increase the friction coefficient to $\mu = 0.51$, the mass will stop at $x_1 = -0.898$, $x_2 = 0.796$, $x_3 = -0.694$, $x_4 = 0.592$, $x_5 = -0.49$, $x_6 = 0.388$, $x_7 = -0.286$, $x_8 = 0.184$, $x_9 = -0.082$, and finally at $x_{10} = -0.02$ entering the zone from the left.

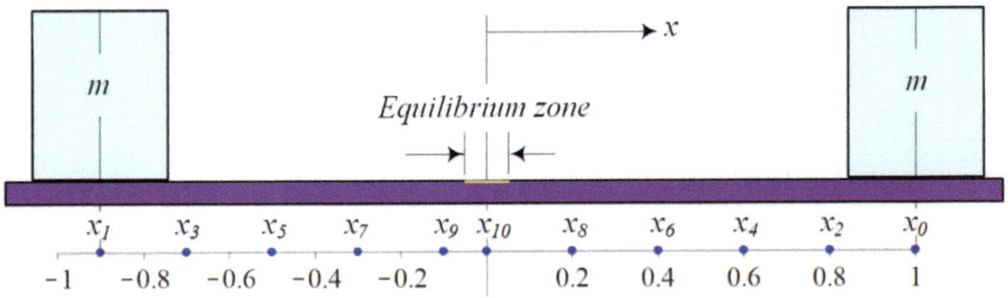


**Fig. 6.15** Illustration of $m$ at the initial position $x_0$ and at the first stop at $x_1$, as well as the location of $x_i$, $i = 1, 2, 3, \cdots, x_{10}$

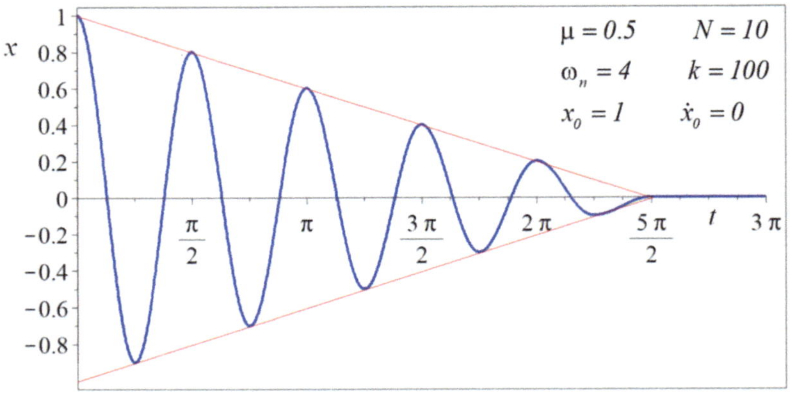


**Fig. 6.16** The time response of the system

## 6.2 ★ Free Vibrations Multi-$DOF$

Analysis of time response of systems with more than one degree of freedom ($DOF$) is very similar to analysis of time response of systems with one $DOF$. It is simple and interesting topic to study. The analysis is very much dependent on natural frequencies and mode shapes of the system. Figure 6.17 illustrates a multi-$DOF$ vibrating system with three oscillating masses.

An $n$ $DOF$ system is expressed by a set of $n$ ordinary differential equations of the second order. The system has as many natural frequencies $\omega_i$ as the degrees of freedom. There is a mode of vibration $\mathbf{u}_i$ associated with each natural frequency. The equations of motion are coupled, and hence, the motion of the masses is a combination of the motions of the individual modes.

### 6.2.1 ★ Undamped Systems

A system shows its natural behavior when it is force free and undamped. Such system will have its basic responses and expresses its natural behavior that is only dependent on initial conditions. We call a system with no external excitation a *free system*, and a system with no damping, an *undamped system.* A linear undamped–free system with constant mass matrix $\mathbf{M}$ and stiffness matrix $\mathbf{K}$ is governed by a set of coupled linear differential equations using $\mathbf{x}$ as the vector of coordinates.

$$\mathbf{M}\ddot{\mathbf{x}} + \mathbf{K}\,\mathbf{x} = \mathbf{0} \tag{6.273}$$

$$\mathbf{x} = \begin{bmatrix} x_1 & x_2 & \cdots & x_n \end{bmatrix}^T \tag{6.274}$$

$$\mathbf{x}(0) = \mathbf{x}_0 \qquad \dot{\mathbf{x}}(0) = \dot{\mathbf{x}}_0 \tag{6.275}$$

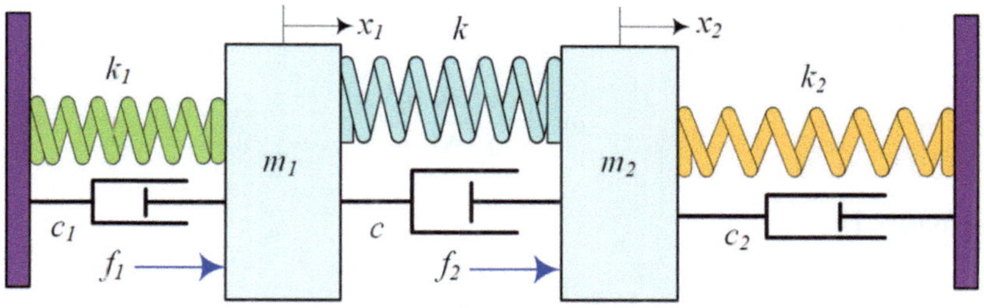


**Fig. 6.17** A multi-$DOF$ vibrating system with two oscillating masses

$$\mathbf{M} = \begin{bmatrix} m_{11} & m_{12} & \cdots & m_{1n} \\ m_{21} & m_{22} & \cdots & m_{2n} \\ \cdots & \cdots & \cdots & \cdots \\ m_{n1} & m_{n2} & \cdots & m_{nn} \end{bmatrix} \tag{6.276}$$

$$\mathbf{K} = \begin{bmatrix} k_{11} & k_{12} & \cdots & k_{1n} \\ k_{21} & k_{22} & \cdots & k_{2n} \\ \cdots & \cdots & \cdots & \cdots \\ k_{n1} & k_{n2} & \cdots & k_{nn} \end{bmatrix} \tag{6.277}$$

The time response of the free system is a linear combination of harmonic functions for each coordinate $x_i$, with $\omega_i$ as the *natural frequencies* and $\mathbf{u}_i$ as the *mode shapes* of the system.

$$\begin{aligned} \mathbf{x} &= \sum_{i=1}^{n} \mathbf{u}_i \left(A_i \sin \omega_i t + B_i \cos \omega_i t\right) \quad i = 1, 2, 3, \cdots, n \\ &= \sum_{i=1}^{n} X_i \mathbf{u}_i \sin\left(\omega_i t + \varphi_i\right) \quad i = 1, 2, 3, \cdots, n \end{aligned} \tag{6.278}$$

$$X_i = \sqrt{A_i^2 + B_i^2} \tag{6.279}$$

$$\tan \varphi_i = \frac{B_i}{A_i} \tag{6.280}$$

The coefficients $A_i$ and $B_i$, or $X_i$ and $\varphi_i$, must be determined from the initial conditions.

The natural frequencies $\omega_i$ are solutions of the characteristic equation of the system

$$\det\left[\mathbf{K} - \omega^2\,\mathbf{M}\right] = 0 \tag{6.281}$$

and the mode shape $\mathbf{u}_i$, corresponding to $\omega_i$, is the solution of the eigenvector equation.

$$\left[\mathbf{K} - \omega_i^2\,\mathbf{M}\right]\,\mathbf{u}_i = 0 \tag{6.282}$$

Mode shapes are orthogonal with respect to the mass $\mathbf{M}$ and stiffness $\mathbf{K}$ matrices.

$$\mathbf{u}_j^T\,\mathbf{M}\,\mathbf{u}_i = 0 \qquad i \neq j \tag{6.283}$$

$$\mathbf{u}_j^T\,\mathbf{K}\,\mathbf{u}_i = 0 \qquad i \neq j \tag{6.284}$$

The orthogonality of mode shapes yields

$$\omega_i^2 = \frac{\mathbf{u}_i^T \mathbf{K} \mathbf{u}_i}{\mathbf{u}_i^T \mathbf{M} \mathbf{u}_i} \tag{6.285}$$

Employing the modal matrix $\mathbf{U}$

$$\mathbf{U} = \begin{bmatrix} \mathbf{u}_1 \ \mathbf{u}_2 \ \mathbf{u}_2 \cdots \mathbf{u}_n \end{bmatrix} \tag{6.286}$$

we can use a linear transformation to change the variables from the generalized coordinates $\mathbf{x}$ to principal coordinates $\mathbf{p}$ and make the equations decoupled.

$$\mathbf{x} = \mathbf{U}\,\mathbf{p} \tag{6.287}$$

$$\mathbf{M}'\ddot{\mathbf{p}} + \mathbf{K}'\mathbf{p} = \mathbf{0} \tag{6.288}$$

$$m_i'\ddot{p}_i + k_i'p_i = 0 \tag{6.289}$$

$$\mathbf{M}' = \begin{bmatrix} m_1' & 0 & 0 & 0 \\ 0 & m_2' & 0 & 0 \\ \vdots & \vdots & \ddots & \vdots \\ 0 & 0 & \cdots & m_n' \end{bmatrix} \qquad \mathbf{K}' = \begin{bmatrix} k_1' & 0 & 0 & 0 \\ 0 & k_2' & 0 & 0 \\ \vdots & \vdots & \ddots & \vdots \\ 0 & 0 & \cdots & k_n' \end{bmatrix} \tag{6.290}$$

The uncoupled equations can be solved individually. The original solution will be found by combining the principal solutions and the transformation (6.287).

***Proof*** If the equations of motion are derived from Lagrange method and the coordinates are all absolute, then all coefficient matrices $\mathbf{M}$, $\mathbf{K}$ are symmetric, $\mathbf{M}^T = \mathbf{M}$, $\mathbf{K}^T = \mathbf{K}$. These matrices can also be shown as $[m] \equiv \mathbf{M}$, $[k] \equiv \mathbf{K}$. Simplifying the equations by eliminating the force and damping terms will provide us with the following set of equations.

$$\mathbf{M}\,\ddot{\mathbf{x}} + \mathbf{K}\,\mathbf{x} = \mathbf{0} \tag{6.291}$$

$$\mathbf{x}(0) = \mathbf{x}_0 \qquad \dot{\mathbf{x}}(0) = \dot{\mathbf{x}}_0 \tag{6.292}$$

The solution of linear differential equations will be exponential function.

$$\mathbf{x} = \mathbf{u}\,e^{i\omega t} \tag{6.293}$$

$$\dot{\mathbf{x}} = i\omega\,\mathbf{x} = i\omega\,\mathbf{u}\,e^{i\omega t} \tag{6.294}$$

$$\ddot{\mathbf{x}} = -\omega^2\mathbf{x} = -\omega^2\,\mathbf{u}\,e^{i\omega t} \tag{6.295}$$

Substituting the exponential functions $\mathbf{x}$ and $\ddot{\mathbf{x}}$ in Equation (6.291) makes a generalized algebraic eigenvalue problem with $\lambda = \omega^2$ as the eigenvalue and $\mathbf{u}$ as the eigenvector.

$$[\mathbf{K} - \lambda\mathbf{M}]\ \mathbf{u} = \mathbf{0} \tag{6.296}$$

$$\lambda = \omega^2 \tag{6.297}$$

Nontrivial solution for $\mathbf{u}$ exists only if determinant of its coefficient is zero.

$$\det[\mathbf{K} - \lambda\mathbf{M}] = |\mathbf{K} - \lambda\mathbf{M}| = 0 \tag{6.298}$$

*Trivial solution* $\mathbf{u} = \mathbf{0}$ indicates the *rest position* of the system and shows no motion.

Determining the constant $\lambda$, such that the set of equations (6.296) provide us with a nontrivial solution, is called the *eigenvalue problem*. Expanding (6.298) makes an algebraic *characteristic equation* that is an $n$th-order equation in terms of $\lambda = \omega^2$ and provides us with $n$ natural frequencies $\omega_i$.

$$a_n\lambda^n + a_{n-1}\lambda^{n-1} + a_{n-2}\lambda^{n-2} + \cdots a_1\lambda + a_0 = 0 \tag{6.299}$$

Having $n$ values for $\omega_i$ indicates that the solution (6.293) is possible for $n$ different frequencies $\omega_i$, $i = 1, 2, 3, \cdots, n$.

Determination of the vectors $\mathbf{u}_i$ to satisfy Equation (6.296) is the *eigenvector problem*. To determine $\mathbf{u}_i$, we substitute $\omega_i$ into Equation (6.296) to find $n$ different $\mathbf{u}_i$.

$$\left[\mathbf{K} - \omega_i^2\mathbf{M}\right]\mathbf{u}_i = 0 \tag{6.300}$$

In mechanical vibrations, the eigenvector $\mathbf{u}_i$ corresponding to the eigenvalue $\omega_i$ is called the *mode shape*.

Equation (6.300) is homogeneous, so if $\mathbf{u}_i$ is a solution, then $a\mathbf{u}_i$, $a \in \mathbb{R}$, is also a solution. Hence, the eigenvectors may be expressed with any length. However, the ratio of any two elements of an eigenvector is unique and, therefore, $\mathbf{u}_i$ has a unique shape. In other words, if one of the elements of $\mathbf{u}_i$ is assigned, the remaining $n-1$ elements are uniquely determined. The shape of an eigenvector $\mathbf{u}_i$ indicates the relative amplitudes of the coordinates of the system at $\omega_i$. Mode shapes are the principal elements of all possible free vibrations of a multi-$DOF$ system.

Having $\omega_i = \pm\sqrt{\lambda}$ and $\mathbf{u}_i$, the solution (6.293) will be determined as

$$\begin{aligned}\mathbf{x}_i &= \mathbf{u}_i\, e^{i\omega_i t} \\ &= \mathbf{u}_i\left(A_i \sin\omega_i t + i B_i \cos\omega_i t\right)\end{aligned} \tag{6.301}$$

or as

$$\begin{aligned}\mathbf{x}_i &= \mathbf{u}_i\, e^{-i\omega_i t} \\ &= \mathbf{u}_i\left(A_i \sin\omega_i t - i B_i \cos\omega_i t\right)\end{aligned} \tag{6.302}$$

Because both the real and imaginary parts of the solutions (6.301) and (6.302) satisfy Equation (6.291), the general and real solution of the equation is a linear combination of all solutions.

$$\mathbf{x} = \sum_{i=1}^{n}\mathbf{u}_i\left(A_i \sin\omega_i t + B_i \cos\omega_i t\right) \qquad i = 1, 2, 3, \cdots, n \tag{6.303}$$

$$= \sum_{i=1}^{n} X_i\mathbf{u}_i \sin\left(\omega_i t + \varphi_i\right) \qquad i = 1, 2, 3, \cdots, n \tag{6.304}$$

Setting $t = 0$ in these expressions and substituting in the initial conditions of (6.292) provide $2n$ equations to determine the constants of integration.

$$\mathbf{x}(0) = \sum_{i=1}^{n}\mathbf{u}_i B_i \tag{6.305}$$

$$\dot{\mathbf{x}}(0) = \sum_{i=1}^{n}\mathbf{u}_i\omega_i A_i \tag{6.306}$$

Using the form of Equations (6.305) and (6.306), the solution of the free vibration problem then becomes

$$\mathbf{x} = \sum_{i=1}^{n}\frac{\mathbf{u}_i\,\mathbf{u}_i^T\,\mathbf{M}}{\mathbf{u}_i^T\,\mathbf{M}\,\mathbf{u}_i}\left(\frac{\dot{\mathbf{x}}(0)}{\omega_i}\sin\omega_i t + \mathbf{x}(0)\cos\omega_i t\right) \tag{6.307}$$

We usually set the natural frequencies $\omega_i$ in numerical order:

$$\omega_1 \le \omega_2 \le \omega_3 \le \cdots \le \omega_n \tag{6.308}$$

We can also multiply Equation (6.291) by $\mathbf{M}^{-1}$ and find the same characteristic equation (6.298).

$$\ddot{\mathbf{x}} + \mathbf{M}^{-1}\mathbf{K}\,\mathbf{x} = \ddot{\mathbf{x}} + \mathbf{A}\,\mathbf{x} = \mathbf{0} \tag{6.309}$$

$$\det\left[\mathbf{A} - \lambda\mathbf{I}\right] = 0 \tag{6.310}$$

$$\mathbf{A} = \mathbf{M}^{-1}\mathbf{K} \tag{6.311}$$

Therefore, determination of the natural frequencies $\omega_i$ is equivalent to determining the eigenvalues $\lambda_i$ of the *characteristic matrix* $\mathbf{A} = \mathbf{M}^{-1}\mathbf{K}$. Alternatively, we can find the eigenvectors $\mathbf{u}_i$ from the matrix $\mathbf{A}$.

$$\left[\mathbf{A} - \lambda\mathbf{I}\right]\mathbf{u}_i = 0 \tag{6.312}$$

Any type of free, transient, or forced response of a linear vibrating system is made by its natural frequencies, mode shapes, and interaction of excitation frequencies. Determination of the natural frequencies $\omega_i$ and their associated mode shapes $\mathbf{u}_i$ is the first step in analysis of multi-$DOF$ vibrating systems. There exists at least one mode shape corresponding to each natural frequency.

Arranging the eigenvalues and mode shapes in the form of the following matrices introduces the modal matrix $\mathbf{U}$ and eigen matrix $\mathbf{\Lambda} \equiv \mathbf{\Omega}^2$,

$$\mathbf{U} = \left[\, \mathbf{u}_1 \; \mathbf{u}_2 \; \mathbf{u}_2 \; \cdots \; \mathbf{u}_n \,\right] \tag{6.313}$$

$$\mathbf{\Lambda} = \begin{bmatrix} \lambda_1 & 0 & \cdots & 0 \\ 0 & \lambda_2 & \cdots & 0 \\ \vdots & \vdots & \ddots & \vdots \\ 0 & 0 & \cdots & \lambda_n \end{bmatrix} = \mathbf{\Omega}^2 = \begin{bmatrix} \omega_1^2 & 0 & \cdots & 0 \\ 0 & \omega_2^2 & \cdots & 0 \\ \vdots & \vdots & \ddots & \vdots \\ 0 & 0 & \cdots & \omega_n^2 \end{bmatrix} \tag{6.314}$$

enables us to collect all equations of eigenvalues

$$\mathbf{K}\,\mathbf{u}_i = \omega_i^2\,\mathbf{M}\,\mathbf{u}_i \tag{6.315}$$

and rewrite them as a compact expression of the complete solution to the eigenvalue problem.

$$\mathbf{K}\,\mathbf{U} = \mathbf{\Omega}^2\,\mathbf{M}\,\mathbf{U} \tag{6.316}$$

The proof of orthogonality conditions of mode shapes with respect to the mass $\mathbf{M}$ and stiffness $\mathbf{K}$ matrices of Equations (6.283) and (6.284) is given in Sect. 8.1.

The orthogonality of the mode shapes with respect to $\mathbf{M}$ and $\mathbf{K}$ implies that the following square matrices are diagonal.

$$\mathbf{U}^T\,\mathbf{K}\,\mathbf{U} = \mathbf{K}' \tag{6.317}$$

$$\mathbf{U}^T\,\mathbf{M}\,\mathbf{U} = \mathbf{M}' \tag{6.318}$$

$$\mathbf{u}_i^T\,\mathbf{K}\,\mathbf{u}_j = \begin{cases} k_i & i = j \\ 0 & i \neq j \end{cases} \tag{6.319}$$

$$\mathbf{u}_i^T\,\mathbf{M}\,\mathbf{u}_j = \begin{cases} m_i & i = j \\ 0 & i \neq j \end{cases} \tag{6.320}$$

$$\mathbf{M}' = \begin{bmatrix} m_1' & 0 & 0 & 0 \\ 0 & m_2' & 0 & 0 \\ \vdots & \vdots & \ddots & \vdots \\ 0 & 0 & \cdots & m_n' \end{bmatrix} \tag{6.321}$$

$$\mathbf{K}' = \begin{bmatrix} k_1' & 0 & 0 & 0 \\ 0 & k_2' & 0 & 0 \\ \vdots & \vdots & \ddots & \vdots \\ 0 & 0 & \cdots & k_n' \end{bmatrix} \tag{6.322}$$

Rearranging the diagonal mass and stiffness matrices makes it easier to show the natural frequencies $\omega_i^2$.

$$\mathbf{\Omega}^2 = \frac{\mathbf{K}'}{\mathbf{M}'} = \frac{\mathbf{U}^T \mathbf{K} \mathbf{U}}{\mathbf{U}^T \mathbf{M} \mathbf{U}} \tag{6.323}$$

$$\omega_i^2 = \frac{\mathbf{u}_i^T \mathbf{K} \mathbf{u}_i}{\mathbf{u}_i^T \mathbf{M} \mathbf{u}_i} \tag{6.324}$$

This is called Rayleigh's quotient. It may also be derived by equating time averages of the maximum value of kinetic energy to the maximum value of potential energy of the system under the assumption that the system is oscillating harmonic at frequency $\omega$ with amplitude ratios $\mathbf{u}$.

The diagonal mass and stiffness matrices make the equations of motion independent in principal coordinates $\mathbf{p}$.

$$\mathbf{x} = \mathbf{U}\,\mathbf{p} \tag{6.325}$$

$$\mathbf{M}'\ddot{\mathbf{p}} + \mathbf{K}'\mathbf{p} = \mathbf{0} \tag{6.326}$$

$$\begin{bmatrix} m_1' & 0 & 0 & 0 \\ 0 & m_2' & 0 & 0 \\ \vdots & \vdots & \ddots & \vdots \\ 0 & 0 & \cdots & m_n' \end{bmatrix} \begin{bmatrix} \ddot{p}_1 \\ \ddot{p}_2 \\ \vdots \\ \ddot{p}_n \end{bmatrix} + \begin{bmatrix} k_1' & 0 & 0 & 0 \\ 0 & k_2' & 0 & 0 \\ \vdots & \vdots & \ddots & \vdots \\ 0 & 0 & \cdots & k_n' \end{bmatrix} \begin{bmatrix} p_1 \\ p_2 \\ \vdots \\ p_n \end{bmatrix} = \begin{bmatrix} 0 \\ 0 \\ \vdots \\ 0 \end{bmatrix} \tag{6.327}$$

Therefore,

$$m_i'\ddot{p}_i + k_i' p_i = 0 \tag{6.328}$$

and hence, the natural frequency of the $i$th equation is

$$\omega_i^2 = \frac{k_i}{m_i} \tag{6.329}$$

and, therefore,

$$\begin{bmatrix} \omega_1^2 & 0 & 0 & 0 \\ 0 & \omega_2^2 & 0 & 0 \\ \vdots & \vdots & \ddots & \vdots \\ 0 & 0 & \cdots & \omega_n^2 \end{bmatrix} = \mathbf{M}'^{-1}\,\mathbf{K}' \tag{6.330}$$

The principal coordinates and decoupled equations suggest another method to solve time response of the multi-$DOF$ systems. ■

**Example 55** A two $DOF$ vibrating system. Study a low $DOF$ vibrating system will show all features of analysis of free vibrations of multi-$DOF$ systems.

Consider free and undamped case of the system in Fig. 6.17.

$$\mathbf{M}\,\ddot{\mathbf{x}} + \mathbf{K}\,\mathbf{x} = \mathbf{0} \tag{6.331}$$

$$\begin{bmatrix} m_1 & 0 \\ 0 & m_2 \end{bmatrix}\begin{bmatrix} \ddot{x}_1 \\ \ddot{x}_2 \end{bmatrix} + \begin{bmatrix} k+k_1 & -k \\ -k & k+k_2 \end{bmatrix}\begin{bmatrix} x_1 \\ x_2 \end{bmatrix} = \begin{bmatrix} 0 \\ 0 \end{bmatrix} \tag{6.332}$$

$$\mathbf{x}(0) = \mathbf{x}_0 = \begin{bmatrix} x_{10} \\ x_{20} \end{bmatrix} \qquad \dot{\mathbf{x}}(0) = \dot{\mathbf{x}}_0 = \begin{bmatrix} \dot{x}_{10} \\ \dot{x}_{20} \end{bmatrix} \tag{6.333}$$

The natural frequencies of the system are the square root of the eigenvalues of **A**-matrix.

$$\begin{aligned} \mathbf{A} = \mathbf{M}^{-1}\mathbf{K} &= \begin{bmatrix} m_1 & 0 \\ 0 & m_2 \end{bmatrix}^{-1}\begin{bmatrix} k+k_1 & -k \\ -k & k+k_2 \end{bmatrix} \\ &= \begin{bmatrix} (k+k_1)/m_1 & -k/m_1 \\ -k/m_2 & (k+k_2)/m_2 \end{bmatrix} \end{aligned} \tag{6.334}$$

$$\det[\mathbf{A} - \lambda\mathbf{I}] = 0 \tag{6.335}$$

$$\lambda^2 - \left(\frac{k+k_1}{m_1} + \frac{k+k_2}{m_2}\right)\lambda + \left(\frac{kk_1 + kk_2 + k_1k_2}{m_1m_2}\right) = 0 \tag{6.336}$$

$$\omega_1 = \sqrt{\lambda_1} = \sqrt{\frac{1}{2}Z_1 - \frac{1}{2}\sqrt{Z_1^2 - 4Z_2}} \tag{6.337}$$

$$\omega_2 = \sqrt{\lambda_2} = \sqrt{\frac{1}{2}Z_1 + \frac{1}{2}\sqrt{Z_1^2 - 4Z_2}} \tag{6.338}$$

$$Z_1 = \frac{k+k_1}{m_1} + \frac{k+k_2}{m_2} \tag{6.339}$$

$$Z_2 = \frac{kk_1 + kk_2 + k_1k_2}{m_1m_2} \tag{6.340}$$

The lower frequency $\omega_1$ is the fundamental and the others are the harmonics. The mode shapes of the system are

$$\mathbf{u}_1 = \begin{bmatrix} u_{11} \\ u_{21} \end{bmatrix} = \begin{bmatrix} \lambda_1 - \frac{k+k_2}{m_2} \\ -k/m_2 \end{bmatrix} = \begin{bmatrix} 1 \\ u_1 \end{bmatrix} \tag{6.341}$$

$$\mathbf{u}_2 = \begin{bmatrix} u_{12} \\ A_{22} \end{bmatrix} = \begin{bmatrix} \lambda_2 - \frac{k+k_2}{m_2} \\ -k/m_2 \end{bmatrix} = \begin{bmatrix} 1 \\ u_2 \end{bmatrix} \tag{6.342}$$

Therefore, the solution of the equations of motion (6.331) can be expressed in a more informative way:

$$\begin{bmatrix} x_1 \\ x_2 \end{bmatrix} = \begin{bmatrix} A_{11} \\ A_{21} \end{bmatrix}\sin(\omega_1 t + \varphi_1) + \begin{bmatrix} A_{12} \\ A_{22} \end{bmatrix}\sin(\omega_2 t + \varphi_2) \tag{6.343}$$

where the $A's$ and $\varphi's$ are constants depending on initial conditions. $A_{ij}$ is the amplitude of $x_i$ at the frequency $\omega_j$. The relative amplitudes of the harmonic components are

$$\frac{A_{11}}{A_{21}} = \frac{k}{k + k_1 - m_1\omega_1^2} = \frac{k + k_2 - m_2\omega_1^2}{k} = \frac{u_{11}}{u_{21}} = \frac{1}{u_1} \tag{6.344}$$

$$\frac{A_{12}}{A_{22}} = \frac{k}{k + k_1 - m_1\omega_2^2} = \frac{k + k_2 - m_2\omega_2^2}{k} = \frac{u_{12}}{u_{22}} = \frac{1}{u_2} \tag{6.345}$$

where $u_i$ are constants, defining the relative amplitudes of $x_1$ and $x_2$ at each of the natural frequencies $\omega_1$ and $\omega_2$. Thus, the solution (6.343) becomes

$$\begin{bmatrix} x_1 \\ x_2 \end{bmatrix} = \begin{bmatrix} 1 \\ u_1 \end{bmatrix} A_{11} \sin(\omega_1 t + \varphi_1) + \begin{bmatrix} 1 \\ u_2 \end{bmatrix} A_{12} \sin(\omega_2 t + \varphi_2) \tag{6.346}$$

There are four constants of integrations, $A_{11}$, $A_{12}$, $\varphi_1$, $\varphi_2$ to be determined from initial conditions. Depending on the initial conditions, the system can vibrate in a principal natural mode or any combination of the mode shapes. The first mode occurs when $A_{12} = 0$,

$$\begin{bmatrix} x_1 \\ x_2 \end{bmatrix} = \begin{bmatrix} 1 \\ u_1 \end{bmatrix} A_{11} \sin(\omega_1 t + \varphi_1) \tag{6.347}$$

and the second mode occurs when $A_{11} = 0$:

$$\begin{bmatrix} x_1 \\ x_2 \end{bmatrix} = \begin{bmatrix} 1 \\ u_2 \end{bmatrix} A_{12} \sin(\omega_2 t + \varphi_2) \tag{6.348}$$

Hence, a principal mode is specified the modal vector at the given natural frequency.

**Example 56** A 2 $DOF$ free vibration. Here is a numerical example of free vibrations of a 2 $DOF$ vibrating system to illustrate its solution graphically in detail.

Consider the system of Fig. 6.17 with the following numerical values:

$$\begin{gathered} m_1 = 1 \qquad m_2 = 2 \\ k_1 = 10 \qquad k = 20 \qquad k_2 = 30 \end{gathered} \tag{6.349}$$

The values of natural frequencies $\omega_1$ and $\omega_2$ are

$$Z_1 = 55 \qquad Z_2 = 550 \tag{6.350}$$

$$\lambda_1 = 13.1386 \qquad \lambda_2 = 41.861 \tag{6.351}$$

$$\omega_1 = \sqrt{\lambda_1} = 3.62 \qquad \omega_2 = \sqrt{\lambda_2} = 6.47 \tag{6.352}$$

$$u_1 = \frac{A_{21}}{A_{11}} = 0.843 \qquad u_2 = \frac{A_{22}}{A_{12}} = -0.593 \tag{6.353}$$

Therefore, the time functions for $x_1$ and $x_2$ are

$$\begin{aligned} \begin{bmatrix} x_1 \\ x_2 \end{bmatrix} &= \begin{bmatrix} 1 \\ u_1 \end{bmatrix} A_{11} \sin(\omega_1 t + \varphi_1) + \begin{bmatrix} 1 \\ u_2 \end{bmatrix} A_{12} \sin(\omega_2 t + \varphi_2) \\ &= \begin{bmatrix} 1 \\ 0.843 \end{bmatrix} A_{11} \sin(\omega_1 t + \varphi_1) + \begin{bmatrix} 1 \\ -0.593 \end{bmatrix} A_{12} \sin(\omega_2 t + \varphi_2) \end{aligned} \tag{6.354}$$

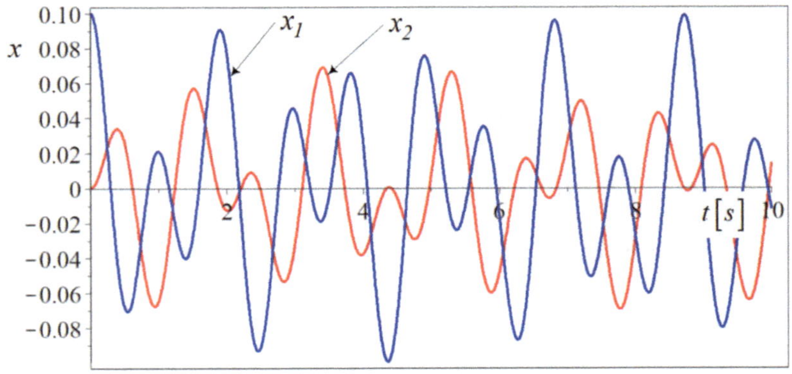


**Fig. 6.18** Free vibrations of a 2 $DOF$ vibrating system

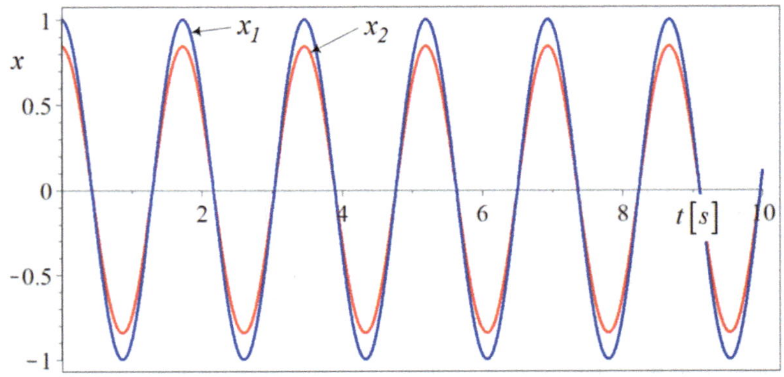


**Fig. 6.19** Free vibrations of a 2 $DOF$ vibrating system at first mode of vibrations

We calculate the constants of integrations for the following initial conditions and calculate the solutions $x_1$ and $x_2$.

$$\begin{bmatrix} x_{10} \\ x_{20} \end{bmatrix} = \begin{bmatrix} 0.1 \\ 0 \end{bmatrix} \qquad \begin{bmatrix} \dot{x}_{10} \\ \dot{x}_{20} \end{bmatrix} = \begin{bmatrix} 0 \\ 0 \end{bmatrix} \tag{6.355}$$

$$A_{11} = 0.041 \qquad A_{12} = 0.058 \tag{6.356}$$

$$\varphi_1 = 1.57\,\text{rad} \qquad \varphi_2 = 1.57\,\text{rad} \tag{6.357}$$

$$x_1 = 0.041\,\sin(3.62t + 1.57) + 0.058\,\sin(6.47t + 1.57) \tag{6.358}$$

$$x_2 = 3.456 \times 10^{-2}\sin(3.62t + 1.57) - 3.439 \times 10^{-2}\sin(6.47t + 1.57) \tag{6.359}$$

Figure 6.18 illustrates the response functions of $x_1$ and $x_2$.

It is more interesting if the initial conditions are set equal to mode shapes of the system. If the initial conditions are

$$\begin{bmatrix} x_{10} \\ x_{20} \end{bmatrix} = \begin{bmatrix} 1 \\ u1 \end{bmatrix} \qquad \begin{bmatrix} \dot{x}_{10} \\ \dot{x}_{20} \end{bmatrix} = \begin{bmatrix} 0 \\ 0 \end{bmatrix} \tag{6.360}$$

then the system will be released from rest when the initial displacement is proportional to the first mode shape. Figure 6.19 illustrates the oscillation of the system at first mode, and Fig. 6.20 illustrates the oscillation of the system at second mode. The higher frequency and opposite phase of oscillation of the second mode compared to the first mode can be recognized in Fig. 6.20.

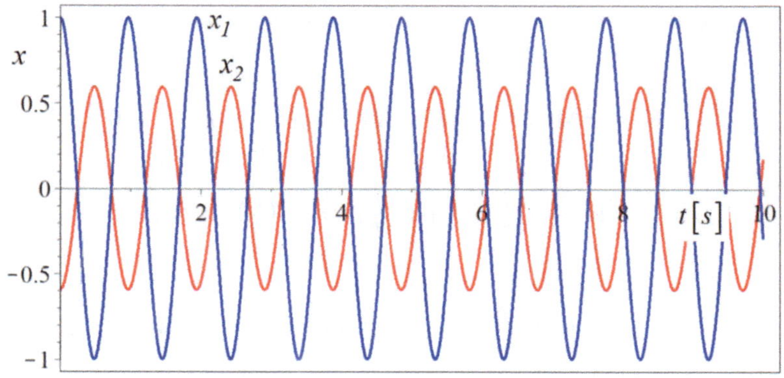


**Fig. 6.20** Free vibrations of a 2 $DOF$ vibrating system at second mode of vibrations

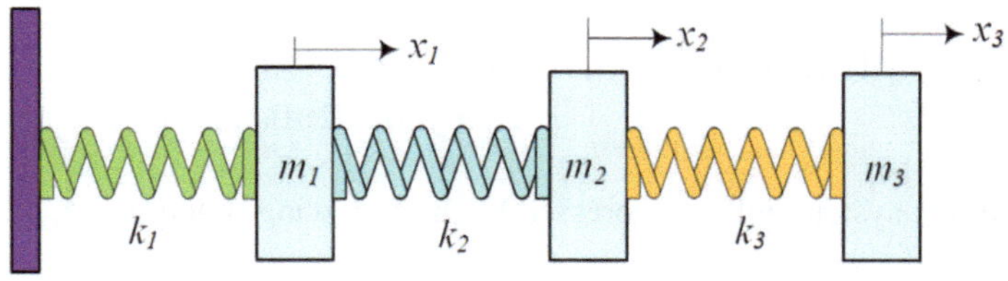


**Fig. 6.21** A 3 $DOF$ free undamped vibrating system

**Example 57** Three degrees of freedom system. A 3 $DOF$ system is a good example to review the method of determining free vibrations of undamped multi-$DOF$ systems.

Consider the 3 $DOF$ vibrating system of Fig. 6.21. The equations of motion of the system are

$$m_1\ddot{x}_1 = -k_1x_1 - k_2(x_1 - x_2) \tag{6.361}$$

$$m_2\ddot{x}_2 = -k_2(x_2 - x_1) - k_3(x_2 - x_3) \tag{6.362}$$

$$m_3\ddot{x}_3 = -k_3(x_3 - x_2) \tag{6.363}$$

or

$$\mathbf{M}\,\ddot{\mathbf{x}} + \mathbf{K}\,\mathbf{x} = \mathbf{0} \tag{6.364}$$

$$\mathbf{M} = \begin{bmatrix} m_1 & 0 & 0 \\ 0 & m_2 & 0 \\ 0 & 0 & m_3 \end{bmatrix} \tag{6.365}$$

$$\mathbf{K} = \begin{bmatrix} k_1 + k_2 & -k_2 & 0 \\ -k_2 & k_2 + k_3 & -k_3 \\ 0 & -k_3 & k_3 \end{bmatrix} \tag{6.366}$$

$$\mathbf{x} = \begin{bmatrix} x_1 & x_2 & x_3 \end{bmatrix}^T \tag{6.367}$$

Determinant $\left[\mathbf{K} - \omega^2\mathbf{M}\right]$ provides us with the characteristic equation of the system determine the natural frequencies.

$$\begin{aligned} &\left|\mathbf{K} - \omega^2\mathbf{M}\right| \\ &= \begin{vmatrix} k_1 + k_2 - m_1\omega^2 & -k_2 & 0 \\ -k_2 & k_2 + k_3 - m_2\omega^2 & -k_3 \\ 0 & -k_3 & k_3 - m_3\omega^2 \end{vmatrix} = 0 \end{aligned} \tag{6.368}$$

Assuming $m_1 = m_2 = m_3 = m$ and $k_1 = k_2 = k_3 = k$ simplifies the analysis and yields

$$\left|\mathbf{K} - \omega^2\mathbf{M}\right| = \begin{vmatrix} 2k - m\omega^2 & -k & 0 \\ -k & 2k - m\omega^2 & -k \\ 0 & -k & k - m\omega^2 \end{vmatrix} = 0 \tag{6.369}$$

$$\omega^6 - 5\left(\tfrac{k}{m}\right)\omega^4 + 6\left(\tfrac{k}{m}\right)^2\omega^2 - \left(\tfrac{k}{m}\right)^3 = 0 \tag{6.370}$$

$$\omega_1^2 = 0.198\frac{k}{m} \qquad \omega_2^2 = 1.55\frac{k}{m}\omega_3^2 = 3.25\frac{k}{m} \tag{6.371}$$

The mode shapes of the simplified system will be

$$\mathbf{U} = \left[\mathbf{u}_1\ \mathbf{u}_2\ \mathbf{u}_3\right] = \begin{bmatrix} 1 & 1 & 1 \\ 1.802 & 0.445 & -1.247 \\ 2.25 & -0.802 & 0.555 \end{bmatrix} \tag{6.372}$$

Therefore, the time response of the system will be expressed by the following functions:

$$\begin{aligned} \mathbf{x} = &\ \mathbf{u}_1 A_{11} \sin(\omega_1 t + \varphi_1) + \mathbf{u}_2 A_{12} \sin(\omega_2 t + \varphi_2) \\ &+ \mathbf{u}_3 A_{13} \sin\left(\omega_3 t + \varphi_3\right) \end{aligned} \tag{6.373}$$

**Example 58** General method to study free vibrations of multi-$DOF$ systems. Here we consider a general two $DOF$ undamped and free system to learn the standard method of determining the response of such systems.

Consider the equations of motion of any statically coupled two degrees of freedom system in the following form:

$$m_1\ddot{x}_1 + k_{11}x_1 + k_{12}x_2 = 0 \tag{6.374}$$

$$m_2\ddot{x}_2 + k_{21}x_1 + k_{22}x_2 = 0 \tag{6.375}$$

$$x_1(0) = x_{10} \qquad x_2(0) = x_{20} \qquad \dot{x}_1(0) = \dot{x}_{10}\dot{x}_2(0) = \dot{x}_{20} \tag{6.376}$$

Let us search for harmonic solutions.

$$x_1 = A_1 \sin(\omega t + \varphi) \qquad x_2 = A_2 \sin(\omega t + \varphi) \tag{6.377}$$

Equations (6.377) are solutions, provided that

$$\left(k_{11} - m_1\omega^2\right)A_1 + k_{12}A_2 = 0 \tag{6.378}$$

$$k_{21}A_1 + \left(k_{22} - m_2\omega^2\right)A_2 = 0 \tag{6.379}$$

Elimination of $A_1/A_2$ and considering a symmetric **K**-matrix with $k_{12} = k_{21}$ lead to the characteristic equation.

$$m_1m_2\omega^4 - (k_{11}m_2 + k_{22}m_1)\,\omega^2 + \left(k_{11}k_{22} - k_{12}^2\right) = 0 \tag{6.380}$$

The solution of characteristic equation is

$$\omega^2 = \frac{1}{2}\left(\frac{k_{11}}{m_1} + \frac{k_{22}}{m_2}\right) \pm \frac{1}{2}\sqrt{\left(\frac{k_{11}}{m_1} - \frac{k_{22}}{m_2}\right)^2 + \frac{4k_{12}^2}{m_1m_2}} \tag{6.381}$$

This equation gives us two real roots for $\omega^2$ as $\omega_1^2$ and $\omega_2^2$ because the term in the square root is always positive. Rewriting the characteristic equation in the following form,

$$\left(\omega^2 - \omega_1^2\right)\left(\omega^2 - \omega_2^2\right) = 0 \tag{6.382}$$

we have

$$\omega_1^2 + \omega_2^2 = \frac{k_{11}m_2 + k_{22}m_1}{m_1 m_2} \tag{6.383}$$

$$\omega_1^2 \omega_2^2 = \frac{k_{11}k_{22} - k_{12}^2}{m_1 m_2} \tag{6.384}$$

that indicated $\omega_1^2$ and $\omega_2^2$ will be positive if $k_{11}k_{22} > k_{12}^2$. Assuming $\omega_1^2 < \omega_2^2$, the minus sign of Equation (6.381) belongs to $\omega_1^2$ and the plus sign indicates $\omega_2^2$.

The ratio $A_1/A_2$ can be calculated from either equations of (6.378) or (6.379). Thus if $u_1$ and $u_2$ are the values of $A_1/A_2$ in the first and second modes, respectively, we have

$$u_1 = \frac{A_1}{A_2} = \frac{m_1\omega_1^2 - k_{11}}{k_{12}} \qquad u_2 = \frac{A_1}{A_2} = \frac{m_1\omega_2^2 - k_{11}}{k_{12}} \tag{6.385}$$

Employing Equations (6.385) and (6.384), we can show that

$$u_1 u_2 = -m_1 m_2 \tag{6.386}$$

Therefore, one of the ratios $u$ is positive and the other negative. That means the masses are vibrating in phase in the lower mode at the natural frequency $\omega_1$, and they are vibrating with a phase difference of 180 deg in the higher mode at the frequency $\omega_2$.

If $A_1$ and $A_2$ are the amplitudes of the masses $m_1$ and $m_2$ in the first and second modes, respectively, we will have

$$x_1 = A_1 \sin(\omega_1 t + \varphi_1) \qquad x_2 = u_1 A_1 \sin(\omega_1 t + \varphi_1) \tag{6.387}$$

$$x_1 = A_2 \sin(\omega_2 t + \varphi_2) \qquad x_2 = u_2 A_2 \sin(\omega_2 t + \varphi_2) \tag{6.388}$$

Therefore, by superposing the normal modes, the general solutions of equations (6.374) and (6.375) are

$$x_1 = A_1 \sin(\omega_1 t + \varphi_1) + A_2 \sin(\omega_2 t + \varphi_2) \tag{6.389}$$

$$x_2 = u_1 A_1 \sin(\omega_1 t + \varphi_1) + u_2 A_2 \sin(\omega_2 t + \varphi_2) \tag{6.390}$$

In the solution functions, $\omega_1$, $\omega_2$, $u_1$, and $u_2$ are fully known from the system parameters and are calculated by Equations (6.381) and (6.385). The unknown amplitudes $A_1$ and $A_2$ and phase angles $\varphi_1$ and $\varphi_2$ are determined from the initial conditions, $x_{10}$, $x_{20}$, $\dot{x}_{10}$, $\dot{x}_{20}$. In general, the initial conditions will be such that the resulting motion has components of all mode shapes. Practically, there is always some damping in the system, causing a slow decay in the vibrations. Every component of the mode shapes will decay at different rates.

**Example 59** Principal coordinate and decoupling the equations of motion. Converting the equations of motion of linear vibrating system into a set of principal coordinates is always possible. The principal coordinate makes the $n$ coupled equations to become $n$ decoupled and independent. The solution of every equation will be simple, and the response of the whole system will be calculated much easier in principal coordinate.

Let us consider the system of Equations (6.374)–(6.376). Combining Equations (6.386) and (6.385) yields

$$m_1 A_1 A_2 + m_2 (u_1 A_1)(u_2 A_2) = 0 \tag{6.391}$$

This is an expression of the orthogonality property of the two mode shapes. We can normalize the mode shapes if the amplitudes $A_1$ and $A_2$ in the first and second modes satisfy

$$m_1 A_1^2 + m_2 (u_1 A_1)^2 = 0 \tag{6.392}$$

$$m_1 A_2^2 + m_2 (u_2 A_2)^2 = 0 \tag{6.393}$$

Let us define the new coordinates, $p_1$ and $p_2$, by the following equations by using $A_1$ and $A_2$ satisfying Equations (6.392) and (6.393).

$$x_1 = A_1 p_1 + A_2 p_2 \tag{6.394}$$

$$x_2 = u_1 A_1 p_1 + u_2 A_2 p_2 \tag{6.395}$$

Substituting $x_1$ and $x_2$ in terms of $p_1$ and $p_2$ in the equations of motion (6.374)–(6.375) converts them into

$$m_1 \left(A_1 \ddot{p}_1 + A_2 \ddot{p}_2\right) + k_{11} \left(A_1 p_1 + A_2 p_2\right) + k_{12} \left(u_1 A_1 p_1 + u_2 A_2 p_2\right) = 0 \tag{6.396}$$

$$m_2 \left(u_1 A_1 \ddot{p}_1 + u_2 A_2 \ddot{p}_2\right) + k_{21} \left(A_1 p_1 + A_2 p_2\right) + k_{22} \left(u_1 A_1 p_1 + u_2 A_2 p_2\right) = 0 \tag{6.397}$$

Multiplying the first and second equations by $A_1$ and $u_1 A_1$, respectively, and adding them up yield

$$\begin{gathered}\left(m_2 A_1^2 u_1^2 + m_1 A_1^2\right) \ddot{p}_1 + \left(m_1 A_1 A_2 + m_2 A_1 A_2 u_1 u_2\right) \ddot{p}_2 \\ + \left(A_1 \left(A_1 k_{11} + A_1 u_1 k_{12}\right) + A_1 u_1 \left(A_1 k_{21} + A_1 u_1 k_{22}\right)\right) p_1 \\ + \left(A_1 \left(A_2 k_{11} + A_2 u_2 k_{12}\right) + A_1 u_1 \left(A_2 k_{21} + A_2 u_2 k_{22}\right)\right) \\ p_2 = 0\end{gathered} \tag{6.398}$$

Rewriting Equation (6.378) in terms of the normalized modes (6.392) for the first mode yields

$$k_{11} A_1 + k_{12}\, u_1 A_1 = m_1 \omega_1^2\, A_1 \tag{6.399}$$

$$k_{21} A_1 + k_{22}\, u_1 A_1 = m_2 \omega_1^2\, u_1 A_1 \tag{6.400}$$

and equations (6.379) in terms of the normalized modes (6.393) for the second mode are

$$k_{11} A_2 + k_{12}\, u_2 A_2 = m_1 \omega_2^2\, A_2 \tag{6.401}$$

$$k_{21} A_2 + k_{22}\, u_2 A_2 = m_2 \omega_2^2\, u_2 A_2 \tag{6.402}$$

Applying equations (6.391), (6.392), (6.393), and (6.399)–(6.402) to the coefficients of Equation (6.399) reduces to

$$\begin{gathered}\ddot{p}_1 + \left(m_1 \omega_1^2\, A_1 + m_2 \omega_1^2\, u_1 A_1\right) p_1 \\ + \left(m_1 \omega_2^2\, A_1 A_2 + m_2 \omega_2^2\, u_1 A_1\, u_2 A_2\right) p_2 = 0\end{gathered} \tag{6.403}$$

Using the normalizing condition and the orthogonal property, it simplifies to

$$\ddot{p}_1 + \omega_1^2\, p_1 = 0 \tag{6.404}$$

Similarly, the second equation of motion (6.375).

$$\ddot{p}_2 + \omega_2^2\, p_2 = 0 \tag{6.405}$$

Equations (6.404) and (6.405) are uncoupled expressions of the equations of motion in principal coordinates $p_1$ and $p_2$. Their solutions are individual simple harmonic functions of time.

$$p_1 = \sin\left(\omega_1 t + \theta_1\right) \tag{6.406}$$

$$p_2 = \sin\left(\omega_2 t + \theta_2\right) \tag{6.407}$$

The general solutions for free vibrations of the Equations (6.374)–(6.376) in coordinates $x_1$ and $x_2$ are obtained by substituting from Equations (6.394)–(6.395).

$$x_1 = A_1 \sin(\omega_1 t + \theta_1) + A_2 \sin(\omega_2 t + \theta_2) \tag{6.408}$$

$$x_2 = u_1 A_1 \sin(\omega_1 t + \theta_1) + u_2 A_2 \sin(\omega_2 t + \theta_2) \tag{6.409}$$

**Example 60** Principal coordinates from expansion theorem. Employing expansion theorem is an alternative method to show there exists a set of principal coordinates that decoupled equations of motion.

Let us begin with the equations of motion of an undamped–free multi-$DOF$ vibrating system.

$$\mathbf{M}\,\ddot{\mathbf{x}} + \mathbf{K}\,\mathbf{x} = \mathbf{0} \tag{6.410}$$

The expansion theorem implies that a general solution can be written as

$$\mathbf{x} = \sum_{i=1}^{n} p_i(t)\ \mathbf{u}_i \tag{6.411}$$

Substituting this solution in the equations of motion yields

$$\mathbf{M}\left(\sum_{i=1}^{n} \ddot{p}_i(t)\ \mathbf{u}_i\right) + \mathbf{K}\left(\sum_{i=1}^{n} p_i(t)\ \mathbf{u}_i\right) = \mathbf{0} \tag{6.412}$$

Inner product of $\mathbf{u}_j$ leads to

$$\sum_{i=1}^{n} \ddot{p}_i(t)\left(\mathbf{u}_j \cdot \mathbf{M}\,\mathbf{u}_i\right) + \sum_{i=1}^{n} p_i(t)\left(\mathbf{u}_j \cdot \mathbf{K}\,\mathbf{u}_i\right) = \mathbf{0} \tag{6.413}$$

$$\sum_{i=1}^{n} \ddot{p}_i(t)\left(\mathbf{u}_j^T\,\mathbf{M}\,\mathbf{u}_i\right) + \sum_{i=1}^{n} p_i(t)\left(\mathbf{u}_j^T\,\mathbf{K}\,\mathbf{u}_i\right) = \mathbf{0} \tag{6.414}$$

$$m_i'\,\ddot{p}_i(t) + k_i'\,p_i(t) = 0 \qquad i = 1, 2, \cdots, n \tag{6.415}$$

Equation (6.411) can be viewed as a linear transformation between the generalized coordinates, $\mathbf{x}$, and the principal coordinates $\mathbf{p}$.

$$\mathbf{x} = \sum_{i=1}^{n} p_i(t)\ \mathbf{u}_i = \mathbf{U}\,\mathbf{p} \tag{6.416}$$

$$\mathbf{p} = \mathbf{U}^{-1}\,\mathbf{x} \tag{6.417}$$

The coefficient matrix is the modal matrix of the system. Because the columns of the modal matrix are linearly independent, the modal matrix is non-singular and the transformations have a one-to-one correspondence.

**Example 61** ★ Eigenvector expansion. Eigenvectors of an $n$ dimensional space are orthogonal, and hence, they make base vectors to decompose any other vectors of the space.

Consider a linear vibrating system with $n$ $DOF$. Every eigenvector $\mathbf{u}_i$ of the system belongs to an $n$ dimensional configuration space and, hence, will have $n$ elements.

$$\mathbf{u}_i = \left[u_{i1}\ u_{i2}\ u_{i2}\ \cdots\ u_{in}\right]^T \tag{6.418}$$

Because eigenvectors are orthogonal,

$$\mathbf{u}_i \cdot \mathbf{u}_i = \begin{cases} |\mathbf{u}_i| \ |\mathbf{u}_i| & i = j \\ 0 & i \neq j \end{cases} \tag{6.419}$$

they can be used as base vectors to express any other vector of the same space. Therefore, any solution $\mathbf{x}$ can be expressed by a linear combination of the eigenvectors. This is called expansion theorem.

$$\mathbf{x} = \sum_{i=1}^{n} C_i \ \mathbf{u}_i \tag{6.420}$$

The initial position $\mathbf{x}_0$ and velocity $\dot{\mathbf{x}}_0$ of the system are vectors of the same space of the eigenvectors of the system. As a result, they can be expressed by the eigenvectors $\mathbf{u}_i$.

$$\mathbf{x}_0 = \sum_{i=1}^{n} a_i \ \mathbf{u}_i \tag{6.421}$$

The coefficients $a_i$ are called the scalar mode multipliers. To calculate $a_i$, let us multiply both sides of (6.421) by $\mathbf{u}_j^T \ \mathbf{M}$. Because of the orthogonality relations of (6.320), all the terms on the right side vanish except the one for which $j = i$.

$$\mathbf{u}_j^T \ \mathbf{M} \ \mathbf{x}_0 = \sum_{i=1}^{n} a_i \ \mathbf{u}_j^T \ \mathbf{M} \ \mathbf{u}_i \qquad a_i = \frac{\mathbf{u}_i^T \ \mathbf{M} \ \mathbf{x}_0}{\mathbf{u}_i^T \ \mathbf{M} \ \mathbf{u}_i} \tag{6.422}$$

Therefore, the vector $\mathbf{x}_0$ can be written in eigenvector expansion form.

$$\mathbf{x}_0 = \sum_{i=1}^{n} \frac{\mathbf{u}_i \ \mathbf{u}_i^T \ \mathbf{M}}{\mathbf{u}_i^T \ \mathbf{M} \ \mathbf{u}_i} \ \mathbf{x}_0 \tag{6.423}$$

The time response of undamped–free vibrating system (6.303) at $t = 0$ makes the initial conditions to be

$$\mathbf{x}(0) = \sum_{i=1}^{n} \mathbf{u}_i B_i \qquad \dot{\mathbf{x}}(0) = \sum_{i=1}^{n} \mathbf{u}_i \omega_i A_i \tag{6.424}$$

The eigenvector expansion of $\mathbf{x}(0)$ and $\dot{\mathbf{x}}(0)$ will be

$$\mathbf{x}(0) = \sum_{i=1}^{n} \frac{\mathbf{u}_i \ \mathbf{u}_i^T \ \mathbf{M}}{\mathbf{u}_i^T \ \mathbf{M} \ \mathbf{u}_i} \ \mathbf{x}(0) \qquad \dot{\mathbf{x}}(0) = \sum_{i=1}^{n} \frac{\mathbf{u}_i \ \mathbf{u}_i^T \ \mathbf{M}}{\mathbf{u}_i^T \ \mathbf{M} \ \mathbf{u}_i} \frac{\dot{\mathbf{x}}(0)}{\omega_i} \tag{6.425}$$

and hence, the solution for general initial conditions can be expressed as follows:

$$\mathbf{x} = \sum_{i=1}^{n} \frac{\mathbf{u}_i \ \mathbf{u}_i^T \ \mathbf{M}}{\mathbf{u}_i^T \ \mathbf{M} \ \mathbf{u}_i} \left( \frac{\dot{\mathbf{x}}(0)}{\omega_i} \sin \omega_i t + \mathbf{x}(0) \cos \omega_i t \right) \tag{6.426}$$

### 6.2.2 ★ Damped Systems

The best way to solve free vibrations of multi-$DOF$ systems with general damping

$$\mathbf{M} \ \ddot{\mathbf{x}} + \mathbf{C} \ \dot{\mathbf{x}} + \mathbf{K} \ \mathbf{x} = \mathbf{0} \tag{6.427}$$

is to convert the $n$ second-order differential equations (6.427) of motion to $2n$ first-order differential equations and develop the exponential solutions.

$$^{+}\mathbf{M}\,\dot{\mathbf{y}} + {}^{+}\mathbf{K}\,\mathbf{y} = \mathbf{0} \tag{6.428}$$

$$\mathbf{y} = \begin{bmatrix} \dot{\mathbf{x}} \\ \mathbf{x} \end{bmatrix} \tag{6.429}$$

$$^{+}\mathbf{M} = \begin{bmatrix} \mathbf{0} & \mathbf{M} \\ \mathbf{M} & \mathbf{C} \end{bmatrix} \tag{6.430}$$

$$^{+}\mathbf{K} = \begin{bmatrix} -\mathbf{M} & \mathbf{0} \\ \mathbf{0} & \mathbf{K} \end{bmatrix} \tag{6.431}$$

A solution to Equation (6.428) is exponential function of time,

$$\mathbf{y} = A\,\mathbf{u}\,e^{-\lambda t} \tag{6.432}$$

where $\lambda$ are the eigenvalues and $\mathbf{u}$ are eigenvectors of $^{+}\mathbf{M}^{-1}\,{}^{+}\mathbf{K}$. Therefore, the complete solution will be a linear combination of all individual solutions. The coefficients $A_i$ will be calculated from initial conditions.

$$\mathbf{y} = \sum_{i=1}^{n} A_i\,\mathbf{u}_i\,e^{-\lambda_i t} \tag{6.433}$$

***Proof*** The general equations of motion of free damped multi-degree-of-freedom ($DOF$) vibrating system are

$$\mathbf{M}\,\ddot{\mathbf{x}} + \mathbf{C}\,\dot{\mathbf{x}} + \mathbf{K}\,\mathbf{x} = \mathbf{0} \tag{6.434}$$

A general method of solution of free vibrations of damped system is to convert the equations into a set of first-order equations. The solution of the resultant set of first-order linear equations will be exponential. To convert the set of second-order equations of motion (6.434) is to define $2n$ new coordinates $\mathbf{y}$ for $\mathbf{x}$ and $\dot{\mathbf{x}}$.

$$\mathbf{y} = \begin{bmatrix} \mathbf{x} \\ \dot{\mathbf{x}} \end{bmatrix} \quad or \quad \mathbf{y} = \begin{bmatrix} \dot{\mathbf{x}} \\ \mathbf{x} \end{bmatrix} \tag{6.435}$$

Employing the new variables $\mathbf{y}$, we can rewrite Equation (6.434) in the following form:

$$\begin{bmatrix} \mathbf{0} & \mathbf{M} \\ \mathbf{M} & \mathbf{C} \end{bmatrix} \begin{bmatrix} \ddot{\mathbf{x}} \\ \dot{\mathbf{x}} \end{bmatrix} + \begin{bmatrix} -\mathbf{M} & \mathbf{0} \\ \mathbf{0} & \mathbf{K} \end{bmatrix} \begin{bmatrix} \dot{\mathbf{x}} \\ \mathbf{x} \end{bmatrix} = \begin{bmatrix} \mathbf{0} \\ \mathbf{0} \end{bmatrix} \tag{6.436}$$

$$^{+}\mathbf{M}\,\dot{\mathbf{y}} + {}^{+}\mathbf{K}\,\mathbf{y} = \mathbf{0} \tag{6.437}$$

The solution of the resultant set of equations will be exponential function of time.

$$\mathbf{y} = \mathbf{u}\,e^{-\lambda t} \tag{6.438}$$

Substituting the solution into the equation yields

$$\gamma\,{}^{+}\mathbf{M}\,\mathbf{u} = {}^{+}\mathbf{K}\,\mathbf{u} \tag{6.439}$$

$$\gamma\,\mathbf{u} = {}^{+}\mathbf{M}^{-1}\,{}^{+}\mathbf{K}\,\mathbf{u} \tag{6.440}$$

Hence, $\lambda$ are the eigenvalues and $\mathbf{u}$ are eigenvectors of $^{+}\mathbf{M}^{-1}\,{}^{+}\mathbf{K}$. The values of $\gamma$ occur in complex conjugate pairs $\gamma, \bar{\gamma}$. Eigenvectors corresponding to complex conjugate eigenvalues are also complex conjugates of one another $\mathbf{u}, \bar{\mathbf{u}}$. Eigenvectors corresponding to eigenvalues that are not complex conjugates satisfy the orthogonality relation.

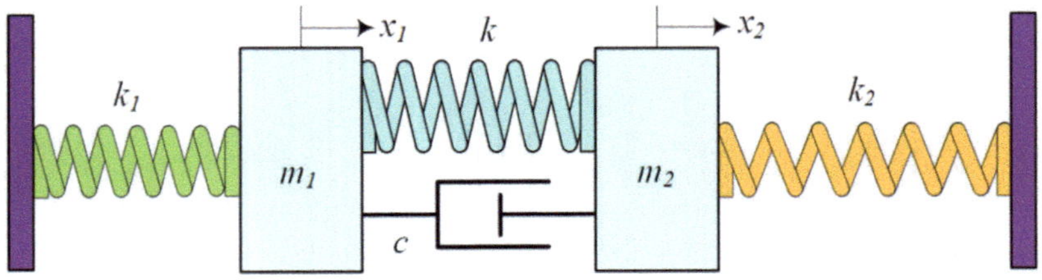


**Fig. 6.22** A 2 $DOF$ damped vibrating system

$$\bar{\mathbf{u}}_i^T \ {}^+\mathbf{M}\,\mathbf{u}_j = \mathbf{0} \tag{6.441}$$

Therefore, the complete solution is a linear combination of all individual solutions with coefficients $A_i$ to be calculated from initial conditions.

$$\mathbf{y} = \sum_{i=1}^{n} A_i \, \mathbf{u}_i \, e^{-\lambda_i t} \tag{6.442}$$

■

**Example 62** A two $DOF$ vibrating system with damping. If damper is not connected to the fixed wall, it will damp the relative displacements between its two ends, but an steady-state oscillation can remain in the system. Here is an example. Figure 6.22 illustrates a 2 $DOF$ damped vibrating system.

$$K = \frac{1}{2}m_1\dot{x}_1^2 + \frac{1}{2}m_2\dot{x}_2^2 \tag{6.443}$$

$$D = \frac{1}{2}c\,(\dot{x}_1 - \dot{x}_2)^2 \tag{6.444}$$

$$P = \frac{1}{2}k_1x_1^2 + \frac{1}{2}k_2x_2^2 + \frac{1}{2}k\,(x_1 - x_2)^2 \tag{6.445}$$

The Lagrange method provides us with the following equations of motion:

$$\frac{d}{dt}\left(\frac{\partial K}{\partial \dot{x}_r}\right) - \frac{\partial K}{\partial x_r} + \frac{\partial D}{\partial \dot{x}_r} + \frac{\partial P}{\partial x_r} = \qquad r = 1, 2 \tag{6.446}$$

$$m_1\ddot{x}_1 + c\,(\dot{x}_1 - \dot{x}_2) + k_1x_1 + k\,(x_1 - x_2) = 0 \tag{6.447}$$

$$m_2\ddot{x}_2 - c\,(\dot{x}_1 - \dot{x}_2) + k_2x_2 - k\,(x_1 - x_2) = 0 \tag{6.448}$$

$$\mathbf{M}\,\ddot{\mathbf{x}} + \mathbf{K}\,\mathbf{x} = \mathbf{0} \tag{6.449}$$

$$\begin{bmatrix} m_1 & 0 \\ 0 & m_2 \end{bmatrix}\begin{bmatrix} \ddot{x}_1 \\ \ddot{x}_2 \end{bmatrix} + \begin{bmatrix} c & -c \\ -c & c \end{bmatrix}\begin{bmatrix} \dot{x}_1 \\ \dot{x}_2 \end{bmatrix} + \begin{bmatrix} k+k_1 & -k \\ -k & k+k_2 \end{bmatrix}\begin{bmatrix} x_1 \\ x_2 \end{bmatrix} = \begin{bmatrix} 0 \\ 0 \end{bmatrix} \tag{6.450}$$

The equivalent set of first-order equations is

$${}^+\mathbf{M}\,\dot{\mathbf{y}} + {}^+\mathbf{K}\,\mathbf{y} = \mathbf{0} \tag{6.451}$$

$$\mathbf{y} = \begin{bmatrix} \dot{x}_1 & \dot{x}_2 & x_1 & x_2 \end{bmatrix}^T \tag{6.452}$$

$$= \begin{bmatrix} y_1 & y_2 & y_3 & y_4 \end{bmatrix}^T \tag{6.453}$$

$$
{}^{+}\mathbf{M} = \begin{bmatrix} \mathbf{0} & \mathbf{M} \\ \mathbf{M} & \mathbf{C} \end{bmatrix} = \begin{bmatrix} 0 & 0 & m_1 & 0 \\ 0 & 0 & 0 & m_2 \\ m_1 & 0 & c & -c \\ 0 & m_2 & -c & c \end{bmatrix} \tag{6.454}
$$

$$
{}^{+}\mathbf{K} = \begin{bmatrix} -\mathbf{M} & \mathbf{0} \\ \mathbf{0} & \mathbf{K} \end{bmatrix} = \begin{bmatrix} -m_1 & 0 & 0 & 0 \\ 0 & -m_2 & 0 & 0 \\ 0 & 0 & k+k_1 & -k \\ 0 & 0 & -k & k+k_2 \end{bmatrix} \tag{6.455}
$$

$$
{}^{+}\mathbf{M}\,\dot{\mathbf{y}} + {}^{+}\mathbf{K}\,\mathbf{y} = \begin{bmatrix} m_1\dot{y}_3 - m_1 y_1 \\ m_2\dot{y}_4 - m_2 y_2 \\ (k+k_1)\,y_3 + c\dot{y}_3 - c\dot{y}_4 - k y_4 + m_1\dot{y}_1 \\ (k+k_2)\,y_4 - c\dot{y}_3 + c\dot{y}_4 - k y_3 + m_2\dot{y}_2 \end{bmatrix} = 0 \tag{6.456}
$$

The natural frequencies of the system are the square root of the eigenvalues of ${}^{+}\mathbf{M}^{-1}\,{}^{+}\mathbf{K}$.

$$
{}^{+}\mathbf{M}^{-1}\,{}^{+}\mathbf{K} = \begin{bmatrix} \frac{c}{m_1} & -\frac{c}{m_1} & \frac{1}{m_1}(k+k_1) & -\frac{k}{m_1} \\ -\frac{c}{m_2} & \frac{c}{m_2} & -\frac{k}{m_2} & \frac{1}{m_2}(k+k_2) \\ -1 & 0 & 0 & 0 \\ 0 & -1 & 0 & 0 \end{bmatrix} \tag{6.457}
$$

Assume nonzero initial conditions and some numerical values for the parameters of the system.

$$
\mathbf{x}(0) = \mathbf{x}_0 = \begin{bmatrix} 0.1 \\ 0 \end{bmatrix} \qquad \dot{\mathbf{x}}(0) = \dot{\mathbf{x}}_0 = \begin{bmatrix} 0 \\ 0 \end{bmatrix} \tag{6.458}
$$

$$
m_1 = 1\,\text{kg} \qquad m_2 = 1\,\text{kg} \qquad c = 100\,\text{N/m/s} \tag{6.459}
$$

$$
k_1 = 100\,\text{kg} \qquad k_2 = 100\,\text{kg} \qquad k = 100\,\text{kg} \tag{6.460}
$$

$$
{}^{+}\mathbf{M}^{-1}\,{}^{+}\mathbf{K} = \begin{bmatrix} 100 & -100 & 200 & -100 \\ -100 & 100 & -100 & 200 \\ -1 & 0 & 0 & 0 \\ 0 & -1 & 0 & 0 \end{bmatrix} \tag{6.461}
$$

$$
\mathbf{\Omega}^2 = \begin{bmatrix} -10i & 0 & 0 & 0 \\ 0 & 10i & 0 & 0 \\ 0 & 0 & 100 - 10\sqrt{97} & 0 \\ 0 & 0 & 0 & 10\sqrt{97} + 100 \end{bmatrix} \tag{6.462}
$$

$$
\mathbf{U} = \begin{bmatrix} 10i & -10i & 100 - 10\sqrt{97} & 10\sqrt{97} + 100 \\ 10i & -10i & 10\sqrt{97} - 100 & -10\sqrt{97} - 100 \\ 1 & 1 & -1 & -1 \\ 1 & 1 & 1 & 1 \end{bmatrix} \tag{6.463}
$$

$$
\mathbf{y} = \sum_{i=1}^{4} A_i\,\mathbf{u}_i\,e^{-\lambda_i t} \tag{6.464}
$$

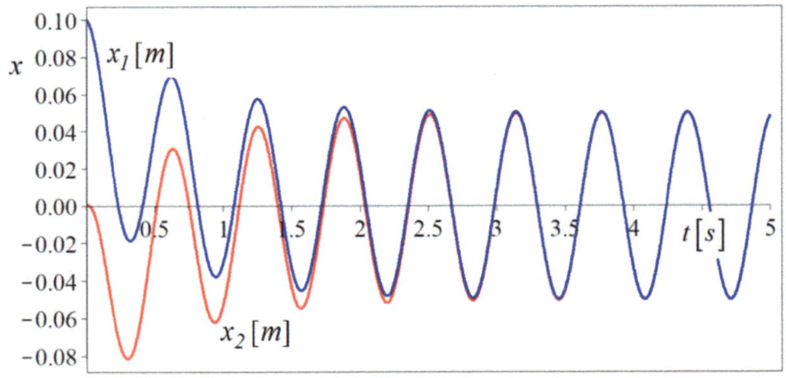


**Fig. 6.23** Displacements of $x_1$ and $x_2$ versus time

$$\mathbf{y} = A_1 \begin{bmatrix} 10i \\ 10i \\ 1 \\ 1 \end{bmatrix} e^{-10i\ t} + A_3 \begin{bmatrix} 100 - 10\sqrt{97} \\ 10\sqrt{97} - 100 \\ -1 \\ 1 \end{bmatrix} e^{(100-10\sqrt{97})t}$$
$$+ A_2 \begin{bmatrix} -10i \\ -10i \\ 1 \\ 1 \end{bmatrix} e^{10i\ t} + A_4 \begin{bmatrix} 10\sqrt{97} + 100 \\ -10\sqrt{97} - 100 \\ -1 \\ 1 \end{bmatrix} e^{(10\sqrt{97}+100)t} \tag{6.465}$$

Employing Euler complex number identity and converting the solution into real functions of time yield

$$y_1 = -\frac{3\sqrt{97}}{388} e^{(-100+10\sqrt{97})t} + \frac{3\sqrt{97}}{388} e^{(-10\sqrt{97}-100)\ t} - \frac{1}{2} \sin 10t \tag{6.466}$$

$$y_2 = \frac{3\sqrt{97}}{388} e^{(10\sqrt{97}-100)t} - \frac{3\sqrt{97}}{388} e^{(-100-10\sqrt{97})\ t} - \frac{1}{2} \sin 10t \tag{6.467}$$

$$y_3 = \left(\frac{1}{40} - \frac{\sqrt{97}}{388}\right) e^{(-100-10\sqrt{97})t} + \left(\frac{1}{40} + \frac{\sqrt{97}}{388}\right) e^{(10\sqrt{97}-100)t} + \frac{1}{20} \cos 10t \tag{6.468}$$

$$y_4 = \left(\frac{\sqrt{97}}{388} - \frac{1}{40}\right) e^{(-100-10\sqrt{97})t} + \left(-\frac{\sqrt{97}}{388} - \frac{1}{40}\right) e^{(10\sqrt{97}-100)t} + \frac{1}{20} \cos 10t \tag{6.469}$$

The time responses of the system for the displacements $x_1$ and $x_2$ are illustrated in Fig. 6.23.

The damper kills the relative motion between $m_1$ and $m_2$, and eventually, both masses move together to the right and left with same amplitude $X_1 = X_2 = 1/20$. The initial energy of the system was

$$E_1 = \frac{1}{2} k_1 x_{01}^2 = \frac{1}{2} (100) (0.1)^2 = \frac{1}{2} \text{ Nm} \tag{6.470}$$

and the final level of energy of the system is

$$E_2 = \frac{1}{2} k_1 X_1^2 + \frac{1}{2} k_2 X_2^2$$
$$= \frac{1}{2} (100) \left(\frac{1}{20}\right)^2 + \frac{1}{2} (100) \left(\frac{1}{20}\right)^2 = \frac{1}{4} \text{ Nm} \tag{6.471}$$

The difference between $E_1$ and $E_2$ is the amount of dissipated energy by the damper.

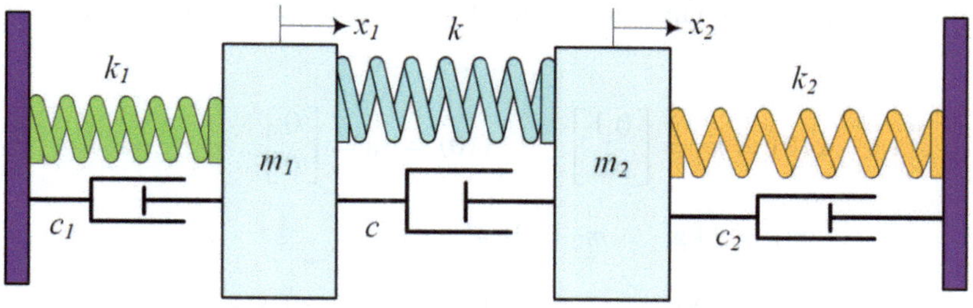


**Fig. 6.24** A multi-$DOF$ vibrating damped system with two oscillating masses

**Example 63** A two $DOF$ damped vibrating system. It is to compare free vibrations of damped and undamped systems.

To investigate the effect of fixed end damping, let us consider free and damped case of the system shown in Fig. 6.24.

$$\mathbf{M}\,\ddot{\mathbf{x}}+\mathbf{C}\,\dot{\mathbf{x}}+\mathbf{K}\,\mathbf{x}=\mathbf{0} \tag{6.472}$$

$$\begin{bmatrix} m_1 & 0 \\ 0 & m_2 \end{bmatrix}\begin{bmatrix} \ddot{x}_1 \\ \ddot{x}_2 \end{bmatrix}+\begin{bmatrix} c+c_1 & -c \\ -c & c+c_2 \end{bmatrix}\begin{bmatrix} \dot{x}_1 \\ \dot{x}_2 \end{bmatrix} \tag{6.473}$$

$$+\begin{bmatrix} k+k_1 & -k \\ -k & k+k_2 \end{bmatrix}\begin{bmatrix} x_1 \\ x_2 \end{bmatrix}=\begin{bmatrix} 0 \\ 0 \end{bmatrix} \tag{6.474}$$

The equivalent set of first-order equations is

$$^{+}\mathbf{M}\,\dot{\mathbf{y}}+\,^{+}\mathbf{K}\,\mathbf{y}=\mathbf{0} \tag{6.475}$$

$$\mathbf{y}=\begin{bmatrix} \dot{x}_1 & \dot{x}_2 & x_1 & x_2 \end{bmatrix}^T \tag{6.476}$$

$$=\begin{bmatrix} y_1 & y_2 & y_3 & y_4 \end{bmatrix}^T \tag{6.477}$$

$$^{+}\mathbf{M}=\begin{bmatrix} \mathbf{0} & \mathbf{M} \\ \mathbf{M} & \mathbf{C} \end{bmatrix}=\begin{bmatrix} 0 & 0 & m_1 & 0 \\ 0 & 0 & 0 & m_2 \\ m_1 & 0 & c+c_1 & -c \\ 0 & m_2 & -c & c+c_2 \end{bmatrix} \tag{6.478}$$

$$^{+}\mathbf{K}=\begin{bmatrix} -\mathbf{M} & \mathbf{0} \\ \mathbf{0} & \mathbf{K} \end{bmatrix}=\begin{bmatrix} -m_1 & 0 & 0 & 0 \\ 0 & -m_2 & 0 & 0 \\ 0 & 0 & k+k_1 & -k \\ 0 & 0 & -k & k+k_2 \end{bmatrix} \tag{6.479}$$

$$^{+}\mathbf{M}\,\dot{\mathbf{y}}+\,^{+}\mathbf{K}\,\mathbf{y}=\begin{bmatrix} m_1\dot{y}_3-m_1y_1 \\ m_2\dot{y}_4-m_2y_2 \\ (k+k_1)\,y_3+(c+c_1)\,\dot{y}_3-c\dot{y}_4-ky_4+m_1\dot{y}_1 \\ (k+k_2)\,y_4-c\dot{y}_3+(c+c_2)\,\dot{y}_4-ky_3+m_2\dot{y}_2 \end{bmatrix}=0 \tag{6.480}$$

The natural frequencies of the system are the square root of the eigenvalues of $^{+}\mathbf{M}^{-1}\,^{+}\mathbf{K}$.

$$^{+}\mathbf{M}^{-1}\,^{+}\mathbf{K}=\begin{bmatrix} \frac{c+c_1}{m_1} & -\frac{c}{m_1} & \frac{k+k_1}{m_1} & -\frac{k}{m_1} \\ -\frac{c}{m_2} & \frac{c+c_2}{m_2} & -\frac{k}{m_2} & \frac{k+k_2}{m_2} \\ -1 & 0 & 0 & 0 \\ 0 & -1 & 0 & 0 \end{bmatrix} \tag{6.481}$$

Nonzero initial conditions and a set of given numerical values for the parameters of the system produce the following eigen and modal matrices:

$$\mathbf{x}(0) = \mathbf{x}_0 = \begin{bmatrix} 0.1 \\ 0 \end{bmatrix} \qquad \dot{\mathbf{x}}(0) = \dot{\mathbf{x}}_0 = \begin{bmatrix} 0 \\ 0 \end{bmatrix} \tag{6.482}$$

$$m_1 = 1\,\text{kg} \qquad m_2 = 1\,\text{kg} \tag{6.483}$$

$$k_1 = 100\,\text{kg} \qquad k_2 = 100\,\text{kg} \qquad k = 100\,\text{kg} \tag{6.484}$$

$$c_1 = 100\,\text{kg} \qquad c_2 = 2\,\text{kg} \qquad c = 2\,\text{kg} \tag{6.485}$$

$$^{+}\mathbf{M}^{-1}\,{}^{+}\mathbf{K} = \begin{bmatrix} 102 & -100 & 200 & -100 \\ -100 & 102 & -100 & 200 \\ -1 & 0 & 0 & 0 \\ 0 & -1 & 0 & 0 \end{bmatrix} \tag{6.486}$$

$$\mathbf{\Omega}^2 = \begin{bmatrix} \sqrt{9901}+101 & 0 & 0 & 0 \\ 0 & 101-\sqrt{9901} & 0 & 0 \\ 0 & 0 & 1-3i\sqrt{11} & 0 \\ 0 & 0 & 0 & 3i\sqrt{11}+1 \end{bmatrix} \tag{6.487}$$

$$\mathbf{U} = \begin{bmatrix} \sqrt{9901}+101 & 101-\sqrt{9901} & 3i\sqrt{11}-1 & -3i\sqrt{11}-1 \\ -\sqrt{9901}-101 & \sqrt{9901}-101 & 3i\sqrt{11}-1 & -3i\sqrt{11}-1 \\ -1 & -1 & 1 & 1 \\ 1 & 1 & 1 & 1 \end{bmatrix} \tag{6.488}$$

The solution of the equations will be

$$\mathbf{y} = \sum_{i=1}^{4} A_i\, \mathbf{u}_i\, e^{-\lambda_i t} \tag{6.489}$$

$$\begin{aligned} y_1 = {} & -0.075374\, e^{-1.4962t} + 0.075374\, e^{-200.5038t} \\ & -0.502519\, e^{-t} \sin 9.95t \end{aligned} \tag{6.490}$$

$$\begin{aligned} y_2 = {} & 0.075374\, e^{-1.4962t} - 0.075374\, e^{-200.5038t} \\ & -0.502519\, e^{-t} \sin 9.95t \end{aligned} \tag{6.491}$$

$$\begin{aligned} y_3 = {} & 0.050376\, e^{-1.4962\,t} - 0.00037592326\, e^{-200.5038\,t} \\ & + \frac{e^{-t}}{20} \cos 9.95t + 0.005025\, e^{-t} \sin 9.95t \end{aligned} \tag{6.492}$$

$$\begin{aligned} y_4 = {} & -0.050376\, e^{-1.4962\,t} + 0.00037592326\, e^{-200.5038\,t} \\ & + \frac{e^{-t}}{20} \cos 9.95t + 0.005025\, e^{-t} \sin 9.95t \end{aligned} \tag{6.493}$$

The time responses of $x_1$ and $x_2$ are illustrated in Fig. 6.25.

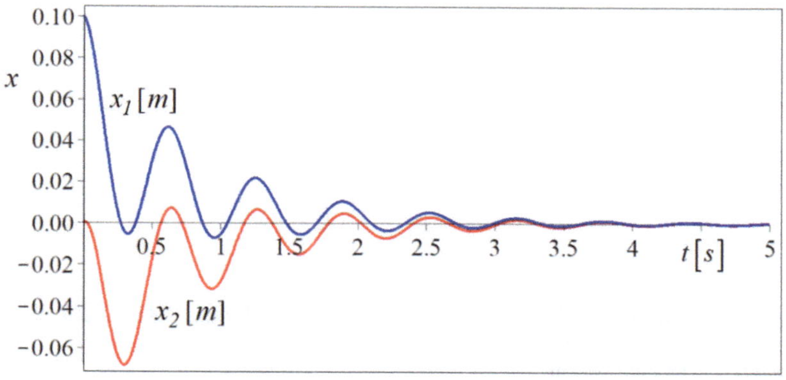


**Fig. 6.25** Displacements of $x_1$ and $x_2$ versus time for the damped two $DOF$ vibrating system

**Example 64** Alternative method for converting damped system into first-order equations. We can use other methods to convert a set of $n$ second-order damped equations into a set of $2n$ first-order equations to be solved by exponential matrix function.

Consider a multi-$DOF$ damped linear vibrating system.

$$\mathbf{M}\,\ddot{\mathbf{x}} + \mathbf{C}\,\dot{\mathbf{x}} + \mathbf{K}\,\mathbf{x} = \mathbf{0} \tag{6.494}$$

Let us define a new $n$ component coordinate vector $\mathbf{z}$, to convert the set of equations into first-order equations.

$$\dot{\mathbf{x}} = \mathbf{z} \tag{6.495}$$

$$\mathbf{M}\,\dot{\mathbf{z}} + \mathbf{C}\,\mathbf{z} + \mathbf{K}\,\mathbf{x} = \mathbf{0} \tag{6.496}$$

$$\dot{\mathbf{z}} = -\mathbf{M}^{-1}\,\mathbf{C}\,\mathbf{z} - \mathbf{M}^{-1}\,\mathbf{K}\,\mathbf{x} \tag{6.497}$$

Now we define a new $2n$ component coordinate vector $\mathbf{y}$

$$\mathbf{y} = \begin{bmatrix} \mathbf{x} \\ \mathbf{z} \end{bmatrix} = \begin{bmatrix} \mathbf{x} \\ \dot{\mathbf{x}} \end{bmatrix} \tag{6.498}$$

and make $2n$ first-order equations

$$\begin{bmatrix} \dot{\mathbf{x}} \\ \dot{\mathbf{z}} \end{bmatrix} = \begin{bmatrix} \mathbf{0} & \mathbf{I} \\ -\mathbf{M}^{-1}\,\mathbf{K} & -\mathbf{M}^{-1}\,\mathbf{C} \end{bmatrix} \begin{bmatrix} \mathbf{x} \\ \mathbf{z} \end{bmatrix} \tag{6.499}$$

It can be written in a compact form,

$$\dot{\mathbf{y}} = \mathbf{A}\,\mathbf{y} \tag{6.500}$$

where

$$\mathbf{A} = \begin{bmatrix} \mathbf{0} & \mathbf{I} \\ -\mathbf{M}^{-1}\,\mathbf{K} & -\mathbf{M}^{-1}\,\mathbf{C} \end{bmatrix} \tag{6.501}$$

The solution of the set of coupled first-order linear homogeneous equations is

$$\mathbf{y} = \sum_{j=1}^{2n} e^{-\lambda_j t} C_j \mathbf{u}_j \tag{6.502}$$

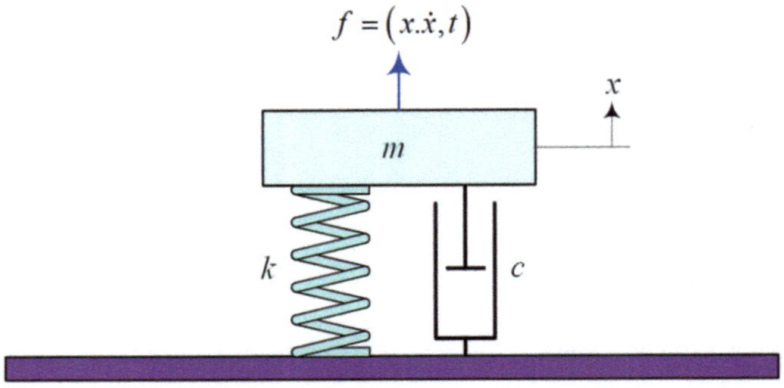


**Fig. 6.26** The model of a one $DOF$ vibrating system

where $\mathbf{u}_j$ are the eigenvectors and $\lambda_j$, $j = 1, 2, \cdots, 2n$, are eigenvalues of the $2n \times 2n$ coefficient matrix $\mathbf{A}$. This solution can equivalently be expressed by

$$\mathbf{y}(t) = e^{-\mathbf{A}t}\mathbf{C} = e^{-\mathbf{A}t}\mathbf{y}_0 \tag{6.503}$$

where

$$\begin{aligned} e^{-\mathbf{A}t} &= \mathbf{I} - \mathbf{A}t + \frac{1}{2!}\mathbf{A}^2t^2 - \frac{1}{3!}\mathbf{A}^3t^3 + \frac{1}{4!}\mathbf{A}^4t^4 - \cdots \\ &= \sum_{j=0}^{\infty}(-1)^j \frac{(\mathbf{A}t)^j}{j!} \end{aligned} \tag{6.504}$$

We may also express the solution of the homogeneous equations (6.730) as

$$\mathbf{x} = e^{-\mathbf{A}t}\mathbf{C} = \mathbf{U}\, e^{-\mathbf{\Lambda}t}\, \mathbf{U}^{-1}\, \mathbf{C} \tag{6.505}$$

where $\mathbf{\Lambda}$ is the associated diagonalized matrix of $\mathbf{A}$, and $\mathbf{U}$ is the modal matrix of $\mathbf{A}$.

$$\mathbf{\Lambda} = \mathbf{U}^{-1}\,\mathbf{A}\,\mathbf{U} = \begin{bmatrix} \lambda_1 & 0 & \cdots & 0 \\ 0 & \lambda_2 & \cdots & 0 \\ \vdots & \vdots & \ddots & \vdots \\ 0 & 0 & \cdots & \lambda_{2n} \end{bmatrix} \tag{6.506}$$

$$\mathbf{U} = \begin{bmatrix} \mathbf{u}_1\; \mathbf{u}_2 \cdots \mathbf{u}_{2n} \end{bmatrix} \tag{6.507}$$

## 6.3 Forced Vibrations One $DOF$

Consider a one degree of freedom ($DOF$) vibrating system as is shown in Fig. 6.26. The equation of motion of the system is a forced linear second-order differential equation along with a set of given initial conditions.

$$m\ddot{x} + c\dot{x} + kx = f(x, \dot{x}, t) \tag{6.508}$$

$$x(0) = x_0 \qquad \dot{x}(0) = \dot{x}_0 \tag{6.509}$$

The coefficients of mass $m$, stiffness $k$, and damping $c$ are assumed constant, although they may be functions of time in more general problems. The solution of such problem, $x = x(t)$, $t > 0$, will be studied in this chapter.

When there is no excitation, $f = 0$, the equation is called *homogeneous*,

$$m\ddot{x} + c\dot{x} + kx = 0 \tag{6.510}$$

otherwise, it is *nonhomogeneous*. The solution of the general equation of motion (6.508) is made up of two parts: $x_h(t)$, which is the *homogeneous solution*, and $x_p(t)$, which is the *particular solution*.

$$x(t) = x_h(t) + x_p(t) \tag{6.511}$$

In mechanical vibrations, the no force case $f = 0$ is called a *free vibration*, and its solution $x_h(t)$ is called the *free vibration response*. The nonhomogeneous case is called *forced vibration*, and its solution is called *forced vibration response*.

The *order* of an equation is the highest number of derivatives in the equation. In this chapter, we study all systems that can be modeled by the second-order differential equation (6.508). A second-order differential equation will have two independent solutions. If $x_1(t)$ and $x_2(t)$ are the two independent solutions of a second-order equation, then its general solution is a linear combination of them with unknown coefficients to be determined by initial conditions of the system:

$$x(t) = a_1 x_1(t) + a_2 x_2(t) \tag{6.512}$$

Determination of homogeneous solution $x_h$ and particular solution $x_p$ is the subject of this chapter.

The equation of motion of forced linear vibrations of a single degree of freedom is Equation (6.513). The solution of equation has two parts: the homogenous solution $x_h$ plus the particular solution $x_p$. The homogenous solution $x_h$ is the solution of free vibrations for $f = 0$, and the particular solution is the special solution of the equation associated to the forcing function $f(x, \dot{x}, t)$.

$$m\ddot{x} + c\dot{x} + kx = f(x, \dot{x}, t) \tag{6.513}$$

$$x(t) = x_h(t) + x_p(t) \tag{6.514}$$

The particular solution $x_p$ of a forced vibrating system is not possible to be found for a general force function $f = f(x, \dot{x}, t)$. However, if the force function is a continuous function of time $f = f(t)$ and is a combination of the following functions:

1. A constant, such as $f = F$
2. A polynomial in $t$, such as $f = a_0 + a_1 t + a_2 t^2 + \cdots + a_n t^n$
3. An exponential function, such as $f = F e^{at}$
4. A harmonic function, such as $f = F_1 \sin \omega t + F_2 \cos \omega t$
5. A transient forcing term $f(t) \neq 0, 0 \leq t \leq t_0$

then, the particular solution $x_p(t)$ of the linear equation of motion has the same form as the forcing term:

1. $x_p(t) =$ a constant, such as $x_p(t) = X$
2. $x_p(t) =$ a polynomial of the same degree, such as $x_p(t) = C_0 + C_1 t + C_2 t^2 + \cdots + C_n t^n$
3. $x_p(t) =$ an exponential function, such as $x_p(t) = C e^{at}$
4. $x_p(t) =$ a harmonic function, such as $x_p(t) = A \sin \omega t + B \cos \omega t$
5. $x_p(t) = \int_0^{t_0} f(t - \tau) h(\tau) d\tau = \int_0^{t_0} h(t - \tau) f(\tau) d\tau$

$$h(t) = \frac{1}{m\omega_n} e^{-\xi \omega_n t} \sin \omega_d t \tag{6.515}$$

$$\omega_d = \omega_n \sqrt{1 - \xi^2} \qquad \xi < 1 \tag{6.516}$$

The coefficients of particular solution are calculated by substituting the solution in the differential equation and solving the resultant algebraic equations to satisfy the differential equation. The coefficients will be functions of the differential equation known coefficients. The coefficients of homogenous solution are functions of the initial conditions. The initial conditions must be applied after the exact particular solution $x_p$ and the general homogenous solution $x_h$ are found and added together to have the general solution $x = x_h + x_p$.

***Proof*** Assuming $x_h(t)$ is the solution of equation (6.510), and $x_p(t)$ is the particular solution of equation (6.508) associated to the excitation $f$, we observe that $x = x_h + x_p$ satisfies $m\ddot{x} + c\dot{x} + kx = f$ and hence is the solution.

$$\begin{aligned} m\ddot{x} + c\dot{x} + kx &= m\left(\ddot{x}_h + \ddot{x}_p\right) + c\left(\dot{x}_h + \dot{x}_p\right) + k\left(x_h + x_p\right) \\ &= \left(m\ddot{x}_h + c\dot{x}_h + kx_h\right) + \left(m\ddot{x}_p + c\dot{x}_p + kx_p\right) \\ &= 0 + f = f \end{aligned} \tag{6.517}$$

Therefore, if $x$ is any solution of the nonhomogeneous equation (6.508), then $x - x_p$ is a solution of the homogeneous equation

$$\begin{aligned} & m\frac{d^2}{dt^2}\left(x - x_p\right) + c\frac{d}{dt}\left(x - x_p\right) + k\left(x - x_p\right) \\ &\quad = m\left(\ddot{x} - \ddot{x}_p\right) + c\left(\dot{x} - \dot{x}_p\right) + k\left(x - x_p\right) \\ &\quad = \left(m\ddot{x} + c\dot{x} + kx\right) - \left(m\ddot{x}_p + c\dot{x}_p + kx_p\right) \\ &\quad = f - f = 0 \end{aligned} \tag{6.518}$$

Hence, we find the solution of the nonhomogeneous equation (6.508) by determining the solution $x_h$ of the associated homogeneous equation (6.510) and add the particular solution $x_p$ of the nonhomogeneous equation associated to the forcing term $f$.

Any solution that satisfies the equation of motion (6.513) is a particular solution $x_p$, and it completes the solution of the equation by adding to the homogenous solution, $x = x_h + x_p$. When the equation is linear, the particular solution would be similar to the forcing term. To find a particular solution $x_p$ of the linear differential equation (6.513), we use the *method of undetermined coefficients* and assume a trial solution $x_p$ of the same form as $f = f(t)$ with unknown coefficients. Substituting $x_p$ in the equation, we determine the coefficients of $x_p$ to satisfy the equation:

1. Constant forcing functions.

$$m\ddot{x} + c\dot{x} + kx = f = F \tag{6.519}$$

   Assuming a constant solution, $x = X$, and substituting the assumed solution into the equation provide us with an algebraic equation to determine the solution.

$$x = X \qquad \dot{x} = 0 \qquad \ddot{x} = 0 \tag{6.520}$$

$$X = \frac{F}{k} \tag{6.521}$$

   The case of a constant force $f = F$ is a special case of the polynomial forcing function.
2. Polynomial forcing functions.

$$f = a_0 + a_1 t + a_2 t^2 + a_3 t^3 + \cdots + a_n t^n \tag{6.522}$$

   Assuming a trial polynomial solution

$$x_p = C_0 + C_1 t + C_2 t^2 + C_3 t^3 + \cdots + C_n t^n \tag{6.523}$$

$$\dot{x}_p = C_1 + 2C_2 t + 3C_3 t^2 + \cdots + nC_n t^{n-1} \tag{6.524}$$

$$\ddot{x}_p = 2C_2 + 6tC_3 + 12C_4 t^2 + \cdots + n\left(n-1\right)C_n t^{n-2} \tag{6.525}$$

   and substituting it in Equation (6.513),

$$m\left(2C_2+6C_3 t+12C_4 t^2+20C_5 t^3+\cdots+n\left(n-1\right)C_n t^{n-2}\right)$$

$$+c\left(C_1+2C_2t+3C_3t^2+4C_4t^3+\cdots+nC_nt^{n-1}\right)$$
$$+k\left(C_0+C_1t+C_2t^2+C_3t^3+\cdots+C_{n-1}t^{n-1}+C_nt^n\right)$$
$$=a_0+a_1t+a_2t^2+a_3t^3+\cdots+a_{n-1}t^{n-1}+a_nt^n \tag{6.526}$$

provides us with a set of algebraic equations to calculate the coefficient $C_i$.

$$kC_0+cC_1+2mC_2=a_0 \tag{6.527}$$
$$kC_1+2cC_2+6mC_3=a_1 \tag{6.528}$$
$$kC_2+3cC_3+12mC_4=a_2 \tag{6.529}$$
$$\vdots=\vdots$$
$$kC_{n-1}+ncC_n=a_{n-1} \tag{6.530}$$
$$kC_n=a_n \tag{6.531}$$

There are $n+1$ known constants $a_0,\cdots,a_n$ and $n+1$ unknowns $C_0,\cdots,C_n$, and we are able to set the equations in matrix form:

$$\begin{bmatrix} k & c & 2m & 0 & \cdots & 0 \\ 0 & k & 2c & 6m & \cdots & 0 \\ 0 & 0 & k & 3c & \cdots & 0 \\ \vdots & \vdots & \vdots & \vdots & \ddots & \vdots \\ 0 & 0 & 0 & \cdots & k & nc \\ 0 & 0 & 0 & \cdots & 0 & k \end{bmatrix} \begin{bmatrix} C_0 \\ C_1 \\ C_2 \\ \vdots \\ C_{n-1} \\ C_n \end{bmatrix} = \begin{bmatrix} a_0 \\ a_1 \\ a_2 \\ \vdots \\ a_{n-1} \\ a_n \end{bmatrix} \tag{6.532}$$

The solution of the set of equation indicates that $C_i, i=0,\cdots,n$, will be determined as functions of $a_0,\cdots,a_n$. Therefore, the trial function (6.523) is a solution of the equation of motion.

3. Exponential forcing function.

$$f=Fe^{at} \tag{6.533}$$

The trial solution would also be an exponential function.

$$x_p=Ce^{at} \qquad x_p=Cae^{at} \qquad x_p=Ca^2e^{at} \tag{6.534}$$

Substituting the solution into Equation (6.513),

$$mCa^2e^{at}+cCae^{at}+kCe^{at}=Fe^{at} \tag{6.535}$$

yields

$$C=\frac{F}{ma^2+ca+k} \tag{6.536}$$

$$x_p=\frac{F}{ma^2+ca+k}e^{at} \tag{6.537}$$

4. Harmonic forcing functions.

$$f=F_1\sin\omega t+F_2\cos\omega t \tag{6.538}$$

The particular solution for a harmonic excitation is also a harmonic function.

$$x_p = A \sin \omega t + B \cos \omega t \tag{6.539}$$

$$\dot{x}_p = \omega A \cos \omega t - \omega B \sin \omega t \tag{6.540}$$

$$\ddot{x}_p = -\omega^2 A \sin \omega t - \omega^2 B \cos \omega t \tag{6.541}$$

Substituting the harmonic solution in the equation of motion makes an algebraic equation.

$$\begin{aligned} -m\omega^2 \left(A \sin \omega t + B \cos \omega t\right) + c\omega \left(A \cos \omega t - B \sin \omega t\right) \\ + k \left(A \sin \omega t + B \cos \omega t\right) = F_1 \sin \omega t + F_2 \cos \omega t \end{aligned} \tag{6.542}$$

The harmonic functions $\sin \omega t$ and $\cos \omega t$ are orthogonal; therefore, their coefficients must be balanced on both sides of the equal sign. Balancing the coefficients of $\sin \omega t$ and $\cos \omega t$ provides us with a set of two equations for $A_1$ and $B_1$.

$$\begin{bmatrix} k - m\omega^2 & -c\omega \\ c\omega & k - m\omega^2 \end{bmatrix} \begin{bmatrix} A \\ B \end{bmatrix} = \begin{bmatrix} F_1 \\ F_2 \end{bmatrix} \tag{6.543}$$

Solving for coefficients $A$ and $B$

$$\begin{aligned} \begin{bmatrix} A \\ B \end{bmatrix} &= \begin{bmatrix} k - m\omega^2 & -c\omega \\ c\omega & k - m\omega^2 \end{bmatrix}^{-1} \begin{bmatrix} F_1 \\ F_2 \end{bmatrix} \\ &= \begin{bmatrix} \dfrac{\left(k - m\omega^2\right) F_1 + c\omega F_2}{\left(k - m\omega^2\right)^2 + c^2\omega^2} \\ \dfrac{\left(k - m\omega^2\right) F_2 - c\omega F_1}{\left(k - m\omega^2\right)^2 + c^2\omega^2} \end{bmatrix} \end{aligned} \tag{6.544}$$

determines the particular solution for the harmonic force function.

$$\begin{aligned} x_p = &\frac{\left(k - m\omega^2\right) F_1 + c\omega F_2}{\left(k - m\omega^2\right)^2 + c^2\omega^2} \sin \omega t \\ &+ \frac{\left(k - m\omega^2\right) F_2 - c\omega F_1}{\left(k - m\omega^2\right)^2 + c^2\omega^2} \cos \omega t \end{aligned} \tag{6.545}$$

5. Transient force.

$$f\left(t\right) \neq 0 \qquad 0 \leq t \leq t_0 \tag{6.546}$$

When the forcing function $f\left(t\right)$ disappears after a while, the system is under *transient excitation*. The solution of the equation of motion under a transient excitation is called *transient response*.

Let us show *the unit impulse force* function $\delta\left(t\right)$ at $t = 0$ as

$$\delta\left(t\right) = \begin{cases} \infty & t = 0 \\ 0 & t \neq 0 \end{cases} \tag{6.547}$$

$$\int_{-\infty}^{\infty} \delta\left(t\right) dt = \int_{0^-}^{0^+} \delta\left(t\right) dt = 1 \tag{6.548}$$

If $\delta\left(t\right)$ is applied at another time, $t = \tau$, its definition would be the same with a shift in time.

$$\int_{-\infty}^{\infty} \delta\left(t - \tau\right) dt = \int_{\tau^-}^{\tau^+} \delta\left(t - \tau\right) dt = 1 \tag{6.549}$$

$$\delta(t) = 0 \qquad t \neq \tau \tag{6.550}$$

The unit impulse force function is also called the *Dirac delta function*. The Dirac function is not similar to traditional functions, and its integrals do not exist in the classical mathematics.

Multiplying any function $g(t)$ by $\delta(t)$ makes the integral (6.548) to provide the value of $g(t)$ at the time $\delta(t)$ applies.

$$\int_{-\infty}^{\infty} g(t)\,\delta(t-\tau)\,dt = g(\tau) \tag{6.551}$$

Assume a unit impulse function is applied on a $(m, c, k)$ vibrating system.

$$m\ddot{x} + c\dot{x} + kx = \delta(t) \tag{6.552}$$

The response of the system to a unit impulse force is shown by $h(t)$.

$$h(t) = \frac{1}{m\omega_d} e^{-\xi\omega_n t} \sin \omega_d t \tag{6.553}$$

If the unit impulse force applied at time $t = \tau$, then the response to a unit impulse force will have a time shift, $h(t-\tau)$, where $h(t-\tau) = 0$ for $t < \tau$.

$$h(t-\tau) = \frac{1}{m\omega_d} e^{-\xi\omega_n (t-\tau)} \sin \omega_d (t-\tau) \tag{6.554}$$

To determine the response of a mass–spring–damper system to a given force $f(t)$, we assume the force function to be replaced by a series of pulses. If the response of the system to a pulse is known, then the response to $f(t)$ will be obtained by summation of the responses to each pulse in the sequence. Increasing the number of pulses approximates them by impulses and converts the summation into the *convolution integral*.

$$\begin{aligned} x(t) &= \int_0^t h(t-\tau)\, f(\tau)\, d\tau \\ &= \frac{1}{m\omega_d} \int_0^t e^{-\xi\omega_n (t-\tau)} f(\tau) \sin \omega_d (t-\tau)\ d\tau \end{aligned} \tag{6.555}$$

This is the response of the system to $f(\tau)$ with zero initial conditions. The convolution integral can also be equivalently written as

$$\begin{aligned} x(t) &= \int_0^t h(\tau)\, f(t-\tau)\, d\tau \\ &= \frac{1}{m\omega_d} \int_0^t e^{-\xi\omega_n \tau} f(t-\tau) \sin \omega_d \tau\ d\tau \end{aligned} \tag{6.556}$$

When the initial conditions are not zero, we add the homogeneous response of the system to the convolution integral.

$$\begin{aligned} x &= e^{-\xi\omega_n t} \left( \frac{\dot{x}_0 + \xi\omega_n x_0}{\omega_d} \sin \omega_d t + x_0 \cos \omega_d t \right) \\ &+ \frac{1}{m\omega_d} \int_0^t e^{-\xi\omega_n (t-\tau)} f(\tau) \sin \omega_d (t-\tau)\ d\tau \end{aligned} \tag{6.557}$$

The unit impulse function was first introduced by the German physicist Gustav Robert Kirchhoff (1824–1887) and then by the British physicist Paul Adrien Maurice Dirac (1902–1984). Any function that satisfies the definition (6.548) is a Dirac delta function. It does not necessarily need to be the limit of a rectangular pulse.

Convolution integral is also called Duhamel integral. Jean Marie Constant Duhamel (1797–1872) was a French mathematician and physicist who proposed a theory related to the transmission of heat in crystal structures. ■

**Example 65** Response to a step input. The response of dynamic systems to step input is the most important response to identify the characteristics of the system. Here is the analytical calculation of the response of a vibrating system to the unit step excitation.

The response of a damped system to a step input is a standard and the most important response by which we examine and compare all vibrating systems. Consider a linear second-order system with zero initial conditions.

$$m\ddot{x} + c\dot{x} + kx = f(t) \tag{6.558}$$

$$x(0) = 0 \qquad \dot{x}(0) = 0 \tag{6.559}$$

The equation may equivalently be transformed to the following form:

$$\ddot{x} + 2\xi\omega_n\dot{x} + \omega_n^2 x = \frac{1}{m} f(t) \qquad \xi < 1 \tag{6.560}$$

A step input is a sudden change of the forcing function $f(t)$ from zero to a constant and steady value. If the value is unity, then the excitation is called unit step input.

$$f(t) = \begin{cases} 1\,\text{N} & t > 0 \\ 0 & t \le 0 \end{cases} \tag{6.561}$$

The response of the system to a unit step input is called the unit step response. Linearity of the equation of motion guarantees that the response to a non-unit step input is proportional to the unit step response. A general step function will have a constant force of magnitude $F$.

$$f(t) = \begin{cases} F\,\text{N} & t > 0 \\ 0 & t \le 0 \end{cases} \tag{6.562}$$

The general solution of Equation (6.558) along with sudden force (6.562) is equal to the sum of the homogeneous and particular solutions of the equation, $x = x_h + x_p$. The homogeneous solution is given by Equation (6.116).

$$x_h = e^{-\xi\omega_n t}\left(A \sin\omega_d t + B\cos\omega_d t\right) \tag{6.563}$$

The particular solution would be a constant function, $x_p = X$, because the forcing term is constant $f(t) = F$. Substituting $x_p = X$ in Equation (6.558) provides us with $x_p$.

$$x_p = X = \frac{F}{m\omega_n^2} = \frac{F}{k} \tag{6.564}$$

Therefore, the general solution of Equation (6.558) is

$$\begin{aligned} x &= x_h + x_p \\ &= e^{-\xi\omega_n t}\left(A\sin\omega_d t + B\cos\omega_d t\right) + \frac{F}{m\omega_n^2} \qquad 0 \le t \end{aligned} \tag{6.565}$$

$$\omega_d = \omega_n\sqrt{1-\xi^2} \tag{6.566}$$

To determine the unknown coefficients of the homogenous solution, the initial conditions of the system must be applied after summation of the homogenous and particular solutions. The zero initial conditions are the best to explore the natural behavior of systems to step inputs. Applying a set of zero initial conditions provides us with two equations for $A$ and $B$

$$B + \frac{F}{m\omega_n^2} = 0 \qquad \omega_d A - \xi\omega_n B = 0 \tag{6.567}$$

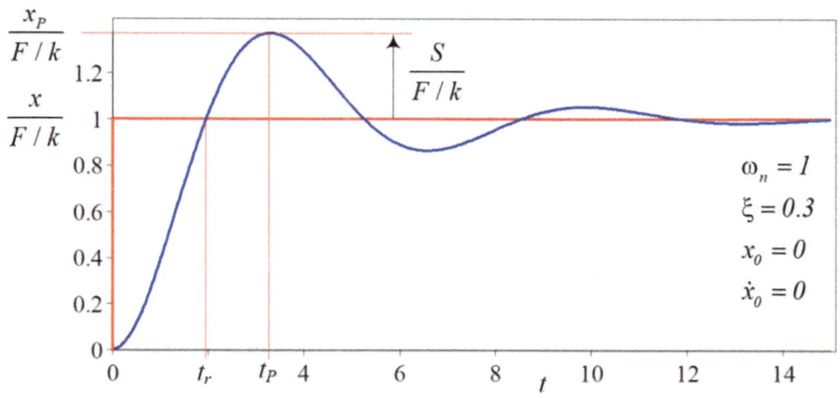


**Fig. 6.27** Response of a one $DOF$ vibrating system to a unit step input

with the following solutions:

$$A = -\frac{\xi F}{m\omega_d \omega_n} = -\frac{\xi \omega_n F}{\omega_d k} \qquad B = -\frac{F}{m\omega_n^2} = -\frac{F}{k} \tag{6.568}$$

Therefore, the step response is

$$\frac{x}{F/k} = 1 - e^{-\xi\omega_n t}\left(\frac{\xi\omega_n}{\omega_d}\sin\omega_d t + \cos\omega_d t\right) \tag{6.569}$$

or equivalently is

$$\frac{x}{F/k} = 1 - \frac{e^{-\xi\omega_n t}}{\sqrt{1-\xi^2}}\sin\left(\omega_d t + \arctan\frac{\sqrt{1-\xi^2}}{\xi}\right) \tag{6.570}$$

When $t \to \infty$, the homogenous equation approaches zero and the step response approaches $x/(F/k) \to 1$. Therefore, the response $x$ approaches the particular solution $x_p = F/k$, which is the static displacement of the system under the constant force $f = F$. Figure 6.27 depicts a step input response for the following numerical values:

$$\xi = 0.3 \qquad \omega_n = 1 \qquad F = 1 \tag{6.571}$$

Figure 6.28 illustrates the effect of the damping ratio $\xi$ on the step input response, and Fig. 6.29 illustrates the effect of the natural frequency $\omega_n$.

**Example 66** Rise time, peak time, overshoot. The characteristics of response of dynamic systems to step input. These characteristics are measurable, and an equivalent second-order dynamic system can be identified based on them.

There are some measurable characteristics for a step response that are being used to identify the system. They are: rise time $t_r$, peak time $t_P$, peak value $x_P$, overshoot $S = x_P - F/k$, and settling time $t_s$. The response of a second-order system with natural frequency $\omega_n$ and damping ratio $\xi$ to a step input with magnitude $F$ from zero initial conditions is

$$\frac{x}{F/k} = 1 - \frac{e^{-\xi\omega_n t}}{\sqrt{1-\xi^2}}\sin\left(\omega_d t + \arctan\frac{\sqrt{1-\xi^2}}{\xi}\right) \tag{6.572}$$

$$= 1 - e^{-\xi\omega_n t}\left(\frac{\xi}{\sqrt{1-\xi^2}}\sin\omega_d t + \cos\omega_d t\right) \tag{6.573}$$

A graphical illustration of this function is shown in Fig. 6.27.

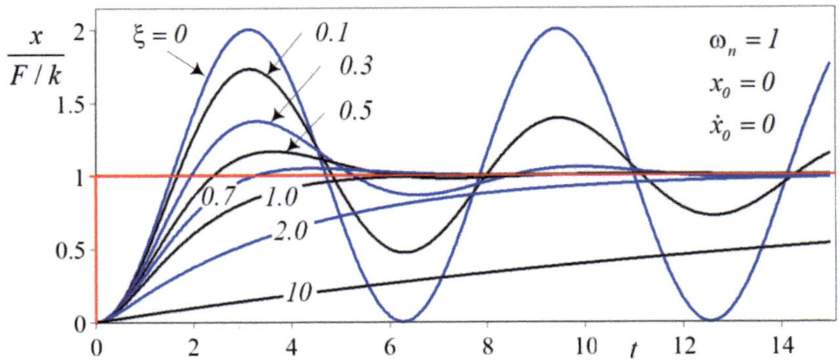


**Fig. 6.28** The effect of damping ratio $\xi$ on step input response

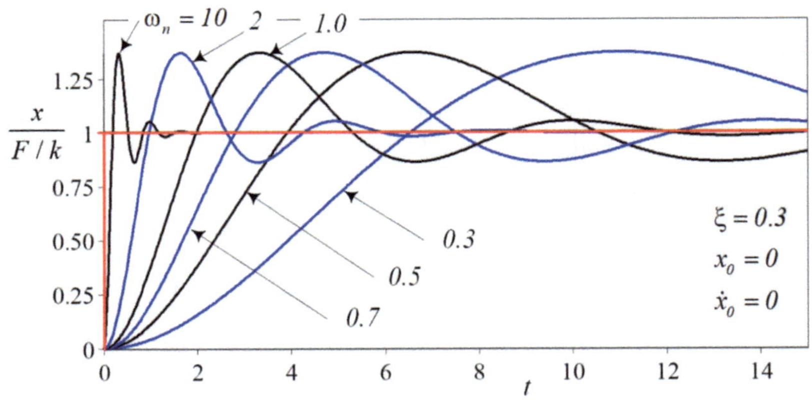


**Fig. 6.29** The effect of natural frequency $\omega_n$ on step input response

The rise time $t_r$ is the first time that the response $x(t)$ reaches the value of the steady-state response of the step input, $F/k$.

$$\frac{F}{k} = \frac{F}{k}\left(1 - e^{-\xi\omega_n t}\left(\frac{\xi}{\sqrt{1-\xi^2}}\sin\omega_d t + \cos\omega_d t\right)\right) \tag{6.574}$$

Because $e^{-\xi\omega_n t} \neq 0$, Equation (6.574) yields

$$\frac{\xi}{\sqrt{1-\xi^2}}\sin\omega_d t + \cos\omega_d t = 0 \tag{6.575}$$

or

$$\tan\omega_d t = -\frac{\sqrt{1-\xi^2}}{\xi} \qquad n = 0, 1, 2, 3, \cdots \tag{6.576}$$

Therefore, the rise time $t_r$ can be calculated.

$$t_r = \frac{1}{\omega_d}\left(\pi - \arctan\frac{\sqrt{1-\xi^2}}{\xi}\right) \tag{6.577}$$

There are also other definitions for rise time. It may also be defined as the inverse of the largest slope of the step response or as the time it takes to pass from 10% to 90% of the steady-state value. Such alternative definitions include the cases of critically or overdamped as well as underdamped systems. However, the definition of rise time as the first time that the response reaches the value of the steady-state value is more applied. In fact, a majority of dynamic systems are designed underdamped.

The peak time $t_P$ is the first time that the response $x(t)$ reaches its maximum value. The times at which $x(t)$ is maximum or minimum are the solutions of the equation $\dot{x} = 0$,

$$\frac{dx}{dt} = \frac{F}{k}\frac{\omega_n}{\sqrt{1-\xi^2}}e^{-\xi\omega_n t}\sin\omega_d t = 0 \tag{6.578}$$

which simplifies to

$$\sin\omega_d t = 0 \tag{6.579}$$

The time of the first maximum is the peak time $t_P$.

$$t_P = \frac{\pi}{\omega_d} \tag{6.580}$$

The value of $x(t)$ at $t_P$ is the peak value $x_P$.

$$\frac{x_P}{F/k} = 1 + e^{-\xi\omega_n\frac{\pi}{\omega_d}} = 1 + e^{-\xi\pi/\sqrt{1-\xi^2}} \tag{6.581}$$

The overshoot $S$ indicates how much the response $x(t)$ exceeds the steady -state response of the step input, $F/k$.

$$S = x_P - \frac{F}{k} = \frac{F}{k}e^{-\xi\pi/\sqrt{1-\xi^2}} \tag{6.582}$$

$$\frac{S}{F/k} = e^{-\xi\pi/\sqrt{1-\xi^2}} \tag{6.583}$$

The value of the overshoot $\frac{S}{F/k}$ is only a function of damping ratio $\xi$ and is always positive $0 < \frac{S}{F/k} < 1$ for underdamped system, $0 < \xi < 1$. It exponentially decreases from $\frac{S}{F/k} = 1$ to $\frac{S}{F/k} = 0$ when $\xi$ increases from zero to one.

$$\lim_{\xi\to 0}\frac{S}{F/k} = 1 \qquad \lim_{\xi\to 1}\frac{S}{F/k} = 0 \tag{6.584}$$

The settling time $t_s$ is four times of the time constant $\tau = 1/(\xi\omega_n)$ of the exponential function $e^{-\xi\omega_n t}$.

$$t_s = \frac{4}{\xi\omega_n} \tag{6.585}$$

There are other definitions for settling time as well. The settling time may also be defined as the required time that the step response $x(t)$ needs to settle within a $\pm n\%$ window of the steady-state value, $F/k$. The value $n = 2$ is commonly used.

$$t_s \approx \frac{\ln\left(n\sqrt{1-\xi^2}\right)}{\xi\omega_n} \tag{6.586}$$

As an example, for a set of sample data,

$$\xi = 0.3 \qquad \omega_n = 1 \qquad F = 1 \tag{6.587}$$

we find the following characteristic values:

$$t_r = 1.966 \tag{6.588}$$

$$t_P = 3.2933 \quad \frac{x_P}{F/k} = 1.3723 \quad \frac{S}{F/k} = 0.3723 \tag{6.589}$$

$$t_s = 13.333 \tag{6.590}$$

On the other hand, measuring only two of the four characteristics, $t_r, t_P, x_P, S, t_s$ would be enough to evaluate $\xi, \omega_n$.

**Example 67** Response to a polynomial and ramp input. Response of a vibrating system to a polynomial force function is an easy problem to solve. Ramp input is an important standard polynomial input to determine laziness of a system.

Let us examine the response of a second-order system to a second-degree polynomial force function.

$$\ddot{x} + 2\xi\omega_n\dot{x} + \omega_n^2 x = \frac{1}{m} f(t) \quad \xi < 1 \tag{6.591}$$

$$f(t) = a_0 + a_1 t + a_2 t^2 \tag{6.592}$$

$$[a_0] = MLT^{-2} \quad [a_1] = MLT^{-3} \quad [a_2] = MLT^{-4} \tag{6.593}$$

The dimensions of the coefficients of the forcing term are shown in Equation (6.593).

Substituting a trial polynomial particular solution $x_p$ into the equation,

$$2b_2 + 2\xi\omega_n(b_1 + 2tb_2) + \omega_n^2\left(b_0 + b_1 t + b_2 t^2\right) = \frac{a_0 + a_1 t + a_2 t^2}{m} \tag{6.594}$$

$$x_p = b_0 + b_1 t + b_2 t^2 \tag{6.595}$$

$$[b_0] = L \qquad [b_1] = LT^{-1} \qquad [b_2] = LT^{-2} \tag{6.596}$$

yields

$$\left(\omega_n^2 b_2 - \frac{a_2}{m}\right)t^2 + \left(b_1\omega_n^2 + 4\xi b_2\omega_n - \frac{a_1}{m}\right)t \tag{6.597}$$

$$+\left(b_0\omega_n^2 + 2\xi b_1\omega_n - \frac{a_0}{m} + 2b_2\right) = 0 \tag{6.598}$$

To have this equation at all times, all coefficients must be zero. Therefore, we will have three equations to determine the coefficients $b_0, b_1, b_2$ of the guessed solution.

$$\omega_n^2 b_2 - \frac{a_2}{m} = 0 \tag{6.599}$$

$$b_1\omega_n^2 + 4\xi b_2\omega_n - \frac{a_1}{m} = 0 \tag{6.600}$$

$$b_0\omega_n^2 + 2\xi b_1\omega_n - \frac{a_0}{m} + 2b_2 = 0 \tag{6.601}$$

The coefficients are

$$b_2 = \frac{a_2}{m\omega_n^2} \tag{6.602}$$

$$b_1 = \frac{1}{m\omega_n^3}(\omega_n a_1 - 4\xi a_2) \tag{6.603}$$

$$b_0 = -\frac{1}{m\omega_n^4}\left(2a_2 - 8\xi^2 a_2 - \omega_n^2 a_0 + 2\xi\omega_n a_1\right) \tag{6.604}$$

Therefore, the general solution of Equation (6.558) is

$$\begin{aligned} x &= x_h+x_p=e^{-\xi\omega_n t}\left(A\sin\omega_d t+B\cos\omega_d t\right)+b_0+b_1t+b_2t^2 \\ &= e^{-\xi\omega_n t}\left(A\sin\omega_d t+B\cos\omega_d t\right) \\ &-\frac{1}{m\omega_n^4}\left(2a_2-8\xi^2a_2-\omega_n^2a_0+2\xi\omega_n a_1\right) \\ &+\frac{1}{m\omega_n^3}\left(\omega_n a_1-4\xi a_2\right)t+\frac{a_2}{m\omega_n^2}t^2 \end{aligned} \tag{6.605}$$

$$\omega_d=\omega_n\sqrt{1-\xi^2} \tag{6.606}$$

If $a_1=0$, $a_2=0$, and $a_0=F$, the solution (6.605) simplifies to the step input response (6.565).

When $a_0=0$, $a_2=0$, the force is proportional to time $t$.

$$f(t)=a_1 t \tag{6.607}$$

Such a forcing term is called the ramp input, and a response to the force is called the ramp response. The ramp input analysis is another standard excitation to identify some characteristics of a dynamic system. The ramp response would be

$$x=x_h+x_p=e^{-\xi\omega_n t}\left(A\sin\omega_d t+B\cos\omega_d t\right)-\frac{2\xi a_1}{m\omega_n^3}+\frac{a_1}{m\omega_n^2}t \tag{6.608}$$

Applying a set of zero initial conditions

$$x(0)=0 \qquad \dot{x}(0)=0 \tag{6.609}$$

provides us with two equations to find the coefficients $A$ and $B$.

$$B-\frac{2\xi a_1}{m\omega_n^3}=0 \qquad \frac{a_1}{m\omega_n^2}-\xi\omega_n B+\omega_d A=0 \tag{6.610}$$

$$A=\frac{\left(2\xi^2-1\right)a_1}{m\omega_n^2\omega_d}=\frac{\left(2\xi^2-1\right)a_1}{k\omega_d} \qquad B=\frac{2\xi a_1}{m\omega_n^3}=\frac{2\xi a_1}{k\omega_n} \tag{6.611}$$

Therefore, the step response to zero initial conditions is calculated.

$$\frac{x\omega_n}{a_1/k}=e^{-\xi\omega_n t}\left(\frac{2\xi^2-1}{\sqrt{1-\xi^2}}\sin\omega_d t+2\xi\cos\omega_d t\right)-2\xi+\omega_n t \tag{6.612}$$

Figure 6.30 depicts a ramp input response for the following numerical values.

$$\xi=0.3 \qquad \omega_n=1 \tag{6.613}$$

The difference between $x\omega_n/(a_1/k)$ and $\omega_n t$ at $t\to\infty$ is called the steady-state error $e_{ss}$.

$$e_{ss}=\lim_{t\to\infty}\left(\frac{x\omega_n}{a_1/k}-\omega_n t\right)=-2\xi \tag{6.614}$$

The steady-state error has a constant value for ramp input, which means $x\omega_n/(a_1/k)$ follows $\omega_n t$ with a constant lag. Figure 6.31 illustrates the effect of the damping ratio $\xi$ on the step input response.

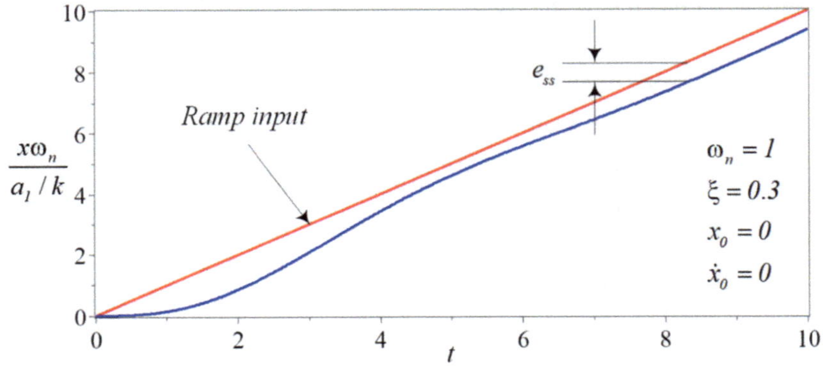


**Fig. 6.30** Response of a one $DOF$ vibrating system to a ramp force input

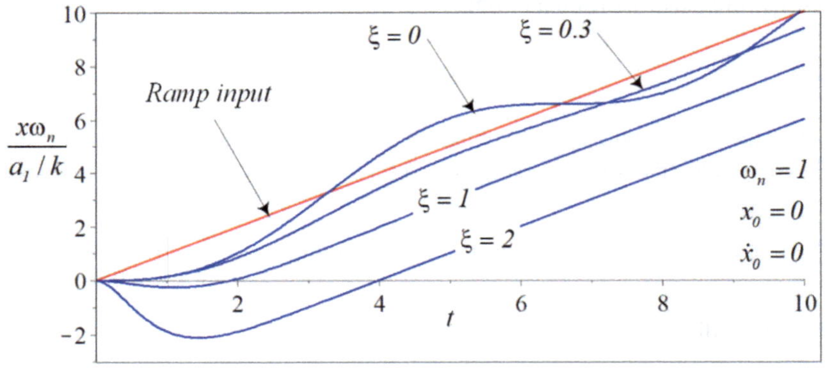


**Fig. 6.31** The effect of damping ratio $\xi$ on ramp input response

**Example 68** Response to a sinusoidal force. The particular solution $x_p$ of a vibrating system to a harmonic forcing term will be a harmonic displacement that must be added to the homogenous solution $x_h$ to make the general solution. The homogenous part is the transient part of the solution that will disappear after a while. Here we study that transient part before the steady-state harmonic response overcomes.

The steady-state response to a harmonic force is a subject of frequency response. However, the system goes through unsteady vibrations before showing its frequency response. Let us discover the unsteady part of the motion of the system. Consider a linear second-order system with the equation of motion

$$m\ddot{x} + c\dot{x} + kx = F\cos\omega t \tag{6.615}$$

$$\ddot{x} + 2\xi\omega_n\dot{x} + \omega_n^2 x = \frac{F}{m}\cos\omega t \qquad \xi < 1 \tag{6.616}$$

The trial particular solution must be a full harmonic function of the same frequency as the excitation with both $\sin\omega t$ and $\cos\omega t$ terms.

$$x_p = C_1\sin\omega t + C_2\cos\omega t \tag{6.617}$$

We substitute the trial solution (6.617) in Equation (6.616), and balancing the harmonic functions on both sides yields

$$C_1\omega_n^2 - 2\xi C_2\omega\omega_n - C_1\omega^2 = 0 \tag{6.618}$$

$$\omega_n^2 C_2 - \omega^2 C_2 - \frac{F}{m} + 2\xi\omega\omega_n C_1 = 0 \tag{6.619}$$

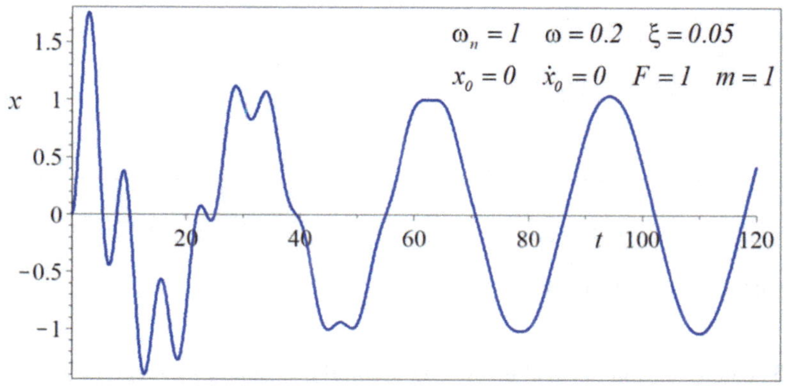


**Fig. 6.32** The transient response of a second-order system to a harmonic excitation

$$\begin{bmatrix} \omega_n^2 - \omega^2 & -2\xi\omega\omega_n \\ 2\xi\omega\omega_n & \omega_n^2 - \omega^2 \end{bmatrix} \begin{bmatrix} C_1 \\ C_2 \end{bmatrix} = \begin{bmatrix} 0 \\ \dfrac{F}{m} \end{bmatrix} \tag{6.620}$$

$$C_1 = \frac{F}{m} \frac{2\xi\omega\omega_n}{\left(\omega^2 - \omega_n^2\right)^2 + 4\xi^2\omega^2\omega_n^2} \tag{6.621}$$

$$C_2 = -\frac{F}{m} \frac{\omega^2 - \omega_n^2}{\left(\omega^2 - \omega_n^2\right)^2 + 4\xi^2\omega^2\omega_n^2} \tag{6.622}$$

Employing the homogenous solution $x_h$ from free vibration in Equation (6.12), the general solution of Equation (6.558) would be

$$x = x_h + x_p = e^{-\xi\omega_n t} \left(A \sin \omega_d t + B \cos \omega_d t\right) + C_1 \sin \omega t + C_2 \cos \omega t \tag{6.623}$$

$$\omega_d = \omega_n \sqrt{1 - \xi^2} \tag{6.624}$$

Assuming a set of zero initial conditions provides us with two equations to determine $A$ and $B$.

$$x(0) = 0 \qquad \dot{x}(0) = 0 \tag{6.625}$$

$$B + C_2 = 0 \qquad \omega C_1 + A\omega_d - B\xi\omega_n = 0 \tag{6.626}$$

$$A = \frac{\xi\omega_n C_2 - \omega C_1}{\omega_d} \qquad B = C_2 \tag{6.627}$$

Therefore, the response to a harmonic force from zero initial condition is as follows:

$$\begin{aligned} x = {} & e^{-\xi\omega_n t} \left( \frac{\xi\omega_n C_2 - \omega C_1}{\omega_d} \sin \omega_d t + C_2 \cos \omega_d t \right) \\ & + C_1 \sin \omega t + C_2 \cos \omega t \end{aligned} \tag{6.628}$$

When $t \to \infty$, the homogeneous solution disappears and the particular harmonic solution remains. Figure 6.32 illustrates the transient response of the system for the following numerical values.

$$\omega_n = 1 \qquad \omega = 0.2 \qquad \xi = 0.05 \qquad F = 1 \qquad m = 1 \tag{6.629}$$

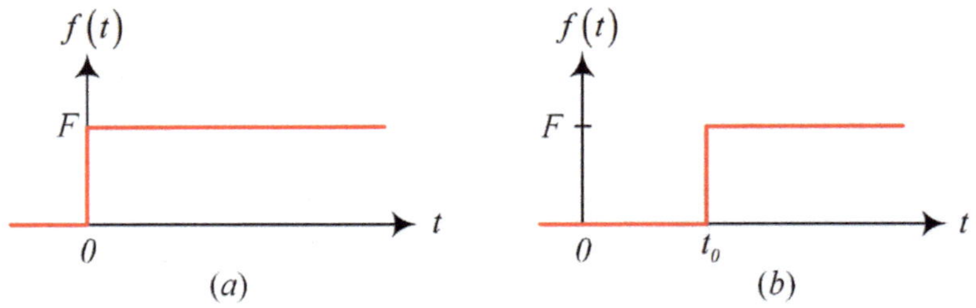


**Fig. 6.33** A step force function: (**a**) at $t = 0$, (**b**) at $t = t_0$

**Example 69** Step input. The convolution integral is a strong method of calculating response of vibrating system to transient forces. It gives us the flexibility to determine the solution to complicated force functions by superposition of solutions to simple forces. Here is to introduce the Heaviside function and show how the solution to a step input can be calculated and shifted in time.

Assume an $(m, k, c)$-system is under the step force input of Fig. 6.33a.

$$m\ddot{x} + c\dot{x} + kx = f(t) \qquad f(t) = F \tag{6.630}$$

A step input is a continuous function, and hence, the method of homogenous plus particular solutions, $x = x_h + x_p$, works well. That method has been used to determine the general solution (6.565), and

$$\begin{aligned} x &= x_h + x_p \\ &= e^{-\xi\omega_n t}(A \sin\omega_d t + B\cos\omega_d t) + \frac{F}{m\omega_n^2} \qquad 0 \le t \end{aligned} \tag{6.631}$$

$$\frac{x}{F/k} = 1 - e^{-\xi\omega_n t}\left(\frac{\xi\omega_n}{\omega_d}\sin\omega_d t + \cos\omega_d t\right) \tag{6.632}$$

$$\omega_d = \omega_n\sqrt{1-\xi^2} \tag{6.633}$$

To determine the response of the system, we may also use the convolution integral (6.555) and derive the same solution.

$$\begin{aligned} x(t) &= \frac{1}{m\omega_d}\int_0^t e^{-\xi\omega_n\tau} f(t-\tau)\sin\omega_d\tau \, d\tau \\ &= \frac{F}{m\omega_d}\int_0^t e^{-\xi\omega_n\tau}\sin\omega_d\tau \, d\tau \\ &= \frac{F}{k}\left(1 - e^{-\xi\omega_n t}\left(\frac{\xi}{\sqrt{1-\xi^2}}\sin\omega_d t + \cos\omega_d t\right)\right) \\ 0 &< t \end{aligned} \tag{6.634}$$

The response of the system to the lagged step input of Fig. 6.33b is the same as (6.634) with a shift $t_0$ in time and valid for $t_0 < t$.

$$\begin{aligned} x(t) &= \frac{1}{m\omega_d}\int_0^t e^{-\xi\omega_n(t-\tau)} f(\tau)\sin\omega_d(t-\tau)\, d\tau \\ &= \int_0^t 0\, d\tau + \frac{F}{m\omega_d}\int_{t_0}^t e^{-\xi\omega_n(t-\tau)}\sin\omega_d(t-\tau)\, d\tau \\ &= \frac{F/m}{\omega_d^2 + \xi^2\omega_n^2} - \frac{Fe^{-\xi\omega_n(t-t_0)}}{m\omega_d}\frac{\omega_d\cos\omega_d(t-t_0) + \xi\omega_n\sin\omega_d(t-t_0)}{\omega_d^2+\xi^2\omega_n^2} \end{aligned}$$

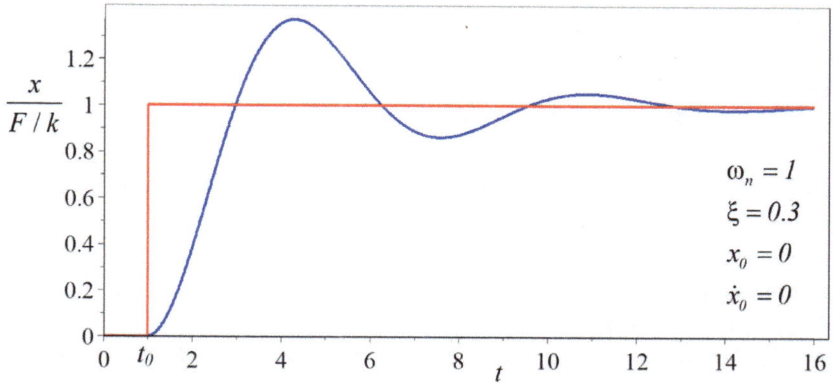


**Fig. 6.34** The response of a second-order system to a step input at $\tau = 1$

$$= \frac{F}{k}\left(1 - e^{-\xi\omega_n(t-t_0)}\left(\frac{\xi\omega_n}{\omega_d}\sin\omega_d(t-t_0) + \cos\omega_d(t-t_0)\right)\right) \tag{6.635}$$

$$t_0 < t$$

However, to express the response by a continuous function starting from $t = 0$, we employ the Heaviside function $H(t - t_0)$ to write the solution to the step input of Fig. 6.33b.

$$x(t) = \frac{F}{k}\left(1 - e^{-\xi\omega_n(t-t_0)}\left(\frac{\xi\sin\omega_d(t-t_0)}{\sqrt{1-\xi^2}} + \cos\omega_d(t-t_0)\right)\right)H(t-t_0) \tag{6.636}$$

$H(t - t_0)$ is the Heaviside function or the unit step function that acts as an on–off switch in mathematical equations.

$$H(t-t_0) = \begin{cases} 1 & t_0 \le t \\ 0 & t < t_0 \end{cases} \tag{6.637}$$

The response to a step input at $t_0 = 1$ is shown in Fig. 6.34.

The British mathematician and physicist Oliver Heaviside (1850–1925) introduced and applied the Heaviside function in telegraphic communications.

## 6.4 ★ Forced Vibrations Multi-$DOF$

Assume that a vibrating system with $n$ degrees of freedom is subjected to given force functions $\mathbf{f}(\mathbf{x}, \dot{\mathbf{x}}, t)$.

$$\mathbf{M}\,\ddot{\mathbf{x}} + \mathbf{C}\,\dot{\mathbf{x}} + \mathbf{K}\,\mathbf{x} = \mathbf{f}(\mathbf{x}, \dot{\mathbf{x}}, t) \tag{6.638}$$

$$\mathbf{x} = \begin{bmatrix} x_1 \; x_2 \; \cdots \; x_n \end{bmatrix}^T \tag{6.639}$$

The complete solution of equation (6.638) has two parts: a homogenous solution $\mathbf{x}_h$, plus a particular solution $\mathbf{x}_p$. The homogenous solution $\mathbf{x}_h$ is the solution of free vibrations for $\mathbf{f} = \mathbf{0}$, and the particular solution is the special solutions of the equations associated to the forcing function $\mathbf{f}(\mathbf{x}, \dot{\mathbf{x}}, t)$. The homogeneous solution depends on system parameters $\mathbf{M}$, $\mathbf{C}$, $\mathbf{K}$, while the particular solution is the response due to the particular form of the excitation $\mathbf{f}(\mathbf{x}, \dot{\mathbf{x}}, t)$.

The particular solution $\mathbf{x}_p$ of a forced vibrating system is not possible to be found for general force functions $\mathbf{f}(\mathbf{x}, \dot{\mathbf{x}}, t)$. However, if the force functions are continuous functions of time $\mathbf{f}(t)$ and are combinations of the following functions:

1. Constant, such as $\mathbf{f} = \mathbf{F}$
2. Polynomials in $t$, such as $\mathbf{f} = \mathbf{a}_0 + \mathbf{a}_1 t + \mathbf{a}_2 t^2 + \cdots + \mathbf{a}_n t^n$
3. Exponential functions, such as $\mathbf{f} = \mathbf{F} e^{at}$
4. Harmonic functions, such as $\mathbf{f} = \mathbf{F}_1 \sin \omega t + \mathbf{F}_2 \cos \omega t$

then, the particular solution $\mathbf{x}_p(t)$ of the linear equation of motion will have the same form as the forcing terms:

1. $\mathbf{x}_p(t) =$ constant, such as $\mathbf{x}_p(t) = \mathbf{X}$
2. $\mathbf{x}_p(t) =$ polynomial of the same degree, such as $\mathbf{x}_p(t) = \mathbf{C}_0 + \mathbf{C}_1 t + \mathbf{C}_2 t^2 + \cdots + \mathbf{C}_n t^n$
3. $\mathbf{x}_p(t) =$ exponential functions, such as $\mathbf{x}_p(t) = \mathbf{C} e^{at}$
4. $\mathbf{x}_p(t) =$ harmonic functions of the same frequency, such as $\mathbf{x}_p(t) = \mathbf{A} \sin \omega t + \mathbf{B} \cos \omega t$

The coefficients of particular solutions are calculated by substituting the solution in the differential equation and solving the resultant algebraic equations to satisfy the differential equations. Every element of the force function $\mathbf{f}$ can be different than the other. The linearity of the equations allows us to use the superposition principle and determine the whole particular solutions $\mathbf{x}_p$ by adding the particular solutions to every individual components of $\mathbf{f}$. The initial conditions will appear in the coefficients of the homogenous solution of the differential equation of motion. The initial conditions must be applied after the particular solution $\mathbf{x}_p$ and the general homogenous solution $\mathbf{x}_h$ are found and added together to have the general solution $\mathbf{x} = \mathbf{x}_h + \mathbf{x}_p$.

The best method to solve forced vibrations of multi-$DOF$ systems with general damping and general forcing functions is to convert the $n$ second -order differential equations (6.638) of motion into $2n$ first-order differential equations and develop the exponential solutions.

$$\dot{\mathbf{y}} + \mathbf{A}\,\mathbf{y} = \mathbf{F} \tag{6.640}$$

$$\mathbf{A} = {}^{+}\mathbf{M}^{-1}\,{}^{+}\mathbf{K} \tag{6.641}$$

$$\mathbf{F} = {}^{+}\mathbf{M}^{-1}\,{}^{+}\mathbf{f} \tag{6.642}$$

$$\mathbf{y} = \begin{bmatrix} \dot{\mathbf{x}} \\ \mathbf{x} \end{bmatrix} \tag{6.643}$$

$${}^{+}\mathbf{M} = \begin{bmatrix} \mathbf{0} & \mathbf{M} \\ \mathbf{M} & \mathbf{C} \end{bmatrix} \tag{6.644}$$

$${}^{+}\mathbf{K} = \begin{bmatrix} -\mathbf{M} & \mathbf{0} \\ \mathbf{0} & \mathbf{K} \end{bmatrix} \tag{6.645}$$

$${}^{+}\mathbf{f} = \begin{bmatrix} \mathbf{0} \\ \mathbf{f} \end{bmatrix} \tag{6.646}$$

The homogenous solution of Equation (6.640) has been developed in previous section on free vibrations of damped systems. The solution of any linear forced vibrating system of any order that can also be expressed by a set of $2n$ coupled first-order linear systems (6.640) is

$$\mathbf{y} = e^{-\mathbf{A}t} \int_0^t e^{\mathbf{A}s}\, \mathbf{F}(s)\; ds + e^{-\mathbf{A}t}\, \mathbf{y}_0 \tag{6.647}$$

$$\mathbf{y} = \begin{bmatrix} y_1 & y_2 & y_3 & \cdots & y_{2n} \end{bmatrix}^T \tag{6.648}$$

$$\mathbf{F} = \left[ F_1 \; F_2 \; F_3 \; \cdots \; F_{2n} \right]^T \tag{6.649}$$

***Proof*** Introducing a set of $2n$ new variables (6.643) will convert the $n$ second-order differential equations (6.638) into $2n$ first-order equations (6.640). The set of equations can always be expressed in matrix form.

$$\dot{\mathbf{y}} + \mathbf{A}\,\mathbf{y} = \mathbf{F} \tag{6.650}$$

$$\mathbf{y} = \left[ y_1 \; y_2 \; y_3 \; \cdots \; y_{2n} \right]^T \tag{6.651}$$

$$\dot{\mathbf{y}} = \left[ \dot{y}_1 \; \dot{y}_2 \; \dot{y}_3 \; \cdots \; \dot{y}_{2n} \right]^T \tag{6.652}$$

To solution of equation (6.650) is a combination of homogenous $\mathbf{x}_h$ and particular $\mathbf{x}_p$ solutions. Multiplying $e^{\mathbf{A}t}$ by both sides of Equation (6.650) yields

$$e^{\mathbf{A}t}\left(\dot{\mathbf{y}} + \mathbf{A}\,\mathbf{y}\right) = \frac{d}{dt}\left(e^{\mathbf{A}t}\,\mathbf{y}\right) = e^{\mathbf{A}t}\,\mathbf{F} \tag{6.653}$$

Integrating this equation yields

$$e^{\mathbf{A}t}\,\mathbf{y} = \int_0^t e^{\mathbf{A}s}\,\mathbf{F}\,ds + \mathbf{y}_0 \tag{6.654}$$

or

$$\mathbf{x} = e^{-\mathbf{A}t}\int_0^t e^{\mathbf{A}s}\,\mathbf{F}\,ds + e^{-\mathbf{A}t}\,\mathbf{y}_0 \tag{6.655}$$

If the forced system is an undamped linear $n$ degree of freedom ($DOF$) vibrating system

$$\mathbf{M}\,\ddot{\mathbf{x}} + \mathbf{K}\,\mathbf{x} = \mathbf{f} \tag{6.656}$$

$$\mathbf{x} = \left[ x_1 \; x_2 \; \cdots \; x_n \right]^T \tag{6.657}$$

then, employing the *principal coordinates* $\mathbf{p}$,

$$\mathbf{x} = \mathbf{U}\,\mathbf{p} \tag{6.658}$$

the equations of motion can be decoupled

$$\mathbf{M}'\,\ddot{\mathbf{p}} + \mathbf{K}'\,\mathbf{p} = \mathbf{F} \tag{6.659}$$

The new mass and stiffness matrices $\mathbf{M}'$, $\mathbf{K}'$, and forcing term $\mathbf{F}$ are diagonal.

$$\mathbf{M}' = \mathbf{U}^T\,\mathbf{M}\,\mathbf{U} = \begin{bmatrix} m_1' & 0 & \cdots & 0 \\ 0 & m_2' & \cdots & 0 \\ \vdots & \vdots & \ddots & \vdots \\ 0 & 0 & \cdots & m_n' \end{bmatrix} \tag{6.660}$$

$$\mathbf{K}' = \mathbf{U}^T\,\mathbf{K}\,\mathbf{U} = \begin{bmatrix} k_1' & 0 & \cdots & 0 \\ 0 & k_2' & \cdots & 0 \\ \vdots & \vdots & \ddots & \vdots \\ 0 & 0 & \cdots & k_n' \end{bmatrix} \tag{6.661}$$

$$\mathbf{F} = \mathbf{U}^T \mathbf{f} = \begin{bmatrix} F_1 \ F_2 \cdots F_n \end{bmatrix}^T \tag{6.662}$$

$$\mathbf{U} = \begin{bmatrix} \mathbf{u}_1 \ \mathbf{u}_2 \ \mathbf{u}_2 \cdots \mathbf{u}_n \end{bmatrix} \tag{6.663}$$

Therefore, the equations of motion are decoupled.

$$m_i' \ \ddot{p}_i + k_i' \ p_i = F_i \qquad i = 1, 2, 3, \cdots, n \tag{6.664}$$

The solution of the decoupled system of equations is addition of homogenous solution $p_h$ and particular solution $p_p$.

$$p_i = p_h + p_p \tag{6.665}$$

$$p_h = A_i \sin \omega_i t + B_i \sin \omega_i t \tag{6.666}$$

The particular solution $p_p$ depends on the force function $F_i$ and will be calculated by the techniques of one $DOF$ forced excitation in previous chapters.

The transformation matrix $\mathbf{U}$ is the *modal matrix* of the system. The columns of the $n \times n$ modal matrix $\mathbf{U}$ are the mode shapes of the system. The principal coordinates $\mathbf{p} = \mathbf{U}^T \mathbf{x}$ are also called the *natural coordinates* of the system. Undamped systems are the base to discover the principal characteristics of a system such as natural frequencies, mode shapes, modal matrix, and principal coordinates.

Having the solutions of the decoupled equations, $p_i(t)$, we can reconstruct the solution in the original coordinates $x_i$ by inverse modal transformation.

$$\mathbf{x} = \mathbf{U} \, \mathbf{p} \tag{6.667}$$

$$\begin{bmatrix} x_1(t) \\ x_2(t) \\ \cdots \\ x_n(t) \end{bmatrix} = \begin{bmatrix} \mathbf{u}_1 \ \mathbf{u}_2 \ \mathbf{u}_2 \cdots \mathbf{u}_n \end{bmatrix} \begin{bmatrix} p_1(t) \\ p_2(t) \\ \cdots \\ p_n(t) \end{bmatrix} \tag{6.668}$$

■

**Example 70** Converting forced system into first-order equations. There are other methods to convert a set of $n$ second-order equations into another set of $2n$ first -order equations. Here is one of them.

Consider a multi-$DOF$ forced linear vibrating system.

$$\mathbf{M} \, \ddot{\mathbf{x}} + \mathbf{C} \, \dot{\mathbf{x}} + \mathbf{K} \, \mathbf{x} = \mathbf{f}(\mathbf{x}, \dot{\mathbf{x}}, t) \tag{6.669}$$

Let us define a new $n$ component coordinate vector $\mathbf{z}$,

$$\dot{\mathbf{x}} = \mathbf{z} \tag{6.670}$$

and convert the set of second-order equations into a set of first-order equations.

$$\mathbf{M} \, \dot{\mathbf{z}} + \mathbf{C} \, \mathbf{z} + \mathbf{K} \, \mathbf{x} = \mathbf{f}(\mathbf{x}, \mathbf{z}, t) \tag{6.671}$$

$$\dot{\mathbf{z}} = \mathbf{M}^{-1} \, \mathbf{f} - \mathbf{M}^{-1} \, \mathbf{C} \, \mathbf{z} - \mathbf{M}^{-1} \, \mathbf{K} \, \mathbf{x} \tag{6.672}$$

Now we define a new $2n$ component coordinate vector $\mathbf{y}$,

$$\mathbf{y} = \begin{bmatrix} \mathbf{x} \\ \mathbf{z} \end{bmatrix} = \begin{bmatrix} \mathbf{x} \\ \dot{\mathbf{x}} \end{bmatrix} \tag{6.673}$$

and make $2n$ first-order equations for the original $n$ second-order equations,

$$\begin{bmatrix} \dot{\mathbf{x}} \\ \dot{\mathbf{z}} \end{bmatrix} = \begin{bmatrix} \mathbf{0} & \mathbf{I} \\ -\mathbf{M}^{-1}\,\mathbf{K} & -\mathbf{M}^{-1}\,\mathbf{C} \end{bmatrix} \begin{bmatrix} \mathbf{x} \\ \mathbf{z} \end{bmatrix} + \begin{bmatrix} \mathbf{0} \\ \mathbf{M}^{-1}\,\mathbf{f} \end{bmatrix} \tag{6.674}$$

or in a compact form,

$$\dot{\mathbf{y}} = \mathbf{A}\,\mathbf{y} + \mathbf{F} \tag{6.675}$$

where

$$\mathbf{A} = \begin{bmatrix} \mathbf{0} & \mathbf{I} \\ -\mathbf{M}^{-1}\,\mathbf{K} & -\mathbf{M}^{-1}\,\mathbf{C} \end{bmatrix} \qquad \mathbf{F} = \begin{bmatrix} \mathbf{0} \\ \mathbf{M}^{-1}\,\mathbf{f} \end{bmatrix} \tag{6.676}$$

**Example 71** General two degrees of freedom system with a harmonic force excitation. It is a practice on determining particular solution of a multi-$DOF$ system when one component of the forcing functions is a harmonic.

Consider a two $DOF$ vibrating system that its motion is expressed by two coupled differential equations. A harmonic forcing function is applied on only one of the coordinates.

$$m_1\ddot{x}_1 + c_{11}\dot{x}_1 + c_{12}\dot{x}_2 + k_{11}x_1 + k_{12}x_2 = F\sin\omega t \tag{6.677}$$

$$m_2\ddot{x}_2 + c_{21}\dot{x}_1 + c_{22}\dot{x}_2 + k_{21}x_1 + k_{22}x_2 = 0 \tag{6.678}$$

Let us only determine the particular solution here. Because the forcing function is harmonic, then the particular responses will be harmonic functions of the same frequency.

$$\mathbf{x}_p(t) = \mathbf{A}\sin\omega t + \mathbf{B}\cos\omega t \tag{6.679}$$

$$x_1 = A_1\sin\omega t + B_1\cos\omega t \tag{6.680}$$

$$x_2 = A_2\sin\omega t + B_2\cos\omega t \tag{6.681}$$

$$\dot{x}_1 = A_1\omega\cos\omega t - B_1\omega\sin\omega t \tag{6.682}$$

$$\dot{x}_2 = A_2\omega\cos\omega t - B_2\omega\sin\omega t \tag{6.683}$$

$$\ddot{x}_1 = -A_1\omega^2\sin\omega t - B_1\omega^2\cos\omega t \tag{6.684}$$

$$\ddot{x}_2 = -A_2\omega^2\sin\omega t - B_2\omega^2\cos\omega t \tag{6.685}$$

Substituting the solution function into the equations of motion provides us with two algebraic equations.

$$\begin{aligned}&\left(A_1k_{11}+A_2\left(k_{12}-m_1\omega^2\right)-\omega\left(B_1c_{11}+B_2c_{12}\right)-F\right)\sin\omega t\\&+\left(B_1k_{11}+B_2\left(k_{12}-m_1\omega^2\right)+\omega\left(A_1c_{11}+A_2c_{12}\right)\right)\cos\omega t=0\end{aligned} \tag{6.686}$$

$$\begin{aligned}&\left(A_1k_{21}+A_2\left(k_{22}-m_2\omega^2\right)-\omega\left(B_1c_{21}+B_2c_{22}\right)\right)\sin\omega t\\&+\left(B_1k_{21}+B_2\left(k_{22}-m_2\right)\omega^2+\omega\left(A_1c_{21}+\omega c_{22}\right)\right)\cos\omega t=0\end{aligned} \tag{6.687}$$

Because $\sin\omega t$ and $\cos\omega t$ are orthogonal functions, the only possible solution for these equations is having all coefficients of $\sin\omega t$ and $\cos\omega t$ to be zero. This will give us a set of algebraic equations to determine the coefficients **A** and **B**.

$$\begin{bmatrix} k_{11} & k_{12}-m_1\omega^2 & -\omega c_{11} & -c_{12}\omega \\ \omega c_{11} & \omega c_{12} & k_{11} & k_{12}-m_1\omega^2 \\ k_{21} & k_{22}-m_2\omega^2 & -\omega c_{21} & -\omega c_{22} \\ \omega c_{21} & \omega c_{22} & k_{21} & k_{22}-m_2\omega^2 \end{bmatrix} \begin{bmatrix} A_1 \\ A_2 \\ B_1 \\ B_2 \end{bmatrix} = \begin{bmatrix} F \\ 0 \\ 0 \\ 0 \end{bmatrix} \tag{6.688}$$

Determining $A_1$, $A_2$, $B_1$, $B_2$ and substituting them in (6.680) and (6.681) provide the particular solutions.

**Example 72** Impulse response. Impulse will only change the initial conditions, and hence, the response of a multi-$DOF$ system will be the transient response to a new set of initial conditions. Here is an example of a 2 $DOF$ undamped system.

Consider statically coupled two degrees of freedom system that an impulse $I$ acting at $m_1$ when the system is at rest.

$$m_1\ddot{x}_1 + k_{11}x_1 + k_{12}x_2 = f_1(t) \tag{6.689}$$

$$m_2\ddot{x}_2 + k_{21}x_1 + k_{22}x_2 = 0 \tag{6.690}$$

$$f_1(t) = I \tag{6.691}$$

The impulse will change the velocity of $m_1$ according to impulse–momentum principle.

$$I = m_1\dot{x}_1 \tag{6.692}$$

The impulse will change the rest initial conditions,

$$x_1(0) = 0 \qquad x_2(0) = 0 \qquad \dot{x}_1(0) = 0 \qquad \dot{x}_2(0) = 0 \tag{6.693}$$

to

$$x_1(0) = 0 \qquad x_2(0) = 0 \qquad \dot{x}_1(0) = \frac{I}{m_1} \qquad \dot{x}_2(0) = 0 \tag{6.694}$$

The solution of free vibration of this system is found as Equations (6.389)–(6.390).

$$x_1 = A_1\sin(\omega_1 t + \varphi_1) + A_2\sin(\omega_2 t + \varphi_2) \tag{6.695}$$

$$x_2 = u_1A_1\sin(\omega_1 t + \varphi_1) + u_2A_2\sin(\omega_2 t + \varphi_2) \tag{6.696}$$

Here we need to determine the unknown amplitudes $A_1$ and $A_2$ and phase angles $\varphi_1$ and $\varphi_2$ from the initial conditions (6.694).

$$A_1\sin\varphi_1 + A_2\sin\varphi_2 = 0 \tag{6.697}$$

$$u_1A_1\sin\varphi_1 + u_2A_2\sin\varphi_2 = 0 \tag{6.698}$$

$$A_1\omega_1\cos\varphi_1 - A_2\omega_2\cos\varphi_2 = \frac{I}{m_1} \tag{6.699}$$

$$u_1A_1\omega_1\cos\varphi_1 - u_2A_2\omega_2\cos\varphi_2 = 0 \tag{6.700}$$

Calculating $A_1$, $A_2$, $\varphi_1$, $\varphi_2$ competes the solutions,

$$x_1 = \frac{I\sin\omega_1 t}{m_1^2\omega_1(1 - u_1/u_2)} + \frac{I\sin\omega_2 t}{m_1^2\omega_2(1 - u_2/u_1)} \tag{6.701}$$

$$x_2 = \frac{u_1 I\sin\omega_1 t}{m_1^2\omega_1(1 - u_1/u_2)} + \frac{u_2 I\sin\omega_2 t}{m_1^2\omega_2(1 - u_2/u_1)} \tag{6.702}$$

where $\omega_1$, $\omega_2$, $u_1$, and $u_2$ are calculated in Equations (6.381) and (6.385).

Considering the force $f_1(t)$ to be divided into a large number of impulses, the response of the masses $m_1$ and $m_2$ will be

$$x_1 = \frac{1}{m_1(u_2 - u_1)}\int_0^t \left(\frac{u_2}{\omega_1}\sin\omega_1(t-\tau) - \frac{u_1}{\omega_2}\sin\omega_2(t-\tau)\right) f_1\,d\tau \tag{6.703}$$

$$x_2 = \frac{u_1u_2}{m_1(u_2 - u_1)}\int_0^t \left(\frac{1}{\omega_1}\sin\omega_1(t-\tau) - \frac{1}{\omega_2}\sin\omega_2(t-\tau)\right) f_1\,d\tau \tag{6.704}$$

The response of the system to forces $f_1(t)$ applied to mass $m_l$ and $f_2(t)$ applied to mass $m_2$, can be formed from Equations (6.703) and (6.704) by manipulation of subscripts and superposition.

The response of the system to a force $f_1(t)$ applied to mass $m_l$ can also be determined by using expression of the equations in principal coordinates. Introducing the change of coordinates of Equations (6.394)–(6.395),

$$x_1 = A_1 p_1 + A_2 p_2 \tag{6.705}$$

$$x_2 = u_1 A_1 p_1 + u_2 A_2 p_2 \tag{6.706}$$

into Equations (6.689)–(6.690), and following the analysis of equations (6.689) to (6.404) yield

$$\ddot{p}_1 + \omega_1^2 p_1 = A_1 f_1 \tag{6.707}$$

$$\ddot{p}_2 + \omega_2^2 p_2 = A_2 f_1 \tag{6.708}$$

where $A_1$ and $A_2$ satisfying the normalizing conditions (6.392)–(6.393). The solutions of equations (6.707)–(6.708) are given in terms of Duhamel's integral as Equation (6.557) for no force and $\xi = 0$.

$$x_1 = \frac{A_1}{\omega_1}\int_0^t f_1(\tau)\sin\omega_1(t-\tau)\, d\tau \tag{6.709}$$

$$x_2 = \frac{A_2}{\omega_2}\int_0^t f_1(\tau)\sin\omega_2(t-\tau)\, d\tau \tag{6.710}$$

Returning to the original coordinates, $x_1$ and $x_2$, the response of masses $m_1$ and $m_2$ will be

$$x_1 = \frac{A_1^2}{\omega_1}\int_0^t \sin\omega_1(t-\tau)\, f_1\, d\tau - \frac{A_2^2}{\omega_2}\int_0^t \sin\omega_2(t-\tau)\, f_1\, d\tau \tag{6.711}$$

$$x_2 = \frac{u_1 A_1^2}{\omega_1}\int_0^t \sin\omega_1(t-\tau)\, f_1\, d\tau - \frac{u_2 A_2^2}{\omega_2}\int_0^t \sin\omega_2(t-\tau)\, f_1\, d\tau \tag{6.712}$$

From (6.392) and (6.393), we have

$$A_1^2 = \frac{1}{m_1 + m_2 u_1^2} \qquad A_2^2 = \frac{1}{m_1 + m_2 u_2^2}, \tag{6.713}$$

and from the relation (6.386)

$$u_1 u_2 = -m_1 m_2 \tag{6.714}$$

we have

$$A_1^2 = \frac{1}{m_1(1 - u_1/u_2)} \qquad A_2^2 = \frac{1}{m_1(1 - u_2/u_1)} \tag{6.715}$$

and hence, the expressions (6.711) and (6.712) are equivalent to (6.703) and (6.704).

**Example 73** General two $DOF$ vibrating system. Study a general low-$DOF$ vibrating system will show all steps and features of analysis of forced multi-$DOF$ systems.

Figure 6.35 illustrates a two $DOF$ vibrating system. Its equations of motion are

$$\mathbf{M}\,\ddot{\mathbf{x}} + \mathbf{C}\,\dot{\mathbf{x}} + \mathbf{K}\,\mathbf{x} = \mathbf{f} \tag{6.716}$$

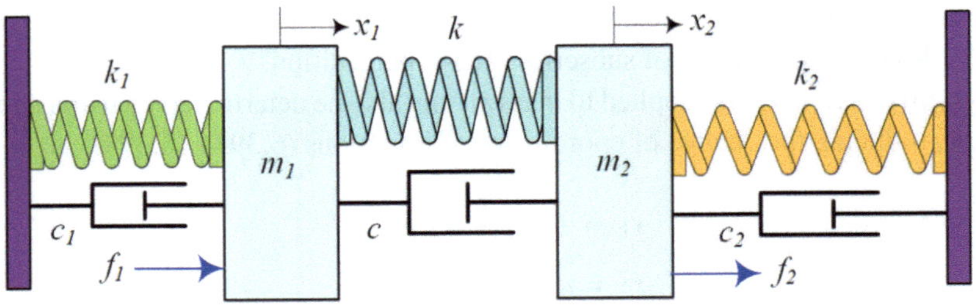


**Fig. 6.35** Two degrees of freedom forced system

$$\begin{bmatrix} m_1 & 0 \\ 0 & m_2 \end{bmatrix}\begin{bmatrix} \ddot{x}_1 \\ \ddot{x}_2 \end{bmatrix} + \begin{bmatrix} c+c_1 & -c \\ -c & c+c_2 \end{bmatrix}\begin{bmatrix} \dot{x}_1 \\ \dot{x}_2 \end{bmatrix} + \begin{bmatrix} k+k_1 & -k \\ -k & k+k_2 \end{bmatrix}\begin{bmatrix} x_1 \\ x_2 \end{bmatrix} = \begin{bmatrix} f_1 \\ f_2 \end{bmatrix} \tag{6.717}$$

Let us define a new set of variables $\mathbf{y}$,

$$\mathbf{y} = \begin{bmatrix} \dot{\mathbf{x}} \\ \mathbf{x} \end{bmatrix} \tag{6.718}$$

to convert the equations of motion into a set of first-order equations.

$$\dot{\mathbf{y}} + \mathbf{A}\,\mathbf{y} = \mathbf{F} \tag{6.719}$$

$$\mathbf{A} = {}^{+}\mathbf{M}^{-1}\,{}^{+}\mathbf{K} = \begin{bmatrix} \frac{c+c_1}{m_1} & -\frac{c}{m_1} & \frac{k+k_1}{m_1} & -\frac{k}{m_1} \\ -\frac{c}{m_2} & \frac{c+c_2}{m_2} & -\frac{k}{m_2} & \frac{k+k_2}{m_2} \\ -1 & 0 & 0 & 0 \\ 0 & -1 & 0 & 0 \end{bmatrix} \tag{6.720}$$

$$\mathbf{F} = {}^{+}\mathbf{M}^{-1}\,{}^{+}\mathbf{f} = \begin{bmatrix} \frac{f_1}{m_1} & \frac{f_2}{m_2} & 0 & 0 \end{bmatrix}^T \tag{6.721}$$

$${}^{+}\mathbf{M} = \begin{bmatrix} \mathbf{0} & \mathbf{M} \\ \mathbf{M} & \mathbf{C} \end{bmatrix} = \begin{bmatrix} 0 & 0 & m_1 & 0 \\ 0 & 0 & 0 & m_2 \\ m_1 & 0 & c+c_1 & -c \\ 0 & m_2 & -c & c+c_2 \end{bmatrix} \tag{6.722}$$

$${}^{+}\mathbf{K} = \begin{bmatrix} -\mathbf{M} & \mathbf{0} \\ \mathbf{0} & \mathbf{K} \end{bmatrix} = \begin{bmatrix} -m_1 & 0 & 0 & 0 \\ 0 & -m_2 & 0 & 0 \\ 0 & 0 & k+k_1 & -k \\ 0 & 0 & -k & k+k_2 \end{bmatrix} \tag{6.723}$$

$${}^{+}\mathbf{f} = \begin{bmatrix} \mathbf{0} \\ \mathbf{f} \end{bmatrix} \tag{6.724}$$

Zero initial conditions and a set of given numerical values for the parameters of the system and a set of force functions produce response shown in Fig. 6.36.

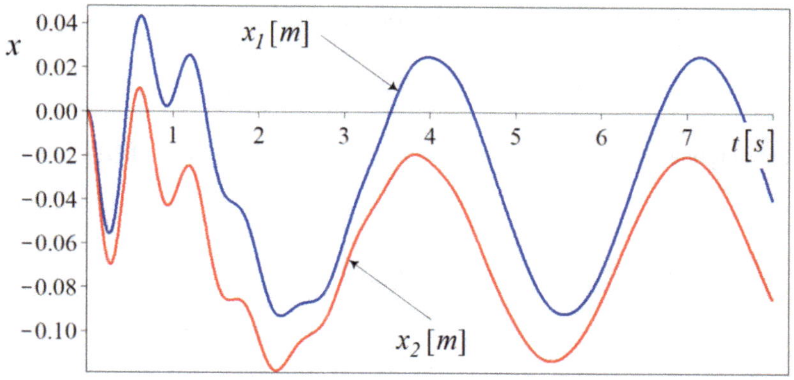


**Fig. 6.36** Forced vibration response of a two $DOF$ system

$$\mathbf{x}(0) = \mathbf{x}_0 = \begin{bmatrix} 0 \\ 0 \end{bmatrix} \qquad \dot{\mathbf{x}}(0) = \dot{\mathbf{x}}_0 = \begin{bmatrix} 0 \\ 0 \end{bmatrix} \tag{6.725}$$

$$m_1 = 1\,\text{kg} \qquad m_2 = 1\,\text{kg} \tag{6.726}$$

$$k_1 = 100\,\text{kg} \qquad k_2 = 100\,\text{kg} \qquad k = 100\,\text{kg} \tag{6.727}$$

$$c_1 = 100\,\text{kg} \qquad c_2 = 2\,\text{kg} \qquad c = 2\,\text{kg} \tag{6.728}$$

$$f_1 = 10 \sin 2t\,\text{N} \qquad f_2 = -10\,\text{N} \tag{6.729}$$

## 6.5 ★ Coupled Systems

Any linear free vibrating system of any order can be expressed by a set of coupled first-order linear differential equations.

$$\dot{\mathbf{x}} + [A]\,\mathbf{x} = \mathbf{0} \tag{6.730}$$

$$[A] = \begin{bmatrix} a_{11} & a_{12} & a_{13} & \cdots & a_{1n} \\ a_{21} & a_{22} & a_{23} & \cdots & a_{2n} \\ \vdots & \vdots & \vdots & \vdots & \vdots \\ a_{n1} & a_{n2} & a_{n3} & \cdots & a_{nn} \end{bmatrix} \tag{6.731}$$

The solution of a set of coupled first-order linear homogeneous equations is

$$\mathbf{x} = \sum_{j=1}^{n} e^{-\lambda_j t} C_j \mathbf{u}_j \tag{6.732}$$

where $\mathbf{u}_j$ are the eigenvectors and $\lambda_j$, $j = 1, 2, \cdots, n$ are eigenvalues of the $n \times n$ coefficient matrix $[A]$. This solution can also be expressed by

$$\mathbf{x}(t) = e^{-[A]t}\mathbf{C} = e^{-[A]t}\mathbf{x}_0 \tag{6.733}$$

where

$$\begin{aligned} e^{-[A]t} &= [I] - [A]\,t + \frac{1}{2!}[A]^2 t^2 - \frac{1}{3!}[A]^3 t^3 + \frac{1}{4!}[A]^4 t^4 - \cdots \\ &= \sum_{n=0}^{\infty} (-1)^n \frac{(\mathbf{A}t)^n}{n!} \end{aligned} \tag{6.734}$$

We may also express the solution of the homogeneous equations (6.730) as

$$\mathbf{x} = e^{-[A]t}\mathbf{C} = [U]\, e^{-[\Lambda]t}\, [U]^{-1}\, \mathbf{C} \tag{6.735}$$

where $[\Lambda]$ is the associated diagonalized matrix of $[A]$, and $[U]$ is the modal matrix of $[A]$.

$$[\Lambda] = [U]^{-1}\, [A]\, [U] = \begin{bmatrix} \lambda_1 & 0 & \cdots & 0 \\ 0 & \lambda_2 & \cdots & 0 \\ \vdots & \vdots & \ddots & \vdots \\ 0 & 0 & \cdots & \lambda_n \end{bmatrix} \tag{6.736}$$

$$[U] = \begin{bmatrix} \mathbf{u}_1 & \mathbf{u}_2 & \cdots & \mathbf{u}_n \end{bmatrix} \tag{6.737}$$

Any linear forced vibrating system of any order can also be expressed by a set of $n$ coupled first-order linear system,

$$\dot{\mathbf{x}} + [A]\,\mathbf{x} = \mathbf{f}\,(t) \tag{6.738}$$

where its solution $\mathbf{x} = \mathbf{x}(t)$ is

$$\mathbf{x} = e^{-[A]t} \int e^{[A]t}\, \mathbf{f}\,(t)\ dt + e^{-[A]t}\mathbf{C} \tag{6.739}$$

$$\mathbf{x} = \begin{bmatrix} x_1 & x_2 & x_3 & \cdots & x_n \end{bmatrix}^T \tag{6.740}$$

$$\mathbf{f} = \begin{bmatrix} f_1 & f_2 & f_3 & \cdots & f_n \end{bmatrix}^T \tag{6.741}$$

$$\mathbf{C} = \begin{bmatrix} C_1 & C_2 & C_3 & \cdots & C_n \end{bmatrix}^T \tag{6.742}$$

Using the modal matrix $[U]$ and its diagonalized matrix $[\Lambda]$, the solution of the forced equations (6.738) can also be expressed as below based on a modal transformation and principal coordinates $\mathbf{p}$,

$$\mathbf{x} = [U]\,\mathbf{p} = [U]\left[e^{-[\Lambda]t} \int e^{[\Lambda]t}\, \mathbf{P}\, dt + e^{-[\Lambda]t}\mathbf{C}\right] \tag{6.743}$$

where

$$\mathbf{p} = [U]^{-1}\, \mathbf{x} \tag{6.744}$$

$$\mathbf{P} = [U]^{-1}\, \mathbf{f}\,(t) \tag{6.745}$$

$$\mathbf{p} = e^{-[\Lambda]t} \int e^{[\Lambda]t}\, \mathbf{P}\, dt + e^{-[\Lambda]t}\mathbf{C} \tag{6.746}$$

***Proof*** Assume the equation of free motion of a vibrating system is expressed by an $n$-order linear differential equation.

$$a_n\,(t)\,\frac{d^n x}{dt^n} + a_{n-1}\,(t)\,\frac{d^{n-1} x}{dt^{n-1}} + \cdots + a_0\,(t)\, x = 0 \tag{6.747}$$

To convert the equation into a set of coupled first-order equations, we introduce $n$ new variables as below.

$$x_1 = x \qquad x_2 = \dot{x} \qquad \cdots \qquad x_n = \frac{d^n x}{dt^n} \tag{6.748}$$

Employing the new set of variables $x_1, x_2, \cdots, x_n$, we make the following set of first-order equations.

$$\dot{x}_1 = x_2 \qquad \dot{x}_2 = x_3 \qquad \cdots \qquad \dot{x}_{n-1} = x_n \tag{6.749}$$

$$\dot{x}_n = -\frac{a_{n-1}(t)\, x_n + a_{n-2}(t)\, x_{n-1} + \cdots + a_0(t)\, x_1}{a_n(t)} \tag{6.750}$$

The set of equations can always be expressed in matrix form.

$$\dot{\mathbf{x}} + [A]\,\mathbf{x} = \mathbf{0} \tag{6.751}$$

$$\mathbf{x} = \begin{bmatrix} x_1 \; x_2 \; x_3 \; \cdots \; x_n \end{bmatrix}^T \tag{6.752}$$

$$\dot{\mathbf{x}} = \begin{bmatrix} \dot{x}_1 \; \dot{x}_2 \; \dot{x}_3 \; \cdots \; \dot{x}_n \end{bmatrix}^T \tag{6.753}$$

$$[A] = \begin{bmatrix} 0 & -1 & 0 & \cdots & 0 \\ 0 & 0 & -1 & \cdots & 0 \\ \vdots & \vdots & \vdots & \vdots & \vdots \\ 0 & 0 & 0 & 0 & -1 \\ \frac{a_0}{a_n} & \frac{a_1}{a_n} & \frac{a_2}{a_n} & \cdots & \frac{a_{n-1}}{a_n} \end{bmatrix} \tag{6.754}$$

To solve the set of homogeneous equations (6.730), let us check a guessed solution,

$$\mathbf{x} = e^{-\lambda t}\mathbf{u} \tag{6.755}$$

where $\mathbf{u}$ is a constant $n \times 1$ vector, and $\lambda$ is a constant value.

$$-\lambda \mathbf{u} e^{-\lambda t} + [A]\,\mathbf{u} e^{-\lambda t} = \mathbf{0} \tag{6.756}$$

Eliminating $e^{-\lambda t}$,

$$[A]\,\mathbf{u} - \lambda \mathbf{u} = 0 \tag{6.757}$$

indicates that the function (6.755) satisfies Equation (6.730) if $\mathbf{u}$ is an eigenvector of $[A]$, and $\lambda$ is an eigenvalue of $[A]$, associated to $\mathbf{u}$. Because the differential equations are linear, the function $\mathbf{x}$ is also a solution.

$$\mathbf{x} = e^{-\lambda t} C\mathbf{u} \tag{6.758}$$

Suppose the $n \times n$ coefficient matrix $[A]$ has eigenvalues $\lambda_j$ and linearly independent eigenvectors $\mathbf{u}_j$, $j = 1, 2, \cdots, n$. Then each function $\mathbf{x}_j = e^{-\lambda_j t}\mathbf{u}_j$ is a solution of Equation (6.730), and the general solution is a linear combination of the $n$ solutions.

$$\mathbf{x} = \sum_{j=1}^{n} e^{-\lambda_j t} C_j \mathbf{u}_j \tag{6.759}$$

To prove Equation (6.733) is the right solution, assume that there is a solution for Equation (6.730) in the form of a time series.

$$\mathbf{x} = \mathbf{b}_0 + \mathbf{b}_1 t + \mathbf{b}_2 t^2 + \mathbf{b}_3 t^3 + \cdots \tag{6.760}$$

Substituting this solution in the equation yields

$$\mathbf{b}_1 + 2\mathbf{b}_2 t + 3\mathbf{b}_3 t^2 + \cdots = -[A]\left(\mathbf{b}_0 + \mathbf{b}_1 t + \mathbf{b}_2 t^2 + \mathbf{b}_3 t^3 + \cdots\right) \tag{6.761}$$

For this equation to hold at any time, we must have

$$\mathbf{b}_1 = -[A]\,\mathbf{b}_0 \tag{6.762}$$

$$\mathbf{b}_2 = -\frac{1}{2}[A]\,\mathbf{b}_1 = \frac{1}{2}[A]^2\,\mathbf{b}_0 \tag{6.763}$$

$$\mathbf{b}_3 = -\frac{1}{3}[A]\,\mathbf{b}_2 = -\frac{1}{3\times 2}[A]^3\,\mathbf{b}_0 = -\frac{1}{3!}[A]^3\mathbf{b}_0 \tag{6.764}$$

$$\vdots$$

The value of $\mathbf{b}_0$ is $\mathbf{x}$ at $t = 0$:

$$\mathbf{b}_0 = \mathbf{x}(0) = \mathbf{x}_0 \tag{6.765}$$

Therefore, we can write the solution as

$$\begin{aligned}\mathbf{x}(t) &= \left([I] - [A]\,t + \frac{1}{2!}[A]^2 t^2 - \frac{1}{3!}[A]^3 t^3 + \frac{1}{4!}[A]^4 t^4 - \cdots\right)\mathbf{x}_0 \\ &= e^{-[A]t}\mathbf{x}_0\end{aligned} \tag{6.766}$$

The matrix $[A]$ turns into a diagonal matrix $[\Lambda]$ when we use the eigenvectors properly. Suppose the $n$ by $n$ matrix $[A]$ has $n$ linearly independent eigenvectors $\mathbf{u}_1, \mathbf{u}_2, \cdots, \mathbf{u}_n$. Put them into columns of an eigenvector matrix $[U]$, and then $[U]^{-1}[A][U]$ will be a diagonal matrix $[\Lambda]$ including the eigenvalues of $[A]$.

$$[\Lambda] = [U]^{-1}[A][U] = \begin{bmatrix} \lambda_1 & 0 & \cdots & 0 \\ 0 & \lambda_2 & \cdots & 0 \\ \vdots & \vdots & \ddots & \vdots \\ 0 & 0 & \cdots & \lambda_n \end{bmatrix} \tag{6.767}$$

$$[A] = [U][\Lambda][U]^{-1} \tag{6.768}$$

To find $[\Lambda]$, we note that when $\mathbf{u}_j$ is an eigenvector, multiplication $[A]$ by $\mathbf{u}_j$ is equal to multiplication $\mathbf{u}_j$ by a number $\lambda_j$.

$$[A]\,\mathbf{u}_j = \lambda_j \mathbf{u}_j \tag{6.769}$$

Let us multiply $[A]$ by its eigenvectors, which are the columns of $[U]$. Then the column number $j$ of $[A][U]$ will be $[A]\,\mathbf{u}_j$, which is equal to $\lambda_j \mathbf{u}_j$. Therefore,

$$\begin{aligned}[A][U] &= [A]\begin{bmatrix}\mathbf{u}_1 & \mathbf{u}_2 & \cdots & \mathbf{u}_n\end{bmatrix} \\ &= \begin{bmatrix}\lambda_1\mathbf{u}_1 & \lambda_2\mathbf{u}_2 & \cdots & \lambda_n\mathbf{u}_n\end{bmatrix}\end{aligned} \tag{6.770}$$

The trick is to split this result into multiplication of $[U]$ and $[\Lambda]$.

$$\begin{aligned}&\begin{bmatrix}\lambda_1\mathbf{u}_1 & \lambda_2\mathbf{u}_2 & \cdots & \lambda_n\mathbf{u}_n\end{bmatrix} \\ &= \begin{bmatrix}\mathbf{u}_1 & \mathbf{u}_2 & \cdots & \mathbf{u}_n\end{bmatrix}\begin{bmatrix} \lambda_1 & 0 & \cdots & 0 \\ 0 & \lambda_2 & \cdots & 0 \\ \vdots & \vdots & \ddots & \vdots \\ 0 & 0 & \cdots & \lambda_n \end{bmatrix}\end{aligned} \tag{6.771}$$

$$= [U][\Lambda] \tag{6.772}$$

Hence, we have

$$[A][U] = [U][\Lambda] \tag{6.773}$$

$$[A] = [U][\Lambda][U]^{-1} \tag{6.774}$$

$$[\Lambda] = [U]^{-1}[A][U] \tag{6.775}$$

Knowing that

$$[A]^2 = [U][\Lambda][U]^{-1}[U][\Lambda][U]^{-1} = [U][\Lambda]^2[U]^{-1} \tag{6.776}$$

and

$$\begin{aligned} e^{[A]} &= [I] + [A]\,t + \frac{1}{2!}[A]^2 t^2 + \frac{1}{3!}[A]^3 t^3 + \frac{1}{4!}[A]^4 t^4 + \cdots \\ &= [I] + [U][\Lambda][U]^{-1} t + \frac{1}{2!}[U][\Lambda][U]^{-1}[U][\Lambda][U]^{-1} t^2 + \cdots \\ &= [U]\left[[I] + [\Lambda]\,t + \frac{1}{2!}[\Lambda]^2 t^2 + \frac{1}{3!}[\Lambda]^3 t^3 + \cdots\right][U]^{-1} \\ &= [U]\,e^{[\Lambda]}[U]^{-1} \end{aligned} \tag{6.777}$$

we have

$$[A]^k = [U][\Lambda]^k[U]^{-1} \qquad k \in \mathbb{N} \tag{6.778}$$

$$e^{[A]} = [U]\,e^{[\Lambda]}[U]^{-1} = e^{[\Lambda]} = \begin{bmatrix} e^{\lambda_1} & 0 & 0 & 0 \\ 0 & e^{\lambda_2} & 0 & 0 \\ 0 & 0 & \ddots & \vdots \\ 0 & 0 & \cdots & e^{\lambda_n} \end{bmatrix} \tag{6.779}$$

Therefore, the solution of Equation (6.730) is

$$\begin{aligned} \mathbf{x}(t) &= e^{-[A]t}\mathbf{x}_0 = [U]\,e^{-[\Lambda]t}[U]^{-1}\mathbf{x}_0 \\ &= e^{-\lambda_1 t}\mathbf{u}_1 + e^{-\lambda_2 t}\mathbf{u}_2 + \cdots + e^{-\lambda_n t}\mathbf{u}_n \end{aligned} \tag{6.780}$$

When the system of equations is forced as Equation (6.738),

$$\dot{\mathbf{x}} + [A]\,\mathbf{x} = \mathbf{f}(t) \tag{6.781}$$

then the solution of the equation will be a combination of homogenous $\mathbf{x}_h$ and particular $\mathbf{x}_p$ solutions. Multiplying $e^{-[A]t}$ by both sides of Equation (6.738) yields

$$e^{[A]t}\left(\dot{\mathbf{x}} + [A]\,\mathbf{x}\right) = \frac{d}{dt}\left(e^{[A]t}\mathbf{x}\right) = e^{[A]t}\,\mathbf{f}(t) \tag{6.782}$$

Integrating this equation yields

$$e^{[A]t}\mathbf{x} = \int e^{[A]t}\,\mathbf{f}(t)\,dt + \mathbf{C} \tag{6.783}$$

or

$$\mathbf{x} = e^{-[A]t} \int e^{[A]t}\, \mathbf{f}(t)\, dt + e^{-[A]t}\mathbf{C} \tag{6.784}$$

■

**Example 74** Solution of a set of two equations. It is to show how a second-order equation will be equal to two first-order equations and checking a set of solutions.

Consider a second-order equation.

$$\ddot{x} - 2 = 0 \tag{6.785}$$

an option is to select two axillary variables $x_1$ and $x_2$ to break the equation into two first-order equations and solve them individually.

$$\dot{x}_1 = 1 \qquad \dot{x}_2 = 2x_1 \tag{6.786}$$

$$x_1 = t + x_1(0) \qquad x_2 = t^2 + x_2(0) \tag{6.787}$$

As another method to break down the equation into a set of first-order equations, we may use different $x_1$ and $x_2$ and solve them.

$$\dot{x}_1 = x_2 \qquad \dot{x}_2 = 2 \tag{6.788}$$

$$x_2 = 2t + x_2(0) \tag{6.789}$$

$$\begin{aligned} x_1 &= \int x_2 dt = \int (2t + x_2(0))\, dt = t^2 + t\, x_2(0) + x_1(0) \\ &= t^2 + t\, \dot{x}_1(0) + x_1(0) \end{aligned} \tag{6.790}$$

A set of second-order systems is expressed as

$$\ddot{\mathbf{x}} + \mathbf{B}\,\dot{\mathbf{x}} + \mathbf{C}\,\mathbf{x} = \mathbf{0} \tag{6.791}$$

This set of equations can be transformed into a set of first-order equations as below.

$$\mathbf{y} = \begin{bmatrix} \mathbf{x} \\ \dot{\mathbf{x}} \end{bmatrix} \qquad \dot{\mathbf{y}} = \begin{bmatrix} \dot{\mathbf{x}} \\ \ddot{\mathbf{x}} \end{bmatrix} \tag{6.792}$$

$$\begin{bmatrix} \dot{\mathbf{x}} \\ \ddot{\mathbf{x}} \end{bmatrix} = \begin{bmatrix} \mathbf{0} & \mathbf{1} \\ -\mathbf{C} & -\mathbf{B} \end{bmatrix} \begin{bmatrix} \mathbf{x} \\ \dot{\mathbf{x}} \end{bmatrix} \tag{6.793}$$

$$\dot{\mathbf{y}} = \mathbf{A}\,\mathbf{y} \tag{6.794}$$

As a practical example, let us examine the vibrating system of Fig. 6.37. It shows two degrees of freedom linear vibrating system that its motion can be expressed by four first-order equations. Newton's second law yields the following coupled second-order ordinary differential equations that describe the motions of the masses.

$$\begin{aligned} &\begin{bmatrix} m_1 & 0 \\ 0 & m_2 \end{bmatrix} \begin{bmatrix} \ddot{y}_1 \\ \ddot{y}_2 \end{bmatrix} + \begin{bmatrix} c_1 + c & -c \\ -c & c_2 + c \end{bmatrix} \begin{bmatrix} \dot{y}_1 \\ \dot{y}_2 \end{bmatrix} \\ &+ \begin{bmatrix} k_1 + k & -k \\ -k & k_2 + k \end{bmatrix} \begin{bmatrix} y_1 \\ y_2 \end{bmatrix} = \begin{bmatrix} f_1 \\ f_2 \end{bmatrix} \end{aligned} \tag{6.795}$$

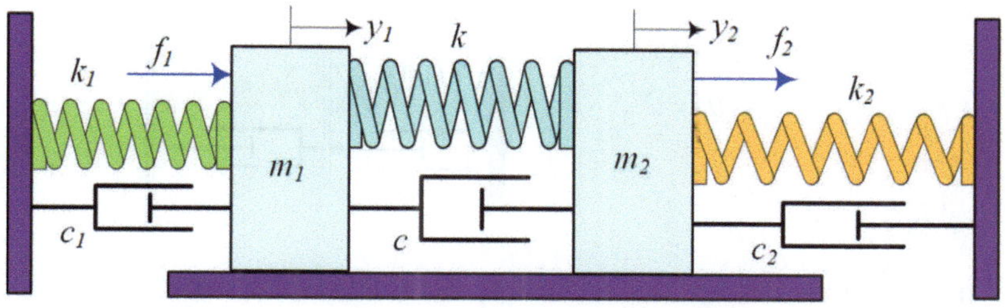


**Fig. 6.37** Two degrees of freedom linear system, expressed by four first-order equations

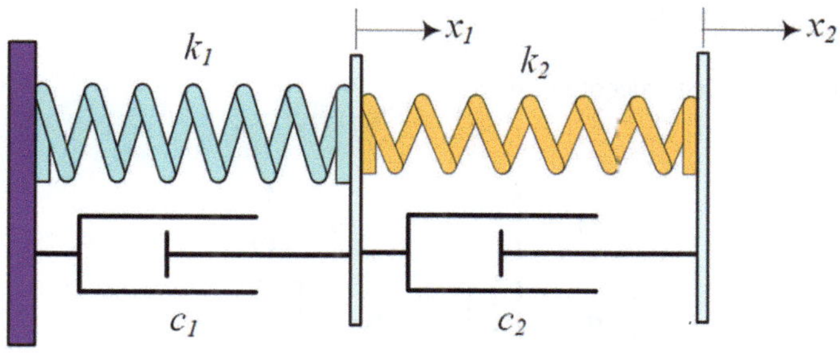


**Fig. 6.38** A series of two first-order systems

$$y_1(0) = 0 \quad y_2(0) = 0 \quad \dot{y}_1(0) = 0 \quad \dot{y}_2(0) = 0 \tag{6.796}$$

Introducing $x_1 = y_1$, $x_2 = \dot{y}_1$, $x_3 = y_2$, $x_4 = \dot{y}_2$, we can express the equations of motion equivalently by a system of first-order ordinary differential equations.

$$\begin{bmatrix} \dot{x}_1 \\ \dot{x}_2 \\ \dot{x}_3 \\ \dot{x}_4 \end{bmatrix} + \begin{bmatrix} 0 & 1 & 0 & 0 \\ \frac{k_1+k}{m_1} & \frac{c_1+c}{m_1} & -\frac{k}{m_1} & -\frac{c}{m_1} \\ 0 & 0 & 0 & 1 \\ -\frac{k}{m_2} & -\frac{c}{m_2} & \frac{k_2+k}{m_2} & \frac{c_2+c}{m_2} \end{bmatrix} \begin{bmatrix} x_1 \\ x_2 \\ x_3 \\ x_4 \end{bmatrix} = \begin{bmatrix} 0 \\ \frac{f_1}{m_1} \\ 0 \\ \frac{f_2}{m_2} \end{bmatrix} \tag{6.797}$$

$$x_1(0) = 0 \qquad x_2(0) = 0 \qquad x_3(0) = 0 \qquad x_4(0) = 0 \tag{6.798}$$

**Example 75** ★ A series of two first-order systems. Free response of connected first-order systems is still similar to a first-order system, without vibrations. As long as there is no mass and no periodic excitations, first-order systems will move from initial conditions to rest position exponentially.

Figure 6.38 illustrates a series of two first-order systems. The equations of motion for the coordinates $x_1$ and $x_2$ are

$$\begin{bmatrix} c_1+c_2 & -c_2 \\ -c_2 & c_2 \end{bmatrix} \begin{bmatrix} \dot{x}_1 \\ \dot{x}_2 \end{bmatrix} + \begin{bmatrix} k_1+k_2 & -k_2 \\ -k_2 & k_2 \end{bmatrix} \begin{bmatrix} x_1 \\ x_2 \end{bmatrix} = 0 \tag{6.799}$$

Multiplying $[c]^{-1}$, we can transform the equations to the standard form of (6.730).

$$\begin{bmatrix} \dot{x}_1 \\ \dot{x}_2 \end{bmatrix} + \begin{bmatrix} \frac{k_1}{c_1} & 0 \\ \frac{k_1}{c_1} - \frac{k_2}{c_2} & \frac{k_2}{c_2} \end{bmatrix} \begin{bmatrix} x_1 \\ x_2 \end{bmatrix} = 0 \tag{6.800}$$

The eigenvalues and eigenvectors of the coefficient matrix are

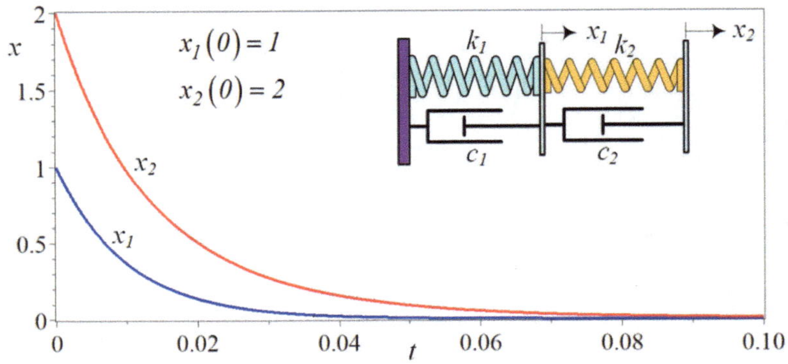


**Fig. 6.39** Time response of the series of two first order of Fig. 6.38

$$\lambda_1 = \frac{k_1}{c_1} \qquad \lambda_2 = \frac{k_2}{c_2} \tag{6.801}$$

$$\mathbf{u}_1 = \begin{bmatrix} 1 \\ 1 \end{bmatrix} \qquad \mathbf{u}_2 = \begin{bmatrix} 0 \\ 1 \end{bmatrix} \tag{6.802}$$

Therefore, the time response of the system is

$$\begin{bmatrix} x_1 \\ x_2 \end{bmatrix} = C_1 e^{-k_1 t/c_1} \begin{bmatrix} 1 \\ 1 \end{bmatrix} + C_2 e^{-k_2 t/c_2} \begin{bmatrix} 0 \\ 1 \end{bmatrix} \tag{6.803}$$

The system has two time constants $\tau_1$, $\tau_2$.

$$\tau_1 = \frac{1}{\lambda_1} = \frac{c_1}{k_1} \qquad \tau_2 = \frac{1}{\lambda_2} = \frac{c_2}{k_2} \tag{6.804}$$

If the initial conditions are

$$x_1(0) = x_{10} \qquad x_2(0) = x_{20} \tag{6.805}$$

then

$$C_1 = x_{10} \qquad C_2 = x_{20} - x_{10} \tag{6.806}$$

Let us assume

$$c_1 = c_2 = 1 \qquad k_1 = 100 \qquad k_2 = 50 \tag{6.807}$$

$$x_1(0) = 1 \qquad x_2(0) = 2 \tag{6.808}$$

and find the response

$$\begin{bmatrix} x_1 \\ x_2 \end{bmatrix} = e^{-100t} \begin{bmatrix} 1 \\ 1 \end{bmatrix} + e^{-50t} \begin{bmatrix} 0 \\ 1 \end{bmatrix} \tag{6.809}$$

Figure 6.39 illustrates the solutions for $t \leq 0.1$.

**Example 76** Homogeneous systems with constant coefficients. Investigating two parametric equations to show the steps toward solution.

Consider the following system of two differential equations with two unknowns.

$$\dot{x}_1 = a_{11}x_1 + a_{12}x_2 \qquad x_1(0) = x_{10} \tag{6.810}$$

$$\dot{x}_2 = a_{21}x_1 + a_{22}x_2 \qquad x_2(0) = x_{20} \tag{6.811}$$

$$\dot{\mathbf{x}} + [A]\,\mathbf{x} = \mathbf{0} \tag{6.812}$$

$$\begin{bmatrix} \dot{x}_1 \\ \dot{x}_2 \end{bmatrix} = \begin{bmatrix} a_{11} & a_{12} \\ a_{21} & a_{22} \end{bmatrix} \begin{bmatrix} x_1 \\ x_2 \end{bmatrix} \tag{6.813}$$

Assuming a set of solution of exponential functions yields

$$x_1 = C_1 e^{st} \qquad x_2 = C_2 e^{st} \tag{6.814}$$

$$C_1 s = a_{11}C_1 + a_{12}C_2 \tag{6.815}$$

$$C_2 s = a_{21}C_1 + a_{22}C_2 \tag{6.816}$$

$$(a_{11} - s)\,C_1 + a_{12}C_2 = 0 \tag{6.817}$$

$$a_{21}C_1 + (a_{22} - s)\,C_2 = 0 \tag{6.818}$$

These equations can be set in matrix form.

$$([A] - s\,[\mathbf{I}])\,\mathbf{C} = \mathbf{0} \tag{6.819}$$

$$\left( \begin{bmatrix} a_{11} - s & a_{12} \\ a_{21} & a_{22} - s \end{bmatrix} - s \begin{bmatrix} 1 & 0 \\ 0 & 1 \end{bmatrix} \right) \begin{bmatrix} C_1 \\ C_2 \end{bmatrix} = \begin{bmatrix} 0 \\ 0 \end{bmatrix} \tag{6.820}$$

Equation (6.819) has nontrivial solutions only if the determinant of the coefficient matrix is zero.

$$|[A] - s\,[\mathbf{I}]| = 0 \tag{6.821}$$

$$(a_{11} - s)\,(a_{22} - s) - a_{12}a_{21} = 0 \tag{6.822}$$

This is the characteristic equation of Equations (6.817) and (6.818), and its roots provide us with the two eigenvalues of $[A]$.

$$s_1, s_2 = \frac{(a_{11} + a_{22}) \pm \sqrt{(a_{11} + a_{22})^2 - 4\,(a_{11}a_{22} - a_{12}a_{21})}}{2} \tag{6.823}$$

The general solution of the differential equations is a linear combination of exponential functions.

$$x_1 = C_{11}e^{s_1 t} + C_{12}e^{s_2 t} \tag{6.824}$$

$$x_2 = C_{21}e^{s_1 t} + C_{22}e^{s_2 t} \tag{6.825}$$

The constant coefficients $C_{11}$, $C_{12}$, $C_{21}$, $C_{22}$, are related to the initial conditions.

$$x_{10} = C_{11} + C_{12} \qquad x_{20} = C_{21} + C_{22} \tag{6.826}$$

Equations (6.817) and (6.818) yield

$$\frac{C_1}{C_2} = \frac{-a_{12}}{a_{11} - s} = \frac{a_{22} - s}{-a_{21}} \tag{6.827}$$

$$(a_{11} - s)\,C_1 + a_{12}C_2 = 0 \tag{6.828}$$

$$a_{21}C_1 + (a_{22} - s)\,C_2 = 0 \tag{6.829}$$

**Example 77** ★Power series of matrix functions. Exponential function of a matrix can be determined by series. Here is the analytic calculations for $e^{\mathbf{A}t}$.

Similar to a power series expansion of a function of a scalar variable, $f(x)$,

$$f(x) = a_0 + a_1 x + a_2 x^2 + a_3 x^3 + \cdots \tag{6.830}$$

we may define power series expansion of a function of a square matrix, $f(\mathbf{A})$,

$$f(\mathbf{A}) = a_0 \mathbf{I} + a_1 \mathbf{A} + a_2 \mathbf{A}^2 + a_3 \mathbf{A}^3 + \cdots \tag{6.831}$$

where $\mathbf{I}$ is the unit matrix of the same order as $\mathbf{A}$. As an example, exponential function of a coefficient matrix $\mathbf{A}$ is

$$e^{\mathbf{A}t} = \mathbf{I} + \mathbf{A}t + \frac{1}{2}\mathbf{A}^2 t^2 + \frac{1}{3!}\mathbf{A}^3 t^3 + \frac{1}{4!}\mathbf{A}^4 t^4 + \cdots \tag{6.832}$$

Assume $\mathbf{A}$ is a square $n \times n$ matrix whose eigenvalues $\lambda_1, \lambda_2, \lambda_3, \cdots, \lambda_n$ are associated to its eigenvectors $\mathbf{u}_1, \mathbf{u}_2, \mathbf{u}_3, \cdots, \mathbf{u}_n$.

$$\mathbf{A}\mathbf{u}_i = \lambda_i \mathbf{u}_i \qquad i = 1, 2, 3, \cdots, n \tag{6.833}$$

$$\mathbf{u}_i = \begin{bmatrix} u_{1i} & u_{2i} & u_{3i} & \ldots & u_{ni} \end{bmatrix}^T \tag{6.834}$$

We may combine all equations of (6.833) and write them in form of an equation,

$$\mathbf{A}\mathbf{U} = \mathbf{\Lambda}\mathbf{U} \tag{6.835}$$

where $\mathbf{\Lambda}$ is a diagonal eigenvalue matrix and $\mathbf{U}$ is a square eigenvector matrix, assuming all eigenvalues are real and distinct,

$$\mathbf{\Lambda} = \begin{bmatrix} \lambda_1 & 0 & 0 & \ldots \\ 0 & \lambda_2 & 0 & \ldots \\ 0 & 0 & \lambda_3 & \ldots \\ \vdots & \vdots & \vdots & \ddots \end{bmatrix} = diag\left(\lambda_1, \lambda_2, \lambda_3, \cdots\right) \tag{6.836}$$

$$\mathbf{U} = \begin{bmatrix} \mathbf{u}_1 \\ \mathbf{u}_2 \\ \mathbf{u}_3 \\ \vdots \end{bmatrix}^T = \begin{bmatrix} u_{11} & u_{21} & u_{31} & \ldots \\ u_{12} & u_{22} & u_{32} & \ldots \\ u_{13} & u_{23} & u_{33} & \ldots \\ \vdots & \vdots & \vdots & \ddots \end{bmatrix} \tag{6.837}$$

and therefore,

$$\mathbf{U}^{-1}\mathbf{A}\mathbf{U} = \mathbf{\Lambda} \tag{6.838}$$

$$\mathbf{A} = \mathbf{U}\mathbf{\Lambda}\mathbf{U}^{-1} \tag{6.839}$$

Now we may have the exponents of $\mathbf{A}$,

$$\mathbf{A}^2 = \left[\mathbf{U}\mathbf{\Lambda}\mathbf{U}^{-1}\right] \cdot \left[\mathbf{U}\mathbf{\Lambda}\mathbf{U}^{-1}\right] = \mathbf{U}\mathbf{\Lambda}^2\mathbf{U}^{-1} \tag{6.840}$$

$$\mathbf{A}^3 = \mathbf{A}^2\mathbf{A} = \left[\mathbf{U}\mathbf{\Lambda}^2\mathbf{U}^{-1}\right] \cdot \left[\mathbf{U}\mathbf{\Lambda}\mathbf{U}^{-1}\right] = \mathbf{U}\mathbf{\Lambda}^3\mathbf{U}^{-1} \tag{6.841}$$

$$\ldots = \ldots$$

$$\mathbf{A}^n = \mathbf{A}^{n-1}\mathbf{A} = \left[\mathbf{U}\mathbf{\Lambda}^{n-1}\mathbf{U}^{-1}\right] \cdot \left[\mathbf{U}\mathbf{\Lambda}\mathbf{U}^{-1}\right] = \mathbf{U}\mathbf{\Lambda}^n\mathbf{U}^{-1} \tag{6.842}$$

where

$$\mathbf{A}^n = \begin{bmatrix} \lambda_1^n & 0 & 0 & \dots \\ 0 & \lambda_2^n & 0 & \dots \\ 0 & 0 & \lambda_3^n & \dots \\ \vdots & \vdots & \vdots & \ddots \end{bmatrix} \tag{6.843}$$

Hence, the power series of a matrix function (6.831) will be

$$\begin{aligned} f(\mathbf{A}) &= a_0\mathbf{I} + a_1\mathbf{A} + a_2\mathbf{A}^2 + a_3\mathbf{A}^3 + \cdots \\ &= a_0\mathbf{U}\mathbf{I}\mathbf{U}^{-1} + a_1\mathbf{U}\mathbf{\Lambda}\mathbf{U}^{-1} + a_2\mathbf{U}\mathbf{\Lambda}^2\mathbf{U}^{-1} \\ &\quad + a_3\mathbf{U}\mathbf{\Lambda}^3\mathbf{U}^{-1} + \cdots \\ &= \mathbf{U}\left[a_0\mathbf{I} + a_1\mathbf{\Lambda} + a_2\mathbf{\Lambda}^2 + a_3\mathbf{\Lambda}^3 + \cdots\right]\mathbf{U}^{-1} \end{aligned} \tag{6.844}$$

where

$$\begin{aligned} \left[a_0\mathbf{I} + a_1\mathbf{\Lambda} + a_2\mathbf{\Lambda}^2 + \cdots\right] = diag(&a_0 + a_1\lambda_1 + a_2\lambda_1^2 \cdots, \\ &a_0 + a_1\lambda_2 + a_2\lambda_2^2 \cdots, \dots \\ &a_0 + a_1\lambda_n + a_2\lambda_n^2 \cdots) \end{aligned} \tag{6.845}$$

As an example, let us consider

$$\mathbf{A} = \begin{bmatrix} 1 & 1 \\ 9 & 1 \end{bmatrix} \tag{6.846}$$

with eigenvalues and eigenvectors of

$$\lambda_1 = -2 \qquad \mathbf{u}_1 = \begin{bmatrix} -1/3 \\ 1 \end{bmatrix} \tag{6.847}$$

$$\lambda_2 = 4 \qquad \mathbf{u}_2 = \begin{bmatrix} 1/3 \\ 1 \end{bmatrix} \tag{6.848}$$

and therefore,

$$\mathbf{\Lambda} = \begin{bmatrix} -2 & 0 \\ 0 & 4 \end{bmatrix} \qquad \mathbf{U} = \begin{bmatrix} \mathbf{u}_1 & \mathbf{u}_2 \end{bmatrix} = \begin{bmatrix} -1/3 & 1/3 \\ 1 & 1 \end{bmatrix} \tag{6.849}$$

and we may check that $\mathbf{A} = \mathbf{U}\mathbf{\Lambda}\mathbf{U}^{-1}$.

$$\begin{aligned} \mathbf{A} = \mathbf{U}\mathbf{\Lambda}\mathbf{U}^{-1} &= \begin{bmatrix} -1/3 & 1/3 \\ 1 & 1 \end{bmatrix} \begin{bmatrix} -2 & 0 \\ 0 & 4 \end{bmatrix} \begin{bmatrix} -1/3 & 1/3 \\ 1 & 1 \end{bmatrix}^{-1} \\ &= \begin{bmatrix} 1 & 1 \\ 9 & 1 \end{bmatrix} \end{aligned} \tag{6.850}$$

Now we may calculate the exponents of **A**,

$$\mathbf{A}^2 = \mathbf{U}\mathbf{\Lambda}^2\mathbf{U}^{-1} = \begin{bmatrix} 10 & 2 \\ 18 & 10 \end{bmatrix} \tag{6.851}$$

$$\mathbf{A}^3 = \mathbf{U}\Lambda^3\mathbf{U}^{-1} = \begin{bmatrix} 28 & 12 \\ 108 & 28 \end{bmatrix} \tag{6.852}$$

$$\cdots$$

to have $e^{\mathbf{A}t}$.

$$\begin{aligned} e^{\mathbf{A}t} &= \mathbf{I} + \mathbf{A}t + \frac{1}{2}\mathbf{A}^2t^2 + \frac{1}{3!}\mathbf{A}^3t^3 + \cdots \\ &= \begin{bmatrix} 1 & 0 \\ 0 & 1 \end{bmatrix} + \begin{bmatrix} 1 & 1 \\ 9 & 1 \end{bmatrix} t + \frac{1}{2}\begin{bmatrix} 10 & 2 \\ 18 & 10 \end{bmatrix} t^2 + \cdots \\ &\simeq \begin{bmatrix} \frac{14}{3}t^3 + 5t^2 + t + 1 & 2t^3 + t^2 + t \\ 18t^3 + 9t^2 + 9t & \frac{14}{3}t^3 + 5t^2 + t + 1 \end{bmatrix} \end{aligned} \tag{6.853}$$

The exact value of $e^{\mathbf{A}t}$ is

$$e^{\mathbf{A}t} = \frac{1}{6}\begin{bmatrix} 3e^{-2t} + 3e^{4t} & e^{4t} - e^{-2t} \\ 9e^{4t} - 9e^{-2t} & 3e^{-2t} + 3e^{4t} \end{bmatrix} \tag{6.854}$$

Knowing that if we have a diagonal matrix, then the exponential function of the matrix would be

$$\exp\begin{bmatrix} a & 0 \\ 0 & b \end{bmatrix} = \begin{bmatrix} e^a & 0 \\ 0 & e^b \end{bmatrix} \tag{6.855}$$

We may also transform a given matrix $\mathbf{A}$ to its Jordan form

$$\begin{bmatrix} 1 & 1 \\ 9 & 1 \end{bmatrix} = \begin{bmatrix} 1 & 1 \\ -3 & 3 \end{bmatrix}\begin{bmatrix} -2 & 0 \\ 0 & 4 \end{bmatrix}\begin{bmatrix} 1 & 1 \\ -3 & 3 \end{bmatrix}^{-1} \tag{6.856}$$

to calculate the exact $e^{\mathbf{A}t}$.

$$\begin{aligned} \exp\begin{bmatrix} 1 & 1 \\ 9 & 1 \end{bmatrix} t &= \begin{bmatrix} 1 & 1 \\ -3 & 3 \end{bmatrix} \exp\begin{bmatrix} -2t & 0 \\ 0 & 4t \end{bmatrix}\begin{bmatrix} 1 & 1 \\ -3 & 3 \end{bmatrix}^{-1} \\ &= \begin{bmatrix} 1 & 1 \\ -3 & 3 \end{bmatrix}\begin{bmatrix} e^{-2t} & 0 \\ 0 & e^{4t} \end{bmatrix}\begin{bmatrix} 1 & 1 \\ -3 & 3 \end{bmatrix}^{-1} \\ &= \frac{1}{6}\begin{bmatrix} 3e^{-2t} + 3e^{4t} & e^{4t} - e^{-2t} \\ 9e^{4t} - 9e^{-2t} & 3e^{-2t} + 3e^{4t} \end{bmatrix} \end{aligned} \tag{6.857}$$

**Example 78** Solving free vibrations of a one $DOF$ second-order system as coupled first-order systems.

Consider a second-order vibrating system.

$$\ddot{x} + 2\xi\omega_n\dot{x} + \omega_n^2 x = 0 \tag{6.858}$$

$$x(0) = x_0 \qquad \dot{x}(0) = \dot{x}_0 \tag{6.859}$$

Let us rewrite the equation by two first-order equations.

$$x_1 = x \qquad x_2 = \dot{x} \tag{6.860}$$

$$\dot{x}_1 = x_2 \tag{6.861}$$

$$\dot{x}_2 = -2\xi\omega_n x_2 - \omega_n^2 x_1 \tag{6.862}$$

This system can be written in matrix form.

$$\begin{bmatrix} \dot{x}_1 \\ \dot{x}_2 \end{bmatrix} = \mathbf{A} \begin{bmatrix} x_1 \\ x_2 \end{bmatrix} \tag{6.863}$$

$$\mathbf{A} = \begin{bmatrix} 0 & 1 \\ -\omega_n^2 & -2\xi\omega_n \end{bmatrix} \tag{6.864}$$

The coefficient matrix of the system has the following eigenvalues and eigenvectors.

$$\lambda_1 = -\xi\omega_n + i\sqrt{1-\xi^2}\omega_n \quad \mathbf{u}_1 = \begin{bmatrix} -\frac{1}{\omega_n}\left(\xi + i\sqrt{1-\xi^2}\right) \\ 1 \end{bmatrix} \tag{6.865}$$

$$\lambda_2 = -\xi\omega_n - i\sqrt{1-\xi^2}\omega_n \quad \mathbf{u}_2 = \begin{bmatrix} -\frac{1}{\omega_n}\left(\xi - i\sqrt{1-\xi^2}\right) \\ 1 \end{bmatrix} \tag{6.866}$$

Therefore, diagonal eigenvalue matrix, $\mathbf{\Lambda}$, and the square eigenvector matrix, $\mathbf{U}$, are

$$\mathbf{\Lambda} = \begin{bmatrix} -\xi\omega_n + i\sqrt{1-\xi^2}\omega_n & 0 \\ 0 & -\xi\omega_n - i\sqrt{1-\xi^2}\omega_n \end{bmatrix} \tag{6.867}$$

$$\mathbf{U} = \begin{bmatrix} -\frac{1}{\omega_n}\left(\xi + i\sqrt{1-\xi^2}\right) & -\frac{1}{\omega_n}\left(\xi - i\sqrt{1-\xi^2}\right) \\ 1 & 1 \end{bmatrix} \tag{6.868}$$

The first four exponents of $\mathbf{A}$ will make a good approximate solution for $e^{\mathbf{A}t}$.

$$\mathbf{A} = \mathbf{U\Lambda U}^{-1} = \begin{bmatrix} 0 & 1 \\ -\omega_n^2 & -2\xi\omega_n \end{bmatrix} \tag{6.869}$$

$$\mathbf{A}^2 = \mathbf{U\Lambda}^2\mathbf{U}^{-1} = \begin{bmatrix} -\omega_n^2 & -2\xi\omega_n \\ 2\xi\omega_n^3 & \omega_n^2\left(4\xi^2-1\right) \end{bmatrix} \tag{6.870}$$

$$\mathbf{A}^3 = \mathbf{U\Lambda}^3\mathbf{U}^{-1} = \begin{bmatrix} 2\xi\omega_n^3 & \omega_n^2\left(4\xi^2-1\right) \\ -\omega_n^4\left(4\xi^2-1\right) & -4\xi\omega_n^3\left(2\xi^2-1\right) \end{bmatrix} \tag{6.871}$$

$$\mathbf{A}^4 = \mathbf{U\Lambda}^4\mathbf{U}^{-1} = \begin{bmatrix} -\omega_n^4\left(4\xi^2-1\right) & -4\xi\omega_n^3\left(2\xi^2-1\right) \\ 4\xi\omega_n^5\left(2\xi^2-1\right) & \omega_n^4\left(16\xi^4-12\xi^2+1\right) \end{bmatrix} \tag{6.872}$$

$$\cdots$$

$$e^{\mathbf{A}t} = \mathbf{I} + \mathbf{A}t + \frac{\mathbf{A}^2t^2}{2} + \frac{\mathbf{A}^3t^3}{3!} + \frac{\mathbf{A}^4t^4}{4!} + \cdots \simeq \begin{bmatrix} r_{11} & r_{12} \\ r_{21} & r_{22} \end{bmatrix} \tag{6.873}$$

$$r_{11} = \left(\frac{1}{24} - \frac{1}{6}\xi^2\right)\omega_n^4t^4 + \frac{1}{3}\xi\omega_n^3t^3 - \frac{1}{2}\omega_n^2t^2 + 1 \tag{6.874}$$

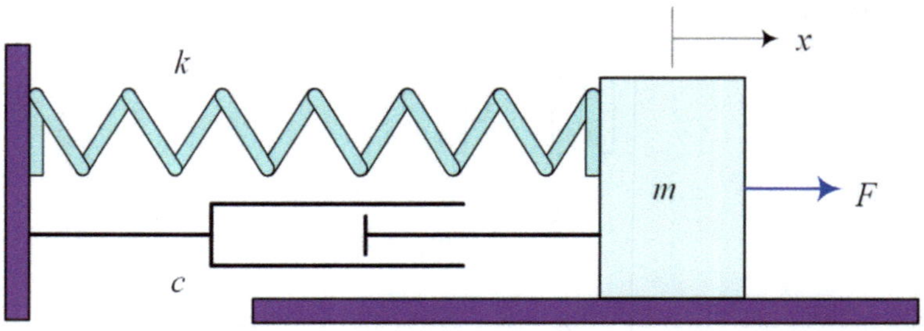


**Fig. 6.40** A constant force on a mass–spring–damper system

$$r_{21} = \frac{1}{6}\left(\left(2\xi^2-1\right)\xi\omega_n^3 t^3+\left(1-4\xi^2\right)\omega_n^2 t^2+6\xi\omega_n t-6\right)\omega_n^2 t \tag{6.875}$$

$$r_{12} = -\frac{1}{6}\left(\left(2\xi^2-1\right)\xi\omega_n^3 t^3+\left(1-4\xi^2\right)\omega_n^2 t^2+6\xi\omega_n t-6\right)t \tag{6.876}$$

$$r_{22} = \left(\frac{2}{3}\xi^4 - \frac{1}{2}\xi^2 + \frac{1}{24}\right)\omega_n^4 t^4 + \left(\frac{2}{3} - \frac{4}{3}\xi^2\right)\xi\omega_n^3 t^3 + \left(2\xi^2 - \frac{1}{2}\right)\omega_n^2 t^2 - 2\xi\omega_n t + 1 \tag{6.877}$$

Alternatively, we can find the Jordan form of $\mathbf{A}$ to its Jordan form to calculate the exact $e^{\mathbf{A}t}$.

$$\mathbf{A} = \begin{bmatrix} 1 & 1 \\ \lambda_{2n} & \lambda_1 \end{bmatrix}\begin{bmatrix} \lambda_2 & 0 \\ 0 & \lambda_1 \end{bmatrix}\begin{bmatrix} 1 & 1 \\ \lambda_{2n} & \lambda_1 \end{bmatrix}^{-1} \tag{6.878}$$

$$\lambda_1 = -\xi\omega_n + i\sqrt{1-\xi^2}\omega_n \qquad \lambda_2 = -\xi\omega_n - i\sqrt{1-\xi^2}\omega_n \tag{6.879}$$

$$\begin{aligned} e^{\mathbf{A}t} &= \begin{bmatrix} 1 & 1 \\ \lambda_2 & \lambda_1 \end{bmatrix}\exp\begin{bmatrix} \lambda_2 t & 0 \\ 0 & \lambda_1 t \end{bmatrix}\begin{bmatrix} 1 & 1 \\ \lambda_2 & \lambda_1 \end{bmatrix}^{-1} \\ &= \begin{bmatrix} 1 & 1 \\ \lambda_2 & \lambda_1 \end{bmatrix}\begin{bmatrix} e^{\lambda_2 t} & 0 \\ 0 & e^{\lambda_1 t} \end{bmatrix}\begin{bmatrix} 1 & 1 \\ \lambda_2 & \lambda_1 \end{bmatrix}^{-1} \\ &= \begin{bmatrix} \dfrac{\lambda_1 e^{\lambda_2 t} - \lambda_2 e^{\lambda_1 t}}{\lambda_1 - \lambda_2} & \dfrac{e^{\lambda_1 t} - e^{\lambda_2 t}}{\lambda_1 - \lambda_2} \\ -\dfrac{e^{\lambda_1 t} - e^{\lambda_2 t}}{\lambda_1 - \lambda_2}\lambda_1\lambda_2 & \dfrac{\lambda_1 e^{\lambda_1 t} - \lambda_2 e^{\lambda_2 t}}{\lambda_1 - \lambda_2} \end{bmatrix} \end{aligned} \tag{6.880}$$

**Example 79** A step force on a first-order system. This is an example of analyzing the forced vibrations of a second-order system as coupled first-order systems.

Consider the system of Fig. 6.40.

$$m\ddot{x} + c\dot{x} + kx = F \tag{6.881}$$

$$x_1(0) = 0 \qquad x_2(0) = 0 \tag{6.882}$$

Introducing two new coordinators $x_1, x_2$, and utilizing the following numerical values,

$$x_1 = x \qquad x_2 = \dot{x} \tag{6.883}$$

$$m = 1 \qquad c = 3 \qquad k = 2 \qquad F = 1 \tag{6.884}$$

we rewrite the equations of motion as a set of first-order equations in terms of $x_1(t)$ and $x_2(t)$.

$$\dot{x}_1 - x_2 = 0 \qquad \dot{x}_2 + 2x_1 + 3x_2 = 1 \tag{6.885}$$

$$\dot{\mathbf{x}} + [A]\,\mathbf{x} = \mathbf{f}(t) \tag{6.886}$$

$$\begin{bmatrix} \dot{x}_1 \\ \dot{x}_2 \end{bmatrix} + \begin{bmatrix} 0 & -1 \\ 2 & 3 \end{bmatrix} \begin{bmatrix} x_1 \\ x_2 \end{bmatrix} = \begin{bmatrix} 0 \\ 1 \end{bmatrix} \tag{6.887}$$

The eigenvalues and modal matrix of the coefficient matrix $[A]$ are

$$\lambda_1 = 1 \qquad \lambda_2 = 2 \qquad [U] = \begin{bmatrix} -1 & -1/2 \\ 1 & 1 \end{bmatrix} \tag{6.888}$$

Therefore,

$$e^{[\Lambda]t} = \begin{bmatrix} e^t & 0 \\ 0 & e^{2t} \end{bmatrix} \tag{6.889}$$

and

$$e^{[A]t} = [U]\,e^{-[\Lambda]t}\,[U]^{-1} = \begin{bmatrix} 2e^t - e^{2t} & e^t - e^{2t} \\ 2e^{2t} - 2e^t & 2e^{2t} - e^t \end{bmatrix} \tag{6.890}$$

$$e^{-[A]t} = \begin{bmatrix} 2e^{-t} - e^{-2t} & e^{-t} - e^{-2t} \\ 2e^{-2t} - 2e^{-t} & 2e^{-2t} - e^{-t} \end{bmatrix} \tag{6.891}$$

The solutions of the equations are

$$\begin{aligned} \mathbf{x} &= e^{-[A]t} \int e^{[A]t}\,\mathbf{f}(t)\,dt + e^{-[A]t}\mathbf{C} \\ &= e^{-[A]t} \int \begin{bmatrix} 2e^t - e^{2t} & e^t - e^{2t} \\ 2e^{2t} - 2e^t & 2e^{2t} - e^t \end{bmatrix} \begin{bmatrix} 0 \\ 1 \end{bmatrix} dt + e^{-[A]t}\mathbf{C} \\ &= \begin{bmatrix} \frac{1}{2} \\ 0 \end{bmatrix} + \begin{bmatrix} 2e^{-t} - e^{-2t} & e^{-t} - e^{-2t} \\ 2e^{-2t} - 2e^{-t} & 2e^{-2t} - e^{-t} \end{bmatrix} \begin{bmatrix} C_1 \\ C_2 \end{bmatrix} \end{aligned} \tag{6.892}$$

Using the initial conditions

$$\begin{bmatrix} 0 \\ 0 \end{bmatrix} = \begin{bmatrix} -\frac{1}{2} \\ 0 \end{bmatrix} + \begin{bmatrix} 1 & 0 \\ 0 & 1 \end{bmatrix} \begin{bmatrix} C_1 \\ C_2 \end{bmatrix} = \begin{bmatrix} C_1 - \frac{1}{2} \\ C_2 \end{bmatrix} \tag{6.893}$$

we calculate the integral constants,

$$C_1 = \frac{1}{2} \qquad C_2 = 0 \tag{6.894}$$

and we determine the time response of the system:

$$\mathbf{x} = \begin{bmatrix} e^{-t} - \frac{1}{2}e^{-2t} + \frac{1}{2} \\ e^{-2t} - e^{-t} \end{bmatrix} \tag{6.895}$$

## 6.6 Summary

There are cases where the behavior of a dynamic system can be modeled by first-order or reducible to first-order differential equations.

$$a\dot{x} + bx = f \tag{6.896}$$

The motion of a first-order system with no external force is the natural motion.

$$\dot{x} + \alpha x = 0 \tag{6.897}$$

The solution of natural motion is

$$x = x\left(0\right)\ e^{-\alpha t} = x_0 e^{-t/\tau} \qquad \tau = \frac{1}{\alpha} \tag{6.898}$$

The most important characteristic of first-order systems is their time constant $\tau$. The response of a system after a period of one time constant $t = \tau$ reaches $e^{-1}$ of its initial value. A time constant is passed when $x$ drops by about %64 of its initial value $x_0$.

$$x = x\left(t + \tau\right) = \frac{x\left(t\right)}{e} \tag{6.899}$$

The natural motion of first-order systems is either exponentially decreasing or increasing function of time, and they do not show vibrations.

The excited case of every linear first-order system is expressed by a full first-order equation.

$$\dot{x} + \alpha x = f\left(t\right) \tag{6.900}$$

The solution $x = x(t)$, for $t > 0$, is

$$x = e^{-\alpha t} \int e^{\alpha t}\ f\left(t\right)\ dt + Ce^{-\alpha t} \tag{6.901}$$

Any linear free vibrating system of any order can be expressed by a set of coupled first-order linear differential equations.

$$\dot{\mathbf{x}} + [A]\,\mathbf{x} = \mathbf{0} \tag{6.902}$$

$$[A] = \begin{bmatrix} a_{11} & a_{12} & a_{13} & \cdots & a_{1n} \\ a_{21} & a_{22} & a_{23} & \cdots & a_{2n} \\ \vdots & \vdots & \vdots & \vdots & \vdots \\ a_{n1} & a_{n2} & a_{n3} & \cdots & a_{nn} \end{bmatrix} \tag{6.903}$$

The solution of the set of coupled first-order linear homogeneous equations is

$$\mathbf{x} = \sum_{j=1}^{n} e^{-\lambda_j t} C_j \mathbf{u}_j \tag{6.904}$$

where $\mathbf{u}_j$ are the eigenvectors and $\lambda_j$, $j = 1, 2, \cdots, n$ are eigenvalues of the $n \times n$ coefficient matrix $[A]$. This solution can also be expressed by

$$\mathbf{x}\left(t\right) = e^{-[A]t}\mathbf{C} = e^{-[A]t}\mathbf{x}_0 \tag{6.905}$$

where

$$e^{-[A]t} = [I] - [A]\,t + \frac{1}{2!}\,[A]^2\,t^2 - \frac{1}{3!}\,[A]^3\,t^3 + \frac{1}{4!}\,[A]^4\,t^4 - \cdots$$
$$= \sum_{n=0}^{\infty} (-1)^n \,\frac{(\mathbf{A}t)^n}{n!} \tag{6.906}$$

The forced vibration of any number of first-order equations can be expressed by a set of $n$ coupled equations

$$\dot{\mathbf{x}} + [A]\,\mathbf{x} = \mathbf{f}\,(t) \tag{6.907}$$

where its solution $\mathbf{x} = \mathbf{x}(t)$ is

$$\mathbf{x} = e^{-[A]t} \int e^{[A]t}\,\mathbf{f}\,(t)\;dt + e^{-[A]t}\mathbf{C} \tag{6.908}$$

$$\mathbf{x} = \left[\, x_1\; x_2\; x_3 \cdots x_n \,\right]^T \tag{6.909}$$

$$\mathbf{f} = \left[\, f_1\; f_2\; f_3 \cdots f_n \,\right]^T \tag{6.910}$$

$$\mathbf{C} = \left[\, C_1\; C_2\; C_3 \cdots C_n \,\right]^T \tag{6.911}$$

## 6.7 Key Symbols

| | |
|---|---|
| $a, b$ | coefficient |
| $\mathbf{b}$ | $n \times 1$ coefficient vector |
| $A, B$ | coefficients of frequency response |
| $A, B, C, D$ | functions of $x$ and $t$ |
| $\mathbf{A}, \mathbf{B}$ | coefficients matrix |
| $[A]$ | coefficient matrix |
| $c$ | damping |
| $C$ | constant, constant of integration |
| $\mathbf{C}$ | vector of constants of integration |
| $e$ | exponential function |
| $E$ | $= K + P$ mechanical energy |
| $f$ | force, function |
| $F$ | force amplitude, force magnitude, constant force |
| $\mathbf{f}, \mathbf{F}$ | $n \times 1$ force set |
| $g$ | functions of displacement |
| $g, \mathbf{g}$ | gravitational acceleration vector |
| $h$ | functions of time, homogeneity factor |
| $i, j$ | dummy index |
| $I$ | impulse |
| $[\mathbf{I}]$ | identity matrix |
| $k$ | stiffness |
| $K$ | kinetic energy |
| $l$ | distance, length |
| $m$ | mass |
| $M, N$ | homogeneous functions of $x$ and $t$ |
| $n$ | number of equations, number of coordinates, $DOF$ |
| $N$ | normal force |

| | |
|---|---|
| $p, q$ | functions of $x$ |
| $\mathbf{p}$ | momentum, principal coordinates set |
| $\mathbf{P}$ | $n \times 1$ principal force set |
| $\mathbf{P}$ | $n \times 1$ principal coordinate |
| $P$ | potential energy |
| $s$ | characteristic value, eigenvalue |
| $t$ | time |
| $u$ | solution of equation, displacement |
| $\mathbf{u}$ | eigenvector |
| $[U]$ | modal matrix |
| $v \equiv \dot{x}$ | velocity |
| $v_0 = \dot{x}_0$ | initial velocity |
| $x, \mathbf{x}$ | displacement, $n \times 1$ displacement set |
| $x_0$ | initial displacement |
| $x_e$ | equilibrium point |
| $X$ | displacement amplitude |
| $y$ | displacement excitation, variable, displacement |
| $z$ | variable |

Greek

| | |
|---|---|
| $\alpha$ | constant, coefficient, inverse of time constant |
| $[\alpha]$ | coefficient matrix |
| $\lambda$ | eigenvalue |
| $[\Lambda]$ | diagonalized coefficient matrix of $[\alpha]$ |
| $s$ | eigenvalue, characteristic value |
| $\omega$ | frequency |
| $\tau$ | time constant |
| $\mu$ | friction coefficient |

Subscript

| | |
|---|---|
| $c$ | critical |
| $I$ | imaginary |
| $h$ | natural solution |
| $n$ | natural |
| $p$ | particular solution |
| $R$ | real |
| $s$ | steady state |
| $v$ | velocity coefficient |
| $x$ | position coefficient |

Symbols

| | |
|---|---|
| $[\ ]$ | matrix |
| $\int$ | integral |
| $\lvert\ \rvert$ | absolute value |
| Hz | Hertz |
| $d$ | differential |
| det | determinant |
| $DOF$ | degree of freedom |
| $FBD$ | free body diagram |
| $\partial$ | partial derivative |
| $\mathcal{L}$ | Lagrangean |
| $O$ | order of magnitude |
| $\nabla$ | gradient |

## References

1. J. Angeles, *Dynamic Response of Linear Mechanical Systems: Modeling, Analysis and Simulation* (Springer, New York, 2011)
2. B. Balachandran, E.B. Magrab, *Vibrations* (Brooks/Cole, Pacific Grove, 2003)
3. M. Del Pedro, P. Pahud, *Vibration Mechanics* (Kluwer Academic Publishers, The Netherland, 1991)
4. J.P. Den Hartog, *Mechanical Vibrations* (McGraw-Hill, New York, 1934)
5. A. Dimarogonas, *Vibration for Engineers* (Prentice Hall, New Jersey, 1996)
6. R.N. Jazar, *Advanced Dynamics: Rigid Body, Multibody, and Aero-Space Applications* (Wiley, New York, 2011)
7. R.N. Jazar, *Advanced Vibrations: A Modern Approach* (Springer, New York, 2013)
8. R.N. Jazar, *Vehicle Dynamics: Theory and Application*, 3rd edn. (Springer, New York, 2017)
9. R.N. Jazar, *Advanced Vibrations: Theory and Application* (Springer, New York, 2023)
10. R.N. Jazar, *Approximation Methods in Science and Engineering* (Springer, New York, 2020)
11. R.N. Jazar, *Perturbation Methods in Science and Engineering* (Springer, New York, 2021)
12. R.N. Jazar, M. Kazemi, S. Borhani, *Mechanical Vibrations* (in Persian) (Ettehad Publications, Tehran, 1992)
13. L. Meirovitch, *Analytical Methods in Vibrations* (Macmillan, New York, 1967)
14. L. Meirovitch, *Principles and Techniques of Vibrations* (Prentice Hall, New Jersey, 1997)
15. L. Meirovitch, *Fundamentals of Vibrations* (McGraw-Hill, New York, 2002)
16. D.E. Newland, E.E. Ungar, *Mechanical Vibration Analysis and Computation* (Wiley, New York, 1989)
17. K. Ogata, *System Dynamics* (Prentice Hall, New Jersey, 2004)
18. Z. Osinski, *Damping of Vibrations* (A. A. Balkema Publishers, Rotterdam, 1998)
19. W.J. Palm, *Mechanical Vibration* (Wiley, New York, 2006)
20. A.G. Piersol, T.L. Paez, *Harris' Shock and Vibration Handbook* (McGraw Hill, New York, 2009)
21. S.S. Rao, *Mechanical Vibrations* (Pearson Prentice Hall, Harlow, 2018)
22. D.J. Rixen, *Mechanical Vibrations: Theory and Application to Structural Dynamics* (Wiley, Chichester, 2015)
23. M. Roseau, *Vibrations in Mechanical Systems* (Springer, Berlin, 1987)
24. J.C. Snowdon, *Vibration and Shock in Damped Mechanical Systems* (Wiley, New York, 1968)
25. S. Timoshenko, *Vibration Problems in Engineering* (D. Van Nostrand Company, New York, 1928)
26. F.S. Tse, I.E. Morse, R.T. Hinkle, *Mechanical Vibrations Theory and Applications* (Allyn and Bacon Inc., Boston, 1978)
27. W. Weaver Jr, S. Timoshenko, D.H. Young, *Vibration Problems in Engineering* (Wiley, New York, 1991)

# Frequency Response

# 7

The harmonic excitation is the most widely used continuous excitation in vehicle vibration applications. Any combination of sin and cos functions is called a harmonic function. Frequency response analysis is simple and straightforward because the response of linear discrete vibrating systems to harmonic excitation is harmonic with proportional magnitude. The magnitudes of harmonic responses are functions of the excitation frequencies and depend on the system parameters. The goal of frequency response analysis is to find and plot the magnitude of the harmonic response versus the excitation frequency. Such a function is called the frequency response and indicates the behavior of the system at different excitation frequencies. Figure 7.1 illustrates the only four types of harmonically excited one-$DOF$ systems.

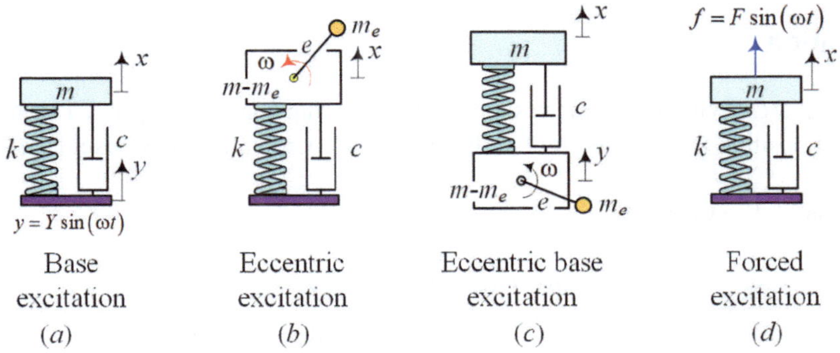


**Fig. 7.1** The four practical types of one-$DOF$ harmonically excited systems: (**a**) base excitation, (**b**) eccentric excitation, (**c**) eccentric base excitation, (**d**) forced excitation

## 7.1 One $DOF$ Systems

The frequency response is the *steady-state* solution of the equation of motion at each frequency, when the system is *harmonically excited*. Steady-state response refers to the amplitude of oscillations after the effect of the initial conditions is exhausted.

A harmonic excitation is any combination of *sinusoidal functions* that is applied to a vibrating system. If the system is linear, then a harmonic excitation generates a harmonic response with a frequency-dependent amplitude. In frequency response analysis, we look for the steady-state amplitude of the oscillation as a function of the excitation frequency.

A large number of vibrating systems can be modeled by a one-$DOF$ mass–spring–damper system. There are four types of harmonically excited one-$DOF$ systems (Fig. 7.1):

1 – Base excitation
2 – Eccentric excitation


R. N. Jazar, H. Marzbani, *Vehicle Vibrations*, https://doi.org/10.1007/978-3-031-43486-0_7

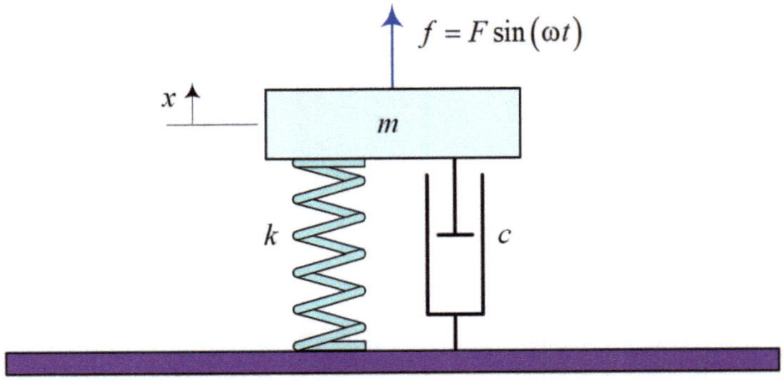


**Fig. 7.2** Harmonically forced excited, single-$DOF$ system

3 – Eccentric base excitation
4 – Forced excitation

*Base excitation* is the most practical model of the forced vibration of mechanical systems, such as vehicles, and structures. *Eccentric excitation* is a model for every type of rotary motor on a suspension, such as an engine on engine mounts. *Eccentric base excitation* is a model of the vibrations of any equipment mounted on an engine. *Forced excitation* has almost no practical application; however, it is the simplest model of forced vibration and has good pedagogical value.

Because of its simplicity, we first examine the frequency response of a harmonically forced vibrating system.

### 7.1.1 Forced Excitation

Figure 7.2 illustrates a one-$DOF$ vibrating mass $m$ supported by a spring $k$ and a damper $c$. The absolute motion of $m$ with respect to its equilibrium position is measured by the displacement coordinate $x$. A sinusoidal excitation force is applied to $m$.

The equation of motion of the system is

$$m\ddot{x} + c\dot{x} + kx = f\,, \tag{7.1}$$

$$f = F\sin\omega t, \tag{7.2}$$

which generates a frequency response expressed by either of the following functions:

$$x = A_1 \sin\omega t + B_1 \cos\omega t\,, \tag{7.3}$$

$$= X\sin\left(\omega t - \varphi_x\right). \tag{7.4}$$

The steady-state response has an *amplitude* $X$ and a *phase* $\varphi_x$:

$$X = \frac{1}{\sqrt{\left(k - m\omega^2\right)^2 + c^2\omega^2}} F\,, \tag{7.5}$$

$$\tan\varphi_x = \frac{c\omega}{k - m\omega^2}. \tag{7.6}$$

We usually express the frequency response $X$ and *phase* $\varphi_x$ in dimensionless form,

$$\frac{X}{F/k} = \frac{1}{\sqrt{\left(1 - r^2\right)^2 + (2\xi r)^2}}, \tag{7.7}$$

$$\varphi_x = \tan^{-1}\frac{2\xi r}{1 - r^2}, \tag{7.8}$$

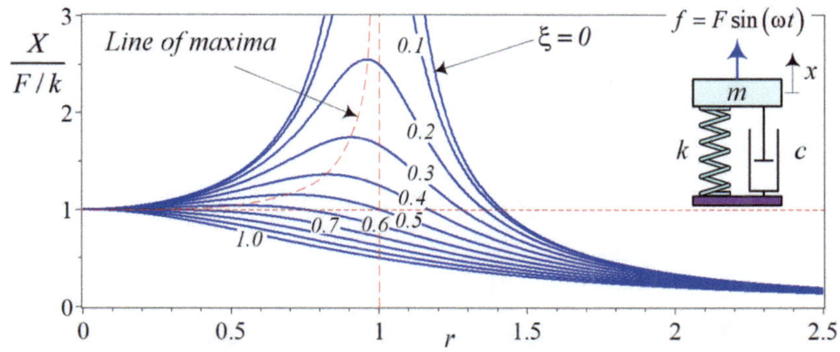


**Fig. 7.3** Position frequency response for $\frac{X}{F/k}$

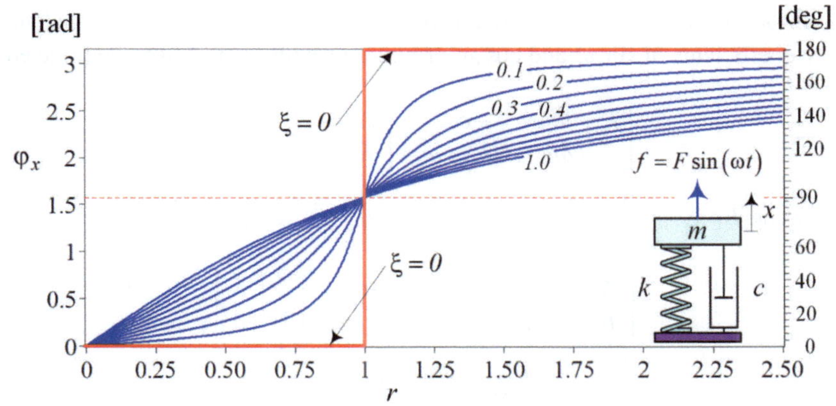


**Fig. 7.4** Frequency response for $\varphi_x$

where we use the *frequency ratio* $r$, *natural frequency* $\omega_n$, and *damping ratio* $\xi$:

$$r = \frac{\omega}{\omega_n}, \tag{7.9}$$

$$\xi = \frac{c}{2\sqrt{km}}, \tag{7.10}$$

$$\omega_n = \sqrt{\frac{k}{m}}. \tag{7.11}$$

The phase $\varphi_x$ indicates the *angular lag* of the response $x$ with respect to the excitation $f$.

The frequency responses for $X$ and $\varphi_x$ as a function of $r$ and $\xi$ are plotted in Figs. 7.3 and 7.4.

***Proof*** Applying Newton's method and using the free body diagram of the system, as shown in Fig. 7.5, generates the equation of motion

$$m\ddot{x} + c\dot{x} + kx = F \sin \omega t. \tag{7.12}$$

The steady-state solution of the linear Eq. (7.12) will be the same function as the excitation with an unknown amplitude and phase. Therefore, the solution can be (7.3) or (7.4). We substitute the solution in the equation of motion to find the amplitude and phase of the response. Let us examine solution (7.3) and obtain the following equation:

$$\begin{gathered} -m\omega^2 (A_1 \sin \omega t + B_1 \cos \omega t) + c\omega (A_1 \cos \omega t - B_1 \sin \omega t) \\ +k (A_1 \sin \omega t + B_1 \cos \omega t) = F \sin \omega t \end{gathered} \tag{7.13}$$

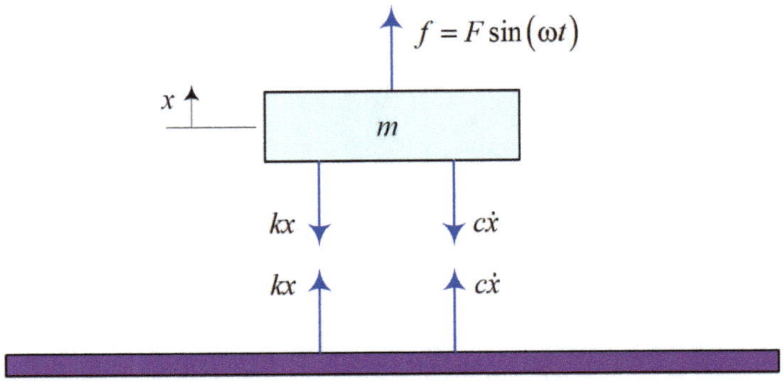


**Fig. 7.5** Free body diagram of harmonically forced excited single-$DOF$ system

The harmonic functions $\sin \omega t$ and $\cos \omega t$ are orthogonal; therefore, their coefficients must be balanced on both sides of the equals sign. Balancing the coefficients of $\sin \omega t$ and $\cos \omega t$ provides us with a set of two algebraic equations for $A_1$ and $B_1$:

$$\begin{bmatrix} k - m\omega^2 & -c\omega \\ c\omega & k - m\omega^2 \end{bmatrix} \begin{bmatrix} A_1 \\ B_1 \end{bmatrix} = \begin{bmatrix} F \\ 0 \end{bmatrix}. \tag{7.14}$$

Solving for coefficients $A_1$ and $B_1$

$$\begin{aligned} \begin{bmatrix} A_1 \\ B_1 \end{bmatrix} &= \begin{bmatrix} k - m\omega^2 & -c\omega \\ c\omega & k - m\omega^2 \end{bmatrix}^{-1} \begin{bmatrix} F \\ 0 \end{bmatrix} \\ &= \begin{bmatrix} \dfrac{k - m\omega^2}{\left(k - m\omega^2\right)^2 + c^2\omega^2} F \\ \dfrac{-c\omega}{\left(k - m\omega^2\right)^2 + c^2\omega^2} F \end{bmatrix} \end{aligned} \tag{7.15}$$

provides the steady-state solution (7.3)

$$x = F \frac{\left(k - m\omega^2\right)}{\left(k - m\omega^2\right)^2 + c^2\omega^2} \sin \omega t - F \frac{c\omega}{\left(k - m\omega^2\right)^2 + c^2\omega^2} \cos \omega t. \tag{7.16}$$

The amplitude $X$ and phase $\varphi_x$ can be found by equating Eqs. (7.3) and (7.4):

$$\begin{aligned} A_1 \sin \omega t + B_1 \cos \omega t &= X \sin\left(\omega t - \varphi_x\right) \\ &= X \cos \varphi_x \sin \omega t - X \sin \varphi_x \cos \omega t, \end{aligned} \tag{7.17}$$

which shows that

$$A_1 = X \cos \varphi_x \qquad B_1 = -X \sin \varphi_x, \tag{7.18}$$

and therefore

$$X = \sqrt{A_1^2 + B_1^2} \qquad \tan \varphi_x = \frac{-B_1}{A_1}. \tag{7.19}$$

Substituting $A_1$ and $B_1$ from (7.15) yields

$$X = \frac{1}{\sqrt{\left(k - m\omega^2\right)^2 + c^2\omega^2}} F, \tag{7.20}$$

$$\tan\varphi_x = \frac{c\omega}{k - m\omega^2}. \tag{7.21}$$

Coefficients $A_1$ and $B_1$ based on $r$ and $\xi$ are

$$\begin{bmatrix} A_1 \\ B_1 \end{bmatrix} = \begin{bmatrix} \frac{F}{k} \frac{1 - r^2}{\left(1 - r^2\right)^2 + (2\xi r)^2} \\ \frac{F}{k} \frac{-2\xi r}{\left(1 - r^2\right)^2 + (2\xi r)^2} \end{bmatrix}. \tag{7.22}$$

The frequency responses of the amplitude $X$ and phase $\varphi_x$ are more useful if we express them by dimensionless parameters $r$ and $\xi$ as expressed in Eqs. (7.7) and (7.8).

When we apply a constant force $f = F$ on $m$, a constant steady-state displacement, $\delta_s$, appears:

$$\delta_s = \frac{F}{k}. \tag{7.23}$$

If we call $\delta_s$ the "*static amplitude*" and $X$ the "*dynamic amplitude*," then $X/\delta_s$ is the ratio of dynamic to static amplitudes. Thus, we can plot $X/(F/k)$ instead of $X$ to make the graph of frequency response dimensionless. We have $X/(F/k) = 1$ at $r = 0$, and $X \to 0$, when $r \to \infty$. However, $X$ takes a high value when $r \to 1$ or $\omega \to \omega_n$. Theoretically, $X \to \infty$ if $\xi = 0$ and $r \to 1$. The frequency domain around the natural frequency is called the *resonance zone*. We reduce the high amplitude of vibration in the resonance zone by introducing damping.

The displacement frequency response function $X = X(\omega)$ is the most important frequency-dependent behavior of vibrating systems. Furthermore, we may use the frequency response of every characteristic of the system that is a function of the excitation frequency, such as velocity frequency response $\dot{X} = \dot{X}(\omega)$ and transmitted force frequency response $f_T = f_T(\omega)$.

In mechanics, there is no practical method to apply a periodic force on an object without attaching a mechanical device and applying a displacement. Hence, the forced vibrating system shown in Fig. 7.2 has no practical application in mechanics. However, it is possible to apply a periodic force on a mass $m$ using a ferromagnetic material and applying an alternating or periodic magnetic force. ■

**Example 80** Forced vibrating system. A sample numerical example of the frequency response of a force excited system to visualize the phase lag between response and excitation.

Consider a mass–spring–damper system with the following numerical values:

$$m = 200\,\text{kg} \qquad k = 10{,}000\,\text{N/m} \qquad c = 100\,\text{N s/m}. \tag{7.24}$$

The natural frequency and damping ratio of the system are

$$\omega_n = \sqrt{\frac{k}{m}} = \sqrt{\frac{10{,}000}{200}} = 7.0711\,\text{rad/s} \approx 1.125\,\text{Hz}, \tag{7.25}$$

$$\xi = \frac{c}{2\sqrt{km}} = \frac{100}{2\sqrt{10{,}000 \times 200}} = 0.0353. \tag{7.26}$$

If a harmonic force $f$ is applied to $m$,

$$f = 100 \sin 10t\,, \tag{7.27}$$

$$r = \frac{\omega}{\omega_n} = \frac{10}{7.0711} = 0.707, \tag{7.28}$$

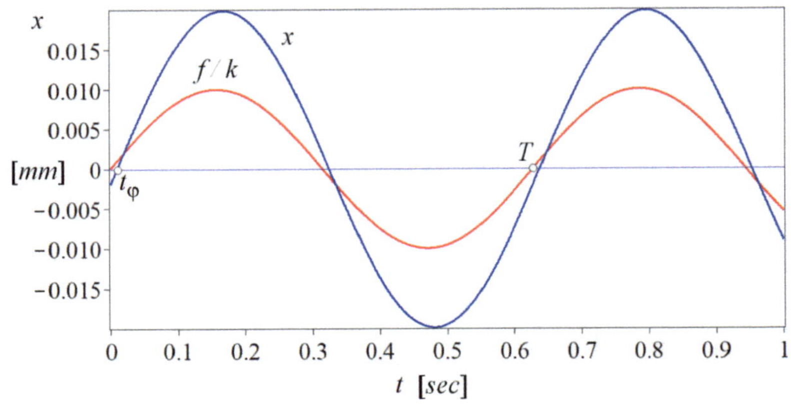


**Fig. 7.6** Effect of phase lag $\varphi_x$ on displacement $x$

then the steady-state amplitude of vibrations of the mass will be

$$X = \frac{F/k}{\sqrt{\left(1-r^2\right)^2 + (2\xi r)^2}} = 0.0199\,\text{m}. \tag{7.29}$$

The phase $\varphi_x$ of the vibration is

$$\varphi_x = \tan^{-1}\frac{2\xi r}{1-r^2} = 0.1\,\text{rad} \approx 5.73\,\text{deg}\,. \tag{7.30}$$

Therefore, the steady-state vibrations of $m$ can be expressed by

$$x = 0.0199\sin(10t - 0.1)\ \text{mm}\,. \tag{7.31}$$

The values of $X$ and $\varphi_x$ may also be found from Figs. 7.3 and 7.4 approximately.

Figure 7.6 illustrates the effect of the phase lag $\varphi_x$ on the steady-state behavior of $x$. Excitation (7.27) starts from $f/k = 0$ at $t = 0$ and returns to its state after the period $T = \frac{2\pi}{\omega} = \frac{2\pi}{10} = 0.6283$ s has elapsed. The response of the system in (7.31) is at $x = -1.9867 \times 10^{-3}$ when $t = 0$, which is $\varphi_x = 0.1$ rad or $t_\varphi = 0.01$ s behind $f$.

**Example 81** Velocity and acceleration frequency responses. Displacement amplitude is the most important frequency response that engineers work with. However, there are applications in which we would need velocity, acceleration, jerk, transmitted force, or other dynamic frequency responses. This example shows how to derive and visualize the velocity and acceleration frequency responses of the forced excitation systems. It also shows the method to derive the frequency response of other variables.

Having the steady-state position frequency response, $x$, we are able to calculate the velocity and acceleration steady-state responses by the derivative:

$$x = A_1\sin\omega t + B_1\cos\omega t = X\sin\left(\omega t - \varphi_x\right), \tag{7.32}$$

$$\begin{aligned}\dot{x} &= A_1\omega\cos\omega t - B_1\omega\sin\omega t = X\omega\cos\left(\omega t - \varphi_x\right)\\ &= \dot{X}\cos\left(\omega t - \varphi_x\right),\end{aligned} \tag{7.33}$$

$$\begin{aligned}\ddot{x} &= -A_1\omega^2\sin\omega t - B_1\omega^2\cos\omega t = -X\omega^2\sin\left(\omega t - \varphi_x\right)\\ &= \ddot{X}\sin\left(\omega t - \varphi_x\right).\end{aligned} \tag{7.34}$$

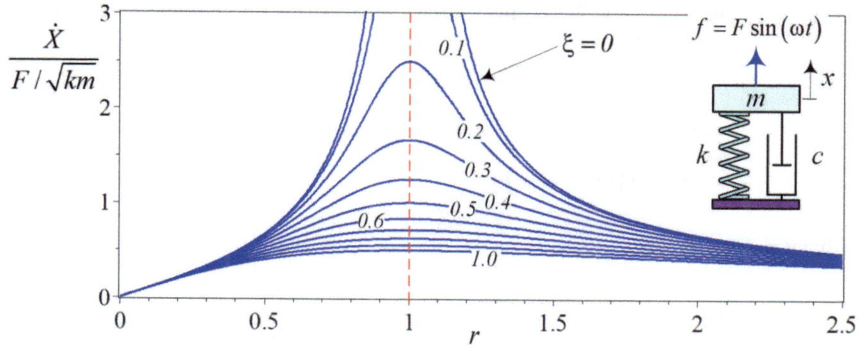


**Fig. 7.7** Velocity frequency response for $\frac{\dot{X}}{F/\sqrt{km}}$

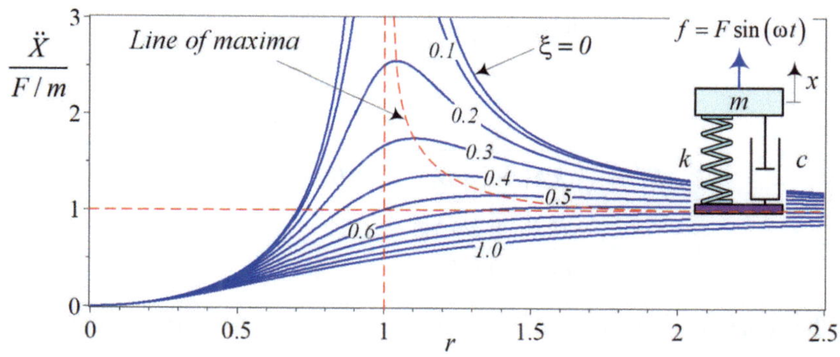


**Fig. 7.8** Acceleration frequency response for $\frac{\ddot{X}}{F/m}$

The amplitude of velocity and acceleration frequency responses are shown by $\dot{X}$ and :$\ddot{X}$

$$\dot{X} = \frac{\omega}{\sqrt{\left(k - m\omega^2\right)^2 + c^2\omega^2}} F, \tag{7.35}$$

$$\ddot{X} = \frac{\omega^2}{\sqrt{\left(k - m\omega^2\right)^2 + c^2\omega^2}} F, \tag{7.36}$$

which can also be written in dimensionless forms:

$$\frac{\dot{X}}{F/\sqrt{km}} = \frac{r}{\sqrt{\left(1 - r^2\right)^2 + (2\xi r)^2}}, \tag{7.37}$$

$$\frac{\ddot{X}}{F/m} = \frac{r^2}{\sqrt{\left(1 - r^2\right)^2 + (2\xi r)^2}}. \tag{7.38}$$

The velocity and acceleration frequency responses (7.37) and (7.38) are plotted in Figs. 7.7 and 7.8.

**Example 82** Force transmitted to base. Force transmitted to the base is an engineering measured date to design foundations. In this example, we calculate the transmitted force frequency response of a forced excited $m$; $c$, $k$ system.

A force excited system transmits a force, $f_T$, to the ground that is equal to the sum of forces in the spring and damper:

$$f_T = f_k + f_c = kx + c\dot{x}. \tag{7.39}$$

Substituting $x$ from (7.3) and $A_1$ and $B_1$ from (7.15) shows that the steady-state response of the transmitted force is

$$\begin{aligned} f_T &= k\,(A_1 \sin \omega t + B_1 \cos \omega t) + c\,\omega\,(A_1 \cos \omega t - B_1 \sin \omega t) \\ &= (kA_1 - c\omega B_1) \sin t\omega + (kB_1 + c\omega A_1) \cos t\omega \\ &= F_T \sin\left(\omega t - \varphi_{F_T}\right). \end{aligned} \tag{7.40}$$

The amplitude $F_T$ and phase $\varphi_{F_T}$ of $f_T$ are

$$\frac{F_T}{F} = \frac{\sqrt{k + c^2\omega^2}}{\sqrt{\left(k - m\omega^2\right)^2 + c^2\omega^2}} = \frac{\sqrt{1 + (2\xi r)^2}}{\sqrt{\left(1 - r^2\right)^2 + (2\xi r)^2}} \tag{7.41}$$

$$\tan \varphi_{F_T} = \frac{c\omega}{k - m\omega^2} = \frac{2\xi r}{1 - r^2} \tag{7.42}$$

because

$$F_T = \sqrt{(kA_1 - c\omega B_1)^2 + (kB_1 + c\omega A_1)^2}, \tag{7.43}$$

$$\tan \varphi_{F_T} = \frac{-\,(kB_1 + c\omega A_1)}{kA_1 - c\omega B_1}. \tag{7.44}$$

It is also possible to use the equation of motion and substitute $x$ from (7.3) to find the transmitted force frequency response of $f_T$:

$$\begin{aligned} f_T &= F \sin \omega t - m\ddot{x} \\ &= F \sin \omega t + m\omega^2 \,(A_1 \sin \omega t + B_1 \cos \omega t) \\ &= \left(mA_1\omega^2 + F\right) \sin t\omega + m\omega^2 B_1 \cos t\omega \\ &= F_T \sin\left(\omega t - \varphi_x\right). \end{aligned} \tag{7.45}$$

The amplitude $F_T$ and phase $\varphi_{F_T}$ would be the same as (7.41) and (7.42) because

$$\begin{aligned} F_T &= \sqrt{\left(mA_1\omega^2 + F\right)^2 + \left(m\omega^2 B_1\right)^2} \\ &= \frac{\sqrt{k + c^2\omega^2}}{\sqrt{\left(k - m\omega^2\right)^2 + c^2\omega^2}}, \end{aligned} \tag{7.46}$$

$$\tan \varphi_{F_T} = \frac{-m\omega^2 B_1}{m\omega^2 A_1 + F} = \frac{c\omega}{k - m\omega^2}. \tag{7.47}$$

The dimensionless transmitted force frequency response $F_T/F$ is plotted in Fig. 7.9, and because $\varphi_{F_T}$ is the same as Eq. (7.8), the graph for $\varphi_{F_T}$ is the same as Fig. 7.8.

**Example 83** Turning point: low and high frequencies. The definition of low and high frequency is based on the frequency ratio at $r = \sqrt{2}$ associated to the node of frequency responses.

The **node** in Fig. 7.9 is a turning point at which the relation between $F_T/F$ and $\xi$ reverses. Let us call the frequencies before the node **low frequency** and those after the node **high frequency**. At low frequency, the transmitted force $F_T$ increases by decreasing $\xi$, and at high frequency $F_T$ increases by increasing $\xi$. The transmitted force $F_T$ is always greater than the applied force $F$ at low frequency, and $F_T$ is always less than the applied force $F$ at high frequency. From a design engineer viewpoint, the suspension of $m$ amplifies the transmitted force $F_T$ with respect to the applied force $F$ at low frequency, then the operating frequency of the system should be at high frequency. The high-frequency zone is the safe zone to operate the system.

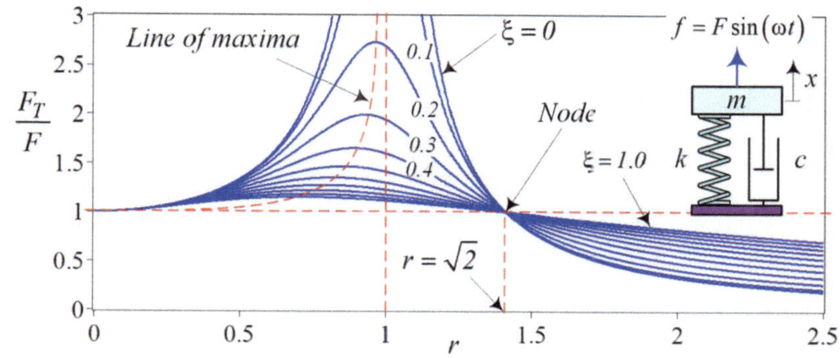


**Fig. 7.9** Frequency response for $\frac{F_T}{F}$

As a numerical example, consider a mass–spring–damper system with

$$m = 1\,\mathrm{kg} \qquad k = 1000\,\mathrm{N/m} \qquad c = 10\,\mathrm{N\,s/m} \tag{7.48}$$

that is under a harmonic forced excitation:

$$f = 10 \sin 2\pi t. \tag{7.49}$$

If the suspension of $m$ is supposed to act as an isolator, then the force $F_T$ transmitted to the ground must be less than the applied force $F = 10\,\mathrm{N}$. To obtain $F_T \le F$, we need

$$\frac{F_T}{F} = \frac{\sqrt{1 + (2\xi r)^2}}{\sqrt{\left(1 - r^2\right)^2 + (2\xi r)^2}} \le 1 \tag{7.50}$$

or

$$\sqrt{1 + (2\xi r)^2} \le \sqrt{\left(1 - r^2\right)^2 + (2\xi r)^2}. \tag{7.51}$$

After simplification, we obtain the following condition:

$$r^2 \ge 2, \tag{7.52}$$

which is consistent with Fig. 7.9.

**Example 84** ★ Frequency response is attractive. Every constant excitation frequency $\omega$ produces a steady-state oscillation at amplitude $X(\omega)$. That amplitude is attractive, which means any disturbance $\delta X$ to move $X$ to $X + \delta X$ will disappear after a while and the system will return to the same oscillating amplitude $X(\omega)$.

Consider a mass–spring–damper system under a harmonic force excitation at a given constant frequency $\omega_0$:

$$m\ddot{x} + c\dot{x} + kx = F \sin \omega_0 t. \tag{7.53}$$

The steady-state response of vibration would be

$$x_o = X_0 \sin\left(\omega_0 t - \varphi_x\right), \tag{7.54}$$

$$X_0 = \frac{F/k}{\sqrt{\left(1 - r^2\right)^2 + (2\xi r)^2}} \qquad r = \frac{\omega_0}{\omega_n}. \tag{7.55}$$

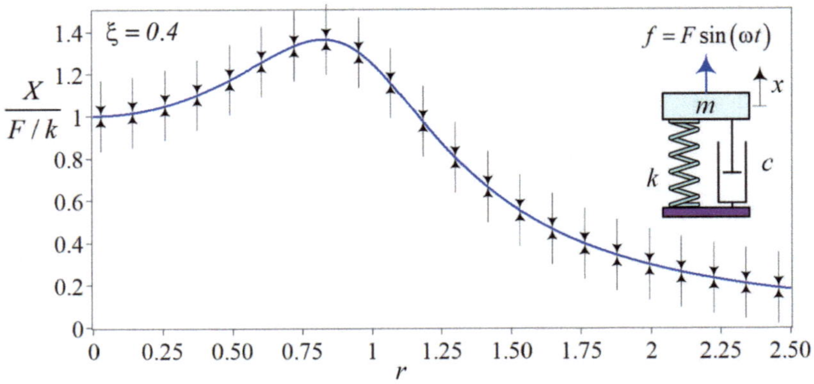


**Fig. 7.10** Stability and attraction characteristic of frequency response

Now assume that because of a sudden disturbance, the amplitude changes to $X_0 + \delta X$ at the same excitation frequency $\omega_0$, and the response of the system changes to $x = x_0 + \delta x$:

$$x = x_0 + \delta x = X \sin\left(\omega t - \varphi_x\right) + \delta x. \tag{7.56}$$

Substituting $x$ in Eq. (7.53)

$$m\ddot{x}_0 + c\dot{x}_0 + kx_0 + m\,\delta\ddot{x} + c\,\delta\dot{x} + k\,\delta x = F\sin\omega_0 t \tag{7.57}$$

and considering that $x_0$ satisfies the equation, we obtain the equation to determine $\delta x$:

$$m\,\delta\ddot{x} + c\,\delta\dot{x} + k\,\delta x = 0. \tag{7.58}$$

The solution of this equation with initial conditions

$$\delta x\,(0) = \delta X \qquad \delta\dot{x}\,(0) = 0 \tag{7.59}$$

is

$$\begin{aligned}\delta x =\; & \frac{1}{2}\frac{-c^2 - c\sqrt{c^2-4km} + 4km}{-c^2+4km}\delta X \exp\left(\frac{c-\sqrt{c^2-4km}}{-2m}t\right) \\ & + \frac{1}{2}\frac{-c^2 + c\sqrt{c^2-4km} + 4km}{-c^2+4km}\delta X \exp\left(\frac{c+\sqrt{c^2-4km}}{-2m}t\right).\end{aligned} \tag{7.60}$$

As long as $c^2 > 4km$, we have

$$\lim_{t\to\infty} \delta x = 0, \tag{7.61}$$

and hence,

$$\lim_{t\to\infty} x = x_0 = X\sin\left(\omega t - \varphi_x\right). \tag{7.62}$$

Therefore, any disturbance to a steady-state vibrating system will disappear after a while, and the system will return to its original steady-state vibration. This is the stability characteristic of frequency response and indicates that frequency response curves are attractive for small-amplitude disturbances, as shown in Fig. 7.10 for $\xi = 0.4$.

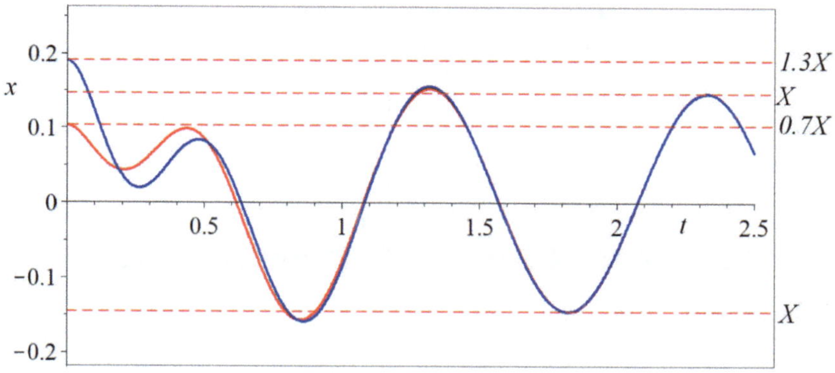


**Fig. 7.11** Attraction process of amplitude to steady-state value after sudden jump to $1.3X$ or $0.7X$

For instance, let us examine a system with

$$m = 1\,\mathrm{kg} \qquad k = 100\,\mathrm{N/m} \qquad c = 5\,\mathrm{N\,s/m}, \tag{7.63}$$

that is under a force $f$

$$f = 10\sin(2\pi t)\ \mathrm{N}. \tag{7.64}$$

The steady-state amplitude of this system will be

$$X = \frac{F}{\sqrt{\left(k - m\omega^2\right)^2 + c^2\omega^2}} = 0.1466\,\mathrm{m}. \tag{7.65}$$

Now suppose that, because of a disturbance, $x$ jumps $\pm 30\%$ to an amplitude of $1.3X$ or $0.7X$. Figure 7.11 illustrates the time history of the motions and shows how the system settles down to $X$ again.

**Example 85** ★ Why do we consider $X/(F/k)$ and $r = \omega/\omega_n$? It is a dimensional analysis of harmonic oscillators. A great advantage of making equations nondimensional is to make them work with the minimum number of parameters and variables. The second advantage is that the solution can be made applicable to all similar systems regardless of their size.

Consider a forced excited, undamped, single-$DOF$ system:

$$m\ddot{x} + kx = F\sin\omega t. \tag{7.66}$$

The involved input, output, and system parameters are $m$, $k$, $F$, $X$, and $t \equiv 1/\omega$. Based on the *Buckingham $\pi$-theorem*, any complete physical relation of a system can be expressed in terms of a set of independent dimensionless products of its $\pi$ terms, which are the products of relevant physical parameters of the system. Because each $\pi$ term is dimensionless, we must have a balance for the dimensions involved:

$$X^a k^b m^c F^d \omega^e = L^0 M^0 T^0. \tag{7.67}$$

Let us substitute the parameters $X, k, m, F, \omega$ by their dimensions of mass $M$, length $L$, and time $T$:

$$L^a \left(\frac{M}{T^2}\right)^b M^c \left(\frac{ML}{T^2}\right)^d \left(\frac{1}{T}\right)^e = L^0 M^0 T^0, \tag{7.68}$$

so we have

$$L^{a+d} M^{b+c+d} T^{-2b-2d-e} = L^0 M^0 T^0, \tag{7.69}$$

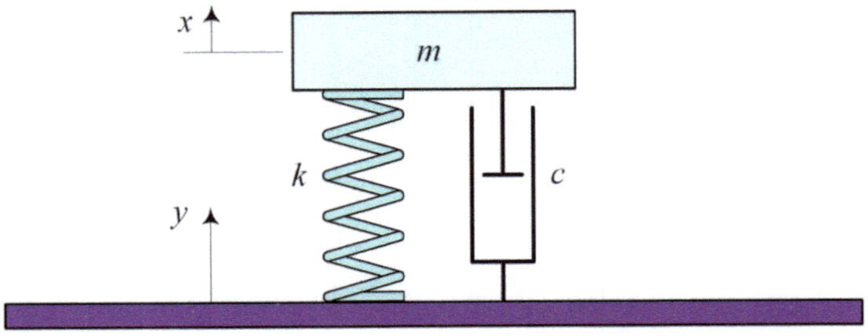


**Fig. 7.12** Harmonically base excited single-$DOF$ system

and hence,

$$a + d = 0, \tag{7.70}$$

$$b + c + d = 0, \tag{7.71}$$

$$-2b - 2d - e = 0. \tag{7.72}$$

These three equations may be solved for any three out of five exponents $a, b, c, d, e$. The solution, obtained by assuming $a$ and $e$ are known, is

$$d = -a, \qquad b = a - \frac{1}{2}e, \qquad c = \frac{1}{2}e. \tag{7.73}$$

Substituting them back, we have

$$X^a k^{a-e/2} m^{e/2} F^{-a} \omega^e = L^0 M^0 T^0, \tag{7.74}$$

which indicates

$$\left(\frac{X}{F/k}\right)^a \left(\frac{\omega}{\sqrt{k/m}}\right)^e = L^0 M^0 T^0. \tag{7.75}$$

The dimensionless $\pi$ terms are

$$\pi_1 = \frac{X}{F/k}, \qquad \pi_2 = \frac{\omega}{\sqrt{k/m}}. \tag{7.76}$$

The $\pi$-theorem states that there is a function $f$ of the $\pi$ terms such that

$$f\left(\frac{X}{F/k}, \frac{\omega}{\sqrt{k/m}}\right) = 0. \tag{7.77}$$

Thus, $X/(F/k)$ is a function of $\omega/\sqrt{k/m}$. If the solution to Eq. (7.66) is expressed based on $\pi_1$ and $\pi_2$, then it will be the solution to any vibrating system that could be modeled by Eq. (7.66).

### 7.1.2 Base Excitation

Figure 7.12 illustrates a one-$DOF$ base excited vibrating system made of a mass $m$ supported by a spring $k$ and a damper $c$. The base excited system is a good model of a vehicle suspension system or any equipment mounted on a vibrating base. The absolute motion of $m$ with respect to its equilibrium position is measured by the coordinate $x$. A sinusoidal excitation motion, $y$, is applied to the base of the suspension that makes the system vibrate:

$$y = Y \sin \omega t. \tag{7.78}$$

The equation of motion for the system can be expressed by any of the following equations for the absolute displacement $x$:

$$m\ddot{x} + c\dot{x} + kx = cY\omega \cos \omega t + kY \sin \omega t \,, \tag{7.79}$$

$$\ddot{x} + 2\xi\omega_n \dot{x} + \omega_n^2 x = 2\xi\omega_n \omega Y \cos \omega t + \omega_n^2 Y \sin \omega t \,, \tag{7.80}$$

$$\omega_n = \sqrt{\frac{k}{m}}, \qquad r = \frac{\omega}{\omega_n}, \qquad \xi = \frac{c}{2\sqrt{km}}. \tag{7.81}$$

The system may also be expressed by any of the following equations for the relative displacement $z$:

$$m\ddot{z} + c\dot{z} + kz = m\omega^2 Y \sin \omega t \,, \tag{7.82}$$

$$\ddot{z} + 2\xi\omega_n \dot{z} + \omega_n^2 z = \omega^2 Y \sin \omega t \,, \tag{7.83}$$

$$z = x - y. \tag{7.84}$$

The equations of motion generate the absolute and relative steady-state responses as

$$x = A_2 \sin \omega t + B_2 \cos \omega t \,, \tag{7.85}$$

$$= X \sin \left(\omega t - \varphi_x\right) \,, \tag{7.86}$$

$$z = A_3 \sin \omega t + B_3 \cos \omega t \,, \tag{7.87}$$

$$= Z \sin \left(\omega t - \varphi_z\right). \tag{7.88}$$

The steady-state response of $x$ has an amplitude $X$ and phase lag $\varphi_x$, while the steady-state response of $z$ has an amplitude $Z$ and phase lag $\varphi_z$:

$$\frac{X}{Y} = \frac{\sqrt{1 + (2\xi r)^2}}{\sqrt{\left(1 - r^2\right)^2 + (2\xi r)^2}}, \tag{7.89}$$

$$\frac{Z}{Y} = \frac{r^2}{\sqrt{\left(1 - r^2\right)^2 + (2\xi r)^2}}, \tag{7.90}$$

$$\varphi_x = \tan^{-1} \frac{2\xi r^3}{1 - r^2 + (2\xi r)^2}, \tag{7.91}$$

$$\varphi_z = \tan^{-1} \frac{2\xi r}{1 - r^2}. \tag{7.92}$$

The phase $\varphi_x$ indicates the *angular lag* of the response $x$ with respect to the excitation $y$ and the phase $\varphi_z$ is the *angular lag* of the relative motion response $z$ with respect to the excitation $y$. The frequency responses for $X$, $Z$, and $\varphi_x$ as a function of $r$ and $\xi$ are plotted in Figs. 7.13, 7.14, and 7.15.

***Proof*** The free body diagram of the system, as shown in Fig. 7.16, generates the equation of motion,

$$m\ddot{x} = -c\left(\dot{x} - \dot{y}\right) - k\left(x - y\right), \tag{7.93}$$

which, after substituting the excitation function (7.78), leads to equation of motion (7.79). Equation (7.79) can be transformed to (7.80) by dividing by $m$ and using definitions (7.81) for natural frequency $\omega_n$, damping ratio $\xi$, and frequency ratio $r$.

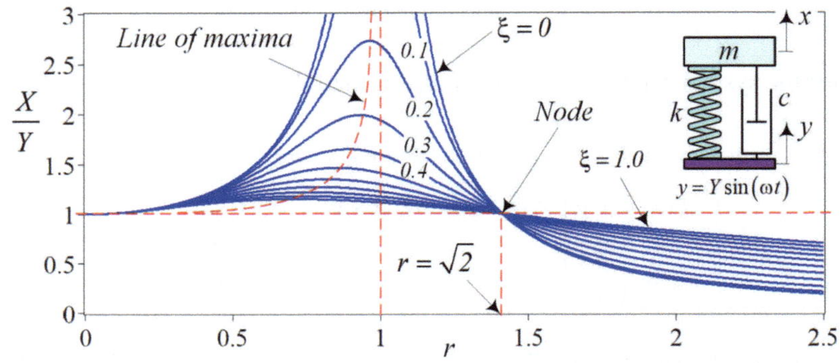


**Fig. 7.13** Position frequency response for $\frac{X}{Y}$

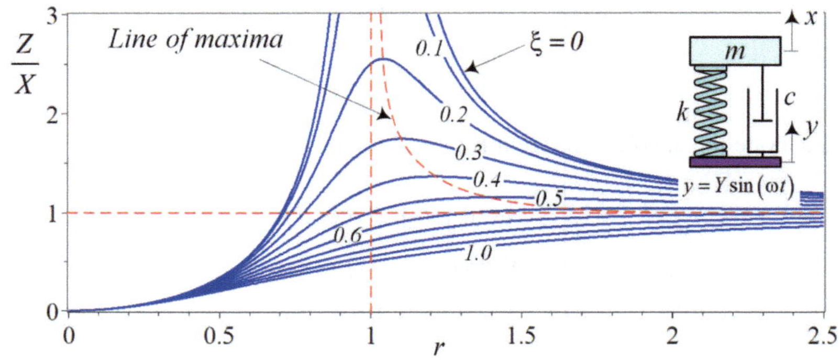


**Fig. 7.14** Frequency response for $\frac{Z}{Y}$

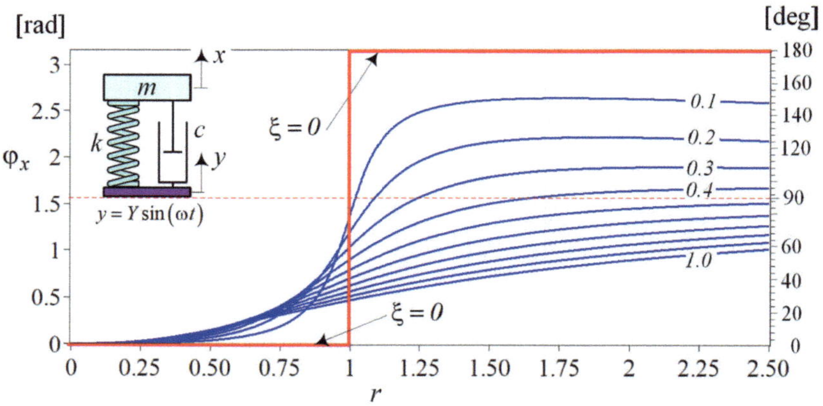


**Fig. 7.15** Frequency response for $\varphi_x$ of base excited one-$DOF$ system

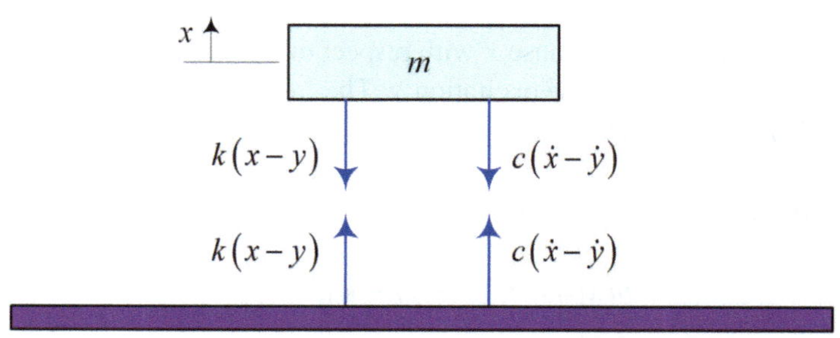


**Fig. 7.16** Harmonically based excited single-$DOF$ system

A practical response for a base excited system is the relative displacement:

$$z = x - y. \tag{7.94}$$

For every mechanical device that is mounted on a suspension, we usually need to determine and control the maximum or minimum distance between the base and the device. Therefore, the relative displacement $z$ is the physical variable to be measured. Taking derivatives from (7.94)

$$\ddot{z} = \ddot{x} - \ddot{y} \tag{7.95}$$

and substituting in (7.93)

$$m\ (\ddot{z} + \ddot{y}) = -c\,\dot{z} - kz \tag{7.96}$$

generates Eqs. (7.82) and (7.83).

The steady-state solution of Eq. (7.79) can be (7.85), or (7.86). To calculate the amplitude and phase of the response, we substitute solution (7.85) into equation of motion (7.79):

$$\begin{aligned} -m\omega^2\ (A_2 \sin \omega t + B_2 \cos \omega t) + c\omega\ (A_2 \cos \omega t - B_2 \sin \omega t) \\ +k\ (A_2 \sin \omega t + B_2 \cos \omega t) = cY\omega \cos \omega t + kY \sin \omega t\ . \end{aligned} \tag{7.97}$$

The coefficients of the functions $\sin \omega t$ and $\cos \omega t$ must balance on both sides of the equation:

$$kA_2 - mA_2\omega^2 - cB_2\omega = Yk\,, \tag{7.98}$$

$$kB_2 - m\omega^2 B_2 + c\omega A_2 = Yc\omega. \tag{7.99}$$

Therefore, we obtain two algebraic equations to calculate $A_2$ and $B_2$:

$$\begin{bmatrix} k - m\omega^2 & -c\omega \\ c\omega & k - m\omega^2 \end{bmatrix} \begin{bmatrix} A_2 \\ B_2 \end{bmatrix} = \begin{bmatrix} Yk \\ Yc\omega \end{bmatrix}. \tag{7.100}$$

Solving for coefficients $A_2$ and $B_2$ provides us with the steady-state solution (7.85):

$$\begin{aligned} \begin{bmatrix} A_2 \\ B_2 \end{bmatrix} &= \begin{bmatrix} k - m\omega^2 & -c\omega \\ c\omega & k - m\omega^2 \end{bmatrix}^{-1} \begin{bmatrix} Yk \\ Yc\omega \end{bmatrix} \\ &= \begin{bmatrix} \dfrac{k\left(k - m\omega^2\right) + c^2\omega^2}{\left(k - m\omega^2\right)^2 + c^2\omega^2} Y \\ \dfrac{c\omega\left(k - m\omega^2\right) - ck\omega}{\left(k - m\omega^2\right)^2 + c^2\omega^2} Y \end{bmatrix}. \end{aligned} \tag{7.101}$$

The amplitude $X$ and phase $\varphi_x$ can be found by

$$X = \sqrt{A_2^2 + B_2^2}, \tag{7.102}$$

$$\tan \varphi_x = \frac{-B_2}{A_2}, \tag{7.103}$$

which, after substituting $A_2$ and $B_2$ from (7.101), will be

$$X = \frac{\sqrt{k^2 + c^2\omega^2}}{\sqrt{\left(k - m\omega^2\right)^2 + c^2\omega^2}} Y, \tag{7.104}$$

$$\tan \varphi_x = \frac{-cm\omega^3}{k\left(k - m\omega^2\right) + c^2\omega^2}. \tag{7.105}$$

More practical expressions for $X$ and $\varphi_x$ are Eqs. (7.89) and (7.91), which can be found by employing $r$ and $\xi$.

To find the relative displacement frequency response (7.90), we substitute Eq. (7.87) in (7.82):

$$-m\omega^2 \left(A_3 \sin \omega t + B_3 \cos \omega t\right) + c\omega \left(A_3 \cos \omega t - B_3 \sin \omega t\right)$$
$$+ k\left(A_3 \sin \omega t + B_3 \cos \omega t\right) = m\omega^2 Y \sin \omega t. \tag{7.106}$$

Balancing the coefficients of $\sin \omega t$ and $\cos \omega t$,

$$kA_3 - mA_3\omega^2 - cB_3\omega = m\omega^2 Y, \tag{7.107}$$

$$kB_3 - m\omega^2 B_3 + c\omega A_3 = 0, \tag{7.108}$$

provides us with two algebraic equations to find $A_3$ and $B_3$:

$$\begin{bmatrix} k - m\omega^2 & -c\omega \\ c\omega & k - m\omega^2 \end{bmatrix} \begin{bmatrix} A_3 \\ B_3 \end{bmatrix} = \begin{bmatrix} m\omega^2 Y \\ 0 \end{bmatrix}. \tag{7.109}$$

Solving for coefficients $A_3$ and $B_3$ we have

$$\begin{bmatrix} A_3 \\ B_3 \end{bmatrix} = \begin{bmatrix} k - m\omega^2 & -c\omega \\ c\omega & k - m\omega^2 \end{bmatrix}^{-1} \begin{bmatrix} m\omega^2 Y \\ 0 \end{bmatrix}$$
$$= \begin{bmatrix} \dfrac{m\omega^2\left(k - m\omega^2\right)}{\left(k - m\omega^2\right)^2 + c^2\omega^2} Y \\ -\dfrac{mc\omega^3}{\left(k - m\omega^2\right)^2 + c^2\omega^2} Y \end{bmatrix}, \tag{7.110}$$

which provides us with the steady-state solution (7.87). The amplitude $Z$ and phase $\varphi_z$ can be found by

$$Z = \sqrt{A_3^2 + B_3^2}, \tag{7.111}$$

$$\tan \varphi_z = \frac{-B_3}{A_3}, \tag{7.112}$$

which, after substituting $A_3$ and $B_3$ from (7.110), leads to the following solutions:

$$Z = \frac{m\omega^2}{\sqrt{\left(k - m\omega^2\right)^2 + c^2\omega^2}} Y, \tag{7.113}$$

$$\tan \varphi_z = \frac{c\omega}{k - m\omega^2}. \tag{7.114}$$

More practical expressions for $Z$ and $\varphi_z$ are Eqs. (7.90) and (7.92) employing $r$ and $\xi$. ■

**Example 86** A base excited system. This is a numerical example to work with equations to understand and compare the orders of magnitude.

Consider a mass–spring–damper system with

$$m = 2\,\mathrm{kg} \qquad k = 10{,}000\,\mathrm{N/m} \qquad c = 100\,\mathrm{N\,s/m}. \tag{7.115}$$

If a harmonic base excitation $y$ is applied to the system,

$$y = 0.02 \sin 140t, \tag{7.116}$$

then the absolute and relative steady-state amplitude of vibrations of the mass, $X$ and $Z$, would be

$$X = \frac{Y\sqrt{1 + (2\xi r)^2}}{\sqrt{\left(1 - r^2\right)^2 + (2\xi r)^2}} = 0.0106\,\mathrm{m}, \tag{7.117}$$

$$Z = \frac{Y r^2}{\sqrt{\left(1 - r^2\right)^2 + (2\xi r)^2}} = 0.0242\,\mathrm{m}, \tag{7.118}$$

because

$$\omega_n = \sqrt{\frac{k}{m}} = 70.71\,\mathrm{rad/s} \approx 11.254\,\mathrm{Hz}, \tag{7.119}$$

$$\xi = \frac{c}{2\sqrt{km}} = 0.353\,, \tag{7.120}$$

$$r = \frac{\omega}{\omega_n} = 1.98. \tag{7.121}$$

The phases $\varphi_x$ and $\varphi_z$ for $x$ and $z$ are

$$\varphi_x = \tan^{-1} \frac{2\xi r^3}{1 - r^2 + (2\xi r)^2} = 1.744\,\mathrm{rad} \approx 49.96\,\mathrm{deg}, \tag{7.122}$$

$$\varphi_z = \tan^{-1} \frac{2\xi r}{1 - r^2} = 2.694\,\mathrm{rad} \approx 77.19\,\mathrm{deg}\,. \tag{7.123}$$

Therefore, the steady-state vibrations of the mass $m$ can be expressed by

$$x = 0.0106 \sin\left(140t - 1.744\right), \tag{7.124}$$

$$z = 0.0242 \sin\left(140t - 2.694\right). \tag{7.125}$$

The excitation $y$ and the steady-state responses $x$ and $z$ are plotted in Fig. 7.17 to illustrate their relative phase and amplitude at $r = 1.98$.

**Example 87** Comparison of frequency responses. Several equations and graphs of different dynamic characteristics of one-$DOF$ harmonically excited systems in dimensionless form are similar. Here are a few examples.

A comparison shows that Eq. (7.90) is equal to Eq. (7.37), and therefore the relative frequency response $\frac{Z}{Y}$ for a base excited system is the same as the acceleration frequency response $\frac{\ddot{X}}{F/m}$ for a force excited system.

Comparing Eqs. (7.89) and (7.41) indicates that the amplitude frequency response of a base excited system, $\frac{X}{Y}$ is the same as the transmitted force frequency response of a harmonically force excited system $\frac{F_T}{F}$. However, the phases of these two responses are different.

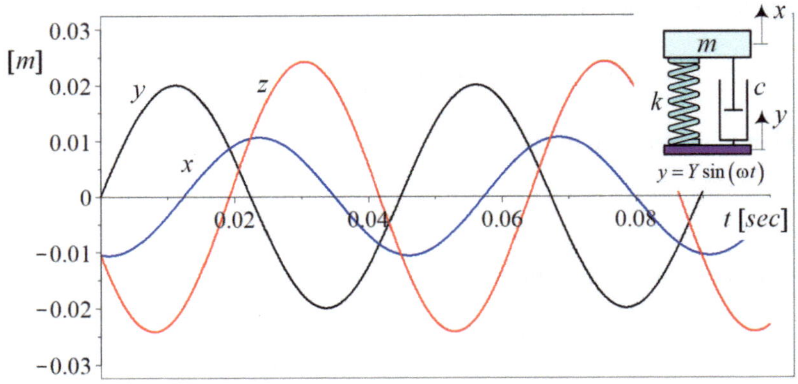


**Fig. 7.17** Excitation $y$ and steady-state responses $x$ and $z$ at $r = 1.98$

**Example 88** Absolute velocity and acceleration of a base excited system. This example shows how to derive and visualize the velocity and acceleration frequency responses of base excited systems.

Having the position frequency response of a base excited system,

$$x = A_2 \sin \omega t + B_2 \cos \omega t = X \sin \left( \omega t - \varphi_x \right), \tag{7.126}$$

we are able to calculate the velocity and acceleration frequency responses:

$$\begin{aligned} \dot{x} &= A_2 \omega \cos \omega t - B_2 \omega \sin \omega t \\ &= X \omega \cos \left( \omega t - \varphi_x \right) = \dot{X} \cos \left( \omega t - \varphi_x \right), \end{aligned} \tag{7.127}$$

$$\begin{aligned} \ddot{x} &= -A_2 \omega^2 \sin \omega t - B_2 \omega^2 \cos \omega t \\ &= -X \omega^2 \sin \left( \omega t - \varphi_x \right) = \ddot{X} \sin \left( \omega t - \varphi_x \right). \end{aligned} \tag{7.128}$$

The amplitude of velocity and acceleration frequency responses, $\dot{X}$ and $\ddot{X}$, are

$$\dot{X} = \frac{\omega \sqrt{k^2 + c^2 \omega^2}}{\sqrt{\left(k - m\omega^2\right)^2 + c^2 \omega^2}} Y, \tag{7.129}$$

$$\ddot{X} = \frac{\omega^2 \sqrt{k^2 + c^2 \omega^2}}{\sqrt{\left(k - m\omega^2\right)^2 + c^2 \omega^2}} Y, \tag{7.130}$$

which can also be written in dimensionless forms:

$$\frac{\dot{X}}{\omega_n Y} = \frac{r \sqrt{1 + (2\xi r)^2}}{\sqrt{\left(1 - r^2\right)^2 + (2\xi r)^2}}, \tag{7.131}$$

$$\frac{\ddot{X}}{\omega_n^2 Y} = \frac{r^2 \sqrt{1 + (2\xi r)^2}}{\sqrt{\left(1 - r^2\right)^2 + (2\xi r)^2}}. \tag{7.132}$$

The velocity and acceleration frequency responses (7.131) and (7.132) are plotted in Figs. 7.18 and 7.19.

There is a **switching point** or **node** in both figures at which the behavior of $\dot{X}$ and $\ddot{X}$ switches. Before the node, $\dot{X}$ and $\ddot{X}$ increase by increasing $\xi$, while they decrease after the node. To find the node, we can find the intersection between frequency response curves for $\xi = 0$ and $\xi = \infty$ because the amplitude at the node is independent of the damping. Let us apply this

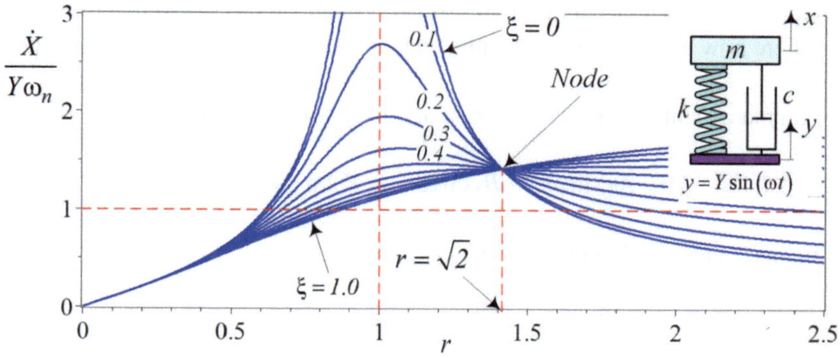


**Fig. 7.18** Velocity frequency response

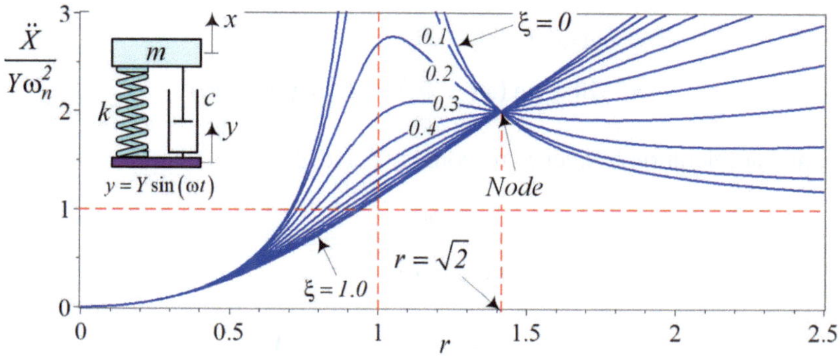


**Fig. 7.19** Acceleration frequency response

method to the acceleration frequency response:

$$\lim_{\xi \to 0} \frac{\ddot{X}}{\omega_n^2 Y} = \pm \frac{r^2}{\left(1 - r^2\right)}, \tag{7.133}$$

$$\lim_{\xi \to \infty} \frac{\ddot{X}}{\omega_n^2 Y} = \pm r^2. \tag{7.134}$$

Therefore, the frequency ratio $r$ at the intersection of these two limits is the solution of

$$r^2 \left(r^2 - 2\right) = 0. \tag{7.135}$$

The nodal frequency response is then equal to

$$r = \sqrt{2}. \tag{7.136}$$

The value of the acceleration frequency response at the node is only a function of $\xi$:

$$\lim_{r \to \sqrt{2}} \frac{\ddot{X}}{\omega_n^2 Y} = \frac{\sqrt{2\xi^2 + 1}}{\sqrt{8\xi^2 + 1}}. \tag{7.137}$$

Applying the same method to the velocity frequency response results in the same nodal frequency ratio, $r = \sqrt{2}$, with a different amplitude:

$$\lim_{r \to \sqrt{2}} \frac{\dot{X}}{\omega_n Y} = \sqrt{2} \frac{\sqrt{2\xi^2 + 1}}{\sqrt{8\xi^2 + 1}}. \tag{7.138}$$

**Example 89** Relative velocity and acceleration of a base excited system. As a way to determine the frequency response of relative variables, we show here how to calculate the relative velocity and relative acceleration frequency responses of a base excited one-$DOF$ system.

We can use the relative displacement frequency response of a base excited system,

$$z = A_3 \sin \omega t + B_3 \cos \omega t = Z \sin \left(\omega t - \varphi_z\right), \tag{7.139}$$

and calculate the relative velocity and acceleration frequency responses:

$$\begin{aligned} \dot{z} &= A_3 \omega \cos \omega t - B_3 \omega \sin \omega t \\ &= Z\omega \cos \left(\omega t - \varphi_z\right) = \dot{Z} \cos \left(\omega t - \varphi_z\right), \end{aligned} \tag{7.140}$$

$$\begin{aligned} \ddot{z} &= -A_3 \omega^2 \sin \omega t - B_3 \omega^2 \cos \omega t \\ &= -Z\omega^2 \sin \left(\omega t - \varphi_z\right) = \ddot{Z} \sin \left(\omega t - \varphi_z\right). \end{aligned} \tag{7.141}$$

The amplitude of velocity and acceleration frequency responses, $\dot{Z}$ and $\ddot{Z}$,

$$\dot{Z} = \frac{m\omega^3}{\sqrt{\left(k - m\omega^2\right)^2 + c^2\omega^2}} Y, \tag{7.142}$$

$$\ddot{Z} = \frac{m\omega^4}{\sqrt{\left(k - m\omega^2\right)^2 + c^2\omega^2}} Y, \tag{7.143}$$

can also be written in dimensionless forms:

$$\frac{\dot{Z}}{\omega_n Y} = \frac{r^3}{\sqrt{\left(1 - r^2\right)^2 + (2\xi r)^2}}, \tag{7.144}$$

$$\frac{\ddot{Z}}{\omega_n^2 Y} = \frac{r^4}{\sqrt{\left(1 - r^2\right)^2 + (2\xi r)^2}}. \tag{7.145}$$

**Example 90** Force transmitted to base of a base excited system. Force transmitted to the base is always an important design factor. Here is the frequency response of the force transmitted to the base of a base excited system.

The force transmitted $f_T$ to the ground by a base excited system is equal to the sum of forces in the spring and damper:

$$f_T = f_k + f_c = k\,(x - y) + c\,(\dot{x} - \dot{y}), \tag{7.146}$$

which, based on equation of motion (7.93), is also equal to

$$f_T = -m\ddot{x}. \tag{7.147}$$

Substituting $\ddot{x}$ from (7.128) and (7.132) shows that the frequency response of the transmitted force is

$$\frac{F_T}{kY} = \frac{\omega^2\sqrt{k^2 + c^2\omega^2}}{\sqrt{\left(k - m\omega^2\right)^2 + c^2\omega^2}} \tag{7.148}$$

$$= \frac{r^2\sqrt{1 + (2\xi r)^2}}{\sqrt{\left(1 - r^2\right)^2 + (2\xi r)^2}}. \tag{7.149}$$

The frequency response of $\frac{F_T}{kY}$ is the same as that shown in Fig. 7.19.

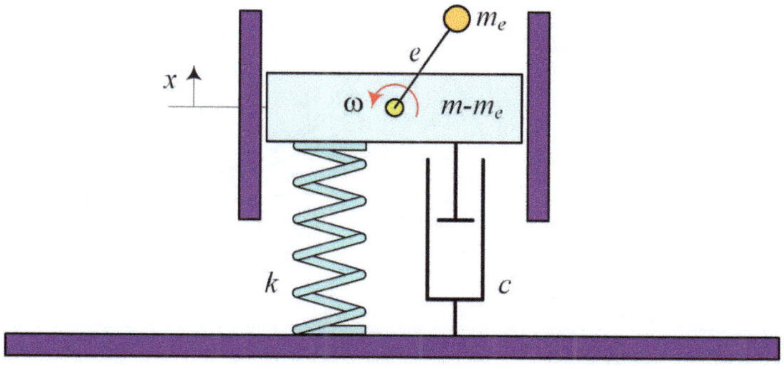


**Fig. 7.20** Accentric excited single-$DOF$ system

### 7.1.3 Eccentric Excitation

Figure 7.20 illustrates a one-$DOF$ eccentric excited vibrating system with a mass $m$ supported by a suspension made of a spring $k$ and a damper $c$. There is an unbalance mass $m_e$ at a distance $e$ that is rotating at an angular velocity $\omega$. An eccentric excited vibrating system is the proper model for the vibration analysis of an engine of a vehicle or any rotary motor that is mounted on a stationary base with flexible suspension.

The absolute motion of $m$ with respect to its equilibrium position is measured by the coordinate $x$. When the lateral motion of $m$ is protected, a harmonic excitation force is applied on $m$ in the $x$-direction and makes the system vibrate.

$$f_x = m_e e\omega^2 \sin \omega t. \tag{7.150}$$

The distance $e$ is called the *eccentricity*, and $m_e$ is called the *eccentric mass*.

The equation of motion of the system is

$$m\ddot{x} + c\dot{x} + kx = m_e e\omega^2 \sin \omega t \tag{7.151}$$

or, equivalently,

$$\ddot{x} + 2\xi\omega_n \dot{x} + \omega_n^2 x = \varepsilon e\omega^2 \sin \omega t, \tag{7.152}$$

$$\omega_n = \sqrt{\frac{k}{m}}, \qquad r = \frac{\omega}{\omega_n}, \qquad \xi = \frac{c}{2\sqrt{km}}, \qquad \varepsilon = \frac{m_e}{m}. \tag{7.153}$$

The absolute displacement response of the system is

$$x = A_4 \sin \omega t + B_4 \cos \omega t \tag{7.154}$$

$$= X \sin\left(\omega t - \varphi_e\right), \tag{7.155}$$

which has an amplitude $X$ and phases $\varphi_e$:

$$\frac{X}{e\varepsilon} = \frac{r^2}{\sqrt{\left(1 - r^2\right)^2 + (2\xi r)^2}}, \tag{7.156}$$

$$\varphi_e = \tan^{-1} \frac{2\xi r}{1 - r^2}. \tag{7.157}$$

The phase $\varphi_e$ indicates the angular lag of the response $x$ with respect to the excitation $m_e e\omega^2 \sin \omega t$. The frequency responses for $X$ and $\varphi_e$ as a function of $r$ and $\xi$ are plotted in Figs. 7.21 and 7.22.

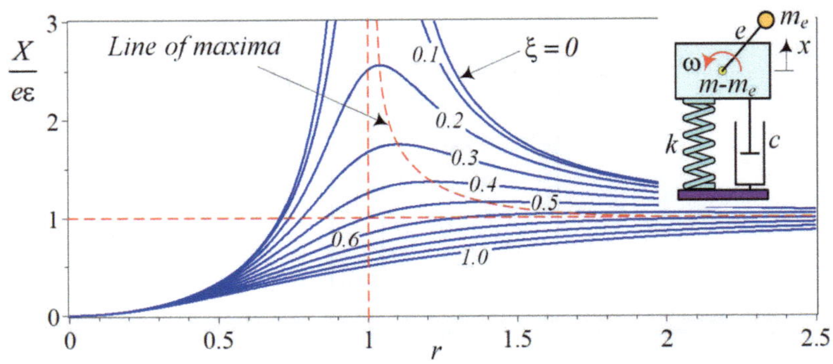


**Fig. 7.21** Position frequency response for $\frac{X}{e\varepsilon}$

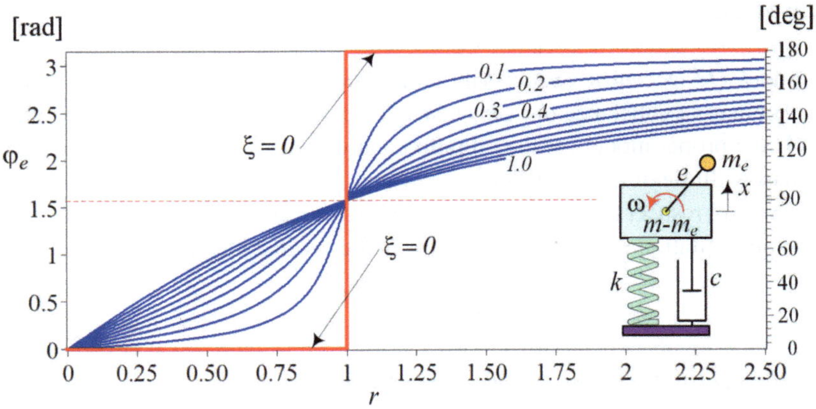


**Fig. 7.22** Frequency response for $\varphi_e$

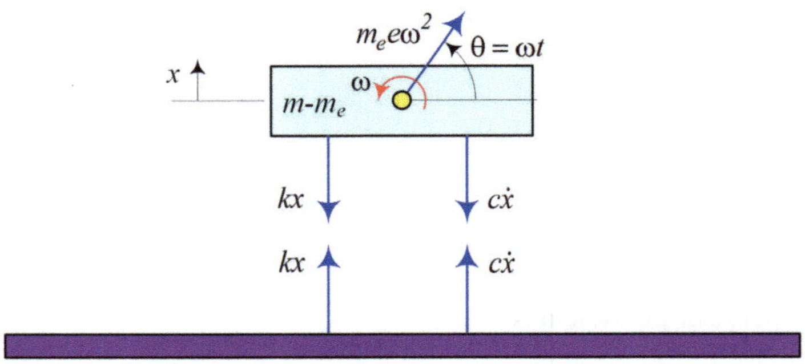


**Fig. 7.23** Free body diagram of eccentric excited single-$DOF$ system

***Proof*** Employing a free body diagram of the system, as shown in Fig. 7.23, and applying Newton's method in the $x$-direction generates the equation of motion

$$m\,\ddot{x} = -c\,\dot{x} - kx + m_e e\omega^2 \sin \omega t. \tag{7.158}$$

Equation (7.151) can be transformed into (7.152) by dividing by $m$ and using the definitions in (7.152) for natural frequency, damping ratio, and frequency ratio. The parameter $\varepsilon = m_e/m$ is the *mass ratio*, the ratio between the eccentric mass $m_e$ and the total mass $m$.

The steady-state solution of Equations (7.151) can be (7.154) or (7.155). To find the amplitude and phase of the response, we substitute solution (7.154) into the equation of motion

$$-m\omega^2 (A_4 \sin \omega t + B_4 \cos \omega t) + c\omega (A_4 \cos \omega t - B_4 \sin \omega t) \\ + k (A_4 \sin \omega t + B_4 \cos \omega t) = m_e e\omega^2 \sin \omega t. \tag{7.159}$$

The coefficients of the functions $\sin\omega t$ and $\cos\omega t$ must be balanced on both sides of the equation:

$$kA_4 - mA_4\omega^2 - cB_4\omega = m_e e\omega^2, \tag{7.160}$$

$$kB_4 - m\omega^2 B_4 + c\omega A_4 = 0. \tag{7.161}$$

Therefore, we find two algebraic equations to calculate $A_4$ and $B_4$:

$$\begin{bmatrix} k-\omega^2 m & -c\omega \\ c\omega & k-\omega^2 m \end{bmatrix}\begin{bmatrix} A_4 \\ B_4 \end{bmatrix} = \begin{bmatrix} e\omega^2 m_e \\ 0 \end{bmatrix}, \tag{7.162}$$

$$\begin{aligned}\begin{bmatrix} A_4 \\ B_4 \end{bmatrix} &= \begin{bmatrix} k-\omega^2 m & -c\omega \\ c\omega & k-\omega^2 m \end{bmatrix}^{-1}\begin{bmatrix} e\omega^2 m_e \\ 0 \end{bmatrix} \\ &= \begin{bmatrix} \dfrac{k-m\omega^2-\omega^2 m_e}{\left(k-\omega^2 m\right)^2+c^2\omega^2} e\omega^2 m_e \\ \dfrac{-c\omega}{\left(k-\omega^2 m\right)^2+c^2\omega^2} e\omega^2 m_e \end{bmatrix}.\end{aligned} \tag{7.163}$$

The solution provides us with steady-state solution (7.154). The amplitude $X$ and phase $\varphi_e$ can be found by

$$X = \sqrt{A_4^2 + B_4^2}, \tag{7.164}$$

$$\tan\varphi_e = \frac{-B_4}{A_4}, \tag{7.165}$$

which, after substituting $A_4$ and $B_4$ from (7.163), yield

$$X = \frac{\omega^2 e m_e}{\sqrt{\left(k-m\omega^2\right)^2+c^2\omega^2}}, \tag{7.166}$$

$$\tan\varphi_e = \frac{c\omega}{k-m\omega^2}. \tag{7.167}$$

A more practical expression for $X$ and $\varphi_e$ are Eqs. (7.156) and (7.157) by employing $r$ and $\xi$.

All rotating machines, such as engines, turbines, generators, and turning machines, have imperfections in their rotating components or have irregular mass distribution, which creates dynamic imbalances. When the unbalanced components rotate, an eccentric load applies to the structure. The load can be decomposed into two perpendicular harmonic forces in the plane of rotation in the lateral and normal directions. If the lateral force component is balanced by a reaction of a support, the normal component provides us with a harmonically variable force with an amplitude depending on the eccentricity $m_e e$. Unbalanced rotating machines are a common source of vibration excitation in electromechanical systems. ■

**Example 91** Eccentric excited system. This is a numerical study of an eccentric excited system to work with equations and compare the orders of magnitude.

Consider an engine with a mass $m$

$$m = 110\,\text{kg} \tag{7.168}$$

that is supported by four engine mounts, each with the following stiffness and damping values:

$$k = 100{,}000\,\text{N/m}, \qquad c = 1000\,\text{Ns/m}. \tag{7.169}$$

The engine runs at

$$\omega = 5000\, rpm \approx 523.6\,\text{rad/s} \approx 83.333\,\text{Hz} \tag{7.170}$$

with the following eccentric parameters:

$$m_e = 0.001\,\text{kg}, \qquad e = 0.12\,\text{m}. \tag{7.171}$$

The natural frequency $\omega_n$, damping ratio $\xi$, mass ratio $\varepsilon$ of the system, and frequency ratio $r$ are

$$\omega_n = \sqrt{\frac{k}{m}} = \sqrt{\frac{100{,}000}{110}} = 30.15\,\text{rad/s} \approx 4.8\,\text{Hz}, \tag{7.172}$$

$$\xi = \frac{c}{2\sqrt{km}} = 0.015, \tag{7.173}$$

$$\varepsilon = \frac{m_e}{m} = \frac{0.001}{110} = 9.0909 \times 10^{-6}, \tag{7.174}$$

$$r = \frac{\omega}{\omega_n} = \frac{523.60}{30.15} = 17.366\,. \tag{7.175}$$

The engine's amplitude of vibration is

$$X = \frac{r^2 e\varepsilon}{\sqrt{\left(1 - r^2\right)^2 + (2\xi r)^2}} = 1.094 \times 10^{-6}\,\text{m}. \tag{7.176}$$

However, if the speed of the engine is at the natural frequency of the system,

$$\omega = 30.15\,\text{rad/s} \approx 4.8\,\text{Hz}, \qquad r = 1, \tag{7.177}$$

then the amplitude of the engine's vibration increases to

$$X = \frac{r^2 e\varepsilon}{\sqrt{\left(1 - r^2\right)^2 + (2\xi r)^2}} = 3.61 \times 10^{-6}\,\text{m}. \tag{7.178}$$

**Example 92** Absolute velocity and acceleration of an eccentric excited system. The velocity and acceleration frequency responses of eccentric excited systems are calculated and visualized in this example.

Using the position frequency response of an eccentric excited system, we can find the velocity and acceleration frequency responses:

$$x = A_4 \sin \omega t + B_4 \cos \omega t = X \sin\left(\omega t - \varphi_e\right), \tag{7.179}$$

$$\begin{aligned} \dot{x} &= A_4\omega \cos \omega t - B_4\omega \sin \omega t \\ &= X\omega \cos\left(\omega t - \varphi_e\right) = \dot{X} \cos\left(\omega t - \varphi_e\right), \end{aligned} \tag{7.180}$$

$$\begin{aligned} \ddot{x} &= -A_4\omega^2 \sin \omega t - B_4\omega^2 \cos \omega t \\ &= -X\omega^2 \sin\left(\omega t - \varphi_e\right) = \ddot{X} \sin\left(\omega t - \varphi_e\right). \end{aligned} \tag{7.181}$$

The amplitude of velocity and acceleration frequency responses, $\dot{X}$ and $\ddot{X}$, are

$$\dot{X} = \frac{\omega^3 e m_e}{\sqrt{\left(k - m\omega^2\right)^2 + c^2\omega^2}}, \tag{7.182}$$

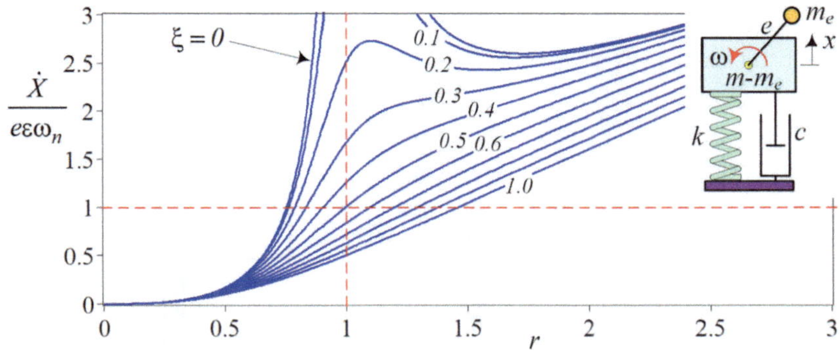


**Fig. 7.24** Velocity frequency response for $\frac{\dot{X}}{e\varepsilon\omega_n}$

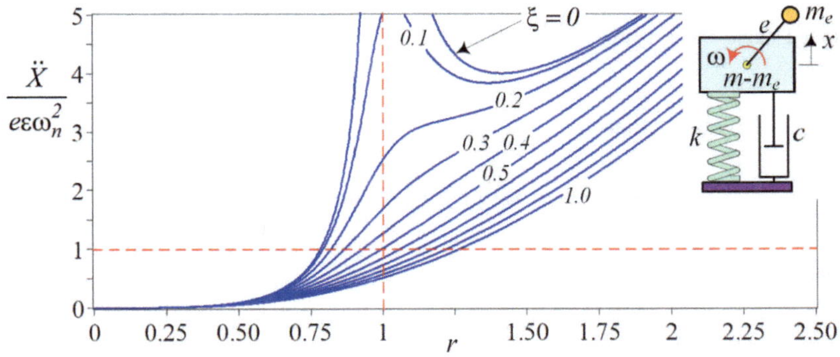


**Fig. 7.25** Acceleration frequency response for $\frac{\ddot{X}}{e\varepsilon\omega_n^2}$

$$\ddot{X} = \frac{\omega^4 e m_e}{\sqrt{\left(k - m\omega^2\right)^2 + c^2\omega^2}}, \tag{7.183}$$

which can be rewritten in dimensionless forms:

$$\frac{\dot{X}}{e\varepsilon\omega_n} = \frac{r^3}{\sqrt{\left(1 - r^2\right)^2 + (2\xi r)^2}}, \tag{7.184}$$

$$\frac{\ddot{X}}{e\varepsilon\omega_n^2} = \frac{r^4}{\sqrt{\left(1 - r^2\right)^2 + (2\xi r)^2}}. \tag{7.185}$$

Figures 7.24 and 7.25 illustrate the velocity and acceleration frequency responses $\frac{\dot{X}}{e\varepsilon\omega_n}$ and $\frac{\ddot{X}}{e\varepsilon\omega_n^2}$, respectively.

**Example 93** Force transmitted to base of eccentric excited system. For an eccentric excited system, the force transmitted to the base foundation is a key factor in designing a foundation or managing the installation of a machine at higher levels of a building. The frequency response of force transmitted to a base is derived as follows.

The force transmitted $f_T$ to the ground by an eccentric excited system is equal to the sum of forces in the spring and damper:

$$f_T = F_T \sin(\omega t - \varphi_T), \tag{7.186}$$

$$f_T = f_k + f_c = kx + c\dot{x}. \tag{7.187}$$

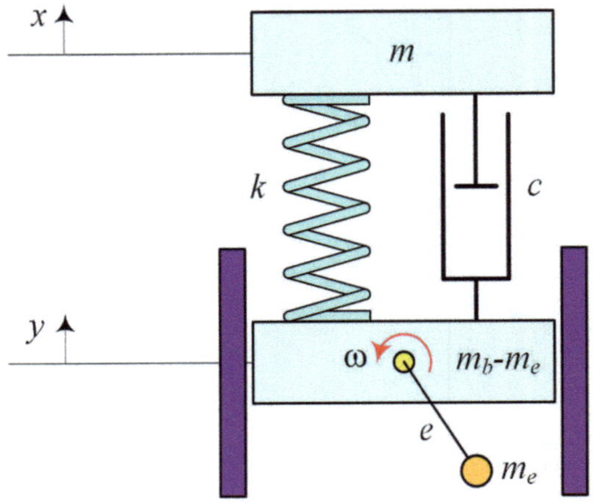


**Fig. 7.26** Eccentric base excited single-$DOF$ system

Substituting $x$ and $\dot{x}$ from (7.154) shows that

$$f_T = (kA_4 - c\omega B_4)\sin t\omega + (kB_4 + c\omega A_4)\cos t\omega; \tag{7.188}$$

therefore, the amplitude of the transmitted force is

$$\begin{aligned} F_T &= \sqrt{(kA_4 - c\omega B_4)^2 + (kB_4 + c\omega A_4)^2} \\ &= e\omega^2 m_e \sqrt{\frac{c^2\omega^2 + k^2}{\left(k - m\omega^2\right)^2 + c^2\omega^2}}. \end{aligned} \tag{7.189}$$

The frequency response of the transmitted force can be simplified to the following applied dimensionless equation:

$$\frac{F_T}{e\omega^2 m_e} = \frac{\sqrt{1 + (2\xi r)^2}}{\sqrt{\left(1 - r^2\right)^2 + (2\xi r)^2}}. \tag{7.190}$$

The graph of $F_T/\left(e\omega^2 m_e\right)$ is the same as that of $X/Y$ in Fig. 7.13 for the absolute displacement frequency response of a base excited system.

### 7.1.4 ★ Eccentric Base Excitation

Figure 7.26 illustrates a one-$DOF$ eccentric base excited vibrating system with a mass $m$ suspended by a spring $k$ and a damper $c$ on a base with mass $m_b$. The base has an unbalance mass $m_e$ at an eccentricity distance $e$ that rotates at an angular velocity $\omega$. The eccentric base excited system is the proper model for the vibration analysis of various types of equipment attached to the engine of a vehicle or any equipment mounted on a rotary motor.

The absolute motion of the mass $m$ and the base are measured by variables $x$ and $y$, respectively. Using the relative motion of $m$ with respect to the base,

$$z = x - y, \tag{7.191}$$

we develop the equation of motion as

$$\frac{mm_b}{m_b + m}\ddot{z} + c\,\dot{z} + kz = \frac{mm_e}{m_b + m}e\omega^2 \sin\omega t \tag{7.192}$$

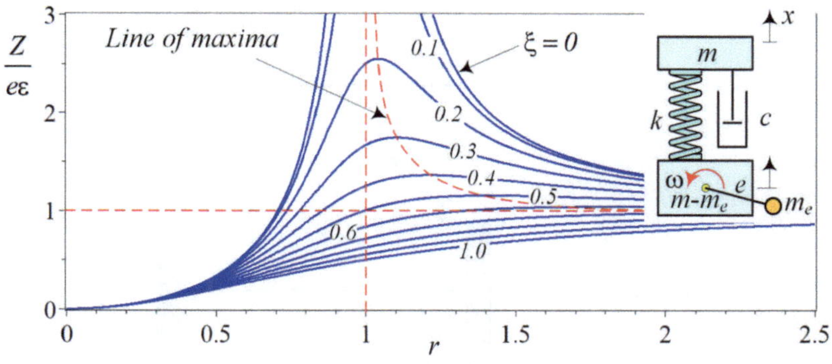


**Fig. 7.27** Position frequency response for $\dfrac{Z}{e\varepsilon}$

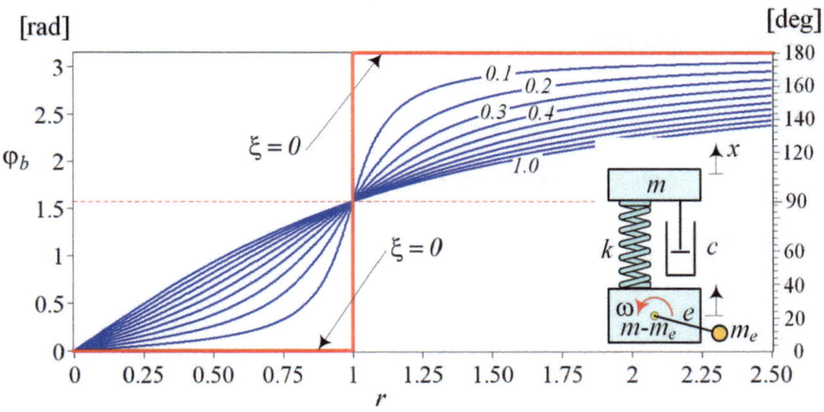


**Fig. 7.28** Frequency response for $\varphi_b$

or, equivalently,

$$\ddot{z} + 2\xi\omega_n\,\dot{z} + \omega_n^2\,z = \varepsilon e\omega^2 \sin\omega t, \tag{7.193}$$

$$\omega_n = \sqrt{\frac{k}{m}}, \qquad r = \frac{\omega}{\omega_n}, \qquad \xi = \frac{c}{2\sqrt{km}}, \qquad \varepsilon = \frac{m_e}{m_b}. \tag{7.194}$$

The relative displacement response of the system is

$$z = A_5 \sin\omega t + B_5 \cos\omega t \tag{7.195}$$

$$= Z \sin\left(\omega t - \varphi_b\right), \tag{7.196}$$

which has an amplitude $Z$ and phase $\varphi_b$:

$$\frac{Z}{e\varepsilon} = \frac{r^2}{\sqrt{\left(1 - r^2\right)^2 + (2\xi r)^2}}, \tag{7.197}$$

$$\varphi_b = \tan^{-1}\frac{2\xi r}{1 - r^2}. \tag{7.198}$$

The frequency responses for $Z$ and $\varphi_b$ as functions of $r$ and $\xi$ are plotted in Figs. 7.27 and 7.28.

***Proof*** The free body diagram in Fig. 7.29, along with Newton's equation in the $x$-direction, will be used to obtain the equations of absolute motion:

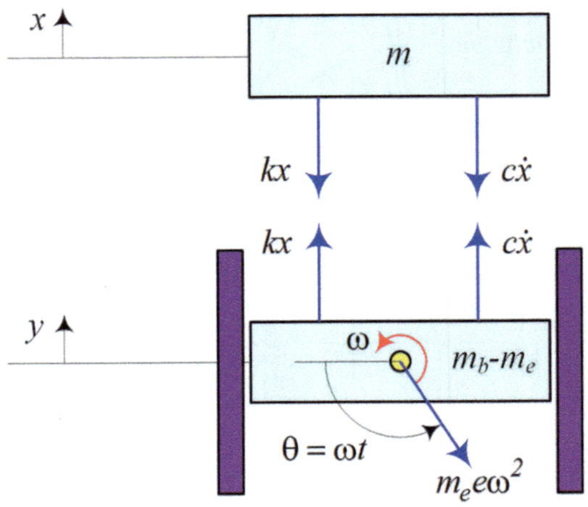


**Fig. 7.29** Free body diagram of eccentric base excited single-$DOF$ system

$$m\ddot{x} = -c(\dot{x} - \dot{y}) - k(x - y), \tag{7.199}$$

$$m_b\ddot{y} = c(\dot{x} - \dot{y}) + k(x - y) - m_e e\omega^2 \sin\omega t. \tag{7.200}$$

Using $z = x - y$ and its derivatives,

$$\dot{z} = \dot{x} - \dot{y}, \tag{7.201}$$

$$\ddot{z} = \ddot{x} - \ddot{y}, \tag{7.202}$$

we combine Eqs. (7.199) and (7.200) to obtain the equation of relative motion:

$$\frac{mm_b}{m_b + m}\ddot{z} + c\,\dot{z} + kz = \frac{mm_e}{m_b + m}e\omega^2 \sin\omega t. \tag{7.203}$$

Equation (7.203) can be transformed into (7.193) if we divide it by $\frac{mm_b}{m_b+m}$ and use the following definitions:

$$\xi = \frac{c}{2\sqrt{k\dfrac{mm_b}{m_b + m}}}, \tag{7.204}$$

$$\omega_n = \sqrt{k\Big/\frac{mm_b}{m_b + m}} = \sqrt{k\frac{m_b + m}{mm_b}}. \tag{7.205}$$

The parameter $\varepsilon = m_e/m_b$ is the *mass ratio* and indicates the ratio between the eccentric mass $m_e$ and the total base mass $m_b$.

The steady-state solutions of Equation (7.193) are (7.195) and (7.196). To find the amplitude and phase of the response, we substitute the solution (7.195) into the equation of motion:

$$\begin{aligned} -\omega^2(A_5 \sin\omega t + B_5 \cos\omega t) + 2\xi\omega_n\omega(A_5 \cos\omega t - B_5 \sin\omega t) \\ + \omega_n^2(A_5 \sin\omega t + B_5 \cos\omega t) = \varepsilon e\omega^2 \sin\omega t \end{aligned}$$

$$. \tag{7.206}$$

The coefficients of the functions $\sin\omega t$ and $\cos\omega t$ must balance on both sides of the equations:

$$\omega_n^2 A_5 - \omega^2 A_5 - 2\xi\omega\omega_n B_5 = \varepsilon\omega^2 e, \tag{7.207}$$

$$2\xi A_5\omega\omega_n - B_5\omega^2 + B_5\omega_n^2 = 0. \tag{7.208}$$

Therefore, we obtain two algebraic equations to calculate $A_5$ and $B_5$:

$$\begin{bmatrix} \omega_n^2 - \omega^2 & -2\xi\omega\omega_n \\ 2\xi\omega\omega_n & \omega_n^2 - \omega^2 \end{bmatrix} \begin{bmatrix} A_5 \\ B_5 \end{bmatrix} = \begin{bmatrix} \varepsilon\omega^2 e \\ 0 \end{bmatrix}. \tag{7.209}$$

Solving them for coefficients $A_5$ and $B_5$ provides us with the steady-state solution (7.195):

$$\begin{bmatrix} A_5 \\ B_5 \end{bmatrix} = \begin{bmatrix} \omega_n^2 - \omega^2 & -2\xi\omega\omega_n \\ 2\xi\omega\omega_n & \omega_n^2 - \omega^2 \end{bmatrix}^{-1} \begin{bmatrix} \varepsilon\omega^2 e \\ 0 \end{bmatrix} = \begin{bmatrix} \dfrac{\omega_n^2 - \omega^2}{\left(\omega_n^2 - \omega^2\right)^2 + (2\xi\omega\omega_n)^2}\varepsilon\omega^2 e \\ \dfrac{-2\xi\omega\omega_n}{\left(\omega_n^2 - \omega^2\right)^2 + (2\xi\omega\omega_n)^2}\varepsilon\omega^2 e \end{bmatrix}. \tag{7.210}$$

The amplitude $Z$ and phase $\varphi_b$ will be

$$X = \sqrt{A_5^2 + B_5^2} \tag{7.211}$$

$$\tan\varphi_b = \frac{-B_5}{A_5} \tag{7.212}$$

which, after inserting $A_5$ and $B_5$ from (7.210), yields

$$Z = \frac{\omega^2 e\varepsilon}{\sqrt{\left(\omega_n^2 - \omega^2\right)^2 + (2\xi\omega\omega_n)^2}}, \tag{7.213}$$

$$\tan\varphi_b = \frac{2\xi\omega\omega_n}{\omega_n^2 - \omega^2}. \tag{7.214}$$

Equations (7.213) and (7.214) can be simplified to the more practical expressions (7.197) and (7.198) by employing $r = \omega/\omega_n$. ■

**Example 94** ★ A base eccentric excited system. A numerical substitution of an eccentric base excited system will show how equations work and the orders of magnitude.

Consider an engine with a mass $m_b$

$$m_b = 110\,\text{kg} \tag{7.215}$$

and an air intake device with a mass $m$ that is mounted on an engine using an elastic mount with the following equivalent stiffness and damping:

$$m = 2\,\text{kg}, \qquad k = 10{,}000\,\text{N/m}, \qquad c = 100\,\text{Ns/m}. \tag{7.216}$$

The engine runs at

$$\omega = 576.0\,rpm \approx 60.302\,\text{rad/s} \approx 9.6\,\text{Hz}, \tag{7.217}$$

with the eccentric parameters $m_e$ and $e$:

$$m_e = 0.001\,\text{kg}, \qquad e = 0.12\,\text{m}. \tag{7.218}$$

The natural frequency $\omega_n$, damping ratio $\xi$, mass ratio $\varepsilon$ of the system, and frequency ratio $r$ are as follows:

$$\omega_n = \sqrt{k\frac{m_b + m}{m m_b}} = 71.35\,\text{rad/s} \approx 11.356\,\text{Hz}, \tag{7.219}$$

$$\xi = \frac{c}{2\sqrt{k\dfrac{m m_b}{m_b + m}}} = 0.357, \tag{7.220}$$

$$\varepsilon = \frac{m_e}{m_b} = 0.5 \times 10^{-3}, \qquad r = \frac{\omega}{\omega_n} = 0.845\,. \tag{7.221}$$

The relative amplitude of the device's vibration is

$$Z = \frac{e\varepsilon r^2}{\sqrt{\left(1 - r^2\right)^2 + (2\xi r)^2}} = 64.225 \times 10^{-6}\,\text{m}. \tag{7.222}$$

**Example 95** ★ Absolute displacement of upper mass of eccentric base excited system. The absolute displacement of the upper mass and relative displacement between the upper and lower masses are the two important measurements for real-world machinery. The derivation of the equation to determine the absolute displacement of the upper mass is as follows.

Equation (7.199), along with solution (7.195), can be used to calculate the displacement frequency response of the upper mass $m$ for an eccentric base excited system, such as that shown in Fig. 7.27:

$$\ddot{x} = -\frac{c}{m}\left(\dot{x} - \dot{y}\right) - \frac{k}{m}\left(x - y\right) = -\frac{c}{m}\dot{z} - \frac{k}{m}z. \tag{7.223}$$

Assuming a steady-state displacement

$$x = A_6 \sin \omega t + B_6 \cos \omega t = X \sin\left(\omega t - \varphi_{bx}\right). \tag{7.224}$$

We have

$$\begin{aligned} -\omega^2\left(A_6 \sin \omega t + B_6 \cos \omega t\right) &= -\frac{c}{m}\dot{z} - \frac{k}{m}z \\ = -\frac{c}{m}\omega\left(A_5 \cos \omega t - B_5 \sin \omega t\right) &- \frac{k}{m}\left(A_5 \sin \omega t + B_5 \cos \omega t\right) \\ = \left(\frac{c}{m}\omega B_5 - \frac{k}{m}A_5\right)\sin t\omega &+ \left(-\frac{k}{m}B_5 - \frac{c}{m}\omega A_5\right)\cos t\omega, \end{aligned} \tag{7.225}$$

and therefore,

$$-\omega^2 A_6 = \frac{c}{m}\omega B_5 - \frac{k}{m}A_5, \tag{7.226}$$

$$-\omega^2 B_6 = -\frac{k}{m}B_5 - \frac{c}{m}\omega A_5. \tag{7.227}$$

Substituting $A_5$ and $B_5$ from (7.210) and using

$$X = \sqrt{A_6^2 + B_6^2} \qquad \tan \varphi_{bx} = \frac{-B_6}{A_6} \tag{7.228}$$

yields

$$A_6 = -\frac{2c\xi\omega^2\omega_n + k\left(\omega_n^2 - \omega^2\right)}{\left(\omega_n^2 - \omega^2\right)^2 + \left(2\xi\omega\omega_n\right)^2}\frac{1}{m}\varepsilon e, \tag{7.229}$$

$$B_6 = \frac{-c\left(\omega_n^2 - \omega^2\right) + 2k\xi\omega_n}{\left(\omega_n^2 - \omega^2\right)^2 + (2\xi\omega\omega_n)^2}\frac{1}{m}\varepsilon\omega e, \tag{7.230}$$

and therefore, the amplitude $X$ of steady-state vibration of the upper mass in an eccentric base excited system is

$$X = \frac{\sqrt{c^2\omega^2 + k^2}}{\sqrt{\left(\omega_n^2 - \omega^2\right)^2 + (2\xi\omega\omega_n)^2}}\frac{\varepsilon}{m}e, \tag{7.231}$$

$$\tan\varphi_{bx} = -\frac{-c\left(\omega_n^2 - \omega^2\right) + 2k\xi\omega_n}{2c\xi\omega^2\omega_n + k\left(\omega_n^2 - \omega^2\right)}\omega. \tag{7.232}$$

## 7.2 Classification of Frequency Responses of One-$DOF$ Systems

A harmonically excited one-$DOF$ system can be one of the four systems shown in Fig. 7.30. Any steady-state response of these systems is equal to one of the following equations, (7.233)–(7.240), and the phase of the motion is equal to one of the Eqs. (7.241)–(7.244):

$$S_0 = \frac{1}{\sqrt{\left(1 - r^2\right)^2 + (2\xi r)^2}}, \tag{7.233}$$

$$S_1 = \frac{r}{\sqrt{\left(1 - r^2\right)^2 + (2\xi r)^2}} = r S_0, \tag{7.234}$$

$$S_2 = \frac{r^2}{\sqrt{\left(1 - r^2\right)^2 + (2\xi r)^2}} = r^2 S_0, \tag{7.235}$$

$$S_3 = \frac{r^3}{\sqrt{\left(1 - r^2\right)^2 + (2\xi r)^2}} = r^3 S_0, \tag{7.236}$$

$$S_4 = \frac{r^4}{\sqrt{\left(1 - r^2\right)^2 + (2\xi r)^2}} = r^4 S_0, \tag{7.237}$$

$$G_0 = \frac{\sqrt{1 + (2\xi r)^2}}{\sqrt{\left(1 - r^2\right)^2 + (2\xi r)^2}}, \tag{7.238}$$

$$G_1 = \frac{r\sqrt{1 + (2\xi r)^2}}{\sqrt{\left(1 - r^2\right)^2 + (2\xi r)^2}} = r G_0, \tag{7.239}$$

$$G_2 = \frac{r^2\sqrt{1 + (2\xi r)^2}}{\sqrt{\left(1 - r^2\right)^2 + (2\xi r)^2}} = r^2 G_0, \tag{7.240}$$

$$\Phi_0 = \tan^{-1}\frac{2\xi r}{1 - r^2} \tag{7.241}$$

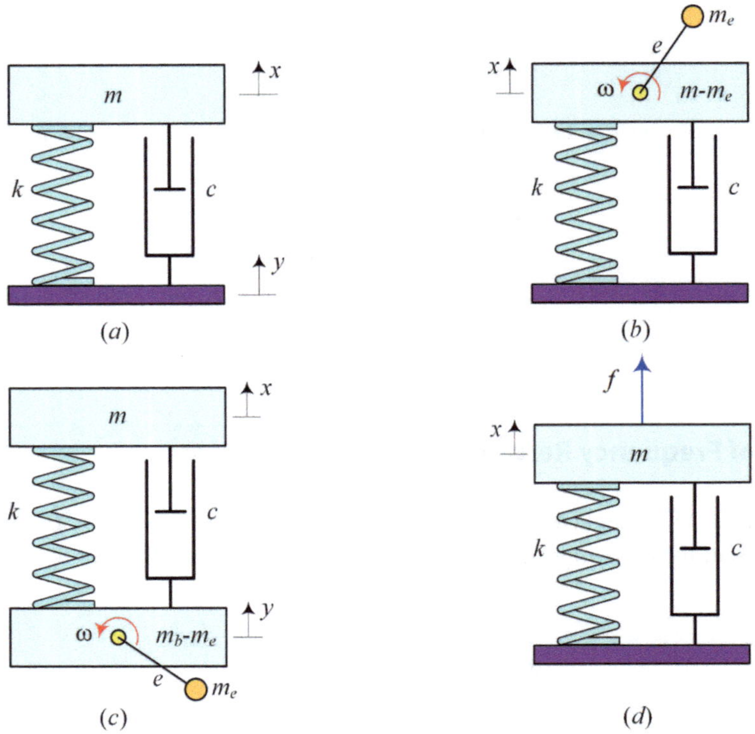


**Fig. 7.30** Four practical types of one-$DOF$ harmonically excited systems: $a$—base excitation, $b$—eccentric excitation, $c$—eccentric base excitation, $d$—forced excitation

$$\Phi_1 = \tan^{-1}\frac{1-r^2}{-2\xi r}, \tag{7.242}$$

$$\Phi_2 = \tan^{-1}\frac{-2\xi r}{1-r^2}, \tag{7.243}$$

$$\Phi_3 = \tan^{-1}\frac{2\xi r^3}{\left(1-r^2\right)^2+(2\xi r)^2}. \tag{7.244}$$

The functions $S_0$ and $G_0$ are the main parts of all kinds of frequency response of single-$DOF$ harmonically excited systems of Fig. 7.30. To have a visual intuition of the behavior of the different responses, we usually plot them as a function of $r$ and use $\xi$ as a parameter. The mass $m$, stiffness $k$, and damping $c$ of the system are fixed; hence, the excitation frequency $\omega$ is the only quasi-static variable. We combine $m$, $k$, $c$, and $\omega$, and we define the two parameters $r$ and $\xi$ to express the frequency responses by only two variable functions.

To classify the frequency responses of one-$DOF$ forced vibration systems, let us show the absolute displacement of the main mass by $x$, displacement excitation of the base by $y$, and relative displacement by $z = x - y$. We indicate the frequency responses related to the systems of Fig. 7.30 by adding a subscript and express their different responses as follows:

1. For a *base excited system*, we use the frequency responses of the relative and absolute motions denoted by $Z_B$, $\dot{Z}_B$, and $\ddot{Z}_B$ for relative displacement, velocity, and acceleration and $X_B$, $\dot{X}_B$, and $\ddot{X}_B$ for absolute displacement, velocity, and acceleration of $m$, and we denote the force transmitted to the base by $F_{T_B}$.
2. For an *eccentric excited system*, we use the frequency responses of the absolute motions $X_E$, $\dot{X}_E$, and $\ddot{X}_E$, and we denote the force transmitted to the base by $F_{T_E}$.
3. For an *eccentric base excited system*, we use the frequency responses of the relative and absolute motions denoted by $Z_R$, $\dot{Z}_R$, and $\ddot{Z}_R$ for relative displacement, velocity, and acceleration and $X_R$, $\dot{X}_R$, and $\ddot{X}_R$ for the absolute displacement, velocity, and acceleration of $m$, and we denote the force transmitted to the base by $F_{T_R}$.
4. For a *force excited system*, we use the frequency responses of the absolute motions $X_F$, $\dot{X}_F$, and $\ddot{X}_F$, and we denote the force transmitted to the base by $F_{T_F}$.

We summarize and label the frequency response of different characteristics of the four systems of Fig. 7.30 by the following nondimensionalized equations:

$$S_0 = \frac{X_F}{F/k},$$

$$S_1 = \frac{\dot{X}_F}{F/\sqrt{km}}, \tag{7.245}$$

$$S_2 = \frac{\ddot{X}_F}{F/m} = \frac{Z_B}{Y} = \frac{X_E}{e\varepsilon_E} = \frac{Z_R}{e\varepsilon_R}, \tag{7.246}$$

$$S_3 = \frac{\dot{Z}_B}{\omega_n Y} = \frac{\dot{X}_E}{e\varepsilon_E \omega_n} = \frac{\dot{Z}_R}{e\varepsilon_R \omega_n}, \tag{7.247}$$

$$S_4 = \frac{\ddot{Z}_B}{\omega_n^2 Y} = \frac{\ddot{X}_E}{e\varepsilon_E \omega_n^2} = \frac{\ddot{Z}_R}{e\varepsilon_R \omega_n^2}, \tag{7.248}$$

$$G_0 = \frac{F_{T_F}}{F} = \frac{X_B}{Y}, \tag{7.249}$$

$$G_1 = \frac{\dot{X}_B}{\omega_n Y}, \tag{7.250}$$

$$G_2 = \frac{\ddot{X}_B}{\omega_n^2 Y} = \frac{F_{T_B}}{kY} = \frac{F_{T_E}}{e\omega_n^2 m_e} = \frac{F_{T_R}}{e\omega_n^2 m_e}\left(1 + \frac{m_b}{m}\right). \tag{7.251}$$

Appendix A visualizes these frequency responses to guide design engineers to evaluate the behavior of the systems. The exact value of the responses should be found from the associated equations.

***Proof*** The equations of motion of a harmonically forced vibrating one-$DOF$ system can always be modeled by a second-order differential equation, where the variable $q$ is a general coordinate to show the absolute displacement $x$ or relative displacement $z = x - y$:

$$m\ddot{q} + c\dot{q} + kq = f(q, \dot{q}, t). \tag{7.252}$$

The forcing term $f(x, \dot{x}, t)$ is a harmonic function, which in the general case can be a combination of $\sin \omega t$ and $\cos \omega t$, where $\omega$ is the excitation frequency:

$$f(q, \dot{q}, t) = a \sin \omega t + b \cos \omega t. \tag{7.253}$$

Depending on the system and the frequency response that we are investigating, the coefficients $a$ and $b$ can be zero, constant, or proportional to $\omega$, $\omega^2$, $\omega^3$, $\omega^4$, $\cdots$, $\omega^n$. To cover every practical harmonically forced vibrating system and derive Equations (7.233)–(7.240), let us assume $a$ and $b$ to be polynomial functions of $\omega$:

$$a = a_0 + a_1\omega + a_2\omega^2, \tag{7.254}$$

$$b = b_0 + b_1\omega + b_2\omega^2. \tag{7.255}$$

We divide the equation of motion (7.252) by $m$ to express it with $\xi$ and $\omega_n$:

$$\ddot{q} + 2\xi\omega_n\dot{q} + \omega_n^2 q = \left(A_0 + A_1\omega + A_2\omega^2\right)\sin \omega t + \left(B_0 + B_1\omega + B_2\omega^2\right)\cos \omega t. \tag{7.256}$$

$$A_0 + A_1\omega + A_2\omega^2 = \frac{1}{m}\left(a_0 + a_1\omega + a_2\omega^2\right), \tag{7.257}$$

$$B_0 + B_1\omega + B_2\omega^2 = \frac{1}{m}\left(b_0 + b_1\omega + b_2\omega^2\right). \tag{7.258}$$

The solution to the equation of motion would be a harmonic response with unknown coefficients:

$$q = A\sin\omega t + B\cos\omega t = Q\sin(\omega t - \varphi). \tag{7.259}$$

To calculate the steady-state amplitude of the response $Q$ and phase $\varphi$, we should substitute solution (7.259) into the equation of motion:

$$\begin{aligned} &-\omega^2(A\sin\omega t + B\cos\omega t) + 2\xi\omega_n\omega(A\cos\omega t - B\sin\omega t) \\ &+\omega_n^2(A\sin\omega t + B\cos\omega t) \\ &= \left(A_0 + A_1\omega + A_2\omega^2\right)\sin\omega t \\ &\quad + \left(B_0 + B_1\omega + B_2\omega^2\right)\cos\omega t\,. \end{aligned} \tag{7.260}$$

The coefficients of the functions $\sin\omega t$ and $\cos\omega t$ must balance on both sides of the equation:

$$\omega_n^2 A - \omega^2 A - 2\xi\omega\omega_n B = A_0 + A_1\omega + A_2\omega^2, \tag{7.261}$$

$$2\xi A\omega\omega_n - B\omega^2 + B\omega_n^2 = B_0 + B_1\omega + B_2\omega^2. \tag{7.262}$$

Therefore, we obtain two algebraic equations to calculate $A$ and $B$:

$$\begin{bmatrix} \omega_n^2 - \omega^2 & -2\xi\omega\omega_n \\ 2\xi\omega\omega_n & \omega_n^2 - \omega^2 \end{bmatrix}\begin{bmatrix} A \\ B \end{bmatrix} = \begin{bmatrix} A_0 + A_1\omega + A_2\omega^2 \\ B_0 + B_1\omega + B_2\omega^2 \end{bmatrix}. \tag{7.263}$$

Solving for coefficients $A$ and $B$ provides us with the steady-state solution amplitude $Q$ and phase $\varphi$:

$$\begin{aligned} \begin{bmatrix} A \\ B \end{bmatrix} &= \begin{bmatrix} \omega_n^2 - \omega^2 & -2\xi\omega\omega_n \\ 2\xi\omega\omega_n & \omega_n^2 - \omega^2 \end{bmatrix}^{-1}\begin{bmatrix} A_0 + A_1\omega + A_2\omega^2 \\ B_0 + B_1\omega + B_2\omega^2 \end{bmatrix} \\ &= \begin{bmatrix} \dfrac{Z_1}{\left(1 - r^2\right)^2 + (2\xi r)^2} \\ \dfrac{Z_2}{\left(1 - r^2\right)^2 + (2\xi r)^2} \end{bmatrix}, \end{aligned} \tag{7.264}$$

$$Z_1 = 2\xi r\frac{1}{\omega_n^2}\left(B_2\omega^2 + B_1\omega + B_0\right) + \frac{1}{\omega_n^2}\left(1 - r^2\right)\left(A_2\omega^2 + A_1\omega + A_0\right), \tag{7.265}$$

$$Z_2 = \frac{1}{\omega_n^2}\left(1 - r^2\right)\left(B_2\omega^2 + B_1\omega + B_0\right) - 2\xi r\frac{1}{\omega_n^2}\left(A_2\omega^2 + A_1\omega + A_0\right), \tag{7.266}$$

$$Q = \sqrt{A^2 + B^2}, \tag{7.267}$$

$$\tan\varphi = \frac{-B}{A}. \tag{7.268}$$

We are able to reproduce any of the steady-state responses $S_i$ and $G_i$ by setting the coefficients $A_0$, $A_1$, $A_2$, $B_0$, $B_1$, and $B_2$ properly. ■

**Example 96** Base excited frequency responses. This is an example to show that the steady-state frequency responses $S_i$ and $G_i$ can be recovered by substituting $A_0$, $A_1$, $A_2$, $B_0$, $B_1$, and $B_2$ from the equation of motion of the system.

A one-$DOF$ base excited vibrating system is shown in Fig. 7.30a. The equation of relative motion $z = x - y$ with a harmonic base excitation $y = Y \sin \omega t$ is

$$\ddot{z} + 2\xi \omega_n \dot{z} + \omega_n^2 z = \omega^2 Y \sin \omega t. \tag{7.269}$$

This equation can be obtained from Eq. (7.256) if $q = z$ and

$$\begin{aligned} A_0 &= 0, & A_1 &= 0, & A_2 &= Y, \\ B_0 &= 0, & B_1 &= 0, & B_2 &= 0. \end{aligned} \tag{7.270}$$

Therefore, the frequency response of the system would be

$$\begin{aligned} Z = Q &= \sqrt{A^2 + B^2} \\ &= \sqrt{\left(\frac{1}{\omega_n^2} \frac{(1-r^2)\, Y\omega^2}{(1-r^2)^2 + (2\xi r)^2}\right)^2 + \left(\frac{1}{\omega_n^2} \frac{-2\xi r Y \omega^2}{(1-r^2)^2 + (2\xi r)^2}\right)^2} \\ &= \frac{r^2}{\sqrt{(1 - r^2)^2 + (2\xi r)^2}} Y. \end{aligned} \tag{7.271}$$

**Example 97** Two-term forcing. Let us use the superposition principle and the general solutions $S_i$ and $G_i$ to determine the frequency response of a system to a multiharmonic forcing function.

Consider a forced vibration system as shown in Fig. 7.30d:

$$m\ddot{x} + c\dot{x} + kx = f, \tag{7.272}$$

to which the following harmonic forcing load is applied:

$$f = F_1 \sin \omega t + F_2 \cos \omega t. \tag{7.273}$$

Employing the parameters

$$\begin{aligned} A_0 &= \frac{F_1}{m}, & A_1 &= 0, & A_2 &= 0, \\ B_0 &= \frac{F_2}{m}, & B_1 &= 0, & B_2 &= 0 \end{aligned} \tag{7.274}$$

provides us with the following frequency response:

$$x = A \sin \omega t + B \cos \omega t = Q \sin (\omega t - \varphi), \tag{7.275}$$

$$A = \frac{1}{m\omega_n^2} \frac{2\xi r F_2 + (1 - r^2) F_1}{(1 - r^2)^2 + (2\xi r)^2} \tag{7.276}$$

$$B = \frac{1}{m\omega_n^2} \frac{(1 - r^2) F_2 - 2\xi r F_1}{(1 - r^2)^2 + (2\xi r)^2}, \tag{7.277}$$

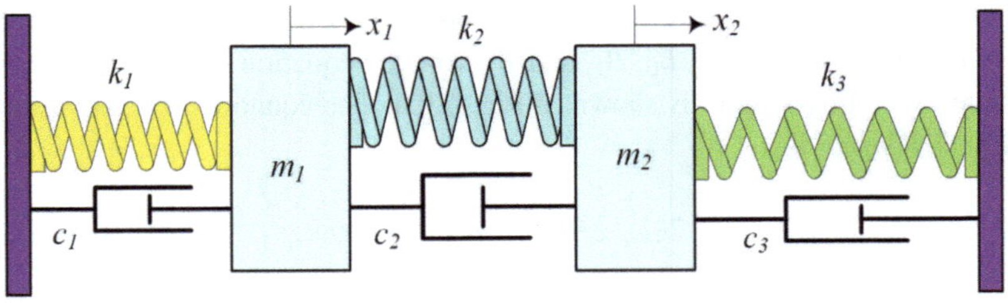


**Fig. 7.31** A two-$DOF$ vibrating system

$$Q = \frac{1}{m\omega_n^2} \frac{\sqrt{F_1^2 + F_2^2}}{\sqrt{\left(1 - r^2\right)^2 + (2\xi r)^2}}. \tag{7.278}$$

$$\varphi = \tan^{-1} \frac{2\xi r F_1 - \left(1 - r^2\right) F_2}{2\xi r F_2 + \left(1 - r^2\right) F_1}. \tag{7.279}$$

Let us also solve the problem using the superposition principle. Two harmonic forces are applied to the system:

$$f_1 = F_1 \sin \omega t\,, \qquad f_2 = F_2 \cos \omega t. \tag{7.280}$$

Employing solution (7.7), we have

$$\frac{X_1}{F_1/k} = \frac{1}{\sqrt{\left(1 - r^2\right)^2 + (2\xi r)^2}}, \tag{7.281}$$

$$\frac{X_2}{F_2/k} = \frac{1}{\sqrt{\left(1 - r^2\right)^2 + (2\xi r)^2}}, \tag{7.282}$$

$$Q = \sqrt{X_1^2 + X_2^2} = \frac{1}{k} \frac{\sqrt{F_1^2 + F_2^2}}{\sqrt{\left(1 - r^2\right)^2 + (2\xi r)^2}}, \tag{7.283}$$

which is consistent with (7.278).

## 7.3 Multi-$DOF$ Systems

Systems with more than one $DOF$, such as that in Fig. 7.31, introduce multiple natural frequencies, mode shapes, and mode interactions. These characteristics make multi-$DOF$ systems different from one-$DOF$ systems. This chapter aims to study the frequency response of multi-$DOF$ systems.

### 7.3.1 Natural Frequency and Mode Shape

Free and undamped vibrations of a system are the basic response of the system and express its natural behavior. We call a system with no external excitation a *free system*, and one with no damping we call an *undamped system*. Using the set of generalized coordinates,

$$\mathbf{x} = \left[ x_1\ x_2\ \cdots\ x_n \right]^T, \tag{7.284}$$

a linear undamped free system is governed by the following set of differential equations:

$$[m]\ddot{\mathbf{x}} + [k]\mathbf{x} = \mathbf{0}. \tag{7.285}$$

As long as the mass matrix $[m]$ and stiffness matrix $[k]$ are constant, the time response of the free system is harmonic:

$$\begin{aligned}\mathbf{x} &= \sum_{i=1}^{n} \mathbf{u}_i \left(A_i \sin \omega_i t + B_i \cos \omega_i t\right) \quad i = 1, 2, 3, \cdots, n \\ &= \sum_{i=1}^{n} C_i \mathbf{u}_i \sin\left(\omega_i t + \varphi_i\right) \quad i = 1, 2, 3, \cdots, n,\end{aligned} \tag{7.286}$$

$$C_i = \sqrt{A_i^2 + B_i^2} \tag{7.287}$$

$$\tan \varphi_i = \frac{B_i}{A_i}, \tag{7.288}$$

where $\omega_i$ are the *natural frequencies* and $\mathbf{u}_i$ are the *mode shapes* of the system. The unknown coefficients $A_i$ and $B_i$, or $C_i$ and $\varphi_i$, must be determined from the initial conditions.

The natural frequencies $\omega_i$ are solutions of the characteristic equation of the system

$$\det\left[[k] - \omega^2 [m]\right] = 0, \tag{7.289}$$

and the mode shape $\mathbf{u}_i$, corresponding to $\omega_i$, is the solution of the eigenvector equation

$$\left[[k] - \omega_i^2 [m]\right] \mathbf{u}_i = 0. \tag{7.290}$$

Mode shapes are orthogonal with respect to the mass $[m]$ and stiffness $[k]$ matrices:

$$\mathbf{u}_j^T [m] \mathbf{u}_i = 0 \qquad i \neq j, \tag{7.291}$$

$$\mathbf{u}_j^T [k] \mathbf{u}_i = 0 \qquad i \neq j. \tag{7.292}$$

Employing the orthogonality property of the mode shapes, we can determine the generalized mass $m_i$ and stiffness $k_i$ associated to $\mathbf{u}_i$:

$$\mathbf{u}_i^T [m] \mathbf{u}_i = m_i, \tag{7.293}$$

$$\mathbf{u}_i^T [k] \mathbf{u}_i = k_i. \tag{7.294}$$

***Proof*** The general equations of motion of any linear multi-$DOF$ vibrating system is

$$[m]\ddot{\mathbf{x}} + [c]\dot{\mathbf{x}} + [k]\mathbf{x} = \mathbf{F}. \tag{7.295}$$

Eliminating the force and damping terms from the general equations of motion provides us with the equations of an undamped free system:

$$[m]\ddot{\mathbf{x}} + [k]\mathbf{x} = \mathbf{0}. \tag{7.296}$$

Let us search for a possible solution of the following form:

$$\mathbf{x} = \mathbf{u}\, q(t), \tag{7.297}$$

$$x_i = u_i\, q(t) \qquad i = 1, 2, 3, \cdots, n, \tag{7.298}$$

$$\mathbf{u} = \left[\mathbf{u}_1\ \mathbf{u}_2 \cdots \mathbf{u}_n\right]^T. \tag{7.299}$$

This solution implies that the amplitude ratio of any two coordinates during motion does not depend on time. Substituting (7.297) into Eq. (7.296)

$$[m]\,\mathbf{u}\,\ddot{q}(t) + [k]\,\mathbf{u}\,q(t) = \mathbf{0} \tag{7.300}$$

and separating the time-dependent terms yields

$$-\frac{\ddot{q}(t)}{q(t)} = [[m]\,\mathbf{u}]^{-1}\,[[k]\,\mathbf{u}] = \frac{\sum_{j=1}^{n} k_{ij}u_j}{\sum_{j=1}^{n} m_{ij}u_j} = \omega^2 \quad i = 1, 2, \cdots, n. \tag{7.301}$$

Because the right-hand side of this equation is time independent and the left-hand side is independent of the index $i$, both sides must be equal to a constant. Let us assume the constant is a positive number $\omega^2$. In that case, Eq. (7.301) can be separated into two equations,

$$\ddot{q}(t) + \omega^2 q(t) = 0 \tag{7.302}$$

and

$$\left[[k] - \omega^2\,[m]\right]\mathbf{u} = 0, \tag{7.303}$$

$$\sum_{j=1}^{n}\left(k_{ij} - \omega^2 m_{ij}\right)u_j = 0 \qquad i = 1, 2, 3, \cdots n. \tag{7.304}$$

The solution of (7.302) is

$$q(t) = \sin(\omega t + \varphi) = \cos\varphi\,\sin\omega t + \sin\varphi\,\cos\omega t, \tag{7.305}$$

which shows that all the coordinates of the system, $x_i$, have a harmonic motion with identical frequency $\omega$ and identical phase angle $\varphi$. The frequency $\omega$ will be determined from Eq. (7.303), which is a set of homogeneous equations for the unknown $\mathbf{u}$.

Both the left and right sides of Eq. (7.301) must be equal to a constant. The sign of the constant $\omega^2$ is dictated by physical considerations. A free and undamped vibrating system is conservative and has a constant mechanical energy, so the amplitude of vibration must remain finite when $t \to \infty$. If the constant is positive, then the response will be harmonic with a constant amplitude; however, if the constant is negative, the response will be hyperbolic with an exponentially increasing amplitude. Therefore, the sign of the constant $\omega^2$ must be positive.

The set of Eqs. (7.303) always has a *trivial solution* $\mathbf{u} = \mathbf{0}$, which is the *rest position* of the system and shows no motion. To have a *nontrivial solution*, the determinant of the coefficient matrix must be zero:

$$\det\left[[k] - \omega^2\,[m]\right] = 0. \tag{7.306}$$

Determining the constant $\omega$, such that the set of Eqs. (7.303) provides us with a nontrivial solution, is called the *eigenvalue problem*. Expanding the determinant (7.306) provides us with the algebraic *characteristic equation*. The characteristic equation is an $n$th-order equation in terms of $\omega^2$ and provides us with $n$ natural frequencies $\omega_i$. We usually set the natural frequencies $\omega_i$ in numerical order:

$$\omega_1 \le \omega_2 \le \omega_3 \le \cdots \le \omega_n. \tag{7.307}$$

Having $n$ values for $\omega$ indicates that the solution (7.305) is possible for $n$ different frequencies $\omega_i$, $i = 1, 2, 3, \cdots, n$.

We may multiply Eq. (7.296) by $[m]^{-1}$,

$$\ddot{\mathbf{x}} + [m]^{-1}[k]\,\mathbf{x} = \ddot{\mathbf{x}} + [A]\,\mathbf{x} = \mathbf{0}, \tag{7.308}$$

and obtain the characteristic equation (7.306) as

$$\det[[A] - \lambda\mathbf{I}] = 0, \tag{7.309}$$

where

$$[A] = [m]^{-1}[k]. \tag{7.310}$$

Therefore, determining the natural frequencies $\omega_i$ would be equivalent to determining the eigenvalues $\lambda_i$ of the *characteristic matrix* $[A] = [m]^{-1}[k]$:

$$\lambda_i = \omega_i^2. \tag{7.311}$$

Determination of the vectors $\mathbf{u}_i$ to satisfy Eq. (7.303) is called the *eigenvector problem*. To determine $\mathbf{u}_i$, we substitute $\omega_i$ into Eq. (7.303) and solve (7.312) to find $n$ different $\mathbf{u}_i$.

$$\left[[k] - \omega_i^2[m]\right]\mathbf{u}_i = 0. \tag{7.312}$$

In mechanical vibrations, the eigenvector $\mathbf{u}_i$ corresponding to the eigenvalue $\omega_i$ is called the *mode shape*. Alternatively, we may find the eigenvectors of the matrix $[A] = [m]^{-1}[k]$ instead of finding the mode shapes from (7.312):

$$[[A] - \lambda_i\mathbf{I}]\,\mathbf{u}_i = 0. \tag{7.313}$$

Equations (7.312) are homogeneous, so if $\mathbf{u}_i$ is a solution, then $a\mathbf{u}_i$, $a \in \mathbb{R}$, is also a solution. Hence, the eigenvectors are not unique and may be expressed by any length. However, the ratio of any two elements of an eigenvector is unique, and therefore $\mathbf{u}_i$ has a unique shape. If one of the elements of $\mathbf{u}_i$ is assigned, then the remaining $n-1$ elements are uniquely determined. The shape of an eigenvector $\mathbf{u}_i$ indicates the relative amplitudes of the coordinates of the system in vibration. Mode shapes are the principal elements of all possible free vibrations of a multi-$DOF$ system.

Because the length of an eigenvector is not uniquely defined, there are many options for expressing $\mathbf{u}_i$. The most common expressions are:

1. normalization,
2. normal form,
3. high-unit,
4. first-unit,
5. last-unit.

In the *normalization* expression, we adjust the length of $\mathbf{u}_i$ such that

$$\mathbf{u}_i^T[m]\,\mathbf{u}_i = 1 \tag{7.314}$$

or

$$\mathbf{u}_i^T[k]\,\mathbf{u}_i = 1 \tag{7.315}$$

and call $\mathbf{u}_i$ a *normal mode* with respect to $[m]$ or $[k]$, respectively.

In the *normal form* expression, we adjust $\mathbf{u}_i$ such that its length has a unity value.

In the *high-unit* expression, we adjust the length of $\mathbf{u}_i$ such that the largest element has a unity value.

In the *first-unit* expression, we adjust the length of $\mathbf{u}_i$ such that the first element has a unity value.

In the *last-unit* expression, we adjust the length of $\mathbf{u}_i$ such that the last element has a unity value.

Let us write Eq. (7.290) for two different natural frequencies $\omega_i, \omega_j, i \neq j$, as

$$[k]\,\mathbf{u}_i = \omega_i^2\,[m]\,\mathbf{u}_i, \tag{7.316}$$

$$[k]\,\mathbf{u}_j = \omega_j^2\,[m]\,\mathbf{u}_j. \tag{7.317}$$

Multiplying $\mathbf{u}_j^T$ by Eq. (7.316) and multiplying the transpose of (7.317) by $\mathbf{u}_i$ yields, respectively,

$$\mathbf{u}_j^T\,[k]\,\mathbf{u}_i = \omega_i^2\,\mathbf{u}_j^T\,[m]\,\mathbf{u}_i, \tag{7.318}$$

$$\mathbf{u}_j^T\,[k]^T\,\mathbf{u}_i = \omega_j^2\,\mathbf{u}_j^T\,[m]^T\,\mathbf{u}_i. \tag{7.319}$$

Assuming $[k]$ and $[m]$ are symmetric matrices, $[k] = [k]^T$, and we can write Eq. (7.319) as

$$\mathbf{u}_j^T\,[k]\,\mathbf{u}_i = \omega_j^2\,\mathbf{u}_j^T\,[m]\,\mathbf{u}_i. \tag{7.320}$$

Subtracting (7.320) from (7.318)

$$\left(\omega_i^2 - \omega_j^2\right)\,\mathbf{u}_j^T\,[m]\,\mathbf{u}_i = 0, \tag{7.321}$$

we conclude that

$$\begin{aligned} \mathbf{u}_j^T\,[m]\,\mathbf{u}_i &= 0 \quad i \neq j, \\ \mathbf{u}_j^T\,[m]\,\mathbf{u}_i &\neq 0 \quad i = j, \end{aligned} \tag{7.322}$$

provided that $\omega_i \neq \omega_j$ and $[m]$ is a positive definite matrix. The mass matrix must be positive definite to guarantee that $\mathbf{u}_i^T\,[m]\,\mathbf{u}_i$ is not equal to zero for the nonzero mode shape $\mathbf{u}_i$. Therefore, we may write Eq. (7.322) as

$$\begin{aligned} \mathbf{u}_j^T\,[m]\,\mathbf{u}_i &= 0 \quad i \neq j, \\ \mathbf{u}_j^T\,[m]\,\mathbf{u}_i &= m_i \quad i = j, \end{aligned} \tag{7.323}$$

where $m_i$ is called the *generalized mass* of the system. We may also write Eq. (7.290) for two different natural frequencies as follows:

$$\frac{1}{\omega_i^2}\,[k]\,\mathbf{u}_i = [m]\,\mathbf{u}_i, \tag{7.324}$$

$$\frac{1}{\omega_j^2}\,[k]\,\mathbf{u}_j = [m]\,\mathbf{u}_j. \tag{7.325}$$

Multiplying $\mathbf{u}_j^T$ by Eq. (7.324) and multiplying the transpose of (7.325) by $\mathbf{u}_i$ and taking advantage of the symmetry of $[k]$ and $[m]$ lead to

$$\frac{1}{\omega_i^2}\mathbf{u}_j^T\,[k]\,\mathbf{u}_i = \mathbf{u}_j^T\,[m]\,\mathbf{u}_i, \tag{7.326}$$

$$\frac{1}{\omega_j^2}\mathbf{u}_j^T\,[k]\,\mathbf{u}_i = \mathbf{u}_j^T\,[m]\,\mathbf{u}_i. \tag{7.327}$$

Subtracting (7.326) from (7.327),

$$\left(\frac{1}{\omega_i^2} - \frac{1}{\omega_j^2}\right)\mathbf{u}_j^T\,[k]\,\mathbf{u}_i = 0, \tag{7.328}$$

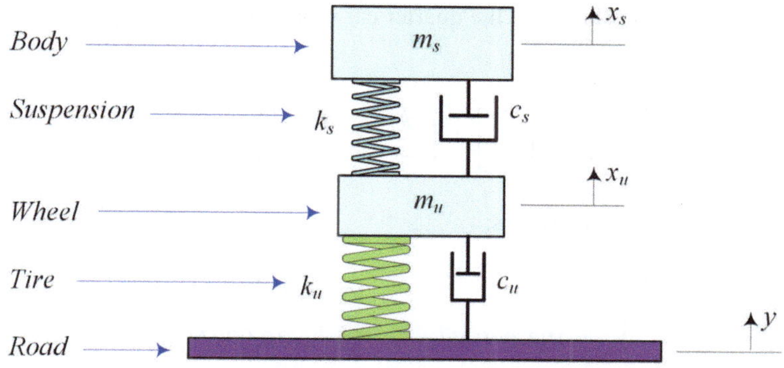


**Fig. 7.32** Quarter car model

yields

$$\begin{aligned} \mathbf{u}_j^T [k] \mathbf{u}_i &= 0 \quad i \neq j, \\ \mathbf{u}_j^T [k] \mathbf{u}_i &= k_i \quad i = j \end{aligned}, \tag{7.329}$$

where $k_i$ is the *generalized stiffness* of the system.

Any type of free, transient, or excited response of a linear vibrating system is dominated by its natural frequencies, mode shapes, and interaction of excitation frequencies. Determination of the natural frequencies $\omega_i$ and their associated mode shapes $\mathbf{u}_i$ is the first step in an analysis of a multi-$DOF$ vibrating system. There exists at least one mode shape corresponding to each natural frequency. If an $n \times n$ matrix $[A]$ has $n$ distinct eigenvalues, then there exist exactly $n$ linearly independent eigenvectors, one associated with each eigenvalue. ■

**Example 98** ★ Quarter car natural frequencies and mode shapes. The quarter model is a great mathematical model for the vertical vibration analysis of vehicles. Here is a mathematical model and a numerical example to calculate natural frequencies and mode shapes of the model.

Figure 7.32 illustrates a quarter car model made of two solid masses, $m_s$ and $m_u$, denoted as *sprung* and *unsprung* masses, respectively. The sprung mass $m_s$ represents 1/4 of the body of the vehicle, and the unsprung mass $m_u$ represents one wheel of the vehicle. A spring of stiffness $k_s$ and a shock absorber with an equivalent viscous damping coefficient $c_s$ support the sprung mass. The unsprung mass $m_u$ is in direct contact with the ground through a spring $k_u$ and a damper $c_u$ representing the tire stiffness and damping.

The governing differential equations of motion for the quarter car model are

$$m_s \ddot{x}_s = -k_s (x_s - x_u) - c_s (\dot{x}_s - \dot{x}_u), \tag{7.330}$$

$$\begin{aligned} m_u \ddot{x}_u &= k_s (x_s - x_u) + c_s (\dot{x}_s - \dot{x}_u) \\ &\quad - k_u (x_u - y) - c_u (\dot{x}_u - \dot{y}), \end{aligned} \tag{7.331}$$

which can be expressed in matrix form:

$$[M] \dot{\mathbf{x}} + [c] \dot{\mathbf{x}} + [k] \mathbf{x} = \mathbf{F}, \tag{7.332}$$

$$\begin{aligned} &\begin{bmatrix} m_s & 0 \\ 0 & m_u \end{bmatrix} \begin{bmatrix} \ddot{x}_s \\ \ddot{x}_u \end{bmatrix} + \begin{bmatrix} c_s & -c_s \\ -c_s & c_s + c_u \end{bmatrix} \begin{bmatrix} \dot{x}_s \\ \dot{x}_u \end{bmatrix} + \\ &\qquad \begin{bmatrix} k_s & -k_s \\ -k_s & k_s + k_u \end{bmatrix} \begin{bmatrix} x_s \\ x_u \end{bmatrix} = \begin{bmatrix} 0 \\ k_u y + c_u \dot{y} \end{bmatrix}. \end{aligned} \tag{7.333}$$

To find the natural frequencies and mode shapes of the quarter car model, we must drop the damping and forcing terms and analyze the following set of equations:

$$\begin{bmatrix} m_s & 0 \\ 0 & m_u \end{bmatrix} \begin{bmatrix} \ddot{x}_s \\ \ddot{x}_u \end{bmatrix} + \begin{bmatrix} k_s & -k_s \\ -k_s & k_s + k_u \end{bmatrix} \begin{bmatrix} x_s \\ x_u \end{bmatrix} = 0. \tag{7.334}$$

Consider a vehicle with the following characteristics:

$$m_s = 375\,\text{kg}, \qquad m_u = 75\,\text{kg},$$
$$k_u = 193{,}000\,\text{N/m}, \qquad k_s = 35{,}000\,\text{N/m}. \tag{7.335}$$

The equations of vibration motion for this vehicle are

$$\begin{bmatrix} 375 & 0 \\ 0 & 75 \end{bmatrix} \begin{bmatrix} \ddot{x}_s \\ \ddot{x}_u \end{bmatrix} + \begin{bmatrix} 35{,}000 & -35{,}000 \\ -35{,}000 & 2.28 \times 10^5 \end{bmatrix} \begin{bmatrix} x_s \\ x_u \end{bmatrix} = 0. \tag{7.336}$$

The natural frequencies of the vehicle can be obtained by solving its characteristic equation:

$$\det\left[[k] - \omega^2 [m]\right] = \det\left[\begin{bmatrix} 35{,}000 & -35{,}000 \\ -35{,}000 & 2.28 \times 10^5 \end{bmatrix} - \omega^2 \begin{bmatrix} 375 & 0 \\ 0 & 75 \end{bmatrix}\right]$$
$$= \det \begin{bmatrix} 35{,}000 - 375\omega^2 & -35{,}000 \\ -35{,}000 & 2.28 \times 10^5 - 75\omega^2 \end{bmatrix}$$
$$= 28{,}125\omega^4 - 8.8125 \times 10^7 \omega^2 + 6.755 \times 10^9, \tag{7.337}$$

$$\omega_1 = 8.8671\,\text{rad/s} \approx 1.41\,\text{Hz}, \tag{7.338}$$

$$\omega_2 = 55.269\,\text{rad/s} \approx 8.79\,\text{Hz}. \tag{7.339}$$

To find the corresponding mode shapes, we use Eq. (7.312):

$$\left[[k] - \omega_1^2 [m]\right] \mathbf{u}_1$$
$$= \left[\begin{bmatrix} 35{,}000 & -35{,}000 \\ -35{,}000 & 2.28 \times 10^5 \end{bmatrix} - 3054.7 \begin{bmatrix} 375 & 0 \\ 0 & 75 \end{bmatrix}\right] \begin{bmatrix} u_{11} \\ u_{12} \end{bmatrix}$$
$$= \begin{bmatrix} -1.1105 \times 10^6 u_{11} - 35{,}000 u_{12} \\ -35{,}000 u_{11} - 1102.5 u_{12} \end{bmatrix} = 0, \tag{7.340}$$

$$\left[[k] - \omega_2^2 [m]\right] \mathbf{u}_2$$
$$= \left[\begin{bmatrix} 35{,}000 & -35{,}000 \\ -35{,}000 & 2.28 \times 10^5 \end{bmatrix} - 78.625 \begin{bmatrix} 375 & 0 \\ 0 & 75 \end{bmatrix}\right] \begin{bmatrix} u_{21} \\ u_{22} \end{bmatrix}$$
$$= \begin{bmatrix} 5515.6 u_{21} - 35{,}000 u_{22} \\ 2.221 \times 10^5 u_{22} - 35{,}000 u_{21} \end{bmatrix} = 0. \tag{7.341}$$

Searching for the first-unit expression of $\mathbf{u}_1$ and $\mathbf{u}_2$ provides us with the following mode shapes:

$$\mathbf{u}_1 = \begin{bmatrix} 1 \\ -3.1729 \times 10^{-3} \end{bmatrix} \qquad \mathbf{u}_2 = \begin{bmatrix} 1 \\ 0.157\,58 \end{bmatrix}. \tag{7.342}$$

Therefore, the free vibrations of the quarter car would be

$$\mathbf{x} = \sum_{i=1}^{n} \mathbf{u}_i \left(A_i \sin \omega_i t + B_i \cos \omega_i t\right) \qquad i = 1, 2, \tag{7.343}$$

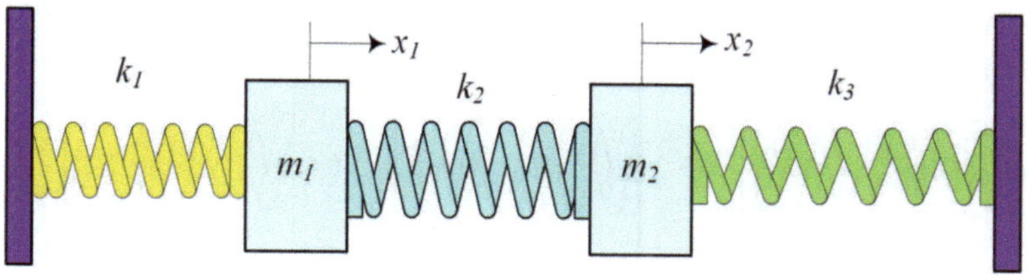


**Fig. 7.33** A two-$DOF$ vibrating system

$$\begin{bmatrix} x_s \\ x_u \end{bmatrix} = \begin{bmatrix} 1 \\ -3.1729 \times 10^{-3} \end{bmatrix} (A_1 \sin 8.8671t + B_1 \cos 8.8671t)$$
$$+ \begin{bmatrix} 1 \\ 0.157\,58 \end{bmatrix} (A_2 \sin 55.269t + B_2 \cos 55.269t)\,. \tag{7.344}$$

**Example 99** Two-$DOF$ vibrating systems, mode shape illustration, nodes, possible free vibrations of multi-$DOF$ systems. This example shows how to derive the equations of motion of multi-$DOF$ vibrating systems using the Lagrange method and how to determine the natural frequencies and mode shapes analytically. It also illustrates geometrically the mode shapes to clarify their meaning. Any possible free vibration of a multi-$DOF$ system will be a linear combination of its mode shapes. An illustration of the mode shapes gives us a vision of how the system will vibrate at each mode.

Figure 7.33 illustrates a two-$DOF$ vibrating system. The kinetic and potential energies of the system are as follows:

$$K = \frac{1}{2} m_1 \dot{x}_1^2 + \frac{1}{2} m_2 \dot{x}_2^2, \tag{7.345}$$

$$P = \frac{1}{2} k_1 x_1^2 + \frac{1}{2} k_2 (x_1 - x_2)^2 + \frac{1}{2} k_3 x_2^2. \tag{7.346}$$

Using the Lagrangean of the system $\mathcal{L} = K - P$ and employing the Lagrange equation,

$$\frac{d}{dt}\left(\frac{\partial \mathcal{L}}{\partial \dot{x}_i}\right) - \frac{\partial \mathcal{L}}{\partial x_i} = 0 \qquad i = 1, 2, \tag{7.347}$$

the equations of motion are

$$m_1 \ddot{x}_1 + k_1 x_1 + k_2 (x_1 - x_2) = 0, \tag{7.348}$$

$$m_2 \ddot{x}_2 - k_2 (x_1 - x_2) + k_3 x_2 = 0. \tag{7.349}$$

We may rearrange the equations in matrix form:

$$\begin{bmatrix} m_1 & 0 \\ 0 & m_2 \end{bmatrix} \begin{bmatrix} \ddot{x}_1 \\ \ddot{x}_2 \end{bmatrix} + \begin{bmatrix} k_1 + k_2 & -k_2 \\ -k_2 & k_2 + k_3 \end{bmatrix} \begin{bmatrix} x_1 \\ x_2 \end{bmatrix} = \begin{bmatrix} 0 \\ 0 \end{bmatrix}. \tag{7.350}$$

The natural frequencies of the system can be found from $[A]$:

$$[A] = \begin{bmatrix} m_1 & 0 \\ 0 & m_2 \end{bmatrix}^{-1} \begin{bmatrix} k_1 + k_2 & -k_2 \\ -k_2 & k_2 + k_3 \end{bmatrix}$$
$$= \begin{bmatrix} \frac{1}{m_1}(k_1 + k_2) & -\frac{k_2}{m_1} \\ -\frac{k_2}{m_2} & \frac{1}{m_2}(k_2 + k_3) \end{bmatrix}, \tag{7.351}$$

which is a symmetric matrix and has the following eigenvalues:

$$\lambda_1 = Z_1 - \sqrt{Z_2} \qquad \lambda_2 = Z_1 + \sqrt{Z_2}, \tag{7.352}$$

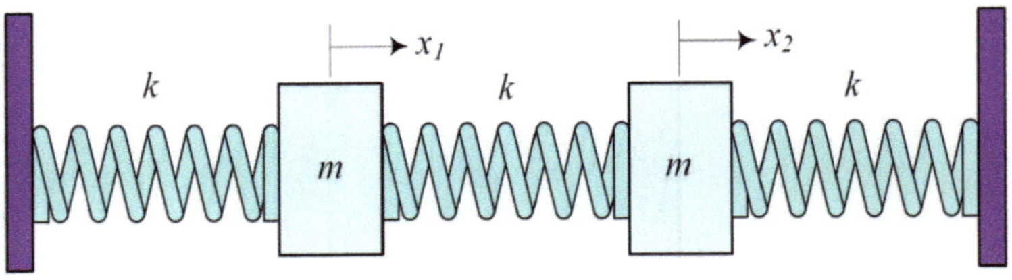


**Fig. 7.34** Symmetrical two-$DOF$ vibrating system with $m_1 = m_2 = m$, $k_1 = k_2 = k_3 = k$

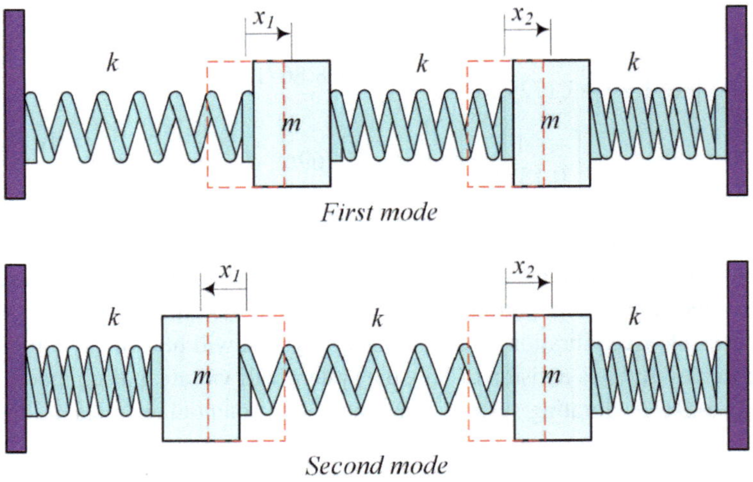


**Fig. 7.35** First- and second-mode shapes of a symmetric two-$DOF$ vibrating system

$$Z_1 = \frac{1}{2}\left(\frac{k_1 + k_2}{m_1} + \frac{k_2 + k_3}{m_2}\right), \tag{7.353}$$

$$Z_2 = \frac{1}{4}\left(\frac{k_1 + k_2}{m_1} - \frac{k_2 + k_3}{m_2}\right)^2 + \frac{k_2^2}{m_1 m_2}. \tag{7.354}$$

In the symmetrical case of Fig. 7.34, where

$$m_1 = m_2 = m, \qquad k_1 = k_2 = k_3 = k, \tag{7.355}$$

the eigenvalues $\lambda_i$ and natural frequencies $\omega_i$ simplify to

$$\lambda_1 = \frac{k}{m}, \qquad \lambda_2 = 3\frac{k}{m}, \tag{7.356}$$

$$\omega_1 = \sqrt{\lambda_1} = \sqrt{\frac{k}{m}}, \qquad \omega_2 = \sqrt{\lambda_2} = \sqrt{3}\sqrt{\frac{k}{m}}, \tag{7.357}$$

of which the eigenvectors, respectively, are

$$\mathbf{u}_1 = \begin{bmatrix} 1 \\ 1 \end{bmatrix}, \qquad \mathbf{u}_2 = \begin{bmatrix} -1 \\ 1 \end{bmatrix}. \tag{7.358}$$

Mode shapes are the principal elements for all possible vibrations of a free system and indicate the relative displacements at each natural frequency. Figure 7.35 illustrates the relative position of the masses in the first and second modes of the two-$DOF$ symmetric system of Fig. 7.34.

To distinguish between masses and springs, let us use the notation of Fig. 7.33. In the first mode, both masses move in the same direction at equal amplitudes. The middle spring $k_2$ has a constant length with no stretch and applies no force on $m_1$ and $m_2$. Therefore, the mass $m_1$ is only under the force of $k_1$, and the mass $m_2$ is only under the force of $k_2$. In the second

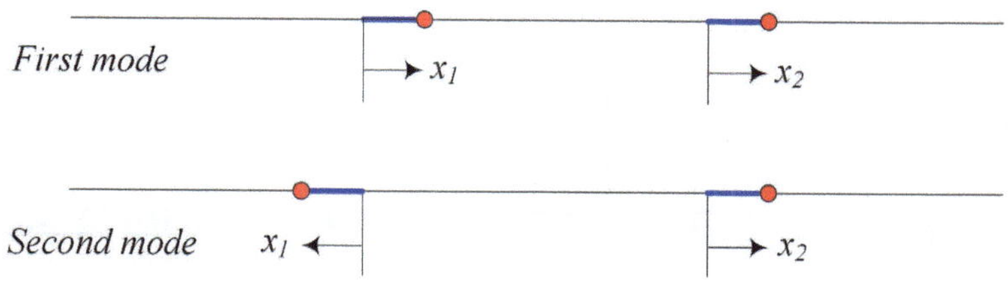


**Fig. 7.36** Physical mode shape illustration

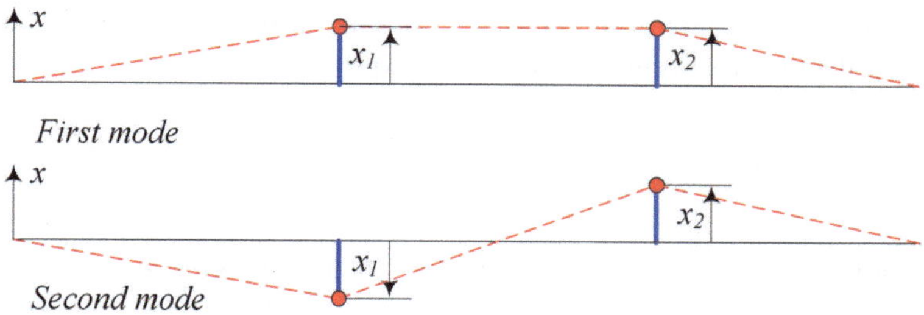


**Fig. 7.37** Geometric mode shape illustration

mode, the masses move in opposite directions with equal amplitudes. The middle spring $k_2$ stretches twice as much as $k_1$ and $k_2$. The midpoint of $k_2$ remains stationary. Thus, $k_2$ applies twice as much force as $k_1$ or $k_3$, and therefore $m_1$ and $m_2$ are under three times as much force as in the first mode.

The motion of the system of the first mode can be expressed by $\mathbf{x}_1$

$$\mathbf{x}_1 = C_1 \mathbf{u}_1 \sin(\omega_1 t + \varphi_1), \tag{7.359}$$

and its motion in the second mode can be expressed by $\mathbf{x}_2$:

$$\mathbf{x}_2 = C_2 \mathbf{u}_2 \sin(\omega_2 t + \varphi_2). \tag{7.360}$$

Any possible free vibration of the system is a linear combination of its two mode shapes, where $C_1$ and $C_2$ are weight factors of the mode shapes:

$$\mathbf{x} = C_1 \mathbf{u}_1 \sin(\omega_1 t + \varphi_1) + C_2 \mathbf{u}_2 \sin(\omega_2 t + \varphi_2). \tag{7.361}$$

To visualize mode shapes, we illustrate each mode shape symbolically by the position of its coordinate with respect to the equilibrium position, as shown physically in Figs. 7.35 and 7.36 or geometrically in Fig. 7.37. The node in the geometric mode shape illustration of Fig. 7.37 indicates that the midpoint of spring $k_2$ is motionless. The motionless points of mode shapes are called nodes. Furthermore, the geometric illustration depicts the relative motion of any point of the springs.

**Example 100** ★ Energy of mode shapes. Free and undamped multi-$DOF$ systems are energy conserved. The initial conditions determine how much energy the system has. The initial conditions and the amount of energy determine the amplitude, mode, and interactions of oscillation modes. This example explains this phenomenon.

Every mode shape is associated to a specific natural frequency. If we order the natural frequencies from lowest to highest, then the system will vibrate slower in lower mode and faster in higher mode shapes. Consider the two-$DOF$ symmetric system of Fig. 7.34 with natural frequencies $\omega_1$ and $\omega_2$ and mode shapes $\mathbf{u}_1$ and $\mathbf{u}_2$:

$$\omega_1 = \sqrt{\frac{k}{m}}, \qquad \omega_2 = \sqrt{3}\sqrt{\frac{k}{m}}, \tag{7.362}$$

$$\mathbf{u}_1 = \begin{bmatrix} 1 \\ 1 \end{bmatrix}, \qquad \mathbf{u}_2 = \begin{bmatrix} -1 \\ 1 \end{bmatrix}. \tag{7.363}$$

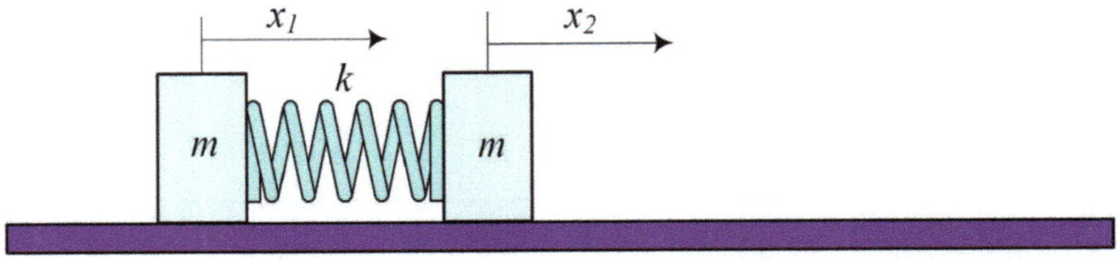


**Fig. 7.38** Two-$DOF$ system with rigid mode

Let us assume the system is oscillating only in the first mode or only in the second mode, both with amplitude $X$. The displacement vector $\mathbf{x}$ of $m_i$ at the first and second modes can be expressed as

$$\mathbf{x}_1 = \begin{bmatrix} X \\ X \end{bmatrix} \sin \omega_1 t, \qquad \mathbf{x}_2 = \begin{bmatrix} -X \\ X \end{bmatrix} \sin \omega_2 t. \tag{7.364}$$

The maximum kinetic and maximum potential energies of the system in the first mode are

$$K_{Max_1} = \frac{1}{2} m\omega_1^2 X^2 + \frac{1}{2} m\omega_1^2 X^2 = m\omega_1^2 X^2, \tag{7.365}$$

$$P_{Max_1} = \frac{1}{2} kX^2 + \frac{1}{2} k (X - X)^2 + \frac{1}{2} kX^2 = kX^2, \tag{7.366}$$

and $K_{Max}$ and $P_{Max}$ in the second mode are

$$K_{Max_2} = \frac{1}{2} m\omega_2^2 X^2 + \frac{1}{2} m\omega_2^2 X^2 = m\omega_2^2 X^2 = 3K_{Max_1}, \tag{7.367}$$

$$P_{Max_2} = \frac{1}{2} kX^2 + \frac{1}{2} k (-X - X)^2 + \frac{1}{2} kX^2 = 3kX^2 = 3P_{Max_1}. \tag{7.368}$$

The system vibrates in the second mode $\sqrt{3}$ times faster than the first mode. Therefore, it passes through the equilibrium position $\sqrt{3}$ times faster in the second mode than the first mode. The amount of required energy to vibrate in the second mode is three times the required energy to vibrate in the first mode at the same amplitude. If the amount of energy is constant, then the amplitude of vibration in the first mode will be $\sqrt{3}$ greater than the amplitude of vibration in the second mode. To make the system vibrate in the first mode, the initial conditions must match those of the first mode, such as $\mathbf{x}(0) = \begin{bmatrix} 1 & 1 \end{bmatrix}$. Similarly, to make the system vibrate in the second mode, the initial conditions must match those of the second mode, such as $\mathbf{x}(0) = \begin{bmatrix} -1 & 1 \end{bmatrix}$. The amount of energy required for the initial conditions in each mode can be calculated and felt by the amount of extension and contraction of the springs (Fig. 7.35).

All real systems have some damping and need an external source of energy to keep the systems vibrating. The level of injection of external energy will depend on the mode and amplitude of vibrations. That is why, when the source of energy is low, every system can vibrate in its first mode.

**Example 101** Zero natural frequency and rigid mode. If a vibrating system is not attached to the ground, it can move with respect to the ground. Such motion is indicated by a zero natural frequency and is called the rigid mode shape. What follows is an introduction to how such zero natural frequency emerges.

When a vibrating system can have any position with respect to a fixed coordinate frame, it has a zero natural frequency, which indicates a rigid motion or rigid mode. For example, consider the two-$DOF$ system in Fig. 7.38 with the following equations of motion:

$$\begin{bmatrix} m_1 & 0 \\ 0 & m_2 \end{bmatrix} \begin{bmatrix} \ddot{x}_1 \\ \ddot{x}_2 \end{bmatrix} + \begin{bmatrix} k & -k \\ -k & k \end{bmatrix} \begin{bmatrix} x_1 \\ x_2 \end{bmatrix} = \begin{bmatrix} 0 \\ 0 \end{bmatrix}. \tag{7.369}$$

The eigenvalues and eigenvectors of the system are as follows:

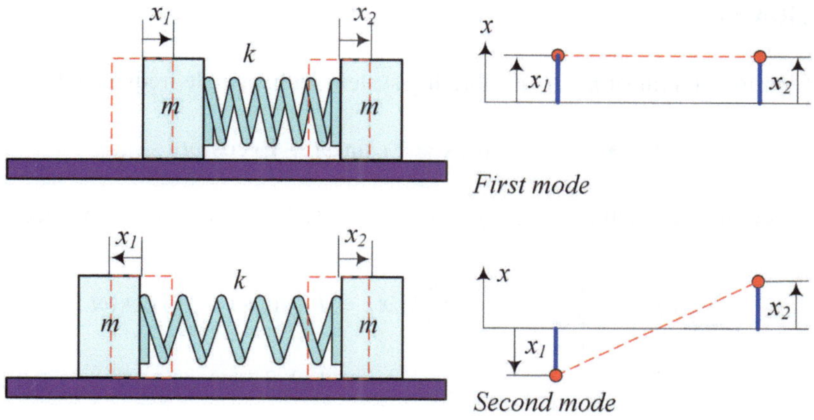


**Fig. 7.39** Mode shapes of two-$DOF$ system with rigid mode

$$\omega_1 = 0, \qquad \omega_2 = \sqrt{k\frac{m_1 + m_2}{m_1 m_2}}, \tag{7.370}$$

$$u_1 = \begin{bmatrix} 1 \\ 1 \end{bmatrix}, \qquad u_2 = \begin{bmatrix} -m_2/m_1 \\ 1 \end{bmatrix}. \tag{7.371}$$

The first mode shape shows a nonvibrating motion of the system, while the second mode indicates a vibrating motion. These modes are illustrated in Fig. 7.39 for $m_1 = m_2$, physically and geometrically. There is a node in the middle of the spring in the second mode of Fig. 7.39. The node will move at the same constant speed as the first mode. The actual motion of the system is determined by the initial conditions of the masses.

The rank of a matrix $[A]$ is the order of the largest nonsingular submatrix of $[A]$. A nonsingular matrix has a nonzero determinant. For an $n$-$DOF$ vibrating system, $[A] = [m]^{-1}[k]$ would be an $n \times n$ matrix. If the rank of $[A]$ is $n'$, then

$$f = n - n' \tag{7.372}$$

indicates the number of rigid modes of the system.

**Example 102** ★ Rigid mode elimination. Mode shapes are coordinate dependent, so we may use different coordinates to derive different mode shapes. Relative coordinates can simplify the equations of motion to hide the rigid body mode.

It is possible to use the relative displacement and eliminate the zero natural frequencies of a system. Consider the system of Fig. 7.38, whose equations of motion are

$$m_1\ddot{x}_1 + kx_1 - kx_2 = 0, \tag{7.373}$$

$$m_2\ddot{x}_2 + kx_2 - kx_1 = 0. \tag{7.374}$$

If we introduce the relative displacement $z$,

$$z = x_1 - x_2, \tag{7.375}$$

then we can subtract the equations to combine them as a single differential equation in terms of $z$:

$$m_1 m_2 \ddot{z} + k(m_1 + m_2)z = 0. \tag{7.376}$$

This equation defines the behavior of relative displacement of the masses. The only natural frequency of this equation is equal to the nonzero natural frequency of the system:

$$\omega_n = \sqrt{k\frac{m_1 + m_2}{m_1 m_2}}. \tag{7.377}$$

### 7.3.2 Harmonic Excitation

The general equations of motion of a linear $n$-$DOF$ vibrating system with a single frequency harmonic excitation is

$$[m]\ddot{\mathbf{x}} + [c]\dot{\mathbf{x}} + [k]\mathbf{x} = \mathbf{F}_1 \sin\omega t + \mathbf{F}_2 \cos\omega t, \tag{7.378}$$

where $\mathbf{F}_1$ and $\mathbf{F}_2$ are the external forces acting on the system. It can also be expressed in expanded form:

$$\sum_{j=1}^{n} m_{ij}\ddot{x}_j + \sum_{j=1}^{n} c_{ij}\dot{x}_j + \sum_{j=1}^{n} k_{ij}x_j = F_{1i}\sin\omega t + F_{2i}\cos\omega t \tag{7.379}$$

$$i = 1, 2, 3, \cdots, n.$$

The steady-state response of the system is

$$\mathbf{x} = \mathbf{A}\sin\omega t + \mathbf{B}\cos\omega t \tag{7.380}$$

or

$$x_i = A_i \sin\omega t + B_i \cos\omega t \tag{7.381}$$

$$= X_i \sin\left(\omega t - \varphi_i\right), \tag{7.382}$$

where

$$\begin{bmatrix} \mathbf{A} \\ \mathbf{B} \end{bmatrix} = \begin{bmatrix} [k] - \omega^2[m] & -\omega[c] \\ \omega[c] & [k] - \omega^2[m] \end{bmatrix}^{-1} \begin{bmatrix} \mathbf{F}_1 \\ \mathbf{F}_2 \end{bmatrix}, \tag{7.383}$$

and the amplitude and phase of the frequency responses are

$$X_i(\omega) = \sqrt{A_i^2 + B_i^2} \qquad i = 1, 2, \cdots, n, \tag{7.384}$$

$$\tan\varphi_i = \frac{-B_i}{A_i}. \tag{7.385}$$

***Proof*** The steady-state solution of the $n$-$DOF$ system (7.378) is a set of $n$ harmonic equations with the same frequency as the excitation:

$$\mathbf{x} = \mathbf{A}\sin\omega t + \mathbf{B}\cos\omega t, \tag{7.386}$$

where $\mathbf{A}$ and $\mathbf{B}$ are $n \times 1$ coefficient matrices. We substitute the solution (7.386) into the equations of motion (7.378) and balance the coefficients of $\cos\omega t$ and $\sin\omega t$:

$$\begin{aligned} &-\omega^2[m](\mathbf{A}\sin\omega t + \mathbf{B}\cos\omega t) + \omega[c](\mathbf{A}\cos\omega t - \mathbf{B}\sin\omega t) \\ &+ [k](\mathbf{A}\sin\omega t + \mathbf{B}\cos\omega t) = \mathbf{F}_1\sin\omega t + \mathbf{F}_2\cos\omega t, \end{aligned} \tag{7.387}$$

$$-\omega^2[m]\mathbf{A} - \omega[c]\mathbf{B} + [k]\mathbf{A} = \mathbf{F}_1, \tag{7.388}$$

$$-\omega^2[m]\mathbf{B} + \omega[c]\mathbf{A} + [k]\mathbf{B} = \mathbf{F}_2. \tag{7.389}$$

These equations provide us with a set of $2n$ algebraic equations to determine the coefficients $\mathbf{A}$ and $\mathbf{B}$:

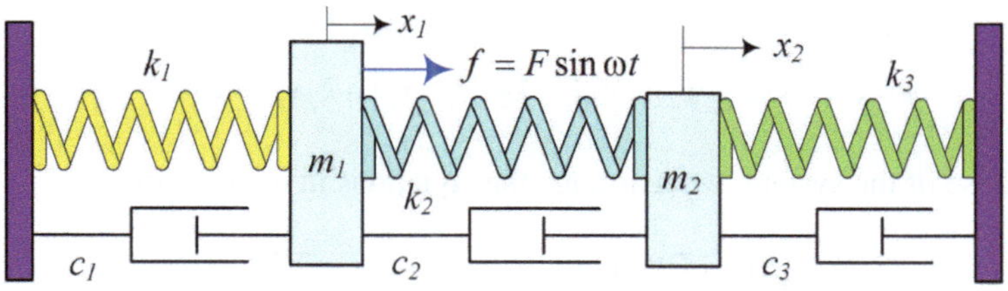


**Fig. 7.40** A two-$DOF$ system under harmonic forced excitation

$$\begin{bmatrix} [k] - \omega^2 [m] & -\omega [c] \\ \omega [c] & [k] - \omega^2 [m] \end{bmatrix} \begin{bmatrix} \mathbf{A} \\ \mathbf{B} \end{bmatrix} = \begin{bmatrix} \mathbf{F}_1 \\ \mathbf{F}_2 \end{bmatrix}, \tag{7.390}$$

where

$$\begin{bmatrix} \mathbf{A} \\ \mathbf{B} \end{bmatrix} = \begin{bmatrix} [k] - \omega^2 [m] & -\omega [c] \\ \omega [c] & [k] - \omega^2 [m] \end{bmatrix}^{-1} \begin{bmatrix} \mathbf{F}_1 \\ \mathbf{F}_2 \end{bmatrix}. \tag{7.391}$$

Given the coefficient matrices **A** and **B**, the steady-state value of the coordinate $x_i$ would be (7.381) or (7.382). Equating (7.381) and (7.382),

$$A_i \sin \omega t + B_i \cos \omega t = X_i \cos \varphi_i \sin \omega t - X_i \sin \varphi_i \cos \omega t, \tag{7.392}$$

we can calculate the amplitude $X_i$ and phase $\varphi_i$ of every coordinate $x_i$:

$$X_i (\omega) = \sqrt{A_i^2 + B_i^2} \qquad i = 1, 2, \cdots, n, \tag{7.393}$$

$$\tan \varphi_i = \frac{-B_i}{A_i} \qquad i = 1, 2, \cdots, n. \tag{7.394}$$

The frequency response $X_i$ is a function of the excitation frequency. The level of $X_i$ at each frequency depends on the damping of the system. In case of no damping, all $X_i$ would approach infinity at $n$ natural frequencies of the system.

Depending on the type of excitation, each element of $\mathbf{F}_1$ and $\mathbf{F}_2$ may be a constant or proportional to $\omega$, $\omega^2$, $\omega^3$, $\cdots$. ■

**Example 103** Forced vibration of a two-$DOF$ system and dimensionless frequency response. It is a detailed example to show how the frequency response of multi-$DOF$ will be determined and what they look like.

Figure 7.40 illustrates a forced excited two-$DOF$ vibrating system. The kinetic and potential energies and the dissipative function of the system are

$$K = \frac{1}{2} m_1 \dot{x}_1^2 + \frac{1}{2} m_2 \dot{x}_2^2, \tag{7.395}$$

$$D = \frac{1}{2} c_1 \dot{x}_1^2 + \frac{1}{2} c_2 (\dot{x}_1 - \dot{x}_2)^2 + \frac{1}{2} c_3 \dot{x}_2^2, \tag{7.396}$$

$$P = \frac{1}{2} k_1 x_1^2 + \frac{1}{2} k_2 (x_1 - x_2)^2 + \frac{1}{2} k_3 x_2^2. \tag{7.397}$$

Employing the Lagrange equation,

$$\frac{d}{dt} \left( \frac{\partial K}{\partial \dot{x}_i} \right) - \frac{\partial K}{\partial x_i} + \frac{\partial D}{\partial \dot{x}_i} + \frac{\partial P}{\partial x_i} = f_i \qquad i = 1, 2, \cdots n, \tag{7.398}$$

the equations of motion will be calculated:

$$m_1\ddot{x}_1+c_1\dot{x}_1+c_2\left(\dot{x}_1-\dot{x}_2\right)+k_1x_1+k_2\left(x_1-x_2\right)=F\sin\omega t, \tag{7.399}$$

$$m_2\ddot{x}_2-c_2\left(\dot{x}_1-\dot{x}_2\right)+c_3\dot{x}_2-k_2\left(x_1-x_2\right)+k_3x_2=0. \tag{7.400}$$

To find the frequency response of the system, we rearrange the equations in matrix form,

$$\begin{bmatrix} m_1 & 0 \\ 0 & m_2 \end{bmatrix}\begin{bmatrix} \ddot{x}_1 \\ \ddot{x}_2 \end{bmatrix}+\begin{bmatrix} c_1+c_2 & -c_2 \\ -c_2 & c_2+c_3 \end{bmatrix}\begin{bmatrix} \dot{x}_1 \\ \dot{x}_2 \end{bmatrix}$$
$$+\begin{bmatrix} k_1+k_2 & -k_2 \\ -k_2 & k_2+k_3 \end{bmatrix}\begin{bmatrix} x_1 \\ x_2 \end{bmatrix}=\begin{bmatrix} F\sin\omega t \\ 0 \end{bmatrix}, \tag{7.401}$$

and substitute a harmonic solution to calculate the submatrices of the coefficient matrix of (7.391):

$$\begin{bmatrix} x_1 \\ x_2 \end{bmatrix}=\begin{bmatrix} A_1 \\ A_2 \end{bmatrix}\sin\omega t+\begin{bmatrix} B_1 \\ B_2 \end{bmatrix}\cos\omega t, \tag{7.402}$$

$$[k]-\omega^2[m]=\begin{bmatrix} -m_1\omega^2+k_1+k_2 & -k_2 \\ -k_2 & -m_2\omega^2+k_2+k_3 \end{bmatrix}, \tag{7.403}$$

$$\omega[c]=\begin{bmatrix} \omega\left(c_1+c_2\right) & -\omega c_2 \\ -\omega c_2 & \omega\left(c_2+c_3\right) \end{bmatrix}. \tag{7.404}$$

Therefore, the set of Eqs. (7.391) to calculate coefficients **A** and **B** are

$$\begin{bmatrix} a_{11} & -k_2 & a_{13} & c_2\omega \\ -k_2 & a_{22} & c_2\omega & a_{24} \\ a_{31} & -c_2\omega & a_{33} & -k_2 \\ -c_2\omega & a_{42} & -k_2 & a_{44} \end{bmatrix}\begin{bmatrix} A_1 \\ A_2 \\ B_1 \\ B_2 \end{bmatrix}=\begin{bmatrix} F \\ 0 \\ 0 \\ 0 \end{bmatrix}, \tag{7.405}$$

$$a_{11}=a_{33}=k_1+k_2-m_1\omega^2, \tag{7.406}$$

$$a_{22}=a_{44}=k_2+k_3-m_2\omega^2, \tag{7.407}$$

$$a_{31}=-a_{13}=\omega\left(c_1+c_2\right), \tag{7.408}$$

$$a_{42}=-a_{24}=\omega\left(c_2+c_3\right), \tag{7.409}$$

where

$$\begin{bmatrix} A_1 \\ A_2 \\ B_1 \\ B_2 \end{bmatrix}=\begin{bmatrix} a_{11} & -k_2 & a_{13} & c_2\omega \\ -k_2 & a_{22} & c_2\omega & a_{24} \\ a_{31} & -c_2\omega & a_{33} & -k_2 \\ -c_2\omega & a_{42} & -k_2 & a_{44} \end{bmatrix}^{-1}\begin{bmatrix} F \\ 0 \\ 0 \\ 0 \end{bmatrix}, \tag{7.410}$$

$$X_1=\sqrt{A_1^2+B_1^2}, \qquad X_2=\sqrt{A_2^2+B_2^2}. \tag{7.411}$$

To calculate $X_1$ and $X_2$ and plot the frequency response curves, we need to examine a numerical example. Let us calculate the frequency response of the system for a set of sample data:

$$m_1=30\,\text{kg}, \qquad m_2=20\,\text{kg},$$
$$c_1=10\,\text{Ns/m}, \qquad c_2=30\,\text{Ns/m}, \qquad c_3=50\,\text{Ns/m},$$

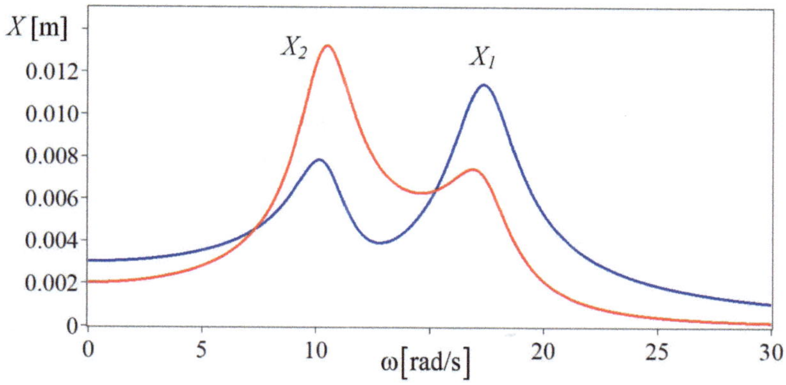


**Fig. 7.41** Frequency responses $X_1$ and $X_2$

$$k_1 = 6000\,\text{N/m}, \qquad k_2 = 2000\,\text{N/m}, \qquad k_3 = 1000\,\text{N/m},$$
$$F = 20\,\text{N}. \tag{7.412}$$

Figure 7.41 illustrates the amplitudes $X_1$ and $X_2$ as functions of excitation frequency $\omega$. Because the system has two $DOF$, it has two natural frequencies and, hence, two resonance zones in which the amplitudes are high. The natural frequencies of the system are

$$\omega_1 = 10.39\,\text{rad/s}, \qquad \omega_2 = 17.57\,\text{rad/s}. \tag{7.413}$$

In case we apply constant forces on $m_1$ and $m_2$, then they will have static displacements $X_{1s}$ and $X_{2s}$. For a better illustration, it is better to make the frequency response equations nondimensional:

$$\begin{bmatrix} k_1 + k_2 & -k_2 \\ -k_2 & k_2 + k_3 \end{bmatrix} \begin{bmatrix} X_{1s} \\ X_{2s} \end{bmatrix} = \begin{bmatrix} F \\ 0 \end{bmatrix}, \tag{7.414}$$

$$\begin{bmatrix} X_{1s} \\ X_{2s} \end{bmatrix} = \frac{F}{k_1 k_2 + k_1 k_3 + k_2 k_3} \begin{bmatrix} k_2 + k_3 \\ k_2 \end{bmatrix}. \tag{7.415}$$

We can make the frequency responses (7.411) dimensionless by dividing the frequency responses by static displacements:

$$\frac{X_1}{X_{1s}} = \frac{\sqrt{A_1^2 + B_1^2}}{X_{1s}}, \qquad \frac{X_2}{X_{2s}} = \frac{\sqrt{A_2^2 + B_2^2}}{X_{2s}}. \tag{7.416}$$

We can plot and illustrate these frequency responses $X_1/X_{1s}$ and $X_2/X_{2s}$; however, the best way to make frequency responses dimensionless is to divide all $X_i$ by only the most important oscillating coordinate. Therefore, frequency response curves will be proportional to the actual displacements, so they can be compared. Let us assume the displacement of $m_1$ is more important to us, and let us make the frequency responses (7.411) dimensionless by dividing the frequency responses by $X_{1s}$:

$$\frac{X_1}{X_{1s}} = \frac{\sqrt{A_1^2 + B_1^2}}{X_{1s}}, \qquad \frac{X_2}{X_{1s}} = \frac{\sqrt{A_2^2 + B_2^2}}{X_{1s}}. \tag{7.417}$$

Figure 7.42 illustrates these frequency responses $X_1/X_{1s}$ and $X_2/X_{1s}$.

**Example 104** Base excited two-$DOF$ system. When the excitation frequency of a vibrating base is in a resonance zone of a system, the amplitude of oscillation of the other components of a multi-$DOF$ can reach a level several times that of the excitation amplitude. What follows is an analysis of such a system.

Figure 7.43 illustrates a two-$DOF$ system with base displacement excitation. Employing the Lagrange method, we derive the equations of motion of the system:

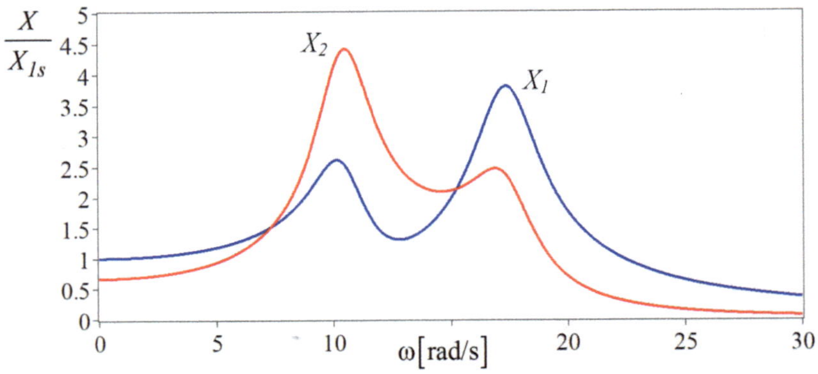


**Fig. 7.42** Dimensionless ferquency responses $X_1/X_{1s}$ and $X_2/X_{1s}$

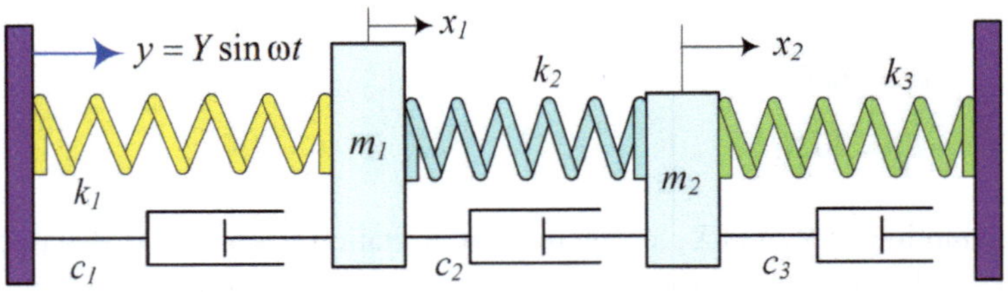


**Fig. 7.43** Two-$DOF$ system with base displacement excitation

$$K = \frac{1}{2}m_1\dot{x}_1^2 + \frac{1}{2}m_2\dot{x}_2^2, \tag{7.418}$$

$$P = \frac{1}{2}k_1\left(y - x_1\right)^2 + \frac{1}{2}k_2\left(x_1 - x_2\right)^2 + \frac{1}{2}k_3x_2^2, \tag{7.419}$$

$$D = \frac{1}{2}c_1\left(\dot{y} - \dot{x}_1\right)^2 + \frac{1}{2}c_2\left(\dot{x}_1 - \dot{x}_2\right)^2 + \frac{1}{2}c_3\dot{x}_2^2, \tag{7.420}$$

$$[m]\,\ddot{\mathbf{x}} + [c]\,\dot{\mathbf{x}} + [k]\,\mathbf{x} = \mathbf{F}, \tag{7.421}$$

$$[m] = \begin{bmatrix} m_1 & 0 \\ 0 & m_2 \end{bmatrix}, \qquad [k] = \begin{bmatrix} k_1 + k_2 & -k_2 \\ -k_2 & k_2 + k_3 \end{bmatrix},$$
$$\mathbf{F} = \begin{bmatrix} k_1 y + c_1\dot{y} \\ 0 \end{bmatrix}, \qquad [c] = \begin{bmatrix} c_1 + k_2 & -c_2 \\ -k_2 & c_2 + c_3 \end{bmatrix}. \tag{7.422}$$

If $y$ is a harmonic displacement excitation, then we have

$$y = Y\sin\omega t, \tag{7.423}$$

$$\mathbf{F} = \begin{bmatrix} k_1 Y\sin\omega t + c_1 Y\omega\cos\omega t \\ 0 \end{bmatrix}. \tag{7.424}$$

Substituting a harmonic solution

$$\begin{bmatrix} x_1 \\ x_2 \end{bmatrix} = \begin{bmatrix} A_1 \\ A_2 \end{bmatrix}\sin\omega t + \begin{bmatrix} B_1 \\ B_2 \end{bmatrix}\cos\omega t \tag{7.425}$$

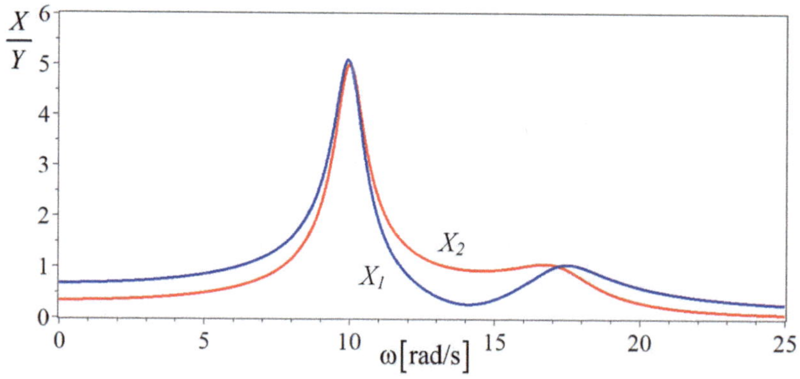


**Fig. 7.44** Frequency response of $m_1$ and $m_2$

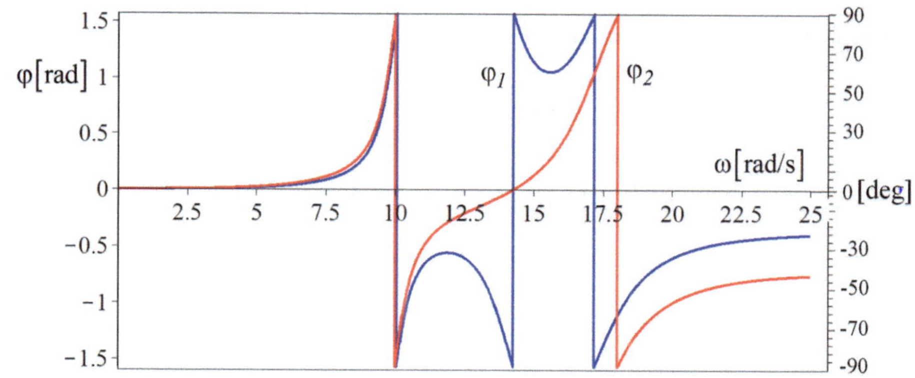


**Fig. 7.45** Phase angle of $X_1$ and $X_2$

provides us with a set of equations to calculate coefficients **A** and **B**:

$$\begin{bmatrix} A_1 \\ A_2 \\ B_1 \\ B_2 \end{bmatrix} = \begin{bmatrix} a_{11} & -k_2 & a_{13} & c_2\omega \\ -k_2 & a_{22} & c_2\omega & a_{24} \\ a_{31} & -c_2\omega & a_{33} & -k_2 \\ -c_2\omega & a_{42} & -k_2 & a_{44} \end{bmatrix}^{-1} \begin{bmatrix} k_1 Y \\ 0 \\ c_1\omega Y \\ 0 \end{bmatrix}, \tag{7.426}$$

$$a_{11} = a_{33} = k_1 + k_2 - m_1\omega^2, \tag{7.427}$$

$$a_{22} = a_{44} = k_2 + k_3 - m_2\omega^2, \tag{7.428}$$

$$a_{31} = -a_{13} = \omega\left(c_1 + c_2\right), \tag{7.429}$$

$$a_{42} = -a_{24} = \omega\left(c_2 + c_3\right). \tag{7.430}$$

Let us examine the frequency response of the system for a set of numerical values:

$$\begin{aligned} m_2 &= 1, \quad m_1 = 1, \quad c_1 = 1, \quad c_2 = 1, \quad c_3 = 1, \\ k_1 &= 100, \quad k_2 = 100, \quad k_3 = 100, \quad Y = 1. \end{aligned} \tag{7.431}$$

Figure 7.44 illustrates the frequency response of $m_1$ and $m_2$, and Fig. 7.45 illustrates the phase angle of $X_1$ and $X_2$.

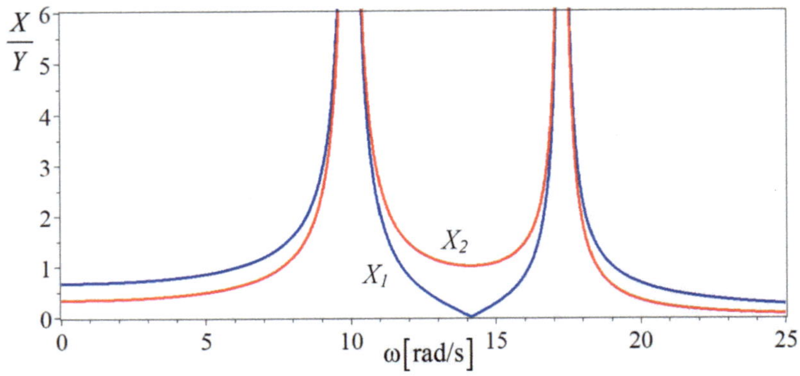


**Fig. 7.46** Absolute value of frequency response of $X_1$ and $X_2$

When there is no damping, the set of equations of the system reduces to

$$\begin{bmatrix} A_1 \\ A_2 \\ B_1 \\ B_2 \end{bmatrix} = \begin{bmatrix} a_{11} & -k_2 & 0 & 0 \\ -k_2 & a_{22} & 0 & 0 \\ 0 & 0 & a_{33} & -k_2 \\ 0 & 0 & -k_2 & a_{44} \end{bmatrix}^{-1} \begin{bmatrix} k_1 Y \\ 0 \\ 0 \\ 0 \end{bmatrix}, \tag{7.432}$$

$$a_{11} = a_{33} = k_1 + k_2 - m_1\omega^2, \tag{7.433}$$

$$a_{22} = a_{44} = k_2 + k_3 - m_2\omega^2, \tag{7.434}$$

and coefficients **A** and **B** will have simpler expressions:

$$\begin{bmatrix} A_1 \\ A_2 \\ B_1 \\ B_2 \end{bmatrix} = \begin{bmatrix} -\dfrac{k_1 a_{22}}{k_2^2 - a_{11}a_{22}} Y \\ -\dfrac{k_1 k_2}{k_2^2 - a_{11}a_{22}} Y \\ 0 \\ 0 \end{bmatrix}. \tag{7.435}$$

The frequency response and phase of the undamped system would be

$$X_1 = \frac{k_1\left(k_2 + k_3 - m_2\omega^2\right)}{\left(k_1 + k_2 - m_1\omega^2\right)\left(k_2 + k_3 - m_2\omega^2\right) - k_2^2} Y, \tag{7.436}$$

$$X_2 = \frac{k_1 k_2}{\left(k_1 + k_2 - m_1\omega^2\right)\left(k_2 + k_3 - m_2\omega^2\right) - k_2^2} Y. \tag{7.437}$$

The dimensionless absolute value of the amplitudes of the undamped system are shown in Fig. 7.46. However, oscillation of the masses may be in phase or out of the phase with respect to the excitons. Figure 7.47 depicts the amplitudes and their sign. The phases of the motion are included in this illustration. All the figures relate to the following data:

$$m_2 = m_1 = 1, \qquad k_1 = k_2 = k_3 = 100, \qquad Y = 1, \tag{7.438}$$

$$X_1 = \frac{m\omega^2 - 2k}{\left(m\omega^2 - 2k\right)^2 - k^2} Y, \qquad X_2 = \frac{k^2}{\left(m\omega^2 - 2k\right)^2 - k^2} Y. \tag{7.439}$$

The natural frequencies of the system for the given value of (7.438) are

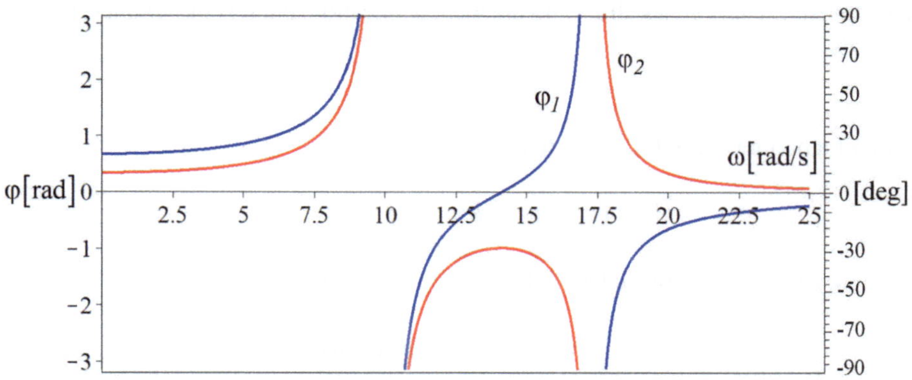


**Fig. 7.47** Frequency response of $X_1$ and $X_2$

$$\omega_1 = \sqrt{Z_1 - \sqrt{Z_2}} \qquad \omega_2 = \sqrt{Z_1 + \sqrt{Z_2}}, \tag{7.440}$$

$$Z_1 = \frac{1}{2}\left(\frac{k_1 + k_2}{m_1} + \frac{k_2 + k_3}{m_2}\right) \tag{7.441}$$

$$Z_2 = \frac{1}{4}\left(\frac{k_1 + k_2}{m_1} - \frac{k_2 + k_3}{m_2}\right)^2 + \frac{k_2^2}{m_1 m_2}, \tag{7.442}$$

$$\omega_1 = \sqrt{300} = 17.321 \qquad \omega_2 = \sqrt{100} = 10. \tag{7.443}$$

The frequency $\omega_0 = \sqrt{200} = 14.142$ is the one at which $X_1 = 0$:

$$\omega_0 = \sqrt{\frac{k_2 + k_3}{m_2}}. \tag{7.444}$$

When $\omega = \omega_0$, regardless of the order of $Y$, the first mass $m_1$ will not move as long as there is no damping in the system:

$$\omega = \omega_0, \qquad X_1 = 0, \qquad X_2 = -\frac{k_1}{k_2} Y. \tag{7.445}$$

At this frequency, the mass $m_2$ acts as a vibration absorber of the system to make $m_1$ motionless.

## 7.4 Summary

Frequency response analysis of one-$DOF$ systems is the most important analysis of vibrating systems in industry. Many vibrating systems can be modeled as a one-$DOF$ mass-spring–damper system. The response of systems to harmonic excitation will be harmonic. The frequency response is the *steady-state* solution of the equation of motion at each frequency when the system is *harmonically excited.* In frequency response analysis, we are interested in the steady-state amplitude of the oscillation as a function of the excitation frequency. The harmonic excitation is the most common excitation in industry.

There are only four types of harmonically excited one-$DOF$ systems:

1 – Base excitation
2 – Eccentric excitation
3 – Eccentric base excitation
4 – Forced excitation

*Base excitation* is the most practical model of the vertical vibration of mechanical systems. *Eccentric excitation* is a model for every type of rotary motor on a suspension. *Eccentric base excitation* is a model for the vibration of any equipment mounted on an engine. *Forced excitation* has almost no practical application, but it is the simplest model for studying forced vibration analysis methods.

**Forced excitation**
The equation of motion of a harmonically forced system is

$$m\ddot{x} + c\dot{x} + kx = F \sin \omega t, \tag{7.446}$$

which generates a frequency response expressed by either of the following functions:

$$x = A_1 \sin \omega t + B_1 \cos \omega t \tag{7.447}$$

$$= X \sin \left(\omega t - \varphi_x\right), \tag{7.448}$$

$$\begin{bmatrix} A_1 \\ B_1 \end{bmatrix} = \begin{bmatrix} \dfrac{F}{k} \dfrac{1 - r^2}{\left(1 - r^2\right)^2 + (2\xi r)^2} \\ \dfrac{F}{k} \dfrac{-2\xi r}{\left(1 - r^2\right)^2 + (2\xi r)^2} \end{bmatrix}. \tag{7.449}$$

The steady-state response has an *amplitude* $X = \sqrt{A^2 + B^2}$ and a *phase* $\varphi_x$,

$$\frac{X}{F/k} = \frac{1}{\sqrt{\left(1 - r^2\right)^2 + (2\xi r)^2}}, \tag{7.450}$$

$$\varphi_x = \tan^{-1} \frac{2\xi r}{1 - r^2}, \tag{7.451}$$

where we use the *frequency ratio* $r$, *natural frequency* $\omega_n$, and *damping ratio* $\xi$:

$$r = \frac{\omega}{\omega_n}, \qquad \xi = \frac{c}{2\sqrt{km}}, \qquad \omega_n = \sqrt{\frac{k}{m}}. \tag{7.452}$$

**Base excitation**
A base excited system is a good model for a vehicle suspension system or any equipment that is mounted on a vibrating base. The equation of motion of such a system can be expressed by either of the following equations for the absolute displacement $x$:

$$m\,\ddot{x} + c\,\dot{x} + kx = cY\omega \cos \omega t + kY \sin \omega t, \tag{7.453}$$

$$\ddot{x} + 2\xi\omega_n\,\dot{x} + \omega_n^2\,x = 2\xi\omega_n\,\omega Y \cos \omega t + \omega_n^2\,Y \sin \omega t, \tag{7.454}$$

$$\omega_n = \sqrt{\frac{k}{m}} \qquad r = \frac{\omega}{\omega_n} \qquad \xi = \frac{c}{2\sqrt{km}} \quad . \tag{7.455}$$

The system can also be expressed by either of the following equations for the relative displacement $z$:

$$m\,\ddot{z} + c\,\dot{z} + kz = m\omega^2\,Y \sin \omega t, \tag{7.456}$$

$$\ddot{z} + 2\xi\omega_n\,\dot{z} + \omega_n^2\,z = \omega^2\,Y \sin \omega t, \tag{7.457}$$

$$z = x - y. \tag{7.458}$$

The equations of motion generate the absolute and relative steady-state and frequency responses as

$$x = A_2 \sin \omega t + B_2 \cos \omega t = X \sin \left(\omega t - \varphi_x\right), \tag{7.459}$$

$$z = A_3 \sin \omega t + B_3 \cos \omega t = Z \sin \left( \omega t - \varphi_z \right) \quad , \tag{7.460}$$

$$\frac{X}{Y} = \frac{\sqrt{1 + (2\xi r)^2}}{\sqrt{\left(1 - r^2\right)^2 + (2\xi r)^2}}, \tag{7.461}$$

$$\frac{Z}{Y} = \frac{r^2}{\sqrt{\left(1 - r^2\right)^2 + (2\xi r)^2}}, \tag{7.462}$$

$$\varphi_x = \tan^{-1} \frac{2\xi r^3}{1 - r^2 + (2\xi r)^2}, \tag{7.463}$$

$$\varphi_z = \tan^{-1} \frac{2\xi r}{1 - r^2}. \tag{7.464}$$

The phase $\varphi_x$ indicates the *angular lag* of the response $x$ with respect to the excitation $y$, and the phase $\varphi_z$ is the *angular lag* of the relative motion response $z$ with respect to the excitation $y$.

**Eccentric excitation**

There is an unbalance mass $m_e$ at an *eccentricity* distance $e$ that rotates at an angular velocity $\omega$. An eccentric excited vibrating system is the proper model for the vibration analysis of an engine of a vehicle or any rotary motor that is mounted on a stationary base with flexible suspension. The equation of motion of such a system is

$$m\,\ddot{x} + c\,\dot{x} + kx = m_e e \omega^2 \sin \omega t \tag{7.465}$$

or, equivalently,

$$\ddot{x} + 2\xi \omega_n \,\dot{x} + \omega_n^2 \,x = \varepsilon e \omega^2 \sin \omega t, \tag{7.466}$$

$$\omega_n = \sqrt{\frac{k}{m}} \qquad r = \frac{\omega}{\omega_n} \qquad \xi = \frac{c}{2\sqrt{km}} \qquad \varepsilon = \frac{m_e}{m}. \tag{7.467}$$

The absolute displacement response of the system is

$$x = A_4 \sin \omega t + B_4 \cos \omega t = X \sin \left( \omega t - \varphi_e \right), \tag{7.468}$$

$$\frac{X}{e\varepsilon} = \frac{r^2}{\sqrt{\left(1 - r^2\right)^2 + (2\xi r)^2}}, \tag{7.469}$$

$$\varphi_e = \tan^{-1} \frac{2\xi r}{1 - r^2}. \tag{7.470}$$

The phase $\varphi_e$ indicates the angular lag of the response $x$ with respect to the excitation $m_e e \omega^2 \sin \omega t$.

**Eccentric base excitation**

The base of a one-$DOF$ eccentric base excited vibrating system has an unbalance mass $m_e$ at an eccentricity distance $e$ that rotates at an angular velocity $\omega$. An eccentric base excited system is the proper model for the vibration analysis of various pieces of equipment attached to the engine of a vehicle or any equipment mounted on a rotary motor.

Using the relative motion $z$ of $m$ with respect to the base, the equation of motion of the system is

$$\frac{mm_b}{m_b+m}\ddot{z}+c\,\dot{z}+kz=\frac{mm_e}{m_b+m}e\omega^2\sin\omega t \tag{7.471}$$

$$z=x-y, \tag{7.472}$$

or, equivalently,

$$\ddot{z}+2\xi\omega_n\,\dot{z}+\omega_n^2\,z=\varepsilon e\omega^2\sin\omega t, \tag{7.473}$$

$$\omega_n=\sqrt{\frac{k}{m}}\quad r=\frac{\omega}{\omega_n}\quad \xi=\frac{c}{2\sqrt{km}}\quad \varepsilon=\frac{m_e}{m_b}. \tag{7.474}$$

The relative displacement response of the system is

$$z=A_5\sin\omega t+B_5\cos\omega t=Z\sin\left(\omega t-\varphi_b\right), \tag{7.475}$$

which has an amplitude $Z$ and phase $\varphi_b$:

$$\frac{Z}{e\varepsilon}=\frac{r^2}{\sqrt{\left(1-r^2\right)^2+(2\xi r)^2}}, \tag{7.476}$$

$$\varphi_b=\tan^{-1}\frac{2\xi r}{1-r^2}. \tag{7.477}$$

All harmonically excited one-$DOF$ systems are one of the aforementioned four systems. All steady-state responses of these systems are equal to one of the following equations:

$$S_0=\frac{1}{\sqrt{\left(1-r^2\right)^2+(2\xi r)^2}}, \tag{7.478}$$

$$S_1=\frac{r}{\sqrt{\left(1-r^2\right)^2+(2\xi r)^2}}=r\,S_0, \tag{7.479}$$

$$S_2=\frac{r^2}{\sqrt{\left(1-r^2\right)^2+(2\xi r)^2}}=r^2\,S_0,$$

$$S_3=\frac{r^3}{\sqrt{\left(1-r^2\right)^2+(2\xi r)^2}}=r^3\,S_0, \tag{7.480}$$

$$S_4=\frac{r^4}{\sqrt{\left(1-r^2\right)^2+(2\xi r)^2}}=r^4\,S_0, \tag{7.481}$$

$$G_0=\frac{\sqrt{1+(2\xi r)^2}}{\sqrt{\left(1-r^2\right)^2+(2\xi r)^2}}, \tag{7.482}$$

$$G_1=\frac{r\sqrt{1+(2\xi r)^2}}{\sqrt{\left(1-r^2\right)^2+(2\xi r)^2}}=r\,G_0, \tag{7.483}$$

$$G_2 = \frac{r^2\sqrt{1+(2\xi r)^2}}{\sqrt{\left(1-r^2\right)^2+(2\xi r)^2}} = r^2 G_0, \tag{7.484}$$

$$\Phi_0 = \tan^{-1}\frac{2\xi r}{1-r^2}, \tag{7.485}$$

$$\Phi_1 = \tan^{-1}\frac{1-r^2}{-2\xi r}, \tag{7.486}$$

$$\Phi_2 = \tan^{-1}\frac{-2\xi r}{1-r^2}, \tag{7.487}$$

$$\Phi_3 = \tan^{-1}\frac{2\xi r^3}{\left(1-r^2\right)^2+(2\xi r)^2}. \tag{7.488}$$

The functions $S_0$ and $G_0$ are the main parts of all kinds of frequency response of single-$DOF$ harmonically excited systems. To classify the frequency responses of one-$DOF$ forced vibration systems, let us show the absolute displacement of the main mass by $x$, the displacement excitation of the base by $y$, and the relative displacement by $z = x - y$. We indicate the frequency responses related to the systems by adding a subscript and express their different responses as follows:

1. For a *base excited system*, we use the frequency responses of the relative and absolute motions denoted by $Z_B$, $\dot{Z}_B$, and $\ddot{Z}_B$ for relative displacement, velocity, and acceleration and $X_B$, $\dot{X}_B$, and $\ddot{X}_B$ for the absolute displacement, velocity, and acceleration of $m$, and we denote the force transmitted to the base by $F_{T_B}$.
2. For an *eccentric excited system*, we use the frequency responses of the absolute motions $X_E$, $\dot{X}_E$, and $\ddot{X}_E$, and we denote the force transmitted to the base by $F_{T_E}$.
3. For an *eccentric base excited system*, we use the frequency responses of the relative and absolute motions denoted by $Z_R$, $\dot{Z}_R$, and $\ddot{Z}_R$ for the relative displacement, velocity, and acceleration and $X_R$, $\dot{X}_R$, and $\ddot{X}_R$ for the absolute displacement, velocity, acceleration of $m$, and we denote the force transmitted to the base by $F_{T_R}$.
4. For a *force excited system*, we use the frequency responses of the absolute motions $X_F$, $\dot{X}_F$, and $\ddot{X}_F$, and we denote the force transmitted to the base by $F_{T_F}$.

We summarize and label the frequency response of different characteristics of the four systems by the following nondimensionalized equations:

$$S_0 = \frac{X_F}{F/k}, \tag{7.489}$$

$$S_1 = \frac{\dot{X}_F}{F/\sqrt{km}}, \tag{7.490}$$

$$S_2 = \frac{\ddot{X}_F}{F/m} = \frac{Z_B}{Y} = \frac{X_E}{e\varepsilon_E} = \frac{Z_R}{e\varepsilon_R}, \tag{7.491}$$

$$S_3 = \frac{\dot{Z}_B}{\omega_n Y} = \frac{\dot{X}_E}{e\varepsilon_E\omega_n} = \frac{\dot{Z}_R}{e\varepsilon_R\omega_n}, \tag{7.492}$$

$$S_4 = \frac{\ddot{Z}_B}{\omega_n^2 Y} = \frac{\ddot{X}_E}{e\varepsilon_E\omega_n^2} = \frac{\ddot{Z}_R}{e\varepsilon_R\omega_n^2}, \tag{7.493}$$

$$G_0 = \frac{F_{T_F}}{F} = \frac{X_B}{Y}, \tag{7.494}$$

$$G_1 = \frac{\dot{X}_B}{\omega_n Y}, \tag{7.495}$$

$$G_2 = \frac{\ddot{X}_B}{\omega_n^2 Y} = \frac{F_{T_B}}{kY} = \frac{F_{T_E}}{e\omega_n^2 m_e} = \frac{F_{T_R}}{e\omega_n^2 m_e}\left(1 + \frac{m_b}{m}\right). \tag{7.496}$$

Figures A.1–A.8 in Appendix A visualize these frequency responses to guide design engineers to evaluate the behavior of systems.

## 7.5 Key Symbols

| | |
|---|---|
| $a \equiv \ddot{x}$ | acceleration |
| $a, b, C$ | Fourier coefficient |
| $a, b, c, d, e$ | dimension exponent |
| $A, B$ | weight factor, coefficients for frequency responses |
| $A, \bar{A}$ | complex conjugate coefficients |
| $c$ | damping |
| $C$ | coefficient |
| $d_1$ | road wave length |
| $d_2$ | road wave amplitude |
| $D$ | dissipation function |
| $e$ | eccentricity arm, exponential function |
| $E$ | mechanical energy, |
| $E_c$ | consumed energy of a damper |
| $f = 1/T$ | cyclic frequency [Hz] |
| $f, F, \mathbf{F}$ | force |
| $f, g$ | function, periodic function |
| $f_c$ | damper force |
| $f_e$ | equivalent force |
| $f_k$ | spring force |
| $f_m$ | required force to move mass $m$ |
| $f_T$ | transmitted force |
| $F$ | amplitude of harmonic force $f = F \sin \omega t$ |
| $F_T$ | amplitude of transmitted force |
| $F_0$ | constant force |
| $g$ | gravitational acceleration, function |
| $g\left(r^2\right)$ | characteristic equation |
| $G$ | general frequency response with node |
| $i, j$ | dummy index |
| $k$ | stiffness |
| $k_e$ | equivalent stiffness |
| $K$ | kinetic energy |
| $l$ | length |
| $L$ | length dimension |
| $\mathcal{L}$ | Lagrangean |
| $m$ | mass, number of octaves |
| $m_b$ | device mass |
| $m_e$ | eccentric mass, equivalent mass |
| $M$ | mass dimension |
| $P$ | power, point |
| $P$ | potential energy |
| $q$ | general coordinate |

| | |
|---|---|
| $Q$ | quality factor, general amplitude |
| $r$ | frequency ratio |
| $r_i,\ i \in N$ | nodal frequency ratio |
| $r_n = \omega_n/\omega_s$ | natural frequency ratio |
| $S$ | general frequency response without node |
| $t$ | time |
| $T$ | period, time dimension |
| $v \equiv \dot{x}$ | velocity |
| $x, y, z, \mathbf{x}$ | displacement |
| $x_0$ | initial displacement |
| $x_s$ | sprung mass displacement |
| $x_u$ | unsprung mass displacement |
| $X$ | steady-state amplitude of $x$ |
| $X_s$ | steady-state amplitude of $x_s$ |
| $X_u$ | steady-state amplitude of $x_u$ |
| $\dot{x}, \dot{y}, \dot{z}$ | velocity, time derivative of $x, y, z$ |
| $\dot{x}_0$ | initial velocity |
| $X, Y, Z$ | amplitude |
| $Y$ | amplitude of base displacement |
| $\dot{y}$ | velocity |
| $z$ | relative displacement |
| $Z$ | amplitude of relative displacement |
| $Z_1, Z_2$ | short notation |

Greek

| | |
|---|---|
| $\alpha$ | angle |
| $\delta$ | difference |
| $\delta_s$ | static deflection |
| $\Delta$ | difference |
| $\gamma$ | angle |
| $\lambda$ | eigenvalue |
| $\Lambda$ | eigenvalue matrix |
| $\varepsilon$ | mass ratio |
| $\varepsilon$ | small coefficient |
| $\theta$ | angular motion |
| $\Pi$ | Buckingham dimensionless term |
| $\rho$ | length mass density |
| $\xi$ | damping ratio |
| $\omega$ | angular frequency |
| $\omega_n$ | natural frequency |
| $\varphi, \Phi$ | phase angle |

| | |
|---|---|
| $\alpha = \omega_s/\omega_u$ | sprung mass ratio |
| $\varepsilon = m_s/m_u$ | sprung mass ratio |
| $\eta = \lvert Z/Y \rvert$ | sprung mass relative frequency response |
| $\mu = \lvert X_s/Y \rvert$ | sprung mass frequency response |
| $\xi = c_s/\left(2\sqrt{k_s m_s}\right)$ | damping ratio |
| $\tau = \lvert X_u/Y \rvert$ | unsprung mass frequency response |
| $\omega = 2\pi f$ | angular frequency [rad/s] |
| $\omega_s = \sqrt{k_s/m_s}$ | sprung mass frequency |
| $\omega_u = \sqrt{k_u/m_u}$ | unsprung mass frequency |

Subscript

| | |
|---|---|
| $i \in N$ | node number |
| $n$ | natural |
| $s$ | sprung |
| $u$ | unsprung |

Symbols

| | |
|---|---|
| $DOF$ | degree of freedom |
| $FBD$ | free body diagram |
| $ODF$ | one degree of freedom |
| Re | real part |
| $RMS$ | root mean square |
| det | determinant |
| Hz | Hertz |
| [ ] | matrix |
| $\int$ | integral |
| $\mid\ \mid$ | absolute value |
| $d$ | differential |
| $\partial$ | partial derivative |
| $\mathcal{L}$ | Lagrangean |
| $O$ | order of magnitude |
| $\nabla$ | gradient |

## References


1. J. Angeles, *Dynamic Response of Linear Mechanical Systems: Modeling, Analysis and Simulation* (Springer, New York, 2011)
2. B. Balachandran, E.B. Magrab, *Vibrations* (Brooks/Cole, Pacific Grove, 2003)
3. C.W. Celia, A.T.F. Nice, K.F. Elliott, *Advanced Mathematics* (Macmillan Education LTD, London, 1982)
4. M. Del Pedro, P. Pahud, *Vibration Mechanics* (Kluwer Academic Publishers, The Netherland, 1991)
5. J.P. Den Hartog, *Mechanical Vibrations* (McGraw-Hill, New York, 1934)
6. A. Dimarogonas, *Vibration for Engineers* (Prentice Hall, New Jersey, 1996)
7. E. Esmailzadeh, R.N. Jazar, Periodic solution of a Mathieu-Duffing type equation. Int. J. Nonlinear Mech. **32**(5), 905–912 (1997)
8. E. Esmailzadeh, R.N. Jazar, Periodic behavior of a cantilever with end mass subjected to harmonic base excitation. Int. J. Nonlinear Mech. **33**(4), 567–577 (1998)
9. E. Esmailzadeh, B. Mehri, R.N. Jazar, Periodic solution of a second order, autonomous, nonlinear system. J. Nonlinear Dyn. **10**(4), 307–316 (1996)
10. K.S. Graham, *Fundamentals of Mechanical Vibrations*, 2nd edn. (McGraw-Hill, New York, 2000)
11. D. Inman, *Engineering Vibrations* (Prentice Hall, New York, 2007)
12. R.N. Jazar, *Advanced Vibrations: A Modern Approach* (Springer, New York, 2013)
13. R.N. Jazar, *Vehicle Dynamics: Theory and Application*, 3rd edn. (Springer, New York, 2017)
14. R.N. Jazar, *Approximation Methods in Science and Engineering* (Springer, New York, 2020)
15. R.N. Jazar, *Perturbation Methods in Science and Engineering* (Springer, New York, 2021)
16. R.N. Jazar, *Advanced Vibrations: Theory and Application* (Springer, New York, 2023)
17. R.N. Jazar, M. Kazemi, S. Borhani, *Mechanical Vibrations* (in Persian) (Ettehad Publications, Tehran, 1992)
18. L. Meirovitch, *Analytical Methods in Vibrations* (Macmillan, New York, 1967)
19. L. Meirovitch, *Principles and Techniques of Vibrations* (Prentice Hall, Ney Jersey, 1997)
20. L. Meirovitch, *Fundamentals of Vibrations* (McGraw-Hill, New York, 2002)
21. K. Ogata, *System Dynamics* (Prentice Hall, Ney Jersey, 2004)
22. Z. Osinski, *Damping of Vibrations* (A. A. Balkema Publishers, Rotterdam, 1998)
23. W.J. Palm, *Mechanical Vibration* (Wiley, New York, 2006)
24. S.S. Rao, *Mechanical Vibrations* (Pearson Prentice Hall, Harlow, 2018)
25. M. Roseau, *Vibrations in Mechanical Systems*(Springer, Berlin, 1987)
26. J.C. Snowdon, *Vibration and Shock in Damped Mechanical Systems* (Wiley, New York, 1968)
27. F.S. Tse, I.E. Morse, R.T. Hinkle, *Mechanical Vibrations Theory and Applications* (Allyn and Bacon, Boston, 1978)

# Part III

# Vehicle Vibrations Analysis

The suspension system of an automobile is a crucial segment of vehicles. The coil and leaf springs are widely used in automobile suspension systems. A well-designed suspension system plays an important role in smooth and jerk-free ride, as well as driving pleasure and motion sickness reduction.

There are five mathematical models for vehicle vibrations: 1/8 vehicle model, quarter car model, bicycle model, half car model, and full car model. The 1/8 vehicle model is a one degree of freedom base excited system in which the vertical vibrations of vehicles will be studied. In this model, wheels are ignored. Considering wheels and stiffness of tires, make the quarter car model to study vertical vibrations of vehicle and suspension.

Vertical vibrations is the most important and the most common vibrating mode of vehicles. The quarter car model is the simplest practical vehicle vibration model to determine main characteristics of vehicle vibrations. Natural frequencies, mode shapes, and frequency response of linear model of quarter car are the base model to determine the rates of spring and damper of vehicle. The quarter car model is capable to be more realistic by considering nonlinear spring and damper. The quarter car model not only is the main system to design and optimize passive and smart suspensions but is also the main system to investigate and optimize semi-active and active suspensions.

The pitch mode of vehicle vibrations is the oscillation of the vehicle about a lateral axis. To study the pitch mode of vehicles, we imaginarily cut the vehicle longitudinally and make bicycle model. The model will show vertical and pitch modes of vehicle vibrations. Compared to vertical vibrations, the pitch oscillations are more uncomfortable. To manage the pitch mode of vibrations, we have two options. Firstly, the ratio of front to rear suspension rates should be optimized, and secondly, we can introduce anti-pitch spring and damper to adjust the pitch mode without changing the vertical mode of oscillations.

The roll mode of vehicle vibrations is the oscillation of the vehicle about the longitudinal axis. To study the roll mode, we cut the vehicle laterally and make the half car model for the front half or the rear half. The model will have vertical and roll modes of vehicle vibrations. Compared to vertical and pitch vibrations, the roll oscillations are the most uncomfortable mode of vehicle vibrations. To study and optimize the roll mode of vibrations, we introduce antiroll spring and damper to adjust the roll mode without changing the vertical mode of oscillations.

Finally, a full car model will be employed to examine the overall vehicle vibration modes and test the optimization results from previous models. Although the optimization process of suspension parameters is done based on frequency response, their effectiveness will be examined by transient responses.

In this part, which is the main mathematical treatment of vehicle vibrations, we will study these models one by one and show how and when to use these models.

# 1/8 Vehicle Model

# 8

The simplest vibrating model of vehicles is considering only vertical vibration of the vehicle while ignoring the flexibility of tires compared to the stiffness of the main springs. Let us cut a vehicle longitudinally and laterally, and we keep one quarter of the vehicle, and then, assume that tires are rigid and massless. This way we model a vehicle with a one degree of freedom (*DOF*) base excited system with the mass $m$ to be one quarter of the mass of the body of vehicle, and suspension spring $k$ and damper $c$ to be equal to $k$ and $c$ of one side of the vehicle. Such model is called 1/8 vehicle model. The 1/8 vehicle model that is shown in Fig. 8.1 is very simple and yet very accurate to optimize the main stiffness and damping of the suspension.

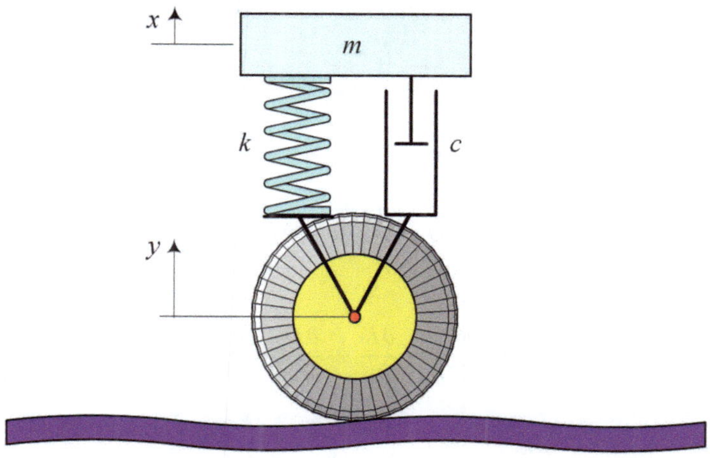


**Fig. 8.1** 1/8 vehicle vibration model as a base excited linear suspension system

In this chapter, we examine a one *DOF*, base excited vibration vehicle as the simplest model for a suspension to separate a mass $m$ from the vibrating base. We also study the optimization method to design the suspension parameters $k$ and $c$, employing the root mean square (*RMS*) optimization method. We develop a design chart to determine the optimal damper and spring for the best vibration isolation.

## 8.1 Mathematical Model

Figure 8.1 illustrates a 1/8 vehicle vibration model as a one *DOF* base excited linear vibrating system. Such system not only can represent a model of vertical vibrations of a vehicle, but it can also model the suspension of any equipment mounted on a vibrating base. As a model of a vehicle, a one-fourth of the mass of the body is modeled as a solid mass $m$ denoted as *sprung mass*. A spring of stiffness $k$ and a shock absorber with viscous damping $c$ support the sprung mass and represent the main suspension of the vehicle. The suspension parameters $k$ and $c$ are the equivalent stiffness and damping for one wheel, measured at one center of the wheel.



R. N. Jazar, H. Marzbani, *Vehicle Vibrations*, https://doi.org/10.1007/978-3-031-43486-0_8

The equation of motion of the system for the absolute displacement $x$ and the relative displacement $z$ are:

$$m\ddot{x} + c\dot{x} + kx = c\dot{y} + ky \tag{8.1}$$

$$m\ddot{z} + c\dot{z} + kz = -m\ddot{y} \tag{8.2}$$

$$z = x - y \tag{8.3}$$

The variable $x$ is the absolute displacement of the body, and $y$ is the absolute displacement of the ground. The equations of motion (8.1) and (8.2), which depend on the three parameters $(m, c, k)$, can be transformed to the following two parameter dependent equations,

$$\ddot{x} + 2\xi\omega_n\dot{x} + \omega_n^2 x = 2\xi\omega_n\dot{y} + \omega_n^2 y \tag{8.4}$$

$$\ddot{z} + 2\xi\omega_n\dot{z} + \omega_n^2 z = -\ddot{y} \tag{8.5}$$

by introducing the *natural frequency* $\omega_n$ and the *damping ratio* $\xi$.

$$\xi = \frac{c}{2\sqrt{km}} \tag{8.6}$$

$$\omega_n = \sqrt{\frac{k}{m}} = 2\pi f_n \tag{8.7}$$

***Proof*** The kinetic energy, potential energy, and dissipation function of the system are

$$K = \frac{1}{2}m\dot{x}^2 \tag{8.8}$$

$$P = \frac{1}{2}k(x - y)^2 \tag{8.9}$$

$$D = \frac{1}{2}c(\dot{x} - \dot{y})^2 \tag{8.10}$$

Employing the Lagrange equation

$$\frac{d}{dt}\left(\frac{\partial K}{\partial \dot{x}}\right) - \frac{\partial K}{\partial x} + \frac{\partial D}{\partial \dot{x}} + \frac{\partial P}{\partial x} = 0 \tag{8.11}$$

we find the equation of motion

$$\frac{d}{dt}(m\dot{x}) + c(\dot{x} - \dot{y}) + k(x - y) = 0 \tag{8.12}$$

which can be simplified to Eq. (8.1). Introducing a relative displacement variable, $z = x - y$, we have

$$\dot{z} = \dot{x} - \dot{y} \tag{8.13}$$

$$\ddot{z} = \ddot{x} - \ddot{y} \tag{8.14}$$

so we can rewrite Eq. (8.12) as

$$m\frac{d}{dt}(\dot{z} + \dot{y}) + c\dot{z} + kz = 0 \tag{8.15}$$

which is equivalent to (8.2).

Dividing Eqs. (8.1) and (8.2) by $m$ and using (8.6) and (8.7) generate the equivalent Eqs. (8.4) and (8.5), respectively.

The function of a vehicle suspension as a vibration isolator is to reduce the magnitude of motion transmitted from a vibrating foundation to the body of the vehicle, or to reduce the magnitude of force transmitted from the vehicle to the road, both in time and frequency domains. In the simplest approach to suspension analysis, the parameters $m$, $k$, and $c$ are considered constant and independent of the excitation frequency or behavior of the vehicle on the road. This assumption is equivalent to considering an infinitely stiff and massive foundation. For rubber isolators, the damping coefficient usually decreases, and the stiffness coefficient increases with excitation frequency. Moreover, neither the road nor the vehicle can be assumed to be infinitely stiff rigid body at high frequencies. ■

**Example 105** 1/8 vehicle model for front and rear parts of a vehicle. Vibration isolator system to model a vehicle will be different for front and rear parts of the vehicle. That is because the mass center of the vehicle is not in the middle and hence, the load on front and rear wheels are unequal, and also the value of $k$ and $c$ are different for the front and rear suspensions.

Consider a vehicle with the following information:

$$car\ mass = 1500\,\text{kg} \quad wheel\ mass = 50\,\text{kg}$$

$$F_{z_1} = 3941.78\,\text{N} \qquad F_{z_2} = 3415.6\,\text{N} \tag{8.16}$$

where $F_{z_1}$ and $F_{z_2}$ are the front and rear wheel loads, respectively. The mass $m$ of the 1/8 vibrating model for the front of the car will be

$$m_1 = \frac{1}{2}\frac{F_{z_1}}{F_{z_1} + F_{z_2}} \times (1500 - 4 \times 50) = 348.25\,\text{kg} \tag{8.17}$$

and the mass for rear of the car will be

$$m_2 = \frac{1}{2}\frac{F_{z_2}}{F_{z_1} + F_{z_2}} \times (1500 - 4 \times 50) = 301.76\,\text{kg} \tag{8.18}$$

The equivalent spring stiffness at the center of the wheel of the rear springs are designed around 20% higher than the front springs. Let us assume that $k_1 = 11\,\text{kN/m}$ and the rear spring is 20% higher. Also let us assume the damping of the front shock absorber to be $c_1 = 610\,\text{N s/m}$ and the damping of the rear damper to be 20% higher. Then the 1/8 model for the front and rear parts will have the following natural frequencies and damping ratios.

$$\xi_1 = \frac{c_1}{2\sqrt{k_1 m_1}} = \frac{610}{2\sqrt{11000 \times 348.25}} = 0.156 \tag{8.19}$$

$$\omega_{n1} = \sqrt{\frac{k_1}{m_1}} = \sqrt{\frac{11000}{348.25}} = 5.62\,\text{rad/s} = 0.894\,\text{Hz} \tag{8.20}$$

$$\xi_2 = \frac{c_1}{2\sqrt{k_2 m_2}} = \frac{610 \times 1.2}{2\sqrt{11000 \times 1.2 \times 301.76}} = 0.184 \tag{8.21}$$

$$\omega_{n2} = \sqrt{\frac{k_2}{m_2}} = \sqrt{\frac{11000 \times 1.2}{301.76}} = 6.614\,\text{rad/s} = 1.053\,\text{Hz} \tag{8.22}$$

**Example 106** Nondimensionalized equations of motion. Converting the equations of motion of mechanical vibrating systems into dimensionless form is the best way to use the results of analysis for all similar systems.

All mechanical vibrating systems work only with three main dimensions of mass $M$, length $L$, and time $T$. To make the equations of motion dimensionless, we need to make the variables of the equations nondimensional by defining new dimensionless variables. There are three variables with length dimension, $x_s$, $x_u$, $y$, and a time variable. Consider the equation of motion of the 1/8 vehicle model in absolute displacement.

$$m\ddot{x} + c\dot{x} + kx = c\dot{y} + ky \tag{8.23}$$

The equation of motion has three variables: $x$, $y$, and $t$. Let us define a characteristic length $x_0$ and a characteristic time $t_0$, to define nondimensionalized displacement $X$, $Y$, and time $\tau$

$$X = \frac{x}{x_0} \qquad Y = \frac{y}{x_0} \qquad \tau = \frac{t}{t_0} \tag{8.24}$$

and replace the actual variables $x$, $y$, and $t$. At the moment we may leave $x_0$ and $t_0$ unknown, although we know that in vibrations, $x_0$ is a constant length such as static deflection or amplitude of displacement excitation, and $t_0$ is inverse of a frequency such as a natural frequency or an excitation frequency.

$$x = x_0 X \qquad y = x_0 Y \qquad t = t_0 \tau \tag{8.25}$$

Using nondimensionalized time $\tau$, we may redefine derivatives.

$$\dot{x} = \frac{dx}{dt} = x_0 \frac{dX}{dt} = x_0 \frac{dX}{d\tau}\frac{d\tau}{dt} = \frac{x_0}{t_0} X' \tag{8.26}$$

$$\ddot{x} = \frac{d}{dt}\frac{dx}{dt} = \frac{x_0}{t_0}\frac{d}{dt} X' = \frac{x_0}{t_0}\frac{d\tau}{dt}\frac{d}{d\tau} X' = \frac{x_0}{t_0^2} X'' \tag{8.27}$$

$$X' = \frac{dX}{d\tau} \qquad X'' = \frac{d^2X}{d\tau^2} \tag{8.28}$$

The equation of motion can now be rewritten as

$$\frac{x_0}{t_0^2} m X'' + \frac{x_0}{t_0} c X' + x_0 k X = \frac{x_0}{t_0} c Y' + x_0 k Y \tag{8.29}$$

After dividing by $m x_0 / t_0^2$ we have

$$X'' + t_0 2\xi\omega_n X' + t_0^2 \omega_n^2 X = 2 t_0 \xi \omega_n Y' + t_0^2 \omega_n^2 Y \tag{8.30}$$

where

$$\omega_n = \sqrt{\frac{k}{m}} \qquad \xi = \frac{c}{2\sqrt{km}} \tag{8.31}$$

It suggests that we use $t_0 = 1/\omega_n$, then the equation will be much simpler.

$$X'' + 2\xi X' + X = 2\xi Y' + Y \tag{8.32}$$

As a fundamental method, consider the general equation of a dependent variable $y$ and independent variable $x$.

$$\frac{d^n y}{dx^n} = f\left(x, y, \frac{dy}{dt}, \frac{d^2y}{dt^2}, \frac{d^3y}{dt^3}, \cdots, \frac{d^{n-1}y}{dx^{n-1}}\right) \tag{8.33}$$

The equation will be transformed into a nondimensional equation

$$\frac{d^n Y}{dX^n} = \frac{x_0^k}{y_0} f\left(x_0 X, y_0 Y, \frac{y_0}{x_0}\frac{dY}{dX}, \frac{y_0}{x_0^2}\frac{d^2Y}{dX^2}, \cdots, \frac{y_0}{x_0^{k-1}}\frac{d^{k-1}Y}{dX^{k-1}}\right) \tag{8.34}$$

by introducing dimensionless variables $Y$ and $X$ using dimensional characteristics $y_0$, $x_0$.

$$Y = \frac{y}{y_0} \qquad X = \frac{x}{x_0} \tag{8.35}$$

## 8.2 Time Response

The time response of the linear 1/8 vehicle model is the solution of the following equations for the absolute displacement $x$, in which $y(t)$ is the road excitation function.

$$m\ddot{x} + c\dot{x} + kx = c\dot{y}(t) + ky(t) \tag{8.36}$$

$$\ddot{x} + 2\xi\omega_n \dot{x} + \omega_n^2 x = 2\xi\omega_n \dot{y} + \omega_n^2 y \tag{8.37}$$

$$\omega_n = \sqrt{\frac{k}{m}} \qquad r = \frac{\omega}{\omega_n} \qquad \xi = \frac{c}{2\sqrt{km}} \tag{8.38}$$

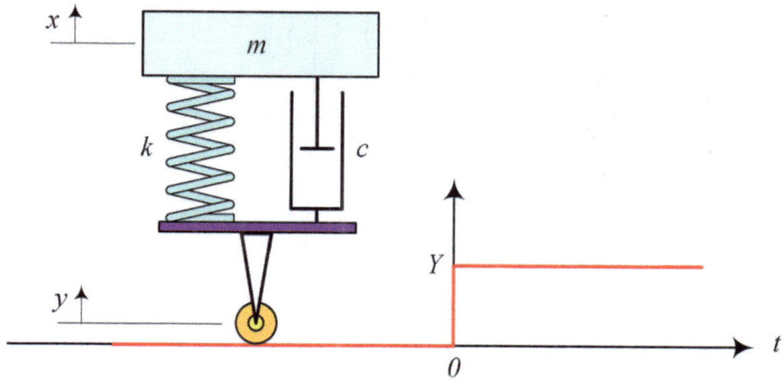


**Fig. 8.2** A vibrating model of a vehicle to have a step displacement under its tire

The equation of motion can also be expressed by either one of the following equations for the relative displacement $z = x - y$.

$$m\ddot{z} + c\dot{z} + kz = m\ddot{y}(t) \tag{8.39}$$

$$\ddot{z} + 2\xi\omega_n\dot{z} + \omega_n^2 z = \ddot{y}(t) \tag{8.40}$$

$$z = x - y \tag{8.41}$$

The road excitation function $y(t)$ is the forcing term that simulates the road displacement. Step input is the simplest and most practical excitation to reveal dynamic characteristics of transient response of the dynamic systems. As a good example of transient response, we determine the vibrating response of vehicles to a step road excitation.

A step input is a sudden change of the excitation function $y(t)$ from zero to a constant steady value $Y$ such as the one shown in Fig. 8.2. If $Y = 1$, then the excitation is called unit step input. The response of the system to a unit step input is called the unit step response. Linearity of the equation of motion guarantees that the response to a non-unit step input is proportional to the unit step response.

$$\begin{aligned} x = {} & \frac{cY}{m\omega_d} e^{-\xi\omega_n t} \sin\omega_d t \\ & + Y\left(1 - e^{-\xi\omega_n t}\left(\frac{\xi\omega_n}{\omega_d}\sin\omega_d t + \cos\omega_d t\right)\right) \end{aligned} \tag{8.42}$$

***Proof*** Let us assume the road displacement, $y(t) = Y$, to be a unit step.

$$m\ddot{x} + c\dot{x} + kx = c\dot{y}(t) + ky(t) \tag{8.43}$$

$$Y = \begin{cases} 1 & t > 0 \\ 0 & t \le 0 \end{cases} \tag{8.44}$$

The solution of the system to a step input is simpler to be treated as a case of constant input.

$$m\ddot{x} + c\dot{x} + kx = kY \qquad x(0) = x_0 \qquad \dot{x}(0) = \dot{x}_0 \tag{8.45}$$

The complete solution is the summation of the homogenous solution $x_h$ and the particular solution $x_p = Y$.

$$x = e^{-\xi\omega_n t}\left(A\sin\omega_d t + B\cos\omega_d t\right) + Y \tag{8.46}$$

Applying the initial conditions provides us with:

$$x_0 = B + Y \qquad \dot{x}_0 = A\omega_d - B\xi\omega_n \tag{8.47}$$

$$A = \frac{1}{\omega_d} \left( \dot{x}_0 - \xi \omega_n (Y - x_0) \right) \tag{8.48}$$

$$B = -\frac{1}{\xi \omega_n} \left( \dot{x}_0 - x_0 + \xi \omega_n (Y - x_0) \right) \tag{8.49}$$

If the initial conditions are zero, then the solution will be

$$\begin{aligned} x &= e^{-\xi \omega_n t} \left( -\frac{\xi \omega_n}{\omega_d} Y \sin \omega_d t - Y \cos \omega_d t \right) + Y \\ &= Y - Y e^{-\xi \omega_n t} \left( \frac{\xi \omega_n}{\omega_d} \sin \omega_d t + \cos \omega_d t \right) \end{aligned} \tag{8.50}$$

The solution is better to be expressed dimensionless.

$$\frac{x}{Y} = 1 - e^{-\xi \omega_n t} \left( \cos \omega_d t + \frac{\xi \omega_n}{\omega_d} \sin \omega_d t \right) \tag{8.51}$$

Having the absolute displacement $x$ and the excitation $y$, the relative displacement response would be

$$z = x - y = Y e^{-\xi \omega_n t} \left( \frac{\xi \omega_n}{\omega_d} \sin \omega_d t + \cos \omega_d t \right) \tag{8.52}$$

Mathematically, the step function is expressed by the on-off switching Heaviside function $H(t)$.

$$H(t - t_0) = \begin{cases} 1 & t > t_0 \\ 0 & t < t_0 \end{cases} \tag{8.53}$$

Therefore,

$$y = Y H(t) \qquad \dot{y} = Y \frac{d}{dt} H = Y \delta(t) \tag{8.54}$$

where the Dirac delta function $\delta(t)$ is the time derivative of the Heaviside function.

Employing the convolution integral, the solution for absolute displacement, $x$, of the body of the vehicle will have two parts due to $\delta(t)$ and $H(t)$.

$$\begin{aligned} x &= cYh(t) + \int_0^t h(\tau) f(t - \tau)\, d\tau \\ &= cYh(t) + kY \int_0^t h(\tau)\, d\tau \end{aligned} \tag{8.55}$$

$$h(t) = \frac{1}{m\omega_d} e^{-\xi \omega_n t} \sin \omega_d t \tag{8.56}$$

The result of convolution integral will be calculated.

$$\begin{aligned} &\frac{1}{m\omega_d} \int_0^t e^{-\xi \omega_n \tau} \sin \omega_d \tau \, d\tau \\ &= \frac{1}{m\omega_d} \left( \frac{\omega_d}{\omega_d^2 + \xi^2 \omega_n^2} - \frac{\omega_d \cos \omega_d t + \xi \omega_n \sin \omega_d t}{\omega_d^2 + \xi^2 \omega_n^2} e^{-\xi \omega_n t} \right) \\ &= \frac{1}{k} \left( 1 - e^{-\xi \omega_n t} \left( \cos \omega_d t + \frac{\xi \omega_n}{\omega_d} \sin \omega_d t \right) \right) \end{aligned} \tag{8.57}$$

Therefore, the displacement function $x$ will be

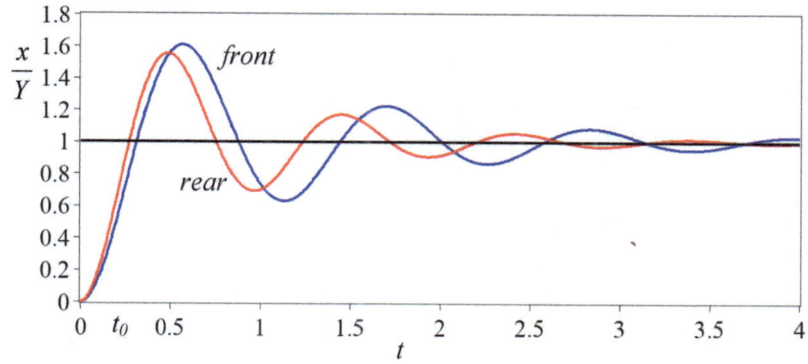


**Fig. 8.3** Time response of the 1/8 vehicle model for the front and rear parts of a vehicle to a step road excitation

$$\begin{aligned} x &= \frac{cY}{m\omega_d} e^{-\xi\omega_n t} \sin \omega_d t + Y \left(1 - e^{-\xi\omega_n t}\left(\cos\omega_d t + \frac{\xi\omega_n}{\omega_d}\sin\omega_d t\right)\right) \\ &\quad + e^{-\xi\omega_n t}\left(A\sin\omega_d t + B\cos\omega_d t\right) \\ &= 2\frac{\xi\omega_n Y}{\omega_d} e^{-\xi\omega_n t} \sin \omega_d t + Y \left(1 - e^{-\xi\omega_n t}\left(\cos\omega_d t + \frac{\xi\omega_n}{\omega_d}\sin\omega_d t\right)\right) \\ &\quad + e^{-\xi\omega_n t}\left(A\sin\omega_d t + B\cos\omega_d t\right) \end{aligned} \tag{8.58}$$

A homogenous response must be added to the solution of the convolution integral because the delta function $\delta(t)$ changes the initial conditions. The coefficients of the homogenous part of the solution will be found to satisfy the zero initial conditions.

$$x_0 = 0 \qquad \dot{x}_0 = 0 \tag{8.59}$$

$$A = -\frac{2\xi\omega_n Y}{\omega_d} \qquad B = 0 \tag{8.60}$$

Substituting $A$ and $B$ determines the solution, matching (8.51)

$$x = Y\left(1 - e^{-\xi\omega_n t}\left(\cos\omega_d t + \frac{\xi\omega_n}{\omega_d}\sin\omega_d t\right)\right) \tag{8.61}$$

Heaviside function was introduced by the English mathematician Oliver Heaviside (1850–1925). The importance and application of the delta function are introduced by British physicist, Paul Adrien Maurice Dirac (1902–1984). ■

**Example 107** Response analysis of the 1/8 vehicle model to step input. Because of importance of step input, we examine the response of the vehicles for different parameters and show them graphically.

Let us assume a vehicle with the following specifications for front and rear 1/8 parts indicated by 1 and 2, respectively.

$$m_1 = 348.25\,\text{kg} \qquad m_2 = 301.76\,\text{kg} \tag{8.62}$$

$$\xi_1 = 0.156 \qquad \xi_2 = 0.184 \tag{8.63}$$

$$\begin{aligned} \omega_{n1} &= 5.62\,\text{rad/s} = 0.894\,\text{Hz} \\ \omega_{n2} &= 6.614\,\text{rad/s} = 1.053\,\text{Hz} \end{aligned} \tag{8.64}$$

Figure 8.3 illustrates the time response of the 1/8 vehicle model for the front and rear parts of the vehicle. Because of different masses, the front and rear 1/8 model of the vehicle have different natural frequencies and damping ratios even if the spring and damper are the same. The rear 1/8 model has higher natural frequency and damping ratio. As a result, the rear model vibrates quicker and damps quicker than the front model.

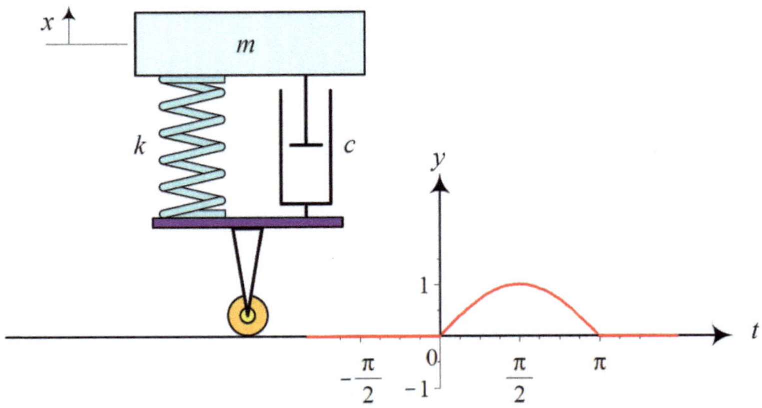


**Fig. 8.4** A one *DOF* system with a half sine transient displacement base excitation

**Example 108** Half-sine transient base excitation. Half-sine is a reasonable mathematical model for bumps on the roads to make vehicles to drive slowly. Here is the analytic calculation of the response of the 1/8 vibrating model of vehicles to a bump.

Figure 8.4 illustrates a 1/8 vibrating model of a vehicle along with a half-sine transient road excitation.

$$y = \begin{cases} Y \sin \omega t & 0 \le t \le t_0 \\ 0 & t_0 < t \end{cases} \qquad t_0 = \frac{\pi}{\omega} \tag{8.65}$$

Let us solve the problem in relative displacement. The equation of motion of the vehicle in relative displacement $z$ is

$$m\ddot{z} + c\dot{z} + kz = -m\ddot{y} \qquad z = x - y \tag{8.66}$$

When the road excitation function $y = y(t)$ is given, then $-m\ddot{y}$ acts as an equivalent external force $f(t)$ in the equation of motion.

The relative displacement response $z(t)$ of the vehicle would be the result of the following convolution integral which is valid only for $0 < t < t_0$.

$$z(t) = \int_0^t h(t - \tau) f(\tau) \, d\tau \tag{8.67}$$

$$h(t) = \frac{1}{m\omega_n} e^{-\xi\omega_n t} \sin \omega_d t \tag{8.68}$$

$$f(t) = -m\ddot{y} = \begin{cases} mY\omega^2 \sin \omega t & 0 < t < t_0 \\ 0 & t_0 < t \end{cases} \qquad t_0 = \frac{\pi}{\omega} \tag{8.69}$$

Therefore,

$$\begin{aligned} z(t) &= \frac{Y\omega^2}{\omega_n} \int_0^t e^{-\xi\omega_n(t-\tau)} \sin \omega_d (t - \tau) \, \sin \omega\tau \, d\tau \\ &= \frac{Y\omega^2 e^{-\xi\omega_n t} \left( \frac{\omega}{\omega_d} \left(\xi^2\omega_n^2 + \omega^2 - \omega_d^2\right) \sin \omega_d t + 2\xi\omega_n\omega \cos \omega_d t \right)}{\left(\xi^2\omega_n^2 + (\omega - \omega_d)^2\right)\left(\xi^2\omega_n^2 + (\omega + \omega_d)^2\right)} \\ &\quad + \frac{Y\omega^2 \left( \left(\xi^2\omega_n^2 - \omega^2 + \omega_d^2\right) \sin \omega t - 2\xi\omega_n\omega \cos \omega t \right)}{\left(\xi^2\omega_n^2 + (\omega - \omega_d)^2\right)\left(\xi^2\omega_n^2 + (\omega + \omega_d)^2\right)} \\ & 0 < t < t_0 \end{aligned} \tag{8.70}$$

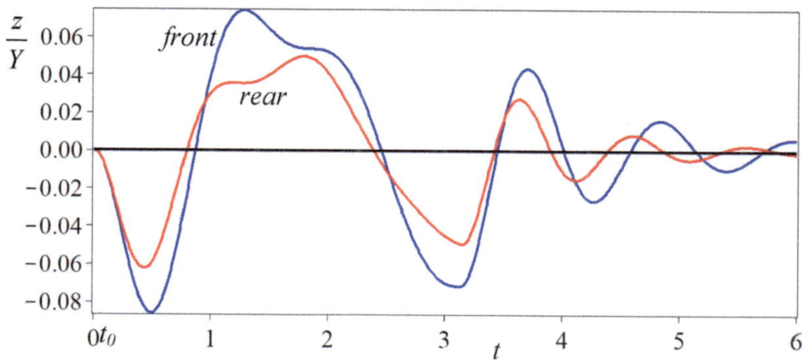


**Fig. 8.5** The response of the 1/8 vehicle model in relative coordinate $z = x - y$, for front and rear parts

The response of the system for $t_0 < t$ will be a free vibrations with initial conditions at the end of the half-sine excitation. The conditions of the system at $t = t_0$ are

$$z(t_0) = 0 \tag{8.71}$$

$$\begin{aligned}\dot{z}(t_0) = & \frac{-Y\omega^3\left(\xi^2\omega_n^2 - \omega^2 + \omega_d^2\right)e^{-\xi\omega_n t_0}}{\left(\xi^2\omega_n^2 + (\omega-\omega_d)^2\right)\left(\xi^2\omega_n^2 + (\omega+\omega_d)^2\right)}\cos\omega_d t_0 \\ & - \frac{Y\omega^3\xi\omega_n\left(\xi^2\omega_n^2 + \omega^2 + \omega_d^2\right)e^{-\xi\omega_n t_0}}{\omega_d\left(\xi^2\omega_n^2 + (\omega-\omega_d)^2\right)\left(\xi^2\omega_n^2 + (\omega+\omega_d)^2\right)}\sin\omega_d t_0 \\ & + Y\omega^3\frac{\left(\xi^2\omega_n^2 - \omega^2 + \omega_d^2\right)\cos\omega t_0 + 2\xi\omega\omega_n\sin\omega t_0}{\left(\xi^2\omega_n^2 + (\omega-\omega_d)^2\right)\left(\xi^2\omega_n^2 + (\omega+\omega_d)^2\right)}\end{aligned} \tag{8.72}$$

and the response of the vehicle after the bump will be a free vibrations.

$$z(t) = e^{-\xi\omega_n t}\left(\frac{\dot{z}(t_0) + \xi\omega_n z(t_0)}{\omega_d}\sin\omega_d t + z(t_0)\cos\omega_d t\right) \tag{8.73}$$

$$t_0 < t$$

To define the response by a continuous function we denote the response in $0 < t < t_0$ by $z_1$, and for $t_0 < t$ by $z_2$. Then, using the Heaviside function, we have

$$z = H(t_0 - t)\, z_1(t) + H(t - t_0)\, z_2(t - t_0) \tag{8.74}$$

Figure 8.5 illustrates the response of the vehicle in relative coordinate $z = x - y$ for front and rear 1/8 model of a vehicle with the following characteristics indicated by 1 and 2, respectively.

$$\xi_1 = 0.156 \qquad \xi_2 = 0.184 \tag{8.75}$$

$$\omega_{n1} = 5.62\,\text{rad/s} \qquad \omega_{n2} = 6.614\,\text{rad/s} \tag{8.76}$$

The absolute displacement $x$ can be calculated analytically.

$$\begin{aligned}x &= z + y \\ &= H(t_0 - t)\, z_1(t) + H(t - t_0)\, z_2(t - t_0) \\ &\quad + H(t - t_0)\, Y\sin\omega t\end{aligned} \tag{8.77}$$

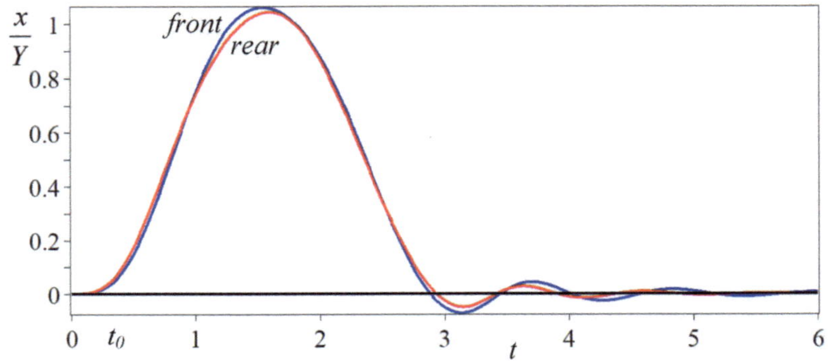


**Fig. 8.6** The response of the 1/8 vehicle model in absolute coordinate $x$, for front and rear parts

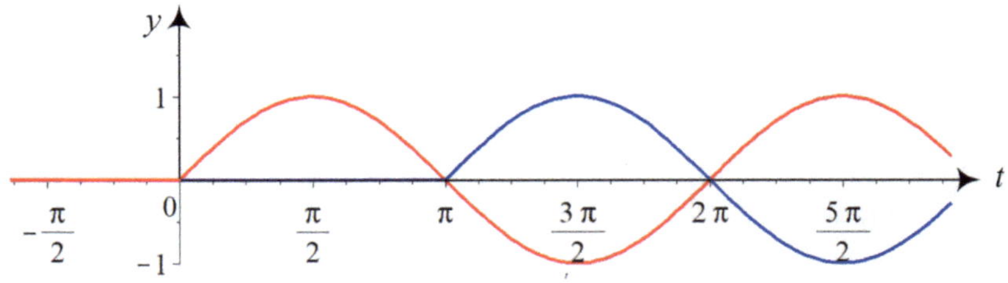


**Fig. 8.7** Making a half-sine excitation by superposition method

Figure 8.6 illustrates the response of the vehicle in absolute coordinate $x$ for the front and rear 1/8 model of the vehicle.

As an alternative method of solution, we may employ superposition method and model the half-sine excitation by summation of two sine functions as is shown in Fig. 8.7.

$$y = \begin{cases} Y \sin \omega t & 0 \le t \\ Y \sin \omega (t - t_0) & t_0 < t \end{cases} \qquad t_0 = \frac{\pi}{\omega} \tag{8.78}$$

$$f(t) = -m\ddot{y} = \begin{cases} mY\omega^2 \sin \omega t & 0 \le t \\ mY\omega^2 \sin \omega (t - t_0) & t_0 < t \end{cases} \tag{8.79}$$

Therefore, the response of the vehicle can be calculated by two convolution integrals.

$$\begin{aligned} z(t) &= \int_0^t h(t-\tau) f(\tau)\, d\tau \\ &= \frac{Y\omega^2}{\omega_n} \int_0^t e^{-\xi\omega_n (t-\tau)} \sin \omega_d (t-\tau) \sin \omega\tau \, d\tau \\ &+ \frac{Y\omega^2}{\omega_n} \int_{t_0}^t e^{-\xi\omega_n (t-\tau)} \sin \omega_d (t-\tau) \sin \omega (\tau - t_0) \, d\tau \end{aligned} \tag{8.80}$$

**Example 109** Nonlinear damper. The transient responses of vehicles to road excitations with ideal practical dampers are different than linear damper assumption. Here is a comparison.

Linear damper is a good model for viscous damping mechanism to express the action of a Newtonian fluid layer between two relatively moving solid surfaces. Practical shock absorbers that we use in automotive industries for damping vibrations have nonlinear behavior. Ideally, they have two damping rates for bound and rebound with bound rate much less than rebound rate. Let us assume a bilinear damper with zero damping in bound and $c$ in rebound as shown in Fig. 8.8 to simulate a theoretically ideal practical shock absorber. Considering $x$ and $y$ to indicate the displacement of the vehicle body and road excitation, respectively, we can define such damping by a Heaviside function $H(\dot{x} - \dot{y})$ and rewrite the equation of motion of the 1/8 vehicle model.

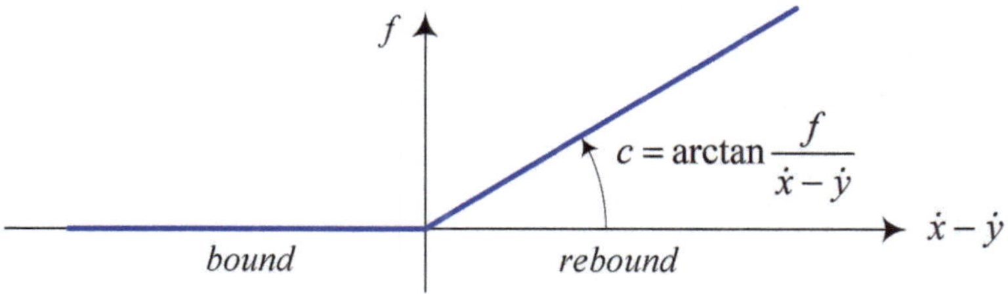


**Fig. 8.8** A bilinear damper with zero damping in bound and $c$ in rebound

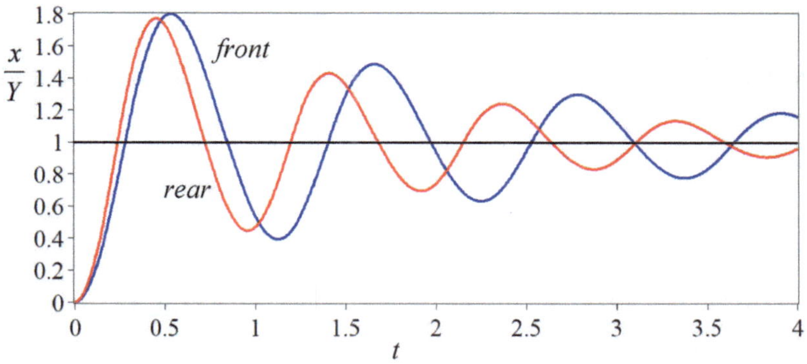


**Fig. 8.9** Time response of a 1/8 vehicle model with a bilinear damping to a unit step road excitation

$$m\ddot{x} + cH(\dot{x} - \dot{y})\,\dot{x} + kx = cH(\dot{x} - \dot{y})\,\dot{y} + ky \tag{8.81}$$

$$m\ddot{z} + cH(\dot{x} - \dot{y})\,\dot{z} + kz = -m\ddot{y} \qquad z = x - y \tag{8.82}$$

To study the effect of bilinear damping on transient vibrations of vehicles, let us consider the front and rear 1/8 models of a vehicle indicated by 1 and 2, respectively.

$$\xi_1 = \frac{c_1}{2\sqrt{km_1}} = 0.156 \qquad \xi_2 = \frac{c_2}{2\sqrt{km_2}} = 0.184 \tag{8.83}$$

$$\omega_{n1} = \sqrt{\frac{k}{m_1}} = 5.62\,\text{rad/s} \qquad \omega_{n2} = \sqrt{\frac{k}{m_2}} = 6.614\,\text{rad/s} \tag{8.84}$$

The response of the vehicle to step input is shown in Fig. 8.9. Comparing the response of the system with a bilinear damping in Fig. 8.9 with the response with a linear damping in Fig. 8.3 shows that having zero damping in bound, make vehicle to vibrate for longer time.

Figures 8.10 and 8.11 illustrate the absolute and relative displacement of the vibrating response of the 1/8 vehicle modes to a half sine bump on the road. They can be compared to Figs. 8.6 and 8.5 for the response of the vehicle with linear damping.

**Example 110** Force analysis. Transmitted force to the vehicle and to the road are two objective functions in optimization of suspension parameters. Here we show how to calculate the transmitted forces in transient vibrations.

The spring and shock absorber make the connection between the vehicle body and the road. All forces that apply to the body and the road are the summation of the force of these elements. Let us show the transmitted force to the body by $f_b$ and the transmitted force to the road by $f_r = -f_b$.

$$f_T = c(\dot{x} - \dot{y}) + k(x - y) \tag{8.85}$$

$$\frac{f_T}{m} = 2\xi\omega_n(\dot{x} - \dot{y}) + \omega_n^2(x - y) \tag{8.86}$$

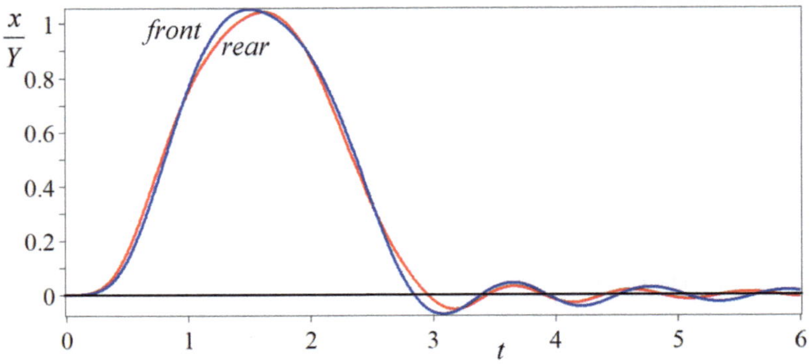


**Fig. 8.10** The response of absolute coordinate $x$ of the 1/8 vehicle model with a bilinear damping for front and rear parts of a vehicle to a road bump

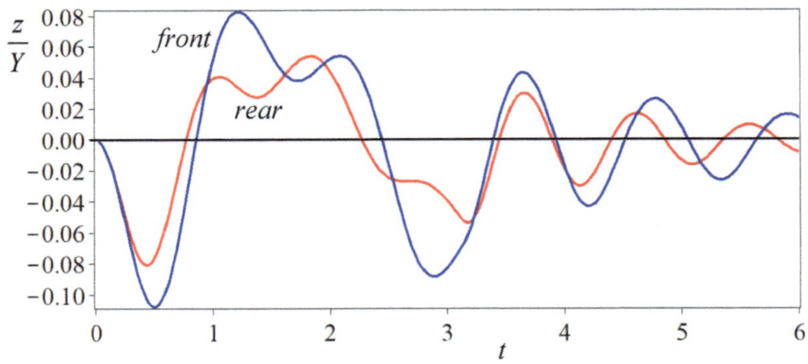


**Fig. 8.11** The response of relative coordinate $z = x - y$ of the 1/8 vehicle model with a bilinear damping for front and rear parts of a vehicle to a road bump

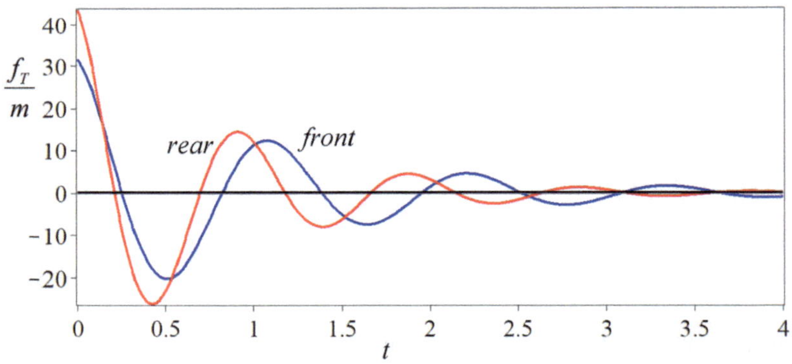


**Fig. 8.12** Illustration of transmitted force $f_T/m$ to a step road input for linear damping and for front and rear 1/8 vehicle model of a vehicle indicated by 1 and 2, respectively

Figure 8.12 illustrates the transmitted acceleration $f_T/m$ to a step road input for linear damping and for the front and rear 1/8 vehicle model of a vehicle with following parameters indicated by 1 and 2, respectively.

$$\xi_1 = 0.156 \qquad \xi_2 = 0.184 \tag{8.87}$$

$$\omega_{n1} = 5.62\,\mathrm{rad/\,s} \qquad \omega_{n2} = 6.614\,\mathrm{rad/\,s} \tag{8.88}$$

If the shock absorber is an ideal bilinear damper, then Fig. 8.13 illustrates the transmitted acceleration $f_T/m$ to the step road input for the same 1/8 vehicle models of a vehicle with the above parameters.

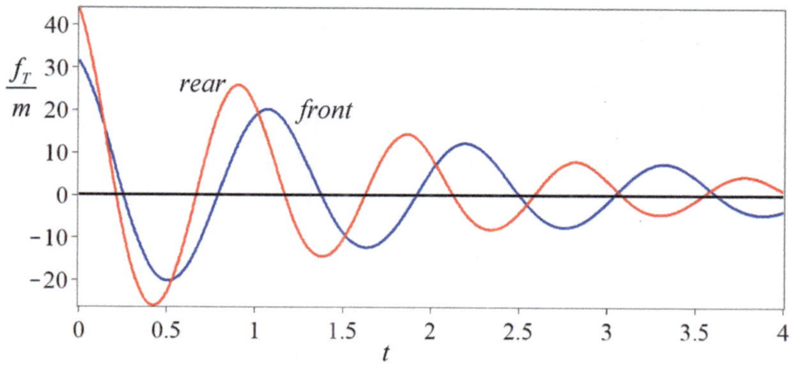


**Fig. 8.13** Illustration of transmitted force $f_T/m$ to a step road input for bilinear damping and for front and rear 1/8 vehicle model of a vehicle indicated by 1 and 2, respectively

## 8.3 Frequency Response

The most important frequency responses of base excited 1/8 vehicle model are absolute displacement $G_0$, relative displacement $S_2$, absolute acceleration $G_2$, and force transmitted to $m$:

$$G_0 = \left|\frac{X}{Y}\right| = \frac{\sqrt{1+(2\xi r)^2}}{\sqrt{\left(1-r^2\right)^2+(2\xi r)^2}} \tag{8.89}$$

$$S_2 = \left|\frac{Z}{Y}\right| = \frac{r^2}{\sqrt{\left(1-r^2\right)^2+(2\xi r)^2}} \tag{8.90}$$

$$G_2 = \left|\frac{\ddot{X}}{Y\omega_n^2}\right| = \frac{r^2\sqrt{1+(2\xi r)^2}}{\sqrt{\left(1-r^2\right)^2+(2\xi r)^2}} \tag{8.91}$$

where

$$r = \frac{\omega}{\omega_n} \qquad \xi = \frac{c}{2\sqrt{km}} \qquad \omega_n = \sqrt{\frac{k}{m}} \tag{8.92}$$

***Proof*** Applying a harmonic road excitation,

$$y = Y \sin \omega t \tag{8.93}$$

the equation of motion of the 1/8 vehicle model in relative displacement (8.5) reduces to

$$\ddot{z} + 2\xi\omega_n \dot{z} + \omega_n^2 z = \omega^2 Y \sin \omega t \tag{8.94}$$

$$z = x - y \tag{8.95}$$

Considering a harmonic solution

$$z = A_3 \sin \omega t + B_3 \cos \omega t \tag{8.96}$$

and substituting it in the equation of motion

$$\begin{aligned}&-A_3\omega^2 \sin \omega t - B_3\omega^2 \cos \omega t + 2\xi\omega_n \left(A_3\omega \cos \omega t - B_3\omega \sin \omega t\right)\\&+\omega_n^2 \left(A_3 \sin \omega t + B_3 \cos \omega t\right) = \omega^2 Y \sin \omega t\end{aligned} \tag{8.97}$$

we find a set of equations to calculate $A_3$ and $B_3$.

$$\begin{bmatrix} \omega_n^2 - \omega^2 & -2\xi\omega\omega_n \\ 2\xi\omega\omega_n & \omega_n^2 - \omega^2 \end{bmatrix} \begin{bmatrix} A_3 \\ B_3 \end{bmatrix} = \begin{bmatrix} Y\omega^2 \\ 0 \end{bmatrix} \tag{8.98}$$

The first row of the set (8.98) is a balance of the coefficients of $\sin\omega t$ in Eq. (8.97), and the second row is a balance of the coefficients of $\cos\omega t$. Therefore, the coefficients $A_3$ and $B_3$ can be found.

$$\begin{aligned} \begin{bmatrix} A_3 \\ B_3 \end{bmatrix} &= \begin{bmatrix} \omega_n^2 - \omega^2 & -2\xi\omega\omega_n \\ 2\xi\omega\omega_n & \omega_n^2 - \omega^2 \end{bmatrix}^{-1} \begin{bmatrix} Y\omega^2 \\ 0 \end{bmatrix} \\ &= \begin{bmatrix} \dfrac{1 - r^2}{\left(1 - r^2\right)^2 + (2\xi r)^2} r^2 Y \\ \dfrac{-2\xi r}{\left(1 - r^2\right)^2 + (2\xi r)^2} r^2 Y \end{bmatrix} \end{aligned} \tag{8.99}$$

The relative displacement amplitude $Z$ provides us with $S_2 = |Z/Y|$ in Eq. (8.90).

$$Z = \sqrt{A_3^2 + B_3^2} \tag{8.100}$$

$$S_2 = \frac{Z}{Y} = \frac{r^2}{\sqrt{\left(1 - r^2\right)^2 + (2\xi r)^2}} \tag{8.101}$$

To find the absolute frequency response $G_0$, we assume

$$x = A_2 \sin\omega t + B_2 \cos\omega t = X \sin\left(\omega t - \varphi_x\right) \tag{8.102}$$

and write

$$z = x - y \tag{8.103}$$

$$A_3 \sin\omega t + B_3 \cos\omega t = A_2 \sin\omega t + B_2 \cos\omega t - Y \sin\omega t \tag{8.104}$$

which shows

$$A_2 = A_3 + Y \tag{8.105}$$

$$B_2 = B_3 \tag{8.106}$$

The absolute displacement amplitude $X$ provides us with $G_0 = |X/Y|$ of Eq. (8.89).

$$X = \sqrt{A_2^2 + B_2^2} = \sqrt{(A_3 + Y)^2 + B_3^2}$$

$$G_0 = \frac{X}{Y} = \frac{\sqrt{1 + (2\xi r)^2}}{\sqrt{\left(1 - r^2\right)^2 + (2\xi r)^2}} \tag{8.107}$$

The absolute acceleration frequency response, $\ddot{X}$, can be found by twice differentiating the displacement frequency response (8.102).

$$\ddot{x} = -X\omega^2 \sin\left(\omega t - \varphi_x\right) = -\ddot{X} \sin\left(\omega t - \varphi_x\right) \tag{8.108}$$

If we denote the amplitude of the absolute acceleration by $\ddot{X}$, then we can define $\ddot{X}$ which provides us with $G_2 = \ddot{X}/\left(\omega_n^2 Y\right)$ as in Eq. (8.91).

$$G_2 = \frac{\ddot{X}}{Y\omega_n^2} = \frac{r^2\sqrt{1+(2\xi r)^2}}{\sqrt{\left(1-r^2\right)^2+(2\xi r)^2}} \tag{8.109}$$

The other important frequency responses of the 1/8 vehicle model are absolute velocity frequency response $\dot{X}/(\omega_n Y)$,

$$G_1 = \frac{\dot{X}}{\omega_n Y} = \frac{r\sqrt{1+(2\xi r)^2}}{\sqrt{\left(1-r^2\right)^2+(2\xi r)^2}} \tag{8.110}$$

relative velocity frequency response $\dot{Z}/(\omega_n Y)$,

$$S_3 = \frac{\dot{Z}}{\omega_n Y} = \frac{r^3}{\sqrt{\left(1-r^2\right)^2+(2\xi r)^2}} \tag{8.111}$$

relative acceleration frequency response $\ddot{Z}/(\omega_n Y)$,

$$S_4 = \frac{\ddot{Z}}{\omega_n^2 Y} = \frac{r^4}{\sqrt{\left(1-r^2\right)^2+(2\xi r)^2}} \tag{8.112}$$

and transmitted force to the base $F_T/(kY)$.

$$G_2 = \frac{F_T}{kY} = \frac{r^2\sqrt{1+(2\xi r)^2}}{\sqrt{\left(1-r^2\right)^2+(2\xi r)^2}} \tag{8.113}$$

The frequency plots of these frequency responses are plotted in Appendix A. These frequency responses show that:

$$G_2 = rG_1 = r^2 G_0 \tag{8.114}$$

$$S_4 = rS_3 = r^2 S_2 \tag{8.115}$$

■

**Example 111** Absolute and relative frequency response curve comparison. Absolute and relative frequency responses are two main characteristics of 1/8 vehicle model systems.

Consider a vehicle suspension with the following data for its front and rear, indicated by 1 and 2.

$$k_1 = 11000\,\text{N/m} \qquad \xi_1 = 0.156 \qquad \omega_{n1} = 5.62\,\text{rad/s} \tag{8.116}$$

$$k_2 = 13200\,\text{N/m} \qquad \xi_2 = 0.171 \qquad \omega_{n2} = 6.156\,\text{rad/s} \tag{8.117}$$

The absolute and relative displacement frequency responses of the systems are shown in Fig. 8.14. The relative displacements start at *zero* and end at *one*, while the absolute displacements start at *one* and end up at *zero*. Higher damping makes the amplitude of the absolute and relative displacement curves of rear to be lower than the front. Higher natural frequency makes the resonant frequency of the rear system to happen at higher frequency than the front.

**Example 112** Relative frequency responses and optimization investigation. Here we study the possibility of optimization of linear suspension by optimization of relative frequency responses and show that there is no optimal suspension.

Employing the frequency responses of the 1/8 vehicle model indicates that the relative frequency responses are independent of damping ratio $\xi$, and only dependent of frequency ratio $r$.

$$\frac{G_2}{G_1} = \frac{S_4}{S_3} = r \qquad \frac{G_2}{G_0} = \frac{S_4}{S_2} = r^2 \tag{8.118}$$

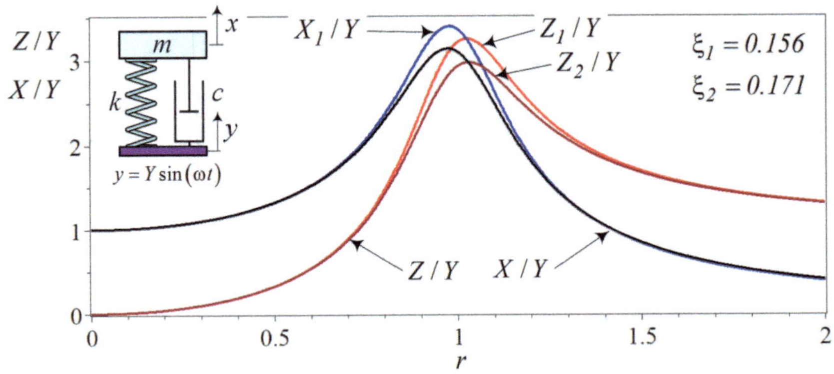


**Fig. 8.14** Absolute and relative displacement frequency responses for a 1/8 vehicle model

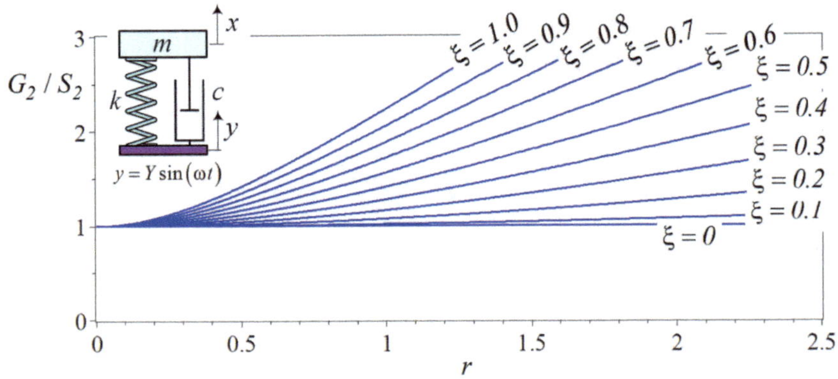


**Fig. 8.15** Graph of the ratio of absolute acceleration with respect to relative displacement, $G_2/S_2$, as a function of frequency ratio $r$ for different damping ratio $\xi$

An acceptable optimization idea might be minimization of transmitted force $G_2$ or absolute acceleration $G_2$ with respect to absolute displacement $G_0$ or relative displacement $S_2$. The ratio of absolute acceleration $G_2$ with respect to absolute displacement $G_0$ is square function of the frequency ratio, $G_2/G_0 = r^2$. The ratio of $G_2/G_0 = r^2$ increases by increasing $r$ and hence, there is no optimal point for $G_2/G_0$. The absolute acceleration will be zero only when absolute displacement is zero at no excitation frequency $r = 0$.

The ratio of absolute acceleration $G_2$ with respect to relative displacement $S_2$ is a function of damping ratio and frequency ratio, $G_2/S_2 = \sqrt{1 + (2\xi r)^2}$.

$$\frac{G_2}{S_2} = \sqrt{1 + (2\xi r)^2}$$

Although $G_2/S_2$ depends on both, $r$ and $\xi$, its behavior is similar to $G_2/G_0$ for any given $\xi$. Figure 8.15 illustrates the behavior.

**Example 113** $G_0 \neq S_2 + 1$. The relative displacement frequency response is not equal to the difference of frequency responses of the relative displacements. Here is to show this fact.

We can find the absolute frequency response $G_0 = |X/Y|$, from the result for $S_2$. The frequency response $S_2$ is

$$S_2 = \frac{Z}{Y} \tag{8.119}$$

However,

$$S_2 \neq \frac{X}{Y} - 1 \neq G_0 - 1 \tag{8.120}$$

because the amplitude of the relative displacement $Z$ is not equal to the amplitude of the absolute displacement $X$ minus the amplitude of the excitation $Y$.

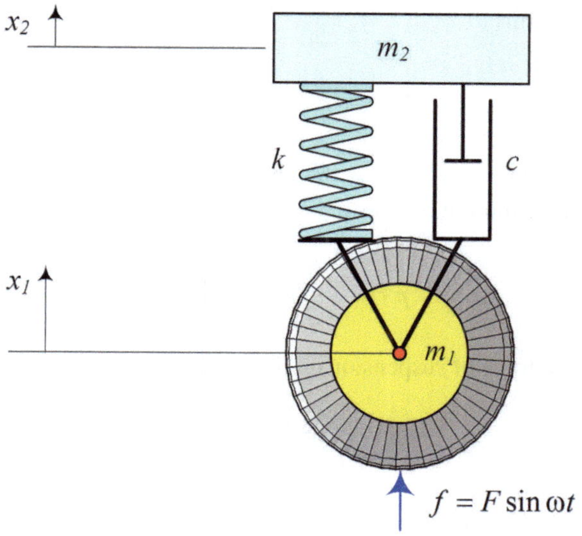


**Fig. 8.16** A base force excited system

$$Z \neq X - Y \tag{8.121}$$

**Example 114** A base force excited system. Vibration isolator is any suspension system separating a mass from a source of vibrations. Here is an example of suspension system between a massive wheel and vehicle body where there is a harmonic force on one of the wheel.

Figure 8.16 illustrates a 1/8 vehicle model with a massive body $m_1$ under a forced excitation along with a secondary system $(m_2, k, c)$. Employing the energies and dissipation function of the system,

$$K = \frac{1}{2}m_1\dot{x}_1^2 + \frac{1}{2}m_2\dot{x}_2^2 \tag{8.122}$$

$$P = \frac{1}{2}k\left(x_1 - x_2\right)^2 \tag{8.123}$$

$$D = \frac{1}{2}c\left(\dot{x}_1 - \dot{x}_2\right)^2 \tag{8.124}$$

the equations of motion of the system will be derived.

$$m_1\ddot{x}_1 + c\left(\dot{x}_1 - \dot{x}_2\right) + k\left(x_1 - x_2\right) = f \tag{8.125}$$

$$m_2\ddot{x}_2 - c\left(\dot{x}_1 - \dot{x}_2\right) - k\left(x_1 - x_2\right) = 0 \tag{8.126}$$

$$f = F\sin\omega t \tag{8.127}$$

Subtracting the equations in relative displacement $z$ yields

$$\frac{m_1 m_2}{m_1 + m_2}\ddot{z} + c\dot{z} + kz = \frac{m_2}{m_1 + m_2}f \tag{8.128}$$

$$z = x_1 - x_2 \tag{8.129}$$

Defining the damping ratio $\xi$ and natural frequency $\omega_n$,

$$\xi = \frac{c}{2\sqrt{k\frac{m_1 m_2}{m_1 + m_2}}} \qquad \omega_n = \sqrt{\frac{k\left(m_1 + m_2\right)}{m_1 m_2}} \tag{8.130}$$

the equation of relative motion transforms to

$$\ddot{z} + \xi \dot{z} + \omega_n z = \frac{1}{m_1} f = \frac{F}{m_1} \sin \omega t \tag{8.131}$$

The equation of relative motion (8.131) is the same as the absolute displacement of the one *DOF* force excitation (7.1). Therefore, the amplitude of the relative motion is $S_0$ as in (7.7),

$$S_0 = \frac{Z}{F/k} = \frac{1}{\sqrt{\left(1 - r^2\right)^2 + (2\xi r)^2}} \tag{8.132}$$

and the force transmitted to $m_2$ or the force of suspension on the body $m_1$ is $G_0$ as in (7.41).

$$F_T = kz + c\dot{z} \tag{8.133}$$

$$G_0 = \frac{F_T}{F} = \frac{\sqrt{1 + (2\xi r)^2}}{\sqrt{\left(1 - r^2\right)^2 + (2\xi r)^2}} \tag{8.134}$$

To determine the frequency response of $m_1$ and $m_2$, we use Eqs. (8.126) and (8.129).

$$\begin{aligned} &\begin{bmatrix} m_1 & 0 \\ 0 & m_2 \end{bmatrix} \begin{bmatrix} \ddot{x}_1 \\ \ddot{x}_2 \end{bmatrix} + \begin{bmatrix} c & -c \\ -c & c \end{bmatrix} \begin{bmatrix} \dot{x}_1 \\ \dot{x}_2 \end{bmatrix} \\ &\quad + \begin{bmatrix} k & -k \\ -k & k \end{bmatrix} \begin{bmatrix} x_1 \\ x_2 \end{bmatrix} = \begin{bmatrix} f \\ 0 \end{bmatrix} \end{aligned} \tag{8.135}$$

Substituting harmonic solutions for $x_1$ and $x_2$,

$$\begin{bmatrix} x_1 \\ x_2 \end{bmatrix} = \begin{bmatrix} A_1 \sin \omega t + B_2 \cos \omega t \\ A_2 \sin \omega t + B_2 \cos \omega t \end{bmatrix} \tag{8.136}$$

we find the coefficients $A_1$, $A_2$, $B_1$, $B_2$ by using (7.391).

$$\begin{aligned} \begin{bmatrix} \mathbf{A} \\ \mathbf{B} \end{bmatrix} &= \begin{bmatrix} A_1 \\ A_2 \\ B_1 \\ B_2 \end{bmatrix} = \begin{bmatrix} [k] - \omega^2 [m] & -\omega [c] \\ \omega [c] & [k] - \omega^2 [m] \end{bmatrix}^{-1} \begin{bmatrix} F \\ 0 \\ 0 \\ 0 \end{bmatrix} \\ &= \begin{bmatrix} k - \omega^2 m_1 & -k & -c\omega & c\omega \\ -k & k - \omega^2 m_2 & c\omega & -c\omega \\ c\omega & -c\omega & k - \omega^2 m_1 & -k \\ -c\omega & c\omega & -k & k - \omega^2 m_2 \end{bmatrix}^{-1} \begin{bmatrix} F \\ 0 \\ 0 \\ 0 \end{bmatrix} \end{aligned} \tag{8.137}$$

The frequency response of $m_1$ and $m_2$ are

$$X_1 = \sqrt{A_1^2 + B_1^2} \qquad X_2 = \sqrt{A_2^2 + B_2^2} \tag{8.138}$$

**Example 115** Principal method for absolute motion $X$. The frequency response of the absolute displacement is calculated here, using principal method of harmonic balance.

To find the absolute frequency response $G_0$, we substitute the base excitation function,

$$y = Y \sin \omega t \tag{8.139}$$

and a harmonic solution for $x$, in Eq. (8.4)

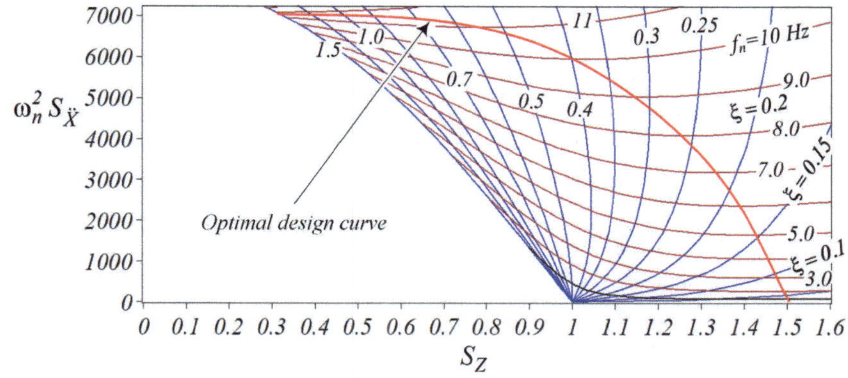


**Fig. 8.17** Design chart for optimal suspension parameters of equipment

$$x = A_2 \sin \omega t + B_2 \cos \omega t \tag{8.140}$$

$$\ddot{x} + 2\xi \omega_n \dot{x} + \omega_n^2 x = 2\xi \omega_n \dot{y} + \omega_n^2 y \tag{8.141}$$

and solve for $X = \sqrt{A_2^2 + B_2^2}$.

$$\begin{aligned} &-\omega^2 A_2 \sin \omega t - \omega^2 B_2 \cos \omega t + 2\xi \omega_n \omega \left(A_2 \cos \omega t - B_2 \sin \omega t\right) \\ &+\omega_n^2 \left(A_2 \sin \omega t + B_2 \cos \omega t\right) = 2\xi \omega_n \omega Y \cos \omega t + \omega_n^2 Y \sin \omega t \end{aligned} \tag{8.142}$$

The set of equations for $A_2$ and $B_2$ from the coefficients of sin and cos results in the following solution:

$$\begin{bmatrix} \omega_n^2 - \omega^2 & -2\xi \omega \omega_n \\ 2\xi \omega \omega_n & \omega_n^2 - \omega^2 \end{bmatrix} \begin{bmatrix} A_2 \\ B_2 \end{bmatrix} = \begin{bmatrix} Y \omega_n^2 \\ 2Y\xi \omega \omega_n \end{bmatrix} \tag{8.143}$$

$$\begin{aligned} \begin{bmatrix} A_2 \\ B_2 \end{bmatrix} &= \begin{bmatrix} \omega_n^2 - \omega^2 & -2\xi \omega \omega_n \\ 2\xi \omega \omega_n & \omega_n^2 - \omega^2 \end{bmatrix}^{-1} \begin{bmatrix} Y \omega_n^2 \\ 2Y\xi \omega \omega_n \end{bmatrix} \\ &= \begin{bmatrix} \dfrac{(2\xi r)^2 - \left(1 - r^2\right)}{\left(1 - r^2\right)^2 + (2\xi r)^2} Y \\ \dfrac{-2\xi r^3}{\left(1 - r^2\right)^2 + (2\xi r)^2} Y \end{bmatrix} \end{aligned} \tag{8.144}$$

Therefore, the amplitude of the absolute displacement $X$ would be the same as (8.107).

## 8.4 ★ Root Mean Square ($RMS$) Optimization

Figure 8.17 is the design chart for optimal suspension parameters of vehicles with harmonic base excited systems. The horizontal axis is the root mean square of the relative displacement frequency response, $S_Z = RMS(S_2)$, and the vertical axis is the root mean square of the absolute acceleration frequency response, $S_{\ddot{X}} = RMS(G_2)$. There are two sets of curves that make the mesh. The first set is the constant natural frequency $f_n$ in Hz, and the second set, which spreads out from $S_Z = 1$, is the constant damping ratio $\xi$. The *optimal design curve*, indicating the optimal suspension parameters, is connecting the minimums of the constant natural frequency curves.

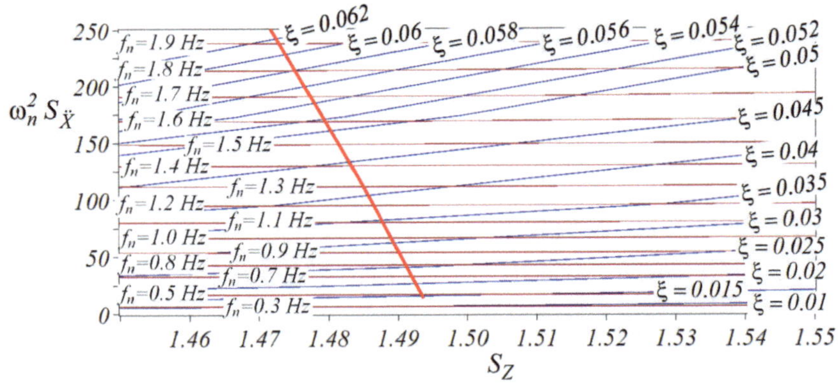


**Fig. 8.18** Design chart for optimal suspension parameters of vehicles

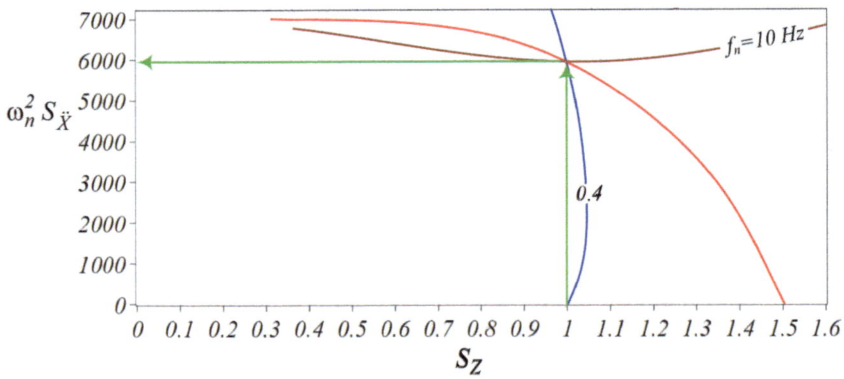


**Fig. 8.19** Application of the design chart for $S_X = 1$ that indicates the optimal values $\xi \approx 0.4$ and $f_n \approx 10\,\text{Hz}$

The main natural frequencies of passenger vehicles are around $f_n = 1\,\text{Hz}$, while the natural frequencies of unsprung masses of vehicles are around $f_n = 10\,\text{Hz}$. Therefore, to employ the design chart of Fig. 8.17 to design the suspension of base excited 1/8 vehicle model, we use the magnified chart shown in the zoomed-in Fig. 8.18.

The optimal design curve is the result of the following optimization strategy:

$$Minimize\ S_{\ddot{X}}\ with\ respect\ to\ S_Z \tag{8.145}$$

which states that the minimum *RMS* of the absolute acceleration with respect to the *RMS* of the relative displacement makes a suspension optimal. Mathematically it is equivalent to the following minimization problem:

$$\frac{\partial S_{\ddot{X}}}{\partial S_Z} = 0 \tag{8.146}$$

$$\frac{\partial^2 S_{\ddot{X}}}{\partial S_Z^2} > 0 \tag{8.147}$$

To use the optimization chart and determine the optimal stiffness $k$ and damping $c$, we begin from an estimated value of $S_X$ on the horizontal axis and draw a vertical line to hit the optimal curve. The intersection point indicates the optimal $f_n$ and $\xi$ for the $S_X$ to have the best vibration isolation.

Figure 8.19 illustrates a sample application for $S_X = 1$, which indicates $\xi \approx 0.4$ and $f_n \approx 10\,\text{Hz}$ as the optimal suspension parameters. The $f_n$, $\xi$, and the mass of the equipment determine the optimal value of $k$ and $c$.

***Proof*** Let us define a working frequency range $0 < f < 20\,\text{Hz}$ to include all road vehicles and denote the *RMS* of $S_2$ and $G_2$ by $S_Z$ and $S_{\ddot{X}}$. In applied vehicle dynamics, we usually measure frequencies in [Hz] instead of [rad/s]; so we develop design calculations based on cyclic frequencies $f$ and $f_n$ in [Hz] and analytic calculations based on angular frequencies $\omega$ and $\omega_n$ in [rad/s].

$$S_Z = RMS(S_2) = \sqrt{\frac{1}{40\pi}\int_0^{40\pi} S_2^2 d\omega} \tag{8.148}$$

$$S_{\ddot{X}} = RMS(G_2) = \sqrt{\frac{1}{40\pi}\int_0^{40\pi} G_2^2 d\omega} \tag{8.149}$$

To calculate $S_Z$ and $S_{\ddot{X}}$ over the working frequency range, we first calculate the integrals of $S_2^2$ and $G_2^2$.

$$\int S_2^2 d\omega = Z_1\omega - \frac{Z_2}{Z_3\sqrt{Z_4}} \arctan\frac{\omega}{\sqrt{Z_4}} + \frac{Z_5}{Z_6\sqrt{Z_7}} \arctan\frac{\omega}{\sqrt{Z_7}} \tag{8.150}$$

$$\omega_n^4 \int G_2^2 d\omega = Z_8\omega + \frac{1}{3}Z_9\omega^3 + \frac{Z_{10}}{Z_{11}\sqrt{Z_{12}}} \arctan\frac{\omega}{\sqrt{Z_{12}}} + \frac{Z_{13}}{Z_{14}\sqrt{Z_{15}}} \arctan\frac{\omega}{\sqrt{Z_{15}}} \tag{8.151}$$

The parameters $Z_1$ through $Z_{15}$ are as follows:

$$Z_1 = 1 \tag{8.152}$$

$$Z_2 = \omega_n^2\left(8\xi^6 - 12\xi^4 + 4\xi^2 - \left(-8\xi^4 + 8\xi^2 - 1\right)\xi\sqrt{1-\xi^2}\right) \tag{8.153}$$

$$Z_3 = -4\xi^2\left(1-\xi^2\right) \tag{8.154}$$

$$Z_4 = \omega_n^2\left(-1 + 2\xi^2 + 2\xi\sqrt{1-\xi^2}\right) \tag{8.155}$$

$$Z_5 = \omega_n^2\left(8\xi^6 - 12\xi^4 + 4\xi^2 - \left(8\xi^4 - 8\xi^2 + 1\right)\xi\sqrt{1-\xi^2}\right) \tag{8.156}$$

$$Z_6 = -4\xi^2\left(1-\xi^2\right) \tag{8.157}$$

$$Z_7 = \omega_n^2\left(-1 + 2\xi^2 - 2\xi\sqrt{1-\xi^2}\right) \tag{8.158}$$

$$Z_8 = \omega_n^4\left(-16\xi^4 + 8\xi^2 + 1\right) \tag{8.159}$$

$$Z_9 = 4\omega_n^2\xi^2 \tag{8.160}$$

$$Z_{10} = \omega_n^6\left(128\xi^{10} - 256\xi^8 + 144\xi^6 - 12\xi^4 - 4\xi^2\right) \tag{8.161}$$

$$-\omega_n^6\left(-128\xi^8 + 192\xi^6 - 64\xi^4 - 4\xi^2 + 1\right)\xi\sqrt{1-\xi^2} \tag{8.162}$$

$$Z_{11} = -4\xi^2\left(1-\xi^2\right) \tag{8.163}$$

$$Z_{12} = \omega_n^2\left(-1 + 2\xi^2 + 2\xi\sqrt{1-\xi^2}\right) \tag{8.164}$$

$$Z_{13} = \omega_n^6\left(128\xi^{10} - 256\xi^8 + 144\xi^6 - 12\xi^4 - 4\xi^2\right) \tag{8.165}$$

$$-\omega_n^6\left(128\xi^8 - 192\xi^6 + 64\xi^4 + 4\xi^2 - 1\right)\xi\sqrt{1-\xi^2} \tag{8.166}$$

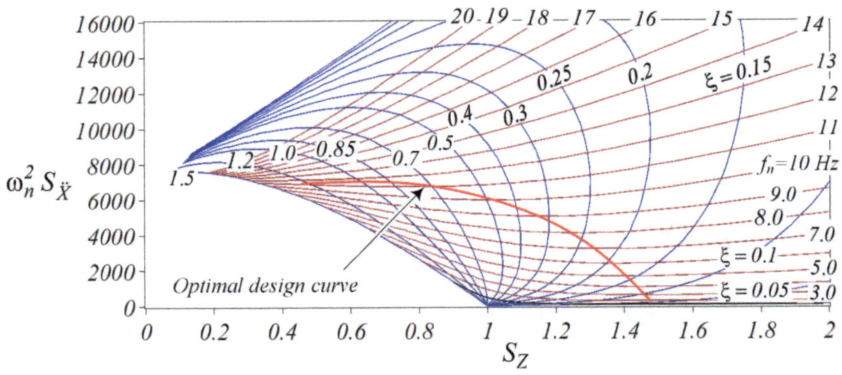


**Fig. 8.20** Behavior of $S_{\ddot{X}}$ with respect to $S_Z$ when $f_n$ and $\xi$ are varied

$$Z_{14} = -4\xi^2\left(1-\xi^2\right) \tag{8.167}$$

$$Z_{15} = \omega_n^2\left(-1+2\xi^2-2\xi\sqrt{1-\xi^2}\right) \tag{8.168}$$

Therefore, $S_Z$ and $S_{\ddot{X}}$ over the frequency range $0 < f < 20$ Hz can be calculated analytically from Eqs. (8.148) and (8.149).

Integrating over the excitation frequency $\omega$, eliminates $\omega$ and shows both $S_{\ddot{X}}$ and $S_Z$ to be dependent on only two parameters $\omega_n$ and $\xi$.

$$S_{\ddot{X}} = S_{\ddot{X}}\left(\omega_n, \xi\right) \tag{8.169}$$

$$S_Z = S_Z\left(\omega_n, \xi\right) \tag{8.170}$$

Therefore, any pair of design parameters $(\omega_n, \xi)$ determines $S_{\ddot{X}}$ and $S_Z$ uniquely. It is also theoretically possible to define $\omega_n$ and $\xi$ as two functions of the variables $S_{\ddot{X}}$ and $S_Z$.

$$\omega_n = \omega_n\left(S_{\ddot{X}}, S_Z\right) \tag{8.171}$$

$$\xi = \xi\left(S_{\ddot{X}}, S_Z\right) \tag{8.172}$$

Hence, we would be able to determine the required $\omega_n$ and $\xi$ for a specific value of $S_{\ddot{X}}$ and $S_Z$.

Using Eqs. (8.169) and (8.170), we draw Fig. 8.20 to illustrate how $S_{\ddot{X}}$ behaves with respect to $S_Z$ when $f_n$ and $\xi$ vary. Keeping $f_n$ constant and varying $\xi$, it is possible to minimize $S_{\ddot{X}}$ with respect to $S_Z$. The minimum points make the optimal curve and determine the best $f_n$ and $\xi$. In other words, for any $f_n$, there is an optimal $\xi$ that minimizes $S_{\ddot{X}}$ with respect to $S_Z$. The key to use the optimal design curve is to adjust, determine, or estimate a value for $S_Z$ or $S_{\ddot{X}}$ and find the associated point on the design curve.

Softening of a suspension decreases the body acceleration; however, it requires a larger room for relative displacement. Due to physical constraints the relative displacement or *wheel travel* is limited, and hence, we must design the suspension to use the available wheel travel as much as possible and decrease the body acceleration as low as possible. Mathematically it is equivalent to (8.146) and (8.147). Figure 8.21 illustrates the optimal curve on a three dimensional graph.

To visualize the optimization principle (8.145), let us plot $\omega_n^2 S_{\ddot{X}}/S_Z$ versus $f_n$ for different values of $\xi$ in Fig. 8.22. It shows that increasing either one of $\xi$ or $f_n$ increases the value of $\omega_n^2 S_{\ddot{X}}/S_Z$. It is equivalent to making the suspension more rigid, which causes an increase in acceleration or decrease in relative displacement. On the contrary, decreasing $\xi$ or $f_n$ decreases the value of $\omega_n^2 S_{\ddot{X}}/S_Z$, which is equivalent to making the suspension softer with less acceleration and more relative displacement.

For minimization of any dynamic variables of vehicle body, such as acceleration, force, jerk, the optimization of linear vehicle suspension approaches to no suspension. To define an optimization condition for linear suspension, we need to consider two variables with opposite behavior. The absolute acceleration of vehicle body is a variable that we wish to minimize for better ride comfort. Reducing suspension spring and damper rates mathematically reduces the body acceleration. On the other hand, reducing suspension rates increases the relative displacement of body with respect to the wheel. Such trade-off

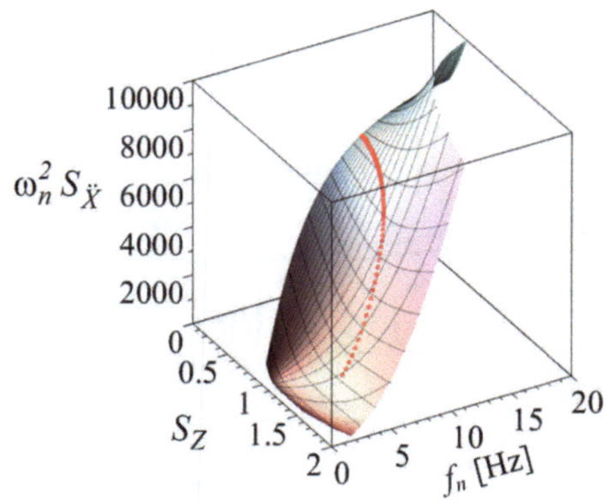


**Fig. 8.21** The optimal curve in a three dimensional surface of $S_{\ddot{X}}$, $S_Z$, and $f_n$

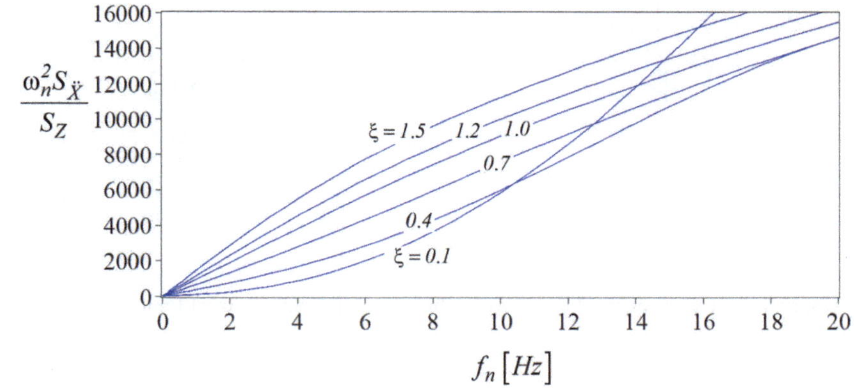


**Fig. 8.22** A plot of ratio $\omega_n^2 S_{\ddot{X}}/S_Z$ versus $f_n$ for different values of $\xi$

makes the optimized suspension rates to be achieved when we reduce the suspension rates until we use the maximum wheel travel.

The magnitude of both, the absolute acceleration and relative displacement, are frequency dependent. To consider them over all frequencies, we calculate *RMS* of absolute acceleration and relative displacement over the important range of frequencies. The *RMS* reduce two scales to be dependent of natural frequency and damping ratio of the vehicle suspension. To determine the optimal suspension parameters, we minimize *RMS* of absolute acceleration with respect to *RMS* of relative displacement while the *RMS* of relative displacement is constraint with the wheel travel of the suspension.

Because $S_{\ddot{X}}$ is minimum on the optimal curve, the sensitivity of acceleration *RMS* with respect to relative displacement *RMS* is minimum at any point on the optimal curve. Therefore, an optimal suspension has the least sensitivity to the available wheel travel variation. If a suspension is optimized for a certain wheel travel, it is still near optimal when the wheel travel changes a little over time or because of weight changes. ■

**Example 116** Wheel travel calculation. The *RMS* optimization design can be well understood by investigating a vehicle suspension. In the theoretical study, we assume the wheel travel to be symmetric, although in practice, it may not be symmetric.

Figure 8.23a illustrates a double A-arm suspension mechanism at its equilibrium position. To limit the motion of the wheel with respect to the body, two stoppers must be employed. There are many possible options for the type and position of the stoppers. Most stoppers are made of stiff rubber balls and mounted somewhere on the chassis or suspension mechanism or both. It is also possible that the bottom up damper acts as a stopper. Figure 8.23a shows an example of stoppers.

The gap sizes $\delta_u$ and $\delta_l$ indicate the upper and lower displacement limits that the suspension mechanism can move. However, the maximum motion of the wheel must be calculated at the center of the wheel. So, we transfer $\delta_u$ and $\delta_l$ to the center of the wheel and show them by $d_u$ and $d_l$.

$$d_u \approx \frac{b_u}{a_u}\delta_u \qquad d_l \approx \frac{b_l}{a_l}\delta_l \tag{8.173}$$

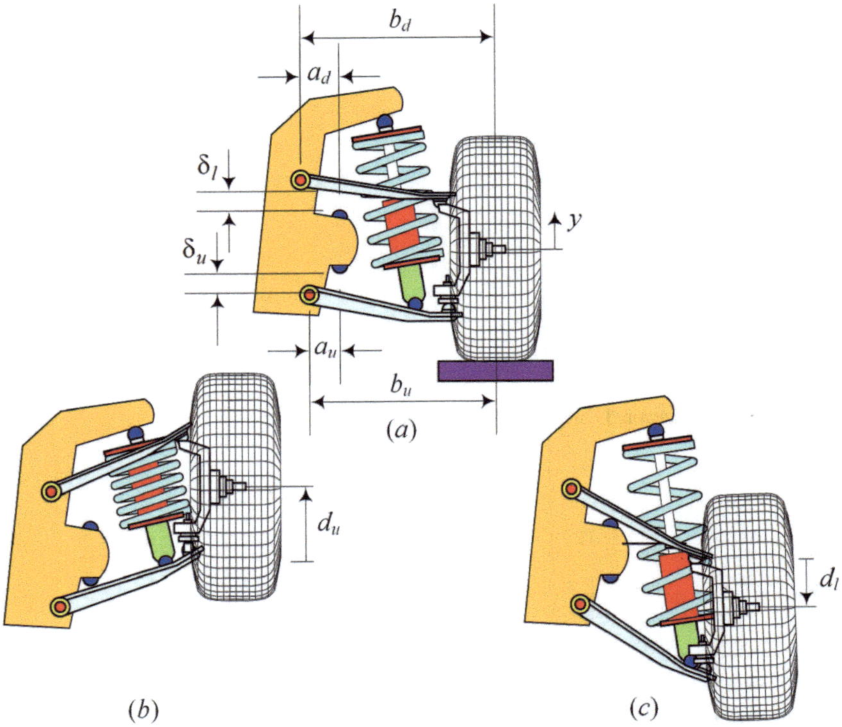


**Fig. 8.23** A double A-arm suspension mechanism at (**a**) equilibrium, (**b**) upper limit, and (**c**) lower limit

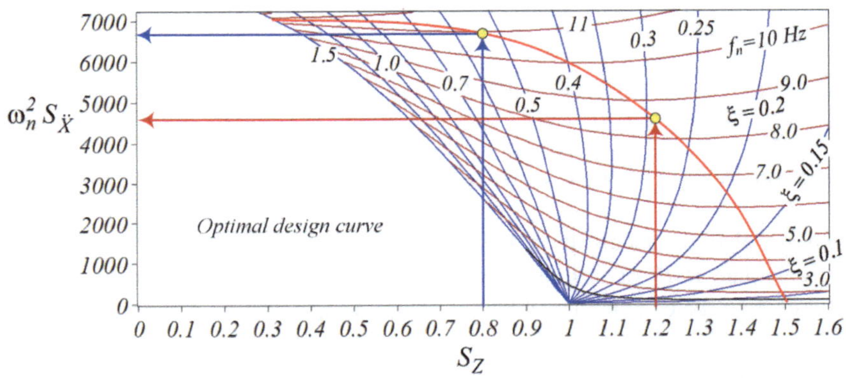


**Fig. 8.24** Comparing two suspensions $A$ and $B$ with $S_{Z_A} = 1.2$ and $S_{Z_B} = 0.8$

Figure 8.23b and c show the mechanism at the upper and lower Limits, respectively. The distance $d_u$ is called the upper wheel travel, and $d_l$ is called the lower wheel travel. The upper wheel travel is important in ride comfort and the lower wheel travel is important for safety. To have better ride comfort, the upper wheel travel should be as high as possible to make the suspension as soft as possible.

Although the upper and lower wheel travels may be different, for practical purposes, we can assume $d_l = d_u$ and design the suspension based on a unique wheel travel. Wheel travel is also called suspension travel, suspension room, and suspension clearance.

**Example 117** Soft and hard suspensions. A practical example to study how to use the design chart for optimal suspension.

Consider two pieces of equipment, $A$ and $B$, under a base excitation with an amplitude $Y = 1$ cm. Equipment $A$ has a suspension travel $d_A = 1.2$ cm and equipment $B$ has $d_B = 0.8$ cm. Let us assume $S_Z = d_u/Y$.

$$S_{Z_A} \approx 1.2 \qquad S_{Z_B} \approx 0.8 \tag{8.174}$$

Using the design chart in Fig. 8.24, the optimal suspensions for $A$ and $B$ are

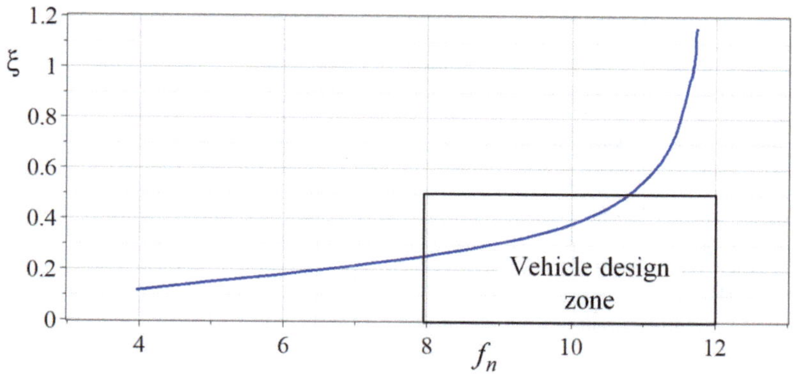


**Fig. 8.25** The optimal curve in the $(f_n, \xi)$-plane for 1/8 vehicle model

$$f_{n_A} \approx 8.53\,\text{Hz} \qquad \xi_A \approx 0.29$$
$$f_{n_B} \approx 10.8\,\text{Hz} \qquad \xi_B \approx 0.56 \tag{8.175}$$

Assuming a mass $m$

$$m = 300\,\text{kg} \tag{8.176}$$

we calculate the optimal springs and dampers as:

$$k_A = \left(2\pi f_{n_A}\right)^2 m = 8.6175 \times 10^5\,\text{N/m} \tag{8.177}$$
$$k_B = \left(2\pi f_{n_B}\right)^2 m = 13.814 \times 10^5\,\text{N/m} \tag{8.178}$$

$$c_A = 2\xi_A\sqrt{k_A m} = 9325.7\,\text{N s/m} \tag{8.179}$$
$$c_B = 2\xi_B\sqrt{k_B m} = 22800\,\text{N s/m} \tag{8.180}$$

Equipment $B$ has a harder suspension compared to equipment $A$. This is because equipment $B$ has less suspension travel, and hence, it has higher acceleration level $\omega_n^2 S_{\ddot{X}}$. Figure 8.24 shows that

$$\omega_n^2 S_{\ddot{X}_A} \approx 4700 \ \ 1/\text{s}^2 \tag{8.181}$$
$$\omega_n^2 S_{\ddot{X}_B} \approx 6650 \ \ 1/\text{s}^2 \tag{8.182}$$

**Example 118** Optimal suspension design curves. The optimal curve can be plotted in different plane using $S_{\ddot{X}_B}$, $S_Z$, $f_n$, $\xi$ as the exes.

The most practical design chart would be the $(f_n, \xi)$-plane in which the engineer can determine the best damping ratio $\xi$ by having the spring rate from the natural frequency $f_n$, or determining the best spring rate $k$ from $f_n$ by having the damping ratio $\xi$. Figure 8.25 illustrates optimal curve in the $(f_n, \xi)$-plane. The optimal curve in $\left(S_{\ddot{X}_B}, S_Z\right)$-plane is shown in Fig. 8.26. Using the *RMS* of the relative displacement $S_Z$ as the input key to pick the optimal $k$ and $c$, Figs. 8.27 and 8.28 illustrate $f_n$ and $\xi$ as functions of $S_Z$.

The behavior of *RMS* of absolute acceleration $S_{\ddot{X}_B}$ and *RMS* of relative displacement $S_Z$ can also be plotted against the optimal natural frequency $f_n$ or optimal damping ratio $\xi$. Figure 8.29 illustrates $S_{\ddot{X}_B}$ against $\xi$ for different $f_n$, and Fig. 8.30 illustrates $S_{\ddot{X}_B}$ versus $f_n$ for different $\xi$. Figure 8.31 illustrates $S_Z$ against $\xi$ for different $f_n$, and Fig. 8.32 illustrates $S_Z$ versus $f_n$ for different $\xi$.

**Example 119** Soft and hard vehicle suspensions. Applying the suspension optimization to vehicles is based on a small part of the design chart due to physical restrictions and limited range of suspension rates. Here is an example for a soft and hard suspensions.

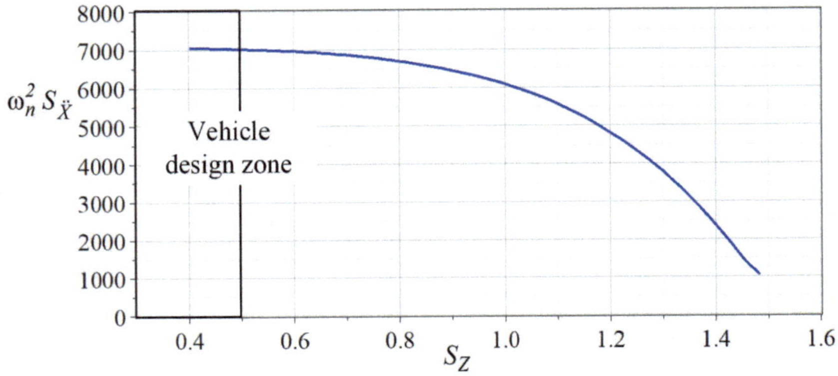


**Fig. 8.26** The optimal curve in the $\left(S_{\ddot{X}_B}, S_Z\right)$-plane for 1/8 vehicle model

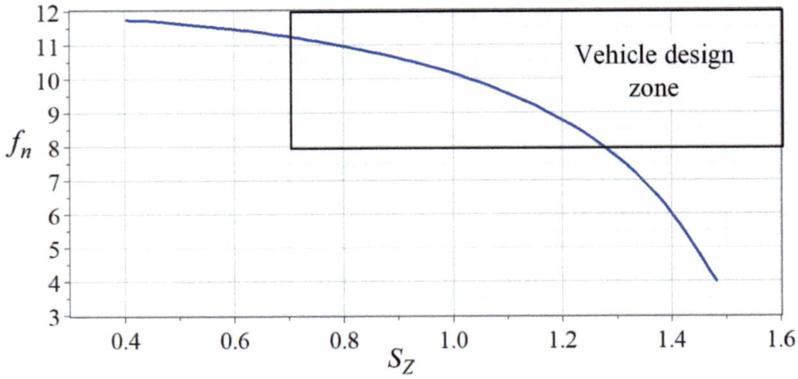


**Fig. 8.27** The optimal curve in the $(f_n, S_Z)$-plane for 1/8 vehicle model

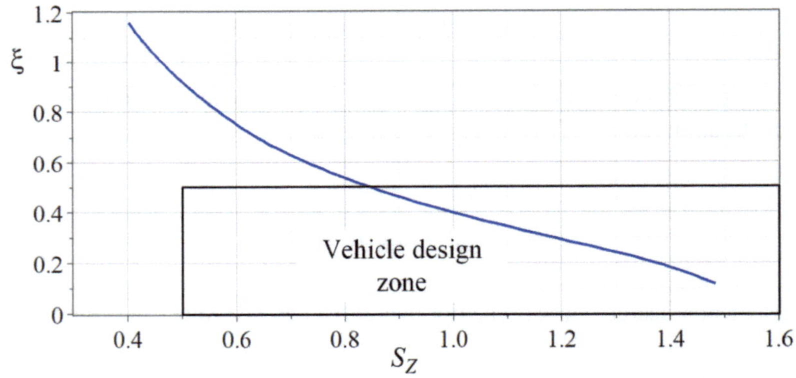


**Fig. 8.28** The optimal curve in the $(\xi, S_Z)$-plane for 1/8 vehicle model

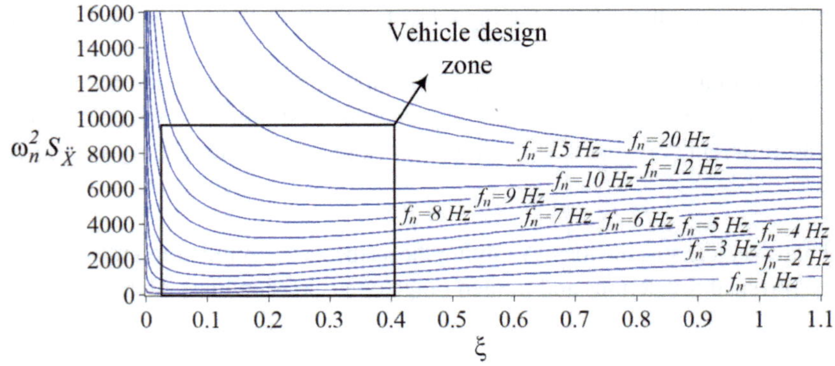


**Fig. 8.29** The absolute acceleration $S_{\ddot{X}_B}$ versus $\xi$ for different $f_n$

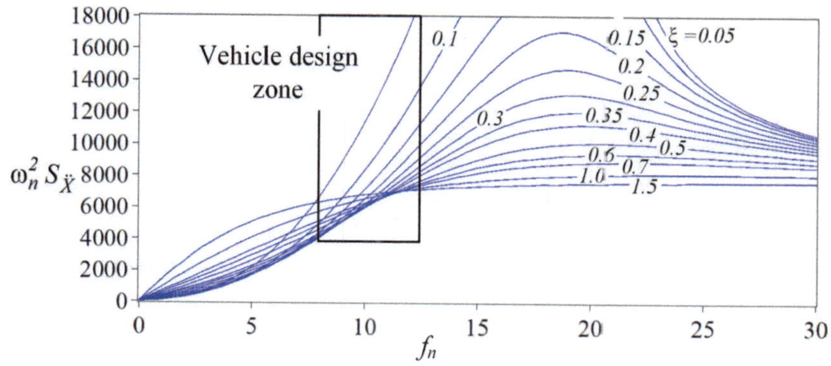


**Fig. 8.30** The absolute acceleration $S_{\ddot{X}_B}$ versus $f_n$ for different $\xi$

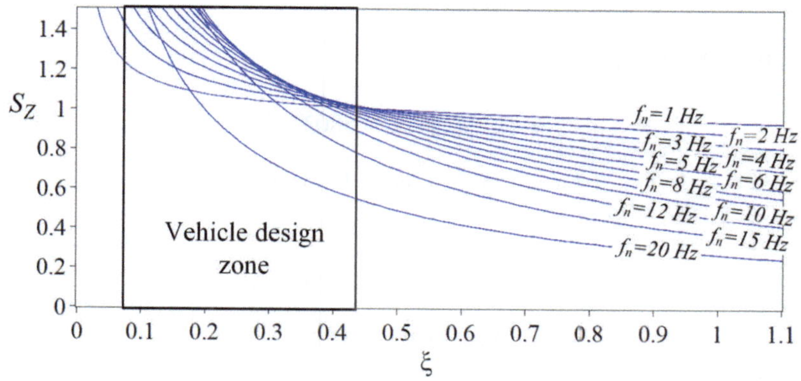


**Fig. 8.31** The relative displacement $S_Z$ versus $\xi$ for different $f_n$

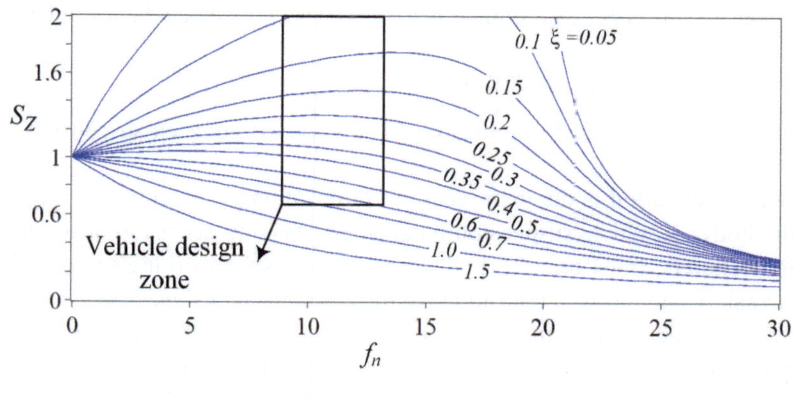


**Fig. 8.32** The relative displacement $S_Z$ versus $f_n$ for different $\xi$

Consider two vehicles $A$ and $B$ that are moving on a bumpy road with an average amplitude $Y = 10\,\text{cm} \approx 3.937\,\text{in}$. Vehicle $A$ has a suspension travel $d_A = 14.77\,\text{cm} \approx 5.816\,\text{in}$ and vehicle $B$ has $d_B = 14.71\,\text{cm} \approx 5.793\,\text{in}$. Let us assume that $S_Z = d_u/Y$.

$$S_{Z_A} = 1.477 \qquad S_{Z_B} = 1.471 \tag{8.183}$$

Using design chart 8.33, the optimal suspensions for vehicles $A$ and $B$ are

$$f_{n_A} \approx 0.7\,\text{Hz} \qquad \xi_A \approx 0.023 \tag{8.184}$$

$$f_{n_B} \approx 1.85\,\text{Hz} \qquad \xi_B \approx 0.06 \tag{8.185}$$

Assuming a mass $m$,

$$m = 300\,\text{kg} \approx 660\,\text{lb} \tag{8.186}$$

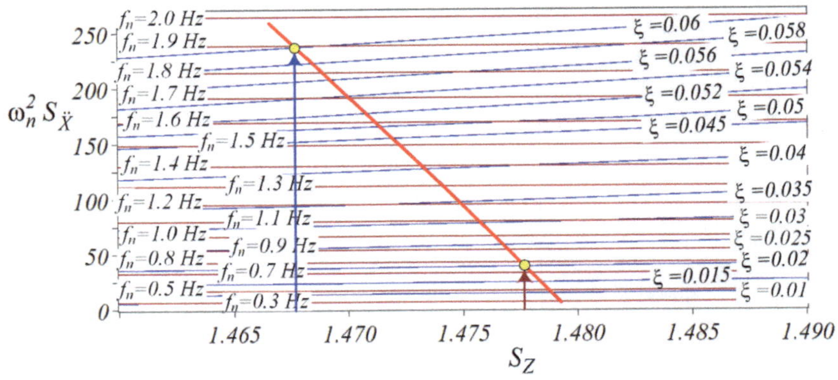

**Fig. 8.33** Comparing two suspensions, $A$ and $B$, with $S_{Z_A} = 1.4772$ and $S_{Z_B} = 1.4714$

we calculate the optimal springs and dampers as:

$$k_A = \left(2\pi f_{n_A}\right)^2 m \approx 5803\,\mathrm{N/m} \tag{8.187}$$

$$k_B = \left(2\pi f_{n_B}\right)^2 m \approx 40534\,\mathrm{N/m} \tag{8.188}$$

$$c_A = 2\xi_A\sqrt{k_A m} \approx 60.7\,\mathrm{N\,s/m} \tag{8.189}$$

$$c_B = 2\xi_B\sqrt{k_B m} \approx 418.5\,\mathrm{N\,s/m} \tag{8.190}$$

These are equivalent dampers and springs at the center of the wheel. The actual value of the suspension parameters depends on the geometry of the suspension mechanism and installment of the spring and damper. Because $k_B > k_A$ and $c_B > c_A$, the suspension of $B$ is harder than $A$. This is because vehicle $B$ has less wheel travel and hence, it has higher acceleration level, $\omega_n^2 S_{\ddot{X}}$. Figure 8.33 shows

$$\omega_n^2 S_{\ddot{X}_B} \approx 220 \;\; 1/\mathrm{s}^2 \qquad \omega_n^2 S_{\ddot{X}_A} \approx 28 \;\; 1/\mathrm{s}^2 \tag{8.191}$$

**Example 120** Suspension trade-off and trivial optimization. The trade-off between suspension travel and acceleration is discussed here.

Reduction of the absolute acceleration is the main goal in optimization problem of suspensions, because it represents the transmitted force to the body. A vehicle suspension reduces the absolute acceleration by increasing deflection of the suspension. The relative deflection is a measure of the clearance known as the working space of the isolator. The clearance should be minimized due to safety and physical constraints in the mechanical design.

There is a trade-off between the acceleration and the relative motion. The ratio of $\omega_n^4 S_{\ddot{X}}$ to $S_Z$ is a monotonically increasing function of $\omega_n$ and $\xi$. Keeping $S_Z$ constant increases $\omega_n^4 S_{\ddot{X}}$ by increasing both $\omega_n$ and $\xi$. However, keeping $\omega_n^4 S_{\ddot{X}}$ constant, we decrease $S_Z$ by increasing $\omega_n$ and $\xi$. Hence, $\omega_n^4 S_{\ddot{X}}$ and $S_Z$ have opposite behaviors. These behaviors show that the suspension of $\omega_n = 0$ and $\xi = 0$ is the trivial and non-practical solution for the best isolation.

**Example 121** $RMS(G_0) \equiv RMS(X/Y)$. Although the absolute displacement is not involved in *RMS* optimization method, the *RMS* of absolute displacement involves in objective functions of many optimization algorithms. Here is the analytic calculation of the function.

For the *RMS* of the absolute displacement, $S_X$, we need the integral of $G_0^2 \equiv (X/Y)^2$, which is determined as follows:

$$\int G_0^2 d\omega = \frac{Z_{16}}{Z_{17}\sqrt{Z_{18}}} \arctan\frac{\omega}{\sqrt{Z_{18}}} + \frac{Z_{19}}{Z_{20}\sqrt{Z_{21}}} \arctan\frac{\omega}{\sqrt{Z_{21}}} \tag{8.192}$$

$$Z_{16} = \omega_n^2\left(-8\xi^6 + 8\xi^4 - \left(8\xi^4 - 4\xi^2 - 1\right)\xi\sqrt{1-\xi^2}\right) \tag{8.193}$$

$$Z_{17} = -4\xi^2 \left(1 - \xi^2\right) \tag{8.194}$$

$$Z_{18} = \omega_n^2 \left(1 - 2\xi^2 - 2\xi\sqrt{1 - \xi^2}\right) \tag{8.195}$$

$$Z_{19} = \omega_n^2 \left(8\xi^6 - 8\xi^4 - \left(8\xi^4 - 4\xi^2 - 1\right)\xi\sqrt{1 - \xi^2}\right) \tag{8.196}$$

$$Z_{20} = Z_{17} = -4\xi^2 \left(1 - \xi^2\right) \tag{8.197}$$

$$Z_{21} = -\omega_n^2 \left(1 - 2\xi^2 - 2\xi\sqrt{1 - \xi^2}\right) \tag{8.198}$$

**Example 122** Alternative optimization methods. Optimization of vibration isolator has a rich background with many different criteria. Here is introducing a couple of other methods than the *RMS* method.

There are various approaches and suggested methods for vehicle suspension and vibration isolator optimization, depending on the application and mathematical modeling. However, there is not a universally accepted method applicable to every application. Every optimization strategy can be transformed to a minimization of a function called the cost function. Considerable attention has been given to minimization of the absolute displacement, known as the main transmissibility. However, for vehicle suspension and vibration isolator, the cost function may include any state variable such as absolute and relative displacements, velocities, accelerations, and jerks.

Constraints determine the domain of acceptable design parameters by dictating an upper and lower limits for $\omega_n$ and $\xi$. For vehicle suspensions, it is generally desired to select $\omega_n$ and $\xi$ such that the absolute acceleration of the system is minimized and the relative displacement does not exceed a prescribed level. The most common optimization strategies are the following.

Minimax the absolute acceleration $S_{\ddot{X}}$ for specified relative displacement $S_{Z_0}$. Specify the allowable relative displacement, and then find the minimax of absolute acceleration:

$$\frac{\partial S_{\ddot{X}}}{\partial \omega_n} = 0 \qquad \frac{\partial S_{\ddot{X}}}{\partial \xi} = 0 \qquad S_Z = S_{Z_0} \tag{8.199}$$

Minimax the relative displacement $S_Z$ for specified absolute acceleration $S_{\ddot{X}_0}$. Specify the allowable absolute acceleration, and then find the minimax relative displacement:

$$\frac{\partial S_Z}{\partial \omega_n} = 0 \qquad \frac{\partial S_Z}{\partial \xi} = 0 \qquad S_{\ddot{X}} = S_{\ddot{X}_0} \tag{8.200}$$

**Example 123** ★ Further application of the design chart. The *RMS* optimization method can be applied to any two characteristics that have a trade-off property. Here is more explanation.

The *RMS* optimization criterion is based on the root mean square of $S_2$ and $G_2$ over a working frequency range.

$$\frac{\partial S_{\ddot{X}}}{\partial S_Z} = 0 \qquad \frac{\partial^2 S_{\ddot{X}}}{\partial S_Z^2} > 0 \tag{8.201}$$

$$S_Z = \sqrt{\frac{1}{40\pi}\int_0^{40\pi} S_2^2 d\omega} \tag{8.202}$$

$$S_{\ddot{X}} = \sqrt{\frac{1}{40\pi}\int_0^{40\pi} G_2 d\omega} \tag{8.203}$$

The design curve is the optimal condition for suspension of a base excited system using the following functions:

$$S_2 = \frac{Z_B}{Y} \qquad G_2 = \frac{\ddot{X}_B}{\omega_n^2 Y} \tag{8.204}$$

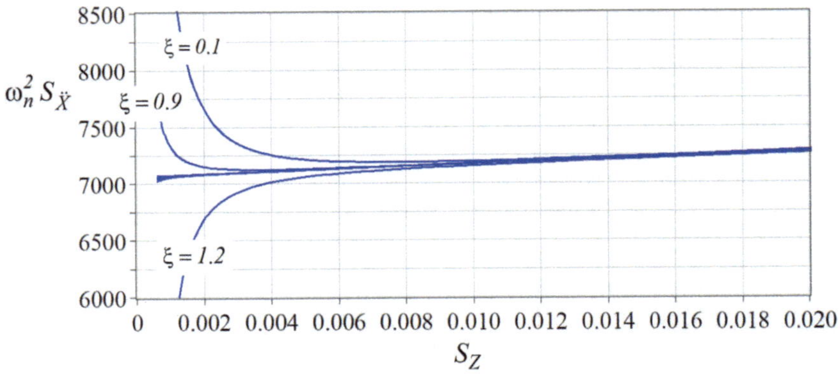


**Fig. 8.34** The curves of $S_{\ddot{X}}$ versus $S_Z$ for different $\xi$ at very low values of $S_Z$

However, because

$$S_2 = \frac{\ddot{X}_F}{F/m} = \frac{Z_B}{Y} = \frac{X_E}{e\varepsilon_E} = \frac{Z_R}{e\varepsilon_R} \tag{8.205}$$

$$G_2 = \frac{\ddot{X}_B}{\omega_n^2 Y} = \frac{F_{T_B}}{kY} = \frac{F_{T_E}}{e\omega^2 m_e} = \frac{F_{T_R}}{e\omega^2 m_e}\left(1 + \frac{m_a}{m}\right) \tag{8.206}$$

the optimal design curve can also be expressed as a minimization condition for any other $G_2$-function with respect to any other $S_2$-function, such as the force transmitted to the base $F_{T_E}/\left(e\omega^2 m_e\right)$ for an eccentric excited system $X_E/(e\varepsilon_E)$. This minimization is equivalent to the optimization of an engine mount.

**Example 124** The graph of $S_{\ddot{X}}$ versus $S_Z$ at low values of $S_Z$.

The optimal curve as well as $(S_{\ddot{X}}, S_Z)$ curves seems to be converging to a particular value for low values of $S_Z$. To evaluate that value, we need to magnify the mesh for very low values of $S_Z$. Figure 8.34 illustrates the curves of $S_{\ddot{X}}$ versus $S_Z$ for different $\xi$ at very low values of $S_Z$. Therefore, all curves of constant $0 < \xi < 0.7$ start from $S_Z = 1$ and end at $\omega_n^2 S_{\ddot{X}_B} \approx 7000 \ 1/\mathrm{s}^2$. This is also where the optimal curve will approach for low values of $S_Z$.

## 8.5 ★ Time Response Optimization

There is no standard method of optimization in time domain. The transient response optimization depends on the type of transient excitation, as well as the definition of the cost function. Figure 8.35 illustrates a 1/8 vehicle model and a unit step displacement.

$$y = \begin{cases} 1 & t > 0 \\ 0 & t \le 0 \end{cases} \tag{8.207}$$

Let us consider a step function as the transient excitation, and the optimization criteria as minimization of the peak value of the acceleration versus peak value of the relative displacement. There is an optimal damping ratio $\xi^\star$ for any natural frequency $f_n$ that provides us with the best transient behavior of a the 1/8 vehicle model. This behavior is shown in the design chart of Fig. 8.36.

$$\xi^\star = 0.4 \tag{8.208}$$

***Proof*** The equation of motion for the base excited one *DOF* system shown in Fig. 8.35 is

$$\ddot{x} + 2\xi\omega_n \dot{x} + \omega_n^2 x = 2\xi\omega_n \dot{y} + \omega_n^2 y \tag{8.209}$$

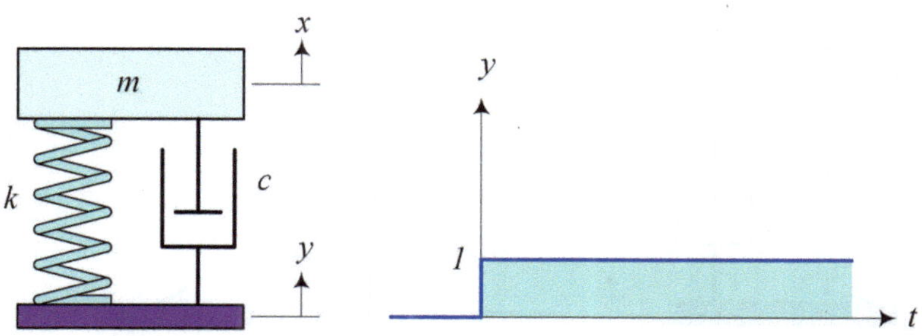


**Fig. 8.35** A 1/8 vehicle model and a unit step displacement base excitation

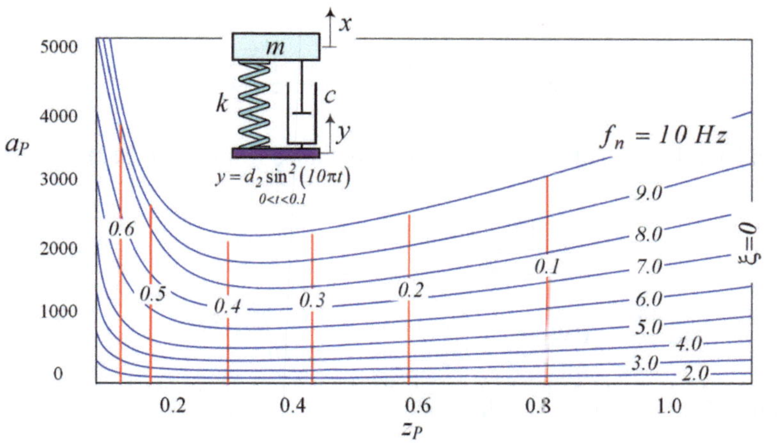


**Fig. 8.36** Peak value of acceleration versus peak value of relative displacement for different $\xi$ and $f_n$

Substituting $y = 1$ in Eq. (8.209) provides us with the following initial value problem to determine the absolute displacement of the mass $m$.

$$\ddot{x} + 2\xi\omega_n\,\dot{x} + \omega_n^2\,x = \omega_n^2 \tag{8.210}$$

$$y(0) = 0 \qquad \dot{y}(0) = 0 \tag{8.211}$$

The solution of the differential equation with zero initial conditions is

$$x = 1 - \frac{1}{2}\frac{C}{i\omega_d}e^{-C\omega_n t} + \frac{1}{2}\frac{\overline{C}}{i\omega_d}e^{-\overline{C}\omega_n t} \tag{8.212}$$

equivalent to

$$x = 1 - e^{-\xi\omega_n t}\left(\frac{\xi\omega_n}{\omega_d}\sin\omega_d t + \cos\omega_d t\right) \tag{8.213}$$

where $C$ and $\overline{C}$ are two complex conjugate numbers.

$$C = -\omega_n\left(\xi + i\sqrt{1-\xi^2}\right) = -\left(\xi\omega_n + i\omega_d\right) \tag{8.214}$$

$$\overline{C} = -\omega_n\left(\xi - i\sqrt{1-\xi^2}\right) = -\left(\xi\omega_n - i\omega_d\right) \tag{8.215}$$

$$\omega_d = \omega_n\sqrt{1-\xi^2} \tag{8.216}$$

Having $x$ and $y = 1$ is enough to calculate the relative displacement $z = x - y$.

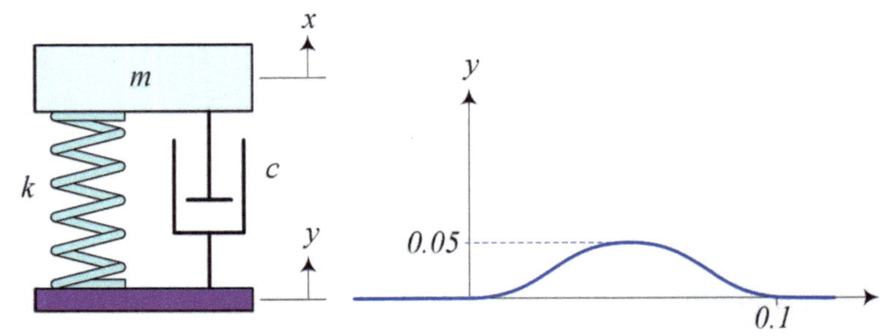


**Fig. 8.37** A base excited one *DOF* system and a sine square bump input

$$z = x - y = -\frac{1}{2}\frac{C}{ib}e^{-A\omega_n t} + \frac{1}{2}\frac{\overline{C}}{ib}e^{-\overline{A}\omega_n t} \tag{8.217}$$

The absolute velocity and acceleration of the mass $m$ can be obtained from Eq. (8.212).

$$\dot{x} = \frac{1}{2}\frac{C^2\omega_n}{ib}e^{-C\omega_n t} - \frac{1}{2}\frac{\overline{C}^2\omega_n}{ib}e^{-\overline{C}\omega_n t} \tag{8.218}$$

$$\ddot{x} = -\frac{1}{2}\frac{C^3\omega_n^2}{ib}e^{-C\omega_n t} + \frac{1}{2}\frac{\overline{C}^3\omega_n^2}{ib}e^{-\overline{C}\omega_n t} \tag{8.219}$$

The peak value of the relative displacement is $z_P$ which occurs when $\dot{z} = 0$ at time $t_1$.

$$z_P = \exp\left(\frac{\arccos\left(2\xi^2 - 1\right)}{\omega_n\sqrt{1-\xi^2}}\right) \tag{8.220}$$

$$t_1 = \frac{-\xi \arccos\left(2\xi^2 - 1\right)}{\sqrt{1-\xi^2}} \tag{8.221}$$

The peak value of the absolute acceleration is $a_P$ which occurs at the beginning of the excitation, $t = 0$, or at the time instant when $\dddot{x} = 0$ at time $t_2$.

$$a_P = \omega_n^2 \exp\left(-\xi\frac{2\arccos\left(2\xi^2 - 1\right) - \pi}{\sqrt{1-\xi^2}}\right) \tag{8.222}$$

$$t_2 = \frac{2\arccos\left(2\xi^2 - 1\right) - \pi}{\omega_n\sqrt{1-\xi^2}} \tag{8.223}$$

Figure 8.36 is a plot for $a_P$ versus $z_P$ for different $\xi$ and $f_n$. The minimum of the curves occur at a damping ratio about $\xi = 0.4$ for every $f_n$. The optimal $\xi$ can be found analytically by finding the minimum point of $a_P$ versus $z_P$. The optimal $\xi$ is the solution of the following transcendental equation which is $\xi = 0.4$.

$$2\xi \arccos\left(2\xi^2 - 1\right) - \pi - 4\xi\sqrt{1-\xi^2} = 0 \tag{8.224}$$

The minimum peak value of the absolute acceleration with respect to the relative displacement is independent of the value of natural frequency $f_n$. ■

**Example 125** ★ Optimal design curve and time response. To examine if the optimal value of $\xi$ based on step input, as well as the optimal suspension based on the optimal curve of *RMS* methods is also working for other inputs, here we study a base exited system on a bump transient input.

To examine transient response of suspensions on the optimal design curve, we compare the response of three suspensions to a road bump simulated by a sine square function as shown in Fig. 8.37.

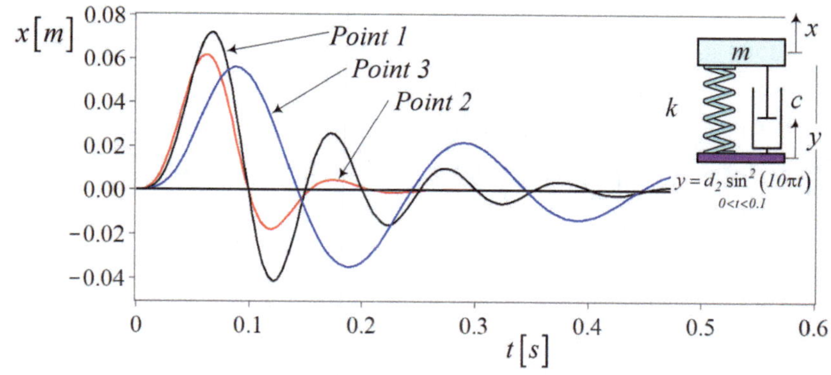


**Fig. 8.38** Absolute displacement time response of the system for three different suspensions

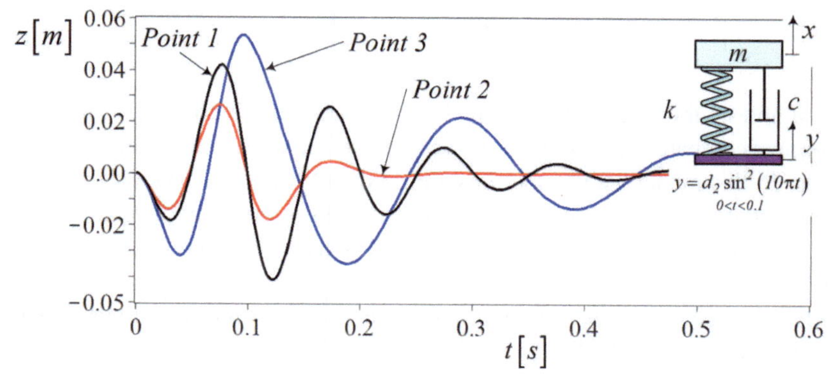


**Fig. 8.39** Relative displacement time response of the system for three different suspensions

$$Point\ P_1 \qquad f_n \approx 10\,\text{Hz} \qquad \xi \approx 0.15 \tag{8.225}$$

$$Point\ P_2 \qquad f_n \approx 10\,\text{Hz} \qquad \xi \approx 0.4 \tag{8.226}$$

$$Point\ P_3 \qquad f_n \approx 5\,\text{Hz} \qquad \xi \approx 0.15 \tag{8.227}$$

Point $P_1$ indicates an off-optimal suspension, and points $P_2$ and $P_3$ are both on the optimal curve of Fig. 8.17. Points $P_2$ and $P_3$ are two alternative optimizations for point $P_1$. Point $P_2$ has $\xi \approx 0.4$ with the same natural frequency as $P_1$, and point $P_3$ has $f_n \approx 5\,\text{Hz}$ with the same damping as point $P_1$.

Figure 8.37 illustrates a base excited one *DOF* system and a sine square bump input.

$$y = \begin{cases} d_2 \sin^2 \dfrac{2\pi v}{d_1} t & 0 < t < 0.1 \\ 0 & t \le 0, t \ge 0.1 \end{cases} \tag{8.228}$$

$$d_2 = 0.05\,\text{m} \qquad v = 10\,\text{m/s} \qquad d_1 = 1\,\text{m} \tag{8.229}$$

The absolute and relative displacement time responses of the system at points 1, 2, and 3 are shown in Figs. 8.38 and 8.39, respectively. The absolute acceleration of $m$ is shown in Fig. 8.40. System 3 has a lower relative displacement peak value and a lower absolute acceleration peak value, but it takes more time to settle down.

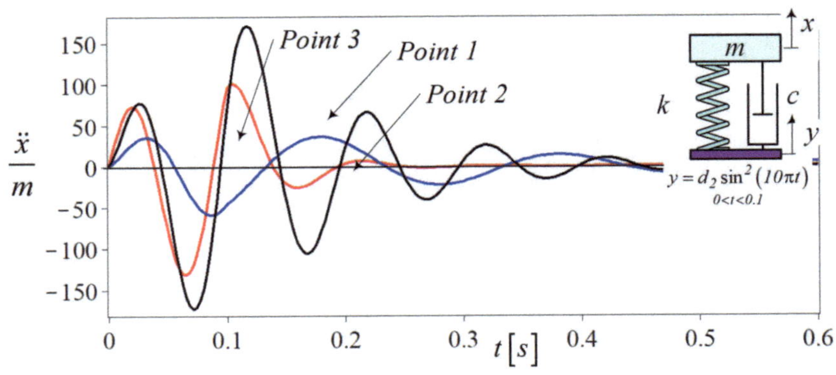


**Fig. 8.40** Absolute acceleration time response of the system for three different suspensions

## 8.6 Summary

1/8 vehicle model is the simplest vibrating model of vehicles that is only considering vertical vibration of the vehicle while ignoring the flexibility of tires compared. The model is made when we cut a vehicle longitudinally and laterally, and we keep one quarter of the vehicle. In this model we assume tires are rigid and massless. This way we model a vehicle with a one DOF base excited system with the mass $m$ to be one quarter of the mass of the body of vehicle, and suspension spring $k$ and damper $c$ to be equal to $k$ and $c$ of one side of the vehicle. The 1/8 vehicle model is very simple and yet very accurate to optimize the main suspension.

The function of a vehicle suspension as a vibration isolator is to reduce the magnitude of motion transmitted from a vibrating foundation to the body of the vehicle. It also reduces the magnitude of force transmitted from the vehicle to the road, both in time and frequency domains.

The equation of motion of the system for the absolute displacement of the body $x$ and the relative displacement $z$ for given road displacement $y$ are

$$m\ddot{x} + c\dot{x} + kx = c\dot{y} + ky \tag{8.230}$$

$$m\ddot{z} + c\dot{z} + kz = -m\ddot{y} \tag{8.231}$$

$$z = x - y \tag{8.232}$$

which can be transformed to the following two parameter dependent equations by introducing the *natural frequency* $\omega_n$ and the *damping ratio* $\xi$.

$$\ddot{x} + 2\xi\omega_n\dot{x} + \omega_n^2 x = 2\xi\omega_n\dot{y} + \omega_n^2 y \tag{8.233}$$

$$\ddot{z} + 2\xi\omega_n\dot{z} + \omega_n^2 z = -\ddot{y} \tag{8.234}$$

$$\xi = \frac{c}{2\sqrt{km}} \qquad \omega_n = \sqrt{\frac{k}{m}} = 2\pi f_n \qquad r = \frac{\omega}{\omega_n} \tag{8.235}$$

The most important frequency responses of base excited 1/8 vehicle model are absolute displacement $G_0$, relative displacement $S_2$, absolute acceleration $G_2$, and force transmitted to $m$.

$$G_0 = \left|\frac{X}{Y}\right| = \frac{\sqrt{1 + (2\xi r)^2}}{\sqrt{\left(1 - r^2\right)^2 + (2\xi r)^2}} \tag{8.236}$$

$$S_2 = \left|\frac{Z}{Y}\right| = \frac{r^2}{\sqrt{\left(1 - r^2\right)^2 + (2\xi r)^2}} \tag{8.237}$$

$$G_2 = \left| \frac{\ddot{X}}{Y\omega_n^2} \right| = \frac{r^2\sqrt{1+(2\xi r)^2}}{\sqrt{\left(1-r^2\right)^2+(2\xi r)^2}} \tag{8.238}$$

The solution of the 1/8 vehicle model to a unit step road displacement $y = 1$ with zero initial conditions is

$$x = 1 - e^{-\xi\omega_n t}\left(\frac{\xi\omega_n}{\omega_d}\sin\omega_d t + \cos\omega_d t\right) \tag{8.239}$$

## 8.7 Key Symbols

| | |
|---|---|
| $a \equiv \ddot{x}$ | acceleration |
| $a, b, C$ | Fourier coefficient |
| $a, b, c, d, e$ | length, dimension exponent |
| $A, B$ | weight factor, coefficients for frequency responses |
| $A, \bar{A}$ | complex conjugate coefficients |
| $c$ | damping |
| $C$ | coefficient, complex coefficient |
| $d_1$ | road wave length |
| $d_2$ | road wave amplitude |
| $D$ | dissipation function |
| $e$ | eccentricity arm, exponential function |
| $E$ | mechanical energy, energy |
| $f = 1/T$ | cyclic frequency [ Hz] |
| $f, F, \mathbf{F}$ | force |
| $f, g$ | function, periodic function |
| $f_c$ | damper force |
| $f_e$ | equivalent force |
| $f_k$ | spring force |
| $f_m$ | required force to move a mass $m$ |
| $f_T$ | transmitted force |
| $F$ | amplitude of the harmonic force $f = F\sin\omega t$ |
| $F_T$ | amplitude of transmitted force |
| $F_0$ | constant force |
| $g$ | gravitational acceleration, function |
| $g\left(r^2\right)$ | characteristic equation |
| $G$ | general frequency response with node |
| $h$ | response to unit impulse |
| $H$ | Heaviside function, complex amplitude |
| $H^*$ | complex conjugate of $H$ |
| $j$ | dummy index |
| $k$ | stiffness |
| $k_e$ | equivalent stiffness |
| $K$ | kinetic energy |
| $l$ | length |
| $L$ | length dimension |
| $\mathcal{L}$ | Lagrangian |
| $m$ | mass, number of octave |
| $m_b$ | device mass |
| $m_e$ | eccentric mass, equivalent mass |
| $M$ | mass dimension |
| $P$ | power, point |
| $P$ | potential energy |

| | |
|---|---|
| $q$ | general coordinate |
| $Q$ | quality factor, general amplitude |
| $r$ | frequency ratio |
| $r_i,\ i \in N$ | nodal frequency ratio |
| $r_n = \omega_n/\omega_s$ | natural frequency ratio |
| $S$ | general frequency response without node |
| $S_u = RMS(u)$ | RMS of $u$ |
| $S_\eta = RMS(\eta)$ | RMS of $\eta$ |
| $t$ | time |
| $T$ | period, time dimension |
| $v \equiv \dot{x}$ | velocity |
| $x,\ y,\ z,\ \mathbf{x}$ | displacement |
| $x_0$ | initial displacement |
| $x_s$ | sprung mass displacement |
| $x_u$ | unsprung mass displacement |
| $X,$ | steady-state amplitude of $x$, dimensionless $x$ |
| $X_s$ | steady-state amplitude of $x_s$ |
| $X_u$ | steady-state amplitude of $x_u$ |
| $\dot{x},\ \dot{y},\ \dot{z}$ | velocity, time derivative of $x,\ y,\ z$ |
| $\dot{x}_0$ | initial velocity |
| $X, Y, Z$ | amplitude |
| $Y$ | amplitude of base displacement, dimensionless $y$ |
| $\dot{y}$ | velocity |
| $z$ | relative displacement, dimensionless $z$ |
| $Z$ | amplitude of relative displacement |
| $Z_1,\ Z_2, \cdots$ | short notation for part of equations |

Greek

| | |
|---|---|
| $\alpha$ | angle |
| $\delta$ | difference |
| $\delta_s$ | static deflection |
| $\Delta$ | difference |
| $\gamma$ | angle |
| $\lambda$ | eigenvalue |
| $\Lambda$ | eigenvalue matrix |
| $\varepsilon$ | mass ratio |
| $\varepsilon$ | small coefficient |
| $\theta$ | angular motion |
| $\Pi$ | Buckingham dimensionless term |
| $\rho$ | length mass density |
| $\xi$ | damping ratio |
| $\omega$ | angular frequency |
| $\omega_n$ | natural frequency |
| $\omega_d$ | damped natural frequency |
| $\varphi,\ \Phi$ | phase angle |
| $\alpha = \omega_s/\omega_u$ | sprung mass ratio |
| $\varepsilon = m_s/m_u$ | sprung mass ratio |
| $\eta = \lvert Z/Y \rvert$ | sprung mass relative frequency response |
| $\mu = \lvert X_s/Y \rvert$ | sprung mass frequency response |
| $\xi = c_s/\left(2\sqrt{k_s m_s}\right)$ | damping ratio |
| $\xi^\star$ | optimal damping ratio |
| $\tau$ | dimensionless $t$ |
| $\tau = \lvert X_u/Y \rvert$ | unsprung mass frequency response |

$\omega = 2\pi f$ angular frequency [rad/s]
$\omega_s = \sqrt{k_s/m_s}$ sprung mass frequency
$\omega_u = \sqrt{k_u/m_u}$ unsprung mass frequency

Subscript
$i \in N$ node number
$n$ natural
$s$ sprung
$u$ unsprung

Symbols
$DOF$ degree of freedom
$FBD$ free body diagram
$ODF$ one degree of freedom
Re real part
$RMS$ root mean square
det determinant
Hz Hertz
[ ] matrix
$\int$ integral
| | absolute value
$d$ differential
$\partial$ partial derivative
$\mathcal{L}$ Lagrangian
$O$ order of magnitude
$\nabla$ gradient

## References


1. R. Alkhatib, R.N. Jazar, M.F. Golnaraghi, Optimal design of passive linear suspension using genetic algorithm. J. Sound Vib. **275**(3–5), 665–691 (2004)
2. M. Akraminia, M. Tatari, M. Fard, R.N. Jazar, Designing active vehicle suspension system using critic-based control strategy. Nonlinear Eng. **4**(3), 141–154 (2015)
3. A. Azizan, M. Fard, M.F. Azari, R.N. Jazar, Effects of vibration on occupant driving performance under simulated driving conditions. Appl. Ergonom. **60**, 348–355 (2017)
4. J. Christopherson, R.N. Jazar, Optimization of classical hydraulic engine mounts based on RMS method. Shock Vib. **12**(2), 119–147 (2005)
5. S. Deshpande, S. Mehta, R.N. Jazar, Jump avoidance conditions for piecewise linear vibration isolator, in *International Design Engineering Technical Conferences and Computers and Information in Engineering Conference, DETC2005-85396* (2008), pp. 1955–1962
6. S. Deshpande, S. Mehta, R.N. Jazar, Sensitivity of jump avoidance condition of piecewise linear vibration isolator to dynamical parameters, in *ASME 2005 International Mechanical Engineering Congress and Exposition, IMECE2005-81853*. Orlando, Florida, USA (2008), pp. 1103–1113
7. R.N. Jazar, R. Alkhatib, M.F. Golnaraghi, Root mean square optimization criterion for vibration behaviour of linear quarter car using analytical methods. Veh. Syst. Dyn. **44**(6), 477–512 (2006)
8. R.N. Jazar, *Advanced Vibrations: A Modern Approach* (Springer, New York, 2013)
9. R.N. Jazar, *Vehicle Dynamics: Theory and Application*, 3rd ed. (Springer, New York, 2017)
10. R.N. Jazar, *Approximation Methods in Science and Engineering* (Springer, New York, 2020)
11. R.N. Jazar, *Perturbation Methods in Science and Engineering* (Springer, New York, 2021)
12. R.N. Jazar, *Advanced Vibrations: Theory and Application* (Springer, New York, 2023)
13. R.N. Jazar, M.F. Golnaraghi, Engine mounts for automotive applications: a survey. Shock Vib. Digest **34**(5), 363–379 (2002)
14. R.N. Jazar, M.F. Golnaraghi, D.A. Swanson, Steady state vibrations of a piecewise linear system, in *ASME 2001 International Design Engineering Technical Conferences and Computers and Information in Engineering Conference, September 9–12, Pittsburgh, Pennsylvania, USA, DETC2001/VIB-21443* (2001), pp. 1197–1206
15. R.N. Jazar, R. Houim, A. Narimani, M.F. Golnaraghi, Frequency response and jump avoidance in a nonlinear passive engine mount. J. Vib. Control **12**(11), 1205–1237 (2006)
16. R.N. Jazar, M. Kazemi, S. Borhani, *Mechanical Vibrations* (in Persian) (Ettehad Publications, Tehran, 1992)
17. H. Marzbani, R.N. Jazar, M. Fard, Hydraulic engine mounts: a survey. J. Vib. Control **20**(10), 1439–1463 (2014)
18. H. Marzbani, R.N. Jazar, A. Khazaei, Smart passive vibration isolation: requirements and unsolved problems. J. Appl. Nonlinear Dyn. **1**(4), 341–386 (2012)

19. L. Meirovitch, *Analytical Methods in Vibrations* (Macmillan, New York, 1967)
20. L. Meirovitch, *Principles and Techniques of Vibrations* (Prentice Hall, Ney Jersey, 1997)
21. L. Meirovitch, *Fundamentals of Vibrations* (McGraw-Hill, New York, 2002)
22. Z. Milovanovic, I. Kovacic, M.J. Brennan, On the displacement transmissibility of a base excited viscously damped nonlinear vibration isolator. J. Vib. Acoust. **131**(5), 054502 (2009)
23. A. Narimani, M.F. Golnaraghi, R.N. Jazar, Frequency response of a piecewise linear vibration isolator. J. Vib. Control **10**(12), 1775–1794 (2004)
24. A. Narimani, R.N. Jazar, M.F. Golnaraghi, Vibration analysis of a piecewise linear system, in *Fourteenth U.S. National Congress of Theoretical and Applied Mechanics*, Blacksburg, Virginia, USA, June 23–28 (2002)
25. K. Ogata, *System Dynamics* (Prentice Hall, New Jersey, 2004)
26. Z. Osinski, *Damping of Vibrations* (A. A. Balkema Publishers, Rotterdam, 1998)
27. W.J. Palm, *Mechanical Vibration* (John Wiley, New York, 2006)
28. O. Pogorilyi, R.N. Jazar, P.M. Trivailo, Challenges in exact response of piecewise linear vibration isolator, in *Nonlinear Approaches in Engineering Applications*, ed. by L. Dai, R. Jazar (Springer, New York, 2015)
29. O. Pogorilyi, P.M. Trivailo, R.N. Jazar, On the piecewise linear exact solution. Nonlinear Eng. **3**(4), 189–196 (2014)
30. M. Roseau, *Vibrations in Mechanical Systems* (Springer, Berlin, 1987)
31. M. Setareh, Application of semi-active tuned mass dampers to base-excited systems. Earthquake Eng. Struct. Dyn. **30**(3), 449–462 (2001)
32. J.C. Snowdon, *Vibration and Shock in Damped Mechanical Systems* (John Wiley, New York, 1968)
33. P. Stahl, R.N. Jazar, Stability analysis of a piecewise nonlinear vibration isolator, in *ASME International Mechanical Engineering Congress, IMECE2005-81856* (2005), pp. 1115–1123
34. P. Stahl, R.N. Jazar, Frequency response analysis of piecewise nonlinear vibration isolator, in *20th ASME Biennial Conference on Mechanical Vibration and Noise*, Long Beach, CA, USA, September 24–28 (2005)
35. F.S. Tse, I.E. Morse, R.T. Hinkle, *Mechanical Vibrations Theory and Applications* (Allyn and Bacon, Boston, 1978)
36. Wiercigroch, M. and Sin, V.W.T., Experimental study of a symmetrical piecewise baseexcited oscillator. J. Appl. Mech. **65**(3), 657–663 (1998)

# Quarter Car Model and Body Bounce Mode

# 9

The most useful and practical vibration model of a vehicle suspension system is the *quarter car model*, shown in Fig. 9.1. Quarter car model is excellent to examine and optimize the body bounce mode of vibrations. This model includes the body bounce $x_s$, wheel hop $x_u$, and the road excitations $y$. We introduce, examine, and optimize the quarter car model in this section.

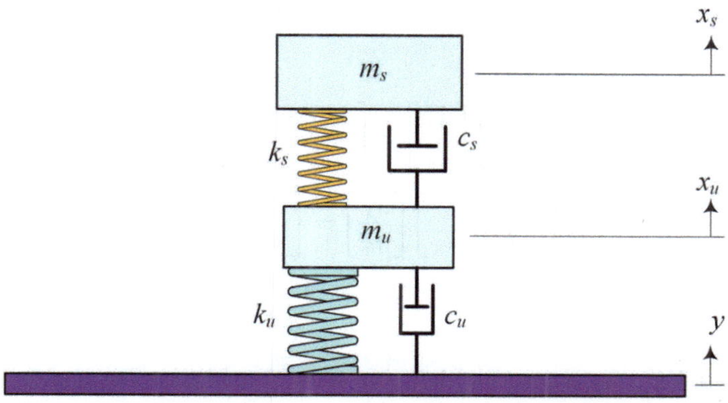


**Fig. 9.1** A quarter car model

## 9.1 Mathematical Model

We investigate the vertical vibrations of a vehicle using a quarter car model made of two solid masses $m_s$ and $m_u$ denoted as *sprung* and *unsprung* masses, respectively. The sprung mass $m_s$ represents $1/4$ of the body of the vehicle, and the unsprung mass $m_u$ represents one wheel of the vehicle. A spring of stiffness $k_s$ and a shock absorber with viscous damping coefficient $c_s$ support the sprung mass and are called the *main suspension*. Ignoring the unsprung damping $c_u$, the unsprung mass $m_u$ is in direct contact with the ground through a spring $k_u$, representing the tire stiffness. The two-mass quarter car model continues to be a useful tool for understanding the bounce mode vibration behavior of vehicle suspensions.

The governing differential equations of motion for the quarter car model, shown in Fig. 9.1, are

$$m_s \ddot{x}_s + c_s (\dot{x}_s - \dot{x}_u) + k_s (x_s - x_u) = 0 \tag{9.1}$$

$$m_u \ddot{x}_u + c_s (\dot{x}_u - \dot{x}_s) + (k_u + k_s) x_u - k_s x_s = k_u y \tag{9.2}$$

***Proof*** The kinetic energy $K$, potential energy $P$, and dissipation function $D$ of the system are

$$K = \frac{1}{2} m_s \dot{x}_s^2 + \frac{1}{2} m_u \dot{x}_u^2 \tag{9.3}$$


R. N. Jazar, H. Marzbani, *Vehicle Vibrations*, https://doi.org/10.1007/978-3-031-43486-0_9

$$P = \frac{1}{2}k_s (x_s - x_u)^2 + \frac{1}{2}k_u (x_u - y)^2 \tag{9.4}$$

$$D = \frac{1}{2}c_s (\dot{x}_s - \dot{x}_u)^2 \tag{9.5}$$

Employing the Lagrange method,

$$\frac{d}{dt}\left(\frac{\partial K}{\partial \dot{x}_s}\right) - \frac{\partial K}{\partial x_s} + \frac{\partial D}{\partial \dot{x}_s} + \frac{\partial P}{\partial x_s} = 0 \tag{9.6}$$

$$\frac{d}{dt}\left(\frac{\partial K}{\partial \dot{x}_u}\right) - \frac{\partial K}{\partial x_u} + \frac{\partial D}{\partial \dot{x}_u} + \frac{\partial P}{\partial x_u} = 0 \tag{9.7}$$

we find the equations of motion

$$m_s \ddot{x}_s = -k_s (x_s - x_u) - c_s (\dot{x}_s - \dot{x}_u) \tag{9.8}$$

$$m_u \ddot{x}_u = k_s (x_s - x_u) + c_s (\dot{x}_s - \dot{x}_u) - k_u (x_u - y) \tag{9.9}$$

which can also be expressed in matrix form.

$$[m]\ddot{\mathbf{x}} + [c]\dot{\mathbf{x}} + [k]\mathbf{x} = \mathbf{F} \tag{9.10}$$

$$\begin{bmatrix} m_s & 0 \\ 0 & m_u \end{bmatrix}\begin{bmatrix} \ddot{x}_s \\ \ddot{x}_u \end{bmatrix} + \begin{bmatrix} c_s & -c_s \\ -c_s & c_s \end{bmatrix}\begin{bmatrix} \dot{x}_s \\ \dot{x}_u \end{bmatrix} + \begin{bmatrix} k_s & -k_s \\ -k_s & k_s + k_u \end{bmatrix}\begin{bmatrix} x_s \\ x_u \end{bmatrix} = \begin{bmatrix} 0 \\ k_u y \end{bmatrix} \tag{9.11}$$

■

**Example 126** ★ Tire damping. We may ignore tire damping to work with simpler equations of motion. Tire damping is much less than the damping of the shock absorber of the main suspension and does not affect much the vibrating behavior of vehicles.

We may add a damper $c_u$ in parallel to $k_u$, as shown in Fig. 9.1, to model possible damping in tires. However, the value of $c_u$ for tires, compared to $c_s$, is very small and hence, we may ignore $c_u$ to simplify the model. Including $c_u$ provides the following equations of motion of quarter car model,

$$m_s \ddot{x}_s = -k_s (x_s - x_u) - c_s (\dot{x}_s - \dot{x}_u) \tag{9.12}$$

$$m_u \ddot{x}_u = k_s (x_s - x_u) + c_s (\dot{x}_s - \dot{x}_u) - c_u (\dot{x}_u - \dot{y}) - k_u (x_u - y) \tag{9.13}$$

$$\begin{bmatrix} m_s & 0 \\ 0 & m_u \end{bmatrix}\begin{bmatrix} \ddot{x}_s \\ \ddot{x}_u \end{bmatrix} + \begin{bmatrix} c_s & -c_s \\ -c_s & c_s + c_u \end{bmatrix}\begin{bmatrix} \dot{x}_s \\ \dot{x}_u \end{bmatrix} + \begin{bmatrix} k_s & -k_s \\ -k_s & k_s + k_u \end{bmatrix}\begin{bmatrix} x_s \\ x_u \end{bmatrix} = \begin{bmatrix} 0 \\ c_u \end{bmatrix}\dot{y} + \begin{bmatrix} 0 \\ k_u \end{bmatrix} y \tag{9.14}$$

because:

$$K = \frac{1}{2}m_s \dot{x}_s^2 + \frac{1}{2}m_u \dot{x}_u^2 \tag{9.15}$$

$$P = \frac{1}{2} k_s (x_s - x_u)^2 + \frac{1}{2} k_u (x_u - y)^2 \tag{9.16}$$

$$D = \frac{1}{2} c_s (\dot{x}_s - \dot{x}_u)^2 + \frac{1}{2} c_u (\dot{x}_u - \dot{y})^2 \tag{9.17}$$

**Example 127** Mathematical model's limitations. Although the quarter car model is the best simple model to examine vibration of vehicles, it has some limitations. There are multiple facts that this model must be expanded to be able to discover.

The quarter car model contains no representation of the geometric effects of the full car and offers no possibility of studying longitudinal and lateral interconnections. However, it contains the most fundamental features of real vehicles and includes a proper representation of the problem of controlling wheel and wheel-body load variations.

In the quarter car model, we assume that tire is always in contact with the ground. This assumption is true at low frequency and low amplitude but might not be true at high frequency or high amplitude of road excitations. A better model must be able to include the possibility of separation between the tire and ground.

Optimal design of two $DOF$ vibration systems, including a quarter car model, is the subject of numerous investigations since the invention of the vibration absorber theory by Frahm in 1909. It seems that the first analytical investigation on the damping properties of two $DOF$ systems is due to the Dutch-American engineer, J. P. Den Hartog (1901–1989).

**Example 128** Nonlinear damping and spring. A good model for practical shock absorbers that we use in automotive has very low damping in bound and a high damping in rebound. Here is the equations of motion of quarter car model with bilinear dampers as shown in Fig. 8.8.

Considering $x_s$ and $x_u$ to indicate the displacement of the vehicle body and wheel movement, respectively, we define such damping by a Heaviside function $H(\dot{x}_s - \dot{x}_u)$ and rewrite the equation of motion of the quarter vehicle model.

$$m_s \ddot{x}_s = -k_s (x_s - x_u) - c_s (\dot{x}_s - \dot{x}_u) H (\dot{x}_s - \dot{x}_u) \tag{9.18}$$

$$m_u \ddot{x}_u = k_s (x_s - x_u) + c_s (\dot{x}_s - \dot{x}_u) H (\dot{x}_s - \dot{x}_u) - k_u (x_u - y) \tag{9.19}$$

The tire stiffness is usually very high and in order of 200,000 N/ m and close to linear. However, the main spring of the suspension, although assumed to be linear, are usually hardening. Such hardening spring make the equations as follows:

$$m_s \ddot{x}_s = -k_s (x_s - x_u)^3 - c_s (\dot{x}_s - \dot{x}_u) H (\dot{x}_s - \dot{x}_u) \tag{9.20}$$

$$m_u \ddot{x}_u = k_s (x_s - x_u)^3 + c_s (\dot{x}_s - \dot{x}_u) H (\dot{x}_s - \dot{x}_u) - k_u (x_u - y) \tag{9.21}$$

**Example 129** Nondimensionalized equations of motion. The best way to study mechanical vibrating systems is to convert their equations of motion into dimensionless form so they can be used for all similar applied systems. Here is the technics to make the quarter car model nondimensionalized.

All mechanical vibrating systems work only with three main dimensions of mass $M$, length $L$, and time $T$. To make the equations of motion dimensionless, we need to make the variables of the equations nondimensional by defining new dimensionless variables. In quarter car model, there are three variables with length dimension, $x_s, x_u, y$, and a time variable $t$.

$$m_s \ddot{x}_s + c_s (\dot{x}_s - \dot{x}_u) + k_s (x_s - x_u) = 0 \tag{9.22}$$

$$m_u \ddot{x}_u + c_s (\dot{x}_u - \dot{x}_s) + (k_u + k_s) x_u - k_s x_s = k_u y \tag{9.23}$$

Let us define a characteristic length $x_0$ and a characteristic time $t_0$, to define nondimensionalized displacement $X_s$, $X_u$, $Y$, and time $\tau$

$$X_s = \frac{x_s}{x_0} \qquad X_u = \frac{x_u}{x_0} \qquad Y = \frac{y}{x_0} \qquad \tau = \frac{t}{t_0} \tag{9.24}$$

and replace the actual variables $x_s$, $x_u$, $y$, and $t$. In vehicle vibrations, $x_0$ is usually a constant length such as static deflection of one of the masses or amplitude of displacement excitation, and $t_0$ is inverse of a frequency such as a natural frequency or an excitation frequency.

$$x_s = x_0 X_s \qquad x_u = x_0 X_u \qquad y = x_0 Y \qquad t = t_0 \tau \tag{9.25}$$

Using nondimensionalized time $\tau$, we may redefine derivatives.

$$\dot{x}_s = \frac{dx_s}{dt} = x_0 \frac{dX_s}{dt} = x_0 \frac{dX_s}{d\tau}\frac{d\tau}{dt} = \frac{x_0}{t_0} X_s' \tag{9.26}$$

$$\ddot{x}_s = \frac{d}{dt}\frac{dx_s}{dt} = \frac{x_0}{t_0}\frac{d}{dt} X_s' = \frac{x_0}{t_0}\frac{d\tau}{dt}\frac{d}{d\tau} X_s' = \frac{x_0}{t_0^2} X_s'' \tag{9.27}$$

$$X_s' = \frac{dX_s}{d\tau} \qquad X_s'' = \frac{d^2 X_s}{d\tau^2} \tag{9.28}$$

The equation of motion can now be rewritten as

$$m_s \frac{x_0}{t_0^2} X_s'' + c_s \frac{x_0}{t_0}\left(X_s' - X_u'\right) + k_s x_0 \left(X_s - X_u\right) = 0 \tag{9.29}$$

$$m_u \frac{x_0}{t_0^2} X_u'' + c_s \frac{x_0}{t_0}\left(X_u' - X_s'\right) + \left(k_u + k_s\right) x_0 X_u - k_s x_0 X_s = k_u x_0 Y \tag{9.30}$$

After dividing by $m_u x_0 / t_0^2$ we have

$$\frac{m_s}{m_u} X_s'' + \frac{c_s}{m_u} t_0 \left(X_s' - X_u'\right) + \frac{k_s}{m_u} t_0^2 \left(X_s - X_u\right) = 0 \tag{9.31}$$

$$X_u'' + \frac{c_s}{m_u} t_0 \left(X_u' - X_s'\right) + \frac{\left(k_u + k_s\right)}{m_u} t_0^2 X_u - \frac{k_s}{m_u} t_0^2 X_s = \frac{k_u}{m_u} t_0^2 Y \tag{9.32}$$

which will be converted to the following equations

$$\varepsilon X_s'' + 2\varepsilon\xi\omega_s t_0 \left(X_s' - X_u'\right) + \varepsilon\omega_s^2 t_0^2 \left(X_s - X_u\right) = 0 \tag{9.33}$$

$$X_u'' + 2\varepsilon\xi\omega_s t_0 \left(X_u' - X_s'\right) + \left(\omega_u^2 + \varepsilon\omega_s^2\right) t_0^2 X_u - \varepsilon\omega_s^2 t_0^2 X_s = \omega_u^2 t_0^2 Y \tag{9.34}$$

by using the following dimensionless parameters.

$$\varepsilon = \frac{m_s}{m_u} \qquad \omega_s = \sqrt{\frac{k_s}{m_s}} \qquad \omega_u = \sqrt{\frac{k_u}{m_u}} \tag{9.35}$$

$$\xi = \frac{c_s}{2 m_s \omega_s} \qquad \alpha = \frac{\omega_u}{\omega_s} \tag{9.36}$$

It suggests that we use $t_0 = 1/\omega_s$, then the equation will be much simpler.

$$X_s'' + 2\xi \left(X_s' - X_u'\right) + \left(X_s - X_u\right) = 0 \tag{9.37}$$

$$X_u'' + 2\varepsilon\xi \left(X_u' - X_s'\right) + \left(\alpha^2 + \varepsilon\right) X_u - \varepsilon X_s = \alpha^2 Y \tag{9.38}$$

## 9.2 Time Response

The time response of the linear quarter car model is the solution of the following equations for the absolute displacement $x_s$, $x_u$, in which $y(t)$ is the road excitation function.

$$m_s \ddot{x}_s + c_s \left(\dot{x}_s - \dot{x}_u\right) + k_s \left(x_s - x_u\right) = 0 \tag{9.39}$$

$$m_u \ddot{x}_u + c_s \left(\dot{x}_u - \dot{x}_s\right) + \left(k_u + k_s\right) x_u - k_s x_s = k_u \; y \tag{9.40}$$

The equation can be transformed to the following more practical form.

$$\ddot{x}_s + 2\xi\omega_s\left(\dot{x}_s - \dot{x}_u\right) + \omega_s^2\left(x_s - x_u\right) = 0 \tag{9.41}$$

$$\ddot{x}_u + 2\varepsilon\xi\omega_s\left(\dot{x}_u - \dot{x}_s\right) + \left(\omega_u^2 + \varepsilon\omega_s^2\right)x_u - \varepsilon\omega_s^2 x_s = \omega_u^2\, y \tag{9.42}$$

$$\varepsilon = \frac{m_s}{m_u} \qquad \omega_s = \sqrt{\frac{k_s}{m_s}} \qquad \omega_u = \sqrt{\frac{k_u}{m_u}} \tag{9.43}$$

$$\xi = \frac{c_s}{2m_s\omega_s} \qquad \alpha = \frac{\omega_u}{\omega_s} \tag{9.44}$$

***Proof*** Dividing the equations of motion (9.39)–(9.40) by $m_u$

$$\frac{m_s}{m_u}\ddot{x}_s + \frac{c_s}{m_u}\left(\dot{x}_s - \dot{x}_u\right) + \frac{k_s}{m_u}\left(x_s - x_u\right) = 0 \tag{9.45}$$

$$\ddot{x}_u + \frac{c_s}{m_u}\left(\dot{x}_u - \dot{x}_s\right) + \frac{\left(k_u + k_s\right)}{m_u}x_u - \frac{k_s}{m_u}x_s = \frac{k_u}{m_u}y \tag{9.46}$$

converts the equations to the following form, depending on four parameters, $\varepsilon, \xi, \omega_s, \omega_u$.

$$\ddot{x}_s + 2\xi\omega_s\left(\dot{x}_s - \dot{x}_u\right) + \omega_s^2\left(x_s - x_u\right) = 0 \tag{9.47}$$

$$\ddot{x}_u + 2\varepsilon\xi\omega_s\left(\dot{x}_u - \dot{x}_s\right) + \left(\omega_u^2 + \varepsilon\omega_s^2\right)x_u - \varepsilon\omega_s^2 x_s = \omega_u^2\, y \tag{9.48}$$

$$\begin{bmatrix} m_s & 0 \\ 0 & m_u \end{bmatrix}\begin{bmatrix} \ddot{x}_s \\ \ddot{x}_u \end{bmatrix} + \begin{bmatrix} 2\xi\omega_s & -2\xi\omega_s \\ -2\varepsilon\xi\omega_s & 2\varepsilon\xi\omega_s \end{bmatrix}\begin{bmatrix} \dot{x}_s \\ \dot{x}_u \end{bmatrix}$$
$$+\begin{bmatrix} \omega_s^2 & -\omega_s^2 \\ -\varepsilon\omega_s^2 & \omega_u^2 + \varepsilon\omega_s^2 \end{bmatrix}\begin{bmatrix} x_s \\ x_u \end{bmatrix} = \begin{bmatrix} 0 \\ \omega_u^2\, y \end{bmatrix} \tag{9.49}$$

The equations of motions are a set of coupled linear forced ordinary deferential equations. The best way to solve them is to transform them to a set of first order equations. Defining four new variables,

$$x_1 = x_u \qquad x_2 = x_s \qquad x_3 = \dot{x}_u \qquad x_4 = \dot{x}_s \tag{9.50}$$

we can transform them into the following set of first order equations.

$$\dot{x}_1 = x_3 \tag{9.51}$$

$$\dot{x}_2 = x_4 \tag{9.52}$$

$$\dot{x}_3 = -2\varepsilon\xi\omega_s\left(x_3 - x_4\right) - \left(\omega_u^2 + \varepsilon\omega_s^2\right)x_1 + \varepsilon\omega_s^2 x_2 + \omega_u^2\, y \tag{9.53}$$

$$\dot{x}_4 = -2\xi\omega_s\left(x_4 - x_3\right) - \omega_s^2\left(x_2 - x_1\right) \tag{9.54}$$

■

**Example 130** Step input. Let us see the reaction of a quarter car to a step road input. The difference of the natural frequency of unsprung and sprung mass will be seen in their response.

Let us assume the road displacement, $y(t) = Y$, to be a unit step displacement.

$$Y = \begin{cases} 1 & t > 0 \\ 0 & t \le 0 \end{cases} \tag{9.55}$$

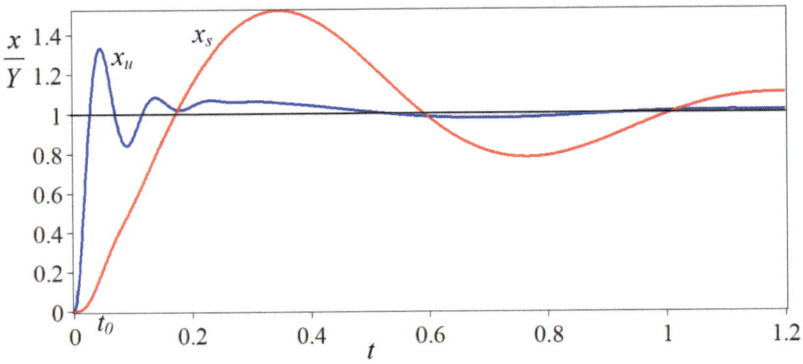


**Fig. 9.2** The reaction of $x_s$ and $x_u$ of a step road input

The equations of motion for the step road input will be

$$\ddot{x}_s + 2\xi\omega_s(\dot{x}_s - \dot{x}_u) + \omega_s^2(x_s - x_u) = 0 \tag{9.56}$$

$$\ddot{x}_u + 2\varepsilon\xi\omega_s(\dot{x}_u - \dot{x}_s) + \left(\omega_u^2 + \varepsilon\omega_s^2\right)x_u - \varepsilon\omega_s^2 x_s = \omega_u^2 Y \tag{9.57}$$

$$x_s(0) = 0 \qquad \dot{x}_s(0) = 0 \qquad x_u(0) = 0 \qquad \dot{x}_u(0) = 0 \tag{9.58}$$

Figure 9.2 illustrates the reaction of $x_s$ and $x_u$ for a vehicle with the following specifications.

$$m_s = 336\,\text{kg} \qquad m_u = 42\,\text{kg} \qquad \varepsilon = \frac{m_s}{m_u} = 8 \tag{9.59}$$

$$k_s = 22{,}000\,\text{N/m} \qquad k_u = 250{,}000\,\text{N/m} \tag{9.60}$$

$$\omega_s = \sqrt{\frac{k_s}{m_s}} = 8.09\,\text{rad/s} \qquad \omega_u = \sqrt{\frac{k_u}{m_u}} = 69\,\text{rad/s} \tag{9.61}$$

$$c = 1630\,\text{N s/m} \qquad \xi = \frac{c}{2m_s\omega_s} = 0.3 \tag{9.62}$$

The response of sprung and unsprung masses are harmonic with their associated natural frequencies. Displacement of the body shows a 50% overshoot and damps slowly in about 1.5 s. Assuming a step of $Y = 10$ cm, there will be a maximum $X_u = 5$ cm displacement of the body with respect to the ground. Displacement of the wheel indicates around 30% overshoot and damps quickly in 0.2 s. Assuming a step of $Y = 10$ cm, there will be a maximum $X_u = 3$ cm displacement of the wheel center with respect to the ground. In case the static deflection of the tire is less than 3 cm, then the tire will separate from the ground before that. Therefore, the assumption of the quarter car model in this chapter is that tire will never separate from the ground. Tire-road separation will be studied in a different chapter in this book.

**Example 131** Nonlinear damper. The transient responses of vehicles to road excitations with bilinear dampers are different than linear damper assumption.

Practical dampers that we use in automotive industries have a low damping in bound and a high damping in rebound. Let us assume the damping in bound phase is zero and investigate the behavior of quarter car mole to road excitations to such an extreme case. Considering $x_s$ and $x_u$ to indicate the displacement of the vehicle body and the wheel, respectively, we can define such damping by a Heaviside function $H(\dot{x}_s - \dot{x}_u)$ and rewrite the equation of motion of the quarter vehicle model with a step road input.

$$c = \begin{cases} c & (\dot{x}_s - \dot{x}_u) > 0 \\ 0 & (\dot{x}_s - \dot{x}_u) \le 0 \end{cases} \tag{9.63}$$

$$c = c\,H(\dot{x}_s - \dot{x}_u) \tag{9.64}$$

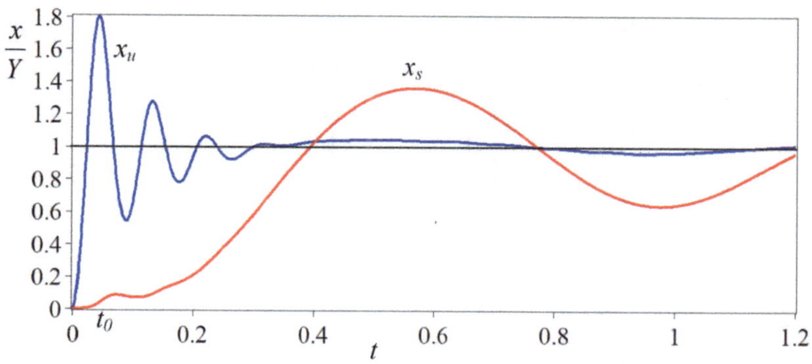


**Fig. 9.3** The reaction of quarter car model, $x_s$ and $x_u$, with zero damping in bound to a step road input

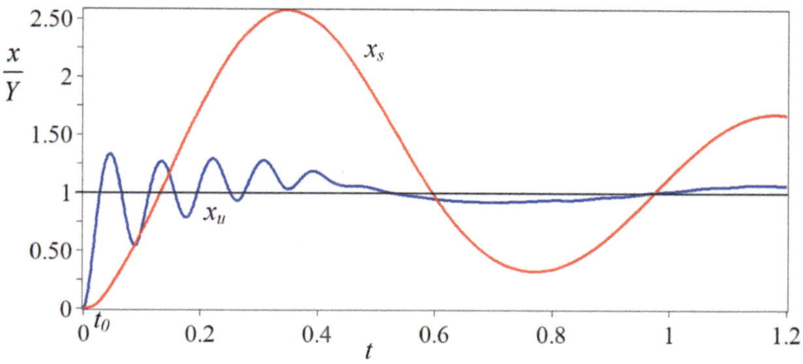


**Fig. 9.4** The reaction of quarter car model, $x_s$ and $x_u$, with zero damping in rebound to a step road input

$$H(t-t_0)=\begin{cases}1 & t>t_0\\ 0 & t<t_0\end{cases} \tag{9.65}$$

$$\ddot{x}_s+2\xi\omega_s\left(\dot{x}_s-\dot{x}_u\right)H\left(\dot{x}_s-\dot{x}_u\right)+\omega_s^2\left(x_s-x_u\right)=0 \tag{9.66}$$

$$\ddot{x}_u+2\varepsilon\xi\omega_s\left(\dot{x}_u-\dot{x}_s\right)H\left(\dot{x}_s-\dot{x}_u\right)+\left(\omega_u^2+\varepsilon\omega_s^2\right)x_u-\varepsilon\omega_s^2x_s=\omega_u^2\,Y \tag{9.67}$$

$$x_s(0)=0 \qquad \dot{x}_s(0)=0 \qquad x_u(0)=0 \qquad \dot{x}_u(0)=0$$

$$Y=\begin{cases}1 & t>0\\ 0 & t\le 0\end{cases} \tag{9.68}$$

To study the effect of bilinear damping on transient vibrations of vehicles, let us consider a quarter vehicle model with the following specifications.

$$\omega_s=8.09\,\mathrm{rad/s} \qquad \omega_u=69\,\mathrm{rad/s} \qquad \xi=\frac{c}{2m_s\omega_s}=0.3 \tag{9.69}$$

The response of the vehicle to step input is shown in Fig. 9.3. Comparing the response of the quarter car model with a bilinear damping, with the response to a linear damping in Fig. 9.2 shows that having zero damping in bound, makes the body to vibrate at lower frequency and for longer time. It also makes the unsprung mass to show higher overshoot with more chance of tire-road separation.

To investigate the engineering point of having dampers with zero bound resistance, let us compare it with Fig. 9.4 that shows the response of the quarter car to a step input with zero damping in rebound. Such an unpractical shock absorber makes the body to show a huge overshoot while decreases the overshoot of the wheel. Therefore, any other type of bilinear damper will show a response between the one in Figs. 9.3 and 9.4.

**Example 132** Bump transient base excitation. Sine square is a good model for speed bumps on the roads to make vehicles to drive slowly. Here is the response of quarter cars to a bump.

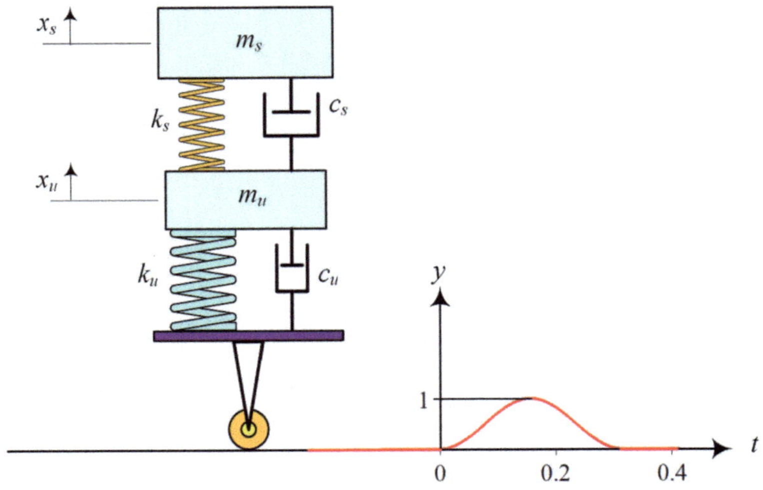


**Fig. 9.5** A quarter car model going over a bump

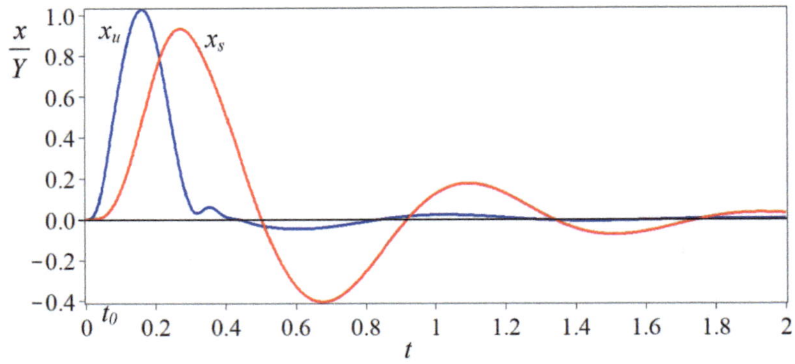


**Fig. 9.6** The response of a quarter vehicle to a bump on the road

Figure 9.5 illustrates a quarter car model and a sine square transient road excitation for a vehicle that is moving with speed $v$ going over a bump with length $w$ and height $h = 1$.

$$y = \begin{cases} Y \sin^2 \omega t & 0 \le t \le t_0 \\ 0 & t_0 < t \end{cases} \qquad \omega = \frac{v\pi}{w} \qquad t_0 = \frac{w}{v} \tag{9.70}$$

The equations of motion of the vehicle in absolute displacements are

$$\ddot{x}_s + 2\xi\omega_s (\dot{x}_s - \dot{x}_u) + \omega_s^2 (x_s - x_u) = 0 \tag{9.71}$$

$$\ddot{x}_u + 2\varepsilon\xi\omega_s (\dot{x}_u - \dot{x}_s) + \left(\omega_u^2 + \varepsilon\omega_s^2\right) x_u - \varepsilon\omega_s^2 x_s = \omega_u^2 \, Y \sin^2 \omega t \tag{9.72}$$

$$x_s(0) = 0 \qquad \dot{x}_s(0) = 0 \qquad x_u(0) = 0 \qquad \dot{x}_u(0) = 0$$

Figure 9.6 illustrates the response of the vehicle with the following characteristics.

$$\omega_s = 8.09 \,\text{rad/s} \qquad \omega_u = 69 \,\text{rad/s} \qquad \xi = \frac{c}{2m_s\omega_s} = 0.3 \tag{9.73}$$

**Example 133** Nonlinear damper and bump on the road. The ideal case of practical dampers has zero resistance in bound phase. Such dampers usually reduce the body displacement.

Assume the damper of a quarter car is bilinear with zero damping in bound and $c$ damping in rebound.

$$c = c\, H\left(\dot{x}_u - \dot{x}_s\right) \tag{9.74}$$

The equations of motion of a quarter car that is moving with speed $v$ going over a bump with length $w$ and height $h = 1$ are

$$y = \begin{cases} Y \sin^2 \omega t & 0 \le t \le t_0 \\ 0 & t_0 < t \end{cases} \qquad \omega = \frac{v\pi}{w} \qquad t_0 = \frac{w}{v} \tag{9.75}$$

$$\ddot{x}_s + 2\xi\omega_s\left(\dot{x}_s - \dot{x}_u\right) H\left(\dot{x}_s - \dot{x}_u\right) + \omega_s^2\left(x_s - x_u\right) = 0 \tag{9.76}$$

$$\ddot{x}_u + 2\varepsilon\xi\omega_s\left(\dot{x}_u - \dot{x}_s\right) H\left(\dot{x}_s - \dot{x}_u\right) + \left(\omega_u^2 + \varepsilon\omega_s^2\right) x_u - \varepsilon\omega_s^2 x_s = \omega_u^2\, y \tag{9.77}$$

$$x_s(0) = 0 \qquad \dot{x}_s(0) = 0 \qquad x_u(0) = 0 \qquad \dot{x}_u(0) = 0$$

Example 133 illustrates the response of the vehicle to the bump on the road for the following characteristics.

$$\omega_s = 8.09\,\mathrm{rad/\,s} \qquad \omega_u = 69\,\mathrm{rad/\,s} \qquad \xi = \frac{c}{2m_s\omega_s} = 0.3 \tag{9.78}$$

Such damping reduces the body displacement significantly.

## 9.3 Frequency Response

To determine the frequency response of the quarter car model, we consider a harmonic base excitation,

$$y = Y \sin \omega t \tag{9.79}$$

and search for a harmonic solution in the form of

$$x_s = A_1 \sin \omega t + B_1 \cos \omega t = X_s \sin\left(\omega t - \varphi_s\right) \tag{9.80}$$

$$x_u = A_2 \sin \omega t + B_2 \cos \omega t = X_u \sin\left(\omega t - \varphi_u\right) \tag{9.81}$$

$$z = x_s - x_u = A_3 \sin \omega t + B_3 \cos \omega t = Z \sin\left(\omega t - \varphi_z\right) \tag{9.82}$$

where $X_s$, $X_u$, and $Z$ are the amplitudes of $x_s$, $x_u$, $x_s - x_u$.

By introducing the following dimensionless characters,

$$\varepsilon = \frac{m_s}{m_u} \tag{9.83}$$

$$\omega_s = \sqrt{\frac{k_s}{m_s}} \tag{9.84}$$

$$\omega_u = \sqrt{\frac{k_u}{m_u}} \tag{9.85}$$

$$\alpha = \frac{\omega_s}{\omega_u} \tag{9.86}$$

$$r = \frac{\omega}{\omega_s} \tag{9.87}$$

$$\xi = \frac{c_s}{2m_s\omega_s} \tag{9.88}$$

we search for the absolute and relative frequency responses:

$$\mu = \left|\frac{X_s}{Y}\right| \tag{9.89}$$

$$\tau = \left|\frac{X_u}{Y}\right| \tag{9.90}$$

$$\eta = \left|\frac{Z}{Y}\right| \tag{9.91}$$

and obtain the following functions:

$$\mu^2 = \frac{\alpha^2\left(4\xi^2 r^2 + \alpha^2\right)}{Z_1^2 + Z_2^2} \tag{9.92}$$

$$\tau^2 = \frac{\alpha^2\left(4\xi^2 r^2 + \alpha^2\right) + r^2\left(r^2 - 2\alpha^2\right)}{Z_1^2 + Z_2^2} \tag{9.93}$$

$$\eta^2 = \frac{r^4}{Z_1^2 + Z_2^2} \tag{9.94}$$

$$Z_1 = \varepsilon r^2 \alpha^2 + \left(\alpha^2 - r^2\right)\left(r^2 - 1\right) \tag{9.95}$$

$$Z_2 = 2\xi r\alpha\left(\varepsilon r^2 + r^2 - 1\right) \tag{9.96}$$

The absolute acceleration of sprung mass and unsprung mass may be defined by the following equations:

$$u = \left|\frac{\ddot{X}_s}{Y\omega_u^2}\right| = r^2\mu \tag{9.97}$$

$$v = \left|\frac{\ddot{X}_u}{Y\omega_u^2}\right| = r^2\tau \tag{9.98}$$

***Proof*** To find the frequency responses, let us apply a harmonic excitation

$$y = Y\sin\omega t \tag{9.99}$$

and assume that the solutions are harmonic functions with unknown coefficients.

$$x_u = A_1 \sin\omega t + B_1 \cos\omega t \tag{9.100}$$

$$x_s = A_2 \sin\omega t + B_2 \cos\omega t \tag{9.101}$$

Substituting the solutions in the equations of motion (9.1)–(9.2) and collecting the coefficients of $\sin\omega t$ and $\cos\omega t$ in both equations provides the following set of algebraic equations for $A_1$, $B_1$, $A_2$, $B_2$:

$$[A]\begin{bmatrix} A_1 \\ A_2 \\ B_1 \\ B_2 \end{bmatrix} = \begin{bmatrix} 0 \\ 0 \\ k_u Y \\ 0 \end{bmatrix} \tag{9.102}$$

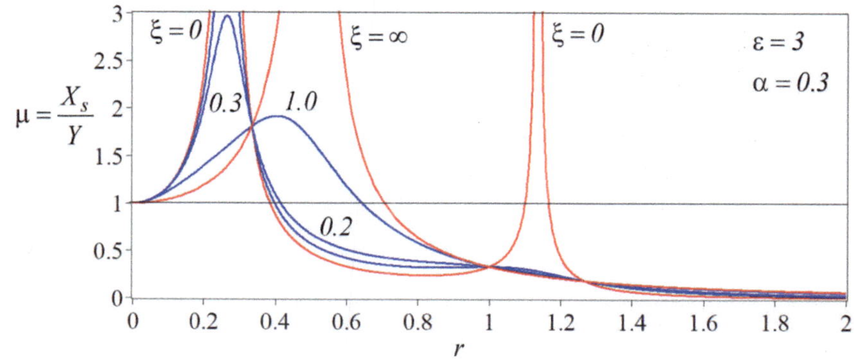


**Fig. 9.7** A sample of the unsprung mass displacement frequency response, $\tau = \left|\frac{X_u}{Y}\right|$

where $[A]$ is the coefficient matrix.

$$[A] = \begin{bmatrix} k_s - m_s\omega^2 & -k_s & -c_s\omega & c_s\omega \\ c_s\omega & -c_s\omega & k_s - m_s\omega^2 & -k_s \\ -k_s & k_s + k_u - m_u\omega^2 & c_s\omega & -c_s\omega \\ -c_s\omega & c_s\omega & -k_s & k_s + k_u - m_u\omega^2 \end{bmatrix} \tag{9.103}$$

The unknowns may be found by matrix inversion

$$\begin{bmatrix} A_1 \\ A_2 \\ B_1 \\ B_2 \end{bmatrix} = [A]^{-1} \begin{bmatrix} 0 \\ 0 \\ k_u Y \\ 0 \end{bmatrix} \tag{9.104}$$

and therefore, the amplitudes $X_s$ and $X_u$ will be found.

$$X_s^2 = A_1^2 + B_1^2 = \frac{k_u}{k_s} \frac{\left(\omega^2 c_s^2 + k_s^2\right)}{Z_5^2 + Z_6^2} Y^2 \tag{9.105}$$

$$X_u^2 = A_2^2 + B_2^2 = \frac{k_u}{k_s} \frac{\left(\omega^4 m_s^2 + \omega^2 c_s^2 - 2\omega^2 k_s m_s + k_s^2\right)}{Z_5^2 + Z_6^2} Y^2 \tag{9.106}$$

$$Z_5 = -\left(\omega^2 \left(k_s m_s + k_s m_u + k_u m_s\right) - k_s k_u - \omega^4 m_s m_u\right) \tag{9.107}$$

$$Z_6 = -\left(\omega^3 \left(c_s m_s + c_s m_u\right) - \omega c_s k_u\right) \tag{9.108}$$

Having $X_s$ and $X_u$ helps us to calculate $z$ and its amplitude $Z$.

$$\begin{aligned} z = x_s - x_u &= (A_1 - A_2) \sin \omega t + (B_1 - B_2) \cos \omega t \\ &= A_3 \sin \omega t + B_3 \cos \omega t = Z \sin\left(\omega t - \varphi_z\right) \end{aligned} \tag{9.109}$$

$$Z^2 = A_3^2 + B_3^2 = \frac{k_u}{k_s} \frac{\omega^4 m_s^2}{Z_5^2 + Z_6^2} Y^2 \tag{9.110}$$

Taking derivative from $X_s$ and $X_u$ provides the acceleration frequency responses $u$ and $v$ for the unsprung and sprung masses. Equations (9.97)–(9.98) express $u$ and $v$.

Using the definitions (9.83)–(9.88), we may transform Eqs. (9.105), (9.106), (9.110) to (9.92), (9.93), (9.94). Figures 9.7, 9.8, 9.9 are samples of the frequency responses $\mu$, $\tau$, and $\eta$ for $\varepsilon = 3$, and $\alpha = 0.2$. ■

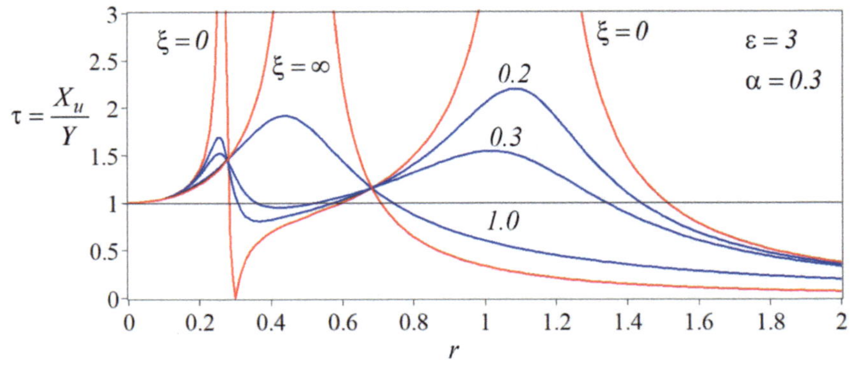


**Fig. 9.8** A sample for the sprung mass displacement frequency response, $\mu = \left|\frac{X_s}{Y}\right|$

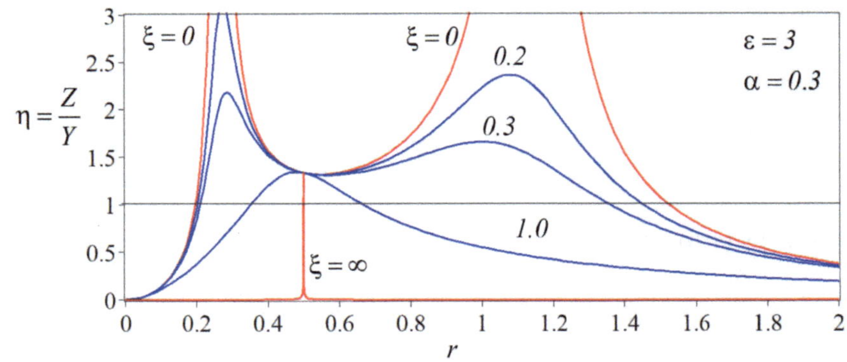


**Fig. 9.9** A sample of the relative displacement frequency response, $\eta = \left|\frac{Z}{Y}\right|$

**Table 9.1** Average value of quarter car parameters

| Parameter | Average | Minimum | Maximum |
|---|---|---|---|
| $\varepsilon = \frac{m_s}{m_u}$ | $5 - 10$ | 2 | 20 |
| $\omega_s = \sqrt{\frac{k_s}{m_s}}$ | $6\,\text{rad/s} \approx 1\,\text{Hz}$ | $1\,\text{rad/s} \approx 0.2\,\text{Hz}$ | $20\,\text{rad/s} \approx 3\,\text{Hz}$ |
| $\omega_u = \sqrt{\frac{k_u}{m_u}}$ | $60\,\text{rad/s} \approx 10\,\text{Hz}$ | $20\,\text{rad/s} \approx 3\,\text{Hz}$ | $120\,\text{rad/s} \approx 20\,\text{Hz}$ |
| $r = \frac{\omega}{\omega_s}$ | $0 - 20$ | 0 | 200 |
| $\alpha = \frac{\omega_s}{\omega_u}$ | 0.1 | 0.01 | 1 |
| $\xi = \frac{c_s}{2m_s\omega_s}$ | 0.5 | 0 | 2 |
| $r_s$ | 1 | 1 | 1 |
| $r_u$ | $1/\alpha$ | $1/\alpha$ | $1/\alpha$ |

**Example 134** Average value of parameters for street cars. It is good to have the numerical range of parameters of the quarter car model for street passenger vehicles as a reference.

Equations (9.92)–(9.94) indicate that the frequency responses $\mu$, $\tau$, and $\eta$ are functions of four parameters: mass ratio $\varepsilon$, damping ratio $\xi$, natural frequency ratio $\alpha$, and excitation frequency ratio $r$. The average, minimum, and maximum of practical values of the characteristic parameters in quarter car model are indicated in Table 9.1.

For quarter car models, usually $m_s > m_u$, and therefore, $\varepsilon > 1$. Typical mass ratio, $\varepsilon$, for vehicles lies in the range 5 to 10, with small cars closer to 5 and large cars near 10. For a street vehicle, the order of magnitude of the stiffness is $k_u > k_s$, so $\omega_u > \omega_s$, and hence, $\alpha < 1$ and $r_u > r_s$.

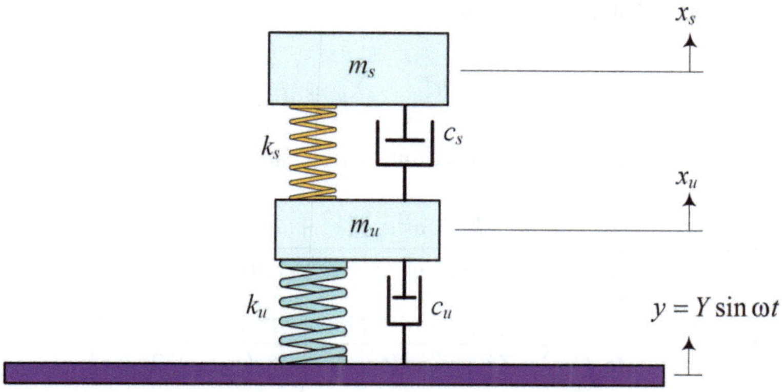


**Fig. 9.10** The quarter car model with harmonic road excitation

**Example 135** Damped tire. It may be interested to have the frequency response of quarter car model including damping in tires. Here is the development of frequency response of the full model shown in Fig. 9.10.

The equations of motion of the system are

$$m_s \ddot{x}_s - c_s (\dot{x}_u - \dot{x}_s) - k_s (x_u - x_s) = 0 \tag{9.111}$$

$$m_u \ddot{x}_u + c_u \dot{x}_u + c_s (\dot{x}_u - \dot{x}_s) + k_u x_u + k_s (x_u - x_s) = k_u y + c_u \dot{y} \tag{9.112}$$

$$y = Y \sin \omega t \tag{9.113}$$

$$\begin{bmatrix} m_s & 0 \\ 0 & m_u \end{bmatrix} \begin{bmatrix} \ddot{x}_s \\ \ddot{x}_u \end{bmatrix} + \begin{bmatrix} c_u & -c_u \\ -c_u & c_s + c_u \end{bmatrix} \begin{bmatrix} \dot{x}_s \\ \dot{x}_u \end{bmatrix} + \begin{bmatrix} k_s & -k_u \\ -k_u & k_s + k_u \end{bmatrix} \begin{bmatrix} x_s \\ x_u \end{bmatrix} = \begin{bmatrix} 0 \\ k_u y + c_u \dot{y} \end{bmatrix} \tag{9.114}$$

$$y = Y \sin \omega t \tag{9.115}$$

Substituting a set of harmonic solutions,

$$x_u = A_1 \sin \omega t + B_1 \cos \omega t \tag{9.116}$$

$$x_s = A_2 \sin \omega t + B_2 \cos \omega t \tag{9.117}$$

and balancing the coefficients of $\sin \omega t$ and $\cos \omega t$ on both sides of the equations provides us with a set of algebraic equations to calculate the coefficients $A_1$, $B_1$, $A_2$, $B_2$.

$$\begin{bmatrix} a_{11} & -k_s & -\omega (c_u + c_s) & \omega c_s \\ -k_s & a_{22} & \omega c_s & -\omega c_s \\ \omega (c_u + c_s) & -\omega c_s & a_{33} & -k_s \\ -\omega c_s & \omega c_s & -k_s & a_{44} \end{bmatrix} \begin{bmatrix} A_1 \\ A_2 \\ B_1 \\ B_2 \end{bmatrix} = \begin{bmatrix} k_u Y \\ 0 \\ c_u \omega Y \\ 0 \end{bmatrix} \tag{9.118}$$

$$a_{11} = a_{33} = k_s + k_u - m_u \omega^2 \tag{9.119}$$

$$a_{22} = a_{44} = k_s - m_s \omega^2 \tag{9.120}$$

The steady-state amplitudes $X_s$ and $X_u$ of $m_s$ and $m_u$ are

$$X_s = \sqrt{A_2^2 + B_2^2} \qquad X_u = \sqrt{A_1^2 + B_1^2} \tag{9.121}$$

and the steady-state amplitudes $X_s$ and $X_u$ for vibrations of $m_s$ and $m_u$ are

$$\left(\frac{X_s}{Y}\right)^2 = \frac{\left(k_u^2 + \omega^2 c_u^2\right)\left(k_s^2 + \omega^2 c_s^2\right)}{Z_7^2 + \omega^2 c_s^2 Z_8^2 + c_u^2 \omega^2 Z_9^2 + \omega^4 c_u c_s \left(c_u c_s + 2\omega^2 m_s^2\right)} \tag{9.122}$$

$$\left(\frac{X_u}{Y}\right)^2 = \frac{\left(k_u^2 + \omega^2 c_u^2\right)\left(\left(k_s^2 + \omega^2 c_s^2\right) + \omega^2 m_s \left(\omega^2 m_s - 2k_s\right)\right)}{Z_7^2 + \omega^2 c_s^2 Z_8^2 + c_u^2 \omega^2 Z_9^2 + \omega^4 c_u c_s \left(c_u c_s + 2\omega^2 m_s^2\right)} \tag{9.123}$$

$$Z_7 = \left(k_u - \omega^2 m_u\right)\left(k_s - \omega^2 m_s\right) - \omega^2 m_s k_s \tag{9.124}$$

$$Z_8 = k_u - \omega^2 m_u - \omega^2 m_s \tag{9.125}$$

$$Z_9 = \left(k_s - \omega^2 m_s\right) \tag{9.126}$$

Introducing the following parameters

$$\varepsilon = \frac{m_s}{m_u} \qquad \xi_u = \frac{c_u}{2m_u\omega_u} \qquad \xi_s = \frac{c_s}{2m_s\omega_s} \tag{9.127}$$

$$\omega_u = \sqrt{\frac{k_u}{m_u}} \qquad \omega_s = \sqrt{\frac{k_s}{m_s}} \qquad \alpha = \frac{\omega_s}{\omega_u} \tag{9.128}$$

$$\mu = \frac{X_u}{Y} \qquad \tau = \frac{X_s}{Y} \qquad r = \frac{\omega}{\omega_u} \tag{9.129}$$

we can transform the frequency responses (9.105) and (9.106) to the following nondimensional forms.

$$\mu^2 = \left(\frac{X_s}{Y}\right)^2 = \frac{\alpha^2\left(4\xi_u^2 r^2 + 1\right)\left(4\xi_s^2 r^2 + \alpha^2\right)}{8\varepsilon\xi_u\xi_s\alpha r^6 + Z_1^2 + Z_2^2 + Z_3} \tag{9.130}$$

$$\tau^2 = \left(\frac{X_u}{Y}\right)^2 = \frac{\left(4\xi_u^2 r^2 + 1\right)\left(\alpha^2\left(4\xi_s^2 r^2 + \alpha^2\right) + Z_4\right)}{8\varepsilon\xi_u\xi_s\alpha r^6 + Z_1^2 + Z_2^2 + Z_3} \tag{9.131}$$

$$Z_1 = \varepsilon r^2\alpha^2 + \left(\alpha^2 - r^2\right)\left(r^2 - 1\right) \tag{9.132}$$

$$Z_2 = 2\xi_s r\alpha\left(\varepsilon r^2 + r^2 - 1\right) \tag{9.133}$$

$$Z_3 = 4\xi_u^2 r^2\left(4\xi_s^2 r^2\alpha^2 + \left(\alpha^2 - r^2\right)^2\right) \tag{9.134}$$

$$Z_4 = r^2\left(r^2 - 2\alpha^2\right) \tag{9.135}$$

The relative displacement frequency response will be calculated by having $x_u$ and $x_s$.

$$\begin{aligned} z = x_s - x_u &= (A_1 - A_2)\sin\omega t + (B_1 - B_2)\cos\omega t \\ &= A_3\sin\omega t + B_3\cos\omega t = Z\sin\left(\omega t - \varphi_z\right) \end{aligned} \tag{9.136}$$

$$Z = \sqrt{(A_1 - A_2)^2 + (B_1 - B_2)^2} \tag{9.137}$$

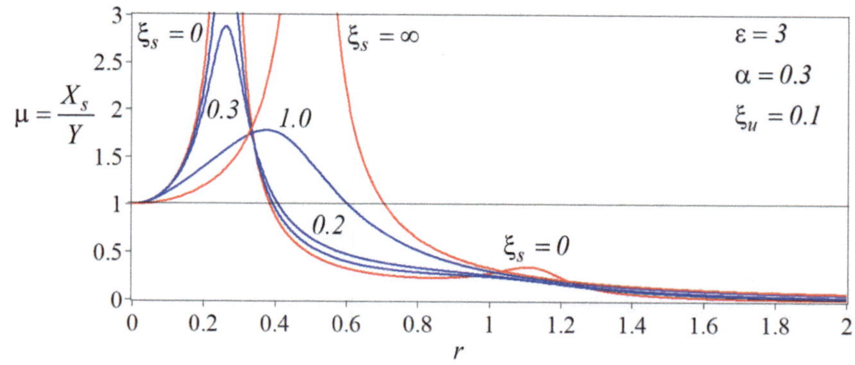


**Fig. 9.11** A sample of the unsprung mass displacement frequency response, $\tau$, with damped tire

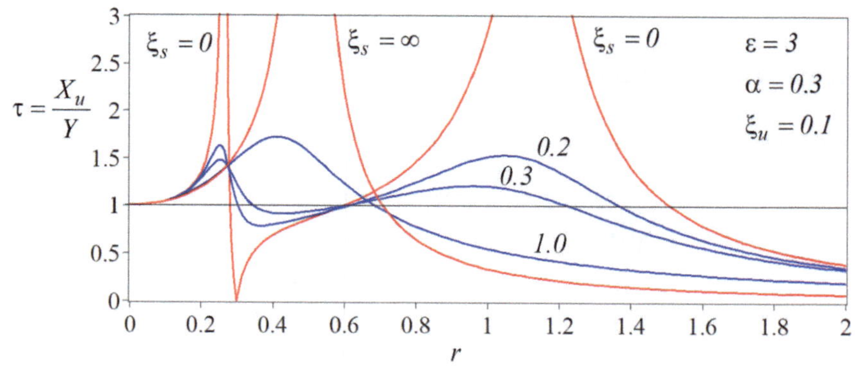


**Fig. 9.12** A sample for the sprung mass displacement frequency response, $\mu$, with damped tire

$$\eta^2 = \left(\frac{Z}{Y}\right)^2 = \frac{r^4\left(4\xi_u^2 r^2 + 1\right)}{8\varepsilon\xi_u\xi_s\alpha r^6 + Z_1^2 + Z_2^2 + Z_3} \tag{9.138}$$

The ratio of the amplitudes of oscillations of $m_u$ and $m_s$ is

$$\frac{X_u^2}{X_s^2} = 1 + \frac{\omega^2 m_s\left(\omega^2 m_s - 2k_s\right)}{\left(k_s^2 + \omega^2 c_s^2\right)} \tag{9.139}$$

$$\frac{\tau^2}{\mu^2} = 1 + \frac{r^2\left(r^2 - 2\alpha^2\right)}{\alpha^2\left(4\xi_s^2 r^2 + \alpha^2\right)} \tag{9.140}$$

The tire parameters $m_u$, $k_u$, $c_u$ do not appear in the amplitude ratio $X_u/X_s$. Therefore, the values of unsprung mass $m_u$, $k_u$, $c_u$ do not affect the ratio of $X_u/X_s$. In the nondimensional form, the ratio of $X_u/X_s$ is a function of $\alpha$, $\xi_s$, $r$, and not a function of $\varepsilon$, $\xi_u$.

The damping ratios of $\xi_u$ and $\xi_s$ are, respectively, proportional to $c_u$ and $c_s$, and they are defined as the damping ratios of the single $DOF$ subsystems. The parameter $\alpha = \omega_s/\omega_u = \sqrt{k_s/(\varepsilon k_u)}$ is the stiffness ratio of the sprung mass to unsprung mass. The frequencies $\omega_u$ and $\omega_s$ are the natural frequencies of the sprung and unsprung subsystems. These are not the natural frequencies of the whole system. The frequency responses $\mu$ and $\tau$ indicate the dynamic amplitude of $m_s$ and $m_u$ compared to the amplitude of the harmonic road excitation. Figures 9.11, 9.12, and 9.13 depict samples for frequency responses of quarter car with some tire damping.

In the quarter car system, we assume that the excitation comes from an ideal source such that none of the characteristics or reactions of the system affects the excitation. The tire is not separable from the road and hence, the source of excitation is assumed to be able to provide infinite energy and power.

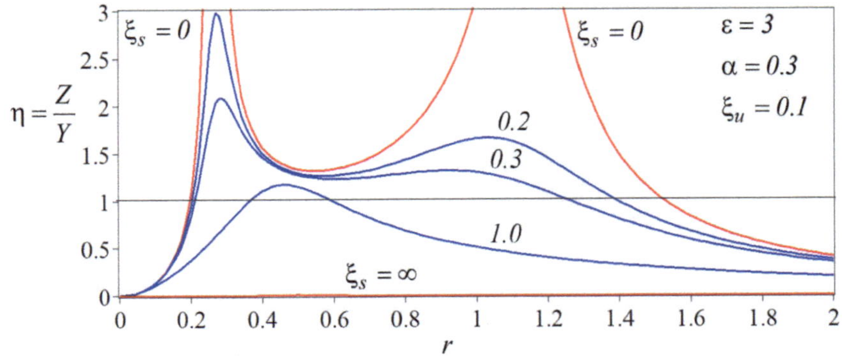


**Fig. 9.13** A sample of the relative displacement frequency response, $\eta$, with damped tire

## 9.4 Natural and Invariant Frequencies

The quarter car system is a two $DOF$ system and therefore it has two natural frequencies $r_{n_1}, r_{n_2}$:

$$r_{n_1} = \frac{\sqrt{2}}{2}\sqrt{\alpha^2 (1+\varepsilon) + 1 - \sqrt{\left(\alpha^2 (1+\varepsilon) + 1\right)^2 - 4\alpha^2}} \tag{9.141}$$

$$r_{n_2} = \frac{\sqrt{2}}{2}\sqrt{\alpha^2 (1+\varepsilon) + 1 + \sqrt{\left(\alpha^2 (1+\varepsilon) + 1\right)^2 - 4\alpha^2}} \tag{9.142}$$

The family of response curves for the frequency response of the sprung mass, $\mu$, are obtained by keeping $\varepsilon$ and $\alpha$ constant, and varying $\xi$. This family has four nodal points, which are at frequencies $r_1, r_2, r_3, r_4$, and at $\mu_1, \mu_2, \mu_3, \mu_4$.

$$\begin{cases} r_1 = 0 & \mu_1 = 1 \\ r_3 = 1 & \mu_3 = \dfrac{1}{\varepsilon} \\ r_2 & \mu_2 = \dfrac{1}{(1+\varepsilon)\, r_2^2 - 1} \\ r_4 & \mu_4 = \dfrac{1}{(1+\varepsilon)\, r_4^2 - 1} \end{cases} \tag{9.143}$$

$$r_2 = \frac{\sqrt{2}}{2}\sqrt{\alpha^2 (1+\varepsilon) + 1 - \sqrt{\left(2\alpha^2 (1+\varepsilon) + 1\right)^2 - 8\alpha^2}} \tag{9.144}$$

$$r_4 = \frac{\sqrt{2}}{2}\sqrt{\alpha^2 (1+\varepsilon) + 1 + \sqrt{\left(2\alpha^2 (1+\varepsilon) + 1\right)^2 - 8\alpha^2}} \tag{9.145}$$

The corresponding transmissivities at $r_2$ and $r_4$ are

$$\mu_2 = \left| \frac{1}{1 - (1+\varepsilon)\, r_2^2} \right| \tag{9.146}$$

$$\mu_4 = \left| \frac{1}{1 - (1+\varepsilon)\, r_4^2} \right| \tag{9.147}$$

The frequencies $r_1, r_2, r_3$ are called *invariant frequencies*, and their corresponding amplitudes are called *invariant amplitudes* because they are not dependent on $\xi$. However, they are dependent on the values of $\varepsilon$ and $\alpha$. The order of magnitude of the natural and invariant frequencies are

$$r_1\,(=0) < r_{n_1} < r_2 < r_3\,(=1) < r_{n_2} < r_4 \tag{9.148}$$

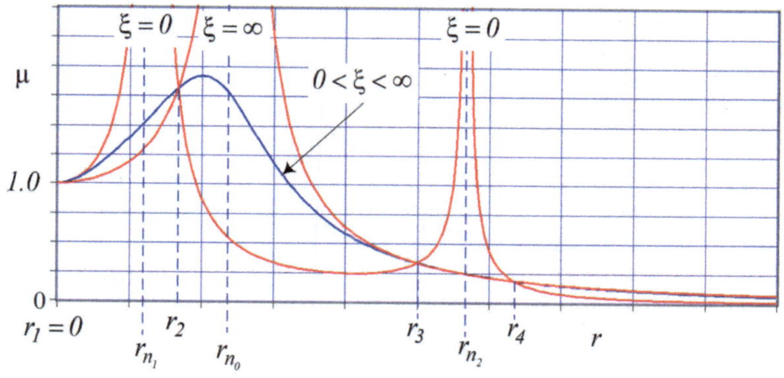


**Fig. 9.14** Schematic illustration of the amplitude $\mu$ versus excitation frequency ratio $r$

The curves for $\mu$ have no other common points except $r_1$, $r_2$, $r_3$. The order of frequencies along with the order of corresponding amplitudes can be used to predict the shape of the frequency response curves of the sprung mass $\mu$. Figure 9.14 shows schematically the shape of the amplitude $\mu$.

***Proof*** The natural or resonant frequencies of a vibrating system are at positions where the amplitude goes to infinity when damping is zero. Hence, the natural frequencies would be the roots of the denominator of the $\mu$ function for $\xi = 0$.

$$\begin{aligned} g\left(r^2\right) &= \left(\alpha^2 - r^2\right)\left(1 - r^2\right) - \varepsilon r^2 \alpha^2 \\ &= r^4 - \left(\alpha^2 + \alpha^2 \varepsilon + 1\right) r^2 + \alpha^2 = 0 \end{aligned} \tag{9.149}$$

The solution of this equation is the natural frequencies given in Eqs. (9.141) and (9.142).

The invariant frequencies are independent of $\xi$, so they can be found by intersecting the $\mu$ curves for $\xi = 0$ and $\xi = \infty$.

$$\lim_{\xi \to 0} \mu^2 = \pm \frac{\alpha^4}{\left(\varepsilon r^2 \alpha^2 + \left(\alpha^2 - r^2\right)\left(r^2 - 1\right)\right)^2} \tag{9.150}$$

$$\lim_{\xi \to \infty} \mu^2 = \pm \frac{1}{\left(\varepsilon r^2 + r^2 - 1\right)^2} \tag{9.151}$$

Therefore, the invariant frequencies, $r_i$, can be determined by solving the following equation:

$$\lim_{\xi \to 0} \mu^2 = \pm \lim_{\xi \to \infty} \mu^2 \tag{9.152}$$

$$\varepsilon r^2 \alpha^2 + \left(\alpha^2 - r^2\right)\left(r^2 - 1\right) = \pm \left(\varepsilon r^2 + r^2 - 1\right) \alpha^2 \tag{9.153}$$

Using the (+) sign, we find $r_1$ and $r_3$ with their corresponding transmissivities $\mu_1$, and $\mu_3$,

$$r_1 = 0 \qquad \mu_1 = 1 \tag{9.154}$$

$$r_3 = 1 \qquad \mu_3 = \frac{1}{\varepsilon} \tag{9.155}$$

and, with the (−) sign, we find the following equation for $r_2$ and $r_4$:

$$-r^4 + \left(2\alpha^2 \varepsilon + 2\alpha^2 + 1\right) r^2 - 2\alpha^2 = 0 \tag{9.156}$$

Equation (9.156) has two real positive roots, $r_2$ and $r_4$,

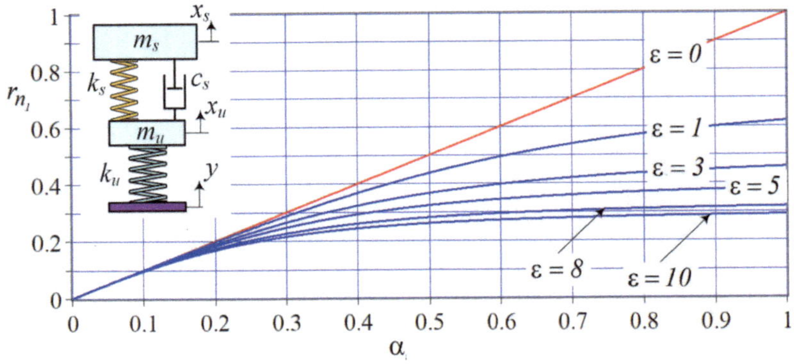


**Fig. 9.15** The natural frequency $r_{n_1}$ as a function of $\varepsilon$ and $\alpha$

$$r_2 = \frac{\sqrt{2}}{2}\sqrt{\alpha^2(1+\varepsilon)+1-\sqrt{\left(2\alpha^2(1+\varepsilon)+1\right)^2-8\alpha^2}} \tag{9.157}$$

$$r_4 = \frac{\sqrt{2}}{2}\sqrt{\alpha^2(1+\varepsilon)+1+\sqrt{\left(2\alpha^2(1+\varepsilon)+1\right)^2-8\alpha^2}} \tag{9.158}$$

with the following relative order of magnitude:

$$r_1\,(=0) < r_2 < r_3\,(=1) < r_4 \tag{9.159}$$

The corresponding amplitudes at $r_2$ and $r_4$ can be found by substituting Eqs. (9.157) and (9.158) in (9.151) as given by (9.146) and (9.147).

It can be checked that

$$\mu_2 > 1 \tag{9.160}$$

$$\mu_4 < 1 \tag{9.161}$$

Using Eq. (9.149), we can evaluate $g\left(r_2^2\right)$, $g\left(r_4^2\right)$, and $g\left(r_3^2\right)$.

$$g\left(r_2^2\right) = (1+\varepsilon)\,r_2^2 - 1 < 0 \tag{9.162}$$

$$g\left(r_4^2\right) = (1+\varepsilon)\,r_4^2 - 1 > 0 \tag{9.163}$$

$$g\left(r_3^2\right) = g\left(\frac{1}{\varepsilon^2}\right) \tag{9.164}$$

therefore, the two positive roots of Eq. (9.149), $r_{n_1}$ and $r_{n_2}\left(> \sqrt{2} > r_2\right)$, have the order of magnitudes as:

$$r_1\,(=0) < r_{n_1} < r_2 < r_3\,(=1) < r_{n_2} < r_4 \tag{9.165}$$

■

**Example 136** Natural frequency variation. Natural frequencies are the first information of a vibrating system we must determine. They show the resonant zones that the excitation frequency must avoid to not damage the system. Here is to imagine how the natural frequencies of the quarter car model depend on the system's parameters.

The natural frequencies $r_{n_1}$ and $r_{n_2}$, as given in Eqs. (9.141) and (9.142), are functions of $\varepsilon$ and $\alpha$. Figures 9.15 and 9.16 illustrate the design charts of variation of the natural frequencies, and Example 136 illustrates $r_{n_2}/r_{n_1}$.

$$r_{n_1} = \frac{\sqrt{2}}{2}\sqrt{\alpha^2(1+\varepsilon)+1-\sqrt{\left(\alpha^2(1+\varepsilon)+1\right)^2-4\alpha^2}} \tag{9.166}$$

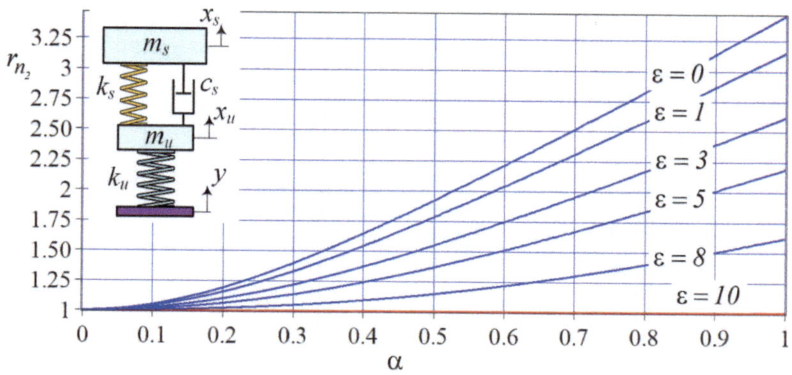


**Fig. 9.16** The natural frequency $r_{n_2}$ as a function of $\varepsilon$ and $\alpha$

$$r_{n_2} = \frac{\sqrt{2}}{2}\sqrt{\alpha^2(1+\varepsilon)+1+\sqrt{\left(\alpha^2(1+\varepsilon)+1\right)^2-4\alpha^2}} \tag{9.167}$$

The first natural frequency $r_{n_1} \leq 1$ increases by increasing the mass ratio $\varepsilon$. $r_{n_1}$ is close to the natural frequency of a 1/8 car model and indicates the **principal natural frequency** of a vehicle. Hence, it is called the **body bounce natural frequency**. The second natural frequency, $r_{n_2}$, begins from 1 for $\alpha = 0$ and goes to higher values when $\alpha$ increases. However, $r_{n_2} \approx 10\,\text{Hz}$ for street cars with acceptable ride comfort. $r_{n_1}$ relates to the unsprung mass and is called the **wheel hop natural frequency**.

Let us consider a numerical example for a quarter car with the following data.

$$m_s = 336\,\text{kg} \qquad m_u = 42\,\text{kg} \qquad \varepsilon = \frac{m_s}{m_u} = 8 \tag{9.168}$$

$$k_s = 22{,}000\,\text{N/m} \qquad k_s = 250{,}000\,\text{N/m} \tag{9.169}$$

$$\omega_s = \sqrt{\frac{k_s}{m_s}} = 8.09\,\text{rad/s} \simeq 1.288\,\text{Hz} \tag{9.170}$$

$$\omega_u = \sqrt{\frac{k_u}{m_u}} = 69\,\text{rad/s} \simeq 10.98\,\text{Hz} \quad \alpha = 0.1172 \tag{9.171}$$

$$c = 1630\,\text{N s/m} \qquad \xi = \frac{c}{2m_s\omega_s} = 0.3 \tag{9.172}$$

The graphs provide us with:

$$r_{n_1} = 0.11118 \qquad r_{n_2} = 1.0541 \tag{9.173}$$

therefore, the natural frequencies are

$$\omega_{n_1} = r_{n_1}\omega_s = 0.899 \qquad \omega_{n_2} = r_{n_2}\omega_s = 8.527 \tag{9.174}$$

For a modern typical passenger car, the values of natural frequencies are around 1 Hz and 10 Hz. The former is due to the bounce of sprung mass and the latter belongs to the unsprung mass. At average speed, bumps with wavelengths much greater than the wheelbase of the vehicle will excite bounce motion of the body. At higher speeds, wavelength of the bumps becomes shorter than a wheelbase and causes heavy vibrations of the unsprung mass. Therefore, when the wheels hit a single bump on the road, the impulse will set the wheels into oscillation at the natural frequency of the unsprung mass around 10 Hz. In turn, for the sprung mass, the excitation will be a vibrating base at the frequency of vibration of the unsprung around 10 Hz. Because the natural frequency of the sprung is approximately 1 Hz, the excellent isolation for sprung mass occurs and the frequency range around 10 Hz that has no essential influence on the sprung discomfort. When the wheel runs over a rough undulating surface, the excitation will consist of a wide range of frequencies. High excitation frequency at 5 Hz to 20 Hz means high frequency input to the sprung mass, which can effectively be isolated. Low frequency excitation, however, will cause resonance in the sprung mass.

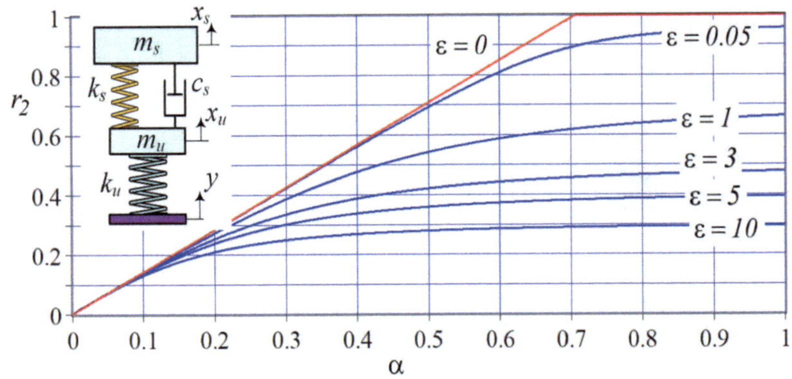


**Fig. 9.17** The nodal frequency $r_2$ as a function of $\varepsilon$ and $\alpha$

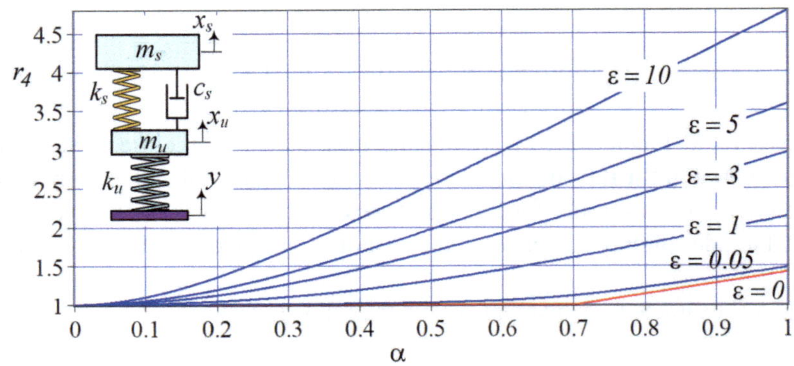


**Fig. 9.18** The nodal frequency $r_4$ as a function of $\varepsilon$ and $\alpha$

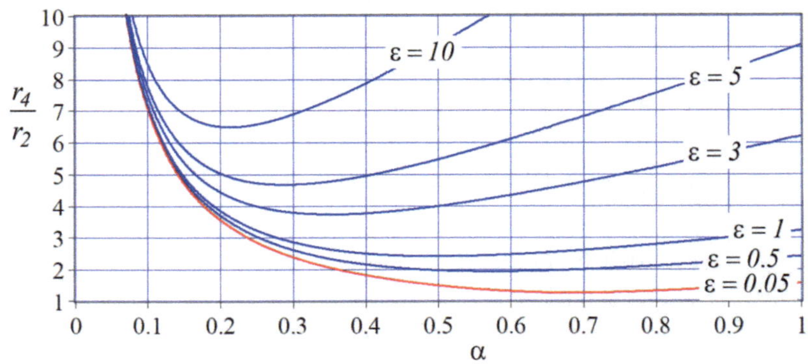


**Fig. 9.19** The nodal frequency ratio $r_4/r_2$ as a function of $\varepsilon$ and $\alpha$

**Example 137** Nodal frequencies and amplitudes. Nodes are the invariant point of frequency response that all curves will pass over, regardless of the value of damping. Therefore, the coordinate of the nodal point is independent of the damping of the system. These fixed point are being used for optimal design.

Figures 9.17 and 9.18 illustrate the nodal frequencies $r_2$ and $r_4$. Figure 9.19 shows that $r_2 < r_4$. Figures 9.20, 9.21 and 9.22 show that $r_2 > r_{n_1}$, $r_3 < r_{n_2}$, $r_4 > r_{n_2}$. These figures help to have good evaluation of the value of nodal and natural frequencies of quarter car model during design stage. Any estimate value out of these figures rings a warning bell to look into the calculations and the design parameters. These graphs also confirm the following order of natural and nodal frequencies.

$$r_1 \, (= 0) < r_{n_1} < r_2 < r_3 \, (= 1) < r_{n_2} < r_4 \tag{9.175}$$

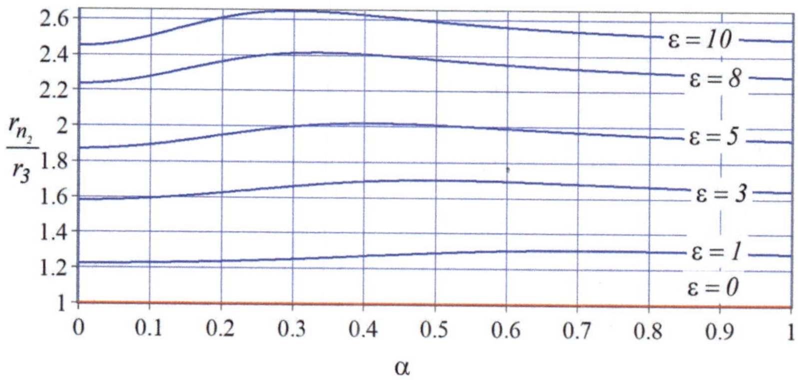


**Fig. 9.20** The nodal frequency ratio $r_2/r_{n_1}$ as a function of $\varepsilon$ and $\alpha$

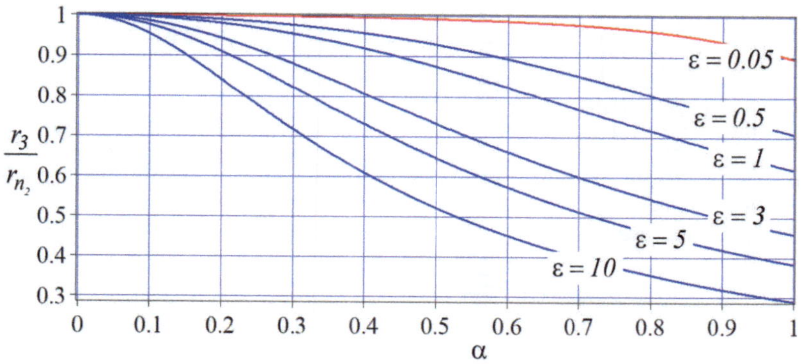


**Fig. 9.21** The nodal frequency ratio $r_3/r_{n_2}$ as a function of $\varepsilon$ and $\alpha$

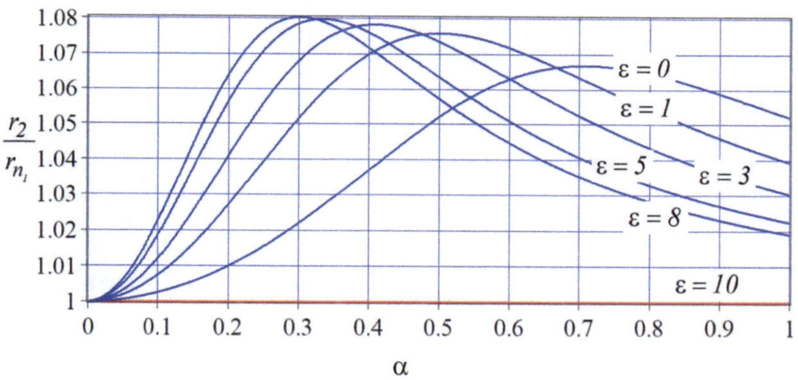


**Fig. 9.22** The nodal frequency ratio $r_4/r_{n_2}$ as a function of $\varepsilon$ and $\alpha$

**Example 138** ★ Nodes' amplitude of the absolute displacement frequency response $\mu$. Nodes are the points at which all frequency response curves will pass over, regardless of the value of the damping. They are important in optimization and analysis of vibrating systems. For the $\mu$ frequency response, there are fur nodes. Here we study $\mu_2$, $\mu_3$, and $\mu_4$.

There are four nodes in the sprung mass absolute displacement frequency response $\mu$ of a quarter car model. The first node is at a trivial point ($r_1 = 0$, $\mu_1 = 1$), which shows that $X_s = Y$ when the excitation frequency is zero. There are also three middle nodes at $(r_2, \mu_2)$, $(r_3, \mu_3)$, and $(r_4, \mu_4)$. Figure 9.23 depicts a magnification of the sprung mass displacement frequency response, $\mu = |X_s/Y|$ around the middle nodes for a sample vehicle. The value of amplitude $\mu_2$ is depicted in Fig. 9.24. The value of $\mu_2$ is always greater than 1. Figure 9.25 illustrates value of amplitude $\mu_3$ as a function of $\varepsilon$ because $\mu_3$ is not dependent on $\alpha$. The graph shows that $\mu_3 > 1$ for the applied range of $\varepsilon$. Figure 9.26 illustrates the amplitude ratio of $\mu_4$.

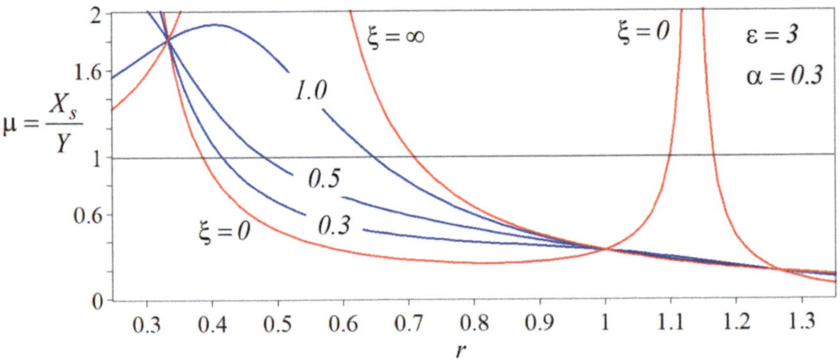


**Fig. 9.23** A magnification around the nodes for the sprung mass displacement frequency response, $\mu = \left|\frac{X_s}{Y}\right|$

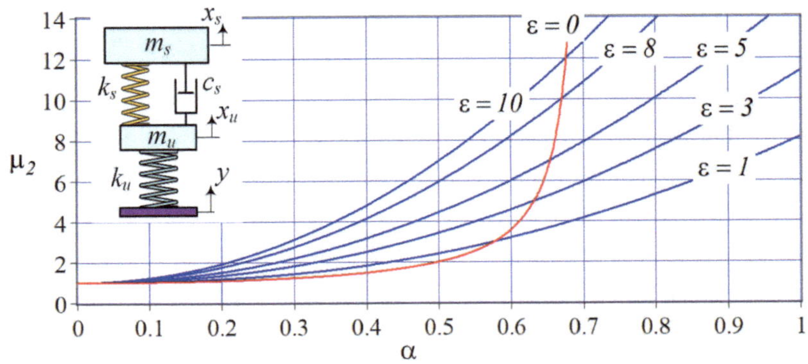


**Fig. 9.24** The value of amplitude $\mu_2$

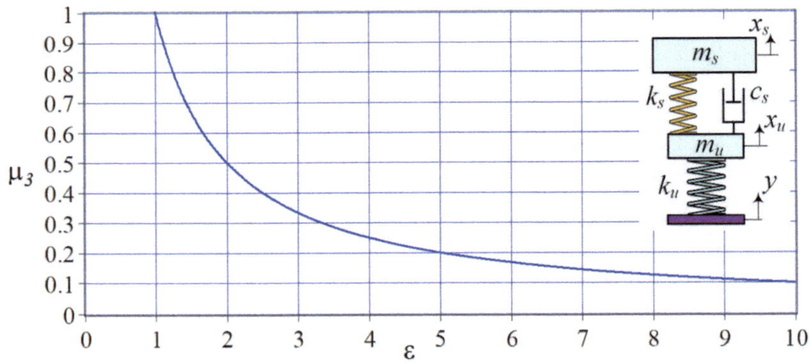


**Fig. 9.25** The value of amplitude $\mu_3$

**Example 139** ★ No Frahm optimal quarter car suspension. Minimization of the amplitude of absolute frequency response $\mu$ is an optimization method to design suspension of vehicles. Frahm suggested a practical minimization method based on nodes of the frequency response to calculate the optimal $k^*$ and $c^*$. Frahm principle is not possible for quarter car model.

Reduction in absolute amplitude is the first attempt for optimization of a vibrating system. If the amplitude frequency response $\mu = \mu(r)$ contains fixed points with respect to some parameters, then using the Frahm method, the optimization process is carried out in two steps:

1. We select the parameters that control the position of the invariant points to equalize the corresponding height at the invariant frequencies and minimize the height of the fixed points as much as possible.
2. We find the remaining parameters such that the maximum amplitude coincides precisely at the invariant points.

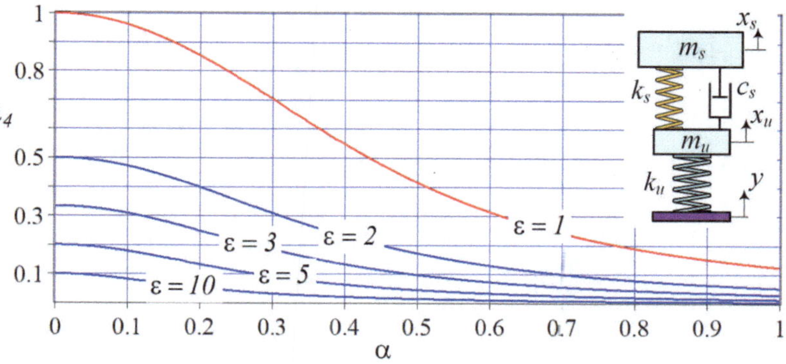


**Fig. 9.26** The value of amplitude $\mu_4$

For a real vehicle quarter car model, the values of mass ratio $\varepsilon$and wheel frequency $\omega_u$ are fixed and we are trying to find the optimum values of $\alpha$ and $\xi$. The parameters $\alpha$ and $\xi$ include the stiffness of the main spring and the damping of the main shock absorber, respectively.

The amplitudes $\mu_i$ at invariant frequencies $r_i$ show that the first invariant point $(r_1 = 0, \mu_1 = 1)$ is always fixed. Figure 9.24 shows that always $\mu_2 > 1$ and Fig. 9.25 shows that always $\mu_3 < 1$. Therefore, the second and third nodes are not possible to be at the same height.

**Example 140** Transmitted force. The transmitted force to the road and to the body are always design factors to construct foundation for a mechanical vibrating device. The frequency response of the transmitted force to the ground and to the body of the quarter car model are calculated and illustrated here.

The transmitted force $f_T$ to the ground and transmitted force $f_B$ to the body are the combined forces of the spring $k_u$ and damper $c_u$.

$$f_T = k_u (x_u - y) + c_u (\dot{x}_u - \dot{y}) \tag{9.176}$$

$$f_B = k_s (x_s - x_u) + c_s (\dot{x}_s - \dot{x}_u) \tag{9.177}$$

$$y = Y \sin \omega t \tag{9.178}$$

We determine the frequency response of $f_T$ by substituting for $x_u$ and $y$.

$$\begin{aligned} f_T &= k_u (A_1 \sin \omega t + B_1 \cos \omega t - Y \sin \omega t) \\ &\quad + c_u \omega (A_1 \cos \omega t - B_1 \sin \omega t - Y \cos \omega t) \\ &= (k_u A_1 - c_u \omega B_1 - k_u Y) \sin \omega t \\ &\quad + (k_u B_1 + c_u \omega A_1 - c_u \omega Y) \cos \omega t \\ &= F_T \sin (\omega t - \varphi_F) \end{aligned} \tag{9.179}$$

The steady-state amplitude of the transmitted force $F_T$ to the road will be transformed to the following form.

$$\left(\frac{F_T}{k_u Y}\right)^2 = \frac{r^4 \left(4\xi_u^2 r^2 + 1\right) \left(\left(\alpha^4 + 4r^2\alpha^2\xi_s^2\right)(1+\varepsilon)^2 - Z_7\right)}{8\varepsilon\xi_u\xi_s\alpha r^6 + Z_1^2 + Z_2^2 + Z_3} \tag{9.180}$$

$$Z_1 = \varepsilon r^2 \alpha^2 + \left(\alpha^2 - r^2\right)\left(r^2 - 1\right) \tag{9.181}$$

$$Z_2 = 2\xi_s r \alpha \left(\varepsilon r^2 + r^2 - 1\right) \tag{9.182}$$

$$Z_3 = 4\xi_u^2 r^2 \left(4\xi_u^2 r^2 \alpha^2 + \left(\alpha^2 - r^2\right)^2\right) \tag{9.183}$$

$$Z_7 = 2r^2\alpha(1+\varepsilon) - r^4 \tag{9.184}$$

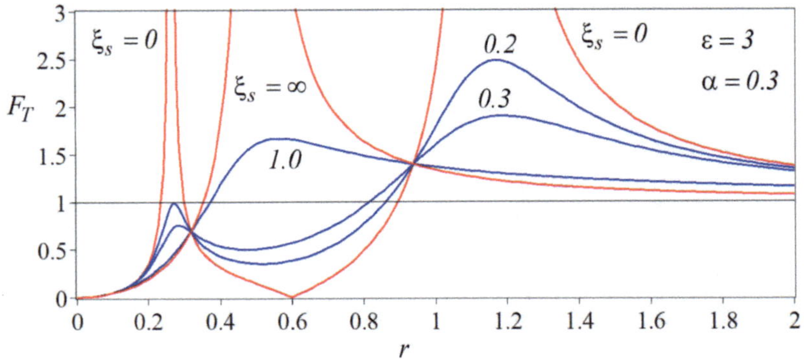


**Fig. 9.27** The transmitted force frequency response $F_T$ to the road for a quarter car

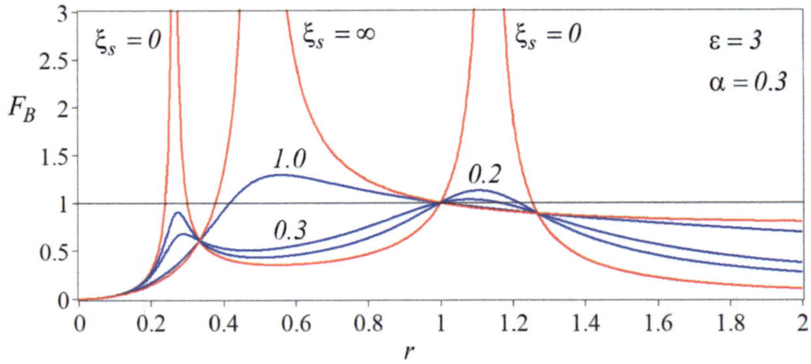


**Fig. 9.28** The transmitted force frequency response $F_B$ to the body for a quarter car

Similarly, the amplitude of the transmitted force $F_B$ to the body will be transformed to the following form.

$$\left(\frac{F_B}{k_u Y}\right)^2 = \frac{\left(4\xi_u^2 r^2 + 1\right)\left(4\xi_s^2 r^2 + \alpha^2\right)}{8\varepsilon\xi_u\xi_s\alpha r^6 + Z_1^2 + Z_2^2 + Z_3} \tag{9.185}$$

Figures 9.27 and 9.28 illustrate an example of $F_T$ and $F_B$.

**Example 141** ★ Invariant frequencies of $\tau$ frequency responses. The frequency response of the unsprung mass, $\tau$, has three invariant nodes. Here we study the $\tau$ and its nodes behavior.

The family of response curves for the frequency response of the unsprung mass, $\tau$, are also obtained by keeping $\varepsilon$ and $\alpha$ constant, and varying $\xi$. This family has three nodal points, which are at frequencies $r_1, r_2, r_3$, and at $\tau_1, \tau_2, \tau_3$.

$$\begin{cases} r_1 = 0 & \tau_1 = 1 \\ r_2 & \tau_2 = \dfrac{1}{1 - (1+\varepsilon)\, r_2^2 \alpha^2} \\ r_3 & \tau_4 = \dfrac{-1}{1 - (1+\varepsilon)\, r_3^2 \alpha^2} \end{cases} \tag{9.186}$$

$$r_2 = \sqrt{\frac{1}{2\alpha^2}\left(1 + 2(1+\varepsilon)\alpha^2 - \sqrt{\left(1 + 2(1+\varepsilon)\alpha^2\right)^2 - 8\alpha^2}\right)} \tag{9.187}$$

$$r_3 = \sqrt{\frac{1}{2\alpha^2}\left(1 + 2(1+\varepsilon)\alpha^2 + \sqrt{\left(1 + 2(1+\varepsilon)\alpha^2\right)^2 - 8\alpha^2}\right)} \tag{9.188}$$

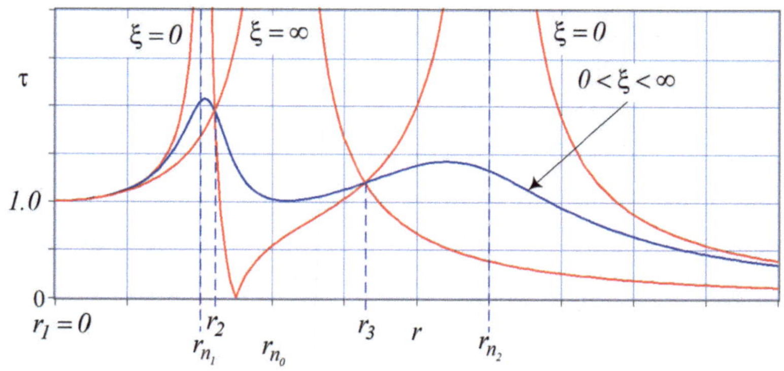


**Fig. 9.29** Schematic illustration of the amplitude $\tau$ versus excitation frequency ratio $r$

The invariant frequencies $r_1$, $r_2$, $r_3$ and their invariant corresponding amplitudes are not dependent on $\xi$. They are only dependent on the values of $\varepsilon$ and $\alpha$. The order of magnitude of the natural and invariant frequencies of $\tau$ are

$$r_1\,(=0) < r_{n_1} < r_2 < r_3 < r_{n_2} \tag{9.189}$$

The curves for $\tau$ have no other common points except $r_1, r_2, r_3$. Figure 9.29 shows schematically the shape of the amplitude $\tau$. The invariant frequencies are independent of $\xi$, so they can be found by intersecting the $\tau$ curves for $\xi = 0$ and $\xi = \infty$.

$$\lim_{\xi\to 0} \tau^2 = \pm \frac{\left(\alpha^2 - r^2\right)^2}{\left(r^2\alpha^2 - \alpha^2 + r^2 - r^4 + r^2\alpha^2\varepsilon\right)^2} \tag{9.190}$$

$$\lim_{\xi\to \infty} \tau^2 = \pm \frac{1}{\left(\varepsilon r^2 + r^2 - 1\right)^2} \tag{9.191}$$

Therefore, the invariant frequencies, $r_i$, can be determined by solving the following equation:

$$\lim_{\xi\to 0} \tau^2 = \pm \lim_{\xi\to \infty} \tau^2 \tag{9.192}$$

$$\varepsilon r^2\alpha^2 + \left(\alpha^2 - r^2\right)\left(r^2 - 1\right) = \pm\left(\varepsilon r^2 + r^2 - 1\right)\left(\alpha^2 - r^2\right) \tag{9.193}$$

Using the (+) sign, we find $r_1$ with their corresponding transmissivities $\mu_1$,

$$r_1 = 0 \qquad \mu_1 = 1 \tag{9.194}$$

and, with the (−) sign, we find the following equation for $r_2$ and $r_3$:

$$(-\varepsilon - 2)\,r^4 + \left(2\alpha^2 + 2\alpha^2\varepsilon + 2\right)r^2 - 2\alpha^2 = 0 \tag{9.195}$$

This equation has two real positive roots, $r_2$ and $r_3$,

$$r_2 = \sqrt{\frac{1}{\varepsilon + 2}\left(\alpha^2 + \alpha^2\varepsilon + 1 - \sqrt{\alpha^4\,(1+\varepsilon)^2 - 2\alpha^2 + 1}\right)} \tag{9.196}$$

$$r_3 = \sqrt{\frac{1}{\varepsilon + 2}\left(\alpha^2 + \alpha^2\varepsilon + 1 + \sqrt{\alpha^4\,(1+\varepsilon)^2 - 2\alpha^2 + 1}\right)} \tag{9.197}$$

with the following relative order of magnitude:

$$r_1\,(=0) < r_2 < \frac{\alpha}{\sqrt{\varepsilon + 2}} < r_3 \tag{9.198}$$

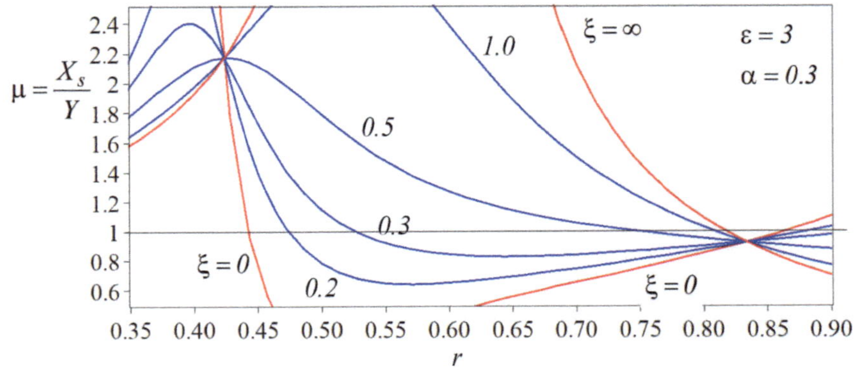


**Fig. 9.30** A magnification around the nodes for the unsprung mass displacement frequency response, $\tau = \left|\frac{X_u}{Y}\right|$

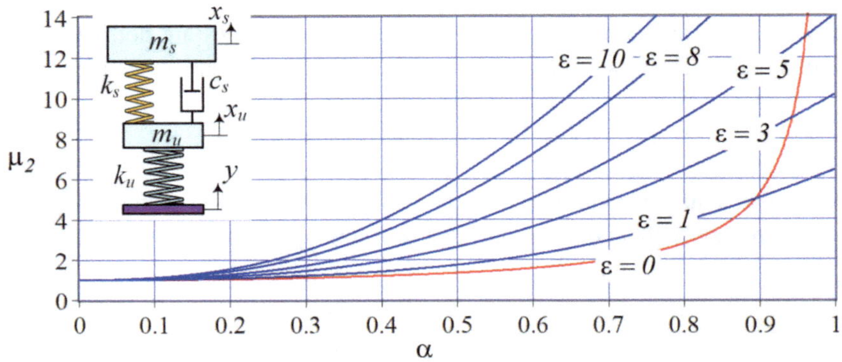


**Fig. 9.31** The value of amplitude $\tau_2$

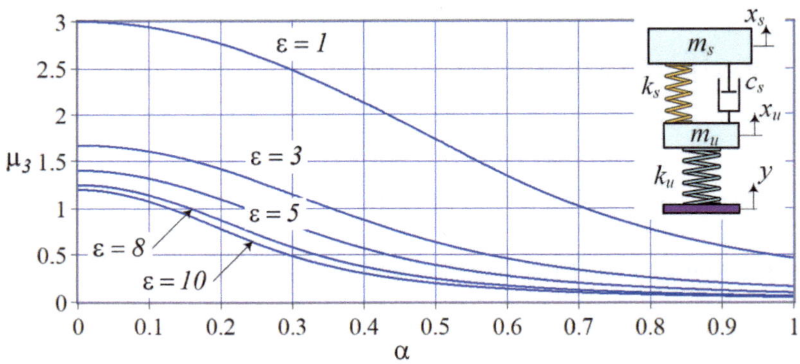


**Fig. 9.32** The value of amplitude $\tau_3$

The corresponding amplitudes of $\tau$ at $r_2$ and $r_3$ can be found by substituting $r_2$ and $r_3$ in (9.191).

$$\tau_2 = \frac{1}{(\varepsilon + 1)\, r_2^2 - 1} \qquad \tau_3 = \frac{1}{(\varepsilon + 1)\, r_3^2 - 1} \tag{9.199}$$

There are three nodes in the unsprung mass absolute displacement frequency response $\tau$ of a quarter car model. The first node is at a trivial point $(r_1 = 0, \tau_1 = 1)$, which shows that $X_u = Y$ when the excitation frequency is zero. There are also two middle nodes at $(r_2, \tau_2)$ and $(r_3, \tau_3)$. Figure 9.30 illustrates a magnification of the unsprung mass displacement frequency response, $\tau = |X_u/Y|$ around the middle nodes for a sample vehicle. The value of amplitude $\tau_2$ is depicted in Fig. 9.31. The value of $\tau_2$ is always greater than 1. Figure 9.32 illustrates value of amplitude $\tau_3$ and it shows that $\tau_3$ can be higher or lower than 1, depending on $\varepsilon$ and $\alpha$. This fact is also shown in Fig. 9.33 which illustrates the amplitude ratio of $\tau_3/\tau_2$. Figure 9.34 illustrates the $\alpha$ and $\varepsilon$ relationship to have $\tau_3 = \tau_2$ in case optimization of the motion of $m_1$ becomes important in some applications.

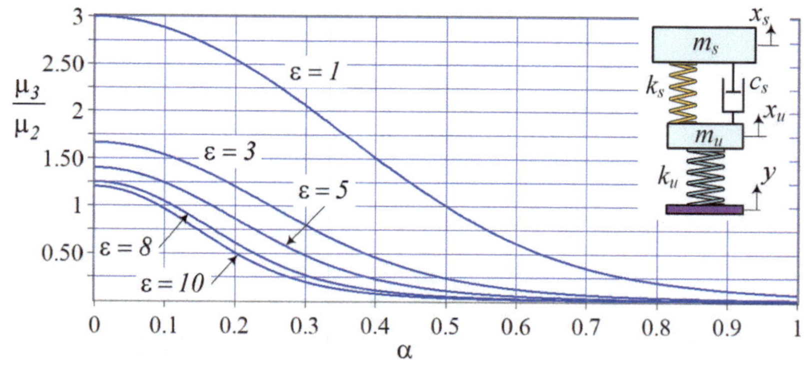


**Fig. 9.33** The amplitude ratio of $\mu_3/\mu_2$

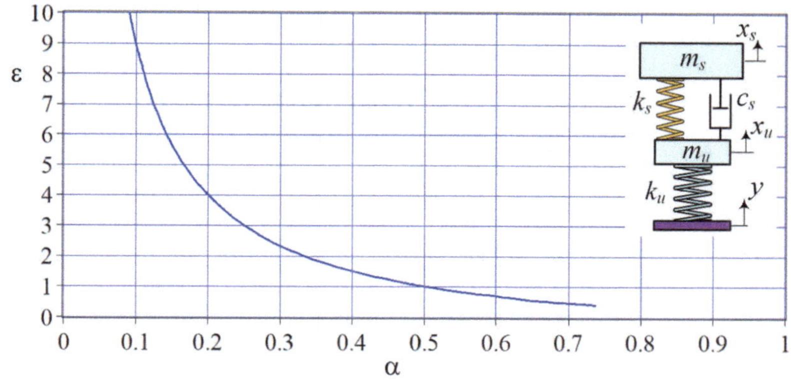


**Fig. 9.34** The relationship between $\varepsilon$ and $\alpha$ for which we have $\mu_3/\mu_2 = 1$

## 9.5 ★ Frahm Theory of Vibration Absorber

Vibration is an unavoidable phenomena in mechanical systems. In this section, we review the Frahm optimization method for vibration absorbers. It is the method to shut down the displacement oscillation of one mass of a multi degree of freedom system under harmonic excitation. Similar to most of the optimization methods for mechanical systems, this theory is also based on frequency response.

Consider a primary system of a mass $m_1$ supported by a spring $k_1$, as is shown in Fig. 9.35. Assume there is a harmonic force $f = F \sin \omega t$ applied on $m_1$. We add a secondary system $(m_2, c_2, k_2)$ to absorb the vibrations of the primary. Such a system is called a *Frahm absorber*, or *Frahm damper*. Adding a vibration absorber to a primary system generates a new system with 2 $DOF$. The goal of having a vibration absorber is to design the suspension of the secondary system $(c_2, k_2)$ to reduce the amplitude of vibration of $m_1$ at a given excitation frequency $\omega$ or in a range of frequencies.

To minimize the steady-state vibration amplitude of $m_1$ when the excitation frequency is variable, we must set $k_2$ at the optimal value $k_2^\star$,

$$k_2^\star = \frac{m_1 m_2}{(m_1 + m_2)^2} k_1 \tag{9.200}$$

and select $c_2$ within the following range.

$$2m_2\omega_1\xi_1^\star < c_2 < 2m_2\omega_1\xi_2^\star \tag{9.201}$$

The optimal damping ratios $\xi_1^\star$ and $\xi_2^\star$ are

$$\xi_1^\star = \sqrt{\frac{-B - \sqrt{B^2 - 4AC}}{2A}} \tag{9.202}$$

$$\xi_2^\star = \sqrt{\frac{-B + \sqrt{B^2 - 4AC}}{2A}} \tag{9.203}$$

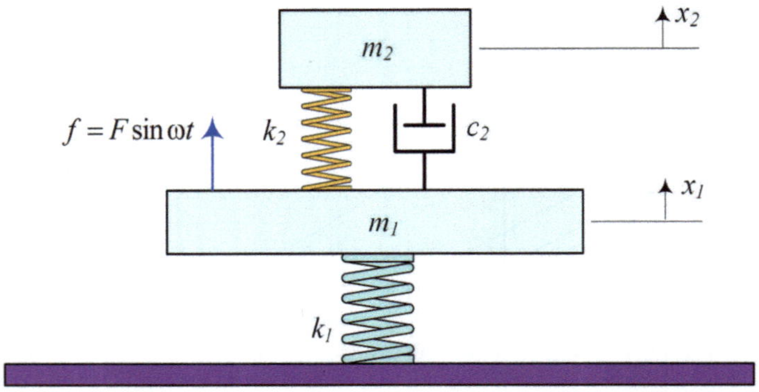


**Fig. 9.35** A secondary vibration absorber system $(m_2, c_2, k_2)$ added to a primary vibrating system $(m_1, k_1)$

$$A = 16Z_8 - 4r^2 (4Z_4 + 8Z_5) \tag{9.204}$$

$$B = 4Z_9 - 4Z_6 r^2 - Z_7 (4Z_4 + 8Z_5) + 4Z_3 Z_8 \tag{9.205}$$

$$C = Z_3 Z_9 - Z_6 Z_7 \tag{9.206}$$

$$Z_3 = 2 \left(r^2 - \alpha^2\right) \tag{9.207}$$

$$Z_4 = \left[r^2 (1 + \varepsilon) - 1\right]^2 \tag{9.208}$$

$$Z_5 = r^2 (1 + \varepsilon) \left[r^2 (1 + \varepsilon) - 1\right] \tag{9.209}$$

$$Z_6 = 2 \left[\varepsilon \alpha^2 r^2 - \left(r^2 - \alpha^2\right) \left(r^2 - 1\right)\right] \times \left[\varepsilon \alpha^2 - \left(r^2 - \alpha^2\right) - \left(r^2 - 1\right)\right] \tag{9.210}$$

$$Z_7 = \left(r^2 - \alpha^2\right)^2 \tag{9.211}$$

$$Z_8 = r^2 \left[r^2 (1 + \varepsilon) - 1\right]^2 \tag{9.212}$$

$$Z_9 = \left[\varepsilon \alpha^2 r^2 - \left(r^2 - 1\right) \left(r^2 - \alpha^2\right)\right]^2 \tag{9.213}$$

***Proof*** The main goal in vibration optimization is to reduce the vibration amplitude of a primary mass, when the system is under a forced vibration. There are two principal methods for decreasing the vibration amplitude of a primary mass: *vibration absorption* and *vibration isolation*. When the suspension of a *primary system* is not easy to optimize, we can add another vibrating system, known as the *vibration absorber* or *secondary system*, to absorb the vibrations of the primary system. The vibration absorber that increases the $DOF$ of the system is an applied method for vibration reduction in frequency domain of systems that have already been designed. The optimization method for the vibration absorber is based on nodal points of the frequency response of the system.

Reduction in absolute displacement amplitude is the first attempt for optimization in most optimal design strategies. If the amplitude frequency response $\mu = X_1/(F/k) = \mu(r)$ contains fixed points with respect to damping ratio $\xi = c_2/(2m_2\omega_1)$, then the optimization process is carried out in two steps:

1. We select the parameter $\alpha = \omega_2/\omega_1$ that controls the position of the nodal points to equalize the corresponding height at the nodal frequencies.
2. We find the damping ratio $\xi = c_2/(2m_2\omega_1)$ such that the maximum amplitude happens at the nodal points.

Let us write the equations of motion for the system shown in Fig. 9.35.

$$m_1\ddot{x}_1 + c_2 (\dot{x}_1 - \dot{x}_2) + k_1 x_1 + k_2 (x_1 - x_2) = F \sin \omega t \tag{9.214}$$

$$m_2\ddot{x}_2 - c_2(\dot{x}_1 - \dot{x}_2) - k_2(x_1 - x_2) = 0 \tag{9.215}$$

$$\begin{bmatrix} m_1 & 0 \\ 0 & m_2 \end{bmatrix}\begin{bmatrix} \ddot{x}_1 \\ \ddot{x}_2 \end{bmatrix} + \begin{bmatrix} c_2 & -c_2 \\ -c_2 & c_2 \end{bmatrix}\begin{bmatrix} \dot{x}_1 \\ \dot{x}_2 \end{bmatrix} + \begin{bmatrix} k_1 + k_2 & -k_2 \\ -k_2 & k_2 \end{bmatrix}\begin{bmatrix} x_1 \\ x_2 \end{bmatrix} = \begin{bmatrix} F \sin \omega t \\ 0 \end{bmatrix} \tag{9.216}$$

Assuming a steady-state condition, and substituting a set of harmonic solutions in the equations of motion, we find the following set of equations for $A_1$, $B_1$, $A_2$, $B_2$.

$$x_1 = A_1 \sin \omega t + B_1 \cos \omega t \tag{9.217}$$

$$x_2 = A_2 \sin \omega t + B_2 \cos \omega t \tag{9.218}$$

$$\begin{bmatrix} a_{11} & -k_2 & -\omega c_2 & \omega c_2 \\ -k_2 & a_{22} & \omega c_2 & -\omega c_2 \\ \omega c_2 & -\omega c_2 & a_{33} & -k_2 \\ -\omega c_2 & \omega c_2 & -k_2 & a_{44} \end{bmatrix}\begin{bmatrix} A_1 \\ A_2 \\ B_1 \\ B_2 \end{bmatrix} = \begin{bmatrix} F \\ 0 \\ 0 \\ 0 \end{bmatrix} \tag{9.219}$$

$$a_{11} = a_{33} = k_2 + k_1 - m_1\omega^2 \tag{9.220}$$

$$a_{22} = a_{44} = k_2 - m_2\omega^2 \tag{9.221}$$

The steady-state amplitudes $X_1$ and $X_2$ for vibration of $m_1$ and $m_2$ are

$$X_1 = \sqrt{A_1^2 + B_1^2} \tag{9.222}$$

$$X_2 = \sqrt{A_2^2 + B_2^2} \tag{9.223}$$

and they are equal to:

$$\left(\frac{X_1}{F}\right)^2 = \frac{\left(k_2 - \omega^2 m_2\right)^2 + \omega^2 c_2^2}{Z_1^2 + \omega^2 c_2^2 Z_2^2} \tag{9.224}$$

$$\left(\frac{X_2}{F}\right)^2 = \frac{k_2^2 + \omega^2 c_2^2}{Z_1^2 + \omega^2 c_2^2 Z_2^2} \tag{9.225}$$

$$Z_1 = \left(k_1 - \omega^2 m_1\right)\left(k_2 - \omega^2 m_2\right) - \omega^2 m_2 k_2 \tag{9.226}$$

$$Z_2 = k_1 - \omega^2 m_1 - \omega^2 m_2 \tag{9.227}$$

Introducing the following parameters

$$\varepsilon = \frac{m_2}{m_1} \tag{9.228}$$

$$\omega_1 = \sqrt{\frac{k_1}{m_1}} \tag{9.229}$$

$$\omega_2 = \sqrt{\frac{k_2}{m_2}} \tag{9.230}$$

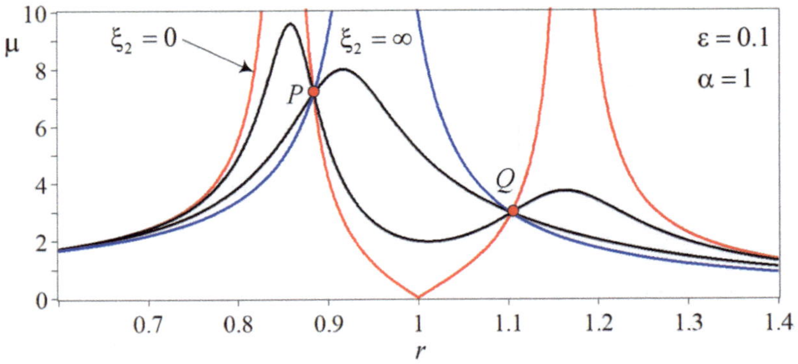


**Fig. 9.36** Frequency response $\mu$ for a set of parameters and different damping ratios

$$\alpha = \frac{\omega_2}{\omega_1} \tag{9.231}$$

$$r = \frac{\omega}{\omega_1} \tag{9.232}$$

$$\xi = \frac{c_2}{2m_2\omega_1} \tag{9.233}$$

$$\mu = \frac{X_1}{F/k_1} \tag{9.234}$$

$$\tau = \frac{X_2}{F/k_1} \tag{9.235}$$

we can rearrange the frequency responses (9.224) and (9.225) to the following dimensionless forms.

$$\mu^2 = \left(\frac{X_1}{F/k_1}\right)^2 = \frac{4\xi_2^2 r^2 + \left(r^2 - \alpha^2\right)^2}{4\xi_2^2 r^2 \left[r^2 (1+\varepsilon) - 1\right]^2 + \left[\varepsilon\alpha^2 r^2 - \left(r^2 - 1\right)\left(r^2 - \alpha^2\right)\right]^2} \tag{9.236}$$

$$\tau^2 = \left(\frac{X_2}{F/k_1}\right)^2 = \frac{4\xi_2^2 r^2 \alpha^2 + \alpha^4}{4\xi_2^2 r^2 \left[r^2 (1+\varepsilon) - 1\right]^2 + \left[\varepsilon\alpha^2 r^2 - \left(r^2 - 1\right)\left(r^2 - \alpha^2\right)\right]^2} \tag{9.237}$$

The parameter $\varepsilon$ is the *mass ratio* of $m_2$ to $m_1$, the parameter $\omega_1$ is the natural frequency of the main system, and $\omega_2$ is the natural frequency of the vibration absorber system. The frequencies $\omega_1$ and $\omega_2$ are not the natural frequencies of the whole system. The parameter $\alpha = \omega_2/\omega_1$ is the natural frequency ratio, $r = \omega/\omega_1$ is the excitation frequency ratio, $\xi = c_2/(2m_2\omega_1)$ is the damping ratio, and $\mu = X_1/(F/k)$ and $\tau = X_2/(F/k)$ are the steady-state amplitudes of the primary and secondary $X_1$, $X_2$ divided by the static deflection $F/k_1$.

Figure 9.36 illustrates the frequency response $\mu$ for

$$\varepsilon = 0.1 \qquad \alpha = 1 \tag{9.238}$$

and damping ratio of

$$\xi = 0 \qquad \xi = 0.2 \qquad \xi = 0.3 \qquad \xi = \infty \tag{9.239}$$

All curves pass through the nodes $P$ and $Q$, independent of the value of damping ratio $\xi$. To find the parameters that control the position of the nodes, we find the intersection points of the curves for $\xi = 0$ and $\xi = \infty$. Setting $\xi = 0$ and $\xi = \infty$ yields

$$\mu^2 = \frac{\left(r^2 - \alpha^2\right)^2}{\left(\varepsilon\alpha^2 r^2 - \left(r^2 - 1\right)\left(r^2 - \alpha^2\right)\right)^2} \tag{9.240}$$

$$\mu^2 = \frac{1}{\left(r^2\left(1+\varepsilon\right) - 1\right)^2} \tag{9.241}$$

When $\xi = 0$, the system is an undamped linear two $DOF$ system with two natural frequencies. The vibration amplitude of the system approaches infinity, $\mu \to \infty$, when the excitation frequency approaches either of the natural frequencies of the system. When $\xi = \infty$, there would be no relative motion between $m_1$ and $m_2$ and the system reduces to an undamped linear one $DOF$ system with one natural frequency:

$$\omega_n = \sqrt{\frac{k_1}{m_1 + m_2}} \tag{9.242}$$

or

$$r_n = \frac{1}{\sqrt{1+\varepsilon}} \tag{9.243}$$

The vibration amplitude of the system approaches infinity, $\mu \to \infty$, when the excitation frequency approaches the natural frequency $\omega \to \omega_n$ or $r \to 1/\left(1+\varepsilon\right)$.

Equating Eqs. (9.240) and (9.241) for $\xi = 0$ and $\xi = \infty$ , we have

$$\frac{\left(r^2 - \alpha^2\right)^2}{\left(\varepsilon\alpha^2 r^2 - \left(r^2 - 1\right)\left(r^2 - \alpha^2\right)\right)^2} = \frac{1}{\left(r^2\left(1+\varepsilon\right) - 1\right)^2} \tag{9.244}$$

which simplifies to:

$$\varepsilon\alpha^2 r^2 - \left(r^2 - 1\right)\left(r^2 - \alpha^2\right) = \pm\left(r^2 - \alpha^2\right)\left(r^2\left(1+\varepsilon\right) - 1\right) \tag{9.245}$$

The negative sign is equivalent to

$$r^4\varepsilon = 0 \tag{9.246}$$

which indicates there is a node at $r = 0$. The plus sign produces a quadratic equation for $r^2$ with two positive solutions $r_1$ and $r_2$ corresponding to nodes $P$ and $Q$.

$$\left(2+\varepsilon\right) r^4 - r^2\left(2 + 2\alpha^2\left(1+\varepsilon\right)\right) + 2\alpha^2 = 0 \tag{9.247}$$

$$r_{1,2}^2 = \frac{1}{\varepsilon + 2}\left(\alpha^2 \pm \sqrt{\left(\varepsilon^2+2\varepsilon+1\right)\alpha^4 - 2\alpha^2 + 1} + \alpha^2\varepsilon + 1\right) \tag{9.248}$$

$$r_1 < r_n < r_2 \tag{9.249}$$

Because the frequency response curves always pass through $P$ and $Q$, the optimal situation would happen when the nodes $P$ and $Q$ have equal height.

$$\mu\left(P\right) = \mu\left(Q\right) \tag{9.250}$$

The value of $\mu^2$ at $P$ and $Q$ are independent of $\xi$. Therefore, we may substitute $r_1$ and $r_2$ in Eq. (9.241) for $u$ corresponding to $\xi = \infty$. However, writing $\mu$ from Eq. (9.241) we see that

$$\mu = \frac{1}{r^2\left(1+\varepsilon\right) - 1} \tag{9.251}$$

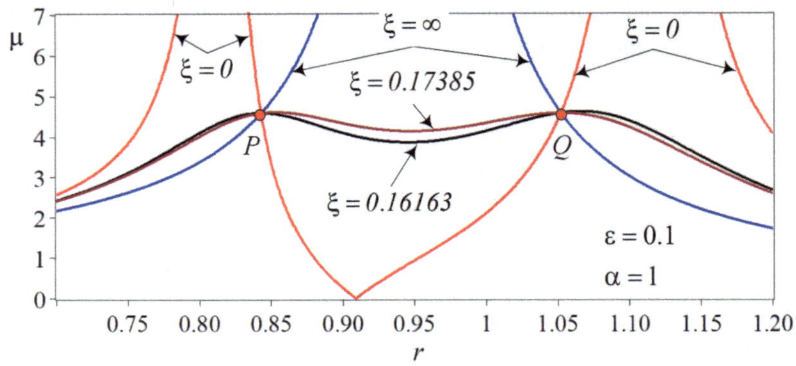


**Fig. 9.37** Optimal damping ratio $\xi$ to have maximum of $\mu$ at $P$ or $Q$

produces a positive number for $r < r_n$ and a negative number for $r > r_n$. Therefore, to equate $\mu(P)$ and $\mu(Q)$ we should solve the following equations

$$\mu(r_1) = -\mu(r_2) \tag{9.252}$$

or

$$\frac{1}{1 - r_1^2(1+\varepsilon)} = \frac{-1}{1 - r_2^2(1+\varepsilon)} \tag{9.253}$$

which simplifies to:

$$r_1^2 + r_2^2 = \frac{2}{1+\varepsilon} \tag{9.254}$$

The sum of the roots using Eq. (9.247) is

$$r_1^2 + r_2^2 = \frac{2 + 2\alpha^2(1+\varepsilon)}{1+\varepsilon} \tag{9.255}$$

and, therefore,

$$\frac{2}{1+\varepsilon} = \frac{2 + 2\alpha^2(1+\varepsilon)}{1+\varepsilon} \tag{9.256}$$

which provides us with the optimal value of $\alpha = \alpha^\star$.

$$\alpha^\star = \frac{1}{1+\varepsilon} = \frac{m_1}{m_1 + m_2} \tag{9.257}$$

Equation (9.257) is the required condition to make the height of the nodes $P$ and $Q$ equal and, hence, provides us with the optimal value of $\alpha$. Having an optimal value for $\alpha^\star$ is equivalent to designing the optimal stiffness $k_2$ for the secondary suspension, because,

$$\alpha = \frac{\omega_2}{\omega_1} = \sqrt{\frac{m_1}{m_2}}\sqrt{\frac{k_2}{k_1}} \tag{9.258}$$

and therefore, Eq. (9.257) provides us with the following condition for optimal stiffness of the vibration absorber, $k_2^\star$.

$$k_2^\star = k_1 \frac{m_1 m_2}{(m_1 + m_2)^2} \tag{9.259}$$

To determine the optimal damping ratio $\xi$, we force $\mu$ to have its maximum at $P$ or $Q$. Having $\mu_{Max}$ at $P$ guarantees that $\mu(r_1)$ is the highest value in a frequency domain around $r_1$, and having $\mu_{Max}$ at $Q$ guarantees that $\mu(r_2)$ is the highest amplitude in a frequency domain around $r_2$. The position of $\mu_{Max}$ is controlled by $\xi$, so we may determine two optimal values of $\xi$ to make $\mu_{Max}$ at $\mu(r_1)$ and at $\mu(r_2)$. An example of this situation is shown in Fig. 9.37.

Using $\alpha^\star$ from (9.257), the nodal frequencies are

$$r_{1,2}^2 = \frac{1}{1+\varepsilon}\left(1 \pm \sqrt{\frac{\varepsilon}{2+\varepsilon}}\right) \tag{9.260}$$

To set the partial derivative $\partial\mu/\partial r$ equal to zero at the nodal frequencies,

$$\left.\frac{\partial\left(\mu^2\right)}{\partial\left(r^2\right)}\right|_{r_1^2}=0 \tag{9.261}$$

$$\left.\frac{\partial\left(\mu^2\right)}{\partial\left(r^2\right)}\right|_{r_2^2}=0 \tag{9.262}$$

let us rewrite $\mu^2$ by the numerator $N(r)$ divided by the denominator $D(r)$ which helps us to find the derivative.

$$\mu^2=\frac{N(r)}{D(r)} \tag{9.263}$$

$$\frac{\partial\mu^2}{\partial r^2}=\frac{1}{D^2}\left(D\frac{\partial N}{\partial r^2}-N\frac{\partial D}{\partial r^2}\right)=\frac{1}{D}\left(\frac{\partial N}{\partial r^2}-\frac{N}{D}\frac{\partial D}{\partial r^2}\right) \tag{9.264}$$

Differentiating yields

$$\frac{\partial N}{\partial r^2}=4\xi^2+Z_3 \tag{9.265}$$

$$\frac{\partial D}{\partial r^2}=4\xi^2 Z_4+8\xi^2 Z_5+Z_6 \tag{9.266}$$

Equations (9.265) and (9.266), along with (9.260), must be substituted in (9.264) to solve for $\xi$. After substitution, the equation $\partial\left(\mu^2\right)/\partial\left(r^2\right)=0$ would be

$$\frac{\partial N}{\partial\left(r^2\right)}-\frac{N}{D}\frac{\partial D}{\partial\left(r^2\right)}=\left(4\xi^2+Z_3\right)\left(4\xi^2 Z_8+Z_9\right)-\left(4\xi^2 r^2+Z_7\right)\left(4\xi^2 Z_4+8\xi^2 Z_5+Z_6\right)=0 \tag{9.267}$$

because:

$$\frac{N}{D}=\frac{4\xi^2 r^2+Z_7}{4\xi^2 Z_8+Z_9} \tag{9.268}$$

Equation (9.267) is a quadratic equation for $\xi^2$

$$\begin{aligned}&\left(16Z_8-4r^2\left(4Z_4+8Z_5\right)\right)\xi^4\\&+\left(4Z_9-4Z_6r^2-Z_7\left(4Z_4+8Z_5\right)+4Z_3Z_8\right)\xi^2\\&+\left(Z_3Z_9-Z_6Z_7\right)=A\left(\xi^2\right)^2+B\xi^2+C=0\end{aligned} \tag{9.269}$$

with the solution

$$\xi^2=\frac{-B\pm\sqrt{B^2-4AC}}{2A} \tag{9.270}$$

where $Z_3$ to $Z_8$ are given in Eqs. (9.207)–(9.213), and $A$, $B$, $C$ are from (9.204)–(9.206).

The positive value of $\xi$ from (9.270) for $r=r_1$ and $r=r_2$ provides us with the limiting values for $\xi_1^\bigstar$ and $\xi_2^\bigstar$. Figure 9.37 illustrates the behavior of $\mu$ for optimal $\alpha$ and $\xi=0,\xi_1^\bigstar,\xi_2^\bigstar,\infty$.

If $\xi=0$, then $\mu=0$ at $r=1$, which shows the amplitude of the primary mass reduces to zero if the natural frequency of the primary and secondary systems are equal to the excitation frequency $r=\alpha=1$. Therefore, the vibration absorber is most effective at $r=\alpha=1$. ■

**Example 142** ★ Optimal spring and damper for $\varepsilon=0.1$. Here is a numerical example to calculate the value of optimal stiffness and damping for a practical application.

Consider a vibration absorber with the following mass ratio.

$$\varepsilon = \frac{m_2}{m_1} = 0.1 \tag{9.271}$$

We determine the optimal frequency ratio $\alpha$ from Eq. (9.257)

$$\alpha^\star = \frac{1}{1+\varepsilon} \approx 0.9091 \tag{9.272}$$

and find the nodal frequencies $r_{1,2}^2$ from (9.260).

$$r_{1,2}^2 = \frac{1}{1+\varepsilon}\left(1 \pm \sqrt{\frac{\varepsilon}{2+\varepsilon}}\right) = 0.71071,\ 1.1075 \tag{9.273}$$

Let us set $r = r_1 = \sqrt{0.71071} \approx 0.843$ and evaluate the parameters $Z_3$ to $Z_9$ from (9.207)–(9.213)

$$\begin{aligned} Z_3 &= -0.231470544 & Z_4 &= 0.0476190476 \\ Z_5 &= -0.1705988426 & Z_6 &= 0.0246326501 \\ Z_7 &= 0.01339465321 & Z_8 &= 0.03384338136 \\ Z_9 &= 0.0006378406298 \end{aligned} \tag{9.274}$$

and then the coefficients $A$, $B$, and $C$ from (9.204)–(9.206).

$$\begin{aligned} A &= 3.879887219 \\ B &= -0.08308086729 \\ C &= -0.0004775871233 \end{aligned} \tag{9.275}$$

The first optimal damping ratio $\xi_1^\star$ will then be calculated.

$$\xi_1^\star = 0.1616320694 \tag{9.276}$$

Using $r = r_2 = \sqrt{1.1075} \approx 1.05236$, we find

$$\begin{aligned} Z_3 &= 0.562049056 & Z_4 &= 0.04761904752 \\ Z_5 &= 0.2658369375 & Z_6 &= -0.375123324 \\ Z_7 &= 0.07897478534 & Z_8 &= 0.05273670508 \\ Z_9 &= 0.003760704084 \end{aligned} \tag{9.277}$$

$$\begin{aligned} A &= -9.421012739 \\ B &= 0.1167823931 \\ C &= 0.005076228579 \end{aligned} \tag{9.278}$$

The second optimal damping ratio $\xi_2^\star$ is then equal to:

$$\xi_2^\star = 0.1738496023 \tag{9.279}$$

Therefore, the optimal $\alpha$ is $\alpha^\star = 0.9091$, and the optimal $\xi$ is within the following range.

$$0.1616320694 < \xi^\star < 0.1738496023 \tag{9.280}$$

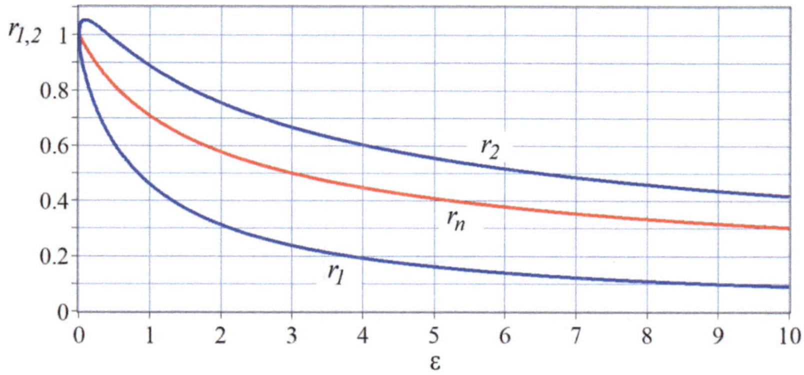


**Fig. 9.38** Behavior of nodal frequencies $r_{1,2}$ as a function of mass ratio $\varepsilon$

**Example 143** ★ The optimal nodal amplitude and optimal $\alpha$ are functions of the mass ratio $\varepsilon$. The equal amplitude at nodal frequencies is a function of mass ratio. The optimal $\alpha$ is also only a function of mass ratio $\varepsilon = m_2/m_1$. A plot of $\alpha$ versus $\varepsilon$ shows how $\alpha$ changes with $\varepsilon$. Practically, we prefer the vibration absorber to be as light as possible, which means $\varepsilon$ to be as low as possible. However, lower $\varepsilon$ produces higher oscillation amplitude. So, it is a trade-off.

Substituting the optimal $\alpha$ from (9.257) in Eq. (9.247),

$$r^4 - \frac{2}{2+\varepsilon} r^2 + \frac{2}{(2+\varepsilon)(1+\varepsilon)^2} = 0 \tag{9.281}$$

provides us with the nodal frequencies $r_{1,2}$.

$$r_{1,2}^2 = \frac{1}{1+\varepsilon}\left(1 \pm \sqrt{\frac{\varepsilon}{2+\varepsilon}}\right) \tag{9.282}$$

Applying $r_{1,2}$ in Eq. (9.251) shows that the equal nodal amplitude $\mu\left(r_{1,2}\right)$ is only a function of mass ratio $\varepsilon$.

$$\mu = \sqrt{\frac{2+\varepsilon}{\varepsilon}} \tag{9.283}$$

The frequencies at nodes for the optimal $\alpha$ are only functions of the mass ratio $\varepsilon$. By selecting $\varepsilon$, the optimal $k_2$ as well as the nodal frequencies are determined. Figure 9.38 illustrates the behavior of $r_{1,2}$ as a function of $\varepsilon$. If $\varepsilon \to 0$, the vibration absorber $m_2$ vanishes, and hence, the system becomes a one $DOF$ primary oscillator. Such a system has only one natural frequency $r_n = 1$. The frequencies $r_{1,2}$ will also approach $r_{1,2} = 1$ by vanishing $m_2$. However, when $\xi \to \infty$ the only natural frequency $r_n$ of the resultant one $DOF$ system is as given in Eq. (9.243).

The nodal frequencies $r_{1,2}$ are always on both sides of the one $DOF$ natural frequency $r_n$, while all of them are decreasing functions of the mass ratio $\varepsilon$.

$$r_1 < r_n < r_2 \tag{9.284}$$

The optimal value of the natural frequency ratio, $\alpha$, is only a function of mass ratio $\varepsilon$, as determined in Eq. (9.257). Figure 9.39 depicts the behavior of $\alpha$ as a function of $\varepsilon$. The value of optimal $\alpha$ and, hence, the value of optimal $k_2$, decreases by increasing $\varepsilon = m_2/m_1$. Therefore, a smaller mass for the vibration absorber needs a softer spring.

**Example 144** ★ Natural frequencies for extreme values of damping. Extreme values of damping are zero and infinity damping. The extreme damping will change the $DOF$ of the system, however, drawing frequency response curves for extreme or any two values of damping determines the nodes of vibrations.

By setting $\xi = 0$ for a given $\varepsilon = 0.1$, we find

$$\mu = \left|\frac{r^2 - 1}{0.1r^2 - \left(r^2 - 1\right)^2}\right| \tag{9.285}$$

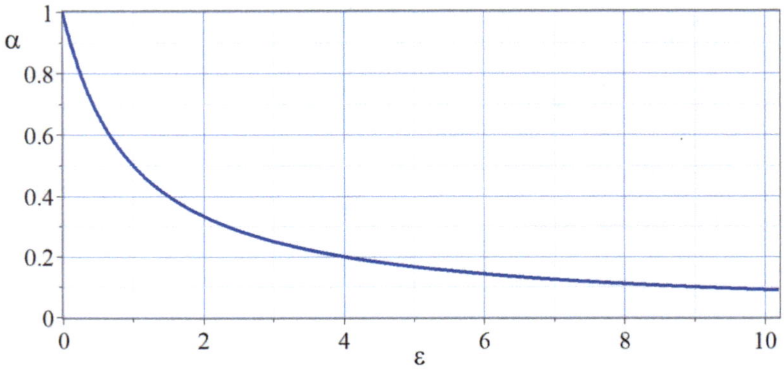


**Fig. 9.39** Optimal value of the natural frequency ratio, $\alpha$, as a function of mass ratio $\varepsilon$

and by setting $\xi = \infty$, we find

$$\mu = \left| \frac{1}{1.1r^2 - 1} \right| \tag{9.286}$$

Having $\xi = 0$ is equivalent to no damping and hence, $\mu$ approaches infinity at the real roots of its denominator, $r_{n_1}$ and $r_{n_2}$, which are the natural frequencies of the system. As an example, the natural frequencies $r_{n_1}$ and $r_{n_2}$, for $\varepsilon = 0.1$, are

$$0.1r^2 - \left(r^2 - 1\right)^2 = 0 \tag{9.287}$$

$$r_{n_1} = 0.85431 \qquad r_{n_2} = 1.1705 \tag{9.288}$$

Having $\xi = \infty$ is equivalent to a rigid connection between $m_1$ and $m_2$. The system would have only one $DOF$ and therefore, $\mu$ approaches infinity at the only root of the denominator, $r_n$

$$1.1r^2 - 1 = 0 \qquad r_n = 0.953 \tag{9.289}$$

where, $r_n$ is always between $r_{n_1}$ and $r_{n_2}$.

$$r_{n_1} < r_n < r_{n_2} \tag{9.290}$$

## 9.6 Summary

The vertical vibrations of vehicles may be modeled by a two-$DOF$ system called *quarter car model*. One-fourth of the body mass, known as *sprung mass*, is suspended by the main suspension of the vehicle $k_s$ and $c_s$. The suspension parameters $k_s$ and $c_s$ are mounted on a wheel of the vehicle, known as *unsprung mass*. The wheel is sitting on the road by a tire with stiffness $k_u$. The governing differential equations of motion for the quarter car model are

$$m_s\ddot{x}_s + c_s\left(\dot{x}_s - \dot{x}_u\right) + k_s\left(x_s - x_u\right) = 0 \tag{9.291}$$

$$m_u\ddot{x}_u + c_s\left(\dot{x}_u - \dot{x}_s\right) + \left(k_u + k_s\right)x_u - k_s x_s = k_u y \tag{9.292}$$

Nondimentionalization showed that the equations of motion and their responses are dependent on only four parameters: the mass ratio $\varepsilon = \frac{m_s}{m_u}$, damping ratio $\xi = \frac{c_s}{2m_s\omega_s}$, systems frequency ratio $\alpha = \frac{\omega_u}{\omega_s}$, and excitation frequency ratio $r = \frac{\omega}{\omega_s}$.

$$\ddot{x}_s + 2\xi\omega_s\left(\dot{x}_s - \dot{x}_u\right) + \omega_s^2\left(x_s - x_u\right) = 0 \tag{9.293}$$

$$\ddot{x}_u + 2\varepsilon\xi\omega_s\left(\dot{x}_u - \dot{x}_s\right) + \left(\omega_u^2 + \varepsilon\omega_s^2\right)x_u - \varepsilon\omega_s^2 x_s = \omega_u^2\, y \tag{9.294}$$

$$\varepsilon = \frac{m_s}{m_u} \qquad \xi = \frac{c_s}{2m_s\omega_s} \qquad \alpha = \frac{\omega_u}{\omega_s} \qquad r = \frac{\omega}{\omega_s} \tag{9.295}$$

$$\omega_s = \sqrt{\frac{k_s}{m_s}} \qquad \omega_u = \sqrt{\frac{k_u}{m_u}} \tag{9.296}$$

Assuming the vehicle is running on a harmonically bumped road

$$y = Y \sin \omega t \tag{9.297}$$

we are able to find the frequency responses of the sprung and unsprung masses, and their relative displacement where $X_s$, $X_u$, and $Z$ are the amplitudes of $x_s$, $x_u$, $z = x_s - x_u$.

$$\mu^2 = \frac{\alpha^2 \left(4\xi^2 r^2 + \alpha^2\right)}{Z_1^2 + Z_2^2} \tag{9.298}$$

$$\tau^2 = \frac{\alpha^2 \left(4\xi^2 r^2 + \alpha^2\right) + r^2 \left(r^2 - 2\alpha^2\right)}{Z_1^2 + Z_2^2} \tag{9.299}$$

$$\eta^2 = \frac{r^4}{Z_1^2 + Z_2^2} \tag{9.300}$$

$$\mu = \left|\frac{X_s}{Y}\right| \qquad \tau = \left|\frac{X_u}{Y}\right| \qquad \eta = \left|\frac{Z}{Y}\right| \tag{9.301}$$

$$Z_1 = \varepsilon r^2 \alpha^2 + \left(\alpha^2 - r^2\right)\left(r^2 - 1\right) \tag{9.302}$$

$$Z_2 = 2\xi r \alpha \left(\varepsilon r^2 + r^2 - 1\right) \tag{9.303}$$

The absolute acceleration of sprung mass and unsprung mass may be defined by the following equations:

$$u = \left|\frac{\ddot{X}_s}{Y\omega_u^2}\right| = r^2 \mu \tag{9.304}$$

$$v = \left|\frac{\ddot{X}_u}{Y\omega_u^2}\right| = r^2 \tau \tag{9.305}$$

The quarter car system is a two $DOF$ system and therefore it has two natural frequencies $r_{n_1}, r_{n_2}$:

$$r_{n_1} = \frac{\sqrt{2}}{2}\sqrt{\alpha^2 (1 + \varepsilon) + 1 - \sqrt{\left(\alpha^2 (1 + \varepsilon) + 1\right)^2 - 4\alpha^2}} \tag{9.306}$$

$$r_{n_2} = \frac{\sqrt{2}}{2}\sqrt{\alpha^2 (1 + \varepsilon) + 1 + \sqrt{\left(\alpha^2 (1 + \varepsilon) + 1\right)^2 - 4\alpha^2}} \tag{9.307}$$

## 9.7 Key Symbols

| | |
|---|---|
| $a, \ddot{x}$ | acceleration |
| $[a], [A]$ | coefficient matrix |
| $[A] = [m]^{-1}[k]$ | coefficient matrix of characteristic equation |
| $c$ | damping |
| $c_e$ | equivalent damping |
| $c_{ij}$ | element of row $i$ and column $j$ of $[c]$ |
| $c_s$ | main suspension damper |
| $[c]$ | damping matrix |
| $D$ | dissipation function, denominator |

| | |
|---|---|
| $e$ | eccentricity arm |
| $E$ | mechanical energy |
| $f$, $\mathbf{F}$ | harmonic force |
| $f = \frac{1}{T}$ | cyclic frequency [Hz] |
| $f_c$ | damper force |
| $f_k$ | spring force |
| $F$ | amplitude of a harmonic force $f$ |
| $F_r$, $Q_r$ | generalized force |
| $g$ | gravitational acceleration |
| $g(r)$ | equation of frequency ratio |
| $H$ | Heaviside function |
| $I$ | mass moment of inertia |
| $\mathbf{I}$ | identity matrix |
| $k$ | stiffness |
| $k_{eq}$ | equivalent stiffness |
| $k_{ij}$ | element of row $i$ and column $j$ of $[k]$ |
| $k_R$ | antiroll bar torsional stiffness |
| $k_s$ | main suspension spring stiffness |
| $k_u$ | tire stiffness |
| $[k]$ | stiffness matrix |
| $[\underline{k}]$ | symmetric stiffness matrix |
| $K$ | kinetic energy |
| $l$ | length |
| $l$ | wheelbase |
| $\mathcal{L}$ | Lagrangian |
| $m$ | mass |
| $m_e$ | eccentric mass |
| $m_{ij}$ | element of row $i$ and column $j$ of $[m]$ |
| $m_s$ | sprung mass |
| $m_u$ | unsprung mass |
| $[m]$ | mass matrix |
| $n$ | number of $DOF$ |
| $N$ | numerator |
| $\mathbf{p}$ | momentum |
| $P$ | potential energy |
| $r = \omega/\omega_s$ | frequency ratio |
| $r_i$, $i \in N$ | nodal frequency ratio |
| $r_n = \omega_n/\omega_s$ | natural frequency ratio |
| $S_u = RMS(u)$ | RMS of $u$ |
| $S_\eta = RMS(\eta)$ | RMS of $\eta$ |
| $t$ | time |
| $T$ | period |
| $u = r^2\alpha^2\mu$ | sprung mass acceleration frequency response |
| $u_{ij}$ | $j$th element of the $i$th mode shape |
| $\mathbf{u}$ | mode shape, eigenvector |
| $\mathbf{u}_i$ | $i$th eigenvector |
| $v = r^2\alpha^2\tau$ | unsprung mass acceleration frequency response |
| $v, \mathbf{v}, \dot{x}, \dot{\mathbf{x}}$ | velocity |
| $W$ | work |
| $x, \mathbf{x}$ | absolute displacement |
| $x_s$ | sprung mass displacement |
| $x_u$ | unsprung mass displacement |
| $X$ | steady-state amplitude of $x$ |

| | |
|---|---|
| $X_s$ | steady-state amplitude of $x_s$ |
| $X_u$ | steady-state amplitude of $x_u$ |
| $y$ | base excitation displacement |
| $Y$ | steady-state amplitude of $y$ |
| $z$ | relative displacement |
| $Z$ | steady-state amplitude of $z$ |
| $Z_i$ | short notation parameter |

Greek

| | |
|---|---|
| $\alpha = \omega_s/\omega_u$ | frequency mass ratio |
| $\alpha^\star$ | optimal damping ratio |
| $\varepsilon = m_s/m_u$ | sprung mass ratio |
| $\eta = \lvert Z/Y \rvert$ | sprung mass relative frequency response |
| $\mu = \lvert X_s/Y \rvert$ | sprung mass frequency response |
| $\xi$ | damping ratio |
| $\xi^\star$ | optimal damping ratio |
| $\tau = \lvert X_u/Y \rvert$ | unsprung mass frequency response |
| $\omega = 2\pi f$ | angular frequency [rad/s] |
| $\omega_i$ | $i$th natural frequency |
| $\omega_s = \sqrt{k_s/m_s}$ | sprung mass frequency |
| $\omega_u = \sqrt{k_u/m_u}$ | unsprung mass frequency |
| $\omega_n$ | natural frequency |
| $\delta$ | deflection |
| $\lambda$ | eigenvalue |
| $\lambda_i$ | $i$th eigenvalue |

Subscript

| | |
|---|---|
| $d$ | driver |
| $i \in N$ | node number |
| $f$ | front |
| $n$ | natural |
| $r$ | rear |
| $s$ | sprung |
| $u$ | unsprung |

Symbols

| | |
|---|---|
| $DOF$ | degree of freedom |
| $FBD$ | free body diagram |
| $ODF$ | one degree of freedom |
| Re | real part |
| $RMS$ | root mean square |
| det | determinant |
| Hz | Hertz |
| [ ] | matrix |
| $\int$ | integral |
| $\lvert\ \rvert$ | absolute value |
| $d$ | differential |
| $\partial$ | partial derivative |
| $\mathcal{L}$ | Lagrangian |
| $O$ | order of magnitude |
| $\nabla$ | gradient |

## References


1. M. Ahmadian, E. Blanchard, Non-dimensionalised closed-form parametric analysis of semi-active vehicle suspensions using a quarter-car model. Veh. Syst. Dyn. **49**(1–2), 219–235 (2011)
2. M. Ahmadian, C.A. Pare, A quarter-car experimental analysis of alternative semiactive control methods. J. Intell. Mater. Syst. Struct. **11**(8), 604–612 (2000)
3. M. Akraminia, M. Tatari, M. Fard, R.N. Jazar, Designing active vehicle suspension system using critic-based control strategy. Nonlinear Eng. **4**(3), 141–154 (2015)
4. R. Alkhatib, R.N. Jazar, M.F. Golnaraghi, Optimal design of passive linear suspension using genetic algorithm. J. Sound vib. **275**(3–5), 665–691 (2004)
5. A. Azizan, M. Fard, M.F. Azari, R.N. Jazar, Effects of vibration on occupant driving performance under simulated driving conditions. Appl. Ergon. **60**, 348–355 (2017)
6. J. Christopherson, R.N. Jazar, Optimization of classical hydraulic engine mounts based on RMS method. Shock Vib. **12**(2), 119–147 (2005)
7. S. Deshpande, S. Mehta, R.N. Jazar, Jump avoidance conditions for piecewise linear vibration isolator, in *International Design Engineering Technical Conferences and Computers and Information in Engineering Conference, DETC2005-85396*, pp. 1955–1962 (2008)
8. S. Deshpande, S. Mehta, R.N. Jazar, Sensitivity of jump avoidance condition of piecewise linear vibration isolator to dynamical parameters, in *ASME 2005 International Mechanical Engineering Congress and Exposition*, November 5–11, Orlando, FL, USA, IMECE2005-81853, pp. 1103–1113 (2008)
9. A. Florin, M.R. Ioan-Cozmin, P. Liliana, Passive suspension modeling using MATLAB, quarter-car model, input signal step type, in *New Technologies and Products in Machine Manufacturing Technologies*, pp. 258–263 (2013)
10. M. Gobbi, G. Mastinu, Analytical description and optimization of the dynamic behavior of passively suspended road vehicles. J. Sound Vib. **245**, 457–481 (2001)
11. A. Hac, I. Youn, Optimal semi-active suspension with preview based on a quarter car model. Trans. ASME J. Vib. Acoust. **114**, 84–92 (1992)
12. R.N. Jazar, R. Alkhatib, M.F. Golnaraghi, Root mean square optimization criterion for vibration behaviour of linear quarter car using analytical methods. Veh. Syst. Dyn. **44**(6), 477–512 (2006)
13. R.N. Jazar, *Advanced Vibrations: A Modern Approach* (Springer, New York, 2013)
14. R.N. Jazar, *Vehicle Dynamics: Theory and Application*, 3rd edn. (Springer, New York, NY, 2017)
15. R.N. Jazar, *Approximation Methods in Science and Engineering* (Springer, New York, 2020)
16. R.N. Jazar, *Perturbation Methods in Science and Engineering* (Springer, New York, 2021)
17. R.N. Jazar, *Advanced Vibrations: Theory and Application* (Springer, New York, 2023)
18. R.N. Jazar, M.F. Golnaraghi, Engine mounts for automotive applications: a survey. Shock Vib. Dig. **34**(5), 363–379 (2002)
19. R.N. Jazar, M.F. Golnaraghi, D.A. Swanson, Steady state vibrations of a piecewise linear system, in *ASME 2001 International Design Engineering Technical Conferences and Computers and Information in Engineering Conference, September 9–12, Pittsburgh, Pennsylvania, USA, DETC2001/VIB-21443*, pp. 1197–1206 (2001)
20. R.N. Jazar, R. Houim, A. Narimani, M.F. Golnaraghi, Frequency response and jump avoidance in a nonlinear passive engine mount. J. Vib. Control **12**(11), 1205–1237 (2006)
21. R.N. Jazar, M. Kazemi, S. Borhani, *Mechanical Vibrations* (in Persian) (Ettehad Publications, Tehran, 1992)
22. H. Marzbani, R.N. Jazar, M. Fard, Hydraulic engine mounts: a survey. J. Vib. Control **20**(10), 1439–1463 (2014)
23. H. Marzbani, R.N. Jazar, A. Khazaei, Smart passive vibration isolation: requirements and unsolved problems. J. Appl. Nonlinear Dyn. **1**(4), 341–386 (2012)
24. L. Meirovitch, *Analytical Methods in Vibrations* (Macmillan, New York, 1967)
25. L. Meirovitch, *Principles and Techniques of Vibrations* (Prentice Hall, Ney Jersey, 1997)
26. Z. Milovanovic, I. Kovacic, M.J. Brennan, On the displacement transmissibility of a base excited viscously damped nonlinear vibration isolator. J. Vib. Acoust. **131**(5), 054502 (2009)
27. A. Narimani, M.F. Golnaraghi, R.N. Jazar, Frequency response of a piecewise linear vibration isolator. J. Vib. Control **10**(12), 1775–1794 (2004)
28. A. Narimani, R.N. Jazar, M.F. Golnaraghi, Vibration analysis of a piecewise linear system, in *Fourteenth U.S. National Congress of Theoretical and Applied Mechanics, Blacksburg, Virginia, USA, June 23–28* (2002)
29. K. Ogata, *System Dynamics* (Prentice Hall, Ney Jersey, 2004)
30. Z. Osinski, *Damping of Vibrations* (A. A. Balkema Publishers, Rotterdam, Netherlands, 1998)
31. W.J. Palm, *Mechanical Vibration* (Wiley, New York, 2006)
32. O. Pogorilyi, R.N. Jazar, P.M. Trivailo, Challenges in exact response of piecewise linear vibration isolator, in *Nonlinear Approaches in Engineering Applications*, ed. by L. Dai, R. Jazar (Springer, New York, 2015)
33. O. Pogorilyi, P.M. Trivailo, R.N. Jazar, On the piecewise linear exact solution. Nonlinear Eng. **3**(4), 189–196 (2014)
34. M. Roseau, *Vibrations in Mechanical Systems* (Springer, Berlin, 1987)
35. M. Setareh, Application of semi-active tuned mass dampers to base-excited systems. Earthq. Eng. Struct. Dyn. **30**(3), 449–462 (2001)
36. R.S. Sharp, D.A. Crolla, Road vehicle suspension system design – a review. Veh. Syst. Dyn. **16**, 167–192 (1987)
37. J.C. Snowdon, *Vibration and Shock in Damped Mechanical Systems* (Wiley, New York, 1968)
38. J.C.N. Smith, Development of a quarter-car test rig for studying road vehicle dynamics, *SAE Technical Paper Series 670139* (1967)
39. P. Stahl, R.N. Jazar, Stability analysis of a piecewise nonlinear vibration isolator, in *ASME International Mechanical Engineering Congress, IMECE2005-81856*, pp. 1115–1123 (2005)
40. P. Stahl, R.N. Jazar, Frequency response analysis of piecewise nonlinear vibration isolator, in *20th ASME Biennial Conference on Mechanical Vibration and Noise, Long Beach, CA, USA, September 24–28* (2005)
41. F.S. Tse, I.E. Morse, R.T. Hinkle, *Mechanical Vibrations Theory and Applications* (Allyn and Bacon Inc., Boston, MA, 1978)
42. G. Verros, S. Natsiavas, G. Stepan, Control and dynamics of quarter-car models with dual-rate damping. J. Vib. Control **6**(7), 1045–1063 (2000)
43. M. Wiercigroch, V.W.T. Sin, Experimental study of a symmetrical piecewise base-excited oscillator. J. Appl. Mech. **65**(3), 657–663 (1998)

# 10 Bicycle Car Vibration Model and Body Pitch Mode

The simplest vehicle model to study the pitch mode of vibrations is the bicycle model in which we longitudinally cut a vehicle and take one side to study. Figure 10.1 illustrates a bicycle vibrating model of a vehicle. This model includes the body bounce $x$, body pitch $\theta$, wheels hop $x_1$ and $x_2$, and the road excitations $y_1$ and $y_2$.

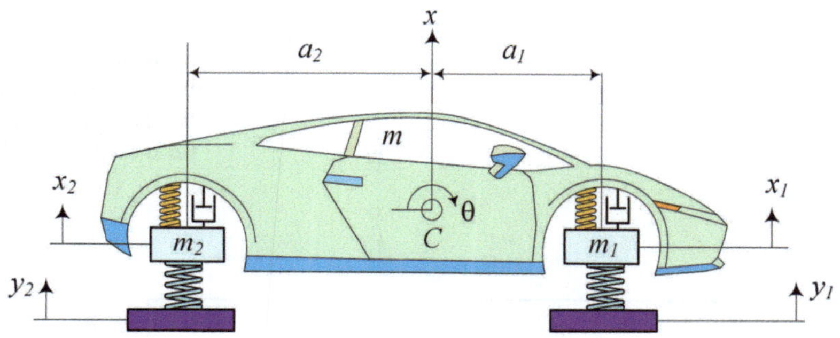


**Fig. 10.1** A bicycle vibrating model of a vehicle

## 10.1 Mathematical Model

The bicycle vehicle model has four degrees of freedom indicated by the coordinates $x$, $\theta$, $x_1$, and $x_2$ along with two input road excitations $y_1$ and $y_2$ as shown in Fig. 10.1. The equations of motion of the bicycle vibrating model of the vehicle are

$$\begin{aligned} & m\ddot{x} + c_1\left(\dot{x} - \dot{x}_1 - a_1\dot{\theta}\right) + c_2\left(\dot{x} - \dot{x}_2 + a_2\dot{\theta}\right) \\ & \quad + k_1\left(x - x_1 - a_1\theta\right) + k_2\left(x - x_2 + a_2\theta\right) = 0 \end{aligned} \tag{10.1}$$

$$\begin{aligned} & I_y\ddot{\theta} - a_1c_1\left(\dot{x} - \dot{x}_1 - a_1\dot{\theta}\right) + a_2c_2\left(\dot{x} - \dot{x}_2 + a_2\dot{\theta}\right) \\ & \quad - a_1k_1\left(x - x_1 - a_1\theta\right) + a_2k_2\left(x - x_2 + a_2\theta\right) \\ & \quad + c_P\left(\dot{\theta} + \frac{\dot{x}_1 - \dot{x}_2}{l}\right) + k_P\left(\theta + \frac{x_1 - x_2}{l}\right) = 0 \end{aligned} \tag{10.2}$$

$$\begin{aligned} & m_1\ddot{x}_1 - c_1\left(\dot{x} - \dot{x}_1 - a_1\dot{\theta}\right) - k_1\left(x - x_1 - a_1\theta\right) \\ & \quad - \frac{c_P}{l}\left(\dot{\theta} + \frac{\dot{x}_1 - \dot{x}_2}{l}\right) - \frac{k_P}{l}\left(\theta + \frac{x_1 - x_2}{l}\right) \\ & \quad + k_{t_1}\left(x_1 - y_1\right) = 0 \end{aligned} \tag{10.3}$$



R. N. Jazar, H. Marzbani, *Vehicle Vibrations*, https://doi.org/10.1007/978-3-031-43486-0_10

**Table 10.1** Parameters of the bicycle vehicle model

| Parameter | Meaning |
|---|---|
| $m$ | Half of body mass [kg] |
| $m_1$ | Mass of a front wheel [kg] |
| $m_2$ | Mass of a rear wheel [kg] |
| $I_y$ | Half of body lateral mass moment [kg m$^2$] |
| $k_1$ | Front spring stiffness rate [N/m] |
| $k_2$ | Rear spring stiffness rate [N/m] |
| $c_1$ | Front shock absorber damping rate [N s/m] |
| $c_2$ | Rear shock absorber damping rate [N s/m] |
| $k_{t_1}$ | Front tire stiffness rate [N/m] |
| $k_{t_2}$ | Rear tire stiffness rate [N/m] |
| $c_{t_1}$ | Front tire damping [N s/m] |
| $c_{t_2}$ | Rear tire damping [N s/m] |
| $k_P$ | Antipitch bar stiffness date [N s m/rad] |
| $c_P$ | Antipitch bar damping date [N m/rad] |
| $a_1$ | Distance of $C$ from front axle [m] |
| $a_2$ | Distance of $C$ from rear axle [m] |
| $l$ | $= a_1 + a_2$, wheelbase of vehicle |

**Table 10.2** Variables of the bicycle vehicle model

| Variable | Meaning |
|---|---|
| $x$ | Body vertical motion coordinate |
| $x_1$ | Front wheel vertical motion coordinate |
| $x_2$ | Rear wheel vertical motion coordinate |
| $\theta$ | Body pitch motion coordinate |
| $y_1$ | Road excitation under the front wheel |
| $y_2$ | Road excitation under the rear wheel |

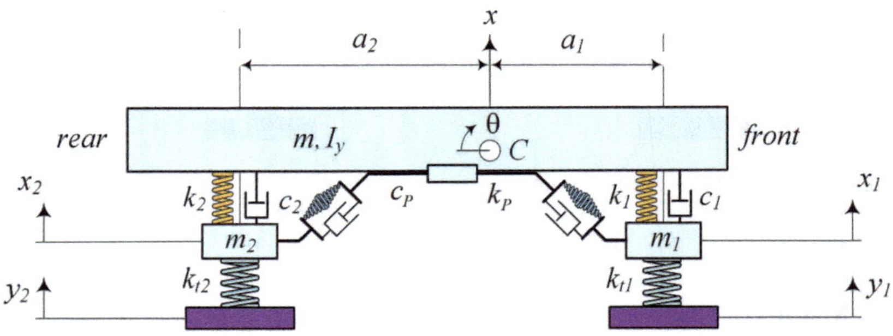


**Fig. 10.2** Bicycle model for a vehicle vibrations

$$m_2\ddot{x}_2 - c_2\left(\dot{x} - \dot{x}_2 + a_2\dot{\theta}\right) - k_2\left(x - x_2 + a_2\theta\right)$$
$$+\frac{c_P}{l}\left(\dot{\theta} + \frac{\dot{x}_1 - \dot{x}_2}{l}\right) + \frac{k_P}{l}\left(\theta + \frac{x_1 - x_2}{l}\right)$$
$$+k_{t_2}\left(x_2 - y_2\right) = 0 \tag{10.4}$$

The definition of the employed parameters and variables are indicated in Tables 10.1 and 10.2.

***Proof*** Figure 10.2 illustrates the mathematical model of the bicycle vehicle system. The body of the vehicle is assumed to be a rigid bar. The bar has a mass $m$, which is half of the total vehicle body mass, and a lateral mass moment $I_y$, which is half of the total vehicle body mass moment. The front and rear wheels have masses $m_1$ and $m_2$, respectively. The stuffiness of front and rear springs are modeled by $k_1$ and $k_2$. The tires stiffness are indicated by parameters $k_{t_1}$ and $k_{t_2}$. The difference in front and rear stuffiness is because the rear spring and tires are usually stiffer than the fronts, although in a simpler model we may assume $k_1 = k_2$ and $k_{t_1} = k_{t_2}$. The front and rear suspensions will have shock absorbers with damping coefficients of $c_1$ and $c_2$. Damping of tires are much smaller than the damping of shock absorbers, and we may ignore the tire damping for simpler calculation.

The vehicle may also have an antipitch bar with a torsional stiffness $k_P$ and torsional damping $c_P$. The antipitch bar provides a torque $-M_P$ proportional to the pitch angle $\theta$ and pitch rate with respect to the wheels' movements.

$$\begin{aligned} M_P &= -k_P \frac{(x - x_1 + a_2\theta) - (x - x_2 - a_1\theta)}{l} \\ &\quad - c_P \frac{\left(\dot{x} - \dot{x}_1 + a_2\dot{\theta}\right) - \left(\dot{x} - \dot{x}_2 - a_1\dot{\theta}\right)}{l} \\ &= -k_P\left(\theta + \frac{x_2 - x_1}{l}\right) - c_P\left(\dot{\theta} + \frac{\dot{x}_2 - \dot{x}_1}{l}\right) \end{aligned} \tag{10.5}$$

To find the equations of motion for the bicycle vehicle vibrating model, we use the Lagrange method. The kinetic $K$ and potential $P$ energies of the system are

$$K = \frac{1}{2}m\dot{x}^2 + \frac{1}{2}m_1\dot{x}_1^2 + \frac{1}{2}m_2\dot{x}_2^2 + \frac{1}{2}I_y\dot{\theta}^2 \tag{10.6}$$

$$\begin{aligned} P &= \frac{1}{2}k_{t_1}(x_1 - y_1)^2 + \frac{1}{2}k_{t_2}(x_2 - y_2)^2 + \frac{1}{2}k_P\left(\theta + \frac{x_1 - x_2}{l}\right)^2 \\ &\quad + \frac{1}{2}k_1(x - x_1 - a_1\theta)^2 + \frac{1}{2}k_2(x - x_2 + a_2\theta) \end{aligned} \tag{10.7}$$

and the dissipation function $D$ is

$$D = \frac{1}{2}c_1\left(\dot{x} - \dot{x}_1 - a_1\dot{\theta}\right)^2 + \frac{1}{2}c_2\left(\dot{x} - \dot{x}_2 + a_2\dot{\theta}\right)^2 + \frac{1}{2}c_P\left(\dot{\theta} + \frac{\dot{x}_1 - \dot{x}_2}{l}\right)^2 \tag{10.8}$$

Applying Lagrange method provides the equations of motion (10.1)–(10.4).

$$\frac{d}{dt}\left(\frac{\partial K}{\partial \dot{q}_r}\right) - \frac{\partial K}{\partial q_r} + \frac{\partial D}{\partial \dot{q}_r} + \frac{\partial P}{\partial q_r} = f_r \qquad r = 1, 2, \cdots 4 \tag{10.9}$$

The set of equations can also be rearranged in a matrix form

$$[m]\,\ddot{\mathbf{x}} + [c]\,\dot{\mathbf{x}} + [k]\,\mathbf{x} = \mathbf{F} \tag{10.10}$$

where

$$\mathbf{x} = \begin{bmatrix} x \\ \theta \\ x_1 \\ x_2 \end{bmatrix} \tag{10.11}$$

$$[m] = \begin{bmatrix} m & 0 & 0 & 0 \\ 0 & I_y & 0 & 0 \\ 0 & 0 & m_1 & 0 \\ 0 & 0 & 0 & m_2 \end{bmatrix} \tag{10.12}$$

$$[c] = \begin{bmatrix} c_1 + c_2 & a_2c_2 - a_1c_1 & -c_1 & -c_2 \\ a_2c_2 - a_1c_1 & c_1a_1^2 + c_2a_2^2 + c_P & a_1c_1 + \frac{c_P}{l} & -a_2c_2 - \frac{c_P}{l} \\ -c_1 & a_1c_1 + \frac{c_P}{l} & c_1 + \frac{c_P}{l^2} & -\frac{c_P}{l^2} \\ -c_2 & -a_2c_2 - \frac{c_P}{l} & -\frac{c_P}{l^2} & c_2 + \frac{c_P}{l^2} \end{bmatrix} \tag{10.13}$$

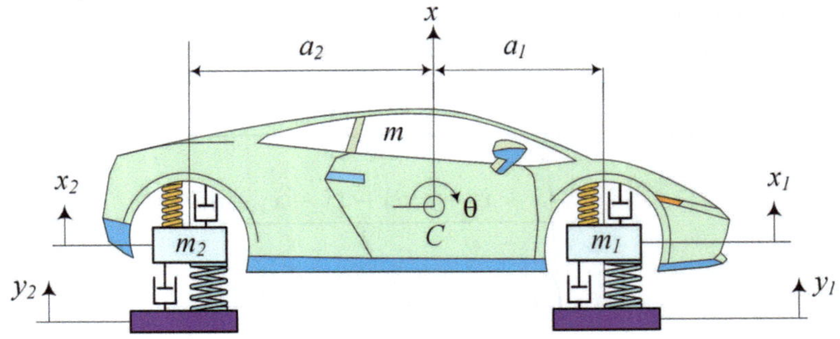


**Fig. 10.3** A bicycle vibrating model of a vehicle with tire damping

$$[k] = \begin{bmatrix} k_1+k_2 & a_2k_2-a_1k_1 & -k_1 & -k_2 \\ a_2k_2-a_1k_1 & k_1a_1^2+k_2a_2^2+k_P & a_1k_1+\frac{k_P}{l} & -a_2k_2-\frac{k_P}{l} \\ -k_1 & a_1k_1+\frac{k_P}{l} & k_1+k_{t_1}+\frac{k_P}{l^2} & -\frac{k_P}{l^2} \\ -k_2 & -a_2k_2-\frac{k_P}{l} & -\frac{k_P}{l^2} & k_2+k_{t_2}+\frac{k_P}{l^2} \end{bmatrix} \tag{10.14}$$

$$\mathbf{F} = \begin{bmatrix} 0 \\ 0 \\ y_1k_{t_1} \\ y_2k_{t_2} \end{bmatrix} \tag{10.15}$$

■

**Example 145** ★ Tire damping. Tire damping is very low and it can be ignored in order to work with simpler equations of motion. Here we see what would be the equations of motion of bicycle vehicle model if we included tire damping.

We may add tire dampers $c_{t_1}$ and $c_{t_2}$ in parallel to tire stiffness $k_{t_1}$ and $k_{t_2}$, as shown in Fig. 10.3, to model possible damping in tires. The value of tire dampers $c_{t_1}$ and $c_{t_2}$ for tires, compared to damping of suspension shock absorbers $c_1$ and $c_2$, are very small. Hence, we may ignore $c_{t_1}$ and $c_{t_2}$ to simplify the model. Including $c_{t_1}$ and $c_{t_2}$ provides the following equations of motion of bicycle car model,

$$[m]\,\ddot{\mathbf{x}}+[c]\,\dot{\mathbf{x}}+[k]\,\mathbf{x}=\mathbf{F} \tag{10.16}$$

where

$$\mathbf{x} = \begin{bmatrix} x \\ \theta \\ x_1 \\ x_2 \end{bmatrix} \tag{10.17}$$

$$[m] = \begin{bmatrix} m & 0 & 0 & 0 \\ 0 & I_y & 0 & 0 \\ 0 & 0 & m_1 & 0 \\ 0 & 0 & 0 & m_2 \end{bmatrix} \tag{10.18}$$

$$[c] = \begin{bmatrix} c_1+c_2 & a_2c_2-a_1c_1 & -c_1 & -c_2 \\ a_2c_2-a_1c_1 & c_1a_1^2+c_2a_2^2+c_P & a_1c_1+\frac{c_P}{l} & -a_2c_2-\frac{c_P}{l} \\ -c_1 & a_1c_1+\frac{c_P}{l} & c_1+\frac{c_P}{l^2}+c_{t_1} & -\frac{c_P}{l^2} \\ -c_2 & -a_2c_2-\frac{c_P}{l} & -\frac{c_P}{l^2} & c_2+\frac{c_P}{l^2}+c_{t_2} \end{bmatrix} \tag{10.19}$$

$$[k] = \begin{bmatrix} k_1+k_2 & a_2k_2 - a_1k_1 & -k_1 & -k_2 \\ a_2k_2 - a_1k_1 & k_1a_1^2+k_2a_2^2+k_P & a_1k_1+\frac{k_P}{l} & -a_2k_2 - \frac{k_P}{l} \\ -k_1 & a_1k_1+\frac{k_P}{l} & k_1+k_{t_1}+\frac{k_P}{l^2} & -\frac{k_P}{l^2} \\ -k_2 & -a_2k_2 - \frac{k_P}{l} & -\frac{k_P}{l^2} & k_2+k_{t_2}+\frac{k_P}{l^2} \end{bmatrix} \tag{10.20}$$

$$\mathbf{F} = \begin{bmatrix} 0 \\ 0 \\ k_{t_1} y_1 + c_{t_1} \dot{y}_1 \\ k_{t_2} y_2 + c_{t_2} \dot{y}_2 \end{bmatrix} \tag{10.21}$$

because:

$$K = \frac{1}{2} m \dot{x}^2 + \frac{1}{2} m_1 \dot{x}_1^2 + \frac{1}{2} m_2 \dot{x}_2^2 + \frac{1}{2} I_y \dot{\theta}^2 \tag{10.22}$$

$$P = \frac{1}{2} k_{t_1} (x_1 - y_1)^2 + \frac{1}{2} k_{t_2} (x_2 - y_2)^2 + \frac{1}{2} k_P \left(\theta + \frac{x_1 - x_2}{l}\right)^2 + \frac{1}{2} k_1 (x - x_1 - a_1\theta)^2 + \frac{1}{2} k_2 (x - x_2 + a_2\theta) \tag{10.23}$$

$$D = \frac{1}{2} c_{t_1} (\dot{x}_1 - \dot{y}_1)^2 \frac{1}{2} c_{t_2} (\dot{x}_2 - \dot{y}_2)^2 + \frac{1}{2} c_P \left(\dot{\theta} + \frac{\dot{x}_1 - \dot{x}_2}{l}\right)^2 + \frac{1}{2} c_1 \left(\dot{x} - \dot{x}_1 - a_1\dot{\theta}\right)^2 + \frac{1}{2} c_2 \left(\dot{x} - \dot{x}_2 + a_2\dot{\theta}\right)^2 \tag{10.24}$$

**Example 146** Mathematical model's limitations. Although the bicycle car model is the best simple model to examine pitch vibrations of vehicles, it has some limitations. There are multiple facts that must be considered.

The bicycle car model contains no component of the geometric effects of the width of a full car and offers no possibility of studying lateral interconnections. However, it contains the most fundamental features of real vehicles and includes a proper representation of the problem of controlling body bounce, body pitch, wheel, and wheel-body load variations. In the bicycle car model, we assume that tire is always in contact with the ground. This assumption is true at low frequency and low amplitude but will not be true at high frequency or high amplitude of road excitations. A better model must be able to include the possibility of separation between the tire and ground.

The main suspension parameters for vertical vibrations of vehicles are designed and optimized by quarter car model. Then, the suspension of the four $DOF$ bicycle vibrating systems will be examined for fine tuning of the suspension parameters by studying the pitch mode of vibrations. Flat ride tuning is a good example of an optimization engineering problem that has been investigated by bicycle car model. We will study the flat ride tuning in another chapter.

**Example 147** Nonlinear damping. Out of the suspension elements, shock absorbers are the most nonlinear element. By an idealization, they are bilinear with very low damping in bound and a high damping in rebound. Here is the equations of motion of bicycle car model with bilinear dampers.

Figure 10.4 illustrates the bilinear damping model of practical shock absorbers in extreme case when there is no damping in bound and a constant damping in rebound, considering $x$ and $y$ as displacements of two ends of the shock absorber. Employing the displacement of the vehicle body and wheels, we can define bilineaar damping by a Heaviside function $H\left(\dot{x} - \dot{x}_1 - a_1\dot{\theta}\right)$ and $H\left(\dot{x} - \dot{x}_2 + a_2\dot{\theta}\right)$ to rewrite the equation of motion of the bicycle vehicle model.

$$m\ddot{x} + c_1 \left(\dot{x} - \dot{x}_1 - a_1\dot{\theta}\right) H\left(\dot{x} - \dot{x}_1 - a_1\dot{\theta}\right) + c_2 \left(\dot{x} - \dot{x}_2 + a_2\dot{\theta}\right) H\left(\dot{x} - \dot{x}_2 + a_2\dot{\theta}\right) + k_1 (x - x_1 - a_1\theta) + k_2 (x - x_2 + a_2\theta) = 0 \tag{10.25}$$

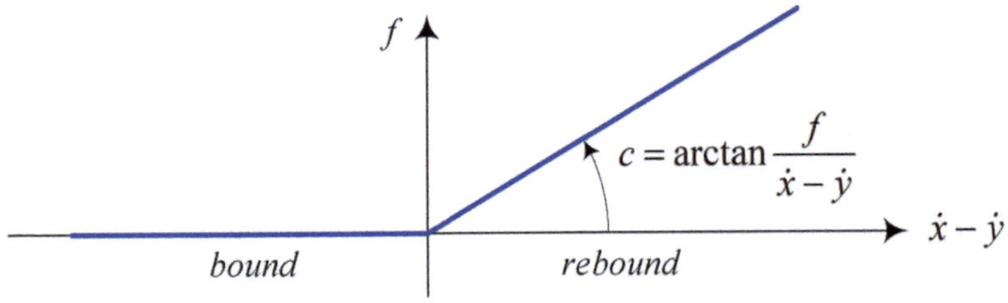


**Fig. 10.4** A bilinear damping model with no damping in bound and a constant damping in rebound

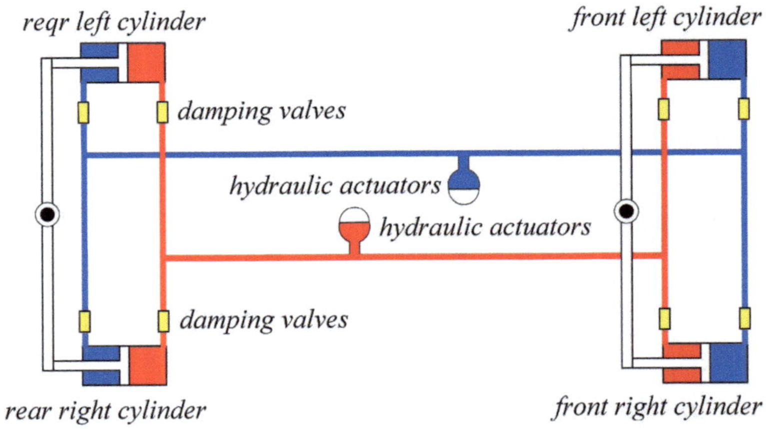


**Fig. 10.5** Illustration of an antipitch hydraulic system

$$I_y\ddot{\theta} - a_1c_1\left(\dot{x} - \dot{x}_1 - a_1\dot{\theta}\right)\, H\left(\dot{x} - \dot{x}_1 - a_1\dot{\theta}\right) + a_2c_2\left(\dot{x} - \dot{x}_2 + a_2\dot{\theta}\right)\, H\left(\dot{x} - \dot{x}_2 + a_2\dot{\theta}\right) \tag{10.26}$$

$$- a_1k_1\left(x - x_1 - a_1\theta\right) + a_2k_2\left(x - x_2 + a_2\theta\right) + c_P\left(\dot{\theta} + \frac{\dot{x}_1 - \dot{x}_2}{l}\right) + k_P\left(\theta + \frac{x_1 - x_2}{l}\right) = 0 \tag{10.27}$$

$$m_1\ddot{x}_1 - c_1\left(\dot{x} - \dot{x}_1 - a_1\dot{\theta}\right)\, H\left(\dot{x} - \dot{x}_1 - a_1\dot{\theta}\right) - k_1\left(x - x_1 - a_1\theta\right) - \frac{c_P}{l}\left(\dot{\theta} + \frac{\dot{x}_1 - \dot{x}_2}{l}\right)$$
$$- \frac{k_P}{l}\left(\theta + \frac{x_1 - x_2}{l}\right) + k_{t_1}\left(x_1 - y_1\right) = 0 \tag{10.28}$$

$$m_2\ddot{x}_2 - c_2\left(\dot{x} - \dot{x}_2 + a_2\dot{\theta}\right)\, H\left(\dot{x} - \dot{x}_2 + a_2\dot{\theta}\right) - k_2\left(x - x_2 + a_2\theta\right) + \frac{c_P}{l}\left(\dot{\theta} + \frac{\dot{x}_1 - \dot{x}_2}{l}\right)$$
$$+ \frac{k_P}{l}\left(\theta + \frac{x_1 - x_2}{l}\right) + k_{t_2}\left(x_2 - y_2\right) = 0 \tag{10.29}$$

**Example 148** Practical antipitch system. Practically, a passive antipitch system can be built by hydraulic circuit system.

To implement an antipitch system, a hydraulic circuit can be utilized, as depicted in Fig. 10.5. The system consists of the following components:

1. Hydraulic Actuators: Each wheel is equipped with a hydraulic actuator that facilitates the movement.
2. Interconnecting Oil Circuits: Two oil circuits interconnect the hydraulic actuators, allowing hydraulic pressure to be transmitted between them.
3. Damping Valves: These valves are integrated into the circuit to consume energy and provide damping. They help dissipate any pitch vibrations that may occur.

4. Gas-Filled Diaphragm Accumulators: Each hydraulic circuit incorporates a gas-filled diaphragm accumulator. These accumulators balance the oil pressure within the circuits.

When the front wheels move upward relative to the vehicle body, the oil circuits prompt the rear wheels to also move upward, effectively counteracting pitch motion and maintaining a level body. The damping valves play a crucial role in dissipating energy and reducing any pitch-related vibrations. A similar reaction occurs when the front wheels move downward, as well as when the rear wheels move upward and downward.

By employing this hydraulic circuit, the antipitch system effectively regulates the movement of the wheels, ensuring pitch control and damping vibrations to enhance the stability and comfort of the vehicle.

**Example 149** Nondimensionalized equations of motion. The nondimensionalized set of equations minimizes the number of parameters and maximizes the generality of the equations. The result of the equations can be applied to infinite number of similar systems whose dimensionless parameters are equal. Here is to derive the bicycle car model nondimensional equations.

All dynamics vibrating systems are involved only with the three main dimensions of mass $M$, length $L$, and time $T$. To make the equations of motion dimensionless, we make the variables of the equations nondimensional we define new dimensionless variables. In bicycle car model, there are five variables with length dimension, $x$, $x_1$, $x_2$, $y_1$, $y_2$, one dimensionless variable $\theta$, and a time variable $t$.

Let us define a characteristic length $x_0$ and a characteristic time $t_0$, to define nondimensionalized displacement $s$, $s_1$, $s_2$, $q_1$, $q_2$, and time $\tau$ and replace the actual variables in the equations of motion.

$$s = \frac{x}{x_0} \qquad s_1 = \frac{x_1}{x_0} \qquad s_2 = \frac{x_2}{x_0}$$
$$q_1 = \frac{y_1}{x_0} \qquad q_2 = \frac{y_2}{x_0} \qquad \tau = \frac{t}{t_0} \tag{10.30}$$

In vehicle vibrations, the characteristic length $x_0$ is usually a constant length such as static deflection of one of the masses or amplitude of displacement excitation, and $t_0$ is usually inverse of a frequency such as a natural frequency or an excitation frequency.

Using nondimensionalized time $\tau$, we redefine derivatives.

$$\dot{x} = \frac{dx}{dt} = x_0 \frac{ds}{dt} = x_0 \frac{ds}{d\tau}\frac{d\tau}{dt} = \frac{x_0}{t_0} s' \tag{10.31}$$

$$\ddot{x} = \frac{d}{dt}\frac{dx_s}{dt} = \frac{x_0}{t_0}\frac{d}{dt}s' = \frac{x_0}{t_0}\frac{d\tau}{dt}\frac{d}{d\tau}s' = \frac{x_0}{t_0^2}s'' \tag{10.32}$$

$$s' = \frac{ds}{d\tau} \qquad s'' = \frac{d^2 s}{d\tau^2} \tag{10.33}$$

For the dimensionless variable $\theta$, we have

$$\dot{\theta} = \frac{1}{t_0}\theta' \qquad \ddot{\theta} = \frac{1}{t_0^2}\theta'' \qquad \theta' = \frac{d\theta}{d\tau} \tag{10.34}$$

The equation of motion can now be rewritten as

$$\frac{m}{t_0^2}s'' + \frac{c_1}{t_0}\left(s' - s_1' - \frac{a_1}{x_0}\theta'\right) + \frac{c_2}{t_0}\left(s' - s_2' + \frac{a_2}{x_0}\theta'\right)$$
$$+k_1\left(s - s_1 - \frac{a_1}{x_0}\theta\right) + k_2\left(s - s_2 + \frac{a_2}{x_0}\theta\right) = 0 \tag{10.35}$$

$$\frac{I_y}{x_0 t_0^2}\theta'' - a_1\frac{c_1}{t_0}\left(s' - s_1' - \frac{a_1}{x_0}\theta'\right) + a_2\frac{c_2}{t_0}\left(s' - s_2' + \frac{a_2}{x_0}\theta'\right)$$
$$-a_1 k_1\left(s - s_1 - \frac{a_1}{x_0}\theta\right) + a_2 k_2\left(s - s_2 + \frac{a_2}{x_0}\theta\right)$$
$$+\frac{c_P}{l t_0}\left(\frac{l}{x_0}\theta' + s_1' - s_2'\right) + \frac{k_P}{l}\left(\frac{l}{x_0}\theta + s_1 - s_2\right) = 0 \tag{10.36}$$

$$\frac{m_1}{t_0^2}s_1'' - \frac{c_1}{t_0}\left(s' - s_1' - \frac{a_1}{x_0}\theta'\right) - k_1\left(s - s_1 - \frac{a_1}{x_0}\theta\right)$$
$$-\frac{c_P}{l^2 t_0}\left(\frac{l}{x_0}\theta' + s_1' - s_2'\right) - \frac{k_P}{l^2}\left(\frac{l}{x_0}\theta + s_1 - s_2\right)$$
$$+k_{t_1}\left(s_1 - q_1\right) = 0 \tag{10.37}$$

$$\frac{m_2}{t_0^2}s_2'' - \frac{c_2}{t_0}\left(s' - s_2' + \frac{a_2}{x_0}\theta'\right) - k_2\left(s - s_2 + \frac{a_2}{x_0}\theta\right)$$
$$+\frac{c_P}{l^2 t_0}\left(\frac{l}{x_0}\theta' + s_1' - s_2'\right) + \frac{k_P}{l^2}\left(\frac{l}{x_0}\theta + s_1 - s_2\right) \tag{10.38}$$
$$+k_{t_2}\left(s_2 - q_2\right) = 0 \tag{10.39}$$

After dividing by $m/t_0^2$ we have

$$s'' + 2\xi_1\omega_1 t_0\left(s' - s_1' - \frac{a_1}{x_0}\theta'\right) + \omega_1^2 t_0^2\left(s - s_1 - \frac{a_1}{x_0}\theta\right)$$
$$+2\xi_2\omega_2 t_0\left(s' - s_2' + \frac{a_2}{x_0}\theta'\right) + \omega_2^2 t_0^2\left(s - s_2 + \frac{a_2}{x_0}\theta\right) = 0 \tag{10.40}$$

$$\frac{I_y}{m x_0}\theta'' - a_1 2\xi_1\omega_1 t_0\left(s' - s_1' - \frac{a_1}{x_0}\theta'\right)$$
$$+a_2 2\xi_2\omega_2 t_0\left(s' - s_2' + \frac{a_2}{x_0}\theta'\right) - a_1\omega_1^2 t_0^2\left(s - s_1 - \frac{a_1}{x_0}\theta\right)$$
$$+a_2\omega_2^2 t_0^2\left(s - s_2 + \frac{a_2}{x_0}\theta\right) + 2\xi_P\omega_P a_1 t_0\left(\frac{l}{x_0}\theta' + s_1' - s_2'\right)$$
$$+\omega_P^2 a_1 t_0^2\left(\frac{l}{x_0}\theta + s_1 - s_2\right) = 0 \tag{10.41}$$

$$\varepsilon_1 s_1'' - 2\xi_1\omega_1 t_0\left(s' - s_1' - \frac{a_1}{x_0}\theta'\right)$$
$$-\omega_1^2 t_0^2\left(s - s_1 - \frac{a_1}{x_0}\theta\right) - 2\xi_P\omega_P\frac{a_1}{l}t_0\left(\frac{l}{x_0}\theta' + s_2' - s_1'\right) \tag{10.42}$$
$$-\omega_P^2\frac{a_1}{l}t_0^2\left(\frac{l}{x_0}\theta + s_1 - s_2\right) + \omega_{t_1}^2 t_0^2\left(s_1 - q_1\right) = 0 \tag{10.43}$$

$$\varepsilon_2 s_2'' - 2\xi_2\omega_2 t_0\left(s' - s_2' + \frac{a_2}{x_0}\theta'\right)$$

$$-\omega_2^2 t_0^2 \left(s - s_2 + \frac{a_2}{x_0}\theta\right) + 2\xi_P \omega_P \frac{a_1}{l} t_0 \left(\frac{l}{x_0}\theta' + s_1' - s_2'\right)$$
$$+\omega_P^2 \frac{a_1}{l} t_0^2 \left(\frac{l}{x_0}\theta + s_1 - s_2\right) + \omega_{t_2}^2 t_0^2 (s_2 - q_2) = 0 \tag{10.44}$$

where

$$\varepsilon_1 = \frac{m_1}{m} \quad \varepsilon_2 = \frac{m_2}{m} \quad \omega_1 = \sqrt{\frac{k_1}{m}} \quad \omega_2 = \sqrt{\frac{k_2}{m}}$$
$$\xi_1 = \frac{c_1}{2m\omega_1} \quad \xi_2 = \frac{c_2}{2m\omega_2} \quad \gamma = \frac{a_1}{a_2} \quad I_0 = \frac{I_y}{m a_1 a_2}$$
$$\xi_P = \frac{c_P}{2m\omega_P} \quad \omega_P = \sqrt{\frac{k_P}{m l a_1}} \quad \omega_{t_1} = \sqrt{\frac{k_{t_1}}{m}} \quad \omega_{t_2} = \sqrt{\frac{k_{t_2}}{m}}$$
$$s = \frac{x}{a_2} \quad s_1 = \frac{x_1}{a_2} \quad s_2 = \frac{x_2}{a_2} \quad q_1 = \frac{y_1}{a_2} \quad q_2 = \frac{y_2}{a_2}$$
$$\alpha = \frac{\omega_2}{\omega_1} \quad \alpha_P = \frac{\omega_P}{\omega_1} \quad \alpha_{t_1} = \frac{\omega_{t_1}}{\omega_1} \quad \alpha_{t_2} = \frac{\omega_{t_2}}{\omega_1} \tag{10.45}$$

Let us define the characteristic length $x_0$ and time $t_0$.

$$x_0 = a_2 \qquad t_0 = \frac{1}{\omega_1} \tag{10.46}$$

These definitions make the equations simpler.

$$s'' + 2\xi_1 \left(s' - s_1' - \gamma\theta'\right) + (s - s_1 - \gamma\theta) + 2\xi_2\alpha \left(s' - s_2' + \theta'\right) + \alpha^2 (s - s_2 + \theta) = 0 \tag{10.47}$$

$$I_0\theta'' - 2\xi_1 \left(s' - s_1' - \gamma\theta'\right) + \frac{2\xi_2}{\gamma}\alpha \left(s' - s_2' + \theta'\right) - (s - s_1 - \gamma\theta) + \frac{\alpha^2}{\gamma} (s - s_2 + \theta)$$
$$+ 2\xi_P\alpha_P \left((1+\gamma)\theta' + s_1' - s_2'\right) + \alpha_P^2 ((1+\gamma)\theta + s_1 - s_2) = 0 \tag{10.48}$$

$$\varepsilon_1 s_1'' - 2\xi_1 \left(s' - s_1' - \gamma\theta'\right) - (s - s_1 - \gamma\theta) - 2\xi_P \frac{\gamma}{1+\gamma}\alpha_P \left((1+\gamma)\theta' + s_1' - s_2'\right)$$
$$- \frac{\gamma}{1+\gamma}\alpha_P^2 ((1+\gamma)\theta + s_1 - s_2) + \alpha_{t_1}^2 (s_1 - q_1) = 0 \tag{10.49}$$

$$\varepsilon_2 s_2'' - 2\xi_2\alpha \left(s' - s_2' + \theta'\right) - \alpha^2 (s - s_2 + \theta) + 2\xi_P \frac{\gamma}{1+\gamma}\alpha_P \left((1+\gamma)\theta' + s_1' - s_2'\right)$$
$$+ \frac{\gamma}{1+\gamma}\alpha_P^2 ((1+\gamma)\theta + s_1 - s_2) + \alpha_{t_2}^2 (s_2 - q_2) = 0 \tag{10.50}$$

Employing the nondimensionalization, we reduced the number of parameters from 14: $m$, $m_1$, $m_2$, $c_1$, $c_2$, $c_P$, $k_1$, $k_2$, $k_P$, $k_{t_1}$, $k_{t_2}$ $I_y$, $a_1$, $a_2$ to 11: $\varepsilon_1$, $\varepsilon_2$, $I_0$, $\xi_1$, $\xi_2$, $\xi_P$, $\alpha$, $\alpha_P$, $\alpha_{t_1}$, $\alpha_{t_2}$, $\gamma$. When the number of parameters are too many, then although nondimensionalization provides us with the minimum number of independent parameters, they can still be too many. However, the parameters can be divided into two groups: constant and adjustable. Considering optimization of the suspension of the bicycle car model, then $c_1$, $c_2$, $c_P$, $k_1$, $k_2$, $k_P$ or $\xi_1$, $\xi_2$, $\xi_P$, $\alpha$, $\alpha_P$ are unknown parameters and the rest of the parameters are assumed to be known.

**Example 150** Mass Moments determination. Mass moment of an object about an axis can be evaluated by measuring period of oscillation of a compound pendulum carrying the object.

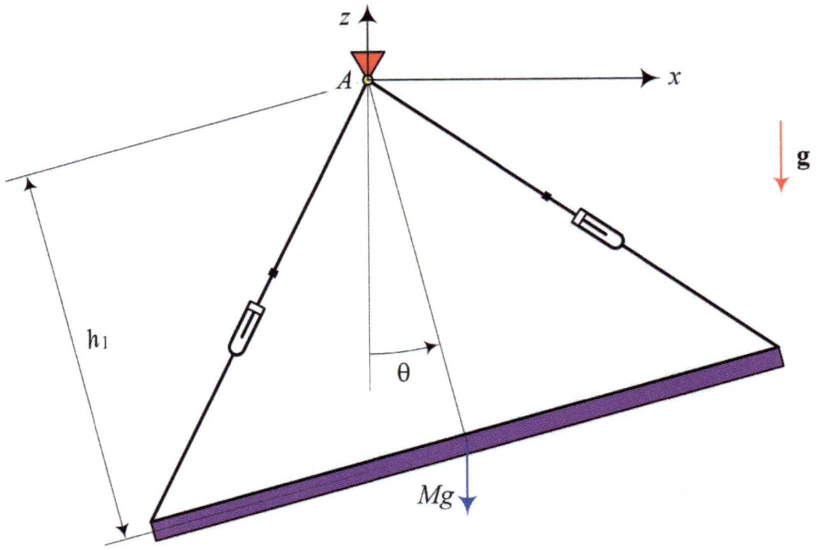


**Fig. 10.6** An oscillating platform hung from point $A$

Mass moments are important characteristics of a vehicle that affect its dynamic behavior. The main principal mass moments of $I_x$, $I_y$, $I_z$ can be calculated by an oscillating experiment. Figure 10.6 illustrates an oscillating platform hung from point $A$. Assume that the platform has a mass $M$ and a mass moment $I_0$ about the pivot point $A$. Ignoring the mass of the cables, we write the Euler equation of motion about point $A$.

$$\sum M_y = I_0 \ddot{\theta} = -Mgh_1 \sin\theta \tag{10.51}$$

$$I_0 \ddot{\theta} + Mgh_1 \sin\theta = 0 \tag{10.52}$$

If the angle of oscillation $\theta$ is very small, then $\sin\theta \approx \theta$ and then Eq. (10.52) simplifies to a linear equation where $\omega_n$ is the natural frequency of the oscillation.

$$\ddot{\theta} + \omega_n^2 \theta = 0 \qquad \omega_n = \sqrt{\frac{Mgh_1}{I_0}} \tag{10.53}$$

We measure $\omega_n$ as the frequency of small oscillations about $A$ when the platform is set free after a small deviation from the vertical equilibrium position. The natural period of oscillation $T_n = 2\pi/\omega_n$ is measured by an average period of a few cycles, and therefore, the mass moment $I_0$ will be determined.

$$I_0 = \frac{1}{4\pi^2} Mgh_1 T_n^2 \tag{10.54}$$

Now consider the swing shown in Fig. 10.7. A car with mass $m$ and mass center $C$ is on the platform such that $C$ is above the mass center of the platform. To find the car's pitch mass moment of inertia $I_y$ about $C$, we apply the Euler equation about point $A$, when the oscillator is deviated from the equilibrium condition.

$$\sum M_y = I_A \ddot{\theta} \tag{10.55}$$

$$-Mgh_1 \sin\theta - mgh_2 \sin\theta = I_0 + I_y + mh_2^2 \tag{10.56}$$

Assuming a very small oscillation, we may use $\sin\theta \approx \theta$ and then Eq. (10.56) reduces to a linear oscillator.

$$\ddot{\theta} + \omega_n^2 \theta = 0 \qquad \omega_n = \sqrt{\frac{(Mh_1 + mh_2)\, g}{I_0 + I_y + mh_2^2}} \tag{10.57}$$

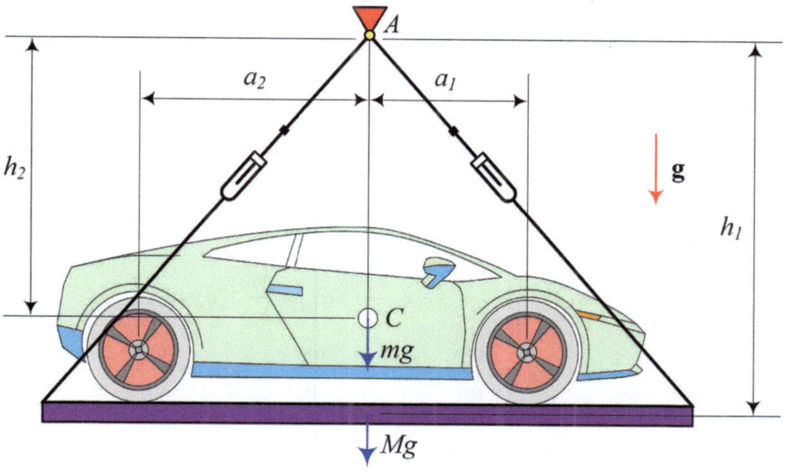


**Fig. 10.7** A car with mass $m$ on an oscillating platform hung from point $A$

Therefore, the pitch mass moment $I_y$ can be calculated by measuring the natural period of oscillation $T_n = 2\pi/\omega_n$.

$$I_y = \frac{1}{4\pi^2}(Mh_1 + mh_2)\, g T_n^2 - I_0 - mh_2^2 \tag{10.58}$$

## 10.2 Time Response

Because of the time lag between the front and rear wheels to hit a bump, the bicycle car model will show pitch oscillations in addition to the vertical vibrations. The time response of the system is the solution of the following equations for the absolute displacement $x$, $\theta$, $x_1$, and $x_2$ or the dimensionless displacements $s = x/a_2$, $\theta$, $s_1 = x_1/a_2$, and $s_2 = x_2/a_2$, along with two input road excitations $y_1$ and $y_2$. The equation can be transformed to the following dimensionless and more practical form.

$$\ddot{s} + 2\xi_1\omega_1\left(\dot{s} - \dot{s}_1 - \gamma\dot{\theta}\right) + 2\xi_2\omega_2\left(\dot{s} - \dot{s}_2 + \dot{\theta}\right) + \omega_1^2(s - s_1 - \gamma\theta) + \omega_2^2(s - s_2 + \theta) = 0 \tag{10.59}$$

$$\begin{aligned} I_0\ddot{\theta} - 2\xi_1\omega_1\left(\dot{s} - \dot{s}_1 - \gamma\dot{\theta}\right) + \frac{2\xi_2\omega_2}{\gamma}\left(\dot{s} - \dot{s}_2 + \dot{\theta}\right) - \omega_1^2(s - s_1 - \gamma\theta) + \frac{\omega_2^2}{\gamma}(s - s_2 + \theta) \\ + 2\xi_P\omega_P\left((1+\gamma)\dot{\theta} + \dot{s}_1 - \dot{s}_2\right) + \omega_P^2\left((1+\gamma)\theta + s_1 - s_2\right) = 0 \end{aligned} \tag{10.60}$$

$$\begin{aligned} \varepsilon_1\ddot{s}_1 - 2\xi_1\omega_1\left(\dot{s} - \dot{s}_1 - \gamma\dot{\theta}\right) - \omega_1^2(s - s_1 - \gamma\theta) - 2\xi_P\omega_1\frac{\gamma}{1+\gamma}\left((1+\gamma)\dot{\theta} + \dot{s}_1 - \dot{s}_2\right) \\ - \omega_P^2\frac{\gamma}{1+\gamma}\left((1+\gamma)\theta + s_1 - s_2\right) + \omega_{t_1}^2(s_1 - q_1) = 0 \end{aligned} \tag{10.61}$$

$$\begin{aligned} \varepsilon_2\ddot{s}_2 - 2\xi_2\omega_1\left(\dot{s} - \dot{s}_2 + \dot{\theta}\right) - \omega_2^2(s - s_2 + \theta) + 2\xi_P\omega_1\frac{\gamma}{1+\gamma}\left((1+\gamma)\dot{\theta} + \dot{s}_1 - \dot{s}_2\right) \\ + \omega_P^2\frac{\gamma}{1+\gamma}\left((1+\gamma)\theta + s_1 - s_2\right) + \omega_{t_2}^2(s_2 - q_2) = 0 \end{aligned} \tag{10.62}$$

$$\varepsilon_1 = \frac{m_1}{m} \qquad \varepsilon_2 = \frac{m_2}{m} \qquad \omega_1 = \sqrt{\frac{k_1}{m}} \qquad \omega_2 = \sqrt{\frac{k_2}{m}}$$
$$\xi_1 = \frac{c_1}{2m\omega_1} \qquad \xi_2 = \frac{c_2}{2m\omega_2} \qquad \gamma = \frac{a_1}{a_2} \qquad I_0 = \frac{I_y}{m a_1 a_2}$$
$$\xi_P = \frac{c_P}{2m\omega_P} \qquad \omega_P = \sqrt{\frac{k_P}{m l a_1}} \qquad \omega_{t_1} = \sqrt{\frac{k_{t_1}}{m}} \qquad \omega_{t_2} = \sqrt{\frac{k_{t_2}}{m}}$$
$$s = \frac{x}{a_2} \qquad s_1 = \frac{x_1}{a_2} \qquad s_2 = \frac{x_2}{a_2} \qquad q_1 = \frac{y_1}{a_2} \qquad q_2 = \frac{y_2}{a_2}$$
$$\alpha = \frac{\omega_2}{\omega_1} \qquad \alpha_P = \frac{\omega_P}{\omega_1} \qquad \alpha_{t_1} = \frac{\omega_{t_1}}{\omega_1} \qquad \alpha_{t_2} = \frac{\omega_{t_2}}{\omega_1} \tag{10.63}$$

***Proof*** Dividing the equations of motion (10.1)–(10.4) by $m$

$$\ddot{x} + \frac{c_1}{m}\left(\dot{x} - \dot{x}_1 - a_1\dot{\theta}\right) + \frac{c_2}{m}\left(\dot{x} - \dot{x}_2 + a_2\dot{\theta}\right)$$
$$+ \frac{k_1}{m}\left(x - x_1 - a_1\theta\right) + \frac{k_2}{m}\left(x - x_2 + a_2\theta\right) = 0 \tag{10.64}$$

$$\frac{I_y}{m}\ddot{\theta} - a_1\frac{c_1}{m}\left(\dot{x} - \dot{x}_1 - a_1\dot{\theta}\right) + a_2\frac{c_2}{m}\left(\dot{x} - \dot{x}_2 + a_2\dot{\theta}\right)$$
$$- a_1\frac{k_1}{m}\left(x - x_1 - a_1\theta\right) + a_2\frac{k_2}{m}\left(x - x_2 + a_2\theta\right)$$
$$+ \frac{c_P}{m}\left(\dot{\theta} + \frac{\dot{x}_1 - \dot{x}_2}{l}\right) + \frac{k_P}{m}\left(\theta + \frac{x_1 - x_2}{l}\right) = 0 \tag{10.65}$$

$$\frac{m_1}{m}\ddot{x}_1 - \frac{c_1}{m}\left(\dot{x} - \dot{x}_1 - a_1\dot{\theta}\right) - \frac{k_1}{m}\left(x - x_1 - a_1\theta\right)$$
$$- \frac{c_P}{ml}\left(\dot{\theta} + \frac{\dot{x}_1 - \dot{x}_2}{l}\right) - \frac{k_P}{ml}\left(\theta + \frac{x_1 - x_2}{l}\right) \tag{10.66}$$
$$+ \frac{k_{t_1}}{m}\left(x_1 - y_1\right) = 0 \tag{10.67}$$

$$\frac{m_2}{m}\ddot{x}_2 - \frac{c_2}{m}\left(\dot{x} - \dot{x}_2 + a_2\dot{\theta}\right) - \frac{k_2}{m}\left(x - x_2 + a_2\theta\right)$$
$$+ \frac{c_P}{ml}\left(\dot{\theta} + \frac{\dot{x}_1 - \dot{x}_2}{l}\right) + \frac{k_P}{ml}\left(\theta + \frac{x_1 - x_2}{l}\right) \tag{10.68}$$
$$+ \frac{k_{t_2}}{m}\left(x_2 - y_2\right) = 0 \tag{10.69}$$

and defining new parameters and variables, we will have the Eqs. (10.59)–(10.62). ■

**Example 151** Transient response of bicycle vehicles model to step input. Step input is a good excitation to examine natural behavior, damping rate, overshoot, and many other sensitivities of dynamic systems. In this example we study the response of bicycle car model to step inputs.

Although there would be a time lag between the front and rear wheels hitting the step, we first assume the step applies to the front and rear wheels at the same time. Knowing that the antipitch stiffness $k_P$ and damping $c_P$ are not common in vehicles, let us assume they are zero. The response of the car would then be the solution of the following equations of motion with respect to the symmetric road step input.

$$[m]\,\ddot{\mathbf{x}} + [c]\,\dot{\mathbf{x}} + [k]\,\mathbf{x} = \mathbf{F} \tag{10.70}$$

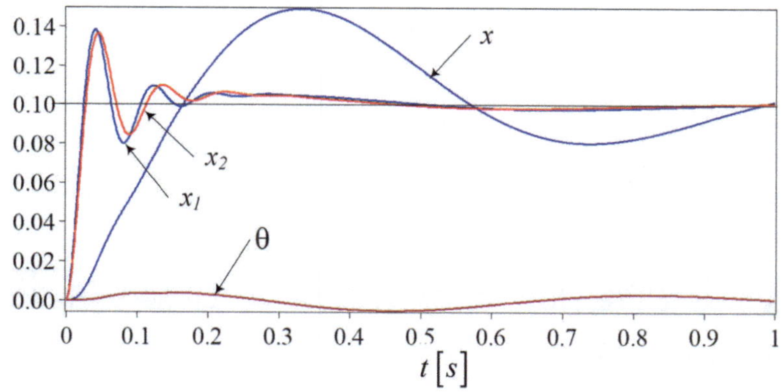


**Fig. 10.8** Body bounce, $x$, body pitch, $\theta$, front wheel, $x_1$, rear wheel, $x_2$, responses of a bicycle vehicle to a symmetric step excitation

$$[m] = \begin{bmatrix} m & 0 & 0 & 0 \\ 0 & I_y & 0 & 0 \\ 0 & 0 & m_1 & 0 \\ 0 & 0 & 0 & m_2 \end{bmatrix} \tag{10.71}$$

$$[c] = \begin{bmatrix} c_1 + c_2 & a_2c_2 - a_1c_1 & -c_1 & -c_2 \\ a_2c_2 - a_1c_1 & c_1a_1^2 + c_2a_2^2 & a_1c_1 & -a_2c_2 \\ -c_1 & a_1c_1 & c_1 & 0 \\ -c_2 & -a_2c_2 & 0 & c_2 \end{bmatrix} \tag{10.72}$$

$$[k] = \begin{bmatrix} k_1 + k_2 & a_2k_2 - a_1k_1 & -k_1 & -k_2 \\ a_2k_2 - a_1k_1 & k_1a_1^2 + k_2a_2^2 & a_1k_1 & -a_2k_2 \\ -k_1 & a_1k_1 & k_1 + k_{t_1} & 0 \\ -k_2 & -a_2k_2 & 0 & k_2 + k_{t_2} \end{bmatrix} \tag{10.73}$$

$$\mathbf{F} = \begin{bmatrix} 0 \\ 0 \\ y_1k_{t_1} \\ y_2k_{t_2} \end{bmatrix} \qquad \mathbf{x} = \begin{bmatrix} x \\ \theta \\ x_1 \\ x_2 \end{bmatrix} \tag{10.74}$$

$$y_1 = \begin{cases} Y & t > 0 \\ 0 & t \le 0 \end{cases} \qquad y_2 = \begin{cases} Y & t > 0 \\ 0 & t \le 0 \end{cases} \tag{10.75}$$

Using a set of nominal numerical values,

$$\begin{aligned} &m = 672\,\text{kg} \qquad m_1 = 42\,\text{kg} \qquad m_2 = 50\,\text{kg} \qquad I_y = 1050\,\text{kg}\,\text{m}^2 \\ &a_1 = 1.4\,\text{m} \qquad a_2 = 1.47\,\text{m} \qquad l = a_1 + a_2 = 2.87\,\text{m} \\ &k_1 = 22{,}000\,\text{N/m} \qquad k_2 = 25{,}000\,\text{N/m} \\ &k_{t_1} = k_{t_2} = 250{,}000\,\text{N/m} \\ &c_1 = 1630\,\text{N}\,\text{s/m} \qquad c_2 = 1900\,\text{N}\,\text{s/m} \end{aligned} \tag{10.76}$$

$$Y = 0.1\,\text{m} \tag{10.77}$$

the response of the vehicle would be as shown in Fig. 10.8.

The front and rear of this vehicle have different dynamic characteristics and that is the reason we have nonzero pitch angle $\theta$. The wheel masses and stiffness of springs are different, as well as different distances of the mass center from the front and rear wheels. The transient inputs reveal the natural behavior of systems. The figure shows that the frequency of vertical oscillation of the body is much less than the vertical oscillation of the front and rear wheels, and the frequency of oscillation of the wheels are close to each other. The pitch frequency is also close to the vertical frequency of the body. The pitch vibration

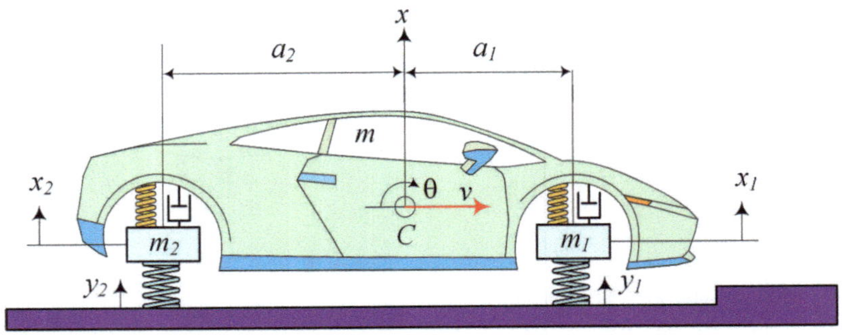


**Fig. 10.9** A bicycle vehicle model going over a step

has only a small amplitude due to symmetry of input and not much difference between front and rear suspension parameters and geometry. The pitch vibration will have higher amplitude when the time lag between front and rear wheels is considered.

Transient response of vehicles is the reaction that vehicles and passengers feel in short time that is usually associated with high acceleration, impact, jerk. On the other hand, the frequency response provides us with steady-state oscillations which is highly dependent on the closeness of excitation frequency and natural frequencies.

The vertical, pitch, and roll vibrations are the three possible vibrations of the body of a full vehicle model. To analyze these modes in the simplest possible models, we use a quarter car for vertical vibrations, a bicycle car for pitch vibrations, and a half car for roll vibrations. The bicycle- and half car models also contain the body bounce. Using these three models we can optimize the three different modes of oscillations of a vehicle. Then, using a full car model, we analyze the interaction of the three modes. Generally speaking, the rotational vibrations of roll and pitch of the body are more uncomfortable than the vertical vibrations.

**Example 152** Bicycle vehicles over a step. Depending on the velocity of the vehicle and the wheelbase there is a time lag between the front and rear wheels to hit the step. Here is the analysis.

Figure 10.9 illustrates a bicycle car model going over a step. The real situation of a vehicle going over a step is to consider the time lag between the front and rear wheels excitations. Consider the vehicle is moving with velocity $v$ and hitting a step of height $Y$. Starting the time from the moment that the front wheel hits the step, we can calculate the time $t_0$ at which the rear wheel reaches the step.

$$t_0 = \frac{l}{v} \tag{10.78}$$

Therefore, the excitation of the rear wheel, $y_2$, would be a lagged step input,

$$y_1 = \begin{cases} Y & t > 0 \\ 0 & t \le 0 \end{cases} \qquad y_2 = \begin{cases} Y\,H\left(t - \dfrac{l}{v}\right) & t > 0 \\ 0 & t \le 0 \end{cases} \tag{10.79}$$

where $H(t - t_0)$ is the Heaviside function.

$$H(t - t_0) = \begin{cases} 1 & t > t_0 \\ 0 & t \le t_0 \end{cases} \tag{10.80}$$

Let us assume a speed of $v$ and the same data as given in Eq. (10.76).

$$Y = 0.1\,\text{m} \qquad l = a_1 + a_2 = 2.87\,\text{m} \qquad v = 10\,\text{m/s} \tag{10.81}$$

The responses of the car are shown in Fig. 10.10.

Examining the time response graphs indicates that the vibration frequencies of the wheels are higher than the vibration frequencies of the body. The wheels also stop vibrating much faster than the body. For the given data, the unpleasant pitch vibration $\theta$ would last longer than the other coordinates, almost as long as the body vertical vibrations, $x$. If the time lag between the front and rear wheels to hit the step, $t_0 = l/v$, is $(2n+1)/2$, $n \in \mathbb{N}$ times of the inverse of natural pitch frequency, the pitch oscillation will have higher amplitude. So, for every vehicle, there would be critical speed that makes the pitch oscillation worse. The pitch natural frequency and the wheelbase of the vehicle determine the critical speed of the vehicle.

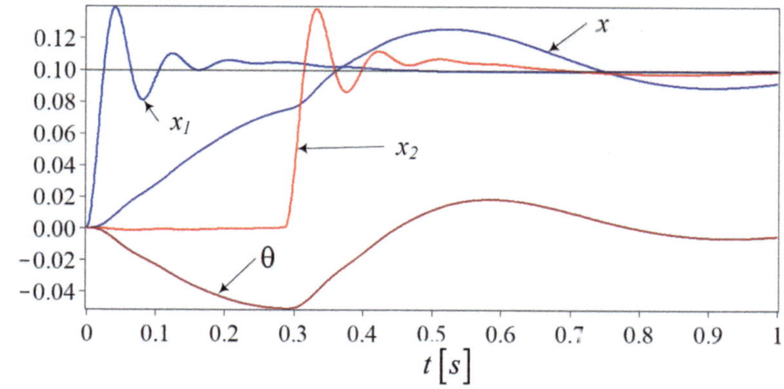


**Fig. 10.10** Body bounce, $x$, body pitch, $\theta$, front wheel, $x_1$, rear wheel, $x_2$, responses of a bicycle vehicle to a lagged step excitation

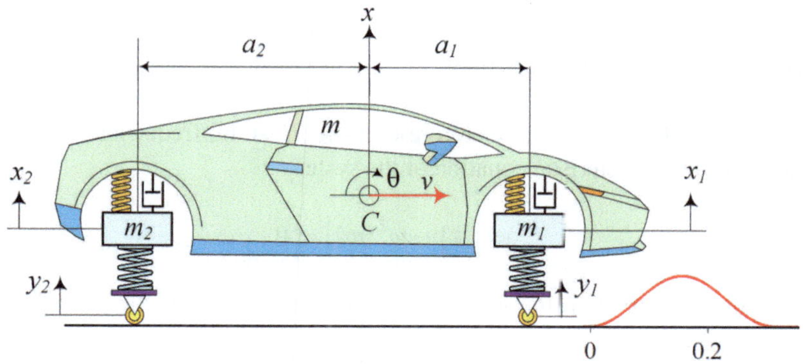


**Fig. 10.11** A bicycle car model and a sine square transient road excitation

**Example 153** Bump on the road and bicycle car. Let us examine the reaction of bicycle vehicle model to a bump on the road. Sine square is a good mathematical model for speed bumps on the roads.

Figure 10.11 illustrates a bicycle car model and a sine square transient road excitation for a vehicle that is moving with speed $v$ going over the bump with length $q$ and height $Y$.

$$y_1 = \begin{cases} Y \sin^2 \omega t & 0 \le t \le \dfrac{q}{v} \\ 0 & \dfrac{q}{v} < t \end{cases} \qquad \omega = \frac{v\pi}{q} \qquad t_0 = \frac{l}{v} \tag{10.82}$$

$$y_2 = \begin{cases} Y \sin^2 \omega \left(t - \dfrac{q}{v}\right) & t_0 \le t \le t_0 + \dfrac{q}{v} \\ 0 & t_0 + \dfrac{q}{v} < t \end{cases} \tag{10.83}$$

$$Y = 0.1\,\text{m} \qquad q = \frac{\pi}{10}\,\text{m} \qquad v = 10\,\text{m/s} \tag{10.84}$$

Figure 10.12 illustrates the response of the vehicle with the following characteristics.

$$m = 672\,\text{kg} \quad m_1 = 42\,\text{kg} \quad m_2 = 50\,\text{kg} \quad I_y = 1050\,\text{kg}\,\text{m}^2$$
$$a_1 = 1.4\,\text{m} \qquad a_2 = 1.47\,\text{m} \qquad l = a_1 + a_2 = 2.87\,\text{m}$$
$$k_1 = 22{,}000\,\text{N/m} \qquad k_2 = 25{,}000\,\text{N/m}$$
$$k_{t_1} = k_{t_2} = 250{,}000\,\text{N/m}$$
$$c_1 = 1630\,\text{N s/m} \qquad c_2 = 1900\,\text{N s/m} \tag{10.85}$$

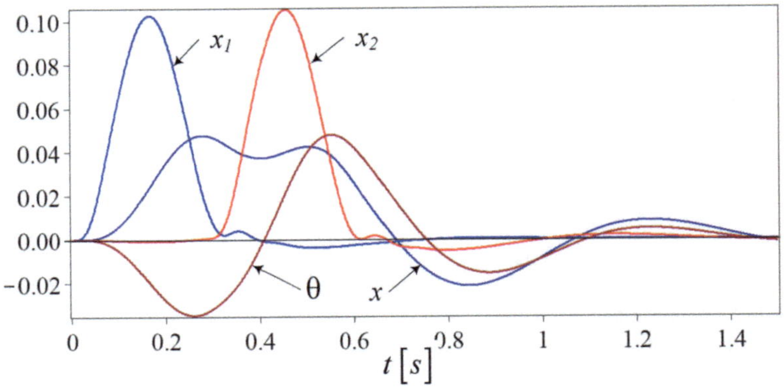


**Fig. 10.12** The response of a bicycle vehicle model to a bump on the road

## 10.3 Natural Frequencies and Mode Shapes

The bicycle car system is a four $DOF$ system and therefore it has four natural frequencies $\omega_{n_1}, \omega_{n_2}, \omega_{n_3}, \omega_{n_4}$. The natural frequencies $\omega_{n_i}$ are solutions of the characteristic equation of the system

$$\det\left[[k] - \omega^2 [m]\right] = 0 \tag{10.86}$$

and the mode shape $\mathbf{u}_i$, corresponding to $\omega_i$, is the solution of the eigenvector equation.

$$\left[[k] - \omega_i^2 [m]\right] \mathbf{u}_i = 0 \tag{10.87}$$

The matrices $[m]$ and $[k]$ are mass and stiffness matrices of the system, respectively.

$$[m] = \begin{bmatrix} m & 0 & 0 & 0 \\ 0 & I_y & 0 & 0 \\ 0 & 0 & m_1 & 0 \\ 0 & 0 & 0 & m_2 \end{bmatrix} \tag{10.88}$$

$$[k] = \begin{bmatrix} k_1+k_2 & a_2k_2 - a_1k_1 & -k_1 & -k_2 \\ a_2k_2 - a_1k_1 & k_1a_1^2+k_2a_2^2+k_P & a_1k_1+\frac{k_P}{l} & -a_2k_2 - \frac{k_P}{l} \\ -k_1 & a_1k_1+\frac{k_P}{l} & k_1+k_{t_1}+\frac{k_P}{l^2} & -\frac{k_P}{l^2} \\ -k_2 & -a_2k_2 - \frac{k_P}{l} & -\frac{k_P}{l^2} & k_2+k_{t_2}+\frac{k_P}{l^2} \end{bmatrix} \tag{10.89}$$

***Proof*** To determine the natural frequencies and mode shapes of a vibrating system, we must set all dampings to zero and eliminate any excitation term. The natural frequencies and mode shapes can equally be determined by calculating the eigenvalues and eigenvectors of $\mathbf{A} = [m]^{-1}[k]$,

$$\mathbf{A} = \begin{bmatrix} A_{11} & -\frac{a_1k_1 - a_2k_2}{m} & -\frac{k_1}{m} & -\frac{k_2}{m} \\ -\frac{a_1k_1 - a_2k_2}{I_y} & A_{22} & \frac{a_1k_1}{I_y}+\frac{k_P}{I_yl} & -\frac{a_2k_2}{I_y} - \frac{k_P}{I_yl} \\ -\frac{k_1}{m_1} & \frac{a_1k_1}{m_1}+\frac{k_P}{m_1l} & A_{33} & -\frac{k_P}{m_1l^2} \\ -\frac{k_2}{m_2} & -\frac{a_2k_2}{m_2} - \frac{k_P}{m_2l} & -\frac{k_P}{m_2l^2} & A_{44} \end{bmatrix}$$

$$A_{11} = \frac{k_1 + k_2}{m} \qquad A_{22} = \frac{k_1 a_1^2 + k_2 a_2^2 + k_P}{I_y} \tag{10.90}$$

$$A_{33} = \frac{k_1 + k_2}{m_1} + \frac{k_P}{m_1 l^2} \qquad A_{44} = \frac{k_1 + k_2}{m_2} + \frac{k_P}{m_2 l^2} \tag{10.91}$$

or equivalently the following expression.

$$\mathbf{A} = \begin{bmatrix} A_{11} & -a_1\omega_1^2 + a_2\omega_2^2 & -\omega_1^2 & -\omega_2^2 \\ -a_1\omega_1^2 + a_2\omega_2^2 & A_{22} & \frac{\omega_1^2 \varepsilon_1}{I_0 a_2} + \frac{k_P}{I_y l} & -\frac{\omega_2^2 \varepsilon_2}{I_0 a_1} - \frac{k_P}{I_y l} \\ -\omega_1^2 & a_1\omega_1^2 + \frac{k_P}{m_1 l} & A_{33} & -\frac{k_P}{m_1 l^2} \\ -\omega_2^2 & -a_2\omega_2^2 - \frac{k_P}{m_2 l} & -\frac{k_P}{m_2 l^2} & A_{44} \end{bmatrix} \tag{10.92}$$

$$A_{11} = \omega_1^2 + \omega_2^2 \qquad A_{22} = \frac{\omega_1^2 \gamma}{I_0} + \frac{\omega_2^2}{I_0 \gamma} + \frac{k_P}{I_y} \tag{10.93}$$

$$A_{33} = \omega_1^2 + \frac{\varepsilon_2}{\varepsilon_1}\omega_2^2 + \frac{k_P}{m_1 l^2} \qquad A_{44} = \frac{\varepsilon_1}{\varepsilon_2}\omega_1^2 + \omega_2^2 + \frac{k_P}{m_2 l^2} \tag{10.94}$$

The analytic expression of the natural frequencies and mode shapes of the bicycle car are too complicated to have practical use. To develop useful equations, we need to separate the parameters to fixed and adjustable. The fixed parameters are usually $a_1, a_2, l, m, m_1, m_2, I_y$, and the adjustable parameters are $k_1, k_2, k_P$. Out of the three adjustable parameters, we adjust the value of $k_1, k_2$, by quarter car model analysis. Hence, $k_P$ is the only parameter we adjust by bicycle vehicle model.

All responses of multi degree of freedom $(DOF)$ systems depend on natural frequencies and mode shapes of the system. Free and transient vibrations always happen in modes of linear combinations of mode shapes. The weight factors of mode shapes are determined by initial conditions. Forced harmonic vibration of multi-$DOF$ systems happen at the excitation frequency and the amplitude of the response depends on mode shapes and closeness of the excitation frequency to natural frequencies. Hence, the first step in study the vibrations of a dynamic system is to determine its natural frequencies and mode shapes. ■

**Example 154** Natural frequencies and mode shapes of a bicycle car model. There are four natural frequencies and mode shapes for bicycle vehicle model. Here is to show how to calculate them.

Consider a vehicle with a heavy solid axle in the rear and independent suspensions in front. The vehicle has the following characteristics.

$$\begin{aligned} &m = 672\,\text{kg} \quad m_1 = 42\,\text{kg} \quad m_2 = 50\,\text{kg} \quad I_y = 1050\,\text{kg m}^2 \\ &a_1 = 1.4\,\text{m} \quad a_2 = 1.47\,\text{m} \quad l = a_1 + a_2 = 2.87\,\text{m} \\ &k_1 = 22{,}000\,\text{N/m} \quad k_2 = 25{,}000\,\text{N/m} \\ &k_{t_1} = k_{t_2} = 250{,}000\,\text{N/m} \quad k_P = 0 \end{aligned} \tag{10.95}$$

The natural frequencies should be found by using the undamped and free vibration equations of motion. The characteristic equation of the system is

$$\begin{aligned} \det[[k] - \lambda[m]] &= \lambda^4 - 12{,}139\lambda^3 + 3.748\,8 \times 10^7\lambda^2 \\ &\quad -5.357\,2 \times 10^9\lambda + 9109 \times 10^{11} \\ &= \omega^8 - 12{,}139\omega^6 + 3.748\,8 \times 10^7\omega^4 \\ &\quad -5.357\,2 \times 10^9\omega^2 + 9109 \times 10^{11} \end{aligned} \tag{10.96}$$

$$\lambda = \omega^2 \tag{10.97}$$

because

$$[m] = \begin{bmatrix} 672 & 0 & 0 & 0 \\ 0 & 1050 & 0 & 0 \\ 0 & 0 & 42 & 0 \\ 0 & 0 & 0 & 50 \end{bmatrix} \tag{10.98}$$

$$[k] = \begin{bmatrix} 47{,}000 & 5950 & -22{,}000 & -25{,}000 \\ 5950 & 97{,}142.5 & 30{,}800 & -36{,}750 \\ -22{,}000 & 30{,}800 & 272{,}000 & 0 \\ -25{,}000 & -36{,}750 & 0 & 275{,}000 \end{bmatrix} \tag{10.99}$$

To find the natural frequencies we may solve the characteristic equation (10.96) or search for eigenvalues of $[A] = [m]^{-1}[k]$.

$$\begin{aligned} [A] &= [m]^{-1}[k] \\ &= \begin{bmatrix} 69.94 & 8.8542 & -32.738 & -37.202 \\ 5.6667 & 92.517 & 29.333 & -35 \\ -523.81 & 733.33 & 6476.2 & 0 \\ -500 & -735 & 0 & 5500 \end{bmatrix} \end{aligned} \tag{10.100}$$

The eigenvalues of $[A]$ are

$$\begin{aligned} \lambda_1 &= 62.197 \qquad \lambda_2 = 86.045 \\ \lambda_3 &= 5508.2 \qquad \lambda_4 = 6482.2 \end{aligned} \tag{10.101}$$

therefore, the natural frequencies of the bicycle car model are

$$\begin{aligned} \omega_1 &= \sqrt{\lambda_1} = 7.88\,\text{rad/s} \approx 1.25\,\text{Hz} \\ \omega_2 &= \sqrt{\lambda_2} = 9.27\,\text{rad/s} \approx 1.47\,\text{Hz} \\ \omega_3 &= \sqrt{\lambda_3} = 74.22\,\text{rad/s} \approx 11.81\,\text{Hz} \\ \omega_4 &= \sqrt{\lambda_4} = 80.51\,\text{rad/s} \approx 12.81\,\text{Hz} \end{aligned} \tag{10.102}$$

The normal form of the mode shapes of the system are

$$\mathbf{u}_1 = \begin{bmatrix} 1.000 \\ -0.211 \\ 0.103 \\ 0.06 \end{bmatrix} \qquad \mathbf{u}_2 = \begin{bmatrix} 0.317 \\ 1.000 \\ -0.081 \\ 0.156 \end{bmatrix} \tag{10.103}$$

$$\mathbf{u}_3 = \begin{bmatrix} 0.0068 \\ 0.0064 \\ -0.0012 \\ -1.000 \end{bmatrix} \qquad \mathbf{u}_4 = \begin{bmatrix} 0.0051 \\ -0.0046 \\ -1.000 \\ 0.000842 \end{bmatrix} \tag{10.104}$$

The biggest element of the third mode shape $\mathbf{u}_3$ belongs to $x_2$. It shows that in the third mode of vibrations at $\omega_3 \approx 11.81$ Hz the rear wheel will have the largest amplitude, while the amplitude of the other components are

$$\Theta = \frac{u_{32}}{u_{34}} = 0.0064 X_2 \tag{10.105}$$

$$X = \frac{u_{31}}{u_{34}} = 0.0068 X_2 \tag{10.106}$$

$$X_1 = \frac{u_{33}}{u_{34}} = -0.0012 X_2 \tag{10.107}$$

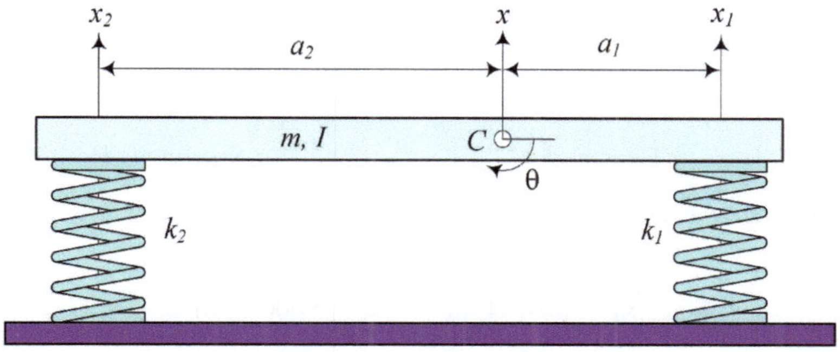


**Fig. 10.13** A beam with mass $m$ and mass moment $I$ and siting on two springs $k_1$ and $k_2$

In this example, the biggest element of the first mode shape $\mathbf{u}_1$ belongs to $x$, the biggest element of the second mode shape $\mathbf{u}_2$ belongs to $\theta$, and the biggest element of the fourth mode shape $\mathbf{u}_4$ belongs to $x_1$. Similar to the fourth mode shape $\mathbf{u}_3$, we can see the relative amplitude of different coordinates at each mode.

Consider a car that starts to move on a bumpy road at a very small acceleration. By increasing the speed, the first resonance occurs at $\omega_1 \approx 1.25$ Hz, at which the bounce vibration is the most observable vibration. The second resonance occurs at $\omega_2 \approx 1.47$ Hz when the pitch vibration of the body is the most observable vibration. The third and fourth resonances at $\omega_3 \approx 11.81$ Hz and $\omega_4 \approx 12.81$ Hz are related to rear and front wheels, respectively.

When the excitation frequency of a multiple-$DOF$ system increases, we will see that observable vibration moves from a coordinate to the others in the order of natural frequencies and associated mode shapes. When the excitation frequency is exactly at a natural frequency, the relative amplitudes of vibration are exactly similar to the associated mode shape. If the excitation frequency is not on a natural frequency, then vibration of the system is a combination of all modes of vibration. However, the weight factor of the closer modes are higher.

**Example 155** Translational and rotational coordinates. A multi-$DOF$ vibrating system may have a combination of translational and rotational coordinates. The analysis of such system mathematically has no difference; however, the mode shapes are more complicated to be interpreted. The smart way to show mode shapes is to convert all coordinates to be translational or rotational.

Consider the two $DOF$ system in Fig. 10.13. A beam with mass $m$ and mass moment $I$ about the mass center $C$ is sitting on two springs $k_1$ and $k_2$. This is a good model to study bicycle car model to determine and compare pitch and bounce vibrations of vehicles. The translational coordinate $x$ of $C$ and the rotational coordinate $\theta$ are the usual generalized coordinates that we use to measure the kinematics of the beam. The equations of motion and the mode shapes of the system are functions of the chosen coordinates.

The kinetic and potential energies of the beam for small motions are

$$K = \frac{1}{2}m\dot{x}^2 + \frac{1}{2}I\dot{\theta}^2 \tag{10.108}$$

$$P = \frac{1}{2}k_2 (x + a_2\theta)^2 + \frac{1}{2}k_1 (x - a_1\theta)^2 \tag{10.109}$$

The energies provide us with the equations of motion.

$$m\ddot{x} + k_2 (x + a_2\theta) + k_1 (x - a_1\theta) = 0 \tag{10.110}$$

$$I\ddot{\theta} + a_2k_2 (x + a_2\theta) - a_1k_1 (x - a_1\theta) = 0 \tag{10.111}$$

$$\begin{bmatrix} m & 0 \\ 0 & I \end{bmatrix} \begin{bmatrix} \ddot{x} \\ \ddot{\theta} \end{bmatrix} + \begin{bmatrix} k_1 + k_2 & a_2k_2 - a_1k_1 \\ a_2k_2 - a_1k_1 & a_2^2k_2 + a_1^2k_1 \end{bmatrix} \begin{bmatrix} x \\ \theta \end{bmatrix} = 0 \tag{10.112}$$

Let us use a set of numeric data:

$$m = 1000\,\text{kg} \qquad I = 200\,\text{kg m}^2$$

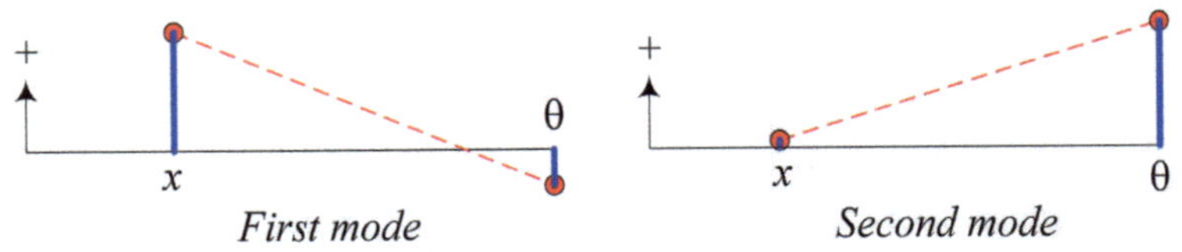


**Fig. 10.14** Mode shapes of the two $DOF$ beam with coordinates $x$, and $\theta$

$$
\begin{aligned}
k_1 &= 8200\,\mathrm{N/m} \qquad & k_2 &= 9000\,\mathrm{N/m} \\
a_1 &= 0.75\,\mathrm{m} \qquad & a_2 &= 1.1\,\mathrm{m}
\end{aligned}
\tag{10.113}
$$

to find the following equations:

$$
\begin{bmatrix} 1000 & 0 \\ 0 & 200 \end{bmatrix} \begin{bmatrix} \ddot{x} \\ \ddot{\theta} \end{bmatrix} + \begin{bmatrix} 17{,}200 & 3750 \\ 3750 & 15{,}503 \end{bmatrix} \begin{bmatrix} x \\ \theta \end{bmatrix} = 0 \tag{10.114}
$$

The natural frequencies and mode shapes of the system are calculated based on eigenvalues and eigenvectors of $[A] = \left[[m]^{-1}[k]\right]$:

$$
\begin{aligned}
[A] &= \begin{bmatrix} 1000 & 0 \\ 0 & 200 \end{bmatrix}^{-1} \begin{bmatrix} 17{,}200 & 3750 \\ 3750 & 15{,}503 \end{bmatrix} \\
&= \begin{bmatrix} 17.2 & 3.75 \\ 18.75 & 77.51 \end{bmatrix}
\end{aligned}
\tag{10.115}
$$

The $\omega_i$ and $\mathbf{u}_i$ are

$$
\omega_1 = 4.007\,\mathrm{rad/s} \qquad \omega_2 = 8.868\,\mathrm{rad/s} \tag{10.116}
$$

$$
\mathbf{u}_1 = \begin{bmatrix} 0.95647 \\ -0.29182 \end{bmatrix} \qquad \mathbf{u}_2 = \begin{bmatrix} 0.06091 \\ 0.99814 \end{bmatrix} \tag{10.117}
$$

Because of the chosen coordinate vector $\begin{bmatrix} x & \theta \end{bmatrix}^T$, the first component of the mode shapes refers to the displacement $x$, and the second components refers to the rotation $\theta$. Assuming the components of the mode shapes are comparable, the mode shapes in Fig. 10.14 show that the displacement $x$ is the most observable motion in the first mode and the rotation $\theta$ in the second mode. However, because the mode shapes are coordinate dependent, when the coordinates are measured in different dimensions, the rotational coordinate would not necessarily get the biggest number in any mode. These mode shapes do not show the configuration of the system.

To compare the mode shapes of a system practically, we recommend to choose the generalized coordinates with the same dimension. Let us employ the coordinates $x_1$ and $x_2$ instead of $x$ and $\theta$, as shown in Fig. 10.13. The energies of the system in this set of coordinates will be

$$
K = \frac{1}{2} m \left( \frac{a_2 \dot{x}_1 + a_1 \dot{x}_2}{a_1 + a_2} \right)^2 + \frac{1}{2} I \left( \frac{\dot{x}_2 - \dot{x}_1}{a_1 + a_2} \right)^2 \tag{10.118}
$$

$$
P = \frac{1}{2} k_2 x_2^2 + \frac{1}{2} k_1 x_1^2 \tag{10.119}
$$

The energies provide us with the following equations of motion.

$$
\begin{bmatrix} \dfrac{m a_2^2 + I}{(a_1 + a_2)^2} & \dfrac{m a_1 a_2 - I}{(a_1 + a_2)^2} \\ \dfrac{m a_1 a_2 - I}{(a_1 + a_2)^2} & \dfrac{m a_1^2 + I}{(a_1 + a_2)^2} \end{bmatrix} \begin{bmatrix} \ddot{x}_1 \\ \ddot{x}_2 \end{bmatrix} + \begin{bmatrix} k_1 & 0 \\ 0 & k_2 \end{bmatrix} \begin{bmatrix} x_1 \\ x_2 \end{bmatrix} = 0 \tag{10.120}
$$

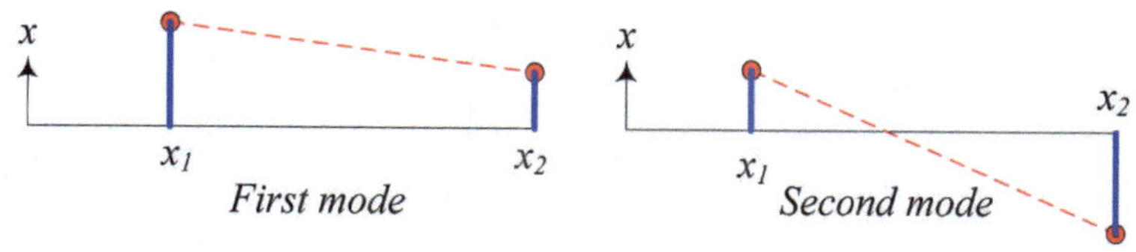


**Fig. 10.15** Mode shapes of the two $DOF$ beam with coordinates $x_1$, and $x_2$

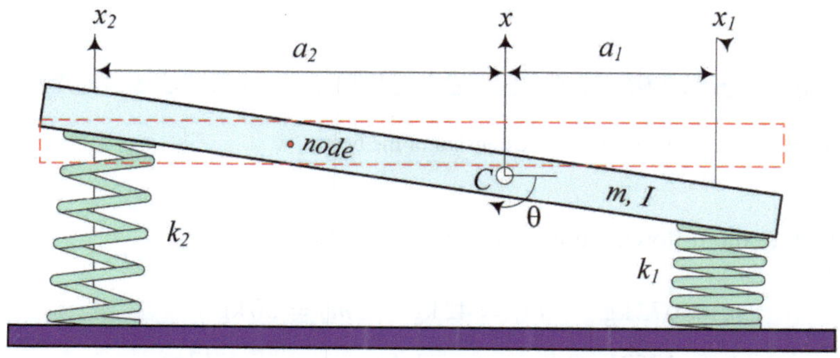


**Fig. 10.16** The second mode shape of the beam with a node on the beam

Using the data (10.113), the equations of motion simplify to

$$\begin{bmatrix} 411.98 & 182.62 \\ 182.62 & 222.79 \end{bmatrix} \begin{bmatrix} \ddot{x}_1 \\ \ddot{x}_2 \end{bmatrix} + \begin{bmatrix} 8200 & 0 \\ 0 & 9000 \end{bmatrix} \begin{bmatrix} x_1 \\ x_2 \end{bmatrix} = 0 \tag{10.121}$$

The natural frequencies and mode shapes of the system in the new coordinates are calculated based on eigenvalues and eigenvectors of new $[A] = \left[[m]^{-1}[k]\right]$:

$$\begin{aligned} [A] &= \begin{bmatrix} 411.98 & 182.62 \\ 182.62 & 222.79 \end{bmatrix}^{-1} \begin{bmatrix} 8200 & 0 \\ 0 & 9000 \end{bmatrix} \\ &= \begin{bmatrix} 31.263 & -28.127 \\ -25.627 & 63.452 \end{bmatrix} \end{aligned} \tag{10.122}$$

The natural frequencies $\omega_i$ will be the same as (10.116) but $\mathbf{u}_i$ will be different.

$$\omega_1 = 4.007 \text{ rad/s} \qquad \omega_2 = 8.868 \text{ rad/s} \tag{10.123}$$

$$\mathbf{u}_1 = \begin{bmatrix} 0.87965 \\ 0.47562 \end{bmatrix} \qquad \mathbf{u}_2 = \begin{bmatrix} 0.51034 \\ -0.85997 \end{bmatrix} \tag{10.124}$$

These mode shapes are illustrated in Fig. 10.15. They well show the configuration of the system. Because the components of the mode shapes have the same dimension, more information can be extracted from them compared to Fig. 10.14. The second mode shape shows a node on the beam. The node is a motionless point of the beam such that the whole beam will rotate about the node as is shown in Fig. 10.16. Although in a real situation the springs will tilt when $\theta \neq 0$, we ignore this fact and we assume the springs remain vertical.

The first mode shape does not show a node on the beam but because the two ends of the beam have different displacements, there is a node on the extension of the centerline of the beam, as is shown in Fig. 10.17.

**Example 156** Effect of the pitch stiffness, $k_P$, on natural frequencies of bicycle vehicle model. The antipitch spring is added to the vehicle suspension system to be able to adjust the pitch frequency. The pitch frequency must be close to the bounce frequency.

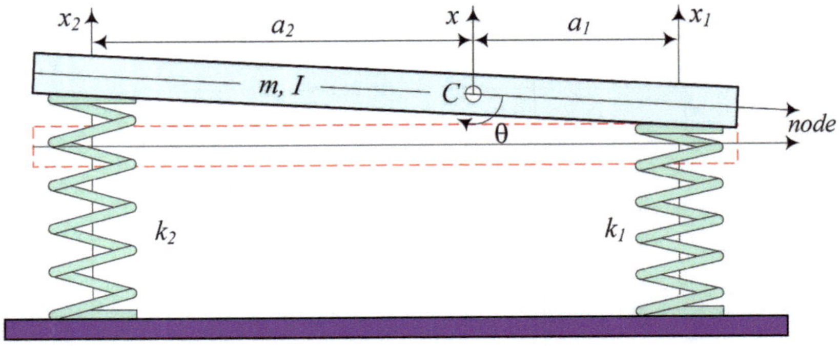


**Fig. 10.17** The first mode shape of the beam with a node on extension of the beam

Let us consider a vehicle with the following nominated data and unknown $k_P$.

$$\begin{aligned} m &= 672\,\text{kg} \qquad m_1 = 42\,\text{kg} \qquad m_2 = 50\,\text{kg} \\ I_y &= 1050\,\text{kg m}^2 \qquad a_1 = 1.4\,\text{m} \qquad a_2 = 1.47\,\text{m} \\ l &= a_1 + a_2 = 2.87\,\text{m} \qquad k_1 = 22{,}000\,\text{N/m} \\ k_2 &= 25{,}000\,\text{N/m} \quad k_{t_1} = k_{t_2} = 250{,}000\,\text{N/m} \end{aligned} \tag{10.125}$$

The mass matrix will be the same as (10.98) while the stiffness matrix of the system will be

$$[k] = \begin{bmatrix} 47{,}000 & 5950 & -22{,}000 & -25{,}000 \\ 5950 & 97{,}142.5 + k_P & 30{,}800 - \dfrac{k_P}{2.87} & -36{,}750 + \dfrac{k_P}{2.87} \\ -22{,}000 & 30{,}800 - \dfrac{k_P}{2.87} & 272{,}000 + \dfrac{k_P}{2.87^2} & -\dfrac{k_P}{2.87^2} \\ -25{,}000 & -36{,}750 + \dfrac{k_P}{2.87} & -\dfrac{k_P}{2.87^2} & 275{,}000 + \dfrac{k_P}{2.87^2} \end{bmatrix} \tag{10.126}$$

The characteristic equation of $[A] = [m]^{-1}[k]$ is a function of $k_P$.

$$\begin{aligned} &\lambda^4 + \left(-6.271 \times 10^{-3}\, k_P - 12{,}139\right)\lambda^3 \\ &\quad + \left(3.581 \times 10^{-11}\, k_P^2 + 44.934\, k_P + 3.748 \times 10^7\right)\lambda^2 \\ &\quad\; + \left(3.089 \times 10^{-13}\, k_P^3 - 3.88 \times 10^{-7}\, k_P^2 - 45{,}570\, k_P - 5.357 \times 10^9\right)\lambda \\ &+ 6.83 \times 10^{-16}\, k_P^4 - 1.238 \times 10^{-9}\, k_P^3 \\ &\qquad -2.253 \times 10^{-3} k_P^2 + 2.724 \times 10^6\, k_P + 1.91 \times 10^{11} = 0 \end{aligned} \tag{10.127}$$

To calculate the natural frequencies $\omega_i$ of the system as functions of $k_P$, we can assign a value to $k_P$ and solve the characteristic equation and determine the value of the eigenvalues $\lambda_i = \omega_i^2$. Repeating this procedure for other values of $k_P$ and recording the natural frequencies will show the frequencies as function of $k_P$. Figure 10.18 illustrates the first two natural frequencies of the vehicle in [ Hz]. When $k_P = 0$, the natural frequencies of the vehicle from the lowest to the highest are

$$\begin{aligned} \omega_1 &\approx 1.25\,\text{Hz} \qquad \omega_2 \approx 1.47\,\text{Hz} \\ \omega_3 &\approx 11.81\,\text{Hz} \qquad \omega_4 \approx 12.81\,\text{Hz} \end{aligned} \tag{10.128}$$

Mode shapes (10.103) are for $\omega_1$ and $\omega_2$.

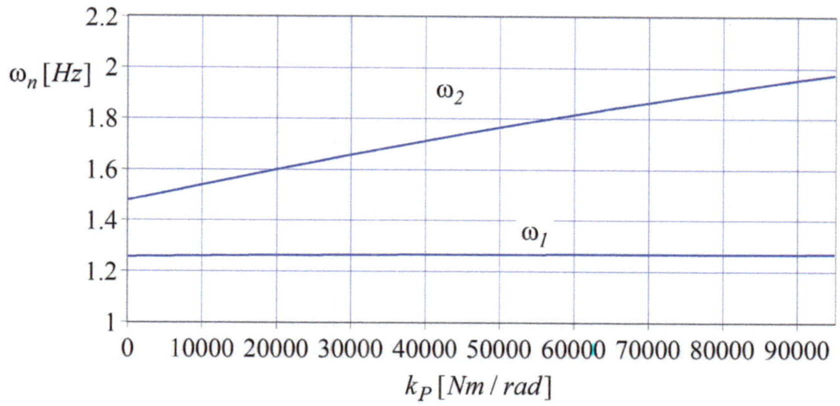


**Fig. 10.18** The first two natural frequencies of a vehicle in versus antipitch stiffness $k_P$ for $I_y = 1050\,\mathrm{kg\,m^2}$

$$\mathbf{u}_1 = \begin{bmatrix} 1.000 \\ -0.211 \\ 0.103 \\ 0.06 \end{bmatrix} \qquad \mathbf{u}_2 = \begin{bmatrix} 0.317 \\ 1.000 \\ -0.081 \\ 0.156 \end{bmatrix} \tag{10.129}$$

They show that $x$ is the dominant displacement at first natural frequency and $\theta$ is the dominant displacement at second natural frequency. Increasing the value of $k_P$ increases $\omega_2$ and has little effect on other natural frequencies.

The reason of introducing $k_P$ is to increase $\omega_2$ to get close to $\omega_1$ and that requires $\omega_2$ to be less than $\omega_1$ initially. The vehicle with data (10.125) has $\omega_2 > \omega_1$ for $k_P = 0$. Therefore introducing $k_P$ is not needed for this vehicle. Let us assume another vehicle with double the value of $I_y$.

$$\begin{aligned} &m = 672\,\mathrm{kg} \qquad m_1 = 42\,\mathrm{kg} \qquad m_2 = 50\,\mathrm{kg} \\ &I_y = 2100\,\mathrm{kg\,m^2} \qquad a_1 = 1.4\,\mathrm{m} \qquad a_2 = 1.47\,\mathrm{m} \\ &l = a_1 + a_2 = 2.87\,\mathrm{m} \qquad k_1 = 22{,}000\,\mathrm{N/m} \\ &k_2 = 25{,}000\,\mathrm{N/m} \qquad k_{t_1} = k_{t_2} = 250{,}000\,\mathrm{N/m} \end{aligned} \tag{10.130}$$

When $k_P = 0$, the natural frequencies of this vehicle from the lowest to the highest are

$$\begin{aligned} &\omega_1 \approx 1.024\,\mathrm{Hz} \qquad \omega_2 \approx 1.28\,\mathrm{Hz} \\ &\omega_3 \approx 11.81\,\mathrm{Hz} \qquad \omega_4 \approx 12.81\,\mathrm{Hz} \end{aligned} \tag{10.131}$$

Mode shapes for $\omega_1$ and $\omega_2$ are

$$\mathbf{u}_1 = \begin{bmatrix} -0.336 \\ 1.0 \\ -0.141 \\ 0.104 \end{bmatrix} \qquad \mathbf{u}_2 = \begin{bmatrix} 1.0 \\ 0.107 \\ 0.069 \\ 0.106 \end{bmatrix} \tag{10.132}$$

They show that with $I_y = 2100$, the vehicle has $\omega_1 < \omega_2$ for $k_P = 0$ and hence the natural frequency at which pitch oscillation is dominant is less that the natural frequency at which the bounce oscillation is dominant. Introducing $k_P$ into the suspension of this vehicle may help to make $\omega_1$ and $\omega_2$ closer. Figure 10.19 illustrates the first two natural frequencies of the vehicle in [Hz]. By increasing the value of $k_P$, the pitch frequency $\omega_1$ increases at first and then saturated, while the bounce frequency $\omega_2$ remains steady at the beginning and increases at higher values. The first two natural frequencies are closest at $k_P \simeq 50{,}000\,\mathrm{N\,m/rad}$.

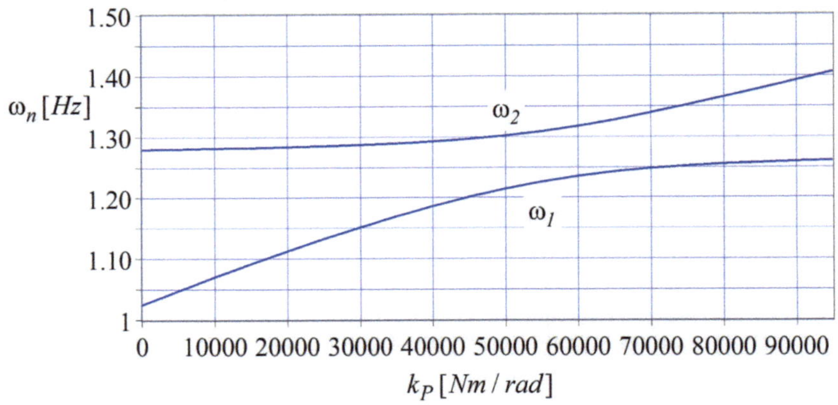


**Fig. 10.19** The first two natural frequencies of a vehicle in versus antipitch stiffness $k_P$ for $I_y = 2100\,\mathrm{kg\,m^2}$

## 10.4 Frequency Response

To determine the frequency response of the bicycle car model, we consider a harmonic base excitation,

$$y_1 = y_2 = Y \sin \omega t \tag{10.133}$$

and substitute a set of harmonic solution into equations of motion.

$$x = A_1 \sin \omega t + B_1 \cos \omega t = X_1 \sin (\omega t - \varphi_1) \tag{10.134}$$

$$\theta = A_2 \sin \omega t + B_2 \cos \omega t = X_2 \sin (\omega t - \varphi_2) \tag{10.135}$$

$$x_1 = A_3 \sin \omega t + B_3 \cos \omega t = X_3 \sin \left(\omega t - \varphi_3\right) \tag{10.136}$$

$$x_2 = A_4 \sin \omega t + B_4 \cos \omega t = X_4 \sin (\omega t - \varphi_4) \tag{10.137}$$

Then the differential equations of motion will convert to a set of algebraic equations to determine the unknown coefficients $A_1$, $B_1$, $A_2$, $B_2$, $A_3$, $B_3$, $A_4$, $B_4$ and calculate the frequency responses $X_1$, $X_2$, $X_3$ and $X_4$ which are the amplitudes of $x$, $\theta$, $x_1$, and $x_2$.

$$X_1 = \sqrt{A_1^2 + B_1^2} \qquad X_2 = \sqrt{A_2^2 + B_2^2}$$
$$X_3 = \sqrt{A_3^2 + B_3^2} \qquad X_4 = \sqrt{A_4^2 + B_4^2} \tag{10.138}$$

***Proof*** The set of equations of the bicycle car model are

$$[m]\,\ddot{\mathbf{x}} + [c]\,\dot{\mathbf{x}} + [k]\,\mathbf{x} = \mathbf{F} \tag{10.139}$$

$$\mathbf{x} = \begin{bmatrix} x \\ \theta \\ x_1 \\ x_2 \end{bmatrix} \tag{10.140}$$

$$[m] = \begin{bmatrix} m & 0 & 0 & 0 \\ 0 & I_y & 0 & 0 \\ 0 & 0 & m_1 & 0 \\ 0 & 0 & 0 & m_2 \end{bmatrix} \tag{10.141}$$

$$[c] = \begin{bmatrix} c_1 + c_2 & a_2c_2 - a_1c_1 & -c_1 & -c_2 \\ a_2c_2 - a_1c_1 & c_1a_1^2 + c_2a_2^2 & a_1c_1 & -a_2c_2 \\ -c_1 & a_1c_1 & c_1 & 0 \\ -c_2 & -a_2c_2 & 0 & c_2 \end{bmatrix} \tag{10.142}$$

$$[k] = \begin{bmatrix} k_1 + k_2 & a_2k_2 - a_1k_1 & -k_1 & -k_2 \\ a_2k_2 - a_1k_1 & k_1a_1^2 + k_2a_2^2 & a_1k_1 & -a_2k_2 \\ -k_1 & a_1k_1 & k_1 + k_{t_1} & 0 \\ -k_2 & -a_2k_2 & 0 & k_2 + k_{t_2} \end{bmatrix} \tag{10.143}$$

$$\mathbf{F} = \begin{bmatrix} 0 \\ 0 \\ y_1k_{t_1} \\ y_2k_{t_2} \end{bmatrix} \tag{10.144}$$

Substituting harmonic solutions in the equations of motion,

$$\begin{bmatrix} x \\ \theta \\ x_1 \\ x_2 \end{bmatrix} = \begin{bmatrix} A_1 \\ A_2 \\ A_3 \\ A_4 \end{bmatrix} \sin \omega t + \begin{bmatrix} B_1 \\ B_2 \\ B_3 \\ B_4 \end{bmatrix} \cos \omega t \tag{10.145}$$

provides us with a set of equations to calculate the coefficients **A** and **B**.

$$\begin{bmatrix} [k] - \omega^2 [m] & -\omega [c] \\ \omega [c] & [k] - \omega^2 [m] \end{bmatrix} \begin{bmatrix} A_1 \\ A_2 \\ A_3 \\ A_4 \\ B_1 \\ B_2 \\ B_3 \\ B_4 \end{bmatrix} = \begin{bmatrix} 0 \\ 0 \\ y_1k_{t_1} \\ y_2k_{t_2} \\ 0 \\ 0 \\ 0 \\ 0 \end{bmatrix} \tag{10.146}$$

The frequency responses are indicated by $X_1$, $X_2$, $X_3$, and $X_4$, where $X_1$ is the body bounce amplitude, $X_2$ is the body pitch amplitude, $X_3$ is the front wheel bounce amplitude, and $X_4$ is the rear wheel bounce amplitude.

Assuming

$$y_1 = Y \sin \omega t \tag{10.147}$$

$$y_2 = Y \sin \omega t \tag{10.148}$$

we calculate the unknown coefficients, and the frequency responses (10.138).

$$\begin{bmatrix} A_1 \\ A_2 \\ A_3 \\ A_4 \\ B_1 \\ B_2 \\ B_3 \\ B_4 \end{bmatrix} = \begin{bmatrix} [k] - \omega^2 [m] & -\omega [c] \\ \omega [c] & [k] - \omega^2 [m] \end{bmatrix}^{-1} \begin{bmatrix} 0 \\ 0 \\ Yk_{t_1} \\ Yk_{t_2} \\ 0 \\ 0 \\ 0 \\ 0 \end{bmatrix} \tag{10.149}$$

■

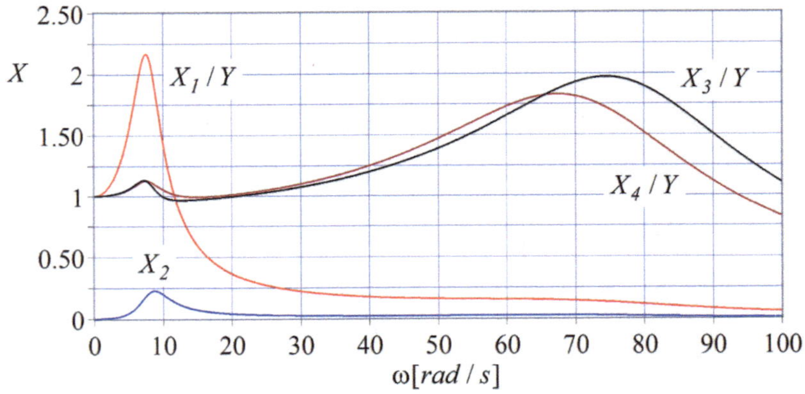


**Fig. 10.20** A sample of frequency response of a bicycle vehicle model

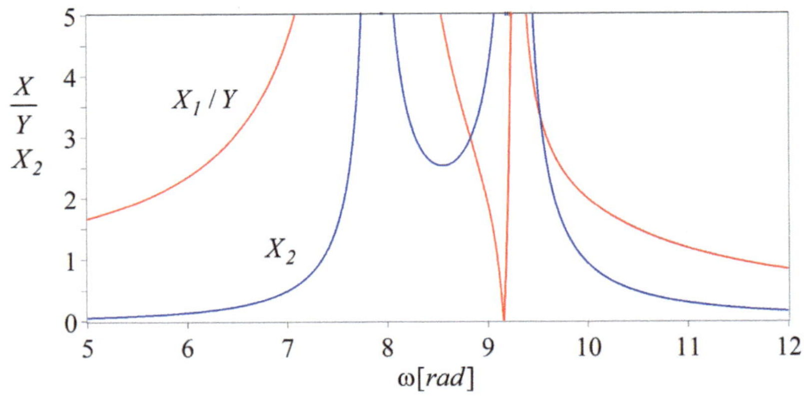


**Fig. 10.21** The frequency response of $x$ and $\theta$ around the first resonance zone for zero damping

**Example 157** Frequency response of a bicycle car model. There are two resonance zones in the frequency response of bicycle car model. A resonance zone around 1 Hz associated with the body bounce and body pitch, and a resonance zone around 10 Hz associated with front and rear wheels bounce vibrations.

Let us consider a bicycle vehicle model with $k_P = 0$ and the following data.

$$\begin{aligned} &m = 672\,\text{kg} \qquad m_1 = 42\,\text{kg} \qquad m_2 = 50\,\text{kg} \\ &I_y = 1050\,\text{kg}\,\text{m}^2 \qquad a_1 = 1.4\,\text{m} \qquad a_2 = 1.47\,\text{m} \\ &l = a_1 + a_2 = 2.87\,\text{m} \qquad k_1 = 22{,}000\,\text{N/m} \\ &k_2 = 25{,}000\,\text{N/m} \quad k_{t_1} = k_{t_2} = 250{,}000\,\text{N/m} \end{aligned} \tag{10.150}$$

To determine frequency response of the vehicle, we apply a harmonic excitation under front and rear wheels with amplitude $Y$ and at a frequency $\omega$.

$$y_1 = y_2 = Y \sin \omega t \tag{10.151}$$

Figure 10.20 illustrates the frequency response of the vehicle. The body bounce is the main kinematics characteristic of vehicle vibrations to be considered in most optimization procedures. The body bounce $x$ goes high at the first resonance zone around 1 Hz and drops significantly for $\omega > 2$ Hz. The body pitch oscillation $\theta$ also has similar behavior with high amplitude around 1 Hz and very low amplitude for $\omega > 2$ Hz. The wheels have a wider range of resonance zone than the body oscillations. The wheels highest amplitude of vibrations appear around 10 Hz. Figure 10.21 illustrates the frequency response of $x$ and $\theta$ around the first resonance zone for zero damping.

## 10.5 Summary

The simplest vehicle model to study the pitch mode of vibrations is the bicycle model in which we longitudinally cut a vehicle and take one side of the vehicle. This model includes the body bounce $x$, body pitch $\theta$, wheels hop $x_1$ and $x_2$, and the road excitations $y_1$ and $y_2$. The equations of motion of the bicycle car model are

$$\begin{aligned} &m\ddot{x} + c_1\left(\dot{x} - \dot{x}_1 - a_1\dot{\theta}\right) + c_2\left(\dot{x} - \dot{x}_2 + a_2\dot{\theta}\right) \\ &\quad + k_1\left(x - x_1 - a_1\theta\right) + k_2\left(x - x_2 + a_2\theta\right) = 0 \end{aligned} \tag{10.152}$$

$$\begin{aligned} &I_y\ddot{\theta} - a_1c_1\left(\dot{x} - \dot{x}_1 - a_1\dot{\theta}\right) + a_2c_2\left(\dot{x} - \dot{x}_2 + a_2\dot{\theta}\right) \\ &\quad - a_1k_1\left(x - x_1 - a_1\theta\right) + a_2k_2\left(x - x_2 + a_2\theta\right) \\ &\quad + c_P\left(\dot{\theta} + \frac{\dot{x}_1 - \dot{x}_2}{l}\right) + k_P\left(\theta + \frac{x_1 - x_2}{l}\right) = 0 \end{aligned} \tag{10.153}$$

$$\begin{aligned} &m_1\ddot{x}_1 - c_1\left(\dot{x} - \dot{x}_1 - a_1\dot{\theta}\right) - k_1\left(x - x_1 - a_1\theta\right) \\ &-\frac{c_P}{l}\left(\dot{\theta} + \frac{\dot{x}_1 - \dot{x}_2}{l}\right) - \frac{k_P}{l}\left(\theta + \frac{x_1 - x_2}{l}\right) + k_{t_1}\left(x_1 - y_1\right) = 0 \end{aligned} \tag{10.154}$$

$$\begin{aligned} &m_2\ddot{x}_2 - c_2\left(\dot{x} - \dot{x}_2 + a_2\dot{\theta}\right) - k_2\left(x - x_2 + a_2\theta\right) \\ &+\frac{c_P}{l}\left(\dot{\theta} + \frac{\dot{x}_1 - \dot{x}_2}{l}\right) + \frac{k_P}{l}\left(\theta + \frac{x_1 - x_2}{l}\right) + k_{t_2}\left(x_2 - y_2\right) = 0 \end{aligned} \tag{10.155}$$

Where the vehicle also has an antipitch bar with a torsional stiffness $k_P$ and torsional damping $c_P$. The $k_P$ is being used to adjust the pitch natural frequency to be as close as possible to the bounce natural frequency.

## 10.6 Key Symbols

| | |
|---|---|
| $a, \ddot{x}$ | acceleration |
| $a_1$ | distance from mass center to front axle |
| $a_2$ | distance from mass center to rear axle |
| $[a], [A]$ | coefficient matrix |
| $[A] = [m]^{-1}[k]$ | coefficient matrix of characteristic equation |
| $c$ | damping |
| $c_e$ | equivalent damping |
| $c_{ij}$ | element of row $i$ and column $j$ of $[c]$ |
| $c_s$ | main suspension damper |
| $[c]$ | damping matrix |
| $[\underline{c}]$ | symmetric damping matrix |
| $C$ | mass center |
| $D$ | dissipation function |
| $E$ | mechanical energy |
| $f$, $\mathbf{F}$ | harmonic force |
| $f = 1/T$ | cyclic frequency [Hz] |
| $f_c$ | damper force |

| | |
|---|---|
| $f_k$ | spring force |
| $F$ | amplitude of a harmonic force $f$ |
| $F_r,\ Q_r$ | generalized force |
| $g$ | gravitational acceleration |
| $h$ | height |
| $H$ | Heaviside function |
| $I$ | mass moment |
| $I_0$ | dimensionless $I$, mass moment about mass center |
| $\mathbf{I}$ | identity matrix |
| $k$ | stiffness |
| $k_{eq}$ | equivalent stiffness |
| $k_{ij}$ | element of row $i$ and column $j$ of $[k]$ |
| $k_R$ | antiroll bar torsional stiffness |
| $k_s$ | main suspension spring stiffness |
| $k_u$ | tire stiffness |
| $[k]$ | stiffness matrix |
| $[\underline{k}]$ | symmetric stiffness matrix |
| $K$ | kinetic energy |
| $l$ | length, wheelbase |
| $\mathcal{L}$ | Lagrangian |
| $m, M$ | mass |
| $m_{ij}$ | element of row $i$ and column $j$ of $[m]$ |
| $[m]$ | mass matrix |
| $[\underline{m}]$ | symmetric mass matrix |
| $M$ | moment, mass |
| $n$ | number of DOF |
| $q$ | dimensionless $y$ |
| $\mathbf{p}$ | momentum |
| $q_i, Q_i$ | generalized force |
| $r$ | excitation frequency ratio |
| $r, R$ | radius |
| $r_i,\ i \in N$ | nodal frequency ratio |
| $r_n = \omega_n/\omega_s$ | natural frequency ratio |
| $s$ | dimensionless $x$ |
| $t$ | time |
| $T$ | period |
| $x$ | absolute displacement |
| $X$ | steady-state amplitude of $x$ |
| $u_{ij}$ | th element of the $i$ th mode shape |
| $\mathbf{u}$ | mode shape, eigenvector |
| $\mathbf{u}_i$ | $i$ th eigenvector |
| $v, \mathbf{v}, \dot{x}, \dot{\mathbf{x}}$ | velocity |
| $V$ | potential energy |
| $w$ | track |
| $y$ | base excitation displacement |
| $Y$ | steady-state amplitude of $y$, step function displacement |
| $z$ | relative displacement |
| $Z$ | steady-state amplitude of $z$ |
| $Z_i$ | short notation parameter |

Greek

| | |
|---|---|
| $\alpha$ | frequency ratio |
| $\varepsilon$ | mass ratio |

| | |
|---|---|
| $\theta$ | pitch angle |
| $\xi$ | damping ratio |
| $\tau$ | dimensionless $t$ |
| $\omega = 2\pi f$ | angular frequency [rad/s] |
| $\omega_i$ | $i$ th natural frequency |
| $\omega_n$ | natural frequency |
| $\delta$ | deflection |
| $\lambda$ | eigenvalue, front/rear mass center ratio |
| $\lambda_i$ | $i$ th eigenvalue |

Subscript

| | |
|---|---|
| $d$ | driver |
| $i \in N$ | node number |
| $f$ | front |
| $n$ | natural |
| $P$ | pitch motion of car |
| $r$ | rear |
| $s$ | sprung |
| $t$ | tire |
| $u$ | unsprung |

Symbols

| | |
|---|---|
| $DOF$ | degree of freedom |
| $FBD$ | free body diagram |
| $ODF$ | one degree of freedom |
| Re | real part |
| $RMS$ | root mean square |
| det | determinant |
| Hz | Hertz |
| [ ] | matrix |
| $\int$ | integral |
| $\mid\ \mid$ | absolute value |
| $d$ | differential |
| $\partial$ | partial derivative |
| $\mathcal{L}$ | Lagrangian |

## References


1. D.J. Cole, D. Cebon, Validation of an articulated vehicle simulation. Vehic. Syst. Dyn. **21**(1), 197–223 (1992)
2. E. Guglielmino, T. Sireteanu, C.W. Stammers, G. Ghita, M. Giuclea, *Semi-active Suspension Control: Improved Vehicle Ride and Road Friendliness* (Springer, London, 2008)
3. N. Jalili, A comparative study and analysis of semi-active vibration-control systems. J. Vib. Acoust. **124**(4), 593–605 (2002)
4. R.N. Jazar, *Advanced Dynamics: Rigid Body, Multibody, and Aerospace Applications* (Wiley, New York, 2011)
5. R.N. Jazar, *Advanced Vibrations: A Modern Approach* (Springer, New York, 2013)
6. R.N. Jazar, *Vehicle Dynamics: Theory and Application*, 3rd edn. (Springer, New York, 2017)
7. R.N. Jazar, *Approximation Methods in Science and Engineering* (Springer, New York, 2020)
8. R.N. Jazar, *Perturbation Methods in Science and Engineering* (Springer, New York, 2021)
9. R.N. Jazar, *Advanced Vibrations: Theory and Application* (Springer, New York, 2023)
10. R.N. Jazar, R. Houim, A. Narimani, M.F. Golnaraghi, Frequency response and jump avoidance in a nonlinear passive engine mount. J. Vibr. Control **12**(11), 1205–1237 (2006)
11. R.N. Jazar, M. Kazemi, S. Borhani, *Mechanical Vibrations, (in Persian)* (Ettehad Publications, Tehran, 1992)
12. N. Louam, D.A. Wilson, R.S. Sharp, Optimal control of a vehicle suspension incorporating the time delay between front and rear wheel inputs. Vehic. Syst. Dyn. **17**(6), 317–336 (1988)
13. H. Marzbani, Flat ride; problems and solutions in vehicle. Nonlinear Eng. **1**(3–4), 101–108 (2012)
14. H. Marzbani, R.N. Jazar, Smart flat ride tuning, in *Nonlinear Approaches in Engineering Applications*, vol. 2. (Springer, New York, 2013), pp. 3–36

15. H. Marzbani, R.N. Jazar, M. Fard, Sustainable flat ride suspension design, in *Sustainable Automotive Technologies 2014: Proceedings of the 6th ICSAT* (Springer International Publishing, Cham, 2015), pp. 41–56
16. H. Marzbani, D. Voquoc, R.N. Jazar, M. Fard, Vehicle flat ride dynamics, in *Intelligent Interactive Multimedia Systems and Services* (Springer International Publishing, Cham, 2017), pp. 278–288
17. L. Meirovitch, *Analytical Methods in Vibrations* (Macmillan, New York, 1967)
18. L. Meirovitch, *Principles and Techniques of Vibrations* (Prentice Hall, Ney Jersey, 1997)
19. L. Meirovitch, *Fundamentals of Vibrations* (McGraw-Hill, New York, 2002)
20. Q. Nguyen, S. Milani, D. Vo, H. Marzbania, R.N. Jazar, Vibration analysis of the bicycle-car model considering tire-road separation. SAE Int. J. Commer. Veh. **15**(2), 111–124 (2022)
21. K. Ogata, *System Dynamics* (Prentice Hall, Ney Jersey, 2004)
22. Z. Osinski, *Damping of Vibrations* (A. A. Balkema Publishers, Rotterdam, 1998)
23. W.J. Palm, *Mechanical Vibration* (Wiley, New York, 2006)
24. S.S. Rao, *Mechanical Vibrations* (Pearson Prentice Hall, Harlow, 2018)
25. M. Roseau, *Vibrations in Mechanical Systems* (Springer, Berlin, 1987)
26. M.A. Salman, A.Y. Lee, N.M. Boustany, Reduced order design of active suspension control. Transactions of the ASME. J. Dyn. Syst. Measure. Control **112**, 604–610 (1990)
27. J.C. Snowdon, *Vibration and Shock in Damped Mechanical Systems* (Wiley, New York, 1968)
28. F.S. Tse, I.E. Morse, R.T. Hinkle, *Mechanical Vibrations Theory and Applications* (Allyn and Bacon, Boston, 1978)
29. H.E. Tseng, D. Hrovat, State of the art survey: active and semi-active suspension control. Vehic. Syst. Dyn. **53**(7), 1034–1062 (2015)

# Half Car Model and Body Roll Mode

# 11

To examine and optimize the roll vibrations of a vehicle, we use the half car vibrating model in which we laterally cut a vehicle and take one half to study. Figure 11.1 illustrates a half car model. This model includes the body bounce $x$, body roll $\varphi$, and wheels hop $x_1$ and $x_2$. There are also two independent road excitations $y_1$ and $y_2$ under the left and right wheels. The roll vibration of a car is the most unpleasant and the most uncomfortable motion of a vehicle.

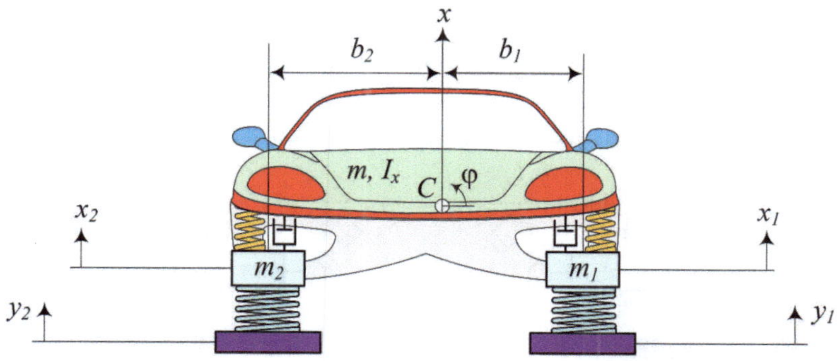


**Fig. 11.1** A half car vibrating model of a vehicle

## 11.1 Mathematical Model

The half car model has four degrees of freedom indicated by the coordinates $x$, $\varphi$, $x_1$, and $x_2$ along with two road excitations $y_1$ and $y_2$ as shown in Fig. 11.1. The equations of motion of the half car model are

$$m\ddot{x} + c(\dot{x} - \dot{x}_1 + b_1\dot{\varphi}) + c(\dot{x} - \dot{x}_2 - b_2\dot{\varphi}) + k(x - x_1 + b_1\varphi) + k(x - x_2 - b_2\varphi) = 0 \tag{11.1}$$

$$I_x\ddot{\varphi} + b_1c(\dot{x} - \dot{x}_1 + b_1\dot{\varphi}) - b_2c(\dot{x} - \dot{x}_2 - b_2\dot{\varphi}) + b_1k(x - x_1 + b_1\varphi) - b_2k(x - x_2 - b_2\varphi) + k_R\varphi = 0 \tag{11.2}$$

$$m_1\ddot{x}_1 - c(\dot{x} - \dot{x}_1 + b_1\dot{\varphi}) + k_t(x_1 - y_1) - k(x - x_1 + b_1\varphi) = 0 \tag{11.3}$$

$$m_2\ddot{x}_2 - c(\dot{x} - \dot{x}_2 - b_2\dot{\varphi}) + k_t(x_2 - y_2) - k(x - x_2 - b_2\varphi) = 0 \tag{11.4}$$


R. N. Jazar, H. Marzbani, *Vehicle Vibrations*, https://doi.org/10.1007/978-3-031-43486-0_11

**Table 11.1** Parameters of the half car model

| Parameter | Meaning |
|---|---|
| $m$ | Half of body mass $\left[\mathrm{kg}\right]$ |
| $m_1$ | Mass of a left wheel $\left[\mathrm{kg}\right]$ |
| $m_2$ | Mass of a right wheel $\left[\mathrm{kg}\right]$ |
| $I_x$ | Half of body longitudinal mass moment $\left[\mathrm{kg\,m^2}\right]$ |
| $k$ | Spring stiffness rate $[\mathrm{N/m}]$ |
| $c$ | Shock absorber damping rate $[\mathrm{N\,s/m}]$ |
| $k_t$ | Tire stiffness rate $[\mathrm{N/m}]$ |
| $c_t$ | Tire damping rate $[\mathrm{N\,s/m}]$ |
| $k_R$ | Antiroll bar stiffness date $[\mathrm{N\,s\,m/rad}]$ |
| $c_R$ | Antiroll bar damping date $[\mathrm{N\,m/rad}]$ |
| $b_1$ | Distance of $C$ from left axle $[\mathrm{m}]$ |
| $b_2$ | Distance of $C$ from right axle $[\mathrm{m}]$ |
| $w$ | $= b_1 + b_2$, track of vehicle |

**Table 11.2** Variables of the half car model

| Variable | Meaning |
|---|---|
| $x$ | Body vertical motion coordinate |
| $x_1$ | Left wheel vertical motion coordinate |
| $x_2$ | Right wheel vertical motion coordinate |
| $\varphi$ | Body roll motion coordinate |
| $y_1$ | Road excitation under the left wheel |
| $y_2$ | Road excitation under the right wheel |

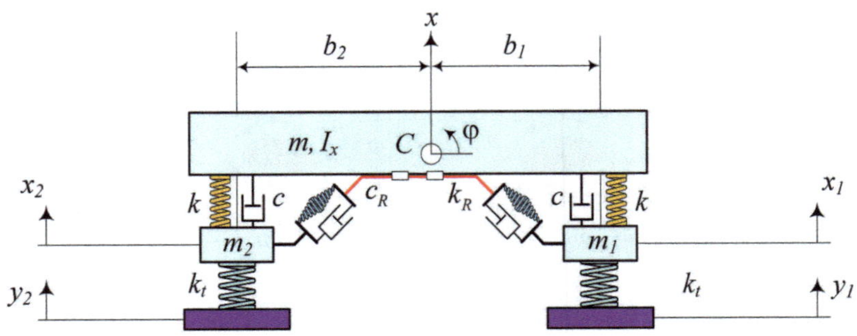


**Fig. 11.2** Half car model for vehicle vibrations

The half car model may be different for the front half and rear half due to different suspensions and vehicle mass distribution. Furthermore, different antiroll bars with different torsional stiffness may be used in the front and rear halves. The definitions of the employed parameters and variables are indicated in Tables 11.1 and 11.2.

***Proof*** Figure 11.2 illustrates the mathematical model of the half car system. The body of the vehicle is assumed to be a rigid bar. The bar has a mass $m$, which is the front or rear half of the total body mass, and a longitudinal mass moment $I_x$, which is half of the total body mass moment. The left and right wheels have masses $m_1$ and $m_2$, respectively, although they are usually equal. The tires' stiffness is indicated by $k_t$. Damping of tires is much smaller than the damping of shock absorbers, so we may ignore the tire damping for simpler calculation. The suspension of the car has stiffness $k$ and damping $c$ for the left and right wheels. It is common to make the suspension of the left and right wheels mirrored. Therefore, their stiffness and damping are equal. However, the half car model has different $k$, $c$, and $k_t$ for front or rear halves.

The vehicle may also have an antiroll bar with a torsional stiffness $k_R$ in front and rear. The antiroll bar provides a torque $M_R$ proportional to the roll angle $\varphi$.

$$\begin{aligned} M_R &= -k_R \frac{(x - x_1 + b_2\varphi) - (x - x_2 - b_1\varphi)}{w} - c_R \frac{(\dot{x} - \dot{x}_1 + b_2\dot{\varphi}) - \left(\dot{x} - \dot{x}_2 - b_1\dot{\theta}\right)}{w} \\ &= -k_R\left(\varphi + \frac{x_2 - x_1}{w}\right) - c_R\left(\dot{\varphi} + \frac{\dot{x}_2 - \dot{x}_1}{w}\right) \end{aligned} \tag{11.5}$$

To find the equations of motion for the half car vibrating model, we use the Lagrange method. The kinetic and potential energies of the system are

$$K = \frac{1}{2}m\dot{x}^2 + \frac{1}{2}m_1\dot{x}_1^2 + \frac{1}{2}m_2\dot{x}_2^2 + \frac{1}{2}I_x\dot{\varphi}^2 \tag{11.6}$$

$$P = \frac{1}{2}k_t\left(x_1 - y_1\right)^2 + \frac{1}{2}k_t\left(x_2 - y_2\right)^2 + \frac{1}{2}k_R\left(\varphi + \frac{x_2 - x_1}{w}\right)^2 + \frac{1}{2}k\left(x - x_1 + b_1\varphi\right)^2 + \frac{1}{2}k\left(x - x_2 - b_2\varphi\right) \tag{11.7}$$

and the dissipation function $D$ is

$$D = \frac{1}{2}c\left(\dot{x} - \dot{x}_1 + b_1\dot{\varphi}\right)^2 + \frac{1}{2}c\left(\dot{x} - \dot{x}_2 - b_2\dot{\varphi}\right)^2 + \frac{1}{2}c_R\left(\dot{\varphi} + \frac{\dot{x}_2 - \dot{x}_1}{w}\right)^2 \tag{11.8}$$

Applying the Lagrange method

$$\frac{d}{dt}\left(\frac{\partial K}{\partial \dot{q}_r}\right) - \frac{\partial K}{\partial q_r} + \frac{\partial D}{\partial \dot{q}_r} + \frac{\partial P}{\partial q_r} = f_r \qquad r = 1, 2, \ldots 4 \tag{11.9}$$

provides us with the following equations of motion (11.1)–(11.4). The set of equations of motion may be rearranged in a matrix form

$$[m]\,\ddot{\mathbf{x}} + [c]\,\dot{\mathbf{x}} + [k]\,\mathbf{x} = \mathbf{F} \tag{11.10}$$

where,

$$\mathbf{x} = \begin{bmatrix} x \\ \varphi \\ x_1 \\ x_2 \end{bmatrix} \tag{11.11}$$

$$[m] = \begin{bmatrix} m & 0 & 0 & 0 \\ 0 & I_x & 0 & 0 \\ 0 & 0 & m_1 & 0 \\ 0 & 0 & 0 & m_2 \end{bmatrix} \tag{11.12}$$

$$[c] = \begin{bmatrix} 2c & c\left(b_1 - b_2\right) & -c & -c \\ c\left(b_1 - b_2\right) & c\left(b_1^2 + b_2^2\right) + c_R & -cb_1 - \frac{c_R}{w} & cb_2 + \frac{c_R}{w} \\ -c & -cb_1 - \frac{c_R}{w} & c + \frac{c_R}{w^2} & -\frac{c_R}{w^2} \\ -c & cb_2 + \frac{c_R}{w} & -\frac{c_R}{w^2} & c + \frac{c_R}{w^2} \end{bmatrix} \tag{11.13}$$

$$[k] = \begin{bmatrix} 2k & kb_1 - kb_2 & -k & -k \\ kb_1 - kb_2 & k\left(b_1^2 + b_2^2\right) + k_R & -kb_1 - \frac{k_R}{w} & kb_2 + \frac{k_R}{w} \\ -k & -kb_1 - \frac{k_R}{w} & k + k_t + \frac{k_R}{w^2} & -\frac{k_R}{w^2} \\ -k & kb_2 + \frac{k_R}{w} & -\frac{k_R}{w^2} & k + k_t + \frac{k_R}{w^2} \end{bmatrix} \tag{11.14}$$

$$\mathbf{F} = \begin{bmatrix} 0 \\ 0 \\ y_1 k_t \\ y_2 k_t \end{bmatrix} \tag{11.15}$$

■

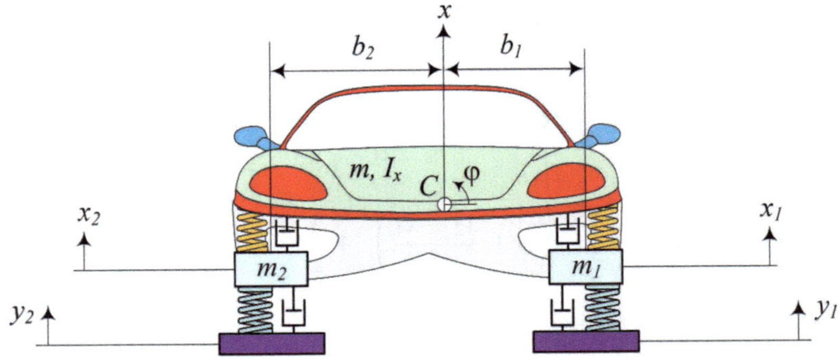


**Fig. 11.3** A half car vibrating model of a vehicle with tire damping

**Example 158** ★ Tire damping. Damping of tires is very low, and it can be ignored to work with simpler equations of motion. Here we derive the equations of motion of half car model including tire damping.

We may add tire dampers $c_t$ in parallel to tire stiffness $k_t$, as shown in Fig. 11.3, to model possible damping in tires. The values of damping $c_t$ for tires, compared to damping of suspension shock absorbers $c$, are very small, and hence, we may ignore $c_t$ to simplify the model. Including $c_t$ provides the following equations of motion of bicycle car model:

$$[m]\,\ddot{\mathbf{x}}+[c]\,\dot{\mathbf{x}}+[k]\,\mathbf{x}=\mathbf{F} \tag{11.16}$$

where

$$\mathbf{x}=\begin{bmatrix} x \\ \varphi \\ x_1 \\ x_2 \end{bmatrix} \tag{11.17}$$

$$[m]=\begin{bmatrix} m & 0 & 0 & 0 \\ 0 & I_x & 0 & 0 \\ 0 & 0 & m_1 & 0 \\ 0 & 0 & 0 & m_2 \end{bmatrix} \tag{11.18}$$

$$[c]=\begin{bmatrix} 2c & c\,(b_1-b_2) & -c & -c \\ c\,(b_1-b_2) & c\left(b_1^2+b_2^2\right)+c_R & -cb_1-\dfrac{c_R}{w} & cb_2+\dfrac{c_R}{w} \\ -c & -cb_1-\dfrac{c_R}{w} & c+c_t+\dfrac{c_R}{w^2} & -\dfrac{c_R}{w^2} \\ -c & cb_2+\dfrac{c_R}{w} & -\dfrac{c_R}{w^2} & c+c_t+\dfrac{c_R}{w^2} \end{bmatrix} \tag{11.19}$$

$$[k]=\begin{bmatrix} 2k & kb_1-kb_2 & -k & -k \\ kb_1-kb_2 & kb_1^2+kb_2^2+k_R & -kb_1-\dfrac{k_R}{w} & kb_2+\dfrac{k_R}{w} \\ -k & -kb_1-\dfrac{k_R}{w} & k+k_t+\dfrac{k_R}{w^2} & -\dfrac{k_R}{w^2} \\ -k & kb_2+\dfrac{k_R}{w} & -\dfrac{k_R}{w^2} & k+k_t+\dfrac{k_R}{w^2} \end{bmatrix} \tag{11.20}$$

$$\mathbf{F}=\begin{bmatrix} 0 \\ 0 \\ k_t y_1+c_t\dot{y}_1 \\ k_t y_2+c_t\dot{y}_2 \end{bmatrix} \tag{11.21}$$

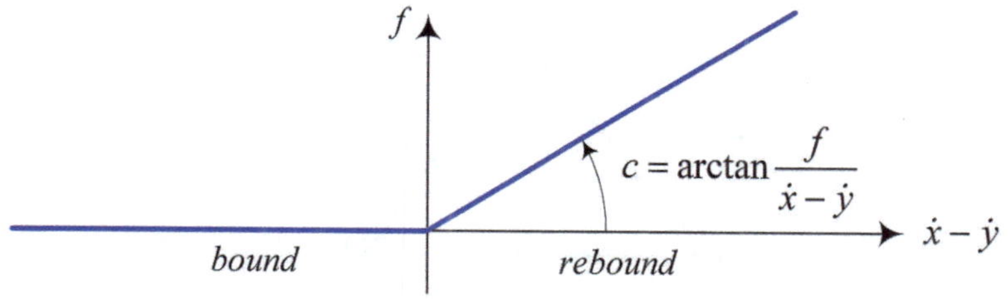


**Fig. 11.4** A bilinear damping model with no damping in bound and a constant damping in rebound

because:

$$K = \frac{1}{2}m\dot{x}^2 + \frac{1}{2}m_1\dot{x}_1^2 + \frac{1}{2}m_2\dot{x}_2^2 + \frac{1}{2}I_x\dot{\varphi}^2 \tag{11.22}$$

$$P = \frac{1}{2}k_t\,(x_1 - y_1)^2 + \frac{1}{2}k_t\,(x_2 - y_2)^2 + \frac{1}{2}k_R\left(\varphi + \frac{x_2 - x_1}{w}\right)^2 + \frac{1}{2}k\,(x - x_1 + b_1\varphi)^2 + \frac{1}{2}k\,(x - x_2 - b_2\varphi) \tag{11.23}$$

$$D = \frac{1}{2}c_t\,(\dot{x}_1 - \dot{y}_1)^2 + \frac{1}{2}c_t\,(\dot{x}_2 - \dot{y}_2)^2 + \frac{1}{2}c_R\left(\dot{\varphi} + \frac{\dot{x}_2 - \dot{x}_1}{w}\right)^2 + \frac{1}{2}c\,(\dot{x} - \dot{x}_1 + b_1\dot{\varphi})^2 + \frac{1}{2}c\,(\dot{x} - \dot{x}_2 - b_2\dot{\varphi})^2 \tag{11.24}$$

**Example 159** Mathematical model's limitations. Half car model is the best and simplest model to examine roll vibrations of vehicles. However, the model has some limitations.

The half car model contains no component of the geometric effects of the length of a full car and offers no possibility to study longitudinal interconnections. However, it contains the most fundamental features of real vehicles and includes a proper representation of the problem of controlling body hup, body roll, wheel, and wheel-body load variations. In the half car model, we assume that tire is always in contact with the ground. This assumption is true only at low frequency and low amplitude but will not be true at high frequency or high amplitude of road excitations. A better model must be able to include the possibility of separation between the tire and ground.

The main suspension parameters for vertical vibrations of vehicles are designed and optimized by the quarter car model. Then, the suspension of the four $DOF$ half car vibrating systems will be examined for fine tuning of the suspension parameters by studying the roll mode of vibrations.

**Example 160** Nonlinear damping. Shock absorbers are the most nonlinear element of the suspension system. Ideally, they are bilinear with very low damping in bound and a high damping in rebound. Here is the equations of motion of half car model with bilinear dampers.

The bilinear damping model of practical shock absorbers is illustrated in Fig. 11.4, considering $x$ and $y$ as displacements of two ends of the shock absorber. At extreme case, there is no damping in bound and a constant damping in rebound. Indicating the displacement of the vehicle body and wheels, we define bilinear damping by a Heaviside function $H\,(\dot{x} - \dot{x}_1 + b_1\dot{\varphi})$ and $H\,(\dot{x} - \dot{x}_2 - b_2\dot{\varphi})$ to rewrite the equations of motion of the half vehicle model.

$$\begin{aligned} &m\ddot{x} + c\,(\dot{x} - \dot{x}_1 + b_1\dot{\varphi})\;H\,(\dot{x} - \dot{x}_1 + b_1\dot{\varphi}) \\ &\quad + c\,(\dot{x} - \dot{x}_2 - b_2\dot{\varphi})\;H\,(\dot{x} - \dot{x}_2 - b_2\dot{\varphi}) \\ &\quad + k\,(x - x_1 + b_1\varphi) + k\,(x - x_2 - b_2\varphi) = 0 \end{aligned} \tag{11.25}$$

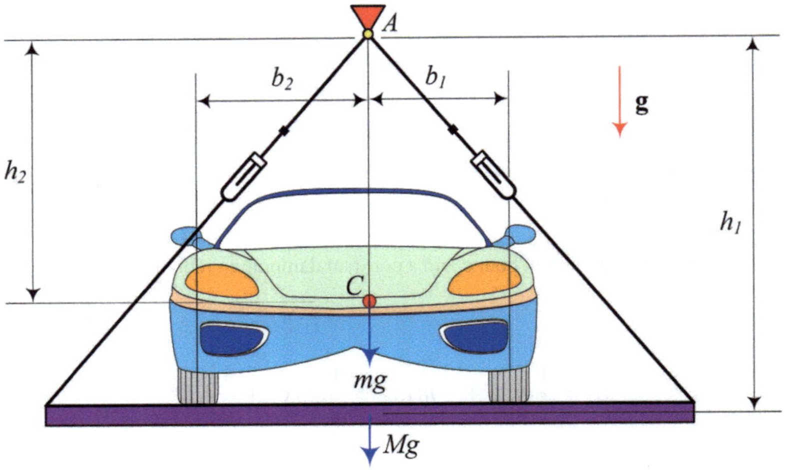


**Fig. 11.5** Roll mass moment $I_x$ measurement using a swinging platform

$$\begin{aligned} &I_x\ddot{\varphi} + b_1 c\left(\dot{x} - \dot{x}_1 + b_1\dot{\varphi}\right) H\left(\dot{x} - \dot{x}_1 + b_1\dot{\varphi}\right) \\ &-b_2 c\left(\dot{x} - \dot{x}_2 - b_2\dot{\varphi}\right) H\left(\dot{x} - \dot{x}_2 - b_2\dot{\varphi}\right) \\ &+b_1 k\left(x - x_1 + b_1\varphi\right) - b_2 k\left(x - x_2 - b_2\varphi\right) + k_R\varphi \\ &+c_R\left(\dot{\varphi} + \frac{\dot{x}_2 - \dot{x}_1}{w}\right) + k_R\left(\varphi + \frac{x_2 - x_1}{w}\right) = 0 \end{aligned} \tag{11.26}$$

$$\begin{aligned} &m_1\ddot{x}_1 - c\left(\dot{x} - \dot{x}_1 + b_1\dot{\varphi}\right) H\left(\dot{x} - \dot{x}_1 + b_1\dot{\varphi}\right) \\ &-k\left(x - x_1 + b_1\varphi\right) - \frac{c_R}{w}\left(\dot{\varphi} + \frac{\dot{x}_2 - \dot{x}_1}{w}\right) \\ &-\frac{k_R}{w}\left(\varphi + \frac{x_2 - x_1}{w}\right) + k_t\left(x_1 - y_1\right) = 0 \end{aligned} \tag{11.27}$$

$$\begin{aligned} &m_2\ddot{x}_2 - c\left(\dot{x} - \dot{x}_2 - b_2\dot{\varphi}\right) H\left(\dot{x} - \dot{x}_2 - b_2\dot{\varphi}\right) \\ &-k\left(x - x_2 - b_2\varphi\right) + \frac{c_R}{w}\left(\dot{\varphi} + \frac{\dot{x}_2 - \dot{x}_1}{w}\right) \\ &+\frac{k_R}{w}\left(\varphi + \frac{x_2 - x_1}{w}\right) + k_t\left(x_2 - y_2\right) = 0 \end{aligned} \tag{11.28}$$

**Example 161** Determination of the roll mass moment $I_x$. Employing equations of motion and natural frequency at low amplitude, we can determine the mass moment of any rigid body including vehicles.

Let us put a car on the platform as is shown in Fig. 11.5. The car with mass $m$ and mass center $C$ is on the platform such that $C$ is above the mass center of the platform. The platform has a mass moment $I_0$ about the pivot point $A$. To find the car's roll mass moment of inertia $I_x$ about $C$, we apply the Euler equation about point $A$.

$$\sum M_x = I_A\,\ddot{\theta} \tag{11.29}$$

$$-Mgh_1 \sin\theta - mgh_2 \sin\theta = I_0 + I_x + mh_2^2 \tag{11.30}$$

Assuming a very small oscillation, we may use $\sin\theta \approx \theta$ and then Equation of motion as a linear oscillator.

$$\ddot{\theta} + \omega_n^2\,\theta = 0 \qquad \omega_n = \sqrt{\frac{(Mh_1 + mh_2)\,g}{I_0 + I_x + mh_2^2}} \tag{11.31}$$

Therefore, the roll mass moment $I_x$ will be calculated by measuring the natural period of oscillation $T_n = 2\pi/\omega_n$.

$$I_x = \frac{1}{4\pi^2}\,(Mh_1 + mh_2)\,gT_n^2 - I_0 - mh_2^2 \tag{11.32}$$

## 11.2 Time Response

Unequal distances of mass center from left and right wheels and asymmetric road excitation between the left and right wheels make a vehicle to show roll oscillations in addition of the vertical vibrations. The time response of the half car model is the solution of the following equations for the absolute displacement $x$,$\varphi$, $x_1$, $x_2$, along with the road excitations $y_1\,(t)$ and $y_2\,(t)$.

$$\begin{aligned} &m\ddot{x} + c\,(\dot{x} - \dot{x}_1 + b_1\dot{\varphi}) + c\,(\dot{x} - \dot{x}_2 - b_2\dot{\varphi}) \\ &+k\,(x - x_1 + b_1\varphi) + k\,(x - x_2 - b_2\varphi) = 0 \end{aligned} \tag{11.33}$$

$$\begin{aligned} &I_x\ddot{\varphi} + b_1c\,(\dot{x} - \dot{x}_1 + b_1\dot{\varphi}) - b_2c\,(\dot{x} - \dot{x}_2 - b_2\dot{\varphi}) \\ &+b_1k\,(x - x_1 + b_1\varphi) - b_2k\,(x - x_2 - b_2\varphi) + k_R\varphi \\ &+c_R\left(\dot{\varphi} + \frac{\dot{x}_2 - \dot{x}_1}{w}\right) + k_R\left(\varphi + \frac{x_2 - x_1}{w}\right) = 0 \end{aligned} \tag{11.34}$$

$$\begin{aligned} &m_1\ddot{x}_1 - c\,(\dot{x} - \dot{x}_1 + b_1\dot{\varphi}) - k\,(x - x_1 + b_1\varphi) \\ &-\frac{c_R}{w}\left(\dot{\varphi} + \frac{\dot{x}_2 - \dot{x}_1}{w}\right) - \frac{k_R}{w}\left(\varphi + \frac{x_2 - x_1}{w}\right) + k_t\,x_1 = k_t\,y_1\,(t) \end{aligned} \tag{11.35}$$

$$\begin{aligned} &m_2\ddot{x}_2 - c\,(\dot{x} - \dot{x}_2 - b_2\dot{\varphi}) - k\,(x - x_2 - b_2\varphi) \\ &+\frac{c_R}{w}\left(\dot{\varphi} + \frac{\dot{x}_2 - \dot{x}_1}{w}\right) + \frac{k_R}{w}\left(\varphi + \frac{x_2 - x_1}{w}\right) + k_t\,x_2 = k_t\,y_2\,(t) \end{aligned} \tag{11.36}$$

***Proof*** The equations of motion for the half car model (11.33)–(11.36) have been driven by using the energy and dissipation functions (11.6)–(11.8) and employing Lagrange equation (11.9).

The set of equations expresses the motion of the four $DOF$ vibration of the half car model. They are coupled equations for four variables: the body bounce $x$, the body roll $\varphi$, and the left and right wheels hop $x_1$, $x_2$. There are also geometric parameters $b_1$, $b_2$, $w = b_1 + b_2$, inertia parameters $m$, $I_x$, stiffness parameters $k$, $k_t$, $k_R$, and damping parameters $c$, $c_R$. The road excitations $y_1$, $y_2$ are the input to the system of equations, and the variables $x$, $\varphi$, $x_1$, $x_2$ are the output of the equations of motion. Out of the parameters $b_1$, $b_2$, $m$, $I_x$, $k$, $k_t$, $k_R$, $c$, $c_R$, the roll stiffness $k_R$ is the parameter we adjust by examining the half car model. ■

**Example 162** Transient response of half car model to step input. Step input is a good excitation to examine natural behavior, damping rate, overshoot, and other dynamic characteristics of dynamic systems. In this example, we study the response of half car model to step inputs.

Let us first assume the step applies to the left and right wheels at the same time. Knowing that the antiroll stiffness $k_R$ and damping $c_R$ are not installed in all vehicles, let us assume they are zero. The response of the car would then be the solution of the following equations of motion with respect to the symmetric road step input.

$$[m]\ddot{\mathbf{x}} + [c]\dot{\mathbf{x}} + [k]\mathbf{x} = \mathbf{F} \tag{11.37}$$

$$\mathbf{x} = \begin{bmatrix} x \\ \varphi \\ x_1 \\ x_2 \end{bmatrix} \tag{11.38}$$

$$[m] = \begin{bmatrix} m & 0 & 0 & 0 \\ 0 & I_x & 0 & 0 \\ 0 & 0 & m_1 & 0 \\ 0 & 0 & 0 & m_2 \end{bmatrix} \tag{11.39}$$

$$[c] = \begin{bmatrix} 2c & c\,(b_1 - b_2) & -c & -c \\ c\,(b_1 - b_2) & c\left(b_1^2 + b_2^2\right) & -cb_1 & cb_2 \\ -c & -cb_1 & c + c_t & 0 \\ -c & cb_2 & 0 & c + c_t \end{bmatrix} \tag{11.40}$$

$$[k] = \begin{bmatrix} 2k & kb_1 - kb_2 & -k & -k \\ kb_1 - kb_2 & kb_1^2 + kb_2^2 & -kb_1 & kb_2 \\ -k & -kb_1 & k + k_t & 0 \\ -k & kb_2 & 0 & k + k_t \end{bmatrix} \tag{11.41}$$

$$\mathbf{F} = \begin{bmatrix} 0 \\ 0 \\ k_t y_1 \\ k_t y_2 \end{bmatrix} \qquad \mathbf{x} = \begin{bmatrix} x \\ \varphi \\ x_1 \\ x_2 \end{bmatrix} \tag{11.42}$$

$$y_1 = \begin{cases} Y & t > 0 \\ 0 & t \le 0 \end{cases} \qquad y_2 = \begin{cases} Y & t > 0 \\ 0 & t \le 0 \end{cases} \tag{11.43}$$

Using a set of nominal numerical values for a front half car model,

$$\begin{aligned} &m = 688.5\,\text{kg} \qquad m_1 = 42\,\text{kg} \qquad m_2 = 42\,\text{kg} \qquad I_x = 900\,\text{kg m}^2 \\ &b_1 = 0.8652\,\text{m} \qquad b_2 = 0.9085\,\text{m} \qquad w = b_1 + b_2 = 1.7737\,\text{m} \\ &k_1 = 22{,}000\,\text{N/m} \qquad k_2 = 22{,}000\,\text{N/m} \qquad k_t = k_t = 250{,}000\,\text{N/m} \\ &c_1 = 1630\,\text{N s/m} \qquad c_2 = 1630\,\text{N s/m} \end{aligned} \tag{11.44}$$

$$Y = 0.1\,\text{m} \tag{11.45}$$

the response of the half vehicle will be as shown in Fig. 11.6.

The left and right sides of this vehicle have different weight distribution due to different distances of the mass center from the left and right wheels and that is the reason we have nonzero roll angle $\varphi$. Transient inputs show the natural behavior of systems. The figure shows that the frequency of vertical oscillation of the body is much less than the vertical oscillation of the left and right wheels, and the frequency of oscillation of the wheels is very close to each other. The roll frequency is also close to the vertical frequency of the body. The roll vibration has only a small amplitude due to symmetry of input and no difference between left and right suspension parameters. The roll vibration will have higher amplitude when the excitation is different for left and right wheels.

**Example 163** Half vehicles over asymmetric step. Depending on the velocity of the vehicle and the wheelbase, there is a time lag between the front and rear wheels to hit the step. Here is the analysis.

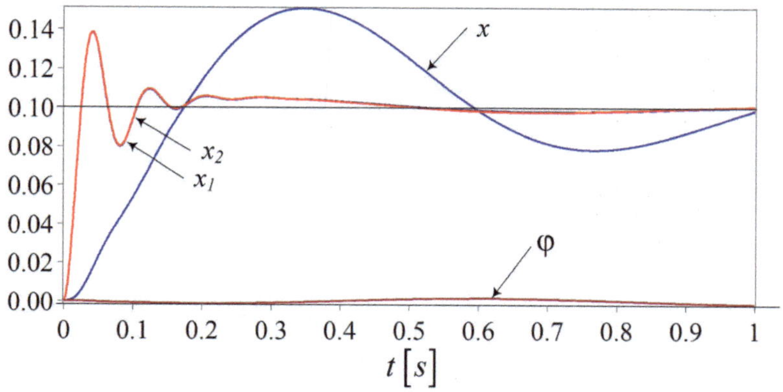


**Fig. 11.6** Body bounce, $x$, body roll, $\varphi$, left wheel, $x_1$, right wheel, $x_2$, responses of a half vehicle to a symmetric step excitation

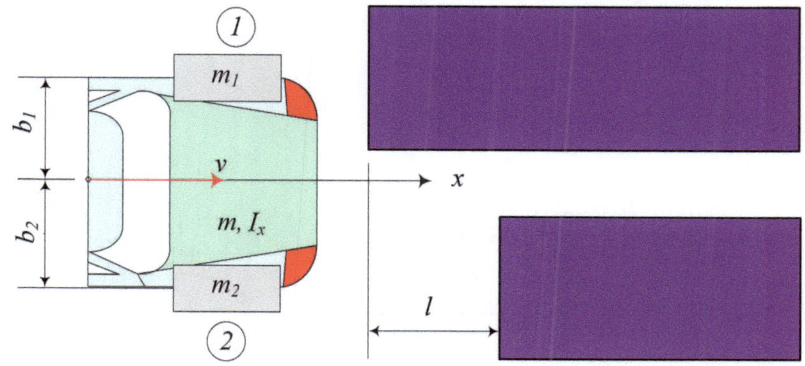


**Fig. 11.7** A half car model going over asymmetric step

Figure 11.7 illustrates a half car model going over an asymmetric step. Consider the vehicle is moving with velocity $v$ and hitting a step of height $Y$. The tire number 1 hits the step first, and after a distance $l$, the right tire hits a step. Starting the time from the moment that the left wheel hits the step, we can calculate the time $t_0$ at which the right wheel reaches the step.

$$t_0 = \frac{l}{v} \tag{11.46}$$

Therefore, the excitation of the right wheel, $y_2$, would be a lagged step input,

$$y_1 = \begin{cases} Y & t > 0 \\ 0 & t \leq 0 \end{cases} \qquad y_2 = \begin{cases} Y\,H\left(t - \frac{l}{v}\right) & t > 0 \\ 0 & t \leq 0 \end{cases} \tag{11.47}$$

where $H(t - t_0)$ is the Heaviside function.

$$H(t - t_0) = \begin{cases} 1 & t > t_0 \\ 0 & t \leq t_0 \end{cases} \tag{11.48}$$

Let us assume a speed of $v$ and the same data as given in Eq. (11.44).

$$Y = 0.1\,\mathrm{m} \qquad 1 = 1\,\mathrm{m} \qquad v = 5\,\mathrm{m/s} \tag{11.49}$$

The responses of the car are shown in Fig. 11.8. The vibration frequencies of the wheels are higher than the vibration frequencies of the body. The vibrations of wheels stop much faster than the body. The unpleasant roll vibration $\varphi$ would last longer than the wheels, almost as long as the body vertical vibrations, $x$.

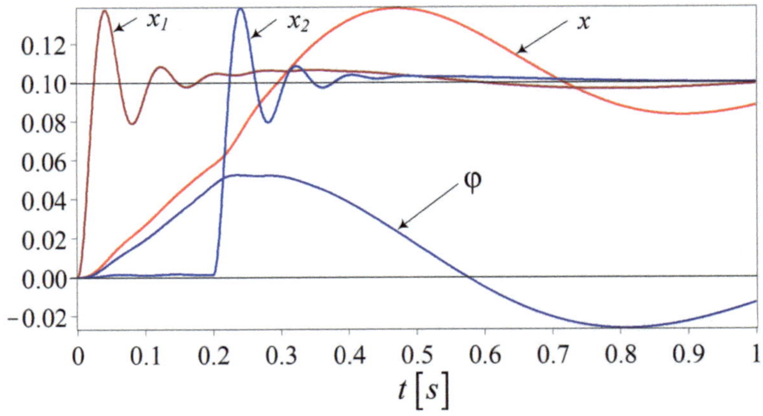


**Fig. 11.8** Response of a half car model going over an asymmetric step

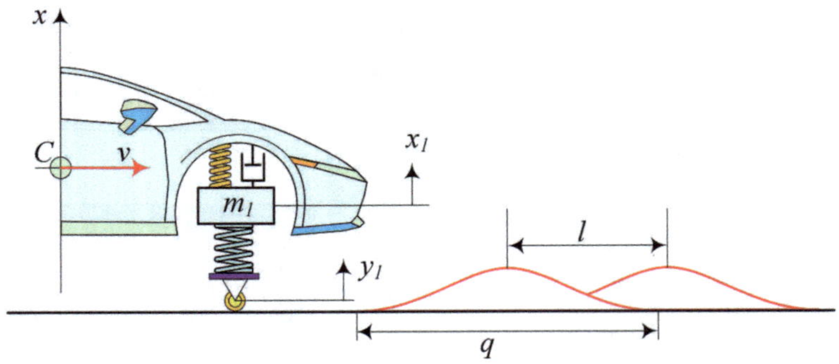


**Fig. 11.9** A half car model and an asymmetric sine square road excitation

**Example 164** Bump on the road and half car. Here we examine the reaction of half vehicle model to a bump on the road. Sine square is a good mathematical model for speed bumps on the roads.

Figure 11.9 illustrates a half car model and an asymmetric sine square road excitation for a vehicle that is moving with speed $v$ over the bump with length $q$ and height $Y$. The bumps hit the left and right wheels with a lag of distance $l$. It is a simulation of going over a bump on the road at an angle, so left and right wheels will hit the bump at different times.

$$y_1 = \begin{cases} Y \sin^2 \omega t & 0 \le t \le t_0 \\ 0 & t_0 < t \end{cases} \qquad \omega = \frac{v\pi}{q} \qquad t_0 = \frac{q}{v} \tag{11.50}$$

$$y_2 = \begin{cases} Y \sin^2 \omega (t - t_0) & \frac{l}{v} \le t \le \frac{l}{v} + t_0 \\ 0 & \frac{l}{v} + t_0 < t \end{cases} \tag{11.51}$$

$$Y = 0.1\,\text{m} \qquad q = \frac{\pi}{10}\,\text{m} \qquad l = 0.5\,\text{m} \qquad v = 10\,\text{m/s} \tag{11.52}$$

Figure 11.10 illustrates the response of the vehicle with the following characteristics:

$$\begin{aligned} & m = 688.5\,\text{kg} \quad m_1 = 42\,\text{kg} \quad m_2 = 42\,\text{kg} \quad I_x = 900\,\text{kg m}^2 \\ & b_1 = 0.8652\,\text{m} \quad b_2 = 0.9085\,\text{m} \quad w = b_1 + b_2 = 1.7737\,\text{m} \\ & k_1 = 22{,}000\,\text{N/m} \qquad k_2 = 22{,}000\,\text{N/m} \\ & k_t = k_t = 250{,}000\,\text{N/m} \\ & c_1 = 1630\,\text{N s/m} \qquad c_2 = 1630\,\text{N s/m} \end{aligned} \tag{11.53}$$

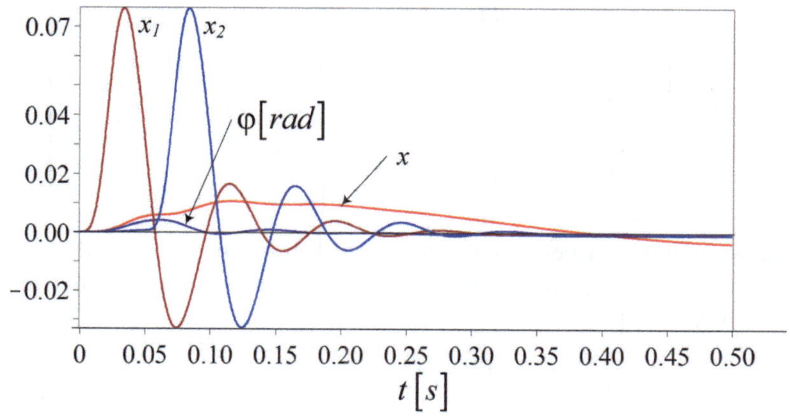


**Fig. 11.10** Body bounce, $x$, body roll, $\varphi$, left wheel, $x_1$, right wheel, $x_2$, responses of a half car to an asymmetric bump excitation

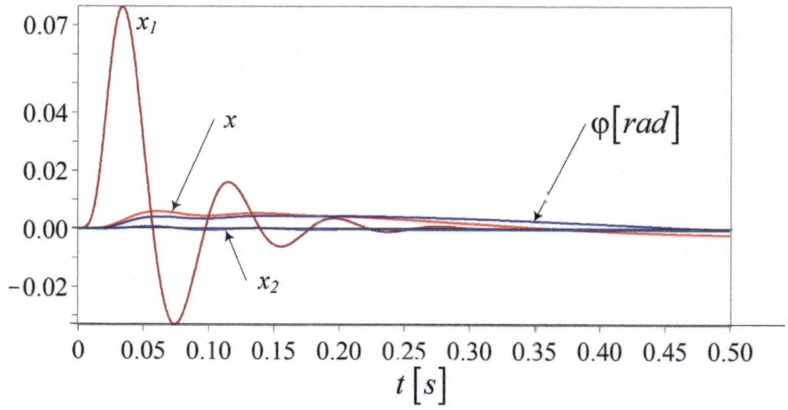


**Fig. 11.11** Body bounce, $x$, body roll, $\varphi$, left wheel, $x_1$, right wheel, $x_2$, responses of a half car to a one side bump excitation

Let us assume that only one side, say left side, of the half car goes over a bump. Figure 11.11 illustrates the reaction of the vehicle to the one side bump. Independent suspension system of the left and right wheels makes the right wheel not to show much oscillation. The vertical and roll oscillations are also very low compared to displacement of the left wheel.

## 11.3 Natural Frequencies and Mode Shapes

The half car vibrating model is a four $DOF$ system, and therefore, it has four natural frequencies $\omega_{n_1}$, $\omega_{n_2}$, $\omega_{n_3}$, $\omega_{n_4}$. The natural frequencies $\omega_{n_i}$ are solutions of the characteristic equation of the system, and the mode shape $\mathbf{u}_i$, corresponding to $\omega_i$, is the solution of the eigenvector equation.

$$\det\left[[k] - \omega^2 [m]\right] = 0 \tag{11.54}$$

$$\left[[k] - \omega_i^2 [m]\right] \mathbf{u}_i = 0 \tag{11.55}$$

The mass matrix $[m]$ and stiffness matrix $[k]$ of the system are

$$[m] = \begin{bmatrix} m & 0 & 0 & 0 \\ 0 & I_x & 0 & 0 \\ 0 & 0 & m_1 & 0 \\ 0 & 0 & 0 & m_2 \end{bmatrix} \tag{11.56}$$

$$[k] = \begin{bmatrix} 2k & kb_1 - kb_2 & -k & -k \\ kb_1 - kb_2 & k\left(b_1^2 + b_2^2\right) + k_R & -kb_1 - \frac{k_R}{w} & kb_2 + \frac{k_R}{w} \\ -k & -kb_1 - \frac{k_R}{w} & k + k_t + \frac{k_R}{w^2} & -\frac{k_R}{w^2} \\ -k & kb_2 + \frac{k_R}{w} & -\frac{k_R}{w^2} & k + k_t + \frac{k_R}{w^2} \end{bmatrix} \tag{11.57}$$

***Proof*** The natural frequencies and mode shapes of any vibrating system will be determined by setting all dampings to zero and eliminating any excitation term. The natural frequencies and mode shapes can also be determined by calculating the eigenvalues and eigenvectors of $\mathbf{A} = [m]^{-1}[k]$.

$$\mathbf{A} = \begin{bmatrix} 2\frac{k}{m} & \frac{kb_1 - kb_2}{m} & -\frac{k}{m} & -\frac{k}{m} \\ \frac{kb_1 - kb_2}{I_x} & \frac{k\left(b_1^2 + b_2^2\right) + k_R}{I_x} & -\frac{kb_1 + \frac{k_R}{w}}{I_x} & \frac{kb_2 + \frac{k_R}{w}}{I_x} \\ -\frac{k}{m_1} & -\frac{kb_1 + \frac{k_R}{w}}{m_1} & \frac{k + k_t + \frac{k_R}{w^2}}{m_1} & -\frac{k_R}{w^2 m_1} \\ -\frac{k}{m_2} & \frac{kb_2 + \frac{k_R}{w}}{m_2} & -\frac{k_R}{w^2 m_2} & \frac{k + k_t + \frac{k_R}{w^2}}{m_2} \end{bmatrix} \tag{11.58}$$

To study the system, we separate the parameters to fixed and adjustable. The fixed parameters are $b_1$, $b_2$, $w$, $m$, $m_1$, $m_2$, $I_x$, and the adjustable parameters are $k_1$, $k_2$, $k_R$. Out of the three adjustable parameters, we adjust the value of $k_1$, $k_2$, by quarter car model analysis and $k_R$ by half car model. ■

**Example 165** Natural frequencies and mode shapes of a half car model. There are four natural frequencies and mode shapes for half car model. This example shows how to calculate them.

Consider a vehicle with the following characteristics:

$$\begin{aligned} &m = 688.5\,\text{kg} \quad m_1 = 42\,\text{kg} \quad m_2 = 42\,\text{kg} \quad I_x = 900\,\text{kg}\,\text{m}^2 \\ &b_1 = 0.8652\,\text{m} \quad b_2 = 0.9085\,\text{m} \quad w = b_1 + b_2 = 1.7737\,\text{m} \\ &k_1 = 22{,}000\,\text{N/m} \qquad k_2 = 22{,}000\,\text{N/m} \\ &k_t = k_t = 250{,}000\,\text{N/m} \\ &c_1 = 1630\,\text{N}\,\text{s/m} \qquad c_2 = 1630\,\text{N}\,\text{s/m} \qquad k_R = 0 \end{aligned} \tag{11.59}$$

The characteristic equation of the system is

$$\begin{aligned} \det\left[[k] - \lambda\,[m]\right] &= \lambda^4 - 13{,}054.76\,\lambda^3 + 4.3216 \times 10^7\,\lambda^2 \\ &\quad -3.976 \times 10^9\,\lambda + 8.7064 \times 10^{10} \\ &= \omega^8 - 13{,}054.76\,\omega^6 + 4.3216 \times 10^7\,\omega^4 \\ &\quad -3.976 \times 10^9\,\omega^2 + 8.7064 \times 10^{10} \end{aligned} \tag{11.60}$$

$$\lambda = \omega^2 \tag{11.61}$$

because

$$[m] = \begin{bmatrix} 688.5 & 0 & 0 & 0 \\ 0 & 900 & 0 & 0 \\ 0 & 0 & 42 & 0 \\ 0 & 0 & 0 & 42 \end{bmatrix} \tag{11.62}$$

$$[k] = \begin{bmatrix} 44{,}000 & -952.6 & -22{,}000 & -22{,}000 \\ -952.6 & 34{,}626.7 & -19{,}034.4 & 19{,}987 \\ -22{,}000 & -19{,}034.4 & 272{,}000 & 0 \\ -22{,}000 & 19{,}987 & 0 & 272{,}000 \end{bmatrix} \tag{11.63}$$

To find the natural frequencies, we may solve the characteristic equation (10.96) or search for eigenvalues of $[A] = [m]^{-1}[k]$. The eigenvalues of $[A]$ are

$$\lambda_1 = 35.292 \qquad \lambda_2 = 58.743$$
$$\lambda_3 = 6479.314 \qquad \lambda_4 = 6481.411 \tag{11.64}$$

therefore, the natural frequencies of the half car model are

$$\omega_1 = \sqrt{\lambda_1} = 5.94\,\mathrm{rad/s} \approx 0.945\,\mathrm{Hz}$$
$$\omega_2 = \sqrt{\lambda_2} = 7.66\,\mathrm{rad/s} \approx 1.22\,\mathrm{Hz}$$
$$\omega_3 = \sqrt{\lambda_3} = 80.49\,\mathrm{rad/s} \approx 12.811\,\mathrm{Hz}$$
$$\omega_4 = \sqrt{\lambda_4} = 80.51\,\mathrm{rad/s} \approx 12.813\,\mathrm{Hz} \tag{11.65}$$

The mode shapes of the system are

$$\mathbf{u}_1 = \begin{bmatrix} 0.05428 \\ 1.0 \\ 0.07477 \\ -0.06947 \end{bmatrix} \qquad \mathbf{u}_2 = \begin{bmatrix} 1.0 \\ -0.04152 \\ 0.07869 \\ 0.0847 \end{bmatrix} \tag{11.66}$$

$$\mathbf{u}_3 = \begin{bmatrix} 0.00035 \\ 0.00648 \\ -1.0 \\ 0.92902 \end{bmatrix} \qquad \mathbf{u}_4 = \begin{bmatrix} 0.0096 \\ -4 \times 10^{-4} \\ -0.92901 \\ -1.0 \end{bmatrix} \tag{11.67}$$

The first mode shape is associated to roll motion at a frequency of about $\omega_1 \approx 1$ Hz. The second mode shape is associated to body bounce at frequency about $\omega_2 \approx 1.2$ Hz. The third and fourth mode shapes are related to left and right wheels, both around 12 Hz. Consider this car that starts to move on a bumpy road at a very small acceleration. By increasing the speed, the first resonance occurs at $\omega_1 \approx 1$ Hz, at which the roll vibration is the most observable vibration. The second resonance occurs at $\omega_2 \approx 1.2$ Hz when the bounce vibration of the body is the most observable vibration. The third and fourth resonances at $\omega_3 \approx 12.8$ Hz are related to rear and front wheels, respectively. When the excitation frequency of a multiple $DOF$ system increases, we will see that observable vibration moves from a coordinate to the others in the order of natural frequencies.

**Example 166** Translational and rotational coordinates. When a multi-$DOF$ vibrating system has a combination of translational and rotational coordinates, the mode shapes are complicated to be interpreted. The smart way is to convert all coordinates to be either translational or rotational.

Figure 11.12 illustrates a two $DOF$ beam with mass $m$ and mass moment $I$ about the mass center $C$, sitting on two springs $k_1 = k_2$. This is a good model to study half car model to determine and compare roll and bounce vibrations of vehicles. The translational coordinate $x$ of $C$ and the rotational coordinate $\varphi$ are the common generalized coordinates that we use to measure the kinematics of the beam. The equations of motion and the mode shapes of the system are functions of the chosen coordinates.

The kinetic and potential energies of the beam for small motions are

$$K = \frac{1}{2}m\dot{x}^2 + \frac{1}{2}I\dot{\varphi}^2 \tag{11.68}$$

$$P = \frac{1}{2}k_2\,(x - b_2\varphi)^2 + \frac{1}{2}k_1\,(x + b_1\varphi)^2 \tag{11.69}$$

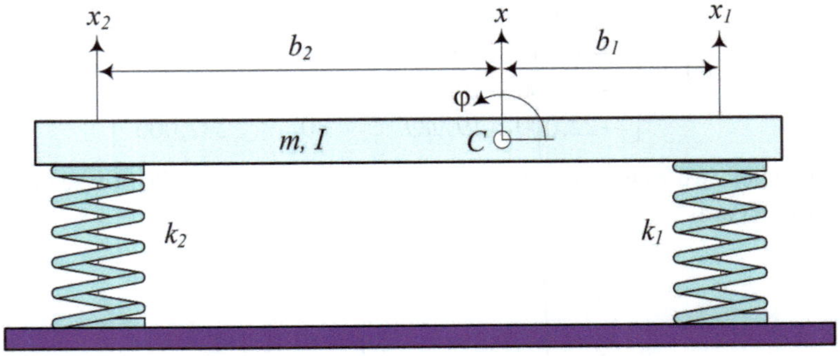


**Fig. 11.12** A beam with mass $m$ and mass moment $I$ siting on two springs $k_1$ and $k_2$

The energies provide us with the equations of motion.

$$m\ddot{x} + k_2 (x - b_2\varphi) + k_1 (x + b_1\varphi) = 0 \tag{11.70}$$

$$I\ddot{\varphi} - b_2k_2 (x - b_2\varphi) + b_1k_1 (x + b_1\varphi) = 0 \tag{11.71}$$

$$\begin{bmatrix} m & 0 \\ 0 & I \end{bmatrix} \begin{bmatrix} \ddot{x} \\ \ddot{\varphi} \end{bmatrix} + \begin{bmatrix} k_1 + k_2 & b_1k_1 - b_2k_2 \\ b_1k_1 - b_2k_2 & b_2^2k_2 + b_1^2k_1 \end{bmatrix} \begin{bmatrix} x \\ \varphi \end{bmatrix} = 0 \tag{11.72}$$

Let us use a set of numeric data:

$$\begin{aligned} m &= 688.5\,\text{kg} & I &= 900\,\text{kg m}^2 \\ k_1 &= 22{,}000\,\text{N/m} & k_2 &= 22{,}000\,\text{N/m} \\ b_1 &= 0.8652\,\text{m} & b_2 &= 0.9085\,\text{m} \end{aligned} \tag{11.73}$$

to find the following equations:

$$\begin{bmatrix} 688.5 & 0 \\ 0 & 900 \end{bmatrix} \begin{bmatrix} \ddot{x} \\ \ddot{\varphi} \end{bmatrix} + \begin{bmatrix} 44{,}000 & -952.6 \\ -952.6 & 34{,}626.75 \end{bmatrix} \begin{bmatrix} x \\ \theta \end{bmatrix} = 0 \tag{11.74}$$

The natural frequencies and mode shapes of the system are calculated based on eigenvalues and eigenvectors of $[A] = \left[[m]^{-1}[k]\right]$:

$$\begin{aligned} [A] &= \begin{bmatrix} 688.5 & 0 \\ 0 & 900 \end{bmatrix}^{-1} \begin{bmatrix} 44{,}000 & -952.6 \\ -952.6 & 34{,}626.75 \end{bmatrix} \\ &= \begin{bmatrix} 63.907 & -1.3836 \\ -1.0584 & 38.474 \end{bmatrix} \end{aligned} \tag{11.75}$$

The $\omega_i$ and $\mathbf{u}_i$ are

$$\omega_1 = 6.198\,\text{rad/s} \qquad \omega_2 = 7.998\,\text{rad/s} \tag{11.76}$$

$$\mathbf{u}_1 = \begin{bmatrix} 5.42 \times 10^{-2} \\ 1 \end{bmatrix} \qquad \mathbf{u}_2 = \begin{bmatrix} 1 \\ -4.148 \times 10^{-2} \end{bmatrix} \tag{11.77}$$

Because of the chosen coordinate vector $\begin{bmatrix} x & \varphi \end{bmatrix}^T$, the first component of the mode shapes refers to the displacement $x$, and the second component refers to the rotation $\varphi$. Therefore, the components of the mode shapes are in different dimensions. The mode shapes are coordinate dependent, and hence, the rotational coordinate would not be able to compare with displacement codominant.

To compare the mode shapes of a system practically, we better choose the generalized coordinates with the same dimension. Let us employ the coordinates $x_1$ and $x_2$ instead of $x$ and $\varphi$, as shown in Fig. 11.12. The energies of the system in this set of coordinates will be

$$K = \frac{1}{2}m\left(\frac{b_2\dot{x}_1 + b_1\dot{x}_2}{b_1 + b_2}\right)^2 + \frac{1}{2}I\left(\frac{\dot{x}_1 - \dot{x}_2}{b_1 + b_2}\right)^2 \tag{11.78}$$

$$P = \frac{1}{2}k_2x_2^2 + \frac{1}{2}k_1x_1^2 \tag{11.79}$$

The energies provide us with the following equations of motion:

$$\begin{bmatrix} \frac{mb_2^2 + I}{(b_1 + b_2)^2} & \frac{mb_1b_2 - I}{(b_1 + b_2)^2} \\ \frac{mb_1b_2 - I}{(b_1 + b_2)^2} & \frac{mb_1^2 + I}{(b_1 + b_2)^2} \end{bmatrix} \begin{bmatrix} \ddot{x}_1 \\ \ddot{x}_2 \end{bmatrix} + \begin{bmatrix} k_1 & 0 \\ 0 & k_2 \end{bmatrix} \begin{bmatrix} x_1 \\ x_2 \end{bmatrix} = 0 \tag{11.80}$$

Using the data (11.73), the equations of motion will be

$$\begin{bmatrix} 466.708 & -114.054 \\ -114.054 & 449.9 \end{bmatrix} \begin{bmatrix} \ddot{x}_1 \\ \ddot{x}_2 \end{bmatrix} + \begin{bmatrix} 22{,}000 & 0 \\ 0 & 22{,}000 \end{bmatrix} \begin{bmatrix} x_1 \\ x_2 \end{bmatrix} = 0 \tag{11.81}$$

The natural frequencies and mode shapes of the system in the new coordinates are calculated based on eigenvalues and eigenvectors of new $[A] = [m]^{-1}[k]$:

$$\begin{aligned} [A] &= \begin{bmatrix} 466.708 & -114.054 \\ -114.054 & 449.9 \end{bmatrix}^{-1} \begin{bmatrix} 22{,}000 & 0 \\ 0 & 22{,}000 \end{bmatrix} \\ &= \begin{bmatrix} 50.252 & 12.739 \\ 12.739 & 52.129 \end{bmatrix} \end{aligned} \tag{11.82}$$

The natural frequencies $\omega_i$ will be the same as (11.76) but $\mathbf{u}_i$ will be different.

$$\omega_1 = 6.198\,\text{rad/s} \qquad \omega_2 = 7.998\,\text{rad/s} \tag{11.83}$$

$$\mathbf{u}_1 = \begin{bmatrix} 1.0 \\ -0.929 \end{bmatrix} \qquad \mathbf{u}_2 = \begin{bmatrix} 0.929 \\ 1.0 \end{bmatrix} \tag{11.84}$$

The components of mode shapes are comparable as they are all in length dimension.

**Example 167** Effect of the roll stiffness, $k_R$, on natural frequencies of half vehicle model. The antiroll spring is added to the vehicle suspension system to adjust the roll frequency. The roll frequency must be close to the bounce frequency.

Let us consider a vehicle with the following nominated data and unknown $k_P$.

$$\begin{aligned} & m = 688.5\,\text{kg} \qquad m_1 = 42\,\text{kg} \qquad m_2 = 42\,\text{kg} \\ & I_x = 900\,\text{kg m}^2 \qquad b_1 = 0.8652\,\text{m} \qquad b_2 = 0.9085\,\text{m} \\ & w = b_1 + b_2 = 1.7737\,\text{m} \qquad k_1 = 22{,}000\,\text{N/m} \\ & k_2 = 22{,}000\,\text{N/m} \qquad k_t = k_t = 250{,}000\,\text{N/m} \end{aligned} \tag{11.85}$$

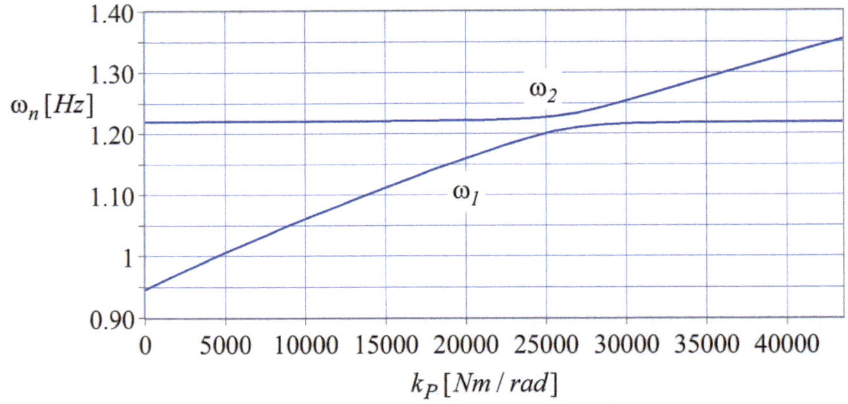


**Fig. 11.13** The first two natural frequencies of a half vehicle versus antiroll stiffness $k_R$

The mass matrix will be the same as (11.62), while the stiffness matrix of the system will be

$$[k] = \begin{bmatrix} 44{,}000 & -952.6 & -22{,}000 & -22{,}000 \\ -952.6 & k_R{+}34{,}626 & -19{,}034{-}\dfrac{k_R}{1.77} & 19{,}987{+}\dfrac{k_R}{1.77} \\ -22{,}000 & -19{,}034{-}\dfrac{k_R}{1.77} & 272{,}000{+}\dfrac{k_R}{1.77^2} & -\dfrac{k_R}{1.77^2} \\ -22{,}000 & 19{,}987{+}\dfrac{k_R}{1.77} & -\dfrac{k_R}{1.77^2} & 272{,}000{+}\dfrac{k_R}{1.77^2} \end{bmatrix} \tag{11.86}$$

The characteristic equation of $[A] = [m]^{-1} [k]$ is a function of $k_R$. To calculate the natural frequencies $\omega_i$ of the system as functions of $k_R$, we assign a value to $k_R$ and solve the characteristic equation and determine the value of the eigenvalues $\lambda_i = \omega_i^2$. Repeating this procedure for other values of $k_R$ and recording the natural frequencies will show the frequencies as function of $k_P$. Figure 10.18 illustrates the first two natural frequencies of the vehicle in [ Hz]. When $k_R = 0$, the natural frequencies of the vehicle from the lowest to the highest are

$$\begin{aligned} \omega_1 &\approx 0.94\,\text{Hz} \qquad \omega_2 \approx 1.22\,\text{Hz} \\ \omega_3 &\approx 12.81\,\text{Hz} \qquad \omega_4 \approx 12.81\,\text{Hz} \end{aligned} \tag{11.87}$$

Figure 11.13 illustrates the first two natural frequencies of the vehicle in [ Hz]. By increasing the value of $k_R$, the roll frequency $\omega_1$ increases at first and then saturates, while the bounce frequency $\omega_2$ remains steady at the beginning and increases at higher values. The first two natural frequencies are closest at $k_R \simeq 27{,}000$ N m/ rad.

## 11.4 Frequency Response

Let us consider a harmonic base excitation to determine the frequency response of the half car model.

$$y_1 = y_2 = Y \sin \omega t \tag{11.88}$$

Then we substitute a set of harmonic solutions with the same excitation frequency $\omega$ into equations of motion.

$$x = A_1 \sin \omega t + B_1 \cos \omega t = X_1 \sin (\omega t - \varphi_1) \tag{11.89}$$

$$\varphi = A_2 \sin \omega t + B_2 \cos \omega t = X_2 \sin (\omega t - \varphi_2) \tag{11.90}$$

$$x_1 = A_3 \sin \omega t + B_3 \cos \omega t = X_3 \sin \left(\omega t - \varphi_3\right) \tag{11.91}$$

$$x_2 = A_4 \sin \omega t + B_4 \cos \omega t = X_4 \sin (\omega t - \varphi_4) \tag{11.92}$$

The differential equations of motion will convert into a set of algebraic equations to determine the unknown coefficients $A_1$, $B_1$, $A_2$, $B_2$, $A_3$, $B_3$, $A_4$, $B_4$ and calculate the frequency responses $X_1$, $X_2$, $X_3$, and $X_4$, which are the amplitudes of $x$, $\varphi$, $x_1$, and $x_2$.

$$X_1 = \sqrt{A_1^2 + B_1^2} \qquad X_2 = \sqrt{A_2^2 + B_2^2}$$

$$X_3 = \sqrt{A_3^2 + B_3^2} \qquad X_4 = \sqrt{A_4^2 + B_4^2} \tag{11.93}$$

***Proof*** Assuming zero damping for the tires, the set of equations of the half car model are

$$[m]\ddot{\mathbf{x}} + [c]\dot{\mathbf{x}} + [k]\mathbf{x} = \mathbf{F} \tag{11.94}$$

$$\mathbf{x} = \begin{bmatrix} x \\ \varphi \\ x_1 \\ x_2 \end{bmatrix} \tag{11.95}$$

$$[m] = \begin{bmatrix} m & 0 & 0 & 0 \\ 0 & I_x & 0 & 0 \\ 0 & 0 & m_1 & 0 \\ 0 & 0 & 0 & m_2 \end{bmatrix} \tag{11.96}$$

$$[c] = \begin{bmatrix} 2c & c(b_1 - b_2) & -c & -c \\ c(b_1 - b_2) & c\left(b_1^2 + b_2^2\right) + c_R & -cb_1 - \frac{c_R}{w} & cb_2 + \frac{c_R}{w} \\ -c & -cb_1 - \frac{c_R}{w} & c + c_t + \frac{c_R}{w^2} & -\frac{c_R}{w^2} \\ -c & cb_2 + \frac{c_R}{w} & -\frac{c_R}{w^2} & c + c_t + \frac{c_R}{w^2} \end{bmatrix} \tag{11.97}$$

$$[k] = \begin{bmatrix} 2k & kb_1 - kb_2 & -k & -k \\ kb_1 - kb_2 & k\left(b_1^2 + b_2^2\right) + k_R & -kb_1 - \frac{k_R}{w} & kb_2 + \frac{k_R}{w} \\ -k & -kb_1 - \frac{k_R}{w} & k + k_t + \frac{k_R}{w^2} & -\frac{k_R}{w^2} \\ -k & kb_2 + \frac{k_R}{w} & -\frac{k_R}{w^2} & k + k_t + \frac{k_R}{w^2} \end{bmatrix} \tag{11.98}$$

$$\mathbf{F} = \begin{bmatrix} 0 \\ 0 \\ y_1 k_t \\ y_2 k_t \end{bmatrix} \tag{11.99}$$

Substituting harmonic solutions in the equations of motion

$$\begin{bmatrix} x \\ \varphi \\ x_1 \\ x_2 \end{bmatrix} = \begin{bmatrix} A_1 \\ A_2 \\ A_3 \\ A_4 \end{bmatrix} \sin\omega t + \begin{bmatrix} B_1 \\ B_2 \\ B_3 \\ B_4 \end{bmatrix} \cos\omega t \tag{11.100}$$

provides us with a set of equations to calculate the coefficients **A** and **B**.

$$\begin{bmatrix} [k] - \omega^2[m] & -\omega[c] \\ \omega[c] & [k] - \omega^2[m] \end{bmatrix} \begin{bmatrix} A_1 \\ A_2 \\ A_3 \\ A_4 \\ B_1 \\ B_2 \\ B_3 \\ B_4 \end{bmatrix} = \begin{bmatrix} 0 \\ 0 \\ y_1 k_t \\ y_2 k_t \\ 0 \\ 0 \\ 0 \\ 0 \end{bmatrix} \tag{11.101}$$

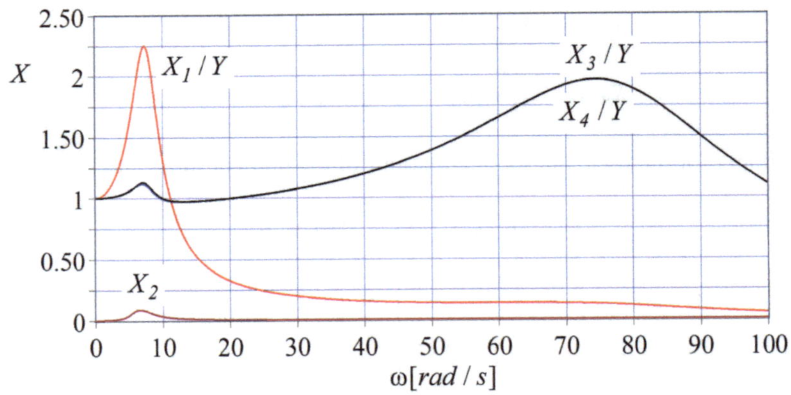


**Fig. 11.14** A sample of frequency response of a half vehicle model

The frequency responses are indicated by $X_1$, $X_2$, $X_3$, and $X_4$, where $X_1$ is the body bounce amplitude, $X_2$ is the body pitch amplitude, $X_3$ is the front wheel bounce amplitude, and $X_4$ is the rear wheel bounce amplitude. Assuming harmonic excitations,

$$y_1 = Y \sin \omega t \tag{11.102}$$

$$y_2 = Y \sin \omega t \tag{11.103}$$

we calculate the unknown coefficients and the frequency responses (11.93).

$$\begin{bmatrix} A_1 \\ A_2 \\ A_3 \\ A_4 \\ B_1 \\ B_2 \\ B_3 \\ B_4 \end{bmatrix} = \begin{bmatrix} [k] - \omega^2 [m] & -\omega [c] \\ \omega [c] & [k] - \omega^2 [m] \end{bmatrix}^{-1} \begin{bmatrix} 0 \\ 0 \\ Yk_t \\ Yk_t \\ 0 \\ 0 \\ 0 \\ 0 \end{bmatrix} \tag{11.104}$$

■

**Example 168** Frequency response of a half car model. There are two resonance zones in the frequency response of half car model. A resonance zone around 1 Hz associated to the body bounce and body roll and a resonance zone around 10 Hz associated to left and right wheels bounce vibrations.

Let us consider a half vehicle model with $k_R = 0$ and the following data.

$$m = 688.5\,\text{kg} \qquad m_1 = 42\,\text{kg} \qquad m_2 = 42\,\text{kg}$$

$$I_x = 900\,\text{kg}\,\text{m}^2 \qquad b_1 = 0.8652\,\text{m} \qquad b_2 = 0.9085\,\text{m}$$

$$w = b_1 + b_2 = 1.7737\,\text{m} \qquad k_1 = 22{,}000\,\text{N/m}$$

$$k_2 = 22{,}000\,\text{N/m} \qquad k_t = k_t = 250{,}000\,\text{N/m}$$

To determine frequency response of the vehicle, we apply a harmonic excitation under front and rear wheels with amplitude $Y$ and at a frequency $\omega$.

$$y_1 = y_2 = Y \sin \omega t \tag{11.105}$$

Figure 11.14 illustrates the frequency response of the vehicle. The amplitude of body bounce $x$ goes high at the first resonance zone around 1 Hz and drops significantly for $\omega > 2$ Hz. The body roll oscillation $\varphi$ has similar behavior with high amplitude around 1 Hz and very low amplitude for $\omega > 2$ Hz. The left and right wheels have a wider range of resonance zones than the body oscillations. The wheels' highest amplitude of vibrations appears around 10 Hz.

## 11.5 Summary

To examine and optimize the roll vibrations of a vehicle, we use the half car vibrating model in which we laterally cut a vehicle and take one half to study. This model includes the body bounce $x$, body roll $\varphi$, and wheels hop $x_1$ and $x_2$. There are also two independent road excitations $y_1$ and $y_2$ under the left and right wheels. The roll vibration of a car is the most unpleasant motion of a vehicle.

The half car model has four degrees of freedom indicated by the coordinates $x, \varphi, x_1$, and $x_2$ along with two road excitations $y_1$ and $y_2$ as shown in Fig. 11.1. The equations of motion of the half car model are

$$m\ddot{x} + c\left(\dot{x} - \dot{x}_1 + b_1\dot{\varphi}\right) + c\left(\dot{x} - \dot{x}_2 - b_2\dot{\varphi}\right)$$
$$+k\left(x - x_1 + b_1\varphi\right) + k\left(x - x_2 - b_2\varphi\right) = 0 \tag{11.106}$$

$$I_x\ddot{\varphi} + b_1c\left(\dot{x} - \dot{x}_1 + b_1\dot{\varphi}\right) - b_2c\left(\dot{x} - \dot{x}_2 - b_2\dot{\varphi}\right)$$
$$+b_1k\left(x - x_1 + b_1\varphi\right) - b_2k\left(x - x_2 - b_2\varphi\right) + k_R\varphi = 0 \tag{11.107}$$

$$m_1\ddot{x}_1 - c\left(\dot{x} - \dot{x}_1 + b_1\dot{\varphi}\right) + k_t\left(x_1 - y_1\right) - k\left(x - x_1 + b_1\varphi\right) = 0 \tag{11.108}$$

$$m_2\ddot{x}_2 - c\left(\dot{x} - \dot{x}_2 - b_2\dot{\varphi}\right) + k_t\left(x_2 - y_2\right) - k\left(x - x_2 - b_2\varphi\right) = 0 \tag{11.109}$$

The vehicles also have an antiroll bar with a torsional stiffness $k_R$. The antiroll bar is being used to adjust the roll natural frequency to be as close as possible to the bounce natural frequency of the vehicle.

## 11.6 Key Symbols

| | |
|---|---|
| $a, \ddot{x}$ | acceleration |
| $[a], [A]$ | coefficient matrix |
| $[A] = [m]^{-1}[k]$ | coefficient matrix of characteristic equation |
| $b_1$ | distance from mass center to left wheel |
| $b_2$ | distance from mass center to right wheel |
| $c$ | damping |
| $c_e$ | equivalent damping |
| $c_{ij}$ | element of row $i$ and column $j$ of $[c]$ |
| $[c]$ | damping matrix |
| $C$ | mass center |
| $D$ | dissipation function |
| $E$ | mechanical energy |
| $f$, $\mathbf{F}$ | force, force vector |
| $f = 1/T$ | cyclic frequency [Hz] |
| $f_c$ | damper force |
| $f_k$ | spring force |
| $F$ | amplitude of a harmonic force $f$ |
| $F_r$, $Q_r$ | generalized force |
| $g$ | gravitational acceleration |
| $h$ | height |

| | |
|---|---|
| $H$ | Heaviside function |
| $I$ | mass moment |
| $\mathbf{I}$ | identity matrix |
| $k$ | stiffness |
| $k_e$ | equivalent stiffness |
| $k_{ij}$ | element of row $i$ and column $j$ of $[k]$ |
| $k_R$ | antiroll bar torsional stiffness |
| $k_s$ | main suspension spring stiffness |
| $k_u$ | tire stiffness |
| $[k]$ | stiffness matrix |
| $K$ | kinetic energy |
| $l$ | length, wheelbase |
| $\mathcal{L}$ | Lagrangean |
| $m, M$ | mass |
| $m_{ij}$ | element of row $i$ and column $j$ of $[m]$ |
| $[m]$ | mass matrix |
| $M$ | moment, mass |
| $P$ | potential energy |
| $r$ | excitation frequency ratio |
| $r, R$ | radius |
| $r_i,\ i \in N$ | nodal frequency ratio |
| $r_n$ | natural frequency ratio |
| $t$ | time |
| $T$ | period |
| $x, y, z$ | body coordinate axes |
| $x$ | body displacement |
| $x_1$ | left wheel displacement |
| $x_2$ | right wheel displacement |
| $X$ | steady-state amplitude of $x$ |
| $u_{ij}$ | $j$th element of the $i$th mode shape |
| $\mathbf{u}$ | mode shape, eigenvector |
| $\mathbf{u}_i$ | $i$th eigenvector |
| $v, \mathbf{v}, \dot{x}, \dot{\mathbf{x}}$ | velocity |
| $w$ | track |
| $W$ | work |
| $y$ | base excitation displacement |
| $Y$ | steady-state amplitude of $y$ |
| $z$ | relative displacement |
| $Z$ | steady-state amplitude of $z$ |
| $Z_i$ | short notation parameter |

Greek

| | |
|---|---|
| $\alpha$ | frequency ratio |
| $\varepsilon$ | mass ratio |
| $\omega = 2\pi f$ | angular frequency [rad/s] |
| $\omega_i$ | $i$th natural frequency |
| $\omega_n$ | natural frequency |
| $\delta$ | deflection |
| $\lambda$ | eigenvalue |
| $\lambda_i$ | $i$th eigenvalue |
| $\varphi$ | roll angle |
| $\xi$ | damping ratio |

Subscript

| | |
|---|---|
| $d$ | driver |
| $i \in N$ | node number |

| | |
|---|---|
| $l$ | left |
| $n$ | natural |
| $r$ | right |
| $R$ | roll motion of car |
| $t$ | tire |
| Symbols | |
| $DOF$ | degree of freedom |
| $FBD$ | free body diagram |
| $ODF$ | one degree of freedom |
| Re | real part |
| $RMS$ | root mean square |
| det | determinant |
| Hz | Hertz |
| [ ] | matrix |
| $\int$ | integral |
| $\vert\ \vert$ | absolute value |
| $d$ | differential |
| $\partial$ | partial derivative |
| $\mathcal{L}$ | Lagrangean |

## References


1. R.N. Jazar, *Advanced Dynamics: Rigid Body, Multibody, and Aerospace Applications* (Wiley, New York, 2011)
2. R.N. Jazar, *Advanced Vibrations: A Modern Approach* (Springer, New York, 2013)
3. R.N. Jazar, *Vehicle Dynamics: Theory and Application*, 3rd edn. (Springer, New York, 2017)
4. R.N. Jazar, *Approximation Methods in Science and Engineering* (Springer, New York, 2020)
5. R.N. Jazar, *Perturbation Methods in Science and Engineering* (Springer, New York, 2021)
6. R.N. Jazar, *Advanced Vibrations: Theory and Application* (Springer, New York, 2023)
7. R.N. Jazar, R. Houim, A. Narimani, M.F. Golnaraghi, Frequency response and jump avoidance in a nonlinear passive engine mount. J. Vibr. Control **12**(11), 1205–1237
8. R.N. Jazar, M. Kazemi, S. Borhani, *Mechanical Vibrations, (in Persian)* (Ettehad Publications, Tehran, 1992)
9. K.C. Le, A. Pieper, Damping of roll vibrations of vehicle suspension. Vehic. Syst. Dyn. **52**(4), 562–579
10. L. Meirovitch, *Analytical Methods in Vibrations* (Macmillan, New York, 1967)
11. L. Meirovitch, *Principles and Techniques of Vibrations* (Prentice Hall, New Jersey, 1997)
12. L. Meirovitch, *Fundamentals of Vibrations* (McGraw-Hill, New York, 2002)
13. K. Ogata, *System Dynamics* (Prentice Hall, New Jersey, 2004)
14. Z. Osinski, *Damping of Vibrations* (A. A. Balkema Publishers, Rotterdam, 1998)
15. W.J. Palm, *Mechanical Vibration* (Wiley, New York, 2006)
16. S.S. Rao, *Mechanical Vibrations* (Pearson Prentice Hall, Harlow, 2018)
17. M. Roseau, *Vibrations in Mechanical Systems* (Springer, Berlin, 1987)
18. J.C. Snowdon, *Vibration and Shock in Damped Mechanical Systems* (Wiley, New York, 1968)
19. F.S. Tse, I.E. Morse, R.T. Hinkle, *Mechanical Vibrations Theory and Applications* (Allyn and Bacon, Boston, 1978)

# 12 Full Car Vibrating Model

The full car model, also known as a general vibrating model of a vehicle, encompasses the complete representation of a vehicle's vibrations, including the bounce, pitch, and roll motions of the body, as well as the vibrations of the wheels. In the case of a four-wheeled vehicle, the full car model incorporates seven degrees of freedom: three for the vehicle body and four for the individual wheels. Figure 12.1 provides a visual depiction of a full car model with four wheels.

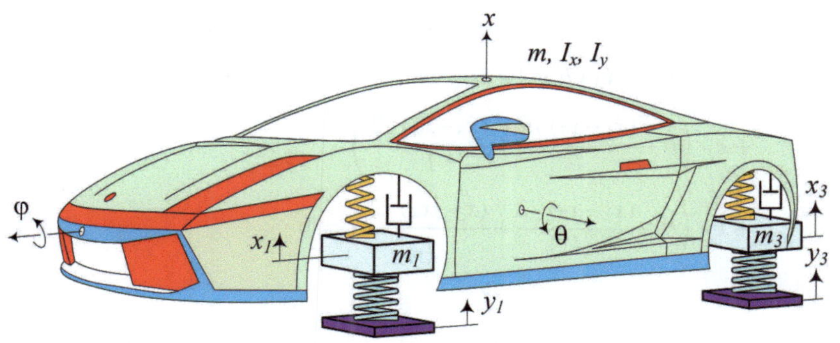


**Fig. 12.1** A full car vibrating model of a vehicle

The full car model is being employed to examine the results of optimization process of the quarter car, bicycle, and half car models and their interactions. Full car vibration analysis plays a crucial role in vehicle design and development, allowing engineers to evaluate and improve the vehicle's suspension design, stiffness and damping characteristics, and other factors that influence ride quality and handling. The final calibration of the primary springs and antiroll bars is accomplished through the utilization of the full car model. This model includes the body bounce $x$, body pitch $\theta$, body roll $\varphi$, and wheels hop $x_1$, $x_2$, $x_3$, and $x_4$. The wheels are indicated by a number, starting from 1 for the front left wheel, number 2 for the front right wheel, then number 3 for the rear left, and number 4 for the rear right wheel. The road excitations $y_1$, $y_2$, $y_3$, and $y_4$ are under the wheels $x_1$, $x_2$, $x_3$, and $x_4$.

## 12.1 Mathematical Model

The seven $DOF$ full car model includes the body bounce $x$, body roll $\varphi$, body pitch $\theta$, and wheels hop $x_1$, $x_2$, $x_3$, and $x_4$. The external excitation of the system is the independent road excitations $y_1$, $y_2$, $y_3$, and $y_4$ under wheels. The equations of motion of the full car model are as follows:

$$m\ddot{x} + c_f\left(\dot{x} - \dot{x}_1 + b_1\dot{\varphi} - a_1\dot{\theta}\right) + c_f\left(\dot{x} - \dot{x}_2 - b_2\dot{\varphi} - a_1\dot{\theta}\right)$$
$$+ c_r\left(\dot{x} - \dot{x}_3 + b_3\dot{\varphi} + a_2\dot{\theta}\right) + c_r\left(\dot{x} - \dot{x}_4 - b_4\dot{\varphi} + a_2\dot{\theta}\right)$$



R. N. Jazar, H. Marzbani, *Vehicle Vibrations*, https://doi.org/10.1007/978-3-031-43486-0_12

$$\begin{aligned}&+k_f\left(x-x_1+b_1\varphi-a_1\theta\right)+k_f\left(x-x_2-b_2\varphi-a_1\theta\right)\\&+k_r\left(x-x_3+b_3\varphi+a_2\theta\right)+k_r\left(x-x_4-b_4\varphi+a_2\theta\right)=0\end{aligned}\tag{12.1}$$

$$\begin{aligned}&I_x\ddot{\varphi}+b_1c_f\left(\dot{x}-\dot{x}_1+b_1\dot{\varphi}-a_1\dot{\theta}\right)-b_2c_f\left(\dot{x}-\dot{x}_2-b_2\dot{\varphi}-a_1\dot{\theta}\right)\\&+b_3c_r\left(\dot{x}-\dot{x}_3+b_3\dot{\varphi}+a_2\dot{\theta}\right)+b_4c_r\left(\dot{x}-\dot{x}_4-b_4\dot{\varphi}+a_2\dot{\theta}\right)\\&+b_1k_f\left(x-x_1+b_1\varphi-a_1\theta\right)-b_2k_f\left(x-x_2-b_2\varphi-a_1\theta\right)\\&+b_3k_r\left(x-x_3+b_3\varphi+a_2\theta\right)+b_4k_r\left(x-x_4-b_4\varphi+a_2\theta\right)\\&+k_{R_f}\left(\varphi-\frac{x_1-x_2}{w_f}\right)+k_{R_r}\left(\varphi-\frac{x_3-x_4}{w_r}\right)\\&+c_{R_f}\left(\dot{\varphi}-\frac{\dot{x}_1-x_2}{w_f}\right)+c_{R_r}\left(\dot{\varphi}-\frac{\dot{x}_3-\dot{x}_4}{w_r}\right)=0\end{aligned}\tag{12.2}$$

$$\begin{aligned}&I_y\ddot{\theta}-a_1c_f\left(\dot{x}-\dot{x}_1+b_1\dot{\varphi}-a_1\dot{\theta}\right)-a_1c_f\left(\dot{x}-\dot{x}_2-b_2\dot{\varphi}-a_1\dot{\theta}\right)\\&+a_2c_r\left(\dot{x}-\dot{x}_3+b_3\dot{\varphi}+a_2\dot{\theta}\right)+a_2c_r\left(\dot{x}-\dot{x}_4-b_4\dot{\varphi}+a_2\dot{\theta}\right)\\&-a_1k_f\left(x-x_1+b_1\varphi-a_1\theta\right)-a_1k_f\left(x-x_2-b_2\varphi-a_1\theta\right)\\&+a_2k_r\left(x-x_3+b_3\varphi+a_2\theta\right)+a_2k_r\left(x-x_4-b_4\varphi+a_2\theta\right)\\&+c_P\left(\dot{\theta}+\frac{\left(\dot{x}_1-\dot{x}_2\right)/2}{l}+\frac{\left(\dot{x}_3-\dot{x}_4\right)/2}{l}\right)\\&+k_P\left(\theta+\frac{\left(x_1-x_2\right)/2}{l}+\frac{\left(x_3-x_4\right)/2}{l}\right)=0\end{aligned}\tag{12.3}$$

$$\begin{aligned}&m_f\ddot{x}_1-c_f\left(\dot{x}-\dot{x}_1+b_1\dot{\varphi}-a_1\dot{\theta}\right)-k_f\left(x-x_1+b_1\varphi-a_1\theta\right)\\&-c_{R_f}\frac{1}{w_f}\left(\dot{\varphi}-\frac{\dot{x}_1-\dot{x}_2}{w_f}\right)-k_{R_f}\frac{1}{w_f}\left(\varphi-\frac{x_1-x_2}{w_f}\right)\\&+\frac{1}{2l}c_P\left(\dot{\theta}+\frac{\left(\dot{x}_1-\dot{x}_2\right)/2}{l}+\frac{\left(\dot{x}_3-\dot{x}_4\right)/2}{l}\right)\\&+\frac{1}{2l}k_P\left(\theta+\frac{\left(x_1-x_2\right)/2}{l}+\frac{\left(x_3-x_4\right)/2}{l}\right)\\&+c_{t_f}\left(\dot{x}_1-\dot{y}_1\right)+k_{t_f}\left(x_1-y_1\right)=0\end{aligned}\tag{12.4}$$

$$\begin{aligned}&m_f\ddot{x}_2-c_f\left(\dot{x}-\dot{x}_2-b_2\dot{\varphi}-a_1\dot{\theta}\right)-k_f\left(x-x_2-b_2\varphi-a_1\theta\right)\\&+c_{R_f}\frac{1}{w_f}\left(\dot{\varphi}-\frac{\dot{x}_1-\dot{x}_2}{w_f}\right)+k_{R_f}\frac{1}{w_f}\left(\varphi-\frac{x_1-x_2}{w_f}\right)\\&-\frac{1}{2l}c_P\left(\dot{\theta}+\frac{\left(\dot{x}_1-\dot{x}_2\right)/2}{l}+\frac{\left(\dot{x}_3-\dot{x}_4\right)/2}{l}\right)\\&-\frac{1}{2l}k_P\left(\theta+\frac{\left(x_1-x_2\right)/2}{l}+\frac{\left(x_3-x_4\right)/2}{l}\right)\\&+c_{t_f}\left(\dot{x}_2-\dot{y}_2\right)+k_{t_f}\left(x_2-y_2\right)=0\end{aligned}\tag{12.5}$$

$$m_r\ddot{x}_3 - c_r\left(\dot{x} - \dot{x}_3 + b_3\dot{\varphi} + a_2\dot{\theta}\right) - k_f\left(x - x_3 + b_3\varphi + a_2\theta\right)$$
$$- c_{R_r}\frac{1}{w_r}\left(\dot{\varphi} - \frac{\dot{x}_3 - \dot{x}_4}{w_f}\right) - k_{R_r}\frac{1}{w_r}\left(\varphi - \frac{x_1 - x_2}{w_f}\right)$$
$$+ \frac{1}{2l}c_P\left(\dot{\theta} + \frac{(\dot{x}_1 - \dot{x}_2)/2}{l} + \frac{(\dot{x}_3 - \dot{x}_4)/2}{l}\right)$$
$$+ \frac{1}{2l}k_P\left(\theta + \frac{(x_1 - x_2)/2}{l} + \frac{(x_3 - x_4)/2}{l}\right)$$
$$+ c_{t_r}(\dot{x}_3 - \dot{y}_3) + k_{t_r}(x_3 - y_3) = 0 \tag{12.6}$$

$$m_r\ddot{x}_4 - c_r\left(\dot{x} - \dot{x}_4 - b_4\dot{\varphi} + a_2\dot{\theta}\right) - k_f\left(x - x_4 - b_4\varphi + a_2\theta\right)$$
$$+ c_{R_r}\frac{1}{w_r}\left(\dot{\varphi} - \frac{\dot{x}_3 - \dot{x}_4}{w_f}\right) - k_{R_r}\frac{1}{w_r}\left(\varphi - \frac{x_1 - x_2}{w_f}\right)$$
$$- \frac{1}{2l}c_P\left(\dot{\theta} + \frac{(\dot{x}_1 - \dot{x}_2)/2}{l} + \frac{(\dot{x}_3 - \dot{x}_4)/2}{l}\right)$$
$$- \frac{1}{2l}k_P\left(\theta + \frac{(x_1 - x_2)/2}{l} + \frac{(x_3 - x_4)/2}{l}\right)$$
$$+ c_{t_r}(\dot{x}_4 - \dot{y}_4) + k_{t_r}(x_4 - y_4) = 0 \tag{12.7}$$

The characteristics of a full car model may be different for the front half and rear half due to different suspensions and vehicle mass distribution. Furthermore, different antiroll bars with different torsional stiffness and damping may be used in the front and rear halves. The definition of the employed parameters and variables in the equations of motion is defined in Tables 12.1 and 12.2.

**Table 12.1** Parameters of the half car model

| Parameter | Meaning |
|---|---|
| $m$ | Mass of the vehicle's body [ kg] |
| $m_f$ | Mass of the front wheel [ kg] |
| $m_r$ | Mass of the rear wheel [ kg] |
| $I_x$ | Body longitudinal mass moment [ kg m$^2$] |
| $I_y$ | Body lateral mass moment [ kg m$^2$] |
| $k_f$ | Front spring stiffness rate [ N/ m] |
| $k_r$ | Rear spring stiffness rate [ N/ m] |
| $c_f$ | Front shock absorber damping rate [ N s/ m] |
| $c_r$ | Rear shock absorber damping rate [ N s/ m] |
| $k_{t_f}$ | Front tire stiffness rate [ N/ m] |
| $k_{t_r}$ | Rear tire stiffness rate [ N/ m] |
| $c_{t_f}$ | Front tire damping rate [ N s/ m] |
| $c_{t_r}$ | Rear tire damping rate [ N s/ m] |
| $k_{R_f}$ | Front antiroll bar stiffness rate [ N s m/ rad] |
| $k_{R_r}$ | Rear antiroll bar stiffness rate [ N s m/ rad] |
| $c_{R_f}$ | Front antiroll damping rate [ N m/ rad] |
| $c_{R_r}$ | Rear antiroll bar damping rate [ N m/ rad] |
| $k_P$ | Antipitch bar stiffness rate [ N s m/ rad] |
| $a_1$ | Longitudinal distance of $C$ from front axle [ m] |
| $a_2$ | Longitudinal distance of $C$ from rear axle [ m] |
| $l$ | $= a_1 + a_2$, wheelbase of vehicle |
| $b_1$ | Lateral distance of $C$ from front left wheel [ m] |
| $b_2$ | Lateral distance of $C$ from front right wheel [ m] |
| $b_3$ | Lateral distance of $C$ from rear left wheel [ m] |
| $b_4$ | Lateral distance of $C$ from rear right wheel [ m] |
| $w_f$ | $= b_1 + b_2$, front track of vehicle |
| $w_r$ | $= b_3 + b_4$, rear track of vehicle |

**Table 12.2** Variables of the half car model

| Variable | Meaning |
|---|---|
| $x$ | Body vertical motion coordinate |
| $x_1$ | Front left wheel vertical motion coordinate |
| $x_2$ | Front right wheel vertical motion coordinate |
| $x_3$ | Rear left wheel vertical motion coordinate |
| $x_4$ | Rear right wheel vertical motion coordinate |
| $\varphi$ | Body roll coordinate |
| $\theta$ | Body pitch coordinate |
| $y_1$ | Road excitation under the front left wheel |
| $y_2$ | Road excitation under the front right wheel |
| $y_3$ | Road excitation under the rear left wheel |
| $y_4$ | Road excitation under the rear right wheel |

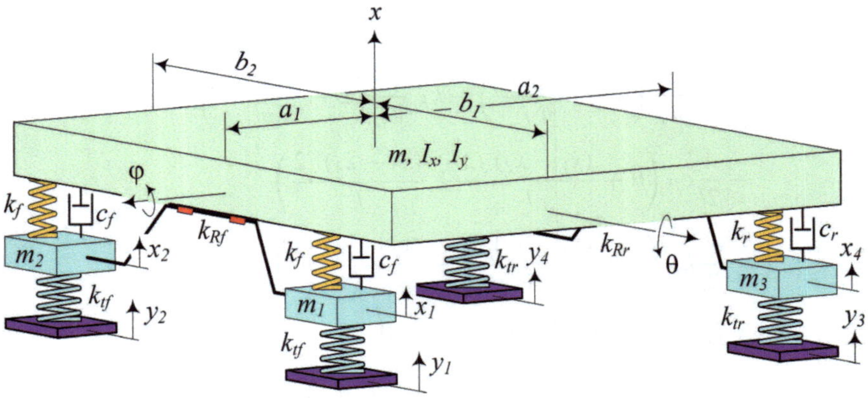


**Fig. 12.2** Full car model for vehicle vibrations

***Proof*** Figure 12.2 depicts the vibrating model of the system. The body of the vehicle is assumed to be a rigid slab. The slab has a mass $m$, which is the total body mass, a longitudinal mass moment $I_x$, and a lateral mass moment $I_y$. The mass moments are only the body mass moments and not the whole vehicle's mass moments. The wheels have masses $m_1, m_2, m_3$, and $m_4$, respectively. It is reasonable to assume front wheels are identical and rear wheels are also identical.

$$m_1 = m_2 = m_f \tag{12.8}$$

$$m_3 = m_4 = m_r \tag{12.9}$$

The front and rear tires' stiffness is indicated by $k_{t_f}$ and $k_{t_r}$, respectively. Because the damping of tires is much smaller than the damping of shock absorbers, we can ignore the tires' damping for simpler calculation.

The suspension of the car has stiffness $k_f$ and damping $c_f$ in the front and stiffness $k_r$ and damping $c_r$ in the rear. It is reasonable to make the suspension of the left and right wheels mirror. So, their stiffness and damping are equal. The vehicle may also have an antiroll bar in the front and in the back, with torsional stiffness $k_{R_f}$ and $k_{R_r}$ and torsional damping $c_{R_f}$ and $c_{R_r}$. The antiroll bar is to make an antiroll moment $M_R$.

$$M_R = -k_{R_f}\left(\varphi - \frac{x_1 - x_2}{w_f}\right) - k_{R_r}\left(\varphi - \frac{x_3 - x_4}{w_r}\right) \tag{12.10}$$

$$- c_{R_f}\left(\dot{\varphi} - \frac{\dot{x}_1 - x_2}{w_f}\right) + c_{R_r}\left(\dot{\varphi} - \frac{\dot{x}_3 - \dot{x}_4}{w_r}\right) \tag{12.11}$$

The vehicle can also have an antipitch bar with torsional stiffness $k_P$ torsional damping $c_P$. The antipitch bar is being used to make an antipitch moment $M_P$.

$$M_P = -k_P\left(\theta + \frac{(x_1 - x_2)/2}{l} + \frac{(x_3 - x_4)/2}{l}\right) \tag{12.12}$$

$$-c_P\left(\dot{\theta} + \frac{(\dot{x}_1 - \dot{x}_2)/2}{l} + \frac{(\dot{x}_3 - \dot{x}_4)/2}{l}\right) \tag{12.13}$$

To find the equations of motion of the full car vibrating model, we use the Lagrange method. The kinetic and potential energies of the system are

$$K = \frac{1}{2}m\dot{x}^2 + \frac{1}{2}I_x\dot{\varphi}^2 + \frac{1}{2}I_y\dot{\theta}^2 + \frac{1}{2}m_f\left(\dot{x}_1^2 + \dot{x}_2^2\right) + \frac{1}{2}m_r\left(\dot{x}_3^2 + \dot{x}_4^2\right) \tag{12.14}$$

$$\begin{aligned}
P = & \frac{1}{2}k_f\left(x - x_1 + b_1\varphi - a_1\theta\right)^2 + \frac{1}{2}k_f\left(x - x_2 - b_2\varphi - a_1\theta\right)^2 \\
& + \frac{1}{2}k_r\left(x - x_3 + b_3\varphi + a_2\theta\right)^2 + \frac{1}{2}k_r\left(x - x_4 - b_4\varphi + a_2\theta\right)^2 \\
& + \frac{1}{2}k_{R_f}\left(\varphi + \frac{x_2 - x_1}{w_f}\right)^2 + \frac{1}{2}k_{R_r}\left(\varphi + \frac{x_4 - x_3}{w_r}\right)^2 \\
& + \frac{1}{2}k_P\left(\theta + \frac{(x_1 + x_3)/2}{l} - \frac{(x_3 + x_4)/2}{l}\right)^2 \\
& + \frac{1}{2}k_{t_f}\left(x_1 - y_1\right)^2 + \frac{1}{2}k_{t_f}\left(x_2 - y_2\right)^2 \\
& + \frac{1}{2}k_{t_r}\left(x_3 - y_3\right)^2 + \frac{1}{2}k_{t_r}\left(x_4 - y_4\right)^2
\end{aligned} \tag{12.15}$$

and the dissipation function is

$$\begin{aligned}
D = & \frac{1}{2}c_f\left(\dot{x} - \dot{x}_1 + b_1\dot{\varphi} - a_1\dot{\theta}\right)^2 + \frac{1}{2}c_f\left(\dot{x} - \dot{x}_2 - b_2\dot{\varphi} - a_1\dot{\theta}\right)^2 \\
& + \frac{1}{2}c_r\left(\dot{x} - \dot{x}_3 + b_3\dot{\varphi} + a_2\dot{\theta}\right)^2 + \frac{1}{2}c_r\left(\dot{x} - \dot{x}_4 - b_4\dot{\varphi} + a_2\dot{\theta}\right)^2 \\
& + \frac{1}{2}c_{R_f}\left(\dot{\varphi} + \frac{\dot{x}_2 - \dot{x}_1}{w_f}\right)^2 + \frac{1}{2}c_{R_r}\left(\dot{\varphi} + \frac{\dot{x}_4 - \dot{x}_3}{w_r}\right)^2 \\
& + \frac{1}{2}c_P\left(\dot{\theta} + \frac{(\dot{x}_1 + \dot{x}_3)/2}{l} - \frac{(\dot{x}_3 + \dot{x}_4)/2}{l}\right)^2 \\
& + \frac{1}{2}c_{t_f}\left(\dot{x}_1 - \dot{y}_1\right)^2 + \frac{1}{2}c_{t_f}\left(\dot{x}_2 - \dot{y}_2\right)^2 \\
& + \frac{1}{2}c_{t_r}\left(\dot{x}_3 - \dot{y}_3\right)^2 + \frac{1}{2}c_{t_r}\left(\dot{x}_4 - \dot{y}_4\right)^2
\end{aligned} \tag{12.16}$$

Applying the Lagrange method provides us with equations of motion (12.1)–(12.7).

$$\frac{d}{dt}\left(\frac{\partial K}{\partial \dot{q}_r}\right) - \frac{\partial K}{\partial q_r} + \frac{\partial D}{\partial \dot{q}_r} + \frac{\partial P}{\partial q_r} = f_r \qquad r = 1, 2, \cdots 7 \tag{12.17}$$

The set of equations of motion may be rearranged in a matrix form

$$[m]\,\ddot{\mathbf{x}} + [c]\,\dot{\mathbf{x}} + [k]\,\mathbf{x} = \mathbf{F} \tag{12.18}$$

where

$$\mathbf{x} = \begin{bmatrix} x & \varphi & \theta & x_1 & x_2 & x_3 & x_4 \end{bmatrix}^T \tag{12.19}$$

$$[m] = \begin{bmatrix} m & 0 & 0 & 0 & 0 & 0 & 0 \\ 0 & I_x & 0 & 0 & 0 & 0 & 0 \\ 0 & 0 & I_y & 0 & 0 & 0 & 0 \\ 0 & 0 & 0 & m_f & 0 & 0 & 0 \\ 0 & 0 & 0 & 0 & m_f & 0 & 0 \\ 0 & 0 & 0 & 0 & 0 & m_r & 0 \\ 0 & 0 & 0 & 0 & 0 & 0 & m_r \end{bmatrix} \tag{12.20}$$

$$[c] = \begin{bmatrix} c_{11} & c_{12} & c_{13} & -c_f & -c_f & -c_r & -c_r \\ c_{21} & c_{22} & c_{23} & c_{24} & c_{25} & c_{26} & c_{27} \\ c_{31} & c_{32} & c_{33} & c_{34} & c_{35} & c_{36} & c_{37} \\ -c_f & c_{42} & c_{43} & c_{44} & c_{45} & \frac{c_P}{4l^2} & -\frac{c_P}{4l^2} \\ -c_f & c_{52} & c_{53} & c_{54} & c_{55} & -\frac{c_P}{4l^2} & \frac{c_P}{4l^2} \\ -c_r & c_{62} & c_{63} & \frac{c_P}{4l^2} & -\frac{c_P}{4l^2} & c_{66} & c_{67} \\ -c_r & c_{72} & c_{73} & -\frac{c_P}{4l^2} & \frac{c_P}{4l^2} & c_{76} & c_{77} \end{bmatrix} \tag{12.21}$$

$$\begin{aligned} &c_{11} = 2c_f + 2c_r \qquad c_{12} = c_{21} = (b_1 - b_2)\, c_f + (b_3 - b_4)\, c_r \\ &c_{13} = c_{31} = 2\left(a_2 c_r - a_1 c_f\right) \\ &c_{22} = \left(b_1^2 + b_2^2\right) c_f + \left(b_3^2 + b_4^2\right) c_r + c_{R_f} + c_{R_r} \end{aligned} \tag{12.22}$$

$$\begin{aligned} &c_{23} = c_{32} = a_1 (b_2 - b_1)\, c_f + a_2 (b_3 - b_4)\, c_r \\ &c_{24} = c_{42} = -b_1 c_f - \frac{c_{R_f}}{w_f} \qquad c_{25} = c_{52} = b_2 c_f + \frac{c_{R_f}}{w_f} \\ &c_{26} = c_{62} = -b_3 c_r - \frac{c_{R_r}}{w_r} \qquad c_{27} = c_{72} = b_4 c_r + \frac{c_{R_r}}{w_r} \end{aligned} \tag{12.23}$$

$$\begin{aligned} &c_{33} = 2a_1^2 c_f + 2a_2^2 c_r + c_P \qquad c_{34} = c_{43} = a_1 c_f + \frac{c_P}{2l} \\ &c_{35} = c_{53} = a_1 c_f - \frac{c_P}{2l} \qquad c_{36} = c_{63} = -a_2 c_r + \frac{c_P}{2l} \\ &c_{37} = c_{73} = -a_2 c_r - \frac{c_P}{2l} \end{aligned} \tag{12.24}$$

$$\begin{aligned} &c_{44} = c_f + \frac{c_{R_f}}{w_f^2} + \frac{c_P}{4l^2} + c_{t_f} \qquad c_{45} = c_{54} = -\frac{c_{R_f}}{w_f^2} - \frac{c_P}{4l^2} \\ &c_{55} = c_f + \frac{c_{R_f}}{w_f^2} + \frac{c_P}{4l^2} + c_{t_f} \qquad c_{66} = c_r + \frac{c_{R_r}}{w_r^2} + \frac{c_P}{4l^2} + c_{t_r} \\ &c_{67} = c_{76} = -\frac{c_{R_r}}{w_r^2} - \frac{c_P}{4l^2} \qquad c_{77} = c_r + \frac{c_{R_r}}{w_r^2} + \frac{c_P}{4l^2} + c_{t_r} \end{aligned} \tag{12.25}$$

$$[k] = \begin{bmatrix} k_{11} & k_{12} & k_{13} & -k_f & -k_f & -k_r & -k_r \\ k_{21} & k_{22} & k_{23} & k_{24} & k_{25} & k_{26} & k_{27} \\ k_{31} & k_{32} & k_{33} & k_{34} & k_{35} & k_{36} & k_{37} \\ -k_f & k_{42} & k_{43} & k_{44} & k_{45} & \frac{k_P}{4l^2} & -\frac{k_P}{4l^2} \\ -k_f & k_{52} & k_{53} & k_{54} & k_{55} & -\frac{k_P}{4l^2} & \frac{k_P}{4l^2} \\ -k_r & k_{62} & k_{63} & \frac{k_P}{4l^2} & -\frac{k_P}{4l^2} & k_{66} & k_{67} \\ -k_r & k_{72} & k_{73} & -\frac{k_P}{4l^2} & \frac{k_P}{4l^2} & k_{76} & k_{77} \end{bmatrix} \quad (12.26)$$

$$\begin{aligned} k_{11} &= 2k_f + 2k_r \qquad k_{12} = k_{21} = (b_1 - b_2)\, k_f + (b_3 - b_4)\, k_r \\ k_{13} &= k_{31} = 2\left(a_2 k_r - a_1 k_f\right) \\ k_{22} &= \left(b_1^2 + b_2^2\right) k_f + \left(b_3^2 + b_4^2\right) k_r + k_{R_f} + k_{R_r} \end{aligned} \quad (12.27)$$

$$\begin{aligned} k_{23} &= k_{32} = a_1\,(b_2 - b_1)\, k_f + a_2\,(b_3 - b_4)\, k_r \\ k_{24} &= k_{42} = -b_1 k_f - \frac{k_{R_f}}{w_f} \qquad k_{25} = k_{52} = b_2 k_f + \frac{k_{R_f}}{w_f} \\ k_{26} &= k_{62} = -b_3 k_r - \frac{k_{R_r}}{w_r} \qquad k_{27} = k_{72} = b_4 k_r + \frac{k_{R_r}}{w_r} \end{aligned} \quad (12.28)$$

$$\begin{aligned} k_{33} &= 2a_1^2 k_f + 2a_2^2 k_r + k_P \qquad k_{34} = k_{43} = a_1 k_f + \frac{k_P}{2l} \\ k_{35} &= k_{53} = a_1 k_f - \frac{k_P}{2l} \qquad k_{36} = k_{63} = -a_2 k_r + \frac{k_P}{2l} \\ k_{37} &= k_{73} = -a_2 k_r - \frac{k_P}{2l} \end{aligned} \quad (12.29)$$

$$\begin{aligned} k_{44} &= k_f + \frac{k_{R_f}}{w_f^2} + \frac{k_P}{4l^2} + k_{t_f} \qquad k_{45} = k_{54} = -\frac{k_{R_f}}{w_f^2} - \frac{k_P}{4l^2} \\ k_{55} &= k_f + \frac{k_{R_f}}{w_f^2} + \frac{k_P}{4l^2} + k_{t_f} \qquad k_{66} = k_r + \frac{k_{R_r}}{w_r^2} + \frac{k_P}{4l^2} + k_{t_r} \\ k_{67} &= k_{76} = -\frac{k_{R_r}}{w_r^2} - \frac{k_P}{4l^2} \qquad k_{77} = k_r + \frac{k_{R_r}}{w_r^2} + \frac{k_P}{4l^2} + k_{t_r} \end{aligned} \quad (12.30)$$

$$\mathbf{F} = \left[0\;0\;0\; k_{t_f}\, y_1\; k_{t_f}\, y_2\; k_{t_r}\, y_3\; k_{t_r}\, y_4\right]^T \quad (12.31)$$

■

**Example 169** Nonlinear damping. Shock absorbers are the only suspension element that is designed nonlinear on purpose. Having a low damping on bound reduces the shock to the vehicle when it hits a bump on the road.

A bilinear damping model for practical shock absorbers is the one that shows two different damping rates in bound and rebound. The rate in bound is supposed to be much lower than the rebound rate. In extreme case, we may assume there is no damping in bound and a constant damping rate $c$ in rebound. Considering $u$ and $v$ as displacements of two ends of the shock absorber, we can define bilinear damping by a Heaviside function $cH\left(\dot{u} - \dot{v}\right)$.

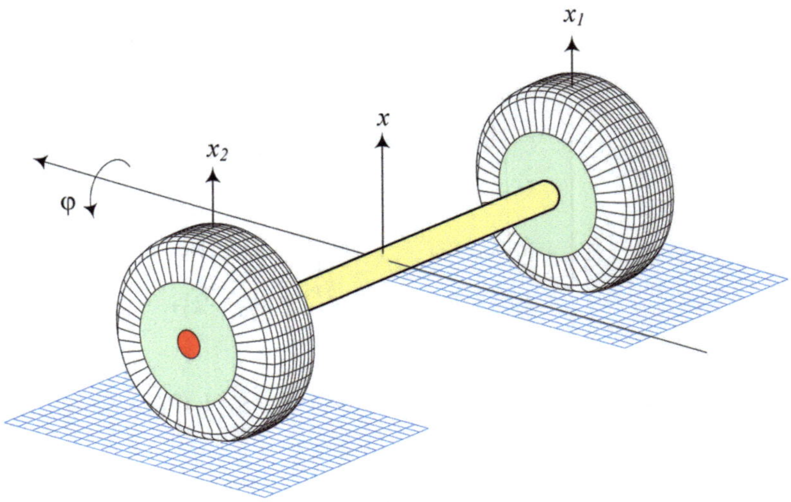


**Fig. 12.3** A solid axle suspension

**Example 170** Solid axle suspension. In case one of having a solid axle in front, in the back, or both, the equations of motion must be changed accordingly.

A solid axle suspension is illustrated in Fig. 12.3. The inertia of the solid axle is expressed by the mass moment $I_A$ and the mass $m_A$. Let us consider a vehicle with independent suspensions for the front wheels and a solid axle suspension in the back. The solid axle suspension makes a change in the kinetic energy of the system by replacing $\frac{1}{2}m_r\left(\dot{x}_3^2+\dot{x}_4^2\right)$ with $\frac{1}{2}m_A\dot{x}_A^2+\frac{1}{2}I_A\dot{\varphi}_A^2$.

$$K=\frac{1}{2}m\dot{x}^2+\frac{1}{2}I_x\dot{\varphi}^2+\frac{1}{2}I_y\dot{\theta}^2+\frac{1}{2}m_f\left(\dot{x}_1^2+\dot{x}_2^2\right)+\frac{1}{2}m_A\dot{x}_A^2+\frac{1}{2}I_A\dot{\varphi}_A^2 \tag{12.32}$$

In the potential energy and dissipation functions, we need to replace the coordinates $x_3$ and $x_4$ with $x_A$ and $\varphi_A$.

$$x_A=\frac{1}{2}\left(x_3+x_4\right) \qquad \varphi_A=\frac{x_3-x_4}{w_r} \tag{12.33}$$

$$x_3=x_A+\frac{1}{2}w_r\varphi_A \qquad x_4=x_A-\frac{1}{2}w_r\varphi_A \tag{12.34}$$

## 12.2 Time Response

The full car model will show pitch oscillations because of time lag between the front and rear axles going over a bump. It shows roll oscillations when there is any asymmetric road excitation between left and right wheels. The time response of the full car system is the solution of the equations of motion (12.1)–(12.7) for the absolute displacement $x$, $\varphi$, $\theta$, $x_1$, $x_2$, $x_3$, and $x_4$ for the road excitation inputs $y_1$, $y_2$, $y_3$, and $y_4$. The equations of motion can be rearranged in the more suitable form for analytic or numeric solutions.

$$\begin{aligned}
&m\ddot{x}+2\left(c_f+c_r\right)\dot{x}+\left(\left(b_1-b_2\right)c_f+\left(b_3-b_4\right)c_r\right)\dot{\varphi}\\
&+2\left(a_2c_r-a_1c_f\right)\dot{\theta}-c_f\dot{x}_1-c_f\dot{x}_2-c_r\dot{x}_3-c_r\dot{x}_4\\
&+2\left(k_f+k_r\right)x+\left(\left(b_1-b_2\right)k_f+\left(b_3-b_4\right)k_r\right)\varphi\\
&-2\left(a_2k_r-a_1k_f\right)\theta-k_fx_1-k_fx_2-k_rx_3-k_rx_4=0
\end{aligned} \tag{12.35}$$

$$
\begin{aligned}
& I_x\ddot{\varphi} + \left((b_1 - b_2)\,c_f + (b_3 - b_4)\,c_r\right)\dot{x} \\
& + \left(\left(b_1^2 + b_2^2\right)c_f + \left(b_3^2 + b_4^2\right)c_r + c_{R_f} + c_{R_r}\right)\dot{\varphi} \\
& + \left(a_1\,(b_2 - b_1)\,c_f + a_2\,(b_3 - b_4)\,c_r\right)\dot{\theta} - \left(b_1 c_f + \frac{c_{R_f}}{w_f}\right)\dot{x}_1 \\
& + \left(b_2 c_f + \frac{c_{R_f}}{w_f}\right)\dot{x}_2 - \left(b_3 c_r + \frac{c_{R_r}}{w_r}\right)\dot{x}_3 + \left(b_4 c_r + \frac{c_{R_r}}{w_r}\right)\dot{x}_4 \\
& + \left((b_1 - b_2)\,k_f + (b_3 - b_4)\,k_r\right)x \\
& + \left(\left(b_1^2 + b_2^2\right)k_f + \left(b_3^2 + b_4^2\right)k_r + k_{R_f} + k_{R_r}\right)\varphi \\
& + \left(a_1\,(b_2 - b_1)\,k_f + a_2\,(b_3 - b_4)\,k_r\right)\theta \\
& - \left(b_1 k_f + \frac{k_{R_f}}{w_f}\right)x_1 + \left(b_2 k_f + \frac{k_{R_f}}{w_f}\right)x_2 \\
& - \left(b_3 k_r + \frac{k_{R_r}}{w_r}\right)x_3 + \left(b_4 k_r + \frac{k_{R_r}}{w_r}\right)x_4 = 0
\end{aligned} \tag{12.36}
$$

$$
\begin{aligned}
& I_y\ddot{\theta} + 2\left(a_2 c_r - a_1 c_f\right)\dot{x} + \left(a_1\,(b_2 - b_1)\,c_f + a_2\,(b_3 - b_4)\,c_r\right)\dot{\varphi} \\
& + \left(2a_1^2 c_f + 2a_2^2 c_r + c_P\right)\dot{\theta} + \left(a_1 c_f + \frac{c_P}{2l}\right)\dot{x}_1 \\
& + \left(a_1 c_f - \frac{c_P}{2l}\right)\dot{x}_2 + \left(-a_2 c_r + \frac{c_P}{2l}\right)\dot{x}_3 - \left(a_2 c_r + \frac{c_P}{2l}\right)\dot{x}_4 \\
& + 2\left(a_2 k_r - a_1 k_f\right)x + \left(a_1\,(b_2 - b_1)\,k_f + a_2\,(b_3 - b_4)\,k_r\right)\varphi \\
& + \left(2a_1^2 k_f + 2a_2^2 k_r + k_P\right)\theta \\
& + \left(a_1 k_f + \frac{k_P}{2l}\right)x_1 + \left(a_1 k_f - \frac{k_P}{2l}\right)x_2 \\
& + \left(-a_2 k_r + \frac{k_P}{2l}\right)x_3 - \left(a_2 k_r + \frac{k_P}{2l}\right)x_4 = 0
\end{aligned} \tag{12.37}
$$

$$
\begin{aligned}
& m_f\ddot{x}_1 - c_f\dot{x} - \left(b_1 c_f + \frac{c_{R_f}}{w_f}\right)\dot{\varphi} + \left(a_1 c_f + \frac{c_P}{2l}\right)\dot{\theta} \\
& + \left(c_f + \frac{c_{R_f}}{w_f^2} + \frac{c_P}{4l^2} + c_{t_f}\right)\dot{x}_1 - \left(\frac{c_{R_f}}{w_f^2} + \frac{c_P}{4l^2}\right)\dot{x}_2 \\
& + \frac{c_P}{4l^2}\dot{x}_3 - \frac{c_P}{4l^2}\dot{x}_4 - k_f\,x - \left(b_1 k_f + \frac{k_{R_f}}{w_f}\right)\varphi \\
& + \left(a_1 k_f + \frac{k_P}{2l}\right)\theta + \left(k_f + \frac{k_{R_f}}{w_f^2} + \frac{k_P}{4l^2} + k_{t_f}\right)x_1 \\
& - \left(\frac{k_{R_f}}{w_f^2} + \frac{k_P}{4l^2}\right)x_2 + \frac{k_P}{4l^2}x_3 - \frac{k_P}{4l^2}x_4 = 0
\end{aligned} \tag{12.38}
$$

$$\begin{aligned}&m_f\ddot{x}_2 - c_f\,\dot{x} + \left(b_2c_f + \frac{c_{R_f}}{w_f}\right)\dot{\varphi} + \left(a_1c_f - \frac{c_P}{2l}\right)\dot{\theta}\\&-\left(\frac{c_{R_f}}{w_f^2} + \frac{c_P}{4l^2}\right)\dot{x}_1 + \left(c_f + \frac{c_{R_f}}{w_f^2} + \frac{c_P}{4l^2} + c_{t_f}\right)\dot{x}_2\\&-\frac{c_P}{4l^2}\dot{x}_3 + \frac{c_P}{4l^2}\dot{x}_4 - k_f\,x + \left(b_2k_f + \frac{k_{R_f}}{w_f}\right)\varphi\\&+\left(a_1k_f - \frac{k_P}{2l}\right)\theta - \left(\frac{k_{R_f}}{w_f^2} + \frac{k_P}{4l^2}\right)x_1\\&+\left(k_f + \frac{k_{R_f}}{w_f^2} + \frac{k_P}{4l^2} + k_{t_f}\right)x_2 - \frac{k_P}{4l^2}x_3 + \frac{k_P}{4l^2}x_4 = 0\end{aligned} \tag{12.39}$$

$$\begin{aligned}&m_r\ddot{x}_3 - c_r\,\dot{x} - \left(b_3c_r + \frac{c_{R_r}}{w_r}\right)\dot{\varphi} - \left(a_2c_r - \frac{c_P}{2l}\right)\dot{\theta}\\&+\frac{c_P}{4l^2}\dot{x}_1 - \frac{c_P}{4l^2}\dot{x}_2 + \left(c_r + \frac{c_{R_r}}{w_r^2} + \frac{c_P}{4l^2} + c_{t_r}\right)\dot{x}_3\\&-\left(\frac{c_{R_r}}{w_r^2} + \frac{c_P}{4l^2}\right)\dot{x}_4 - k_r\,x - \left(b_3k_r + \frac{k_{R_r}}{w_r}\right)\varphi\\&-\left(a_2k_r - \frac{k_P}{2l}\right)\theta + \frac{k_P}{4l^2}x_1 - \frac{k_P}{4l^2}x_2\\&+\left(k_r + \frac{k_{R_r}}{w_r^2} + \frac{k_P}{4l^2} + k_{t_r}\right)x_3 - \left(\frac{k_{R_r}}{w_r^2} + \frac{k_P}{4l^2}\right)x_4 = 0\end{aligned} \tag{12.40}$$

$$\begin{aligned}&m_r\ddot{x}_4 - c_r\dot{x} + \left(b_4c_r + \frac{c_{R_r}}{w_r}\right)\dot{\varphi} - \left(a_2c_r + \frac{c_P}{2l}\right)\dot{\theta} - \frac{c_P}{4l^2}\dot{x}_1\\&+\frac{c_P}{4l^2}\dot{x}_2 - \left(\frac{c_{R_r}}{w_r^2} + \frac{c_P}{4l^2}\right)\dot{x}_3 + \left(c_r + \frac{c_{R_r}}{w_r^2} + \frac{c_P}{4l^2} + c_{t_r}\right)\dot{x}_4\\&-k_r\,x + \left(b_4k_r + \frac{k_{R_r}}{w_r}\right)\varphi - \left(a_2k_r + \frac{k_P}{2l}\right)\theta\\&-\frac{c_P}{4l^2}x_1 + \frac{c_P}{4l^2}x_2 - \left(\frac{k_{R_r}}{w_r^2} + \frac{k_P}{4l^2}\right)x_3 + k_r\,x_4 = 0\end{aligned} \tag{12.41}$$

***Proof*** The equations of motion (12.35)–(12.41) are an expanded form of (12.18). ■

**Example 171** Transient response of full car model to step input. A symmetric and similar road excitation to all wheels simultaneously will show pitch and roll oscillation if there is any asymmetry between front and rear and between left and right.

Let us assume a step input applies to all wheels of a full car model at the same time.

$$y_1 = y_2 = y_3 = y_4 = \begin{cases} Y & t > 0 \\ 0 & t \le 0 \end{cases} \tag{12.42}$$

Using a set of nominal values,

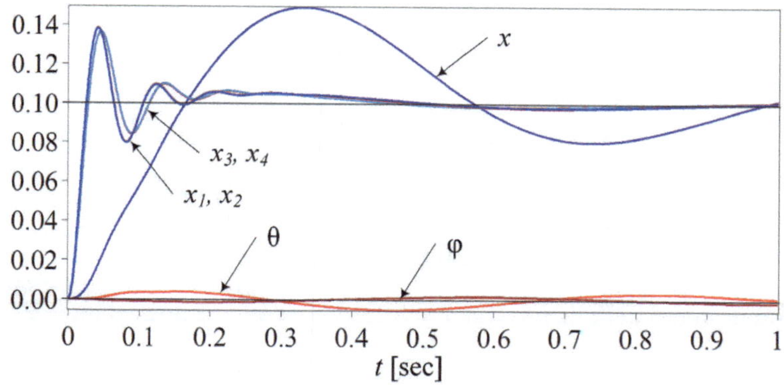


**Fig. 12.4** Body bounce, $x$, body pitch, $\theta$, body roll, $\varphi$, and wheels' bounces, $x_1$, $x_2$, $x_3$, and $x_4$, responses of a full vehicle model to a step road excitation

$$
\begin{aligned}
& m = 1344\,\mathrm{kg} \quad m_1 = 42\,\mathrm{kg} \quad m_2 = 42\,\mathrm{kg} \quad m_3 = 50\,\mathrm{kg} \\
& m_4 = 50\,\mathrm{kg} \quad I_x = 1800\,\mathrm{kg\,m^2} \quad I_y = 2100\,\mathrm{kg\,m^2} \\
& a_1 = 1.4\,\mathrm{m} \quad a_2 = 1.47\,\mathrm{m} \quad l = a_1 + a_2 = 2.87\,\mathrm{m} \\
& b_1 = 0.8652\,\mathrm{m} \quad b_2 = 0.9085\,\mathrm{m} \quad w_f = b_1 + b_2 = 1.7737\,\mathrm{m} \\
& b_3 = 0.8852\,\mathrm{m} \quad b_4 = 0.9285\,\mathrm{m} \quad w_r = b_3 + b_4 = 1.8137\,\mathrm{m} \\
& k_1 = k_2 = 22000\,\mathrm{N/m} \quad k_3 = k_4 = 25000\,\mathrm{N/m} \\
& k_{t_1} = k_{t_2} = k_{t_3} = k_{t_4} = 250000\,\mathrm{N/m} \\
& c_1 = c_2 = 1630\,\mathrm{N\,s/m} \quad c_3 = c_4 = 1900\,\mathrm{N\,s/m} \\
& c_{t_f} = c_{t_r} = c_{R_f} = c_{R_r} = c_P = 0 \quad k_{R_f} = k_{R_r} = k_P = 0
\end{aligned} \tag{12.43}
$$

$$
Y = 0.1\,\mathrm{m} \tag{12.44}
$$

the response of the vehicle is shown in Fig. 12.4.

The front and rear of the vehicle have close but a little different dynamic characteristic and that is the reason a nonzero pitch oscillation $\theta$ appears. The left and right of the vehicle are very similar, and hence, there is not much roll oscillation $\varphi$ for symmetric road excitation. The figure indicates that the frequency of vertical oscillation of the body is much less than the frequency of oscillation of the front and rear wheels. Also, the frequency of oscillation of the wheels is close to each other. The pitch frequency is also close to the vertical frequency of the body. The vertical, pitch, and roll vibrations are the three modes of vibrations of the body of a full vehicle model. Employing the full car model, we analyze the interaction of the three modes.

**Example 172** Full car going over a step. Depending on the velocity of the vehicle and the wheelbase of the vehicle, there will be a time lag between the front and rear wheels to hit the step. Here is the analysis of such a full car model.

Figure 12.5 illustrates a full car model going over a step. Assume the vehicle is moving with velocity $v$ and hitting a step of height $Y$. Starting the time from the moment that the front wheel hits the step, we can calculate the time $t_0$ at which the rear wheel reaches the step.

$$
t_0 = \frac{l}{v} \tag{12.45}
$$

Therefore, the excitation of the rear wheels, $y_3$ and $y_4$, would be a lagged step input, defined by the Heaviside function $H(t - t_0)$.

$$
y_1 = y_2 = \begin{cases} Y & t > 0 \\ 0 & t \le 0 \end{cases} \tag{12.46}
$$

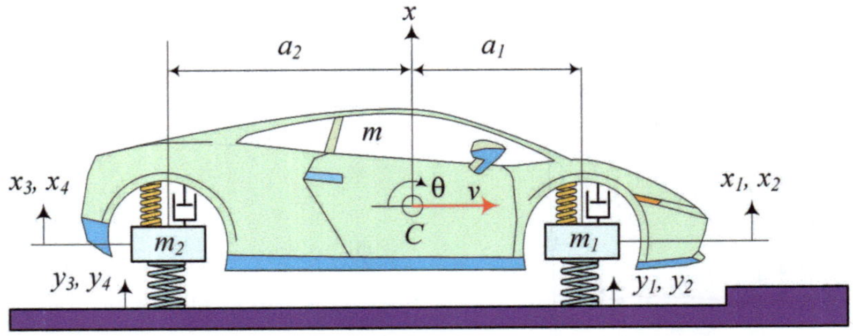


**Fig. 12.5** A full vehicle model going over a step

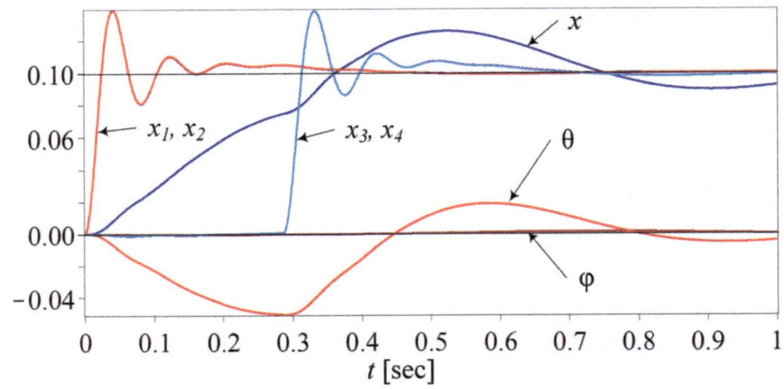


**Fig. 12.6** Body bounce, $x$, body pitch, $\theta$, front wheel, $x_1$, and rear wheel, $x_2$, responses of a bicycle vehicle to a lagged step excitation

$$y_3 = y_4 = \begin{cases} Y\,H\left(t - \dfrac{l}{v}\right) & t > 0 \\ 0 & t \le 0 \end{cases} \tag{12.47}$$

$$H(t - t_0) = \begin{cases} 1 & t > t_0 \\ 0 & t \le t_0 \end{cases} \tag{12.48}$$

Let us employ the same data as given in (12.43) along with the following velocity and height of the step:

$$Y = 0.1\,\text{m} \qquad l = 2.87\,\text{m} \qquad v = 10\,\text{m/s} \tag{12.49}$$

The responses of the car are shown in Fig. 12.6.

For the given data, the wheels stop vibrating much faster than the body, and the pitch vibration $\theta$ would last as long as the body vertical vibrations, $x$. If the time lag between the front and rear wheels to hit the step, $t_0 = l/v$, is $(2n + 1)/2$, $n \in \mathbb{N}$ times of the inverse of natural pitch frequency, the pitch oscillation will have higher amplitude. So, for every vehicle, there would be critical speed that makes the pitch oscillation stronger. The pitch natural frequency and the wheelbase of the vehicle determine the critical speed of the vehicle.

**Example 173** Bump on one side of the road and full car model. Let us examine the reaction of a full vehicle model to a one side bump on the road. Sine square is being used as a mathematical model for speed bumps on the road.

Figure 12.7 illustrates a full car model and a sine square bump on one side of the road. The vehicle is moving at speed $v$ and only the right side of it goes over the bump with length $q$ and height $Y$.

$$y_1 = y_3 = 0 \tag{12.50}$$

$$y_2 = \begin{cases} Y \sin^2 \omega t & 0 \le t \le \dfrac{q}{v} \\ 0 & \dfrac{q}{v} < t \end{cases} \qquad \omega = \frac{v\pi}{q} \qquad t_0 = \frac{l}{v} \tag{12.51}$$

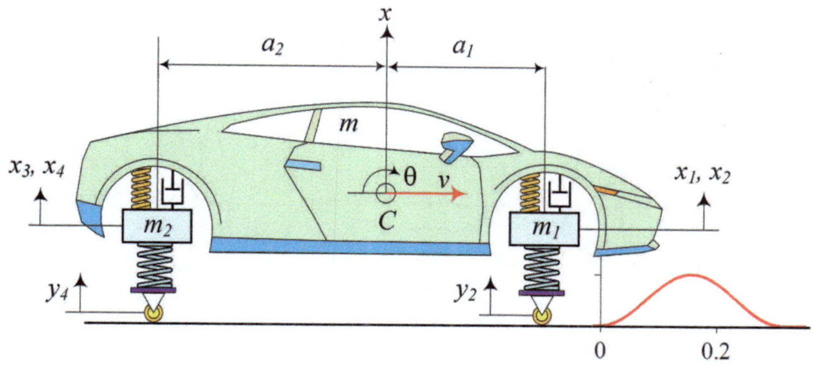


**Fig. 12.7** A bicycle car model and a sine square transient road excitation

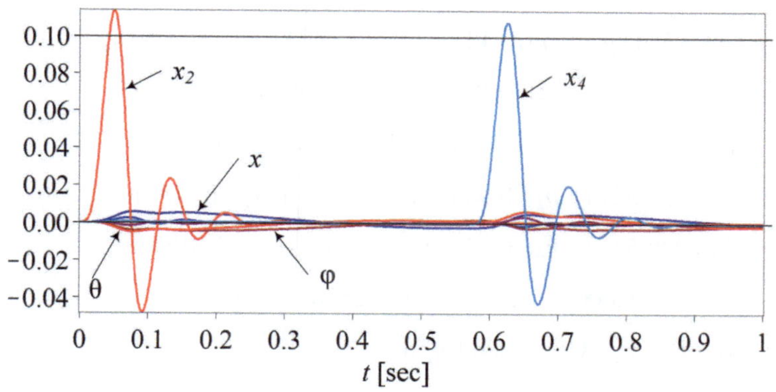


**Fig. 12.8** The response of a full vehicle model to a bump on the road under the right sides wheels

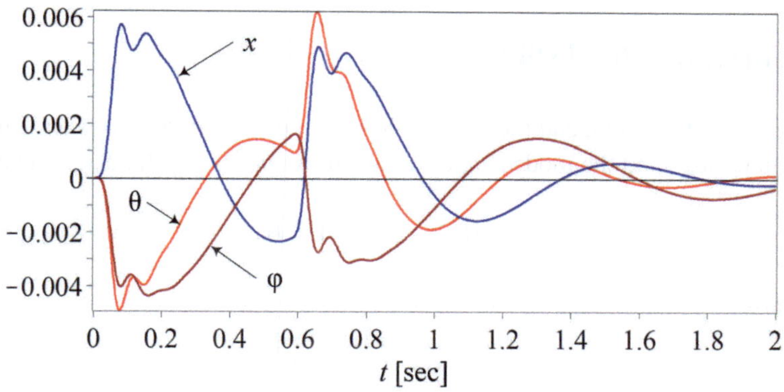


**Fig. 12.9** The response of the body of a full vehicle model to a bump on the road under the right sides wheels

$$y_4 = \begin{cases} Y \sin^2 \omega \left( t - \frac{q}{v} \right) & t_0 \le t \le t_0 + \frac{q}{v} \\ 0 & t_0 + \frac{q}{v} < t \end{cases} \tag{12.52}$$

$$Y = 0.1\,\mathrm{m} \qquad q = \frac{\pi}{10}\,\mathrm{m} \qquad v = 10\,\mathrm{m/s} \tag{12.53}$$

Figure 12.8 illustrates the response of the vehicle with the following characteristics. All displacements are small compared to the movement of the right side wheels, $x_2$ and $x_4$. Figure 12.9 illustrates the body vibrations $x$, $\theta$, and $\varphi$, and Fig. 12.10 depicts vibrations of the left side wheels $x_1$ and $x_3$.

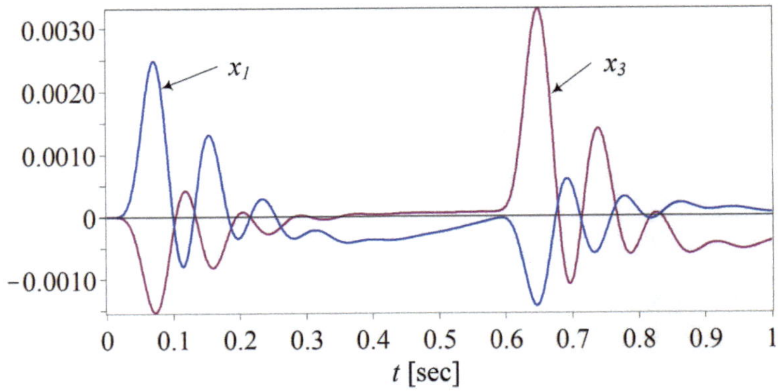


**Fig. 12.10** The response of the left side wheels of a full vehicle model to a bump on the road under the right sides wheels

$$
\begin{aligned}
&m{=}1344\,\text{kg} \qquad m_1{=}42\,\text{kg} \qquad m_2{=}42\,\text{kg} \qquad m_3{=}50\,\text{kg} \\
&m_4{=}50\,\text{kg} \qquad I_x{=}1800\,\text{kg m}^2 \qquad I_y{=}2100\,\text{kg m}^2 \\
&a_1{=}1.4\,\text{m} \qquad a_2{=}1.47\,\text{m} \qquad l{=}a_1+a_2{=}2.87\,\text{m} \\
&b_1{=}0.8652\,\text{m} \quad b_2{=}0.9085\,\text{m} \quad w_f{=}b_1+b_2{=}1.7737\,\text{m} \\
&b_3{=}0.8852\,\text{m} \quad b_4{=}0.9285\,\text{m} \quad w_r{=}b_3+b_4{=}1.8137\,\text{m} \\
&k_1{=}k_2{=}22000\,\text{N/m} \qquad k_3{=}k_4{=}25000\,\text{N/m} \\
&k_{t_1}{=}k_{t_2}{=}k_{t_3}{=}k_{t_4}{=}250000\,\text{N/m} \\
&c_1{=}c_2{=}1630\,\text{N s/m} \qquad c_3{=}c_4{=}1900\,\text{N s/m} \\
&c_{t_f}{=}c_{t_r}{=}c_{R_f}{=}c_{R_r}{=}c_P{=}0 \qquad k_{R_f}{=}k_{R_r}{=}k_P{=}0
\end{aligned}
\tag{12.54}
$$

## 12.3 Natural Frequencies and Mode Shapes

The full car system is an eight $DOF$ system, and therefore, it has eight natural frequencies $\omega_{n_1}$, $\omega_{n_2}$, $\omega_{n_3}$, $\omega_{n_4}$, $\omega_{n_5}$, $\omega_{n_6}$, $\omega_{n_7}$, and $\omega_{n_8}$. The natural frequencies $\omega_{n_i}$, and the mode shapes $\mathbf{u}_i$, are solutions of the characteristic equation of the system

$$
\det\left[[k] - \omega^2 [m]\right] = 0 \tag{12.55}
$$

and the eigenvector equation

$$
\left[[k] - \omega_i^2 [m]\right] \mathbf{u}_i = 0 \tag{12.56}
$$

$$
\mathbf{A} = [m]^{-1} [k]
$$

The $\omega_{n_i}$ and $\mathbf{u}_i$ can also be calculated by determining the eigenvalues and eigenvectors of the $\mathbf{A}$-matrix. The matrices $[m]$ and $[k]$ are mass and stiffness matrices of the system, respectively.

$$
[m] = \begin{bmatrix}
m & 0 & 0 & 0 & 0 & 0 & 0 \\
0 & I_x & 0 & 0 & 0 & 0 & 0 \\
0 & 0 & I_y & 0 & 0 & 0 & 0 \\
0 & 0 & 0 & m_f & 0 & 0 & 0 \\
0 & 0 & 0 & 0 & m_f & 0 & 0 \\
0 & 0 & 0 & 0 & 0 & m_r & 0 \\
0 & 0 & 0 & 0 & 0 & 0 & m_r
\end{bmatrix} \tag{12.57}
$$

$$[k] = \begin{bmatrix} k_{11} & k_{12} & k_{13} & -k_f & -k_f & -k_r & -k_r \\ k_{21} & k_{22} & k_{23} & k_{24} & k_{25} & k_{26} & k_{27} \\ k_{31} & k_{32} & k_{33} & k_{34} & k_{35} & k_{36} & k_{37} \\ -k_f & k_{42} & k_{43} & k_{44} & k_{45} & \frac{k_P}{4l^2} & -\frac{k_P}{4l^2} \\ -k_f & k_{52} & k_{53} & k_{54} & k_{55} & -\frac{k_P}{4l^2} & \frac{k_P}{4l^2} \\ -k_r & k_{62} & k_{63} & \frac{k_P}{4l^2} & -\frac{k_P}{4l^2} & k_{66} & k_{67} \\ -k_r & k_{72} & k_{73} & -\frac{k_P}{4l^2} & \frac{k_P}{4l^2} & k_{76} & k_{77} \end{bmatrix} \tag{12.58}$$

$$\begin{aligned} & k_{11} = 2k_f + 2k_r \quad k_{12} = k_{21} = (b_1 - b_2) k_f + (b_3 - b_4) k_r \\ & k_{13} = k_{31} = 2 \left(a_2 k_r - a_1 k_f\right) \\ & k_{22} = \left(b_1^2 + b_2^2\right) k_f + \left(b_3^2 + b_4^2\right) k_r + k_{R_f} + k_{R_r} \end{aligned} \tag{12.59}$$

$$\begin{aligned} & k_{23} = k_{32} = a_1 (b_2 - b_1) k_f + a_2 (b_3 - b_4) k_r \\ & k_{24} = k_{42} = -b_1 k_f - \frac{k_{R_f}}{w_f} \qquad k_{25} = k_{52} = b_2 k_f + \frac{k_{R_f}}{w_f} \\ & k_{26} = k_{62} = -b_3 k_r - \frac{k_{R_r}}{w_r} \qquad k_{27} = k_{72} = b_4 k_r + \frac{k_{R_r}}{w_r} \end{aligned} \tag{12.60}$$

$$\begin{aligned} & k_{33} = 2a_1^2 k_f + 2a_2^2 k_r + k_P \qquad k_{34} = k_{43} = a_1 k_f + \frac{k_P}{2l} \\ & k_{35} = k_{53} = a_1 k_f - \frac{k_P}{2l} \qquad k_{36} = k_{63} = -a_2 k_r + \frac{k_P}{2l} \\ & k_{37} = k_{73} = -a_2 k_r - \frac{k_P}{2l} \end{aligned} \tag{12.61}$$

$$\begin{aligned} & k_{44} = k_f + \frac{k_{R_f}}{w_f^2} + \frac{k_P}{4l^2} + k_{t_f} \qquad k_{45} = k_{54} = -\frac{k_{R_f}}{w_f^2} - \frac{k_P}{4l^2} \\ & k_{55} = k_f + \frac{k_{R_f}}{w_f^2} + \frac{k_P}{4l^2} + k_{t_f} \qquad k_{66} = k_r + \frac{k_{R_r}}{w_r^2} + \frac{k_P}{4l^2} + k_{t_r} \\ & k_{67} = k_{76} = -\frac{k_{R_r}}{w_r^2} - \frac{k_P}{4l^2} \qquad k_{77} = k_r + \frac{k_{R_r}}{w_r^2} + \frac{k_P}{4l^2} + k_{t_r} \end{aligned} \tag{12.62}$$

The order of coordinates in the matrix form of equations of motion is

$$\mathbf{x} = \left[ x \ \varphi \ \theta \ x_1 \ x_2 \ x_3 \ x_4 \right]^T \tag{12.63}$$

***Proof*** Natural frequencies and mode shapes of a vibrating system are calculated only based on inertia $[m]$ and stiffness $[k]$ of the system. The natural frequencies and mode shapes are eigenvalues and eigenvectors of $\mathbf{A} = [m]^{-1} [k]$.

$$
\mathbf{A} = \begin{bmatrix}
\frac{k_{11}}{m} & \frac{k_{12}}{m} & \frac{k_{13}}{m} & \frac{k_{14}}{m} & \frac{k_{15}}{m} & \frac{k_{16}}{m} & \frac{k_{17}}{m} \\
\frac{k_{21}}{I_x} & \frac{k_{22}}{I_x} & \frac{k_{23}}{I_x} & \frac{k_{24}}{I_x} & \frac{k_{25}}{I_x} & \frac{k_{26}}{I_x} & \frac{k_{27}}{I_x} \\
\frac{k_{31}}{I_y} & \frac{k_{32}}{I_y} & \frac{k_{33}}{I_y} & \frac{k_{34}}{I_y} & \frac{k_{35}}{I_y} & \frac{k_{36}}{I_y} & \frac{k_{37}}{I_y} \\
\frac{k_{41}}{m_f} & \frac{k_{42}}{m_f} & \frac{k_{43}}{m_f} & \frac{k_{44}}{m_f} & \frac{k_{45}}{m_f} & \frac{k_{46}}{m_f} & \frac{k_{47}}{m_f} \\
\frac{k_{51}}{m_f} & \frac{k_{52}}{m_f} & \frac{k_{43}}{m_f} & \frac{k_{54}}{m_f} & \frac{k_{55}}{m_f} & \frac{k_{56}}{m_f} & \frac{k_{57}}{m_f} \\
\frac{k_{61}}{m_r} & \frac{k_{62}}{m_r} & \frac{k_{53}}{m_r} & \frac{k_{64}}{m_r} & \frac{k_{65}}{m_r} & \frac{k_{66}}{m_r} & \frac{k_{67}}{m_r} \\
\frac{k_{71}}{m_r} & \frac{k_{72}}{m_r} & \frac{k_{63}}{m_r} & \frac{k_{74}}{m_r} & \frac{k_{75}}{m_r} & \frac{k_{76}}{m_r} & \frac{k_{77}}{m_r}
\end{bmatrix} \tag{12.64}
$$

To study and design the natural frequencies and mode shapes of a full car model, let us separate the parameters to fixed and adjustable. The fixed parameters are geometric parameters $a_1$, $a_2$, $b_1$, $b_2$, $b_3$, $b_4$ and inertia parameters $m$, $m_f$, $m_r$, $I_x$, $I_y$, and the adjustable parameters are stiffness rates $k_f$, $k_r$, $k_{R_f}$, $k_{R_r}$, $k_{t_f}$, $k_{t_r}$, $k_{R_r}$, $k_P$ and damping rates $c_f$, $c_r$, $c_{t_f}$, $c_{t_r}$, $c_{R_f}$, $c_{R_r}$. All stiffness and damping rates are designed by quarter car, bicycle car, and half car models. The full car model will be employed to examine and fine-tune the adjustable parameters.

The natural frequencies and mode shapes are the first and most important characteristics of a vibrating system to determine. All responses of multi-degree of freedom ($DOF$) systems depend on natural frequencies and mode shapes of the system. Free and transient vibrations happen in modes of linear combinations of mode shapes, where the weight factors of the mode shapes are determined by initial conditions. Forced harmonic vibrations of multi-$DOF$ systems happen at the excitation frequency and the amplitude of the response depends on the mode shapes and closeness of the excitation frequency to the natural frequencies. Hence, the first step in studying the vibrations of a dynamic system is to determine its natural frequencies and mode shapes. ■

**Example 174** Natural frequencies and mode shapes of a full car model. Here we determine the natural frequencies and visualize the mode shapes of a nominated full car model.

Consider a vehicle with the following characteristics:

$$
\begin{aligned}
&m = 1344\,\text{kg} \quad m_1 = 42\,\text{kg} \quad m_2 = 42\,\text{kg} \quad m_3 = 50\,\text{kg} \\
&m_4 = 50\,\text{kg} \quad I_x = 1800\,\text{kg m}^2 \quad I_y = 2100\,\text{kg m}^2 \\
&a_1 = 1.4\,\text{m} \quad a_2 = 1.47\,\text{m} \quad l = a_1 + a_2 = 2.87\,\text{m} \\
&b_1 = 0.8652\,\text{m} \quad b_2 = 0.9085\,\text{m} \quad w_f = b_1 + b_2 = 1.7737\,\text{m} \\
&b_3 = 0.8852\,\text{m} \quad b_4 = 0.9285\,\text{m} \quad w_r = b_3 + b_4 = 1.8137\,\text{m} \\
&k_1 = k_2 = 22000\,\text{N/m} \quad k_3 = k_4 = 25000\,\text{N/m} \\
&k_{t_1} = k_{t_2} = k_{t_3} = k_{t_4} = 250000\,\text{N/m} \\
&c_1 = c_2 = 1630\,\text{N s/m} \quad c_3 = c_4 = 1900\,\text{N s/m} \\
&c_{t_f} = c_{t_r} = c_{R_f} = c_{R_r} = c_P = 0 \quad k_{R_f} = k_{R_r} = k_P = 0
\end{aligned} \tag{12.65}
$$

Employing $[A] = [m]^{-1}[k]$ and solving the associated eigenvalue and eigenvector problems, we find the following natural frequencies and mode shapes for the full car model:

$$
\begin{aligned}
&\omega_1 = 0.985\,\text{Hz} \quad \omega_2 = 1.255\,\text{Hz} \quad \omega_3 = 1.476\,\text{Hz} \\
&\omega_4 = 11.805\,\text{Hz} \quad \omega_5 = 11.812\,\text{Hz} \quad \omega_6 = 12.809\,\text{Hz} \\
&\omega_7 = 12.813\,\text{Hz}
\end{aligned} \tag{12.66}
$$

$$\mathbf{x} = \left[ x \ \varphi \ \theta \ x_1 \ x_2 \ x_3 \ x_4 \right]^T \tag{12.67}$$

$$\mathbf{u}_1 = \begin{bmatrix} 0.0553 \\ 1 \\ -0.0035 \\ 0.0753 \\ -0.0690 \\ 0.0856 \\ -0.0804 \end{bmatrix} \quad \mathbf{u}_2 = \begin{bmatrix} 1 \\ -0.0422 \\ -0.2185 \\ 0.1036 \\ 0.1097 \\ 0.0589 \\ 0.0660 \end{bmatrix} \tag{12.68}$$

$$\mathbf{u}_3 = \begin{bmatrix} 0.3407 \\ -0.0099 \\ 1 \\ -0.0875 \\ -0.0861 \\ 0.1664 \\ 0.1681 \end{bmatrix} \quad \mathbf{u}_4 = \begin{bmatrix} -0.5554 \times 10^{-4} \\ -0.4566 \times 10^{-2} \\ -0.5284 \times 10^{-4} \\ -0.2114 \times 10^{-2} \\ 0.2240 \times 10^{-2} \\ 1 \\ -0.9835 \end{bmatrix} \tag{12.69}$$

$$\mathbf{u}_5 = \begin{bmatrix} -0.6802 \\ 0.1484 \times 10^{-3} \\ -0.641 \times 10^{-2} \\ 0.1244 \times 10^{-2} \\ 0.1102 \times 10^{-2} \\ 0.9835 \\ 1 \end{bmatrix} \quad \mathbf{u}_6 = \begin{bmatrix} -0.4257 \times 10^{-4} \\ -0.3345 \times 10^{-2} \\ 0.3955 \times 10^{-4} \\ 1 \\ -0.9829 \\ 0.1506 \times 10^{-2} \\ -0.1596 \times 10^{-2} \end{bmatrix} \tag{12.70}$$

$$\mathbf{u}_7 = \begin{bmatrix} -0.5051 \times 10^{-2} \\ 0.1112 \times 10^{-3} \\ 0.4551 \times 10^{-2} \\ 0.9828 \\ 1 \\ -0.8848 \times 10^{-3} \\ -0.7822 \times 10^{-2} \end{bmatrix} \tag{12.71}$$

A visual illustration of the mode shapes is shown in Fig. 12.11. The largest element of the mode shapes $\mathbf{u}_1$ to $\mathbf{u}_7$ is $\varphi$, $x$, $\theta$, $x_3$, $x_4$, $x_1$, $x_2$, respectively. These figures depict the relative amplitude of each coordinate of the full car model at a natural frequency.

The natural frequencies of a full car can be separated in two classes. The first class is the natural frequencies of the body: body bounce $x$, body roll $\varphi$, and body pitch $\theta$. Body-related natural frequencies are always around 1 Hz. The second class is the natural frequencies of the wheels' bounce $x_1$, $x_2$, $x_3$, and $x_4$. Wheel-related natural frequencies are always around 10 Hz.

In this example, the third natural frequency is higher than the first and second natural frequencies. As a general rule, designers prefer all the body frequencies be close to each other around 1 Hz and all the wheels' natural frequencies also be close to each other around 10 Hz. In this way, when the excitation frequencies, such as road induced vibrations, are away from 1 Hz, then all of the body modes can be avoided altogether. A distributed natural frequency is not suitable. Here the third mode shape, associated with the third natural frequency, indicates that body pitch is the dominant vibration. Hence, the pitch stiffness must be reduced or the pitch inertia must be increased in order to reduce the third natural frequency.

As a trial and error experiment, let us assume that $I_y$ can be increased from $I_y = 2100\,\mathrm{kg\,m^2}$ to $I_y = 2500\,\mathrm{kg\,m^2}$. The natural frequencies for the new $I_y$ will be

$$\begin{aligned} &\omega_1 = 0.985\,\mathrm{Hz} \quad \omega_2 = 1.241\,\mathrm{Hz} \quad \omega_3 = 1.368\,\mathrm{Hz} \\ &\omega_4 = 11.805\,\mathrm{Hz} \quad \omega_5 = 11.811\,\mathrm{Hz} \quad \omega_6 = 12.809\,\mathrm{Hz} \\ &\omega_7 = 12.813\,\mathrm{Hz} \end{aligned} \tag{12.72}$$

while the mode shapes remain in the same order.

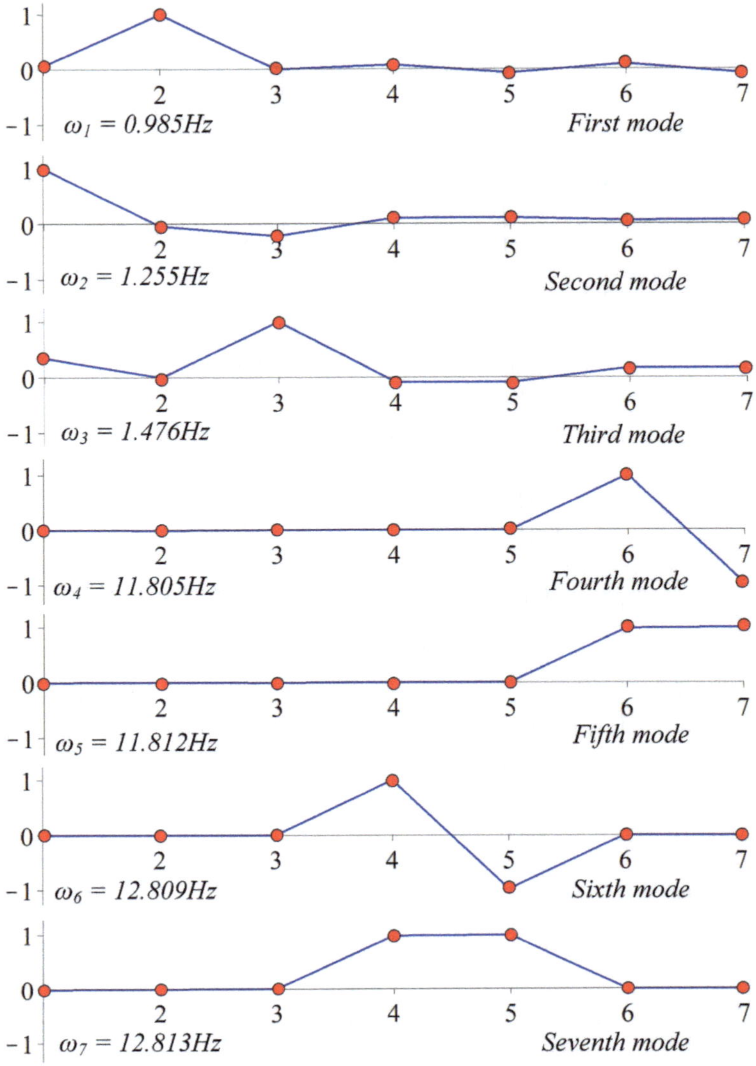


**Fig. 12.11** Vibration mode shape of a full car model

## 12.4 Frequency Response

To determine the frequency response of the half car model, we consider a harmonic base excitation under all wheels

$$y_1 = y_2 = y_3 = y_4 = Y \sin \omega t \tag{12.73}$$

and substitute a set of harmonic solutions into equations of motion

$$x = A_1 \sin \omega t + B_1 \cos \omega t = X_1 \sin (\omega t - \varphi_1) \tag{12.74}$$

$$\varphi = A_2 \sin \omega t + B_2 \cos \omega t = X_2 \sin (\omega t - \varphi_2) \tag{12.75}$$

$$\theta = A_3 \sin \omega t + B_3 \cos \omega t = X_3 \sin \left(\omega t - \varphi_3\right) \tag{12.76}$$

$$x_1 = A_4 \sin \omega t + B_4 \cos \omega t = X_4 \sin (\omega t - \varphi_4) \tag{12.77}$$

$$x_2 = A_5 \sin \omega t + B_5 \cos \omega t = X_5 \sin \left(\omega t - \varphi_5\right) \tag{12.78}$$

$$x_3 = A_6 \sin \omega t + B_6 \cos \omega t = X_6 \sin \left(\omega t - \varphi_6\right) \tag{12.79}$$

$$x_4 = A_7 \sin \omega t + B_7 \cos \omega t = X_7 \sin \left(\omega t - \varphi_7\right) \tag{12.80}$$

This way, the differential equations of motion will convert to a set of algebraic equations to determine the unknown coefficients $A_1$, $B_1$, $A_2$, $B_2$, $A_3$, $B_3$, $A_4$, $B_4$, $A_5$, $B_5$, $A_6$, $B_6$, $A_7$, and $B_7$ and calculate the frequency responses $X_1$, $X_2$, $X_3$, $X_4$, $X_5$, $X_6$, and $X_7$. The frequency responses are the amplitudes of vibrations of the coordinates $x$, $\varphi$, $\theta$, $x_1$, $x_1$, $x_2$, $x_3$, and $x_4$.

$$\begin{aligned} &X_1 = \sqrt{A_1^2 + B_1^2} \qquad X_2 = \sqrt{A_2^2 + B_2^2} \qquad X_3 = \sqrt{A_3^2 + B_3^2} \\ &X_4 = \sqrt{A_4^2 + B_4^2} \qquad X_5 = \sqrt{A_5^2 + B_5^2} \\ &X_6 = \sqrt{A_6^2 + B_6^2} \qquad X_7 = \sqrt{A_7^2 + B_7^2} \end{aligned} \tag{12.81}$$

***Proof*** Assuming the left and right tires are similar, the set of equations of motion of the full car model are

$$[m]\,\ddot{\mathbf{x}} + [c]\,\dot{\mathbf{x}} + [k]\,\mathbf{x} = \mathbf{F} \tag{12.82}$$

$$\mathbf{x} = \left[\, x \;\; \varphi \;\; \theta \;\; x_1 \;\; x_2 \;\; x_3 \;\; x_4 \,\right]^T \tag{12.83}$$

$$[m] = \begin{bmatrix} m & 0 & 0 & 0 & 0 & 0 & 0 \\ 0 & I_x & 0 & 0 & 0 & 0 & 0 \\ 0 & 0 & I_y & 0 & 0 & 0 & 0 \\ 0 & 0 & 0 & m_f & 0 & 0 & 0 \\ 0 & 0 & 0 & 0 & m_f & 0 & 0 \\ 0 & 0 & 0 & 0 & 0 & m_r & 0 \\ 0 & 0 & 0 & 0 & 0 & 0 & m_r \end{bmatrix} \tag{12.84}$$

$$[c] = \begin{bmatrix} c_{11} & c_{12} & c_{13} & -c_f & -c_f & -c_r & -c_r \\ c_{21} & c_{22} & c_{23} & c_{24} & c_{25} & c_{26} & c_{27} \\ c_{31} & c_{32} & c_{33} & c_{34} & c_{35} & c_{36} & c_{37} \\ -c_f & c_{42} & c_{43} & c_{44} & c_{45} & \dfrac{c_P}{4l^2} & -\dfrac{c_P}{4l^2} \\ -c_f & c_{52} & c_{53} & c_{54} & c_{55} & -\dfrac{c_P}{4l^2} & \dfrac{c_P}{4l^2} \\ -c_r & c_{62} & c_{63} & \dfrac{c_P}{4l^2} & -\dfrac{c_P}{4l^2} & c_{66} & c_{67} \\ -c_r & c_{72} & c_{73} & -\dfrac{c_P}{4l^2} & \dfrac{c_P}{4l^2} & c_{76} & c_{77} \end{bmatrix} \tag{12.85}$$

$$\begin{aligned} &c_{11} = 2c_f + 2c_r \qquad c_{12} = c_{21} = (b_1 - b_2)\, c_f + (b_3 - b_4)\, c_r \\ &c_{13} = c_{31} = 2\left(a_2 c_r - a_1 c_f\right) \\ &c_{22} = \left(b_1^2 + b_2^2\right) c_f + \left(b_3^2 + b_4^2\right) c_r + c_{R_f} + c_{R_r} \end{aligned} \tag{12.86}$$

$$\begin{aligned} &c_{23} = c_{32} = a_1\,(b_2 - b_1)\, c_f + a_2\,(b_3 - b_4)\, c_r \\ &c_{24} = c_{42} = -b_1 c_f - \frac{c_{R_f}}{w_f} \qquad c_{25} = c_{52} = b_2 c_f + \frac{c_{R_f}}{w_f} \\ &c_{26} = c_{62} = -b_3 c_r - \frac{c_{R_r}}{w_r} \qquad c_{27} = c_{72} = b_4 c_r + \frac{c_{R_r}}{w_r} \end{aligned} \tag{12.87}$$

$$c_{33} = 2a_1^2 c_f + 2a_2^2 c_r + c_P \qquad c_{34} = c_{43} = a_1 c_f + \frac{c_P}{2l}$$
$$c_{35} = c_{53} = a_1 c_f - \frac{c_P}{2l} \qquad c_{36} = c_{63} = -a_2 c_r + \frac{c_P}{2l}$$
$$c_{37} = c_{73} = -a_2 c_r - \frac{c_P}{2l} \tag{12.88}$$

$$c_{44} = c_f + \frac{c_{R_f}}{w_f^2} + \frac{c_P}{4l^2} + c_{t_f} \qquad c_{45} = c_{54} = -\frac{c_{R_f}}{w_f^2} - \frac{c_P}{4l^2}$$
$$c_{55} = c_f + \frac{c_{R_f}}{w_f^2} + \frac{c_P}{4l^2} + c_{t_f} \qquad c_{66} = c_r + \frac{c_{R_r}}{w_r^2} + \frac{c_P}{4l^2} + c_{t_r}$$
$$c_{67} = c_{76} = -\frac{c_{R_r}}{w_r^2} - \frac{c_P}{4l^2} \qquad c_{77} = c_r + \frac{c_{R_r}}{w_r^2} + \frac{c_P}{4l^2} + c_{t_r} \tag{12.89}$$

$$[k] = \begin{bmatrix} k_{11} & k_{12} & k_{13} & -k_f & -k_f & -k_r & -k_r \\ k_{21} & k_{22} & k_{23} & k_{24} & k_{25} & k_{26} & k_{27} \\ k_{31} & k_{32} & k_{33} & k_{34} & k_{35} & k_{36} & k_{37} \\ -k_f & k_{42} & k_{43} & k_{44} & k_{45} & \frac{k_P}{4l^2} & -\frac{k_P}{4l^2} \\ -k_f & k_{52} & k_{53} & k_{54} & k_{55} & -\frac{k_P}{4l^2} & \frac{k_P}{4l^2} \\ -k_r & k_{62} & k_{63} & \frac{k_P}{4l^2} & -\frac{k_P}{4l^2} & k_{66} & k_{67} \\ -k_r & k_{72} & k_{73} & -\frac{k_P}{4l^2} & \frac{k_P}{4l^2} & k_{76} & k_{77} \end{bmatrix} \tag{12.90}$$

$$k_{11} = 2k_f + 2k_r \qquad k_{12} = k_{21} = (b_1 - b_2)\, k_f + (b_3 - b_4)\, k_r$$
$$k_{13} = k_{31} = 2\left(a_2 k_r - a_1 k_f\right)$$
$$k_{22} = \left(b_1^2 + b_2^2\right) k_f + \left(b_3^2 + b_4^2\right) k_r + k_{R_f} + k_{R_r} \tag{12.91}$$

$$k_{23} = k_{32} = a_1 (b_2 - b_1)\, k_f + a_2 (b_3 - b_4)\, k_r$$
$$k_{24} = k_{42} = -b_1 k_f - \frac{k_{R_f}}{w_f} \qquad k_{25} = k_{52} = b_2 k_f + \frac{k_{R_f}}{w_f}$$
$$k_{26} = k_{62} = -b_3 k_r - \frac{k_{R_r}}{w_r} \qquad k_{27} = k_{72} = b_4 k_r + \frac{k_{R_r}}{w_r} \tag{12.92}$$

$$k_{33} = 2a_1^2 k_f + 2a_2^2 k_r + k_P \qquad k_{34} = k_{43} = a_1 k_f + \frac{k_P}{2l}$$
$$k_{35} = k_{53} = a_1 k_f - \frac{k_P}{2l} \qquad k_{36} = k_{63} = -a_2 k_r + \frac{k_P}{2l}$$
$$k_{37} = k_{73} = -a_2 k_r - \frac{k_P}{2l} \tag{12.93}$$

$$k_{44} = k_f + \frac{k_{R_f}}{w_f^2} + \frac{k_P}{4l^2} + k_{t_f} \qquad k_{45} = k_{54} = -\frac{k_{R_f}}{w_f^2} - \frac{k_P}{4l^2}$$

$$k_{55} = k_f + \frac{k_{R_f}}{w_f^2} + \frac{k_P}{4l^2} + k_{t_f} \qquad k_{66} = k_r + \frac{k_{R_r}}{w_r^2} + \frac{k_P}{4l^2} + k_{t_r}$$

$$k_{67} = k_{76} = -\frac{k_{R_r}}{w_r^2} - \frac{k_P}{4l^2} \qquad k_{77} = k_r + \frac{k_{R_r}}{w_r^2} + \frac{k_P}{4l^2} + k_{t_r} \tag{12.94}$$

$$\mathbf{F} = \left[0\;0\;0\;k_{t_f}\,y_1\;k_{t_f}\,y_2\;k_{t_r}\,y_3\;k_{t_r}\,y_4\right]^T \tag{12.95}$$

Substituting the harmonic solutions in the equations of motion

$$\mathbf{x} = \mathbf{A}\sin\omega t + \mathbf{B}\cos\omega t \tag{12.96}$$

$$\begin{bmatrix} x \\ \varphi \\ \theta \\ x_1 \\ x_2 \\ x_3 \\ x_4 \end{bmatrix} = \begin{bmatrix} A_1 \\ A_2 \\ A_3 \\ A_4 \\ A_5 \\ A_6 \\ A_7 \end{bmatrix} \sin\omega t + \begin{bmatrix} B_1 \\ B_2 \\ B_3 \\ B_4 \\ B_5 \\ B_6 \\ B_7 \end{bmatrix} \cos\omega t \tag{12.97}$$

provides us with a set of algebraic equations to calculate the coefficients **A** and **B**.

$$\begin{bmatrix} [k] - \omega^2[m] & -\omega[c] \\ \omega[c] & [k] - \omega^2[m] \end{bmatrix} \begin{bmatrix} A_1 \\ A_2 \\ A_3 \\ A_4 \\ A_5 \\ A_6 \\ A_7 \\ B_1 \\ B_2 \\ B_3 \\ B_4 \\ B_5 \\ B_6 \\ B_7 \end{bmatrix} = \begin{bmatrix} 0 \\ 0 \\ 0 \\ k_{t_f}\,y_1 \\ k_{t_f}\,y_2 \\ k_{t_r}\,y_3 \\ k_{t_r}\,y_4 \\ 0 \\ 0 \\ 0 \\ 0 \\ 0 \\ 0 \\ 0 \end{bmatrix} \tag{12.98}$$

$$X_6 = \sqrt{A_6 + B_6} \qquad X_7 = \sqrt{A_7 + B_7}$$

The frequency responses are indicated by $X_1$, $X_2$, $X_3$, $X_4$, $X_5$, $X_6$, and $X_7$, where $X_1$ is the body bounce amplitude, $X_2$ is the body roll amplitude, $X_3$ is the body pitch amplitude, $X_4$ is the front left wheel bounce amplitude, $X_5$ is the front right wheel bounce amplitude, $X_6$ is the rear left wheel bounce amplitude, and $X_7$ is the rear right wheel bounce amplitude. Assuming

$$y_1 = Y\sin\omega t \qquad y_2 = Y\sin\omega t$$

$$y_3 = Y\sin\omega t \qquad y_4 = Y\sin\omega t \tag{12.99}$$

we calculate the unknown coefficients and the frequency responses (13.2).

$$\begin{bmatrix} \mathbf{A} \\ \mathbf{B} \end{bmatrix} = \begin{bmatrix} [k] - \omega^2[m] & -\omega[c] \\ \omega[c] & [k] - \omega^2[m] \end{bmatrix}^{-1} \begin{bmatrix} \mathbf{F} \\ \mathbf{0} \end{bmatrix} \tag{12.100}$$

■

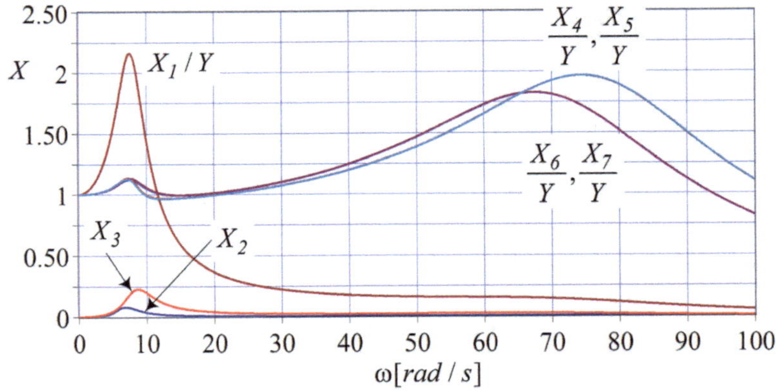


**Fig. 12.12** A sample of frequency response of a full vehicle model

**Example 175** Frequency response of a full car model. Every passenger vehicle has two resonance zones in their frequency response. A resonance zone around 1 Hz associated with the body bounce, roll, and body pitch, and a resonance zone around 10 Hz associated with front and rear wheels bounce vibrations.

Consider a full vehicle model with $k_P = 0$ and the following data:

$$
\begin{aligned}
&m = 1344\,\mathrm{kg} \qquad m_1 = 42\,\mathrm{kg} \qquad m_2 = 42\,\mathrm{kg} \\
&m_3 = 50\,\mathrm{kg} \\
&m_4 = 50\,\mathrm{kg} \qquad I_x = 1800\,\mathrm{kg\,m^2} \qquad I_y = 2100\,\mathrm{kg\,m^2} \\
&a_1 = 1.4\,\mathrm{m} \qquad a_2 = 1.47\,\mathrm{m} \qquad l = a_1 + a_2 = 2.87\,\mathrm{m} \\
&b_1 = 0.8652\,\mathrm{m} \qquad b_2 = 0.9085\,\mathrm{m} \\
&w_f = b_1 + b_2 = 1.7737\,\mathrm{m} \\
&b_3 = 0.8852\,\mathrm{m} \qquad b_4 = 0.9285\,\mathrm{m} \\
&w_r = b_3 + b_4 = 1.8137\,\mathrm{m} \\
&k_1 = k_2 = 22000\,\mathrm{N/\,m} \qquad k_3 = k_4 = 25000\,\mathrm{N/\,m} \\
&k_{t_1} = k_{t_2} = k_{t_3} = k_{t_4} = 250000\,\mathrm{N/\,m} \\
&c_1 = c_2 = 1630\,\mathrm{N\,s/\,m} \qquad c_3 = c_4 = 1900\,\mathrm{N\,s/\,m} \\
&c_{t_f} = c_{t_r} = c_{R_f} = c_{R_r} = c_P = 0 \qquad k_{R_f} = k_{R_r} = k_P = 0
\end{aligned}
\tag{12.101}
$$

To determine the frequency response of the vehicle, we apply a harmonic excitation under all four wheels with amplitude $Y$ at a frequency $\omega$.

$$y_1 = y_2 = y_3 = y_4 = Y \sin \omega t \tag{12.102}$$

Figure 12.12 illustrates the frequency response of the vehicle. The body bounce $x$ goes high at the first resonance zone around 1 Hz and drops significantly for $\omega > 2\,\mathrm{Hz}$. The body roll $\varphi$ and body pitch oscillations $\theta$ also have similar behavior with high amplitude around 1 Hz and very low amplitude for $\omega > 2\,\mathrm{Hz}$. The wheels have a wider range of resonance zone. The wheel's highest amplitude of vibrations is around 10 Hz. Figure 12.13 illustrates the frequency responses of $\varphi$ and $\theta$ in terms of deg and Hz.

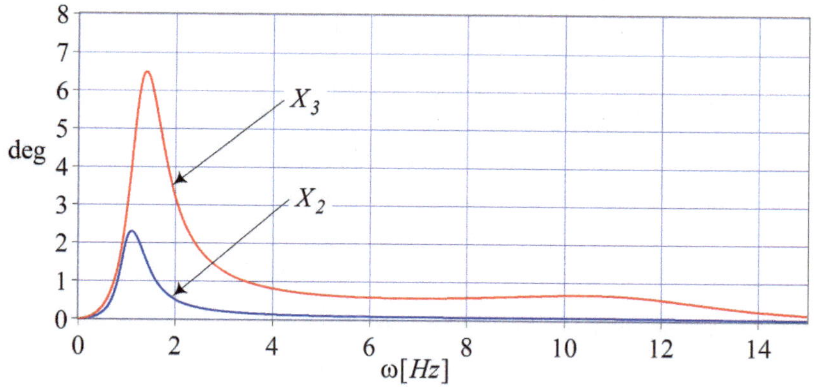


**Fig. 12.13** The frequency response of roll $\varphi \equiv X_2$ and pitch $\theta \equiv X_3$ in deg in terms of Hz

## 12.5 Summary

The full car vibrating model of a vehicle encompasses the complete representation of a vehicle's vibrations, including the bounce, pitch, and roll motions of the body, as well as the vibrations of the wheels. A four-wheeled vehicle includes the body bounce $x$, body pitch $\theta$, body roll $\varphi$, and wheels hop $x_1$, $x_2$, $x_3$, and $x_4$. The wheels are indicated by a number, starting from 1 for the front left wheel, number 2 for the front right wheel, then number 3 for the rear left, and number 4 for the rear right wheel. The road excitations $y_1$, $y_2$, $y_3$, and $y_4$ are under the wheels $x_1$, $x_2$, $x_3$, and $x_4$.

The equations of motion of the full car model are

$$\begin{aligned} & m\ddot{x} + c_f\left(\dot{x} - \dot{x}_1 + b_1\dot{\varphi} - a_1\dot{\theta}\right) + c_f\left(\dot{x} - \dot{x}_2 - b_2\dot{\varphi} - a_1\dot{\theta}\right) \\ & + c_r\left(\dot{x} - \dot{x}_3 + b_3\dot{\varphi} + a_2\dot{\theta}\right) + c_r\left(\dot{x} - \dot{x}_4 - b_4\dot{\varphi} + a_2\dot{\theta}\right) \\ & + k_f\left(x - x_1 + b_1\varphi - a_1\theta\right) + k_f\left(x - x_2 - b_2\varphi - a_1\theta\right) \\ & + k_r\left(x - x_3 + b_3\varphi + a_2\theta\right) + k_r\left(x - x_4 - b_4\varphi + a_2\theta\right) = 0 \end{aligned} \tag{12.103}$$

$$\begin{aligned} & I_x\ddot{\varphi} + b_1c_f\left(\dot{x} - \dot{x}_1 + b_1\dot{\varphi} - a_1\dot{\theta}\right) - b_2c_f\left(\dot{x} - \dot{x}_2 - b_2\dot{\varphi} - a_1\dot{\theta}\right) \\ & + b_3c_r\left(\dot{x} - \dot{x}_3 + b_3\dot{\varphi} + a_2\dot{\theta}\right) + b_4c_r\left(\dot{x} - \dot{x}_4 - b_4\dot{\varphi} + a_2\dot{\theta}\right) \\ & + b_1k_f\left(x - x_1 + b_1\varphi - a_1\theta\right) - b_2k_f\left(x - x_2 - b_2\varphi - a_1\theta\right) \\ & + b_3k_r\left(x - x_3 + b_3\varphi + a_2\theta\right) + b_4k_r\left(x - x_4 - b_4\varphi + a_2\theta\right) \\ & + k_{R_f}\left(\varphi - \frac{x_1 - x_2}{w_f}\right) + k_{R_r}\left(\varphi - \frac{x_3 - x_4}{w_r}\right) \\ & + c_{R_f}\left(\dot{\varphi} - \frac{\dot{x}_1 - \dot{x}_2}{w_f}\right) + c_{R_r}\left(\dot{\varphi} - \frac{\dot{x}_3 - \dot{x}_4}{w_r}\right) = 0 \end{aligned} \tag{12.104}$$

$$\begin{aligned} & I_y\ddot{\theta} - a_1c_f\left(\dot{x} - \dot{x}_1 + b_1\dot{\varphi} - a_1\dot{\theta}\right) - a_1c_f\left(\dot{x} - \dot{x}_2 - b_2\dot{\varphi} - a_1\dot{\theta}\right) \\ & + a_2c_r\left(\dot{x} - \dot{x}_3 + b_3\dot{\varphi} + a_2\dot{\theta}\right) + a_2c_r\left(\dot{x} - \dot{x}_4 - b_4\dot{\varphi} + a_2\dot{\theta}\right) \\ & - a_1k_f\left(x - x_1 + b_1\varphi - a_1\theta\right) - a_1k_f\left(x - x_2 - b_2\varphi - a_1\theta\right) \\ & + a_2k_r\left(x - x_3 + b_3\varphi + a_2\theta\right) + a_2k_r\left(x - x_4 - b_4\varphi + a_2\theta\right) \\ & + c_P\left(\dot{\theta} + \frac{\left(\dot{x}_1 - \dot{x}_2\right)/2}{l} + \frac{\left(\dot{x}_3 - \dot{x}_4\right)/2}{l}\right) \\ & + k_P\left(\theta + \frac{\left(x_1 - x_2\right)/2}{l} + \frac{\left(x_3 - x_4\right)/2}{l}\right) = 0 \end{aligned} \tag{12.105}$$

$$\begin{aligned}
&m_f\ddot{x}_1 - c_f\left(\dot{x} - \dot{x}_1 + b_1\dot{\varphi} - a_1\dot{\theta}\right) - k_f\left(x - x_1 + b_1\varphi - a_1\theta\right) \\
&- c_{R_f}\frac{1}{w_f}\left(\dot{\varphi} - \frac{\dot{x}_1 - \dot{x}_2}{w_f}\right) - k_{R_f}\frac{1}{w_f}\left(\varphi - \frac{x_1 - x_2}{w_f}\right) \\
&+ \frac{1}{2l}c_P\left(\dot{\theta} + \frac{(\dot{x}_1 - \dot{x}_2)/2}{l} + \frac{(\dot{x}_3 - \dot{x}_4)/2}{l}\right) \\
&+ \frac{1}{2l}k_P\left(\theta + \frac{(x_1 - x_2)/2}{l} + \frac{(x_3 - x_4)/2}{l}\right) \\
&+ c_{t_f}\left(\dot{x}_1 - \dot{y}_1\right) + k_{t_f}\left(x_1 - y_1\right) = 0
\end{aligned} \tag{12.106}$$

$$\begin{aligned}
&m_f\ddot{x}_2 - c_f\left(\dot{x} - \dot{x}_2 - b_2\dot{\varphi} - a_1\dot{\theta}\right) - k_f\left(x - x_2 - b_2\varphi - a_1\theta\right) \\
&+ c_{R_f}\frac{1}{w_f}\left(\dot{\varphi} - \frac{\dot{x}_1 - \dot{x}_2}{w_f}\right) + k_{R_f}\frac{1}{w_f}\left(\varphi - \frac{x_1 - x_2}{w_f}\right) \\
&- \frac{1}{2l}c_P\left(\dot{\theta} + \frac{(\dot{x}_1 - \dot{x}_2)/2}{l} + \frac{(\dot{x}_3 - \dot{x}_4)/2}{l}\right) \\
&- \frac{1}{2l}k_P\left(\theta + \frac{(x_1 - x_2)/2}{l} + \frac{(x_3 - x_4)/2}{l}\right) \\
&+ c_{t_f}\left(\dot{x}_2 - \dot{y}_2\right) + k_{t_f}\left(x_2 - y_2\right) = 0
\end{aligned} \tag{12.107}$$

$$\begin{aligned}
&m_r\ddot{x}_3 - c_r\left(\dot{x} - \dot{x}_3 + b_3\dot{\varphi} + a_2\dot{\theta}\right) - k_f\left(x - x_3 + b_3\varphi + a_2\theta\right) \\
&- c_{R_r}\frac{1}{w_r}\left(\dot{\varphi} - \frac{\dot{x}_3 - \dot{x}_4}{w_f}\right) - k_{R_r}\frac{1}{w_r}\left(\varphi - \frac{x_1 - x_2}{w_f}\right) \\
&+ \frac{1}{2l}c_P\left(\dot{\theta} + \frac{(\dot{x}_1 - \dot{x}_2)/2}{l} + \frac{(\dot{x}_3 - \dot{x}_4)/2}{l}\right) \\
&+ \frac{1}{2l}k_P\left(\theta + \frac{(x_1 - x_2)/2}{l} + \frac{(x_3 - x_4)/2}{l}\right) \\
&+ c_{t_r}\left(\dot{x}_3 - \dot{y}_3\right) + k_{t_r}\left(x_3 - y_3\right) = 0
\end{aligned} \tag{12.108}$$

$$\begin{aligned}
&m_r\ddot{x}_4 - c_r\left(\dot{x} - \dot{x}_4 - b_4\dot{\varphi} + a_2\dot{\theta}\right) - k_f\left(x - x_4 - b_4\varphi + a_2\theta\right) \\
&+ c_{R_r}\frac{1}{w_r}\left(\dot{\varphi} - \frac{\dot{x}_3 - \dot{x}_4}{w_f}\right) - k_{R_r}\frac{1}{w_r}\left(\varphi - \frac{x_1 - x_2}{w_f}\right) \\
&- \frac{1}{2l}c_P\left(\dot{\theta} + \frac{(\dot{x}_1 - \dot{x}_2)/2}{l} + \frac{(\dot{x}_3 - \dot{x}_4)/2}{l}\right) \\
&- \frac{1}{2l}k_P\left(\theta + \frac{(x_1 - x_2)/2}{l} + \frac{(x_3 - x_4)/2}{l}\right) \\
&+ c_{t_r}\left(\dot{x}_4 - \dot{y}_4\right) + k_{t_r}\left(x_4 - y_4\right) = 0
\end{aligned} \tag{12.109}$$

The vehicle also has an antiroll bar in the front and in the back, with torsional stiffness $k_{R_f}$ and $k_{R_r}$ and torsional damping $c_{R_f}$ and $c_{R_r}$. It also has an antipitch bar with torsional stiffness $k_P$ torsional damping $c_P$.

The natural frequencies and mode shapes are the first and most important characteristics of a vibrating system to determine. All responses of multi-degree of freedom ($DOF$) systems depend on natural frequencies and mode shapes of the system. Free and transient vibrations happen in modes of linear combinations of mode shapes, where the weight factors of the mode shapes are determined by initial conditions. Forced harmonic vibrations of multi-$DOF$ systems happen at the excitation frequency, and the amplitude of the response depends on the mode shapes and closeness of the excitation frequency to the natural frequencies. Hence, the first step in studying the vibrations of a dynamic system is to determine its natural frequencies and mode shapes. The $\omega_{n_i}$ and $\mathbf{u}_i$ can also be calculated by determining the eigenvalues and eigenvectors of the $\mathbf{A}$-matrix. The matrices $[m]$ and $[k]$ are mass and stiffness matrices of the system, respectively.

$$\mathbf{A} = [m]^{-1} [k] \tag{12.110}$$

## 12.6 Key Symbols

| | |
|---|---|
| $a \equiv \ddot{x}$ | acceleration |
| $a_1, a_2$ | longitudinal distance of mass center from front and rear axles |
| $A$ | coefficient matrix |
| $\mathbf{A} = [m]^{-1}[k]$ | characteristic matrix |
| $A, B$ | weight factor, coefficients for frequency responses |
| $\mathbf{A}, \mathbf{B}$ | coefficient vectors |
| $b_i$ | lateral distance of mass center from wheel number i |
| $c$ | damping |
| $c_{ij}$ | row i, column j of matrix $[c]$ |
| $C$ | coefficient |
| $d_1$ | road wave length |
| $d_2$ | road wave amplitude |
| $D$ | dissipation function |
| $e$ | exponential function |
| $E$ | mechanical energy |
| $f = 1/T$ | cyclic frequency [ Hz] |
| $f, F, \mathbf{F}$ | force |
| $f, g$ | function, periodic function |
| $f_c$ | damper force |
| $f_e$ | equivalent force |
| $f_k$ | spring force |
| $f_m$ | required force to move a mass m |
| $f_T$ | transmitted force |
| $F$ | amplitude of the harmonic force $f = F \sin \omega t$ |
| $F_T$ | amplitude of transmitted force |
| $F_0$ | constant force |
| $g$ | gravitational acceleration, function |
| $G$ | global coordinate, general frequency response with node |
| $H$ | Heaviside function |
| $I$ | mass moment |
| $i, j$ | dummy index |
| $k$ | stiffness |
| $k_{ij}$ | row i, column j of matrix $[k]$ |
| $k_e$ | equivalent stiffness |
| $K$ | kinetic energy |
| $l$ | length, wheelbase |
| L | length dimension |
| $\mathcal{L}$ | Lagrangean |
| $m, M$ | mass |
| $m_{ij}$ | row i, column j of matrix $[m]$ |
| $M$ | moment, mass dimension |
| $P$ | potential energy |
| $q$ | general coordinate |

| | |
|---|---|
| $r$ | frequency ratio |
| $r_i,\ i \in N$ | nodal frequency ratio |
| $r_n$ | natural frequency ratio |
| $s$ | exponent, characteristic value |
| $S$ | general frequency response without node |
| $t$ | time |
| $T$ | period, time dimension |
| $u_{ij}$ | $j$th element of the $i$th mode shape |
| $\mathbf{u}$ | mode shape, eigenvector |
| $\mathbf{u}_i$ | $i$th eigenvector |
| $v \equiv \dot{x}$ | velocity |
| $w$ | track of vehicle |
| $x, y, z$ | body coordinate axes |
| $x, y, z,\ \mathbf{x}$ | displacement |
| $X, Y, Z$ | amplitude |
| $\dot{x}, \dot{y}, \dot{z}$ | velocity, time derivative of $x, y, z$ |
| $y_i$ | road excitation under wheel number $i$ |
| $Y_i$ | amplitude of road displacement |
| $z$ | relative displacement |
| $Z$ | amplitude of relative displacement |
| $Z_i$ | short notation of part of functions |

Greek

| | |
|---|---|
| $\alpha$ | angle, frequency ratio |
| $\delta$ | difference |
| $\delta_s$ | static deflection |
| $\Delta$ | difference |
| $\gamma$ | angle |
| $\lambda$ | eigenvalue |
| $\Lambda$ | eigenvalue matrix |
| $\xi$ | damping ratio |
| $\varepsilon$ | mass ratio |
| $\theta$ | pitch angular motion |
| $\varphi$ | roll angular motion |
| $\rho$ | length mass density |
| $\omega$ | angular frequency |
| $\omega_n$ | natural frequency |
| $\omega = 2\pi f$ | angular frequency [rad/s] |
| $\Omega$ | frequency |

Subscript

| | |
|---|---|
| $i \in N$ | node number, wheel number |
| $n$ | natural |
| $f$ | front |
| $l$ | left |
| $P$ | pitch motion |
| $r$ | rear, right |
| $R$ | roll motion |
| $t$ | tire |

Symbols

| | |
|---|---|
| $DOF$ | degree of freedom |
| $FBD$ | free body diagram |

| | |
|---|---|
| $ODF$ | one degree of freedom |
| Re | real part |
| $RMS$ | root mean square |
| det | determinant |
| Hz | Hertz |
| [ ] | matrix |
| $\int$ | integral |
| $\mid\,\mid$ | absolute value |
| $d$ | differential |
| $\partial$ | partial derivative |
| $\mathcal{L}$ | Lagrangean |

## References


1. H. Akcay, S. Turkay, Influence of tire damping on mixed $H_2/H\infty$ synthesis of half - car active suspensions. J. Sound Vib. **322**, 15–28 (2009)
2. A.A. Aly, F.A. Salem, Vehicle suspension systems control: a review. Int. J. Control Autom. Syst. **2**(2), 46–54 (2013)
3. R. Darus, Y.M. Sam, Modeling and control active suspension system for a full car model, in *5th IEEE International Colloquium on Signal Processing & Its Applications* (2009), pp. 13–18
4. E. Esmailzadeh, F. Fahimi, Optimal adaptive active suspensions for a full car model. Veh. Syst. Dyn. **27**(2), 89–107 (1997)
5. D. Hrovat, Survey of advanced suspension developments and related optimal control applications. Automatica **33**(10), 178–817 (1997)
6. M. Issa, A. Samn, Passive vehicle suspension system optimization using Harris Hawk Optimization algorithm. Math. Comput. Simul. **191**, 328–345 (2022)
7. R.N. Jazar, *Advanced Dynamics: Rigid Body, Multibody, and Aerospace Applications* (Wiley, New York, 2011)
8. R.N. Jazar, *Advanced Vibrations: A Modern Approach* (Springer, New York, 2013)
9. R.N. Jazar, *Vehicle Dynamics: Theory and Application*, 3rd edn. (Springer, New York, 2017)
10. R.N. Jazar, *Approximation Methods in Science and Engineering* (Springer, New York, 2020)
11. R.N. Jazar *Perturbation Methods in Science and Engineering* (Springer, New York, 2021)
12. R.N. Jazar, *Advanced Vibrations: Theory and Application* (Springer, New York, 2023)
13. R.N. Jazar, R. Houim, A. Narimani, M.F. Golnaraghi, Frequency response and jump avoidance in a nonlinear passive engine mount. J. Vib. Control **12**(11), 1205–1237 (2006)
14. R.N. Jazar, M. Kazemi, S. Borhani, *Mechanical Vibrations* (in Persian) (Ettehad Publications, Tehran, 1992)
15. C. Kim, P.I. Ro, An accurate full car ride model using model reducing techniques. J. Mech. Des. **124**(4), 697–705 (2002)
16. R. Krtolica, D. Hrovat, Optimal active suspension control based on a half-car model: an analytical solution. IEEE Trans. Autom. Control **37**(4), 528–532 (1992)
17. I.W. Lee, D.O. Kim, G.H. Jung, Natural frequency and mode shape sensitivities of damped systems: Part I, distinct natural frequencies. J. Sound Vib. **223**(3), 399–412 (1999)
18. L. Meirovitch, *Analytical Methods in Vibrations* (Macmillan, New York, 1967)
19. L. Meirovitch, *Principles and Techniques of Vibrations* (Prentice Hall, Ney Jersey, 1997)
20. L. Meirovitch, *Fundamentals of Vibrations* (McGraw-Hill, New York, 2002)
21. S.Y. Moon, W.H. Kwon, Genetic-based fuzzy control for half-car active suspension systems. Int. J. Syst. Sci. **29**(7), 699–710 (1998)
22. T.A. Nosseir, S.M. Dickinson, The free vibration of a model of a car body. J. Sound Vib. **15**(2), 257–268 (1971)
23. Z. Osinski, *Damping of Vibrations* (A. A. Balkema Publishers, Rotterdam, 1998)
24. S. Palli, R.C. Sharma, P.D. Rao, Dynamic behaviour of a 7 DoF passenger car model. Int. J. Veh. Struct. Syst. **9**(1), 57 (2017)
25. W.J Palm, *Mechanical Vibration* (John Wiley, New York, 2006)
26. S.S. Rao, *Mechanical Vibrations* (Pearson Prentice Hall, Harlow, 2018)
27. M. Roseau, *Vibrations in Mechanical Systems* (Springer, Berlin, 1987)
28. R.S. Sharp, D.A. Crolla, Road vehicle suspension system design-a review. Veh. Syst. Dyn. **16**(3), 167–192 (1987)
29. A. Shirahatti, P.S.S. Prasad, P. Panzade, M.M. Kulkarni, Optimal design of passenger car suspension for ride and road holding. J. Brazil. Soc. Mech. Sci. Eng. **30**, 66–76 (2008)
30. J.C. Snowdon, *Vibration and Shock in Damped Mechanical Systems* (John Wiley, New York, 1968)
31. F.S. Tse, I.E. Morse, R.T. Hinkle, *Mechanical Vibrations Theory and Applications* (Allyn and Bacon, Boston, 1978)
32. C. Wei, H. Taghavifar, A novel approach to energy harvesting from vehicle suspension system: Half-vehicle model. Energy **134**, 279–288 (2017)
33. N. Zhang, W.A. Smith, J. Jeyakumaran, Hydraulically interconnected vehicle suspension: background and modelling. Veh. Syst. Dyn. **48**(1), 17–40 (2010)

# Part IV

# Advanced Vehicle Vibration Problems

The linear analysis of vehicle vibrations provides a valuable means to delve into the fundamental characteristics of vehicle vibration. By investigating essential parameters such as mass moment ($I$), mass ($m$), stiffness ($k$), and damping ($c$), we can gain valuable insights into how changes in these variables influence the behavior of the vehicle. This analysis offers a robust comprehension of the overall effects and consequences involved. Nevertheless, there are situations that demand going beyond the confines of linear analysis and incorporating nonlinear models to examine specific scenarios. Two noteworthy examples of such scenarios are flat-ride tuning and tire-road separations. In the upcoming chapters, our focus will be directed toward exploring these particular cases in greater detail.

Flat-ride tuning poses distinct challenges that go beyond the scope of linear analysis. Achieving optimal performance and fine-tuning flat rides necessitates a comprehensive grasp of the ride's dynamics. By incorporating nonlinear modeling techniques, we can effectively capture the intricate interactions among different ride elements, leading to an elevated and enhanced experience for riders.

Additionally, when examining tire-road separations, linear analysis may not provide a complete picture. Nonlinear modeling becomes crucial in understanding the complex dynamics that arise when tires interact with uneven or imperfect road surfaces. By considering these nonlinear effects, we can better comprehend tire behavior and optimize vehicle performance in real-world road conditions.

In summary, although linear analysis provides valuable insights into vehicle vibrations, there are specific situations, such as flat-ride tuning and tire-road separations, where nonlinear modeling becomes crucial. By delving into these scenarios, we can enhance our understanding and make informed choices to optimize ride experiences and improve overall vehicle performance.

# 13 Flat Ride Tuning

Flat ride refers to a phenomenon where the uncomfortable pitching oscillation of a vehicle transforms into a more tolerable bouncing oscillation when the vehicle encounters a bump while moving forward. Through experimentation, Maurice Olley identified and established two conditions for achieving a flat ride:

1. The pitch radius of gyration should be equal to the product of the distances between the mass centers of the front and rear wheels of the car ($r^2 = a_1 a_2$). This condition allows the car to be viewed as two separate, uncoupled mass–spring systems, front and rear suspensions as shown in Fig. 13.1.
2. The rear suspension should have a stiffness approximately 20% higher than that of the front suspension.

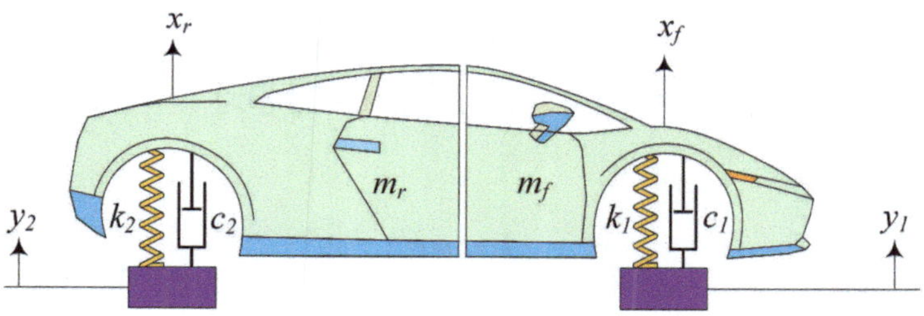


**Fig. 13.1** Uncoupled mass–spring systems of front and rear suspensions of a vehicle

In this chapter, we will analytically examine the conditions required for a flat ride and provide design charts that satisfy these requirements. However, the linear models upon which these conditions were initially based need to be modified to incorporate the nonlinear characteristics of shock absorbers in practical scenarios. By considering the nonlinear properties of shock absorbers, we can refine the conditions for achieving a flat ride, ensuring that the design and performance of the suspensions are optimized accordingly. In summary, this chapter will delve into the analytical review of flat ride conditions, presenting design charts that fulfill the necessary requirements. It will also address the need for adjustments to account for the nonlinear behavior of shock absorbers.

The transmission of road-induced forces to a car in straight motion initially affects the front wheels and, with a time delay, reaches the rear wheels. To eliminate pitch oscillations of the vehicle and allow the car to transition into a bouncing motion, it was generally advised that the natural frequency of the front suspension should be lower than that of the rear. This way, the rear part of the vehicle oscillates at a faster rate, catching up with the front and ultimately transitioning the car into a bounce, which is then damped to reduce vibrations. This concept, known as *flat ride tuning*.

Maurice Olley (1889–1983), a prominent figure in vehicle dynamics, conducted experiments in the 1930s using a modified car that allowed for variation of the pitch mass moment. Despite the subjective nature of the ride evaluations, Olley's guidelines have been regarded as reliable rules of thumb, even in the context of modern cars. These guidelines, commonly referred to as Olley's Flat Ride, remain significant in designing vehicles with improved ride quality.



R. N. Jazar, H. Marzbani, *Vehicle Vibrations*, https://doi.org/10.1007/978-3-031-43486-0_13

The concept of Olley's Flat Ride, when considering only one aspect and without taking into account other prerequisites, can be summarized as follows:

*The front suspension of a vehicle should have a stiffness approximately* 30% *lower than that of the rear suspension.*

The concept of the uncoupling condition, a vital requirement for achieving a flat ride, was initially introduced in 1923. To determine this condition, a bicycle car model has been employed, which helps establish the bounce and pitch centers of the model when placed on the springs. Once the uncoupling condition is satisfied, it allows us to treat the front and rear spring systems of the vehicle as two separate one degree of freedom systems. This simplification enables a more focused analysis and understanding of the vehicle's dynamics.

In this chapter, we employ analytical methods to investigate the flat ride conditions that have been adhered to and implemented by car manufacturers' designers since their initial introduction. This chapter offers a more dependable scientific and mathematical approach to understanding the design criteria for achieving a flat ride in vehicle dynamic studies.

## 13.1 Uncoupling the Bicycle Car

The simplest mathematical model of bicycle car is a beam supported by two springs, as shown in Fig. 13.2. The equations of motion of the model are

$$\begin{bmatrix} \dfrac{ma_2^2+I}{(a_1+a_2)^2} & \dfrac{ma_1a_2-I}{(a_1+a_2)^2} \\ \dfrac{ma_1a_2-I}{(a_1+a_2)^2} & \dfrac{ma_1^2+I}{(a_1+a_2)^2} \end{bmatrix} \begin{bmatrix} \ddot{x}_1 \\ \ddot{x}_2 \end{bmatrix} + \begin{bmatrix} k_1 & 0 \\ 0 & k_2 \end{bmatrix} \begin{bmatrix} x_1 \\ x_2 \end{bmatrix} = 0 \tag{13.1}$$

The equations can be converted into the following form:

$$\begin{bmatrix} \alpha+\gamma & 1-\alpha \\ 1-\alpha & \alpha+\dfrac{1}{\gamma} \end{bmatrix} \begin{bmatrix} \ddot{x}_1 \\ \ddot{x}_2 \end{bmatrix} + \begin{bmatrix} \Omega_1^2 & 0 \\ 0 & \Omega_2^2 \end{bmatrix} \begin{bmatrix} x_1 \\ x_2 \end{bmatrix} = 0 \tag{13.2}$$

by defining these parameters:

$$I = mr^2 \tag{13.3}$$

$$\Omega_1^2 = \frac{k_1}{m}\beta \tag{13.4}$$

$$\Omega_2^2 = \frac{k_2}{m}\beta \tag{13.5}$$

$$\beta = \frac{l^2}{a_1a_2} \tag{13.6}$$

$$\alpha = \frac{r^2}{a_1a_2} \tag{13.7}$$

$$\gamma = \frac{a_2}{a_1} \tag{13.8}$$

$$l = a_1 + a_2 \tag{13.9}$$

Equation (13.2) will be decoupled if

$$\alpha = \frac{r^2}{a_1a_2} = 1 \tag{13.10}$$

where $a_1$ is the distance of mass center $C$ to the center of the front wheel and $a_2$ is the distance of $C$ to the rear wheel. This condition specifies that the pitch radius of gyration, $r$, should be equal to the product of the distances from the mass center, $C$, to the front and rear axles. By setting $\alpha = 1$, the locations of the two vibration modes correspond to the front and rear axles. Consequently, the excitation experienced by the front wheel will not affect the body at the rear axle and vice versa. In essence, for such a car, the front and rear sections operate independently, acting as separate entities as illustrated in Fig. 13.1.

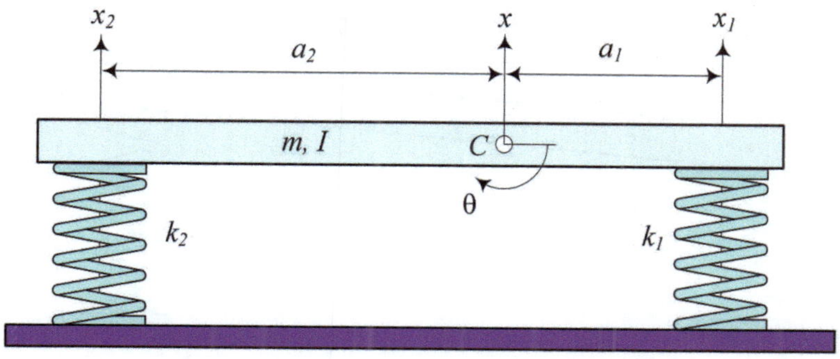


**Fig. 13.2** The bicycle model of a car is a beam with mass $m$ and mass moment $I$, sitting on two springs $k_1$ and $k_2$

***Proof*** Let us consider the two degrees of freedom ($DOF$) system of Fig. 13.2. The system consists of a beam with mass $m$ and a mass moment $I$ about the mass center $C$. It is supported by two springs, $k_1$ and $k_2$. To analyze the system, we utilize the translational coordinate $x$ of point $C$ and the rotational coordinate $\theta$, which are commonly employed as generalized coordinates to describe the kinematics of the beam. The equations of motion and the mode shapes of the system are dependent on the chosen coordinates. By considering these coordinates, we can establish the equations that govern the motion of the system and determine the mode shapes associated with different vibrational patterns.

Employing $x$ and $\theta$ as the generalized coordinates, the energies of the beam for small motions provide us with the following free vibration equations of motion:

$$K = \frac{1}{2}m\dot{x}^2 + \frac{1}{2}I\dot{\theta}^2 \tag{13.11}$$

$$P = \frac{1}{2}k_2 (x + a_2\theta)^2 + \frac{1}{2}k_1 (x - a_1\theta)^2 \tag{13.12}$$

$$\begin{bmatrix} m & 0 \\ 0 & I \end{bmatrix} \begin{bmatrix} \ddot{x} \\ \ddot{\theta} \end{bmatrix} + \begin{bmatrix} k_1 + k_2 & a_2k_2 - a_1k_1 \\ a_2k_2 - a_1k_1 & a_2^2k_2 + a_1^2k_1 \end{bmatrix} \begin{bmatrix} x \\ \theta \end{bmatrix} = 0 \tag{13.13}$$

The dimension of them compares the mode shapes of the system practically, and we employ the coordinates $x_1$ and $x_2$ instead of $x$ and $\theta$, as shown in Fig. 13.2.

$$x = \frac{a_2x_1 + a_1x_2}{a_1 + a_2} \tag{13.14}$$

$$\theta = \frac{x_2 - x_1}{a_1 + a_2} \tag{13.15}$$

$$x_1 = x + a_1\theta \tag{13.16}$$

$$x_2 = x - a_2\theta \tag{13.17}$$

The equations of motion of the system in the new set of coordinates are

$$\begin{bmatrix} \dfrac{ma_2^2 + I}{(a_1 + a_2)^2} & \dfrac{ma_1a_2 - I}{(a_1 + a_2)^2} \\ \dfrac{ma_1a_2 - I}{(a_1 + a_2)^2} & \dfrac{ma_1^2 + I}{(a_1 + a_2)^2} \end{bmatrix} \begin{bmatrix} \ddot{x}_1 \\ \ddot{x}_2 \end{bmatrix} + \begin{bmatrix} k_1 & 0 \\ 0 & k_2 \end{bmatrix} \begin{bmatrix} x_1 \\ x_2 \end{bmatrix} = 0 \tag{13.18}$$

because

$$K = \frac{1}{2}m\dot{x}^2 \left(\frac{a_2\dot{x}_1 + a_1\dot{x}_2}{a_1 + a_2}\right)^2 + \frac{1}{2}I \left(\frac{\dot{x}_2 - \dot{x}_1}{a_1 + a_2}\right)^2 \tag{13.19}$$

$$P = \frac{1}{2}k_1x_1^2 + \frac{1}{2}k_2x_2^2 \tag{13.20}$$

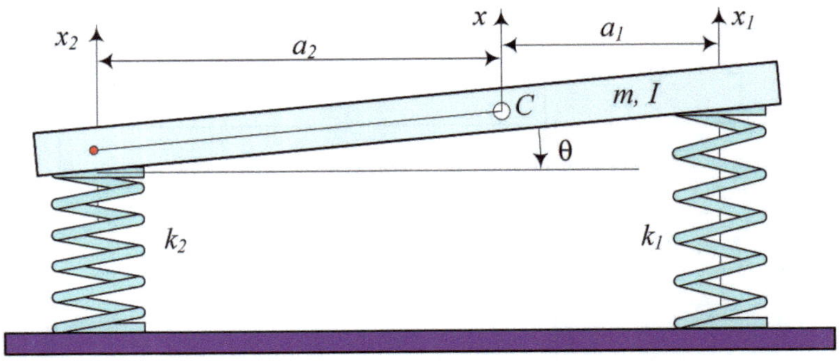


**Fig. 13.3** The first mode shape for $\alpha = 1$ shows that only the front suspension is oscillating and the node is on the rear suspension

Employing the parameters (13.3)–(13.9), we can rewrite the equations of motion.

$$\begin{bmatrix} \alpha+\gamma & 1-\alpha \\ 1-\alpha & \alpha+\frac{1}{\gamma} \end{bmatrix} \begin{bmatrix} \ddot{x}_1 \\ \ddot{x}_2 \end{bmatrix} + \begin{bmatrix} \Omega_1^2 & 0 \\ 0 & \Omega_2^2 \end{bmatrix} \begin{bmatrix} x_1 \\ x_2 \end{bmatrix} = 0 \tag{13.21}$$

The equations will be decoupled if $\alpha = 1$.

$$\alpha = 1 \tag{13.22}$$

$$\begin{bmatrix} 1+\gamma & 0 \\ 0 & 1+\dfrac{1}{\gamma} \end{bmatrix} \begin{bmatrix} \ddot{x}_1 \\ \ddot{x}_2 \end{bmatrix} + \begin{bmatrix} \Omega_1^2 & 0 \\ 0 & \Omega_2^2 \end{bmatrix} \begin{bmatrix} x_1 \\ x_2 \end{bmatrix} = 0 \tag{13.23}$$

$$\mathbf{M}\,\ddot{\mathbf{x}} + \mathbf{K}\,\mathbf{x} = 0 \tag{13.24}$$

$$\mathbf{x} = \begin{bmatrix} x_1 & x_2 \end{bmatrix}^T \tag{13.25}$$

■

**Example 176** The natural frequencies and mode shapes of decoupled bicycle car model. Every one of the mode shapes of decoupled bicycle car has a node. The nodes of mode shapes coincide with the rear and front axles of the car.

The natural frequencies $\omega_i$ and mode shapes $u_i$ of the decoupled bicycle system are calculated from eigenvalues and eigenvectors of the matrix $\mathbf{A} = \mathbf{M}^{-1}\,\mathbf{K}$.

$$\omega_1^2 = \frac{1}{\gamma+1}\Omega_1^2 = \frac{l}{a_2}\frac{k_1}{m} \qquad u_1 = \begin{bmatrix} 1 \\ 0 \end{bmatrix} \tag{13.26}$$

$$\omega_2^2 = \frac{\gamma}{\gamma+1}\Omega_2^2 = \frac{l}{a_1}\frac{k_2}{m} \qquad u_2 = \begin{bmatrix} 0 \\ 1 \end{bmatrix} \tag{13.27}$$

The mode shapes indicate that the nodes of oscillation in the first and second modes are at the rear and front suspensions, respectively. Figures 13.3 and 13.4 illustrate the mode shapes of the bicycle car model for $\alpha = 1$.

**Example 177** Illustration of decoupled front and rear sections of bicycle car. The decoupling condition, expressed as $r^2 = a_1 a_2$, can be interpreted as indicating that the bicycle car model can be viewed as comprising two distinct and independent parts: the front and rear sections.

The decoupling condition $\alpha = 1$ yields

$$r^2 = a_1 a_2 \tag{13.28}$$

which indicates that the pitch radius of gyration, $r$, must be equal to the multiplication of the distance of the mass canter $C$ from the front and rear axles. Therefore, by setting $\alpha = 1$, the nodes of the two modes of vibrations appear to be at the front

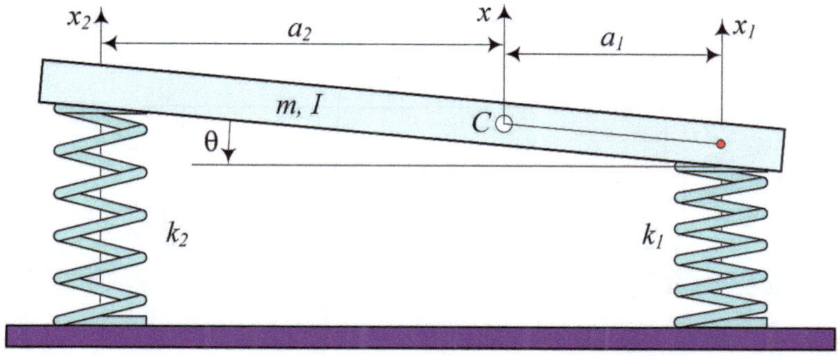


**Fig. 13.4** The second mode shape for $\alpha = 1$ shows that only the rear suspension is oscillating and the node is on the front suspension

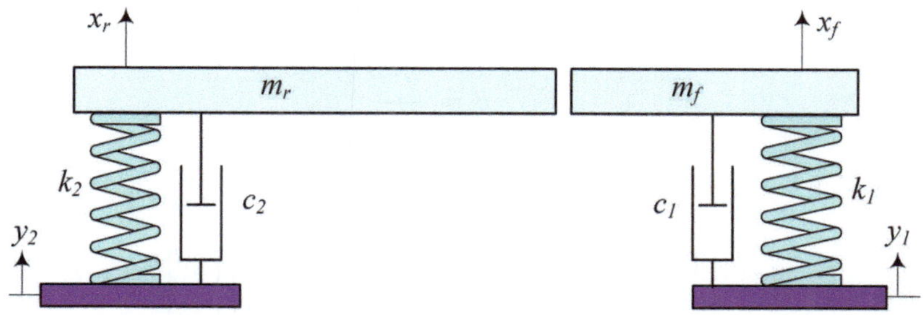


**Fig. 13.5** Car bicycle model after decoupling

and rear axles. As a result, the front wheel excitation will not alter the body at the rear axle and vice versa. For such a car, the front and rear parts of the car act independently. The decoupling condition $\alpha = 1$ allows us to break the initial two $DOF$ system into two independent one $DOF$ systems as illustrated in Fig. 13.5, where

$$m_r = m\frac{a_1}{l} = m\varepsilon \tag{13.29}$$

$$m_f = m\frac{a_2}{l} = m(1-\varepsilon) \tag{13.30}$$

$$\varepsilon = \frac{a_1}{l} \tag{13.31}$$

The equations of motion of the independent front and rear sections will be

$$m(1-\varepsilon)\ddot{x}_1 + c_1\dot{x}_1 + k_1x_1 = k_1y_1 + c_1\dot{y}_1 \tag{13.32}$$

$$m\varepsilon\ddot{x}_2 + c_2\dot{x}_2 + k_2x_2 = k_2y_2 + c_2\dot{y}_2 \tag{13.33}$$

The decoupling condition of the system will not necessarily decouple the general damped system. However, if there is no antipitch spring or antipitch damping to connect front and rear suspensions, then equations of motion of the damped system

$$\begin{aligned} &\begin{bmatrix} \alpha+\gamma & 1-\alpha \\ 1-\alpha & \alpha+\frac{1}{\gamma} \end{bmatrix} \begin{bmatrix} \ddot{x}_1 \\ \ddot{x}_2 \end{bmatrix} + \begin{bmatrix} 2\xi_1\Omega_1 & 0 \\ 0 & 2\xi_2\Omega_2 \end{bmatrix} \begin{bmatrix} \dot{x}_1 \\ \dot{x}_2 \end{bmatrix} \\ &+ \begin{bmatrix} \Omega_1^2 & 0 \\ 0 & \Omega_2^2 \end{bmatrix} \begin{bmatrix} x_1 \\ x_2 \end{bmatrix} \\ &= \begin{bmatrix} 2\xi_1\Omega_1 & 0 \\ 0 & 2\xi_2\Omega_2 \end{bmatrix} \begin{bmatrix} \dot{y}_1 \\ \dot{y}_2 \end{bmatrix} + \begin{bmatrix} \Omega_1^2 & 0 \\ 0 & \Omega_2^2 \end{bmatrix} \begin{bmatrix} y_1 \\ y_2 \end{bmatrix} \end{aligned} \tag{13.34}$$

$$2\xi_1\Omega_1 = \frac{c_1}{m}\beta \tag{13.35}$$

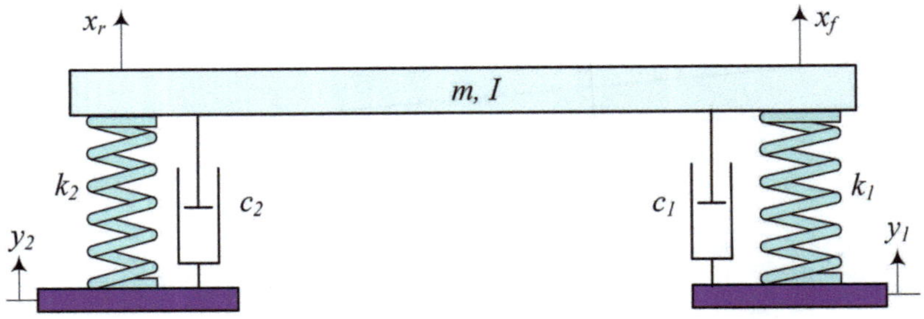


**Fig. 13.6** Bicycle car model on front and rear suspensions

$$2\xi_2\Omega_2 = \frac{c_2}{m}\beta \tag{13.36}$$

$$\beta = \frac{l^2}{a_1 a_2} \tag{13.37}$$

will also be decoupled for $\alpha = 1$.

$$\begin{bmatrix} 1+\gamma & 0 \\ 0 & 1+\frac{1}{\gamma} \end{bmatrix}\begin{bmatrix} \ddot{x}_1 \\ \ddot{x}_2 \end{bmatrix} + \begin{bmatrix} 2\xi_1\Omega_1 & 0 \\ 0 & 2\xi_2\Omega_2 \end{bmatrix}\begin{bmatrix} \dot{x}_1 \\ \dot{x}_2 \end{bmatrix} + \begin{bmatrix} \Omega_1^2 & 0 \\ 0 & \Omega_2^2 \end{bmatrix}\begin{bmatrix} x_1 \\ x_2 \end{bmatrix} = \begin{bmatrix} 2\xi_1\Omega_1 & 0 \\ 0 & 2\xi_2\Omega_2 \end{bmatrix}\begin{bmatrix} \dot{y}_1 \\ \dot{y}_2 \end{bmatrix} + \begin{bmatrix} \Omega_1^2 & 0 \\ 0 & \Omega_2^2 \end{bmatrix}\begin{bmatrix} y_1 \\ y_2 \end{bmatrix} \tag{13.38}$$

**Example 178** Numerical example and decoupling behavior. To compare the vibrating behavior of a vehicle with decoupling condition, we study a vehicle with conditions $\alpha > 1$, $\alpha = 1$, and $\alpha < 1$.

To examine the effect of the decoupling condition and having independent front and rear models of a car, let us compare the responses of the model of Fig. 13.6 using for different $\alpha$. The equations of motion of the system are

$$\begin{bmatrix} \frac{ma_2^2+I}{(a_1+a_2)^2} & \frac{ma_1a_2-I}{(a_1+a_2)^2} \\ \frac{ma_1a_2-I}{(a_1+a_2)^2} & \frac{ma_1^2+I}{(a_1+a_2)^2} \end{bmatrix}\begin{bmatrix} \ddot{x}_1 \\ \ddot{x}_2 \end{bmatrix} + \begin{bmatrix} k_1 & 0 \\ 0 & k_2 \end{bmatrix}\begin{bmatrix} x_1 \\ x_2 \end{bmatrix} + \begin{bmatrix} c_1 & 0 \\ 0 & c_2 \end{bmatrix}\begin{bmatrix} \dot{x}_1 \\ \dot{x}_2 \end{bmatrix} \tag{13.39}$$

$$= \begin{bmatrix} c_1 & 0 \\ 0 & c_2 \end{bmatrix}\begin{bmatrix} \dot{y}_1 \\ \dot{y}_2 \end{bmatrix} + \begin{bmatrix} k_1 & 0 \\ 0 & k_2 \end{bmatrix}\begin{bmatrix} y_1 \\ y_2 \end{bmatrix} \tag{13.40}$$

with after employing the parameters (13.3)–(13.9) will be converted into Eq. (13.34).

Consider a car with the following characteristics:

$$\begin{aligned} m &= 672\,\text{kg} \quad a_1 = 1.4\,\text{m} \quad a_2 = 1.47\,\text{m} \quad l = 2.87\,\text{m} \\ k_1 &= 22000\,\text{N/m} \quad k_2 = 25000\,\text{N/m} \\ c_1 &= 1630\,\text{Ns/m} \quad c_2 = 1900\,\text{N s/m} \end{aligned} \tag{13.41}$$

$$\beta = \frac{l^2}{a_1 a_2} = 4.00238 \qquad \gamma = \frac{a_2}{a_1} = 1.05 \tag{13.42}$$

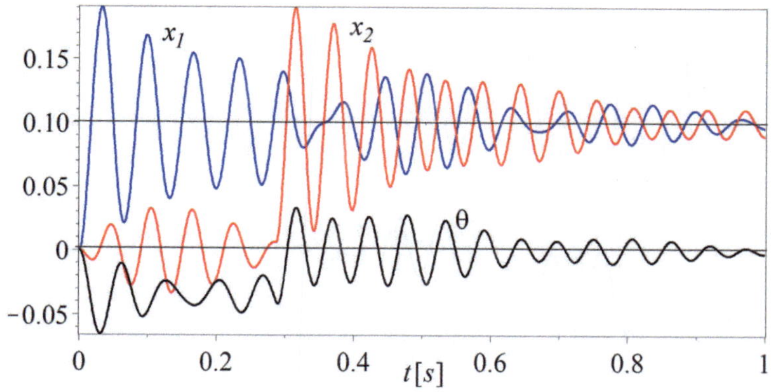


**Fig. 13.7** Case 1: Oscillations of the car with $\alpha < 1$

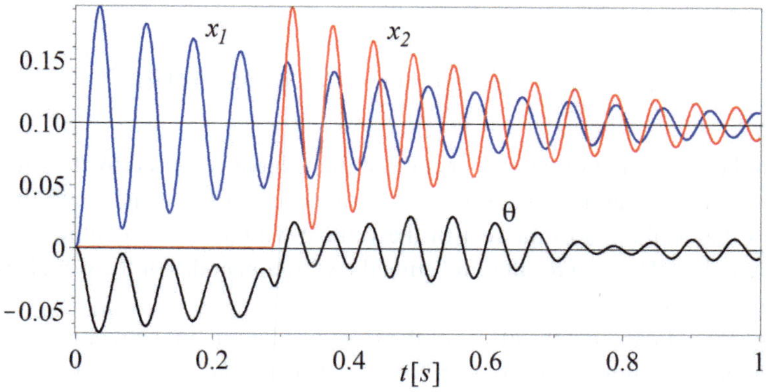


**Fig. 13.8** Case 2: Oscillations of the car with $\alpha = 1$

$$\Omega_1^2 = \frac{k_1}{m}\beta = 131.03 \qquad \Omega_2^2 = \frac{k_2}{m}\beta = 148.89 \tag{13.43}$$

$$\alpha = \frac{r^2}{a_1 a_2} \qquad I = mr^2 \tag{13.44}$$

The car goes over a step. The time lag between the rear and front wheels is expressed by the Heaviside function $H(x)$.

$$y_1 = 0.1\,\text{m} \qquad y_1 = 0.1H\left(t - \frac{l}{v}\right)\,\text{m} \tag{13.45}$$

$$v = 10\,\text{m/s} \tag{13.46}$$

$$H(t-\tau) = \begin{cases} 0 & t \le \tau \\ 1 & t > \tau \end{cases}$$

Let us examine the vibrating response of the car for different values of $\alpha$ which varies. Figure 13.7 depicts the oscillations of the front $x_1$, rear $x_2$, and pitch $\theta = (x_2 - x_1)/l$ for $\alpha = 0.8 < 1$. Figure 13.8 shows $x_1$, $x_2$, and $\theta$ for $\alpha = 1$, and Fig. 13.9 shows $x_1$, $x_2$, and $\theta$ for $\alpha = 1.2 > 1$.

**Example 179** Case 1: $\alpha < 1$

$$\alpha = 0.8 \qquad r = 1.283\,\text{m} \qquad I_y = 1106.38\,\text{kg}\,\text{m}^2 \tag{13.47}$$

Case 2: $\alpha = 1$

$$\alpha = 1 \qquad r = 1.434\,\text{m} \qquad I_y = 1382.97\,\text{kg}\,\text{m}^2 \tag{13.48}$$

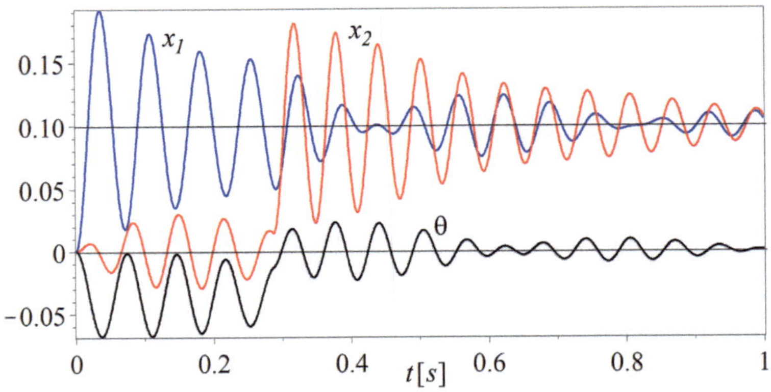


**Fig. 13.9** Case 3: Oscillations of the car with $\alpha > 1$

Case 3: $\alpha > 1$

$$\alpha = 1.2 \qquad r = 1.571\,\text{m} \qquad I_y = 1659.57\,\text{kg m}^2 \tag{13.49}$$

**Example 180** The delay in the oscillation of the rear section of the vehicle is caused by the time lag between the front and rear wheels hitting the step. The time lag is a function of the wheelbase of the vehicle and the traveling speed shown by $\tau = l/v$. Figure 13.8 shows the oscillation of the rear of the vehicle with $\alpha = 1$ and shows that the front wheel does not affect the oscillation of the rear part. That is a result of locating the vibration nodes of the vehicle at the rear and front wheels, respectively, the uncoupling condition.

However, Figs. 13.7 and 13.9 indicate that the effect of the front oscillation on the rear is observable in the form of small amplitude oscillations. Smaller pitch motion in Fig. 13.8 for $\alpha = 1$ makes it obvious that an uncoupled system provides a more comfortable ride compared to the pitch angle oscillation $\theta$, in Figs. 13.7 and 13.9.

## 13.2 Near Flat Ride Suspension

The optimal values of damping ratio $\xi$ and spring ratio $k$ to provide flat ride are

$$\xi = 1 \tag{13.50}$$

$$k = \frac{Z_1}{Z_2\tau^2 + Z_3\tau + Z_4} \tag{13.51}$$

$$Z_1 = -9\varepsilon\pi^2 m(-1 + \xi_1^2) \tag{13.52}$$

$$Z_2 = (\xi^2\xi_1^4 - \xi^2\xi_1^2 + 1 - \xi_1^2)k_1 \tag{13.53}$$

$$Z_3 = 6\pi m(\xi^2\xi_1^2 - \varepsilon\xi^2\xi_1^2 - 1 + \varepsilon)\sqrt{-\frac{k_1(1 - \xi_1^2)}{m(\varepsilon - 1)}} \tag{13.54}$$

$$Z_4 = -9\pi^2 m(\varepsilon\xi^2\xi_1^2 - \varepsilon + 1 - \xi^2\xi_1^2) \tag{13.55}$$

where parameters $k$ and $\xi$ are the ratio of the rear/front spring rates and damping ratios, respectively.

$$\xi = \frac{\xi_2}{\xi_1} = \frac{c_1}{c_2}\frac{\varepsilon}{1 - \varepsilon} \tag{13.56}$$

$$k = \frac{k_2}{k_1} = \frac{k_1}{k_2}\frac{\varepsilon}{1 - \varepsilon} \tag{13.57}$$

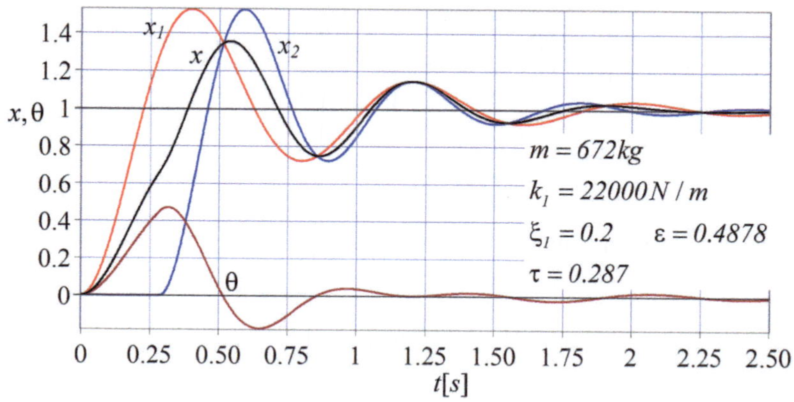


**Fig. 13.10** Response of the front and rear suspensions of a near at ride car to a unit step

$$\xi_1 = \frac{c_1}{m(1-\varepsilon)} \tag{13.58}$$

$$\xi_2 = \frac{c_2}{m\varepsilon} \tag{13.59}$$

$$\varepsilon = \frac{a_1}{l} \tag{13.60}$$

The flat ride optimal values of $k$ and $\xi$ are based on the equations of vibrating motion of the bicycle vehicle that can be written in the following:

$$\ddot{x}_1 + 2\xi_1\dot{x}_1 + \frac{k_1}{m(1-\varepsilon)}x_1 = 2\xi_1\dot{y}_1 + \frac{k_1}{m(1-\varepsilon)}y_1 \tag{13.61}$$

$$\ddot{x}_2 + 2\xi\xi_1\dot{x}_2 + \frac{kk_1}{m\varepsilon}x_2 = 2\xi\xi_1 y_2 + \frac{kk_1}{m\varepsilon}y_2 \tag{13.62}$$

***Proof*** Assuming decoupling condition $\alpha = 1$ is applied, the equations of motion of bicycle car are

$$m(1-\varepsilon)\ddot{x}_1 + c_1\dot{x}_1 + k_1x_1 = k_1y_1 + c_1\dot{y}_1 \tag{13.63}$$

$$m\varepsilon\ddot{x}_2 + c_2\dot{x}_2 + k_2x_2 = k_2y_2 + c_2\dot{y}_2 \tag{13.64}$$

$$\varepsilon = a_1/l \tag{13.65}$$

Dividing the first equation by $m\,(1-\varepsilon)$ and the second equation by $m\varepsilon$ provides us with

$$\ddot{x}_1 + 2\xi_1\dot{x}_1 + \frac{k_1}{m\,(1-\varepsilon)}x_1 = 2\xi_1\dot{y}_1 + \frac{k_1}{m\,(1-\varepsilon)}y_1 \tag{13.66}$$

$$\ddot{x}_2 + 2\xi\xi_1\dot{x}_2 + \frac{kk_1}{m\varepsilon}x_2 = 2\xi\xi_1 y_2 + \frac{kk_1}{m\varepsilon}y_2 \tag{13.67}$$

To achieve a flat ride, it is necessary that the rear section oscillates faster to catch up with the front system at a reasonable time. At that time both systems must be at the same amplitude.

To design the rear suspension parameters for a given set of front parameters to achieve a reasonable flat ride condition, we rephrase the problem by introducing a **near flat ride** condition. The rear suspension must have proper parameters and higher frequency to reach the same amplitude as the front suspension at a reasonable time before the oscillations die out. At that point in time, the vehicle will be level. Subsequently, the system's oscillation subsides before the pitch mode appears noticeably once more.

A near flat ride situation is shown in Fig. 13.10, for a car going over a unit step.

Solving the equations of motion (13.61) and (13.62) for a unit step input,

$$y_1 = \begin{cases} Y \ t > 0 \\ 0 \ t \leq 0 \end{cases} \qquad y_2 = \begin{cases} Y \ H\left(t - \frac{l}{v}\right) & t > 0 \\ 0 & t \leq 0 \end{cases} \tag{13.68}$$

$$H(t - \tau) = \begin{cases} 1 \ t > \tau \\ 0 \ t \leq \tau \end{cases} \qquad Y = 1 \quad \tau = \frac{l}{v} \tag{13.69}$$

where $H(t - \tau)$ is the Heaviside function, yields

$$x_1 = 1 - e^{\frac{-\xi_1 \omega t}{\sqrt{1-\varepsilon}}} (\cos \sqrt{\frac{1 - \xi_1^2}{1 - \varepsilon}} \omega t$$
$$+ \frac{\xi_1}{\sqrt{1 - \xi_1^2}} \sin \sqrt{\frac{1 - \xi_1^2}{1 - \varepsilon}} \omega t) \tag{13.70}$$
$$t \geq 0$$

$$x_2 = \begin{cases} 0 & t < \tau \\ 1 - e^{-\xi \xi_1 \sqrt{\frac{k}{\varepsilon}} \omega (t - \tau)} \left( \cos \sqrt{\frac{k(1 - \xi^2 \xi_1^2)}{\varepsilon}} \omega (t - \tau) \right. & \\ \left. + \frac{\xi \xi_1}{\sqrt{1 - \xi^2 \xi_1^2}} \sin \sqrt{\frac{k(1 - \xi^2 \xi_1^2)}{\varepsilon}} \omega (t - \tau) \right) & t \geq \tau \end{cases} \tag{13.71}$$

where

$$\omega = \sqrt{\frac{k_1}{m}} \tag{13.72}$$

Both solutions indicate damped fluctuating functions started at different times. Let us search for the conditions that both systems reach their third equal peak amplitude at the same time. The third peak means one and a half cycle of oscillations.

The third peak of $x_1$ happens at the time:

$$t_{p_1} = \frac{3\pi \sqrt{1 - \varepsilon}}{\omega \sqrt{1 - \xi_1^2}} \tag{13.73}$$

where its displacement is

$$x_1 = 1 + exp\left( \frac{-3\pi \xi_1}{\sqrt{1 - \xi_1^2}} \right) \tag{13.74}$$

The third peak of $x_2$ happens at

$$t_{p_2} = \tau + \frac{3\pi \sqrt{\frac{k}{\varepsilon}}}{\omega \sqrt{1 - \xi^2 \xi_1^2}} \tag{13.75}$$

where its displacement is

$$x_2 = 1 + exp(\frac{-3\pi \xi \xi_1}{\sqrt{1 - \xi^2 \xi_1^2}}) \tag{13.76}$$

The conditions that $x_1$ and $x_2$ meet after one and a half oscillations are

$$x_1 = x_2 \tag{13.77}$$

$$t_{p_1} = t_{p_2} \tag{13.78}$$

which yield

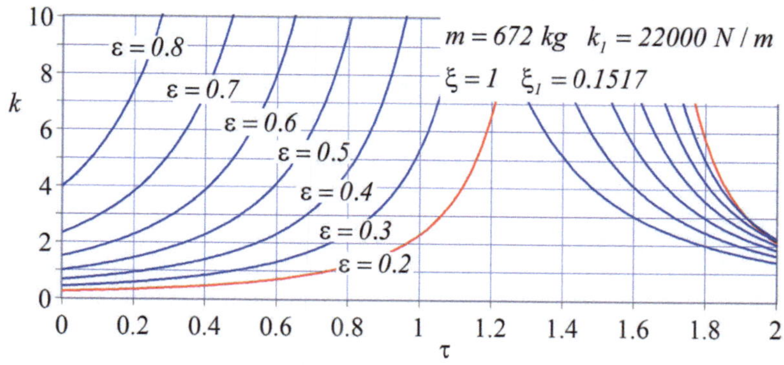


**Fig. 13.11** Illustrates the plot of $k$ versus $\tau$, for different $\varepsilon$ from 0.2 to 0.8

$$1 + exp\left(\frac{-3\pi\xi_1}{\sqrt{1-\xi_1^2}}\right) = 1 + exp\left(\frac{-3\pi\xi\xi_1}{\sqrt{1-\xi^2\xi_1^2}}\right) \tag{13.79}$$

$$\frac{3\pi\sqrt{1-\varepsilon}}{\omega\sqrt{1-\xi_1^2}} = \tau + \frac{3\pi\sqrt{\frac{k}{\varepsilon}}}{\omega\sqrt{1-\xi^2\xi_1^2}} \tag{13.80}$$

Equation (13.79) is independent of the spring ratio $k$ and can be solved for damping ratio $\xi$.

$$\xi = 1 \tag{13.81}$$

Therefore, if the damping coefficients of the front and rear suspensions are equal, it suggests that the displacements at the third peak of motion for both the front and rear sections of the car will be identical.

Solving Eq. (13.80) for spring ratio $k$ yields

$$k = \frac{Z_1}{Z_2\tau^2 + Z_3\tau + Z_4} \tag{13.82}$$

The time lag $\tau = l/v$ between the front and rear suspensions is a function of the wheelbase, $l$, and speed of the vehicle, $v$. Soon after the rear wheels hit a step, the vehicle is at the worst condition of pitching. Maurice Olley (1889 – 1983) experimentally determined a recommendation for the optimum frequency ratio of the front and rear sections of cars. The suggestion for American cars and roads of 1950$s$ was to have the natural frequency of the front approximately 80% of the natural frequency of the rear suspension. ■

**Example 181** Design chart for $k$. Equations (13.51)–(13.55) can be used to develop a design chart to determine $k$ for flat ride of a given car.

Using a set of nominal values,

$$\begin{aligned} &m = 672\,\text{kg} \quad a_1 = 1.4\,\text{m} \quad a_2 = 1.47\,\text{m} \quad l = 2.87\,\text{m} \\ &k_1 = 22000\,\text{N/m} \quad c_1 = 1630\,\text{N s/m} \end{aligned} \tag{13.83}$$

$$\xi_1 = \frac{c_1}{2}\sqrt{\frac{1-\varepsilon}{k_1 m}} = \frac{1630}{2}\sqrt{\frac{1-\frac{1.4}{2.87}}{22000\times 672}} = 0.1517 \tag{13.84}$$

Figure 13.11 illustrates the graph of $k$ of Eq. (13.51) versus $\tau$, for different values of $\varepsilon$. Using this graph, we can determine the value of spring ratio $k = k_2/k_1$ for a given $\varepsilon = a_1/l$ and $\tau = l/v$. The values of $\varepsilon$ and $l$ are fixed for a given vehicle, and hence, the graph determines the required $k$ at speed $v$.

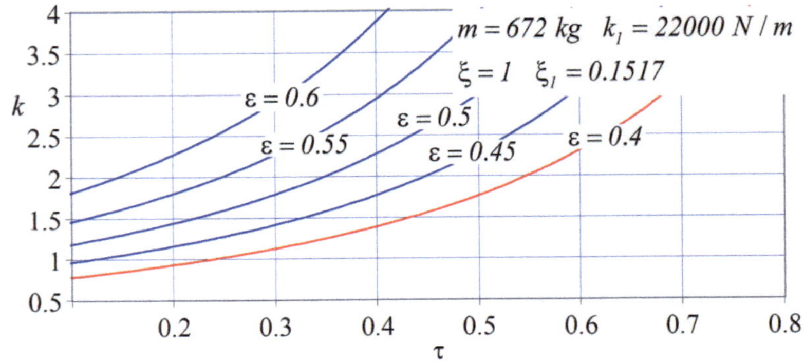


**Fig. 13.12** Plots of $k$ versus $\tau$ for $\xi_1 = 0.1517$ and different $\varepsilon$ in the domain $0.1 < \tau < 1$

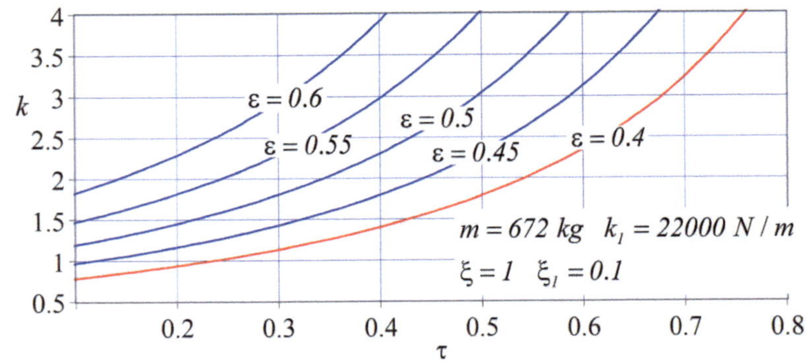


**Fig. 13.13** Plots of $k$ versus $\tau$ for $\xi_1 = 0.1$ and different $\varepsilon$ in the domain $0.1 < \tau < 1$

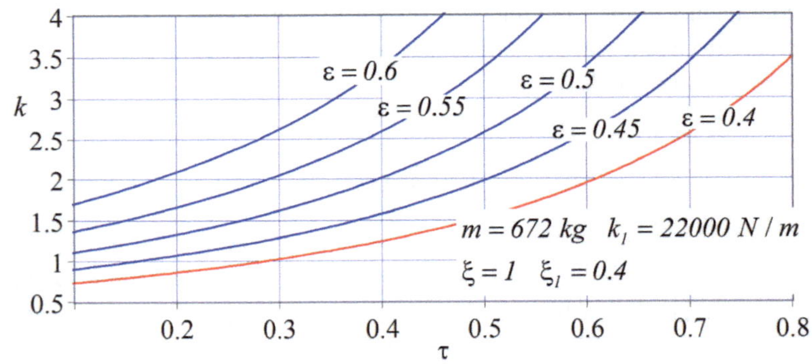


**Fig. 13.14** Plots of $k$ versus $\tau$ for $\xi_1 = 0.4$ and different $\varepsilon$ in the domain $0.1 < \tau < 1$

For the existing designs of normal street vehicles, only a very small section of the horizontal axis of Fig. 13.11 should be investigated. The minimum wheelbase for street cars is usually around 2400 millimeters (94.5 inches) for compact cars or smaller vehicles. On the other hand, the maximum wheelbase for street cars can go up to 4800 millimeters (189 inches) for large vehicles such as full-size sedans, SUVs, or trucks. Let us assume that the speed of a car, which goes over a step and is expected to show a flat ride, is between 4 m/s and 20 m/s. Therefore, the practical domain of the time lag between the front and rear wheels would roughly be $0.1 < \tau < 1.0$.

Furthermore, the mass centers of street cars are normally in the front half of the wheelbase in order to provide understeer condition. Considering $0.4 < \varepsilon < 0.6$ will cover all street and sports cars. Figure 13.12 shows how $k$ varies with $\tau$ for $\xi_1 = 0.1517$ and different $\varepsilon$ to provide a near flat ride. For any $\varepsilon$, the required stiffness ratio $k$ increases by increasing $\tau$. Increasing $\tau$ means reducing speed. Therefore, the ratio of rear to front stuffiness $k$ increases when the speed of the car decreases. It is because the frequency of the rear section must get higher to catch up with the oscillations of the front section.

In order to create comprehensive design charts that encompass a wide range of vehicles, we can generate several plots of $k$ versus $\tau$, utilizing different values of $\xi_1$ as the parameter while incorporating $\varepsilon$. Figures 13.13, 13.14, and 13.15 depict similar information of Fig. 13.12 for $\xi_1 = 0.1$, $\xi_1 = 0.4$, and $\xi_1 = 0.8$, respectively.

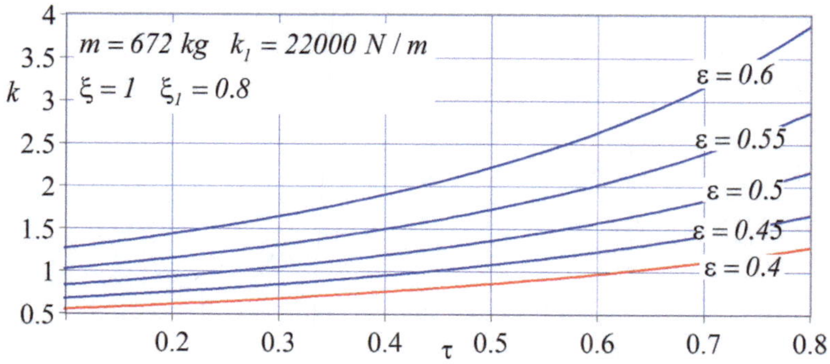


**Fig. 13.15** Plots of $k$ versus $\tau$ for $\xi_1 = 0.8$ and different $\varepsilon$ in the domain $0.1 < \tau < 1$

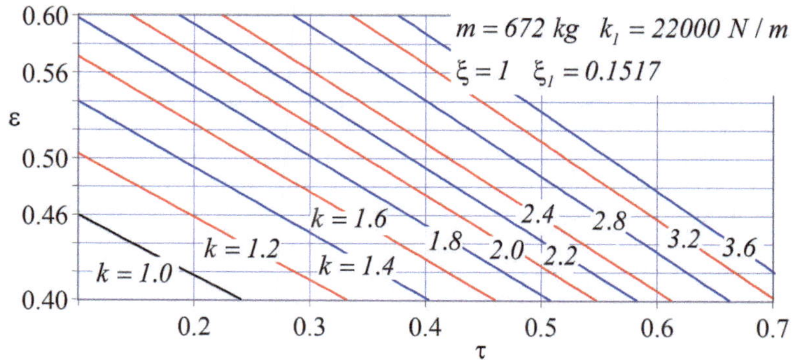


**Fig. 13.16** $\varepsilon$ versus $\tau$ for different spring ratios $k$

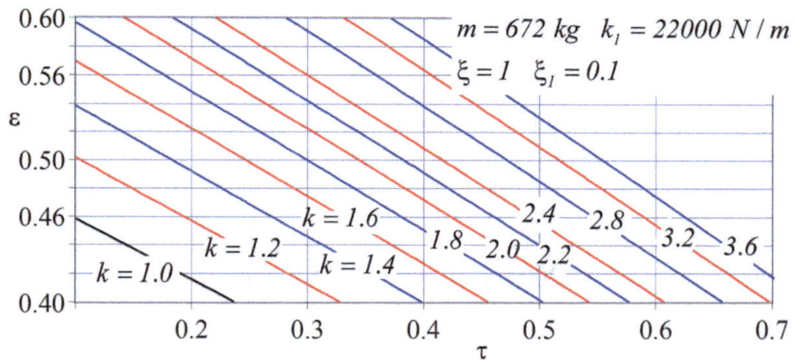


**Fig. 13.17** Near flat ride design chart of $\varepsilon$ versus $\tau$ for different spring ratios $k$ for $\xi_1 = 0.1$

**Example 182** Flat ride design chart in plane $(\varepsilon, \tau)$. The design chart for optimal $k$ to have flat ride is a function of $\varepsilon$ and $\tau$ for a given $\xi_1$. Therefore, we may plot two of the three parameters $k$, $\varepsilon$, $\tau$ in a plane and use the third one as a parameter. Here is the design chart in plane $(\varepsilon, \tau)$.

Figure 13.16 depicts the relationship between $\varepsilon$ and $\tau$ for various values of $k$. It demonstrates how the longitudinal mass center of a car should vary with velocity in order to achieve a nearly flat ride, assuming a constant value of $k$. The parameter $\varepsilon$ exhibits a decreasing trend with respect to $\tau$, indicating that as the car's speed increases, the mass center should move closer to the front axle.

Ideally, the curves in Fig. 13.16 should be horizontal, indicating a fixed mass center, in order to achieve a near flat ride at any speed. However, it is important to note that a perfectly flat ride is unattainable, and even achieving a near flat ride is not feasible with linear suspensions alone.

In order to create comprehensive design charts that encompass a wide range of vehicles, we generate several plots of $\varepsilon$ versus $\tau$ for different $\xi_1$ values, using $k$ as the parameter. Figures 13.17, 13.18, and 13.19 present analogous information to Fig. 13.16 but with varying values of $\xi_1$. The figures display the relationship between $\varepsilon$ and $\tau$ for $\xi_1 = 0.1$, $\xi_1 = 0.4$, and $\xi_1 = 0.8$, respectively.

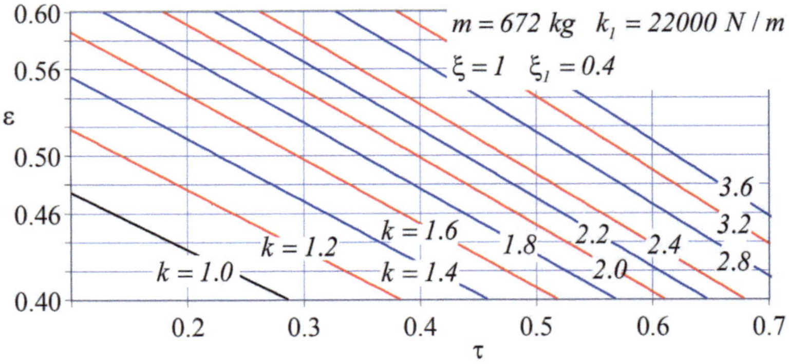


**Fig. 13.18** Near flat ride design chart of $\varepsilon$ versus $\tau$ for different spring ratios $k$ for $\xi_1 = 0.4$

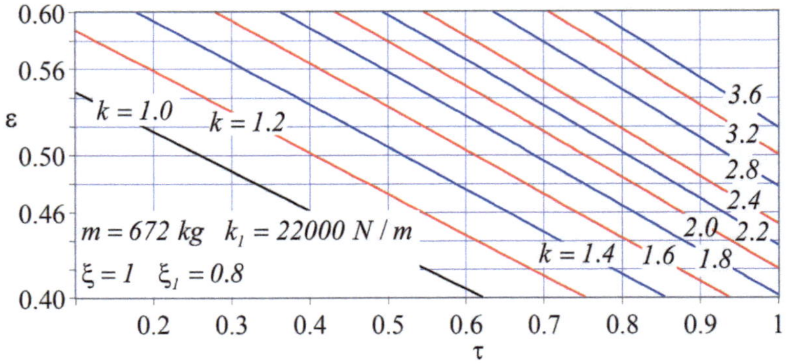


**Fig. 13.19** Near flat ride design chart of $\varepsilon$ versus $\tau$ for different spring ratios $k$ for $\xi_1 = 0.8$

**Example 183** There is no passive flat ride design. Smart active suspension is needed to provide a flat ride. Here is to illustrate how design charts will be used by a smart suspension system to provide a near flat ride.

$$m = 672\,\text{kg} \qquad a_1 = 1.4\,\text{m} \qquad l = 2.87\,\text{m}$$
$$k_1 = 22000\,\text{N/m} \qquad c_1 = 1630\,\text{N s/m} \tag{13.85}$$

$$\varepsilon = \frac{a_1}{l} = 0.4878 \qquad \xi_1 = \frac{c_1}{2}\sqrt{\frac{1-\varepsilon}{k_1 m}} = 0.1517 \tag{13.86}$$

Let us assume having a vehicle with smart rear suspension to change the stiffness of springs in response of traveling speed of the vehicle. As an example, the average length of a sedan vehicle is taken equal to $l = 2.87\,\text{m}$ with a normal weight distribution of a front differential vehicle 1.47/1.4 heavier at the front. Using the given information, some other values can be calculated as: $a_1 = 1470\,\text{mm}$ and $a_2 = 1470\,\text{mm}$, which yields to $\varepsilon = a_1/l = 0.4878$. Assuming that the front suspension of the car has a damping of $\xi_1 = 0.1517$, the diagram in Fig. 13.20 can be used by the smart suspension to provide a near flat ride at different speeds by switching the spring rate to the required value.

The horizontal line in Fig. 13.20 is showing the values that the spring should be switching to as the traveling speed of the vehicle changes. To use the same chart for vehicles with passive suspension, a target speed must be chosen by the designer as the ideal speed for the vehicle to have a near flat ride. The point in Fig. 13.20 is indicating that for a car with passive suspension, with a wheelbase of 2.87 m, and a traveling speed of 13.5 km/h, and $\tau = 0.212$ to get a near flat ride, a stiffness ratio of $k = 1.4$ is needed.

Figure 13.21 illustrates curves of variation $k$ for constant $\varepsilon$ as a function of speed. The line associated to $\varepsilon = 0.4878$ indicates how $k$ must change with speed of the vehicle.

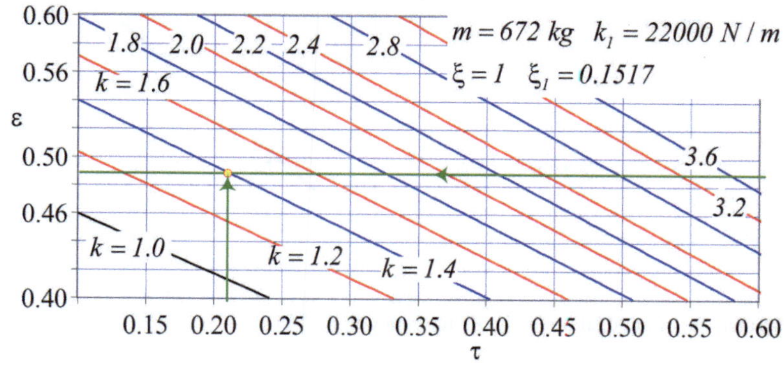


**Fig. 13.20** Design chart for a smart suspension with a linear damper

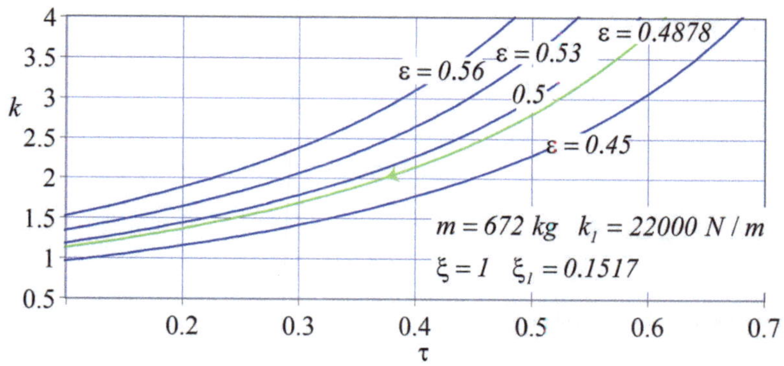


**Fig. 13.21** Illustration of variation $k$ for constant $\varepsilon$ as a function of time lag between front and rear axles to hit a step, $\tau = l/v$

## 13.3 Nonlinear Damper

In practical automotive applications, the dampers used typically exhibit low damping during compression (bound phase) and high damping during extension (rebound phase). To explore the effects of this extreme case, let us assume zero damping during the bound phase and analyze the behavior of a vehicle subjected to flat ride condition.

Employing a bilinear damper as depicted in Fig. 13.22, the damping of the damper will mathematically be expressed by the following equations:

$$c_1 = c_1 \, H(\dot{x}_1 - \dot{y}_1) \tag{13.87}$$

$$c_2 = c_2 \, H(\dot{x}_2 - \dot{y}_2) \tag{13.88}$$

$$H(t - t_0) = \begin{cases} 1 & t > t_0 \\ 0 & t < t_0 \end{cases} \tag{13.89}$$

and we are able to investigate the flat ride conditions for the extreme case of nonlinear dampers, using the following differential equations for the front and rear sections of the vehicle.

$$\ddot{x}_1 + 2\xi_1 (\dot{x}_1 - \dot{y}_1) \; H(\dot{x}_1 - \dot{y}_1) + \frac{\omega^2}{(1 - \varepsilon)} (x_1 - y_1) = 0 \tag{13.90}$$

$$\ddot{x}_2 + 2\xi\xi_1 (\dot{x}_2 - \dot{y}_2) \; H(\dot{x}_2 - \dot{y}_2) + \frac{k}{\varepsilon} \omega^2 (x_2 - y_2) = 0 \tag{13.91}$$

$$\omega = \sqrt{\frac{k_1}{m}} \tag{13.92}$$

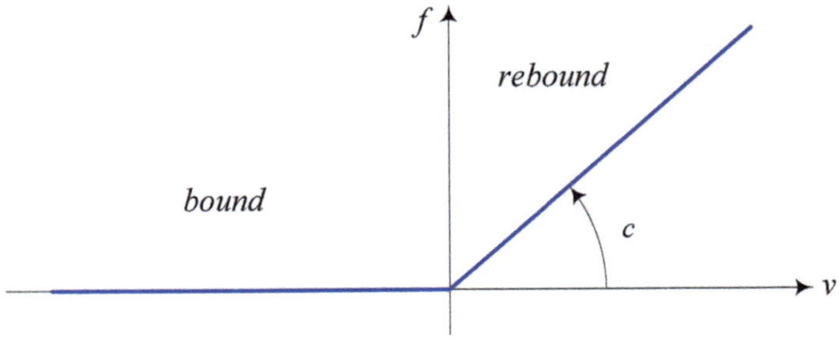


**Fig. 13.22** Force–velocity characteristic of an ideal nonlinear damper model

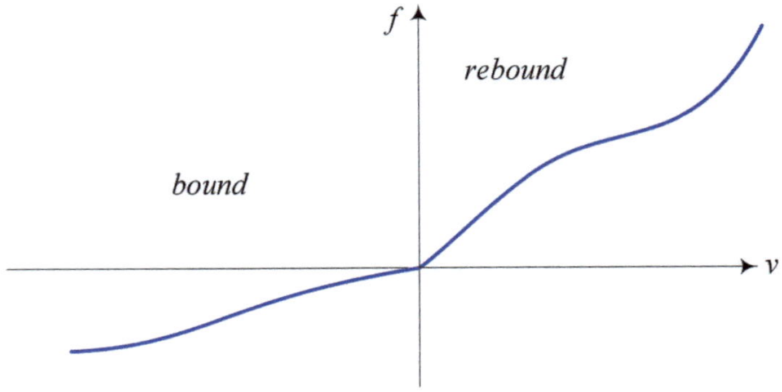


**Fig. 13.23** The force–velocity characteristics of a real damper

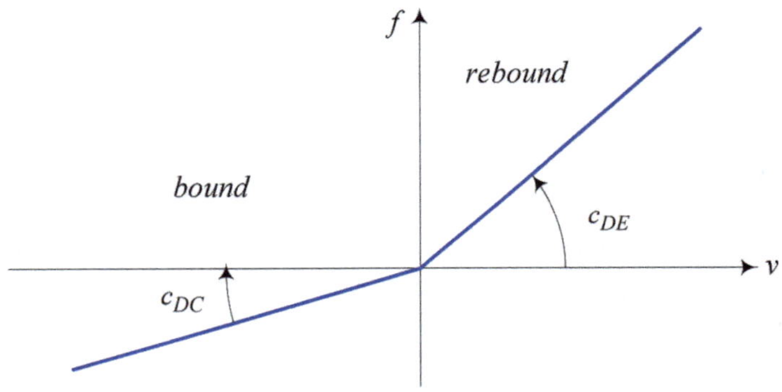


**Fig. 13.24** Force–velocity characteristics of a bilinear damper

***Proof*** The force–velocity characteristics of a real shock absorber can exhibit a high level of complexity, as illustrated in Fig. 13.23. While it is possible to approximate the intricate behavior using a mathematical function, performing analytical calculations can be challenging and may not provide substantial design insights.

The simplest linear viscous damper model is usually used for linear analytical calculation where $c$ is the damping coefficient of the damper and $f$ is the damper force.

$$f = c\, v \tag{13.93}$$

The bound and rebound forces of a practical damper exhibit asymmetry, meaning that the force–velocity characteristics diagram is not symmetrical. In practical terms, a shock absorber tends to compress more easily than it decompresses. This is due to the fact that during the rebound phase, where the damper extends back, it utilizes the stored energy in the spring. With high compression damping, there will not be enough spring compression to accumulate sufficient potential energy.

To achieve a more accurate and realistic analysis of dampers, it is recommended to utilize bilinear dampers. These dampers resemble linear dampers but have distinct coefficients for each direction. By employing such damper models, a more reliable and closer-to-reality response can be obtained. The behavior of these damper models is illustrated in Fig. 13.24 and expressed by

$$f = \begin{cases} c_E\, v & Extension \\ c_C\, v & Compression \end{cases} \tag{13.94}$$

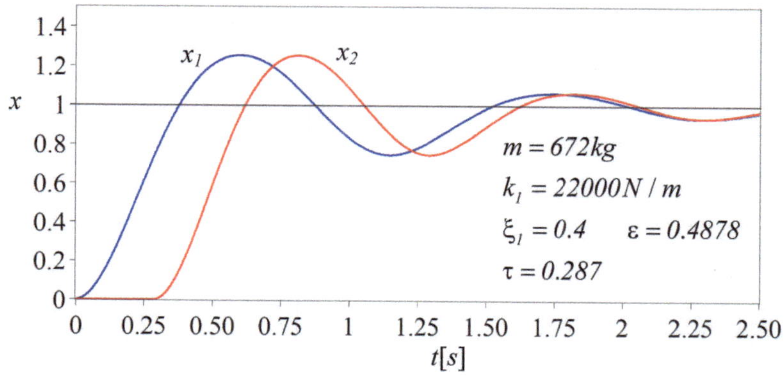


**Fig. 13.25** Response of the front and rear suspensions of a near flat ride vehicle with ideal bilinear damper to a unit step

where $c_E$ is the damping coefficient when damper is extended and $c_C$ is the damping coefficient when the damper is compressed.

An ideal dual behavior damper is characterized by its ability to exhibit no damping in compression, while effectively damping motion during extension.

$$c = c\, H\left(v\right) \tag{13.95}$$

Such dual ideal damping behavior is illustrated in Fig. 13.22.

Applying the decoupling condition

$$\alpha = \frac{r^2}{a_1 a_2} = 1 \tag{13.96}$$

the equations of motion of the front and rear sections of the vehicle are expressed by the following differential equations, where the parameters are defined in (13.56)–(13.60).

$$\ddot{x}_1 + 2\xi_1\left(\dot{x}_1 - \dot{y}_1\right) + \frac{\omega^2}{\left(1-\varepsilon\right)}\left(x_1 - y_1\right) = 0 \tag{13.97}$$

$$\ddot{x}_2 + 2\xi\xi_1\left(\dot{x}_2 - \dot{y}_2\right) + \frac{k}{\varepsilon}\omega^2\left(x_2 - y_2\right) = 0 \tag{13.98}$$

Substituting the linear damping with the bilinear model of Eqs. (13.87) and (13.88), the equations of motion will be converted into (13.90) and (13.91). ■

**Example 184** Linear and bilinear damping comparison. To show the effect of bilinear damping on flat ride tuning, we can solve the equations of motion for both models and analyze the results.

Applying a unit step and considering the time lag between the front and rear wheels, the road excitation under front and rear wheels will be expressed by the following functions:

$$y_1 = \begin{cases} Y & t > 0 \\ 0 & t \le 0 \end{cases} \qquad y_2 = \begin{cases} Y\, H\left(t - \frac{l}{v}\right) & t > 0 \\ 0 & t \le 0 \end{cases} \tag{13.99}$$

$$Y = 1 \qquad \tau = \frac{l}{v} \tag{13.100}$$

Figure 13.25 illustrates the behavior of the vehicle equipped with an ideal bilinear damper when going over a unit step. Figure 13.26 illustrates the behavior of the same vehicle if the damper was linear. Comparison of the behavior of the vehicle for linear and nonlinear dampers is plotted in Figs. 13.27 and 13.28 for front and rear suspensions, respectively. The linear damper dissipates energy in both, bound and rebound cycles, while the nonlinear damper dissipates energy only in rebound cycle. Therefore, the linear damper is more effective in energy dissipation and damps the system faster.

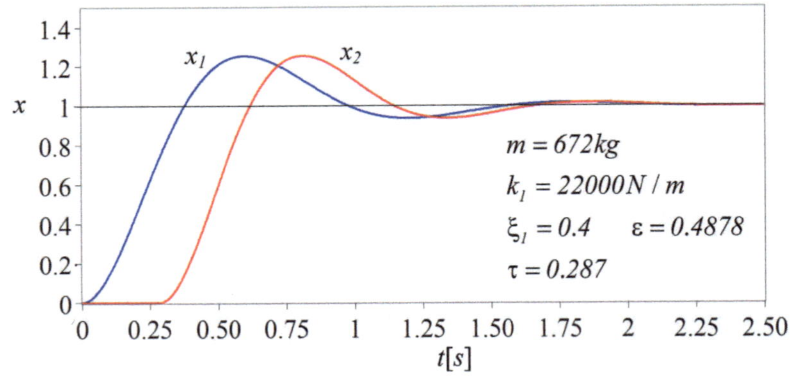


**Fig. 13.26** Response of the front and rear suspensions of a near flat ride vehicle with linear damper to a unit step

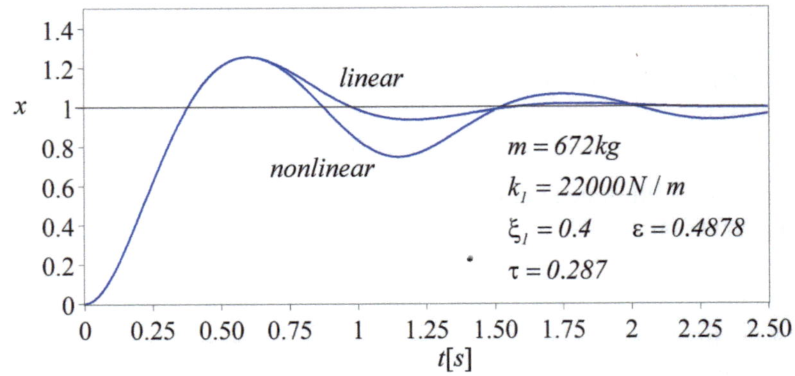


**Fig. 13.27** Comparison of the behavior of the front suspensions of the car for linear and nonlinear dampers

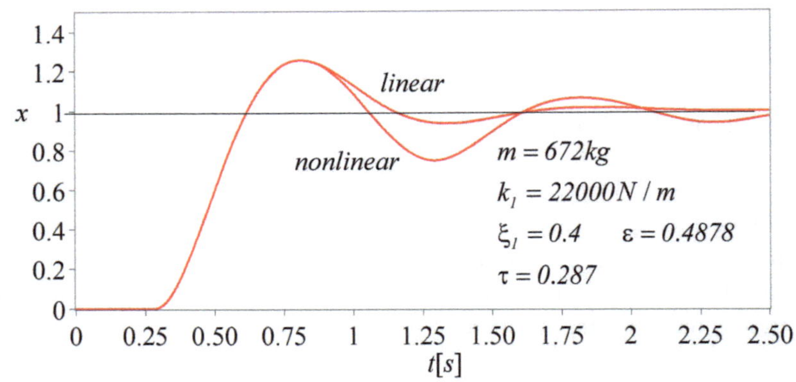


**Fig. 13.28** Comparison of the behavior of the rear suspensions of the car for linear and nonlinear dampers

**Example 185** Wrong and opposite damper. To compare the behavior of a near flat ride tuned vehicle with nonoptimal nonlinear damper, we also analyze the vibration behavior of a vehicle to dampers with zero resistance in rebound and constant resistance in bound cycles.

A wrong design of dampers is the nonlinear one shown in Fig. 13.29. The damper has a constant damping $c$ in bound and zero damping in rebound. The response of the vehicle to the opposite nonlinear damper is illustrated in Fig. 13.30. Comparison of the linear and nonlinear dampers for the front and rear suspensions is plotted in Figs. 13.31 and 13.32, respectively.

The high overshoot because of the wrong damper indicates how effective will be design of a proper optimal damper in dissipating energy while limiting the displacement of masses.

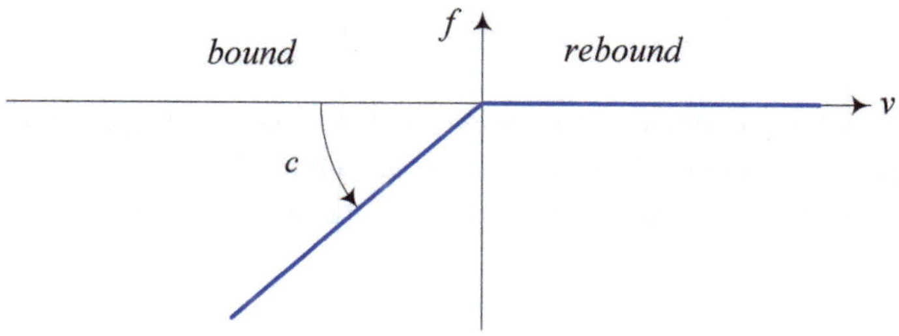


**Fig. 13.29** Force–velocity characteristic of an opposite cycle nonlinear damper model

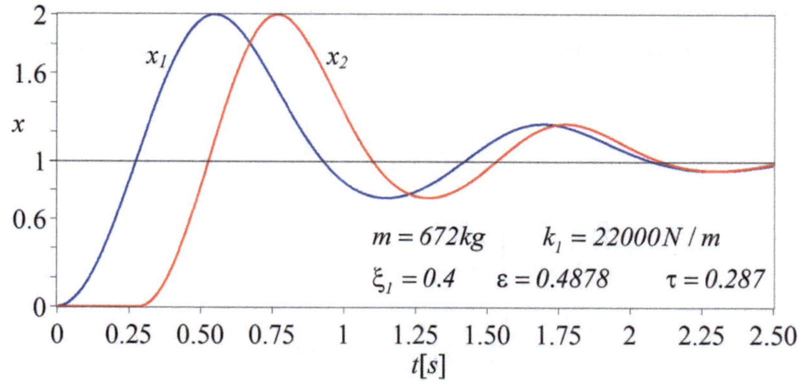


**Fig. 13.30** Response of the front and rear suspensions of a near flat ride car with ideal nonlinear damper to a unit step

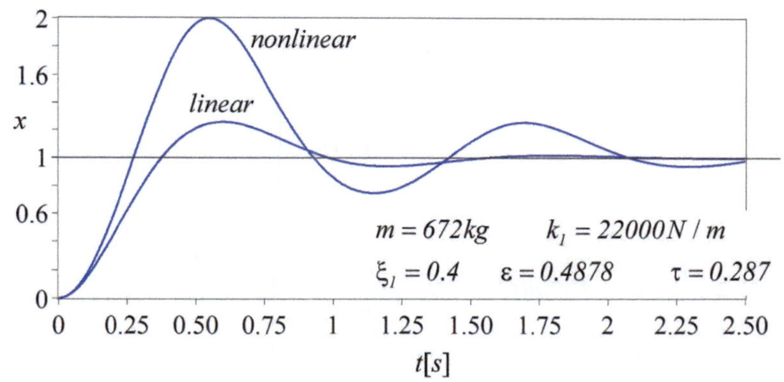


**Fig. 13.31** Comparison of the behavior of the front suspensions of the car for linear and opposite cycle nonlinear dampers

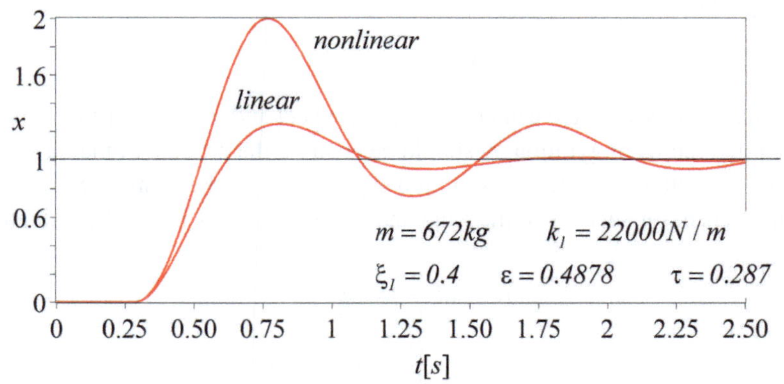


**Fig. 13.32** Comparison of the behavior of the rear suspensions of the car for linear and opposite cycle nonlinear dampers

## 13.4 Summary

Flat ride refers to transforming pitching oscillation of a vehicle into bouncing oscillation. Through experimentation, Maurice Olley identified and established two recommendations for achieving a flat ride:

1. The pitch radius of gyration should be equal to the product of the distances between the mass centers of the front and rear wheels of the car.

$$r^2 = a_1 a_2$$

This condition allows the car to be viewed as two uncoupled mass–spring systems of front and rear suspensions. This recommendation is analytically proven to be correct.

2. The rear suspension should have a stiffness approximately 20% higher than that of the front suspension. This recommendation is analytically proven in this chapter to be velocity dependent.

Modeling a vehicle as a beam on two suspensions of front and rear, the equations of motion can be converted into the following form:

$$\begin{aligned}&\begin{bmatrix}\alpha+\gamma & 1-\alpha\\ 1-\alpha & \alpha+\frac{1}{\gamma}\end{bmatrix}\begin{bmatrix}\ddot{x}_1\\ \ddot{x}_2\end{bmatrix}+\begin{bmatrix}2\xi_1\Omega_1 & 0\\ 0 & 2\xi_2\Omega_2\end{bmatrix}\begin{bmatrix}\dot{x}_1\\ \dot{x}_2\end{bmatrix}\\ &+\begin{bmatrix}\Omega_1^2 & 0\\ 0 & \Omega_2^2\end{bmatrix}\begin{bmatrix}x_1\\ x_2\end{bmatrix}\\ &=\begin{bmatrix}2\xi_1\Omega_1 & 0\\ 0 & 2\xi_2\Omega_2\end{bmatrix}\begin{bmatrix}\dot{y}_1\\ \dot{y}_2\end{bmatrix}+\begin{bmatrix}\Omega_1^2 & 0\\ 0 & \Omega_2^2\end{bmatrix}\begin{bmatrix}y_1\\ y_2\end{bmatrix}\end{aligned} \tag{13.101}$$

$$2\xi_1\Omega_1 = \frac{c_1}{m}\beta \tag{13.102}$$

$$2\xi_2\Omega_2 = \frac{c_2}{m}\beta \tag{13.103}$$

$$\beta = \frac{l^2}{a_1 a_2} \tag{13.104}$$

by defining these parameters:

$$I = mr^2 \qquad \alpha = \frac{r^2}{a_1 a_2} \qquad \beta = \frac{l^2}{a_1 a_2} \tag{13.105}$$

$$2\xi_1\Omega_1 = \frac{c_1}{m}\beta \qquad 2\xi_2\Omega_2 = \frac{c_2}{m}\beta \qquad \gamma = \frac{a_2}{a_1} \tag{13.106}$$

$$l = a_1 + a_2 \tag{13.107}$$

The equations will be decoupled if

$$\alpha = \frac{r^2}{a_1 a_2} = 1 \tag{13.108}$$

where $a_1$ is the distance of mass center $C$ to the center of the front wheel and $a_2$ is the distance of $C$ to the rear wheel. This condition specifies that the pitch radius of gyration, $r$, should be equal to the product of the distances from the mass center, $C$, to the front and rear axles. By setting $\alpha = 1$, the excitation experienced by the front wheel will not affect the body at the rear axle and vice versa. In essence, for such a car, the front and rear sections operate independently.

$$m_r = m\frac{a_1}{l} = m\varepsilon \tag{13.109}$$

$$m_f = m\frac{a_2}{l} = m(1-\varepsilon) \tag{13.110}$$

$$\varepsilon = \frac{a_1}{l} \tag{13.111}$$

The optimal values of damping ratio $\xi$ and spring ratio $k$ to provide a near flat ride are

$$\xi = 1 \tag{13.112}$$

$$k = \frac{Z_1}{Z_2\tau^2 + Z_3\tau + Z_4} \tag{13.113}$$

$$Z_1 = -9\varepsilon\pi^2 m(-1 + \xi_1^2) \tag{13.114}$$

$$Z_2 = (\xi^2\xi_1^4 - \xi^2\xi_1^2 + 1 - \xi_1^2)k_1 \tag{13.115}$$

$$Z_3 = 6\pi m(\xi^2\xi_1^2 - \varepsilon\xi^2\xi_1^2 - 1 + \varepsilon)\sqrt{-\frac{k_1(1 - \xi_1^2)}{m(\varepsilon - 1)}} \tag{13.116}$$

$$Z_4 = -9\pi^2 m(\varepsilon\xi^2\xi_1^2 - \varepsilon + 1 - \xi^2\xi_1^2) \tag{13.117}$$

where parameters $k$ and $\xi$ are the ratio of the rear/front spring rates and damping ratios, respectively.

$$\xi = \frac{\xi_2}{\xi_1} = \frac{c_1}{c_2}\frac{\varepsilon}{1 - \varepsilon} \tag{13.118}$$

$$k = \frac{k_2}{k_1} = \frac{k_1}{k_2}\frac{\varepsilon}{1 - \varepsilon} \tag{13.119}$$

$$\xi_1 = \frac{c_1}{m(1 - \varepsilon)} \qquad \xi_2 = \frac{c_2}{m\varepsilon} \tag{13.120}$$

## 13.5 Key Symbols

| | |
|---|---|
| $a_1$ | distance from the front wheel to the mass center C |
| $a_2$ | distance from the rear wheel to the mass center C |
| $c$ | damping coefficient |
| $c_E$ | damping coefficient in extension |
| $c_C$ | damping coefficient in compression |
| $C$ | mass center |
| $\mathbf{C}$ | damping matrix |
| $f, F, \mathbf{F}$ | force |
| $F_D$ | damping force |
| $H$ | Heaviside function |
| $I = mr^2$ | mass moment |
| $k = k_2/k_1$ | rear to front stiffness ratio |
| $k_r = k_1$ | rear spring rate |
| $k_f = k_2$ | front spring rate |
| $k_t, k_T$ | tire stiffness |
| $K$ | kinetic energy |
| $\mathbf{K}$ | stiffness matrix |
| $l = a_1 + a_2$ | wheelbase |
| $m$ | mass |
| $m_f = m_1$ | front mass section |
| $m_r = m_2$ | rear mass section |
| $\mathbf{M}$ | mass matrix |
| $P$ | potential energy |
| $r$ | radius of gyration |
| $t$ | time |

$t_{p_1}$ time of the third peak of the front section
$t_{p_2}$ time of the third peak of the rear section
$T$ period, time dimension
$\mathbf{u}_1$ first mode shape of the system
$\mathbf{u}_2$ second mode shape of the system
$\mathbf{u}$ mode shape vector
$v \equiv \dot{x}$ velocity
$x, y, z, \mathbf{x}$ displacement
$x_0$ initial displacement
$x_1$ front wheel displacement
$x_2$ rear wheel displacement
$\dot{x}_1$ vertical movement of the front wheel
$\dot{x}_2$ vertical movement of the rear wheel
$X$ steady-state amplitude of x
$X_1$ steady-state amplitude of $x_1$
$X_2$ steady-state amplitude of $x_2$
$\dot{x}, \dot{y}, \dot{z}$ velocity, time derivative of x,y,z
$\dot{x}_0$ initial velocity
$y$ road displacement
$y_1$ road input to the front wheel
$y_2$ road input to the rear wheel
$Y$ amplitude of road displacement
$z$ relative displacement
$Z$ short notation expression

Greek
$\alpha$ $= r^2/(a_1 a_2)$ nondimensional radius of gyration parameter
$\beta$ $= l^2/(a_1 a_2)$ Nondimensional wheelbase parameter
$\gamma$ $= a_2/a_1$ mass center distance ratio
$\theta$ pitch angle
$\delta_s$ static deflection
$\Omega_1$ $= \sqrt{k_1\beta/m}$ nondimensional frequency parameter
$\Omega_2$ $= \sqrt{k_2\beta/m}$ nondimensional frequency parameter
$\omega_n$ natural frequency
$\varepsilon$ $= a_1/l$ nondimensional length parameter
$\xi$ $= \xi_1/\xi_2$ front to rear damping ratio
$\xi_1$ $= c_1\beta/m$ front damping ratio
$\xi_2$ $= c_2\beta/m$ rear damping ratio
$\tau$ $= l/v$ time lag
$\omega$ $= 2\pi f$ angular frequency [rad/s]
$\omega_1$ $= \sqrt{\frac{l}{a_2}\frac{k_1}{m}}$ front natural frequency
$\omega_2$ $= \sqrt{\frac{l}{a_1}\frac{k_2}{m}}$ rear natural frequency

Subscript
$f$ front
$i \in N$ node number
$n$ natural
$P$ peak
$r$ rear
$y$ lateral axis

Symbols

$H$ Heaviside function

$\mathcal{L}$ Lagrangean

$DOF$ degree of freedom

$FBD$ free body diagram

## References


1. P. Barak, Magic numbers in design of suspensions for passenger cars, in *SAE Transactions* (1991), pp. 1698–1733
2. A. Best, Vehicle ride-stages in comprehension. Phys. Technol. **15**(4), 205 (2002)
3. D. Cao, S. Rakheja, C.Y. Su, Heavy vehicle pitch dynamics and suspension tuning. Part I: unconnected suspension. Veh. Syst. Dyn. **46**(10), 931–953 (2008)
4. D.A. Crolla, R.P. King, Olley's "Flat ride" revisited. Veh. Syst. Dyn. **33**(supl), 762–774 (1999)
5. L. Dai, R.N. Jazar (Eds.), *Nonlinear Approaches in Engineering Applications*. Springer, New York, Chapter 1: Smart Flat Ride Tuning (Springer, New York, 2013)
6. J. Dixon, *The Shock Absorber Handbook* (Wiley, New York, 2008)
7. R.N. Jazar, *Advanced Dynamics: Rigid Body, Multibody, and Aerospace Applications* (Wiley, New York, 2011)
8. R.N. Jazar, *Advanced Vibrations: A Modern Approach* (Springer, New York, 2013)
9. R.N. Jazar, *Vehicle Dynamics: Theory and Application*, 3rd ed. (Springer, New York, 2017)
10. R.N. Jazar, *Approximation Methods in Science and Engineering* (Springer, New York, 2020)
11. R.N. Jazar, *Perturbation Methods in Science and Engineering* (Springer, New York, 2021)
12. R.N. Jazar, *Advanced Vibrations: Theory and Application* (Springer, New York, 2023)
13. R.N. Jazar, R. Houim, A. Narimani, M.F. Golnaraghi, Frequency response and jump avoidance in a nonlinear passive engine mount. J. Vib. Control **12**(11), 1205–1237 (2006)
14. R.N. Jazar, M. Kazemi, S. Borhani, *Mechanical Vibrations* (in Persian) (Ettehad Publications, Tehran, 1992)
15. P. Lugner, *Horizontal Motion of Automobiles: Theoretical and Practical Investigations. Dynamics of High-Speed Vehicles* (Springer, Wien, 1982)
16. P. Lugner (Ed.), *Vehicle Dynamics of Modern Passenger Cars* (Springer, New York, 2019)
17. H. Marzbani, Flat ride; problems and solutions in vehicle. Nonlinear Eng. **1**(3–4), 101–108 (2012)
18. H. Marzbani, R.N. Jazar, Smart flat ride tuning, in *Nonlinear Approaches in Engineering Applications*, vol. 2. (Springer, New York, 2013), pp. 3–36
19. H. Marzbani, R.N. Jazar, M. Fard, Sustainable flat ride suspension design, in *Sustainable Automotive Technologies* 2014: Proceedings of the 6th ICSAT. (Springer, Berlin, 2015), pp. 41–56
20. H. Marzbani, D. Voquoc, R.N. Jazar, M. Fard, Vehicle flat ride dynamics, in *Intelligent Interactive Multimedia Systems and Services 2017* (pp. 278–288) (Springer, Cham, 2017)
21. L. Meirovitch, *Analytical Methods in Vibrations* (Macmillan, New York, 1967)
22. L. Meirovitch, *Principles and Techniques of Vibrations* (Prentice Hall, New Jersey, 1997)
23. L. Meirovitch, *Fundamentals of Vibrations* (McGraw-Hill, New York, 2002)
24. W.F. Milliken, D.L. Milliken, M. Olley, *Chassis Design* (Professional Engineering Publishing, 2002)
25. A. Odhams, D. Cebon, An analysis of ride coupling in automobile suspensions. Proc. Inst. Mech. Eng. D: J. Autom. Eng. **220**(8), 1041–61 (2006)
26. K. Ogata, *System Dynamics* (Prentice Hall, New Jersey, 2004)
27. M. Olley, 'Independent wheel suspensionâ€"its whys and wherefores'. Soc. Autom. Eng J. (1934)
28. M. Olley, National influences on American passenger car design. Proc. Inst. Autom. Eng. **32**(2), 509–72 (1938)
29. M. Olley, Road manners of the modern car. Proc. Inst. Autom. Eng. **41**(1), 147–82 (1946)
30. Z. Osinski, *Damping of Vibrations* (A. A. Balkema Publishers, Rotterdam, 1998)
31. W.J. Palm, *Mechanical Vibration* (John Wiley, New York, 2006)
32. C.R. Paton, Ride controls and calibration, in *SAE Transactions* (1938), pp. 313–324
33. H.S. Rowell, J.J. Guest, Proc. Inst. Automobile Engineers, 18 (1923), p. 455
34. S.S. Rao, *Mechanical Vibrations* (Pearson Prentice Hall, Harlow, 2018)
35. M. Roseau, *Vibrations in Mechanical Systems* (Springer, Berlin, 1987)
36. L. Segel, An overview of developments in road-vehicle dynamics: past, present and future, in *Proc. IMechE Conference on Vehicle Ride and Handling* (1993), pp. 1–12
37. R. Sharp, Wheelbase filtering and automobile suspension tuning for minimizing motions in pitch. Proc. Inst. Mech. Eng. D: J. Autom. Eng. **216**(12), 933–46 (2002)
38. M. Simic, Flat ride over the gravel road, in *2017 Australian and New Zealand Control Conference (ANZCC)* (IEEE, 2017), pp. 151–154
39. J.C. Snowdon, *Vibration and Shock in Damped Mechanical Systems* (John Wiley, New York, 1968)
40. T. Sun, Y. Zhang, P. Barak, 4-$DOF$ vehicle ride model (No. 2002-01-1580). SAE Technical Paper (2002)
41. S. Yang, Y. Lu, S. Li, An overview on vehicle dynamics. Int. J. Dyn. Control **1**, 385–395 (2013)

# 14 Tire–Road Separation

Tire–road separation is an infrequent and extraordinary occurrence that arises during a vehicle's ride and handling. It typically manifests as a transient phenomenon that quickly resolves itself. The intricate nature of this dynamic system necessitates the use of two sets of equations to simulate the conditions under which the tire is in contact with the road and when it is not. These equations must be configured to maintain consistent states during the transition between the two sets.

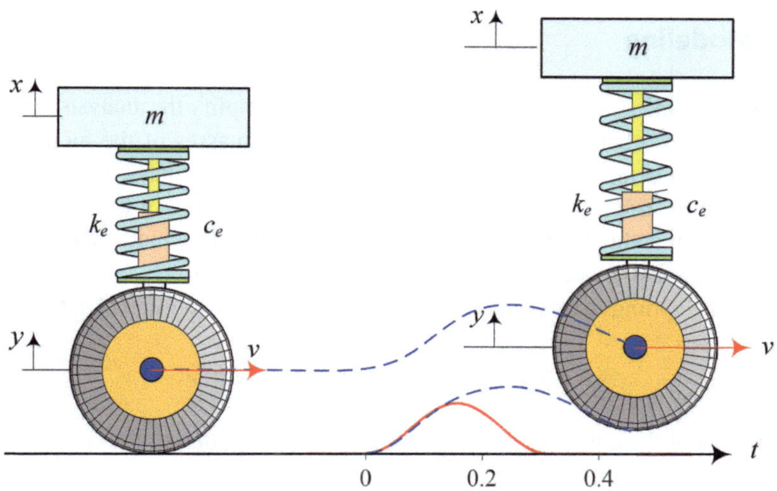


**Fig. 14.1** Illustration of tire–road separation after going over a bump

Figure 14.1 illustrates tire–road separation after going over a bump. Tire–road separation refers to the gap or space that happens between a vehicle's tires and the surface of the road. This separation is an essential aspect of vehicle dynamics and plays a crucial role in maintaining traction, ride, handling, and overall safety while driving. In the context of vehicle ride, tire–road separation refers to the phenomenon where the tire loses contact with the road surface momentarily. This separation can occur due to various factors, such as bumps or uneven road surfaces, that cause the tire to momentarily lift off the ground.

Tire–road separation during vehicle ride can have several effects.

The first effect has to do with suspension response. When a tire loses contact with the road, the suspension system plays a crucial role in absorbing the impact and mitigating the effects of the separation. The suspension components, such as springs and dampers, work to maintain stability and control during the transient period.

Ride comfort is also affected by tire–road separation, which can result in a jolt or sudden change in the vehicle's motion, leading to a temporary decrease in ride comfort. The abrupt loss of contact and subsequent reestablishment of contact with the road surface can transmit vibrations and shocks to vehicle occupants.

In addition, tire–road separation affects traction and handling performance. During separation, tires cannot provide the necessary grip and traction with the road surface. As a result, the vehicle's handling will be compromised momentarily until the tires regain contact with the road surface.



R. N. Jazar, H. Marzbani, *Vehicle Vibrations*, https://doi.org/10.1007/978-3-031-43486-0_14

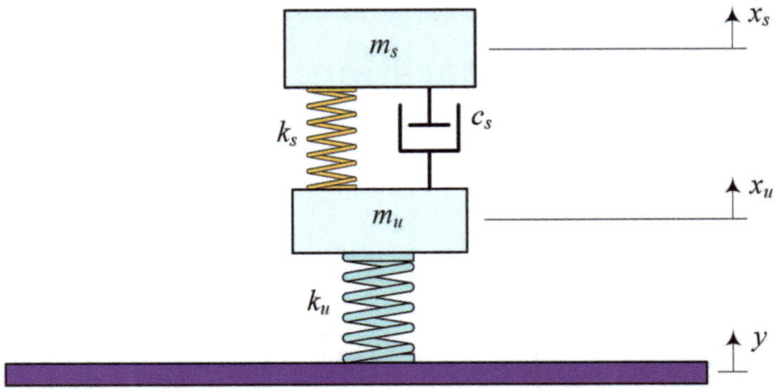


**Fig. 14.2** Quarter car model

Finally, maintaining tire–road contact is crucial for vehicle stability and safety. The friction between tires and the road provides the necessary forces for braking, accelerating, and cornering. Momentary separation of tires from the road can temporarily disrupt these forces, impacting vehicle stability and, potentially, safety.

Automotive engineers aim to design vehicles with suspension systems and tires with features that minimize the occurrences and effects of tire–road separation during normal travel conditions.

## 14.1 Mathematical Modeling

Figure 14.2 illustrates the mathematical model of a linear quarter car. To simplify the analysis, we ignore tire damping versus the damping of the main shock absorber of the suspension system. The masses of the model represented by $m_s$ and $m_u$ indicate sprung mass, which is a quarter of the mass of the body, and unsprung mass, which is the mass of one wheel and its components. The stiffness of the tire is $k_u$, and the stiffness of the main spring of the suspension system is $k_s$. The only damping mechanism of the system is the main shock absorber of the suspension, indicated by $c_s$. The displacements of the masses are indicated by $x_s$ and $x_u$, and the displaceent of the road is shown by $y$.

The equation of motion of the sprung mass is independent of tire–road separation and is always expressed as

$$X_s'' + 2\xi \left(X_s' - X_u'\right) + (X_s - X_u) = 0; \tag{14.1}$$

however, the equation of motion of the unsprung mass will be different, considering tire–road separation:

$$\begin{aligned} X_u'' + 2\varepsilon\xi \left(X_u' - X_s'\right) + \varepsilon (X_u - X_s) + \alpha^2 X_u = \alpha^2 Y \\ X_u - Y \geq 1, \end{aligned} \tag{14.2}$$

$$\begin{aligned} X_u'' + 2\varepsilon\xi \left(X_u' - X_s'\right) + \varepsilon (X_u - X_s) + \alpha^2 = 0 \\ X_u - Y < 1, \end{aligned} \tag{14.3}$$

where the following variables and parameters were used to make the equations dimensionless:

$$\begin{aligned} &\alpha = \frac{\omega_u}{\omega_s}, \qquad \xi = \frac{c_s}{2m_s\omega_s}, \qquad \varepsilon = \frac{m_s}{m_u}, \\ &\omega_s = \sqrt{\frac{k_s}{m_s}}, \qquad \omega_u = \sqrt{\frac{k_u}{m_u}}, \\ &x_0 = x_T = \frac{(m_u + m_s)\, g}{k_u}, \qquad t_0 = \frac{1}{\omega_s} \\ &Y = \frac{y}{x_0}, \qquad X_u = \frac{x_u}{x_0}, \qquad X_s = \frac{x_s}{x_0}, \qquad \tau = \frac{t}{t_0}, \\ &X_T = \frac{x_T}{x_0} = 1, \qquad X_R = \frac{x_R}{x_0} = n. \end{aligned} \tag{14.4}$$

***Proof*** Denoting by $x_T$ the compression displacement of the tire, the separation condition is

$$x_u - y \geq x_T \qquad separation, \tag{14.5}$$

$$x_u - y < x_T \qquad in\ contact. \tag{14.6}$$

The equation of motion of the sprung mass is independent of tire–road contact:

$$m_s \ddot{x}_s + c_s (\dot{x}_s - \dot{x}_u) + k_s (x_s - x_u) = 0. \tag{14.7}$$

However, the equation of motion of the unsprung mass depends on tire–road contact. In the contact state ($x_u - y < x_T$), we have

$$m_u \ddot{x}_u + c_s (\dot{x}_u - \dot{x}_s) + (k_u + k_s) x_u - k_s x_s = k_u y, \tag{14.8}$$

$$x_u - y < x_T, \tag{14.9}$$

where $y = y(t)$ is a given function. In a free-fall state ($x_u - y \geq x_T$), we have

$$m_u \ddot{x}_u + c_s (\dot{x}_u - \dot{x}_s) + k_s (x_u - x_s) + (m_u + m_s) g = 0, \tag{14.10}$$

$$x_u - y \geq x_T. \tag{14.11}$$

Two practical limitations regarding displacements need to be considered. The first constraint relates to the unsprung mass, which will not move downward beyond $n$ times, typically three to five times, of the static displacement of the tire $x_T$:

$$y - x_u \leq n\, x_T, \tag{14.12}$$

$$x_T = \frac{(m_u + m_s) g}{k_u}. \tag{14.13}$$

Second, there is a limit on wheel travel $x_R$, the relative motion of sprung and unsprung masses:

$$|x_u - x_s| \leq x_R, \tag{14.14}$$

$$x_R = n x_T = n \frac{(m_u + m_s) g}{k_u}. \tag{14.15}$$

Let us define the following characteristic parameters for displacement and time, to make the variables dimensionless:

$$x_0 = x_T = \frac{(m_u + m_s) g}{k_u}, \qquad t_0 = \frac{1}{\omega_s}, \tag{14.16}$$

$$X_u = \frac{x_u}{x_0}, \qquad X_s = \frac{x_s}{x_0}, \qquad \tau = \omega_s t. \tag{14.17}$$

Therefore, the derivatives will be

$$\dot{x}_u = \frac{dx_u}{dt} = \omega_s x_0 X'_u, \qquad X'_u = \frac{dX_u}{d\tau}, \tag{14.18}$$

$$\dot{x}_s = \frac{dx_s}{dt} = \omega_s x_0 X'_s, \qquad X'_s = \frac{dX_s}{d\tau}, \tag{14.19}$$

$$\ddot{x}_u = \frac{dX'_u}{dt} = \omega_s^2 x_0 X''_u, \qquad X''_u = \frac{d^2 X_u}{d^2\tau}, \tag{14.20}$$

$$\ddot{x}_s = \frac{dX'_s}{d\tau} = \omega_s^2 x_0 X''_s, \qquad X''_s = \frac{d^2 X_s}{d\tau^2}. \tag{14.21}$$

Following the introduction of the dimensionless variables of (14.4) the conditions of tire–road separation will be

$$\begin{aligned} X_u - Y &\geq 1 \quad separation, \\ X_u - Y &< 1 \quad in\ contact, \end{aligned} \tag{14.22}$$

and the condition preventing the relative motion of the unsprung mass and the ground will be

$$Y - X_u \leq n, \tag{14.23}$$

while the limit of wheel travel will be

$$|X_u - X_s| \leq X_R, \tag{14.24}$$

$$X_R = \frac{x_R}{x_0} = n\frac{x_T}{x_0}. \tag{14.25}$$

The dimensionless variables and parameters (14.4) provide us with a new set of equations of motion:

$$X''_s + 2\xi\left(X'_s - X'_u\right) + (X_s - X_u) = 0, \tag{14.26}$$

$$X''_u + 2\varepsilon\xi\left(X'_u - X'_s\right) + \varepsilon\left(X_u - X_s\right) + \alpha^2 X_u = \alpha^2 Y, \tag{14.27}$$

$$X_u - Y \geq 1,$$

$$X''_s + 2\xi\left(X'_s - X'_u\right) + (X_s - X_u) = 0, \tag{14.28}$$

$$X''_u + 2\varepsilon\xi\left(X'_u - X'_s\right) + \varepsilon\left(X_u - X_s\right) + \alpha^2 = 0, \tag{14.29}$$

$$X_u - Y < 1.$$

■

**Example 186** Quarter car model and tire–road separation. Historically, the linear quarter car model, assuming no tire–road separation, has been extensively used to determine natural frequencies and mode shapes and optimize suspension parameters. Researchers have employed various objective functions and algorithms for optimization purposes. However, all time and frequency response analyses, as well as optimization objective functions, are based on permanent tire–road contact. Here we investigate the vertical vibration behavior of vehicles in the event of tire–road separation.

A two-degree-of-freedom ($DOF$) quarter car model is used to analyze the vertical vibrations of vehicles. The quarter car model captures the fundamental aspects of the problem and effectively represents variations in wheel and wheel–body loads. Therefore, it serves as a valuable tool for studying the ride behavior of vehicles and simulating suspension functionality. The quarter car model is represented in the form of a model of two masses connected to each other through a spring and a damper and connected to the ground through a spring. Suspension springs are characterized by a suspension equivalent stiffness. The equivalent stiffness includes the stiffness of springs, rubber mounts, and compliance in linkages. A damper is characterized by its viscous damping coefficient, so the force across the linear damper is proportional to its velocity, in which the damper could be expanded or contracted.

The assumption that the tire is always in contact with the ground is true at low frequencies and low amplitudes but not at high frequencies and high amplitudes. The solution for the separation period by a differential equation system is complex and requires a numerical simulation to evaluate the response. Once loss of contact happens, the system will return to the road only as a result of gravity. Fly times of the wheel have been measured to validate the quality of the suspension system, where shorter fly times represent better accuracy and increase the control over road perturbations. Free flight will occur if the unsprung mass is displaced a distance greater than the relaxed radial dimension of a tire. Therefore, the suspension system has only two states, contact and free fall. The two states are different physically and mathematically, so the differential equations describing a system's motion would be discontinuous or piecewise. It has also been suggested that tires act as springs in the compression state, but they cannot be extended anymore as there are no physical mechanisms by which the ground could pull the tires.

**Example 187** Nonlinear damper. Shock absobers that we use in the automotive field have very low damping in bound and a high damping in rebound. Here are the equations of motion of a quarter car model with tire–road separation using bilinear dampers.

Employing $x_s$, $x_u$, and $y$ to denote the displacement of the vehicle body, wheel movement, and road displacement respectively, we define a bilinear damping by a Heaviside function $H(\dot{x}_s - \dot{x}_u)$ and rewrite the equation of motion of the quarter vehicle model. The equation of motion of the unsprung mass is not affected by tire–road separation. Considering nonlinear damping, its equation of motion will be

$$m_s\ddot{x}_s + c_s(\dot{x}_s - \dot{x}_u) H(\dot{x}_s - \dot{x}_u) + k_s(x_s - x_u) = 0. \tag{14.30}$$

The equation of motion of the unsprung mass depends on tire–road contact. In the contact state ($x_u - y < x_T$), we have

$$m_u\ddot{x}_u + c_s(\dot{x}_u - \dot{x}_s) H(\dot{x}_s - \dot{x}_u),$$

$$+ (k_u + k_s) x_u - k_s x_s = k_u y, \tag{14.31}$$

$$x_u - y < x_T. \tag{14.32}$$

In free-fall state ($x_u - y \geq x_T$), we have

$$m_u\ddot{x}_u + c_s(\dot{x}_u - \dot{x}_s) H(\dot{x}_s - \dot{x}_u),$$

$$+ k_s(x_u - x_s) + (m_u + m_s) g = 0, \tag{14.33}$$

$$x_u - y \geq x_T. \tag{14.34}$$

## 14.2 Linear Frequency Response

In frequency response analysis, we assume that the road excitation $y$ is a single frequency harmonic input, $y = Y\cos\omega t$. While the excitation amplitude $Y$ and frequency $\omega$ are low, no separation happens. In conducting a frequency response analysis of a quarter car, it will be possible to determine analytically the boundary of contact and separation in the plane of $Y$ and $\omega$ for a given set of system parameters. The boundary curve separates the plane into two regions. The goal is to determine the effect of any parameter of the system on the location of the separation curve, trying to increase the ratio of the contact/separation areas. In the separation area of the frequency response plane, we can define a variable to indicate the fraction of time that the system is in free fall to the time the tire is in touch with the ground in a cycle. The separation area of the plane of $Y$ and $\omega$ will then be divided into different fly time fraction areas.

### 14.2.1 In-Contact Dynamics

Assuming the road excitation is a harmonic function of time at frequency $\omega$,

$$y = y_0\cos\omega t, \tag{14.35}$$

$$Y = \frac{y_0 \cos \omega t}{x_0} = Y_0 \cos r\tau, \tag{14.36}$$

the frequency response of the quarter car system in dimensionless variables will be

$$X_s^2 = A_1^2 + B_1^2 = \frac{4r^2\xi^2 + 1}{Z_1^2 + Z_2^2}\alpha^4 Y_0^2, \tag{14.37}$$

$$X_u^2 = A_2^2 + B_2^2 = \frac{4\xi^2 r^2 + r^2\left(r^2 - 2\right) + 1}{Z_1^2 + Z_2^2}\alpha^4 Y_0^2, \tag{14.38}$$

where

$$Z_1 = r^2\left(r^2 - \alpha^2\right) + \left(\alpha^2 - (1+\varepsilon)\, r^2\right), \tag{14.39}$$

$$Z_2 = 2r\xi\left(\alpha^2 - (1+\varepsilon)\, r^2\right), \tag{14.40}$$

$$\begin{aligned}
&\alpha = \frac{\omega_u}{\omega_s}, \qquad r = \frac{\omega}{\omega_s}, \qquad \xi = \frac{c_s}{2m_s\omega_s}, \qquad \varepsilon = \frac{m_s}{m_u}, \\
&Y = \frac{y}{x_0}, \qquad \omega_s = \sqrt{\frac{k_s}{m_s}}, \qquad \omega_u = \sqrt{\frac{k_u}{m_u}}, \\
&X_T = \frac{x_T}{x_0} = 1, \qquad X_R = \frac{x_R}{x_0} = n, \\
&x_0 = x_T = \frac{(m_u + m_s)\, g}{k_u}, \qquad t_0 = \frac{1}{\omega_s}.
\end{aligned} \tag{14.41}$$

***Proof*** Assuming a harmonic solution for $X_s$ and $X_u$,

$$X_s = A_1 \sin r\tau + B_1 \cos r\tau, \tag{14.42}$$

$$X_u = A_2 \sin r\tau + B_2 \cos r\tau, \tag{14.43}$$

$$X_s' = A_1 r \cos r\tau - B_1 r \sin r\tau, \tag{14.44}$$

$$X_u' = A_2 r \cos r\tau - B_2 r \sin r\tau, \tag{14.45}$$

$$X_s'' = -A_1 r^2 \sin r\tau - B_1 r^2 \cos r\tau, \tag{14.46}$$

$$X_u'' = -A_2 r^2 \sin r\tau - B_2 r^2 \cos r\tau, \tag{14.47}$$

and substituting in the in-contact equations of motion provides us with a set of two algebraic equations:

$$X_s'' + 2\xi\left(X_s' - X_u'\right) + (X_s - X_u) = 0, \tag{14.48}$$

$$X_u'' + 2\varepsilon\xi\left(X_u' - X_s'\right) + \varepsilon\,(X_u - X_s) + \alpha^2 X_u = \alpha^2 Y_0 \cos r\tau, \tag{14.49}$$

$$\begin{aligned}
&\left(A_1 - A_2 - 2\xi\,(rB_1 - rB_2) - r^2 A_1\right)\sin r\tau \\
&\quad + \left(B_1 - B_2 + 2\xi\,(rA_1 - rA_2) - r^2 B_1\right)\cos r\tau = 0,
\end{aligned} \tag{14.50}$$

$$\begin{aligned}
&\left(\alpha^2 A_2 - r^2 A_2 - \varepsilon\,(A_1 - A_2) + 2\xi\varepsilon\,(rB_1 - rB_2)\right)\sin r\tau \\
&\quad + \left(\alpha^2 B_2 - r^2 B_2 - \varepsilon\,(B_1 - B_2) - 2\xi\varepsilon\,(rA_1 - rA_2)\right)\cos r\tau = \alpha^2 Y_0 \cos r\tau.
\end{aligned} \tag{14.51}$$

A harmonic balance of the equations makes four algebraic equations to determine the unknown coefficients of the coefficients $A_1$, $B_1$, $A_2$, and $B_2$:

$$[A]\begin{bmatrix} A_1 \\ B_1 \\ A_2 \\ B_2 \end{bmatrix} = \begin{bmatrix} 0 \\ 0 \\ 0 \\ \alpha^2 Y_0 \end{bmatrix}, \tag{14.52}$$

$$[A] = \begin{bmatrix} 1-r^2 & -2r\xi & -1 & 2r\xi \\ 2r\xi & 1-r^2 & -2r\xi & -1 \\ -\varepsilon & 2r\xi\varepsilon & \alpha^2-r^2+\varepsilon & -2r\xi\varepsilon \\ -2r\xi\varepsilon & -\varepsilon & 2r\xi\varepsilon & \alpha^2-r^2+\varepsilon \end{bmatrix}. \tag{14.53}$$

The solutions are as follows:

$$A_1 = \frac{2r^3\xi\alpha^2\left(\alpha^2-r^2\right)}{Z_1^2+Z_2^2}Y_0, \tag{14.54}$$

$$B_1 = \frac{r^2\left(r^2-\alpha^2\right)-\left(4r^2\xi^2+1\right)\left(r^2\left(1+\varepsilon\right)-\alpha^2\right)}{Z_1^2+Z_2^2}\alpha^2 Y_0, \tag{14.55}$$

$$A_2 = \frac{2r^5\alpha^2\xi\varepsilon}{Z_1^2+Z_2^2}Y_0, \tag{14.56}$$

$$B_2 = \frac{r^2\left(r^2\left(\alpha^2-r^2+\varepsilon+2\right)-2\alpha^2\right)}{Z_1^2+Z_2^2}\alpha^2 Y_0 - \frac{\left(4r^2\xi^2+1\right)\left(r^2\left(1+\varepsilon\right)-\alpha^2\right)}{Z_1^2+Z_2^2}\alpha^2 Y_0, \tag{14.57}$$

which cause the frequency responses of the system $X_s$ and $X_u$ to be

$$X_s = \sqrt{A_1^2+B_1^2}, \tag{14.58}$$

$$X_u = \sqrt{A_2^2+B_2^2}, \tag{14.59}$$

$$X_s^2 = A_1^2+B_1^2 = \frac{4r^2\xi^2+1}{Z_1^2+Z_2^2}\alpha^4 Y_0^2, \tag{14.60}$$

$$X_u^2 = A_2^2+B_2^2 = \frac{4\xi^2r^2+r^2\left(r^2-2\right)+1}{Z_1^2+Z_2^2}\alpha^4 Y_0^2, \tag{14.61}$$

$$Z_1 = r^2\left(r^2-\alpha^2\right)+\left(\alpha^2-\left(1+\varepsilon\right)r^2\right), \tag{14.62}$$

$$Z_2 = 2r\xi\left(\alpha^2-\left(1+\varepsilon\right)r^2\right). \tag{14.63}$$

The linearity of the system caused the frequency response equations to be proportional to the dimensionless amplitude $Y_0$. Therefore, we may draw the in-contact frequency responses using new dimensionless variables $X_s/Y_0$ and $X_u/Y_0$:

$$\frac{X_s}{Y_0} = \frac{x_s}{y_0}, \qquad \frac{X_u}{Y_0} = \frac{x_u}{y_0}. \tag{14.64}$$

■

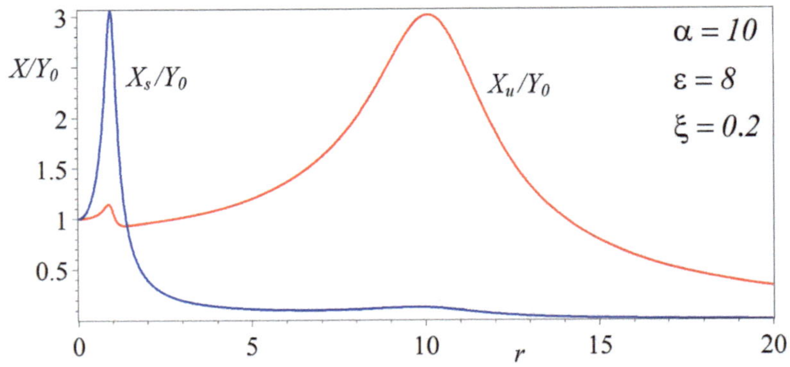


**Fig. 14.3** Frequency responses of system for a set of nominal values

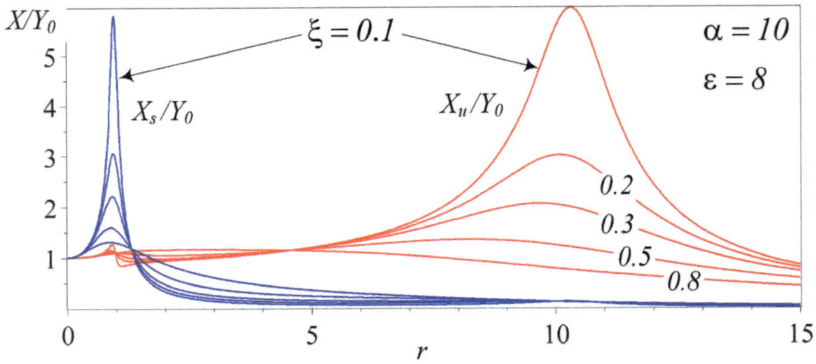


**Fig. 14.4** Frequency response of $X_s/Y_0$ and $X_u/Y_0$ for different damping ratios

**Example 188** Sprung and unsprung mass frequency responses. The graphical illustration of the frequency responses shows the steady-state behavior of a system when the frequency of excitation varies.

An example of the frequency responses of $X_s/Y_0$ and $X_u/Y_0$ is shown in Fig. 14.3 for a nominal quarter car:

$$\alpha = 10, \qquad r = \frac{\omega}{6}, \qquad \xi = 0.2, \qquad \varepsilon = 8. \tag{14.65}$$

These frequency response curves are based on no tire–road separation. Hence, assuming a tire will always be in contact with the ground, the frequency response of the sprung mass, $X_s/Y_0$, has a sharp high amplitude around the first natural frequency of the system and then drops to zero rapidly. On the other hand, the frequency response of the unsprung mass, $X_u/Y_0$, has a wide range over all excitation frequencies, with a higher amplitude around the second natural frequency of the system.

**Example 189** Effect of damping on frequency response. The frequency response of a dynamic system is sensitive to the value of damping in the system. To examine the frequency response behavior of the quarter car, we draw the frequency response curve for different values of damping ratio.

Figure 14.4 illustrates the frequency response of $X_s/Y_0$ and $X_u/Y_0$ for different damping ratios to study the effect of damping ratio on the response of a quarter car. Increasing the damping ratio diminishes the amplitude of both sprung and unsprung masses. It shows that for the first narrow resonance zone around the first natural frequency, the amplitude of the sprung mass is higher than that of the unsprung mass for any value of damping ratio. Also, for all frequencies after the first resonance zone, the amplitude of the unsprung mass is higher than that of the sprung mass.

### 14.2.2 Separation Limit

To determine the boundary of tire–road separation in the frequency response domain, we define a relative displacement variable $x_t$ between the unsprung mass and the road:

$$x_t = x_u - y_0 \cos r\tau. \tag{14.66}$$

As long as the relative displacement of the unsprung mass and road is less than the static deflection of the tire, we will have tire–road contact. The dimensionless frequency response of $x_t$ will be $X_t$:

$$X_t = \sqrt{A_3^2 + B_3^2}, \tag{14.67}$$

$$X_t^2 = A_3^2 + B_3^2$$
$$= \frac{4\xi^2 r^2 (\varepsilon + 1)^2 + \left(r^2 - 1\right)^2 + \varepsilon^2 - 2\varepsilon\left(r^2 - 1\right)}{Z_1^2 + Z_2^2} r^4 Y_0^2, \tag{14.68}$$

$$Z_1 = r^2\left(r^2\alpha^2 - 1\right) + \left(1 - (1 + \varepsilon) r^2\alpha^2\right), \tag{14.69}$$

$$Z_2 = 2r\xi\left(1 - (1 + \varepsilon) r^2\alpha^2\right). \tag{14.70}$$

The relative frequency response of $X_t/Y_0$ will show the zone of contact, $X_t/Y_0 < 1$, and tire–road separation, $X_t/Y_0 > 1$:

$$\frac{X_t}{Y_0} = \frac{x_t}{y_0} < 1 \qquad \text{in contact} \tag{14.71}$$

$$\frac{X_t}{Y_0} = \frac{x_t}{y_0} > 1 \qquad \text{tire–road separation.} \tag{14.72}$$

***Proof*** The boundary where tire–road contact is lost will happen at those points where the relative displacement of the unsprung mass and the road excitation reaches the static deflection of the tire. To determine the boundary in the frequency domain, we need to define a relative displacement variable between unsprung mass and the road:

$$x_t = x_u - y_0 \cos r\tau. \tag{14.73}$$

The relative displacement of the unsprung mass and road excitation would be less than the deflection of the tire to have tire and road in contact. Hence, road separation happens when $X_t = 1$ in the frequency domain, separating the regions of contact $X_t \le 1$ and separation $X_t > 1$:

$$X_t = X_u - \frac{y}{x_0} = X_u - Y_0 \cos r\tau, \tag{14.74}$$

$$x_0 = x_T = \frac{(m_u + m_s)\, g}{k_u}. \tag{14.75}$$

Considering the dimensionless relative displacement variable $X_t$, the in-contact set of equations of motion will be

$$X_s'' + 2\xi\left(X_s' - X_t' + Y_0 r \sin r\tau\right) + \left(X_s - X_t - Y_0 \cos r\tau\right) = 0, \tag{14.76}$$

$$X_t'' - Y_0 r^2 \cos r\tau + 2\varepsilon\xi\left(X_t' - Y_0 r \sin r\tau - X_s'\right)$$
$$+\varepsilon\left(X_t + Y_0 \cos r\tau - X_s\right)$$
$$+\alpha^2\left(X_t + Y_0 \cos r\tau\right) = \alpha^2 Y_0 \cos r\tau. \tag{14.77}$$

Assuming harmonic solutions for the variables,

$$X_s = A_1 \sin r\tau + B_1 \cos r\tau, \tag{14.78}$$

$$X_t = A_3 \sin r\tau + B_3 \cos r\tau, \tag{14.79}$$

$$X_s' = A_1 r \cos r\tau - B_1 r \sin r\tau, \tag{14.80}$$

$$X_t' = A_3 r \cos r\tau - B_3 r \sin r\tau, \tag{14.81}$$

$$X_s'' = -A_1 r^2 \sin r\tau - B_1 r^2 \cos r\tau, \tag{14.82}$$

$$X_t'' = -A_3 r^2 \sin r\tau - B_3 r^2 \cos r\tau, \tag{14.83}$$

we have

$$\begin{aligned} &\left(A_1 - A_3 - r^2 A_1 + 2\xi \left(r B_3 - r B_1 + r Y_0\right)\right) \sin r\tau \\ &+ \left(B_1 - B_3 - Y_0 + 2\xi \left(r A_1 - r A_3\right) - r^2 B_1\right) \cos r\tau = \cos r\tau, \end{aligned} \tag{14.84}$$

$$\begin{aligned} &\left(\left(\alpha^2 - r^2\right) A_3 - \varepsilon \left(A_1 - A_3\right) + 2\xi\varepsilon r \left(B_1 - B_3 - Y_0\right)\right) \sin r\tau \\ &+ \left(\varepsilon \left(-B_1 + B_3 + Y_0\right) + \left(\alpha^2 - r^2\right) B_3 - 2\xi\varepsilon r \left(A_1 - A_3\right)\right) \cos r\tau = Y_0 r^2 \cos r\tau. \end{aligned} \tag{14.85}$$

A harmonic balance of the equations provides us with four algebraic equations to determine the unknown coefficients of the harmonic solutions:

$$[A] \begin{bmatrix} A_1 \\ B_1 \\ A_3 \\ B_3 \end{bmatrix} = \begin{bmatrix} -2r\xi Y_0 \\ Y_0 \\ 2r\xi\varepsilon Y_0 \\ -\varepsilon Y_0 \end{bmatrix}, \tag{14.86}$$

$$[A] = \begin{bmatrix} 1 - r^2 & -2r\xi & -1 & 2r\xi \\ 2r\xi & 1 - r^2 & -2r\xi & -1 \\ -\varepsilon & 2\xi r\varepsilon & \alpha^2 - r^2 + \varepsilon & -2\xi r\varepsilon \\ -2\xi r\varepsilon & -\varepsilon & 2\xi r\varepsilon & \alpha^2 - r^2 + \varepsilon \end{bmatrix}. \tag{14.87}$$

The solutions yield the frequency responses of the system:

$$X_s^2 = A_1^2 + B_1^2 = \frac{4r^2\xi^2 + 1}{Z_1^2 + Z_2^2} \alpha^4 Y_0^2, \tag{14.88}$$

$$X_t^2 = A_3^2 + B_3^2 = \frac{4\xi^2 r^2 (\varepsilon + 1)^2 + \left(r^2 - 1\right)^2 + \varepsilon^2 - 2\varepsilon \left(r^2 - 1\right)}{Z_1^2 + Z_2^2} r^4 Y_0^2, \tag{14.89}$$

$$Z_1 = r^2 \left(r^2\alpha^2 - 1\right) + \left(1 - (1 + \varepsilon) r^2 \alpha^2\right), \tag{14.90}$$

$$Z_2 = 2r\xi \left(1 - (1 + \varepsilon) r^2 \alpha^2\right). \tag{14.91}$$

The frequency response of the in-contact system would be

$$X_s = \sqrt{A_1^2 + B_1^2}, \tag{14.92}$$

$$X_t = \sqrt{A_3^2 + B_3^2}. \tag{14.93}$$

We may draw the in-contact frequency responses employing new dimensionless variables $X_s/Y_0$ and $X_t/Y_0$:

$$\frac{X_s}{Y_0} = \frac{x_s}{y_0}, \qquad \frac{X_t}{Y_0} = \frac{x_t}{y_0}. \tag{14.94}$$

■

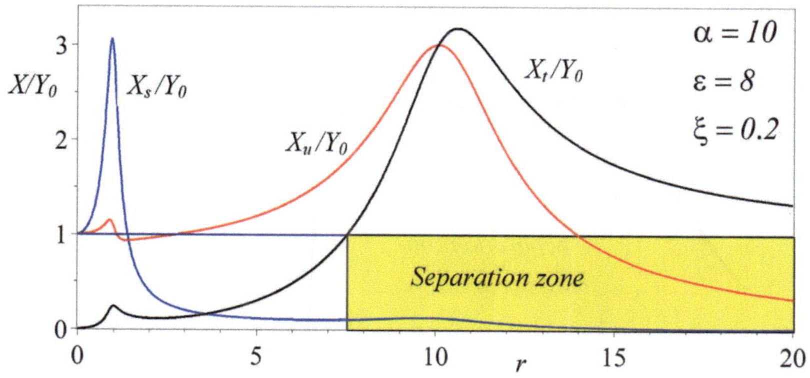


**Fig. 14.5** Relative and absolute displacement frequency responses of sprung mass of quarter car

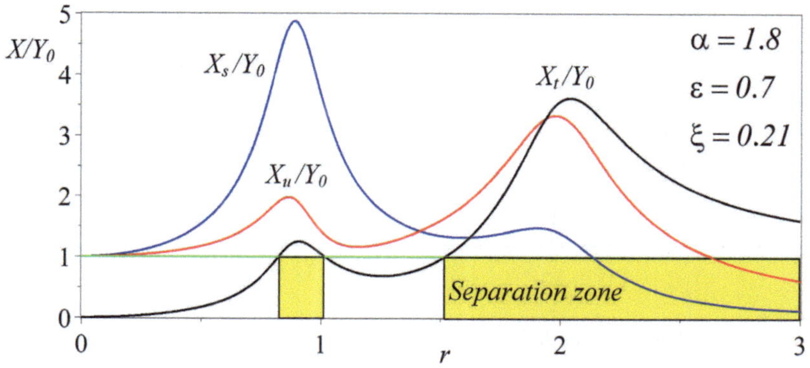


**Fig. 14.6** Relative and absolute displacement frequency responses of two-$DOF$ vibrating system similar to a quarter car

**Example 190** The tire–road separation limit and validity of the frequency response of a system can be shown graphically in the frequency domain.

Figure 14.5 illustrates the relative displacement frequency response $X_t$ as well as the absolute displacement frequency responses $X_s$ and $X_u$ for a quarter car. For the set of nominated numerical values, the curve of $X_t$ hits the horizontal line $X/Y_0 = 1$ at a frequency ratio around $r = 7.536$, indicating the limit of in-contact range. Therefore, the tire will not adhere to the ground for $r > 7.536$, and hence, not all the calculated frequency responses are valid for higher frequencies. The linearity of the system determines that for the given nominal values, $r \leq 7.536$ is the zone of correct frequency response based on no tire–road separation. The developed quarter car model can be used for any base excited two-$DOF$ system with possible separation from the ground. Depending on the value of $\alpha$, $\varepsilon$, and $\xi$, the frequency response of the system may have two separation zones. Figure 14.6 illustrates an example of a system with two separation zones.

**Example 191** Separation function. The frequency response $X_t/Y_0$ can be used to define a separation limit function in the frequency domain $X_t/Y_0 = 1$. The equation can be plotted implicitly on a plane to determine the impossible and possible area as a separation limit curve.

The relative motion frequency response of unsprung mass to road excitation of Eq. (14.89) will show the limit of tire–road contact when $X_t/Y_0 = 1$ or $X_t = Y_0$:

$$\frac{X_t}{Y_0} = r^2 \sqrt{\frac{4\xi^2 r^2 (\varepsilon + 1)^2 + \left(r^2 - 1\right)^2 + \varepsilon^2 - 2\varepsilon \left(r^2 - 1\right)}{Z_1^2 + Z_2^2}} = 1, \tag{14.95}$$

$$Z_1 = r^2 \left(r^2\alpha^2 - 1\right) + \left(1 - (1 + \varepsilon) r^2\alpha^2\right), \tag{14.96}$$

$$Z_2 = 2r\xi \left(1 - (1 + \varepsilon) r^2\alpha^2\right). \tag{14.97}$$

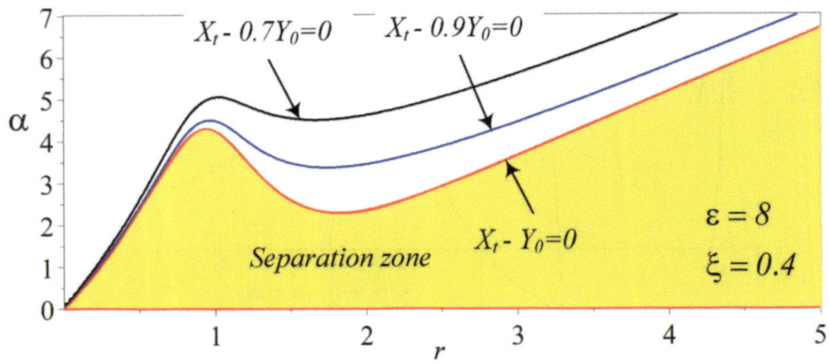


**Fig. 14.7** Illustration of separation function $X_t/Y_0 - 1 = 0$, along with two in-contact functions $X_t/Y_0 - 0.9 = 0$ and $X_t/Y_0 - 0.7 = 0$

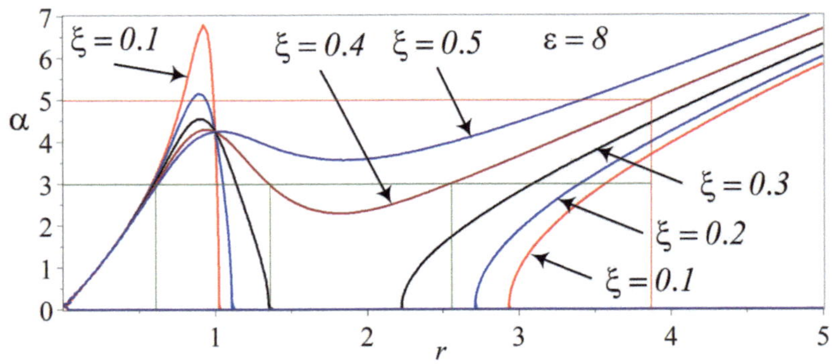


**Fig. 14.8** Effect of $\alpha$ on in-contact limit frequency for different $\xi$ and $\varepsilon = 8$

The equation $X_t/Y_0 = 1$, at which tire–road contact is lost, is an implicit function of $\varepsilon$, $\alpha$, $\xi$, and $r$. The mass ratio $\varepsilon$ is fixed for a given vehicle, and hence, the equation reduces to have three parameters $\alpha$, $\xi$, and $r$. Now we are able to plot any of them versus another for a fixed value of the third one. Let us assume $\xi$ has a given value. The implicit plot of $X_t/Y_0 = 1$ for $\alpha$ versus $r$ shows the locus of the tire–road separation boundary. Every point on the curve determines the value of $\alpha$ and $r$ at which separation occurs. Hence, the curve separates the plane into two sections. The section above indicates the in-contact zone $X_t/Y_0 < 1$, and the section below indicates the separation zone $X_t/Y_0 > 1$. Figure 14.7 illustrates the separation function $X_t/Y_0 - 1 = 0$ along with two in-contact functions $X_t/Y_0 - 0.9 = 0$ and $X_t/Y_0 - 0.7 = 0$. Below the curve $X_t/Y_0 - 1 = 0$ is the separation zone, and no point under the curve can show a frequency response.

**Example 192** Effect of $\alpha$ and $\xi$ on tire–road separation function. The stiffness ratio $\alpha$ and damping ratio $\xi$ are suspension charactristics and subject to optimization for a given $\varepsilon$. Here we investigate their effects on the tire–road separation function.

Let us fix $\varepsilon = 8$ and determine the effect of $\alpha$ and $\xi$ on the limit of tire–road separation. Figure 14.8 illustrates the implicit graphs of $X_t/Y_0 - 1 = 0$ for different $\xi$.

For a given $\xi$, say, $\xi = 0.4$, a high value of $\alpha$, say, $\alpha = 5$, is depicted by a horizontal line $\alpha = 5$, starting from $r = 0$; it extends to higher frequencies until it hits the separation limit curve associated to $\xi = 0.4$. The hit point indicates a frequency, $r = 3.88$, that is the maximum in-contact excitation freqency beyond which the tire loses contact with the road. Interestingly, at lower values of $\alpha$, say, $\alpha = 3$, there would be three intersecting points, $r_1 = 0.6129$, $r_2 = 1.3543$, and $r_3 = 2.5554$, indicating four regions: in-contact region for $r < r_1$, separation region for $r_1 < r < r_2$, in-contact region for $r_2 < r < r_3$, and separation region for $r > r_3$.

Figure 14.9 depicts the relative frequency response $X_t/Y_0$, as well as the intersection points with $X_t/Y_0 = 1$, to determine the in-contact and separation regions. The frequency response cure is correct only for $X_t/Y_0 < 1$, as there would be tire–road separation for $X_t/Y_0 > 1$, and the frequency response would be different. Figures 14.10, 14.11 and 14.12 show the effect sof $\alpha$ on the in-contact limit frequencies for different $\xi$ and for $\varepsilon = 12$, $\varepsilon = 10$, and $\varepsilon = 5$.

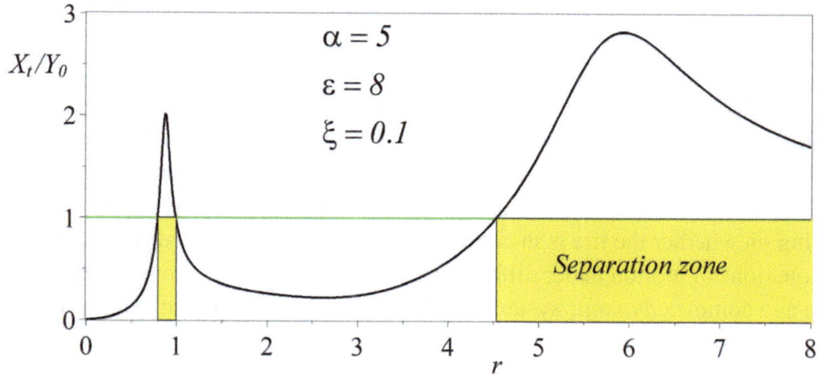


**Fig. 14.9** Relative frequency response $X_t/Y_0$, as well as intersection points with $X_t/Y_0 = 1$, to determine in-contact and separation regions

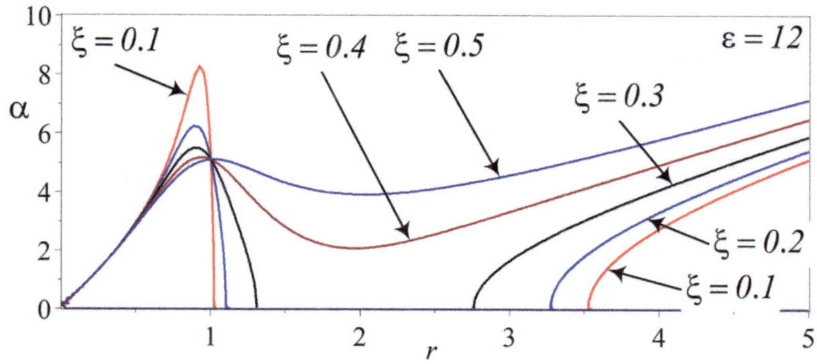


**Fig. 14.10** Effect of $\alpha$ on in-contact limit frequency for different $\xi$ and $\varepsilon = 12$

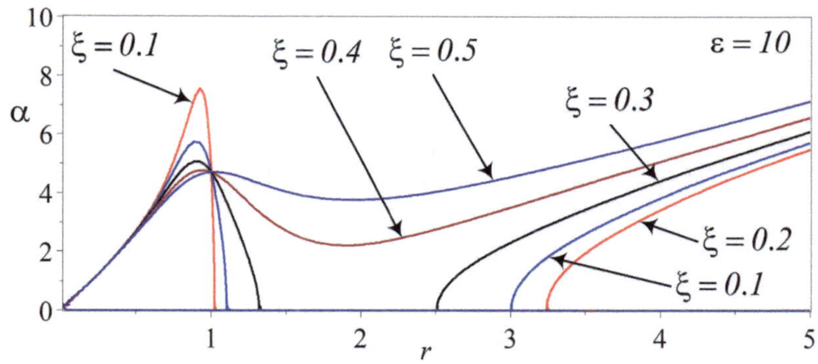


**Fig. 14.11** Effect of $\alpha$ on in-contact limit frequency for different $\xi$ and $\varepsilon = 10$

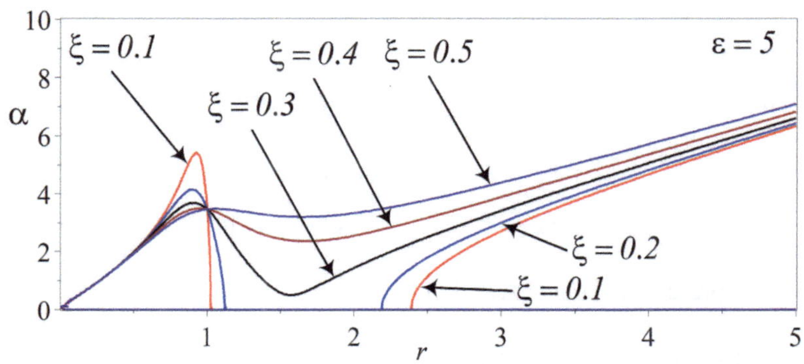


**Fig. 14.12** Effect of $\alpha$ on in-contact limit frequency for different $\xi$ and $\varepsilon = 5$

## 14.3 Time Response

Numerical integration is employed as a method for investigating the transient response of vehicles traversing a road bump to identify potential jump occurrences. Additionally, it is utilized to examine the steady-state response of a vehicle traveling over a series of sine wave-shaped roads. The magnitude of the steady-state response determines the frequency response of the vehicle during instances of tire–road separation. The equations of motion for a quarter car, accounting for tire–road separation, vary depending on whether the tire is in contact with the road or separated from it. This type of dynamic system, encompassing distinct equations of motion under different circumstances such as geometry, time, dynamic components, and kinematics, is referred to as a complex dynamic system. Analysis of complex dynamic systems is interesting and challenging.

### 14.3.1 Transient Response

When a vehicle goes over a bump on the road, a tire may jump over the bump and lose contact with the road. The nondimensional equations of motion of a quarter car considering tire–road separation are different for in-contact and separation phases. When a tire is in contact with the road, the equations of motion of a quarter car are

$$X_s'' + 2\xi\left(X_s' - X_u'\right) + (X_s - X_u) = 0, \tag{14.98}$$

$$X_u'' + 2\varepsilon\xi\left(X_u' - X_s'\right) + \varepsilon\left(X_u - X_s\right) + \alpha^2 X_u = \alpha^2 Y \tag{14.99}$$

$$X_u - Y < 1,$$

and when the tire is separated from the road and the quarter car is in free fall, the equations of motion of quarter car are as follows:

$$X_s'' + 2\xi\left(X_s' - X_u'\right) + (X_s - X_u) = 0, \tag{14.100}$$

$$X_u'' + 2\varepsilon\xi\left(X_u' - X_s'\right) + \varepsilon\left(X_u - X_s\right) + \alpha^2 = 0, \tag{14.101}$$

$$X_u - Y \geq 1,$$

$$\begin{aligned}
&\alpha = \frac{\omega_u}{\omega_s}, \qquad \xi = \frac{c_s}{2m_s\omega_s}, \qquad \varepsilon = \frac{m_s}{m_u}, \\
&\omega_s = \sqrt{\frac{k_s}{m_s}}, \qquad \omega_u = \sqrt{\frac{k_u}{m_u}}, \\
&x_0 = x_T = \frac{(m_u + m_s)\, g}{k_u}, \qquad t_0 = \frac{1}{\omega_s}, \\
&Y = \frac{y}{x_0}, \qquad X_u = \frac{x_u}{x_0}, \qquad X_s = \frac{x_s}{x_0}, \qquad \tau = \frac{t}{t_0}, \\
&X_T = \frac{x_T}{x_0} = 1.
\end{aligned} \tag{14.102}$$

The tire–road separation state will be indicated by the following inequalities:

$$X_u - Y \geq 1 \qquad separation. \tag{14.103}$$

$$X_u - Y < 1 \qquad in\ contact, \tag{14.104}$$

where the function $Y = Y(t)$ represents the road excitation.

***Proof*** Let us substitute variables and parameters (14.102)

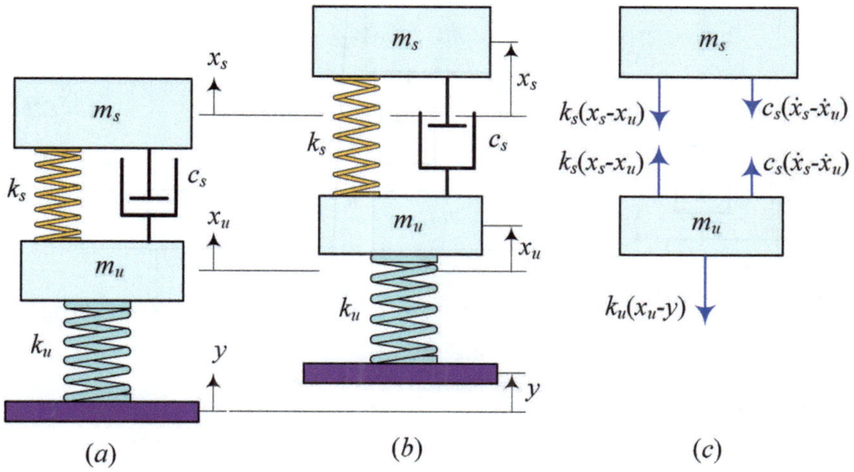


**Fig. 14.13** Free body diagram of quarter car while it is in contact with the road

$$
\begin{aligned}
X_s'' &= \frac{\ddot{x}_s}{\omega_s^2 x_0}, \quad X_s' = \frac{\dot{x}_s}{\omega_s x_0}, \quad X_s = \frac{x_s}{x_0}, \\
X_u'' &= \frac{\ddot{x}_u}{\omega_s^2 x_0}, \quad X_u' = \frac{\dot{x}_u}{\omega_s x_0}, \quad X_u = \frac{x_u}{x_0}, \quad Y = \frac{y}{x_0}, \\
2\xi &= \frac{c_s}{m_s \omega_s}, \quad \omega_s^2 = \frac{k_s}{m_s}, \quad \varepsilon = \frac{m_s}{m_u}, \\
\alpha &= \frac{\omega_u}{\omega_s}, \quad \omega_u^2 = \frac{k_u}{m_u}, \quad x_0 = x_T = \frac{(m_u + m_s)\, g}{k_u},
\end{aligned}
\tag{14.105}
$$

into equation of motion of sprung mass (14.98),

$$
\frac{\ddot{x}_s}{\omega_s^2 x_0} + \frac{c_s}{m_s \omega_s}\left(\frac{\dot{x}_s}{\omega_s x_0} - \frac{\dot{x}_u}{\omega_s x_0}\right) + \left(\frac{x_s}{x_0} - \frac{x_u}{x_0}\right) = 0,
$$

to derive the equation in real variable space

$$
m_s \ddot{x}_s = -c_s\left(\dot{x}_s - \dot{x}_u\right) - k_s\left(x_s - x_u\right). \tag{14.106}
$$

Let us do the same to the equation of motion of the unsprung mass

$$
\begin{aligned}
&\frac{\ddot{x}_u}{\omega_s^2 x_0} + \frac{m_s}{m_u}\frac{c_s}{m_s \omega_s}\left(\frac{\dot{x}_u}{\omega_s x_0} - \frac{\dot{x}_s}{\omega_s x_0}\right) \\
&\qquad + \frac{m_s}{m_u}\left(\frac{x_u}{x_0} - \frac{x_s}{x_0}\right) + \frac{\omega_u^2}{\omega_s^2}\frac{x_u}{x_0} = \frac{\omega_u^2}{\omega_s^2}\frac{y}{x_0}
\end{aligned}
\tag{14.107}
$$

to derive the equation in real variable space:

$$
m_u \ddot{x}_u = c_s\left(\dot{x}_s - \dot{x}_u\right) + k_s\left(x_s - x_u\right) - k_u\left(x_u - y\right) \tag{14.108}
$$

$$
x_u - y < x_T \qquad \textit{in contact}. \tag{14.109}
$$

Equations (14.106) and (14.108) match the free body diagram of a quarter car model shown in Fig. 14.13.

The equation of motion of unsprung mass in tire–road separation mode will be the same as the equation in tire–road contact mode. After substitution of variables and parameters (14.102), the equation of motion of unsprung mass (14.101) in separation mode will be

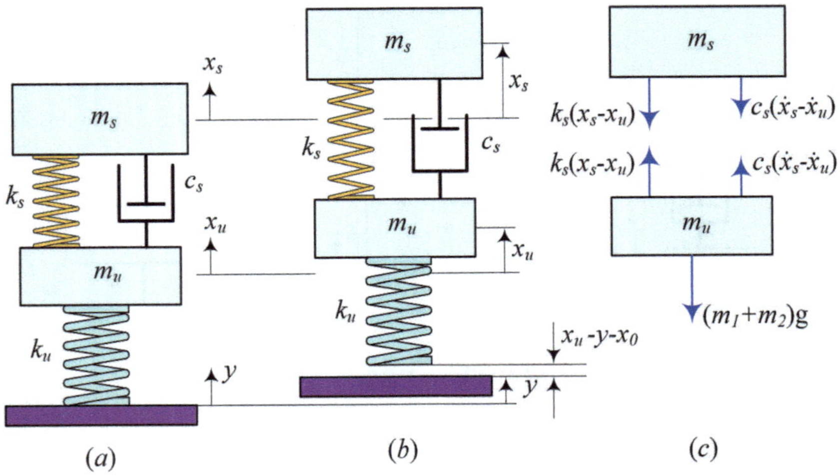


**Fig. 14.14** Free body diagram of quarter car in tire–road separation mode

$$\frac{\ddot{x}_u}{\omega_s^2 x_0} + \frac{m_s}{m_u}\frac{c_s}{m_s\omega_s}\left(\frac{\dot{x}_u}{\omega_s x_0} - \frac{\dot{x}_s}{\omega_s x_0}\right) + \frac{m_s}{m_u}\left(\frac{x_u}{x_0} - \frac{x_s}{x_0}\right) + \frac{\omega_u^2}{\omega_s^2} = 0, \tag{14.110}$$

which will be simplified to the following equations in real variable space:

$$X_s'' + 2\xi\left(X_s' - X_u'\right) + (X_s - X_u) = 0, \tag{14.111}$$

$$X_u'' + 2\varepsilon\xi\left(X_u' - X_s'\right) + \varepsilon\,(X_u - X_s) + \alpha^2 = 0, \tag{14.112}$$

$$m_u\ddot{x}_u = c_s\,(\dot{x}_s - \dot{x}_u) + k_s\,(x_s - x_u) - k_u x_0, \tag{14.113}$$

$$x_u - y \geq x_T \qquad separation. \tag{14.114}$$

Equations (14.106) and (14.113) match the free body diagram of a quarter car in separation mode shown in Fig. 14.14. ■

**Example 193** Tire–road separation indicator. To detect tire–road separation in the time domain, we define a zero-one mathematical function to act as a switch to turn to one when a tire separates from the road and remains zero when the tire is in contact with the road.

Using the nondimensional form of the equations of motion, the tire–road separation condition will be indicated by the following inequalities:

$$X_u - Y \geq 1 \qquad separation, \tag{14.115}$$

$$X_u - Y < 1 \qquad in\ contact. \tag{14.116}$$

Let us define an indicator function $S$ to detect tire–road separation during vehicle motion:

$$S = H\,(X_u - Y - 1) = \begin{cases} 1 & X_u - Y \geq 1, \\ 0 & X_u - Y < 1. \end{cases} \tag{14.117}$$

To show how the indicator works, let us run a quarter car over a half sign bump on the road at a speed at which tire–road separation happens. We model a bump on the road by a sine-square function, $y = y_0 \sin^2 \omega t$:

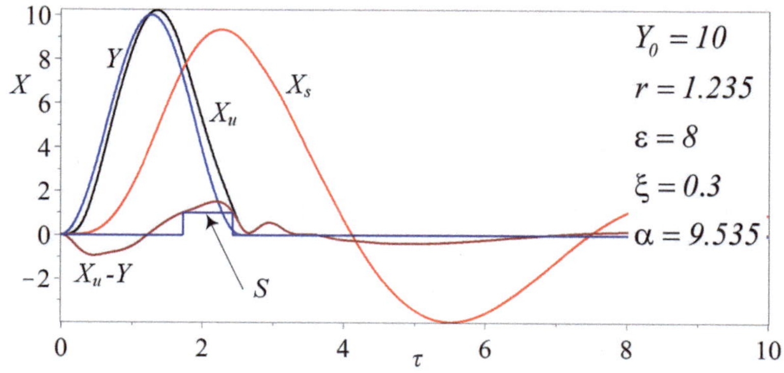


**Fig. 14.15** Response of quarter car going over bump and experiencing road separation

$$y = \begin{cases} y_0 \sin^2 \omega t & 0 \le t \le \dfrac{\pi}{\omega}, \\ 0 & \dfrac{\pi}{\omega} < t, \end{cases} \tag{14.118}$$

$$\omega = \frac{\pi v}{w}, \tag{14.119}$$

or, in dimensionless form,

$$Y = \begin{cases} Y_0 \sin^2 r\tau & 0 \le \tau \le \dfrac{\pi}{r}, \\ 0 & \dfrac{\pi}{r} < \tau, \end{cases} \tag{14.120}$$

$$r = \frac{\pi v}{w\omega_s}, \tag{14.121}$$

$$r = \frac{\omega}{\omega_s}, \quad Y = \frac{y}{x_0}, \quad Y_0 = \frac{y_0}{x_0}, \quad t_0 = \frac{1}{\omega_s}, \tag{14.122}$$

$$x_0 = x_T = \frac{(m_u + m_s)\, g}{k_u}, \qquad \tau = \frac{t}{t_0} = \omega_s t. \tag{14.123}$$

Figure 14.15 depicts the response of a quarter car going over a bump and experiencing road separation. The tire jumps over the bump and for a period of time leaves the ground. The separation time is detected and shown by the indicator function $S$.

**Example 194** Velocity and tire–road separation. For a given bump on the road, the speed of a vehicle will be the cause of tire–road separation. Here we show the dynamic response of a quarter car moving over a bump at different velocities to detect the speed at which separation happens.

Let us consider a quarter car with the following parameters:

$$\begin{aligned} &m_s = 336\,\text{kg}, \quad m_u = 42\,\text{kg}, \quad c = 1630\,\text{N s/m}, \\ &k_s = 22{,}000\,\text{N/m}, \quad k_u = 250{,}000\,\text{N/m}, \\ &\omega_s = \sqrt{\frac{k_s}{m_s}} = 8.09\,\text{rad/s}, \quad \omega_u = \sqrt{\frac{k_u}{m_u}} = 77\,\text{rad/s}. \end{aligned} \tag{14.124}$$

The dimensionless parameters of Eqs. (14.98)–(14.101) will be

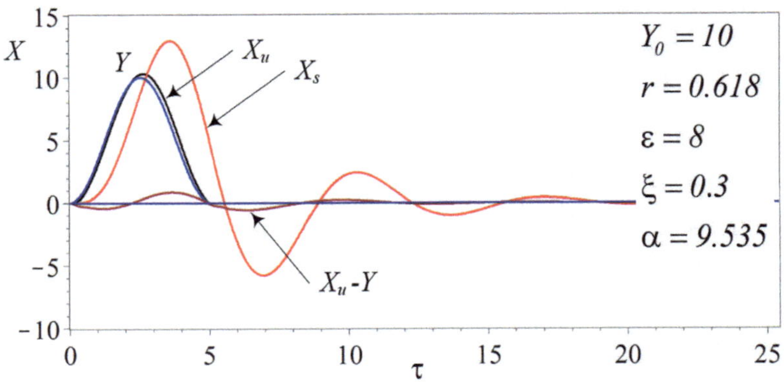


**Fig. 14.16** Response of quarter car at very low velocity at which no tire–road separation occurs

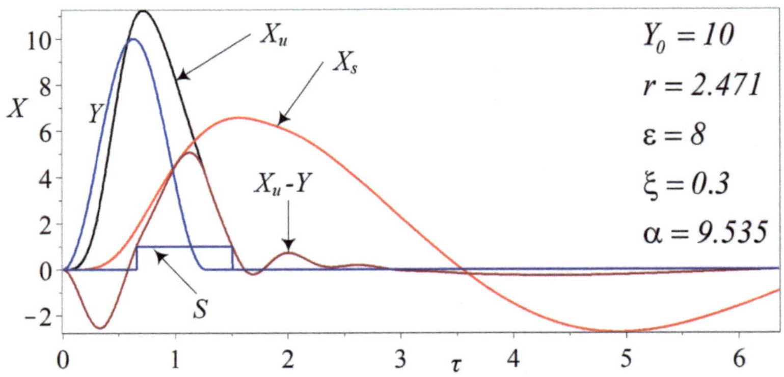


**Fig. 14.17** Response of quarter car at velocity at which tire–road separation occurs

$$\varepsilon = \frac{m_s}{m_u} = 8, \qquad \xi = \frac{c}{2m_s\omega_s} = 0.3, \qquad \alpha = \frac{\omega_u}{\omega_s} = 9.535. \tag{14.125}$$

The static tire deflection is

$$x_T = \frac{(m_u + m_s)\,g}{k_u} = 1.48 \times 10^{-2}\,\text{m}. \tag{14.126}$$

The bump is expressed by function (14.120) with the follwing data:

$$Y = \frac{y}{x_0}, \qquad Y_0 = \frac{y_0}{x_0} = 10, \qquad w = \frac{\pi}{10}\,\text{m}, \tag{14.127}$$

$$r = \frac{\omega}{\omega_s} = \frac{\pi v}{w\omega_s} = 1.236v, \tag{14.128}$$

$$\tau = \frac{t}{t_0} = \omega_s t, \qquad t_0 = \frac{1}{\omega_s} = 0.123\,\text{s}, \tag{14.129}$$

$$Y = \begin{cases} Y_0 \sin^2 1.236v\tau & 0 \le \tau \le \dfrac{\pi}{1.236v}, \\ 0 & \dfrac{\pi}{1.236v} < \tau. \end{cases} \tag{14.130}$$

Figure 14.16 illustrates the response of the quarter car at a low velocity for which no tire–road separation occurs. Figures 14.17 and 14.18 show the quarter car's response at higher velocity where tire–road separation occurred.

**Example 195** Bottom of tire motion. Using the displacement of the unsprung mass, road, and static tire deflection, we are able to calculate the displacement of the point under the tire.

In this chapter, we have assumed that the tire is point size, ignoring tire size. Therefore, the point of contact of the tire with the ground is always at the bottom of the tire. Measuring the displacement of the unsprung mass by $X_u$, the road by

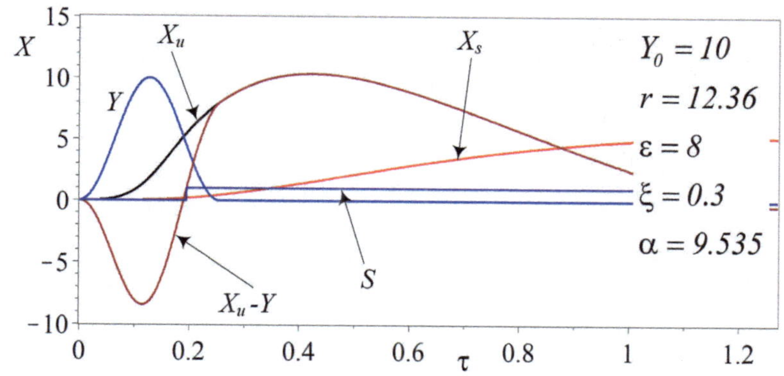


**Fig. 14.18** Response of quarter car at velocity at which tire–road separation occurs

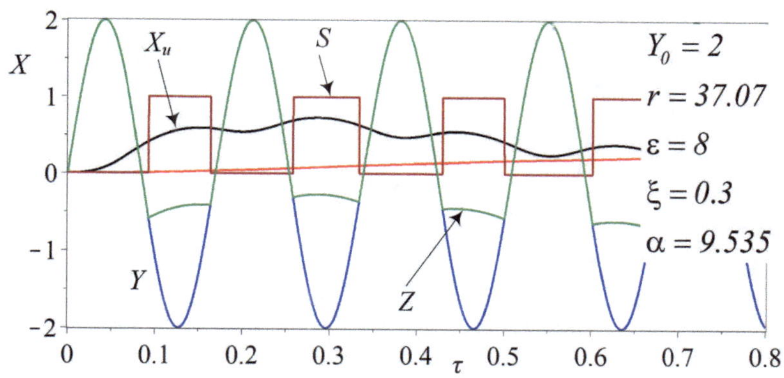


**Fig. 14.19** Illustration of bottom of tire displacement $Z$ as well as $X_u$, $S$, and the road for a quarter car on a harmonic wavy road

$Y$, and the static tire deflection by 1, all dimensionless, we can define a variable $Z$ to indicate the displacement of the point under the tire with respect to the road. This indicates the gap size between tire and road when separation happens:

$$Z = Y + (X_u - Y - 1) H (X_u - Y - 1) \tag{14.131}$$

Figure 14.19 illustrates the variable $Z$ as well as $X_u$ and separation indicator $S$, and the road for a wavy road.

$$Y = Y_0 \sin r\tau = 2 \sin 12.36\tau. \tag{14.132}$$

**Example 196** Jump over a ramp. The mathematical model is capable of detecting and predicting vehicle dynamics when the tire becomes separated from the road. As a complicated example, we simulated a car jumping over a ramp.

Consider a car moving with velocity $v$ and hits a ramp section on the road of length $L$ and height $H$, with the following mathematical expression:

$$y = \frac{H}{L}x = \frac{H}{L}vt, \qquad t \le \frac{L}{v}. \tag{14.133}$$

Employing the charactersistic time $t_0$ and length $x_0$

$$x_0 = x_T = \frac{(m_u + m_s) g}{k_u}, \qquad t_0 = \frac{1}{\omega_s}, \tag{14.134}$$

the road function (14.133) will have a dimensionless expression:

$$Y = \begin{cases} \frac{H}{L}\frac{v}{\omega_s x_0}\tau & \tau \le \frac{L}{v}\omega_s, \\ 0 & \tau > \frac{L}{v}\omega_s, \end{cases} \tag{14.135}$$

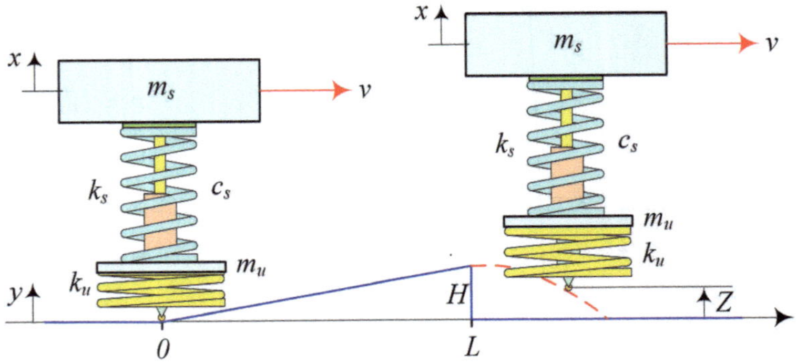


**Fig. 14.20** A ramp and a car jumping over the ram following tire–road separation

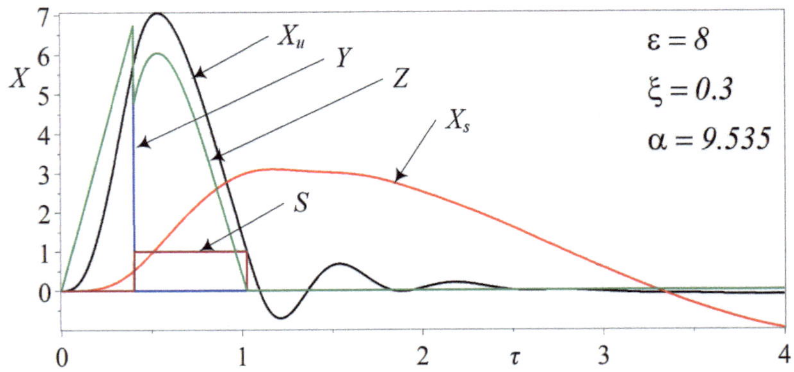


**Fig. 14.21** Response of quarter car jumping over ram on road

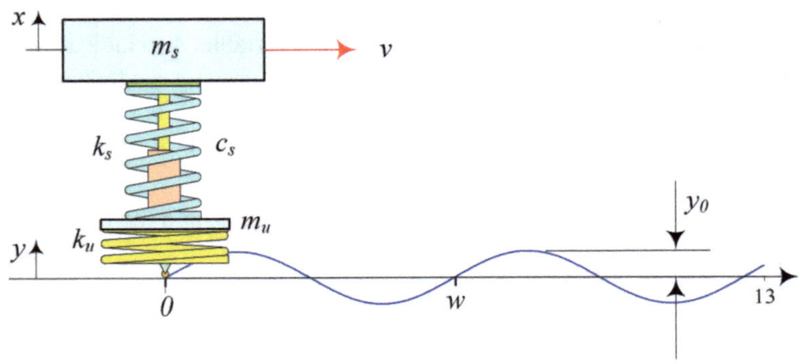


The ramp and the car jumping over the ram are shown in Fig. 14.20.

Let us move the car with velocity $v = 10\,\mathrm{m/s}$ over a ramp of length $L = 0.5\,\mathrm{m}$ and maximum height $H = 0.1\,\mathrm{m}$. Figure 14.21 illustrates the response of the car. The interesting curve is $Z$, the bottom of the tire. It follows the ramp to the tip at which the tire contraction is instantly released. Then $Z$ follows the motion of the unsprung mass $X_u$, while the tire is separated from the road and the car is airborne. The tire touches the ground after a distance and follows the ground. The unsprung mass keeps moving down until the tire compresses again and raises the car up. The sprung mass has a much lower frequency of vibration, and its motion is less than $X_u$ because $k_u$ and $k_s$ absorb the road displacement. The separation indicator $S$ shows the tire–road separation happens right after the cliff until the tire touches the ground after a distance.

### 14.3.2 Harmonic Response

To determine the frequency response of the system in separation zones, we need to numerically solve the equations in the time domain for a given harmonic input. To investigate the time response of the quarter car to the harmonic road excitation, a harmonic function with constant amplitude and undetermined frequency is assumed:

$$Y = Y_0 \sin r\tau = Y_0 \sin \frac{2\pi v}{w\omega_s}\tau. \tag{14.136}$$

***Proof*** Consider a harmonically bumpy road as illustrated in Fig. 14.22. The road will act as a base excitation to a quarter car model. The car is moving with horizontal velocity $v$. The amplitude of the road wave is shown by $y_0$, and one cycle of the wave with length $w$ will take $T = w/v$ time to pass. Using the relation between period and frequency, $\omega = 2\pi/T$, the road wave will be expressed mathematically as

$$y = \sin \omega t = \sin \frac{2\pi v}{w} t. \tag{14.137}$$

Employing the characteristic parameters for displacement $x_0$ and time $t_0$ in (14.4), to make the variables dimensionless, we will have

$$t = t_0 \tau = \frac{\tau}{\omega_s}, \tag{14.138}$$

$$y = x_0 Y, \tag{14.139}$$

$$r = \frac{\omega}{\omega_s}, \tag{14.140}$$

$$Y = Y_0 \sin r\tau. \tag{14.141}$$

The equations of motion to be solved are (14.98)–(14.101) with the same tire–road separation conditions. ■

**Example 197** Response to harmonically bumpy road. A harmonic excitation is supposed to have a steady-state periodic response in a dynamic system. Here we have a quarter car as a complex dynamic system whose steady-state response is the issue as tire–road separation happens.

Consider a quarter car with the parameters

$$\varepsilon = \frac{m_s}{m_u} = 8, \quad \xi = \frac{c}{2m_s\omega_s} = 0.3, \quad \alpha = \frac{\omega_u}{\omega_s} = 9.535 \tag{14.142}$$

and a harmonic road excitation function:

$$Y = Y_0 \sin r\tau = 2 \sin 12.36\tau. \tag{14.143}$$

This is be a road with the following characteristics:

$$w = \frac{\pi}{5}\,\text{m}, \qquad v = 10\,\text{m/s},$$
$$x_T = 1.48\,\text{cm}, \qquad t_0 = 0.12358\,\text{s}. \tag{14.144}$$

Figure 14.22 illustrates the response of a quarter car moving over a bumpy road with zero initial conditions. As a result, we will have a transient part of the time response before the steady-state response settles. The separation indicator $S$ shows that the car tire loses contact with the road in every cycle of the road wave. Since not all variables are showing steady-state behavior, the response is still in transient mode.

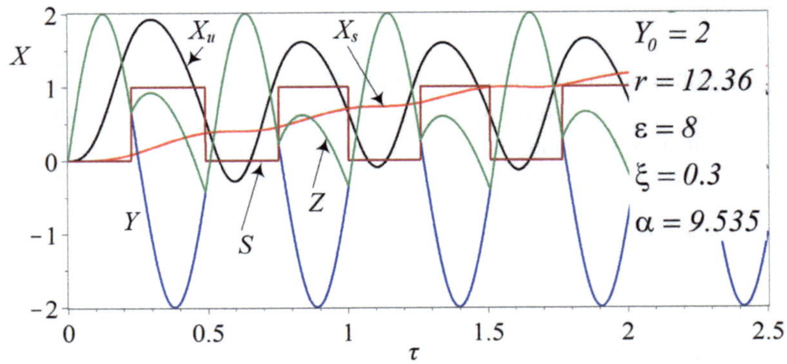


**Fig. 14.22** A harmonic bumpy road

To visually determine whether a system has reached a steady-state response, it is necessary to observe responses over an extended period of time. Examining the sprung mass displacement shown in Fig. 14.22 reveals that the response has not yet reached a steady state. Figure 14.23 shows time response of the sprung mass for a longer period of time, indicating that the transient part of the response eventually passes as long as there is damping in the system.

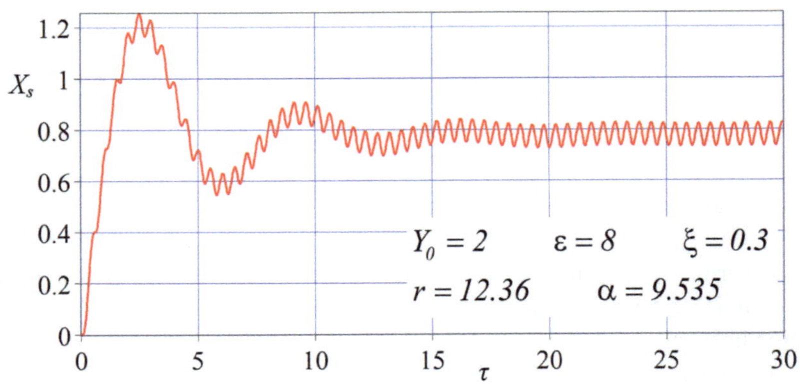


**Example 198** Jump over peak and land on another bump. One issue would be whether a car could jump over a bump and land on another bump, and whether the developed mathematical equations of motion could detect such jumps and road separations.

Let us consider a high-amplitude wavy road and an unknown velocity $v$ of the vehicle:

$$m_s = 336\,\text{kg}, \quad m_u = 42\,\text{kg}, \quad c = 1630\,\text{N s/m},$$
$$k_s = 22{,}000\,\text{N/m}, \quad k_u = 250{,}000\,\text{N/m}, \tag{14.145}$$
$$\omega_s = \sqrt{\frac{k_s}{m_s}} = 8.09\,\text{rad/s}, \quad \omega_u = \sqrt{\frac{k_u}{m_u}} = 77\,\text{rad/s},$$

$$w = \frac{\pi}{5}\,\text{m}, \quad x_T = 1.48\,\text{cm}, \quad \frac{w}{x_0} = 42.36, \tag{14.146}$$

$$t_0 = 0.12358\,\text{s}, \quad T = \frac{w}{v} = \frac{0.62832}{v}\,\text{s}, \tag{14.147}$$

$$Y = Y_0 \sin r\tau = 15 \sin 2.472 v\tau, \tag{14.148}$$

$$\varepsilon = \frac{m_s}{m_u} = 8, \quad \xi = \frac{c}{2m_s\omega_s} = 0.3, \quad \alpha = \frac{\omega_u}{\omega_s} = 9.535. \tag{14.149}$$

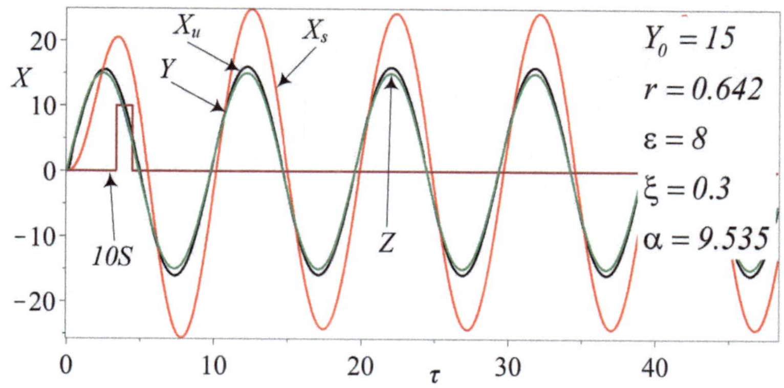


**Fig. 14.23** Response of quarter car to harmonic bumpy road at speed to obtain $r = 0.64263$

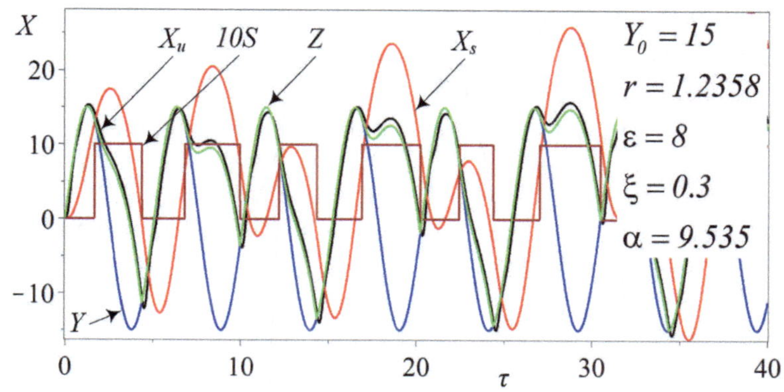


**Fig. 14.24** Time response of the sprung mass for a longer period of time

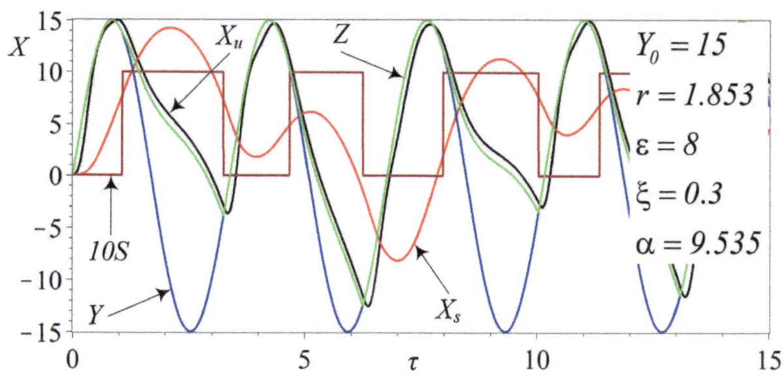


**Fig. 14.25** Response of quarter car to harmonic wavy road at a speed to obtain $r = 1.853$

The response of the car depends on the value of the velocity $v$, which determines the value of the frequency ratio $r$ and the characteristic time $t_0$. To examine the complexity of the possible response of the car, we determine the response at different velocities. Figure 14.23 depicts the response of the quarter car to a harmonic wave with dimensionless length $w/x_0 = 42.36$ and amplitude $Y_0 = y_0/x_0 = 15$ at a velocity that yields a frequency ratio of $r = 0.64263$. The time history of the system is shown for $\tau = 50$, which is equal to $t = 0.12358 \times 50 = 6.179$ s, and this is usually not long enough for the system to settle a steady-state response. However, it is long enough to answer our question and visualize how complicated the response could be.

By increasing the velocity $v$, the frequency ratio $r$ increases, and Figs. 14.24, 14.25, 14.26, 14.27, 14.28 and 14.29 depict the response of the quarter car at higher velocities. Increasing the speed increases the separation time until in Fig. 14.29, for $r = 9.886$, the car jumps over a peak and lands on another bump to answer in the affirmative the question posed in this example.

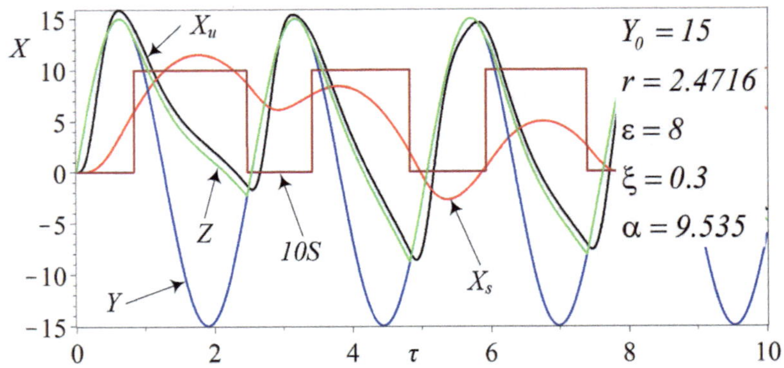


**Fig. 14.26** Response of quarter car to harmonic wavy road at a speed to obtain $r = 2.4716$

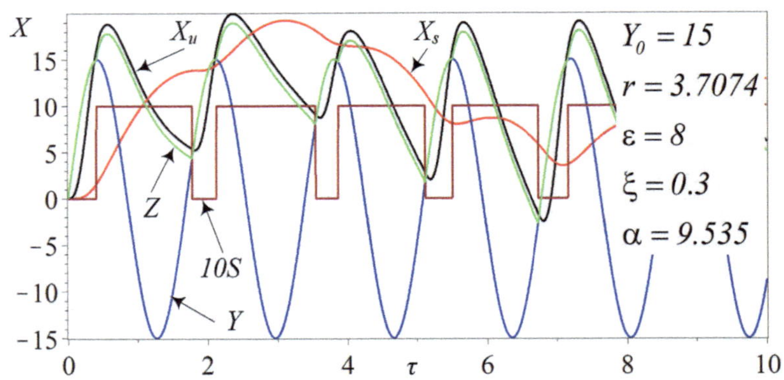


**Fig. 14.27** Response of quarter car to harmonic wavy road at a speed to obtain $r = 3.7074$

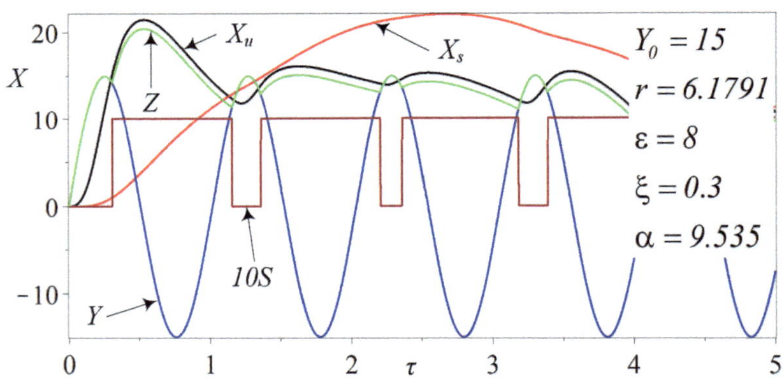


**Fig. 14.28** Response of quarter car to harmonic wavy road at a speed to obtain $r = 6.1791$

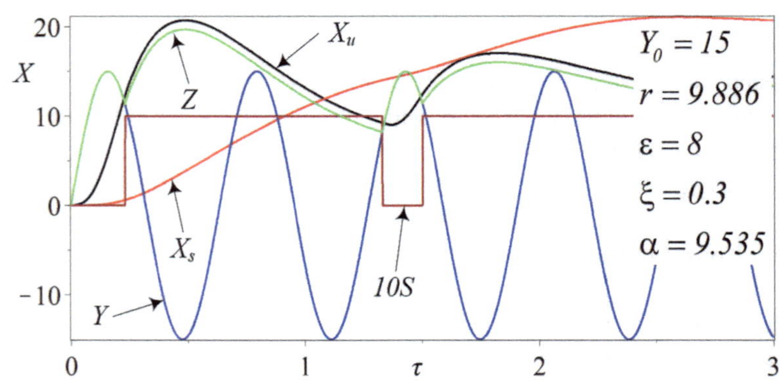


**Fig. 14.29** Response of quarter car to harmonic wavy road at a speed to obtain $r = 9.886$

## 14.4 Separation Frequency Response

Assuming damping of the system will remove the effects of the initial conditions and eventually set the system at steady-state periodic response, the amplitude of the oscillation will be recorded for the given excitation frequency, and the frequency response will be determined. The system is a complex dynamic system for which an exact analytic solution is not achievable. Hence, numerical integration is an approximate alternative solution method.

The equations of motion for in-contact and separation conditions are as follows. The equations must be solved numerically until a steady-state response is reached and the amplitude $X_s$ detect for a given frequency ratio $r$. The plot of $X_s$ versus $r$ will show the frequency response:

$$X_s'' + 2\xi\left(X_s' - X_u'\right) + (X_s - X_u) = 0, \tag{14.150}$$

$$X_u'' + 2\varepsilon\xi\left(X_u' - X_s'\right) + \varepsilon\left(X_u - X_s\right) + \alpha^2 X_u = \alpha^2 Y_0 \sin r\tau, \tag{14.151}$$

$$X_u - Y < 1,$$

$$X_s'' + 2\xi\left(X_s' - X_u'\right) + (X_s - X_u) = 0, \tag{14.152}$$

$$X_u'' + 2\varepsilon\xi\left(X_u' - X_s'\right) + \varepsilon\left(X_u - X_s\right) + \alpha^2 = 0, \tag{14.153}$$

$$X_u - Y \geq 1,$$

$$\alpha = \frac{\omega_u}{\omega_s}, \qquad \xi = \frac{c_s}{2m_s\omega_s}, \qquad \varepsilon = \frac{m_s}{m_u},$$

$$\omega_s = \sqrt{\frac{k_s}{m_s}}, \qquad \omega_u = \sqrt{\frac{k_u}{m_u}},$$

$$x_0 = x_T = \frac{(m_u + m_s)\,g}{k_u}, \qquad t_0 = \frac{1}{\omega_s}, \tag{14.154}$$

$$Y = \frac{y}{x_0}, \qquad X_u = \frac{x_u}{x_0}, \qquad X_s = \frac{x_s}{x_0}, \qquad \tau = \frac{t}{t_0},$$

$$X_T = \frac{x_T}{x_0} = 1,$$

***Proof*** The frequency response is a mathematical representation of how a system responds to different frequencies. It describes the relationship between the input excitation frequency and the output amplitude and phase. To obtain the frequency response from a time-domain solution of a lineaar system, multiple methods can be used, such as operational transformations, to determine the frequency response analytically. However, there is no method to determine the exact frequency response of nonlinear or complex dynamic systems. Perturbation methods are able to provide approximate frequency responses for weak nonlinearity.

When there is damping in a system, the system will technically approach a steady-state periodic response for a harmonic excitation from any initial conditions. Hence, if the time solution is calculated long enough, and the amplitude of the steady state is captured, the frequency response can be determined. Figure 14.30 illustrates an example of a quarter car time response to a harmonic road excitation. For the given data, it seems the response approaches its steady state by $\tau > 100$. Capturing the amplitude from such responses for different frequencies could provide us with a good approximation of the frequency response.

There are several shortcomings in calculating a frequency response from a time response. The most important ones are as follows: (1) Theoretically it takes to infinity for a system to settle at its steady-state solution. Practically, it takes a long time to obtain steady-state conditions, especially for low damped systems. (2) Superposition will not be applied to nonlinear and complex systems. Therefore, the output will not be proportional to the input, and hence, the response depends on the value of the input signal, in this case, the amplitude of the wavy road. (3) The positive and negative peak values at steady state may not be equal. In that case, the system will have two frequency responses for positive

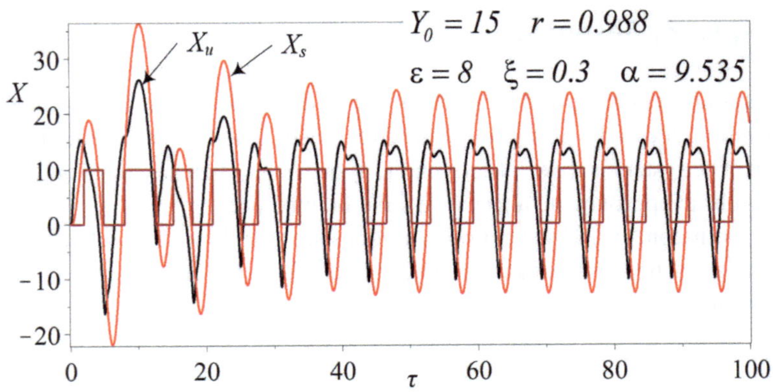


**Fig. 14.30** Example of quarter car time response to harmonic road excitation

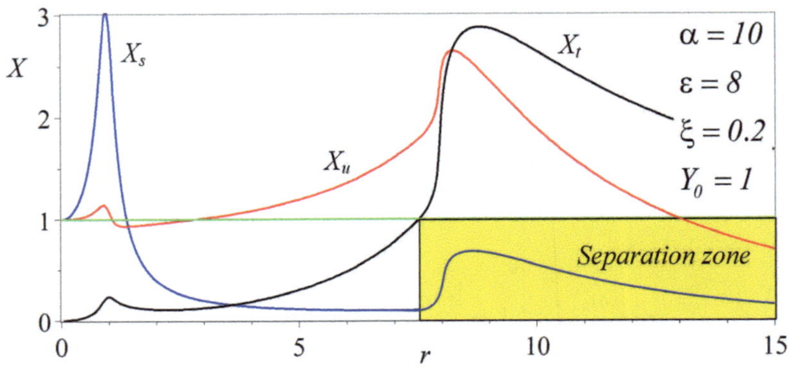


**Fig. 14.31** Frequency response of quarter car with tire–road separation range of frequency

and negative amplitudes. (4) The frequency response of nonlinear systems usually shows a jump phenomenon, which indicates there will be a range of frequencies for which there will be three amplitudes, two stable and one unstable in between the two stable ones. At the limits of the frequency range, a jump will happen between the higher and lower amplitude depending on whether the excitation frequency is increasing or decreasing. (5) A nonlinear system might have a frequency island, which is not easy to detect from a time response, especially if we do not know the frequency island exists. ■

**Example 199** Frequency response of quarter car with tire–road separation. To examine how the numerical calculation of a frequency response for a quarter car with tire–road separation works, here we calculate frequency response of a sample quarter car.

Figure 14.31 illustrates the numerically calculated frequency response of a quarter car. The tire–road separation occurs for the frequency above where the point $X_t$ crosses the horizontal line of $X = Y_0$. The curve shows that at the lower range of frequencies where tire–road separation happens, the amplitudes are very sensitive to the frequency ratio $r$. For this car, the tire–road separation happens only in the second resonance zone; the tire will not leave the ground in the first resonance zone.

**Example 200** Linear and nonlinear frequency response comparison. Plotting the linear frequency response and frequency response with tire–road separation is a good tool to understand how tire–road separation affects the frequency response of quarter cars.

A comparsion of the frequency responses of tire–road separation and tire–road in contact reveals two facts. First, the numerical and analytic calculations must match when there is no tire–road separation. This compatibility will show how the numerical method is close to an exact response. Second, such a comparison shows how the frequency responses differ in the separation zone. The difference answers the question of whether the in-contact assumption can be accepted for the separation zone.

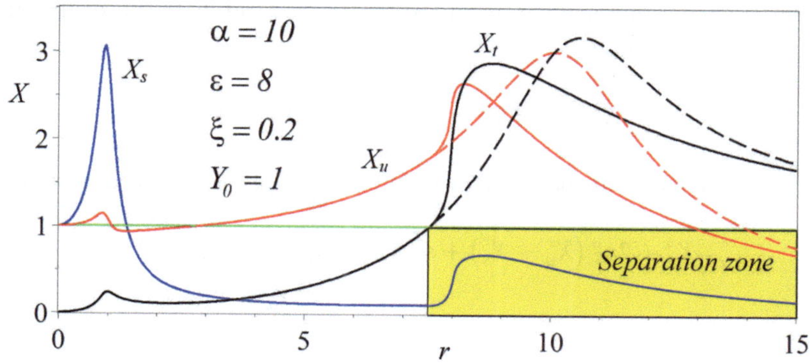


**Fig. 14.32** Dashed lines are analytic frequency responses based on the assumption that the tire is in contact with the road, and the solid lines are the numerical calculation of the frequency response with separation

Figure 14.32 illustrates the analytic frequency response of a quarter car by assuming the tire will never separate from the road (dashed lines) and the numerical calculation of the frequency response when the tire can separate from the road (solid lines). The frequency responses are identical before the excitation frequency enters the separation zone, in which the real frequency response of the quarter car differs significantly from the analytic solution for no tire–road separation.

## 14.5 Summary

Tire–road separation is an extraordinary occurrence that arises during the analysis of a vehicle's ride behavior. It typically manifests as a transient phenomenon that quickly resolves itself. The intricate nature of this dynamic system necessitates the use of two sets of equations to simulate the conditions when the tire is in contact with the road and when it is not. These equations must be configured to maintain consistent states during the transition between the two sets.

To study tire–road separation, we use a quarter car model in which $m_s$ and $m_u$ indicate sprung and unsprung masses. The sprung mass is a quarter of the mass of the body, whereas the unsprung mass is the mass of one wheel and its components. The stiffness of the tire is $k_u$, and the stiffness of the main spring of the suspension system is $k_s$. The only damping mechanism of the system is the main shock absorber of the suspension system, indicated by $c_s$. The displacement of the masses is indicated by $x_s$ and $x_u$, and the displaceent of the road is shown by $y$.

Indicating $x_T$ as the compression displacement of the tire, the separation condition is

$$x_u - y \geq x_T \qquad \textit{separation}, \tag{14.155}$$

$$x_u - y < x_T \qquad \textit{in contact}. \tag{14.156}$$

The equation of motion of the sprung mass is independent of tire–road contact.

$$m_s \ddot{x}_s + c_s (\dot{x}_s - \dot{x}_u) + k_s (x_s - x_u) = 0. \tag{14.157}$$

However, the equation of motion of the unsprung mass depends on tire–road contact. In contact state $(x_u - y < x_T)$ we have

$$m_u \ddot{x}_u + c_s (\dot{x}_u - \dot{x}_s) + (k_u + k_s) x_u - k_s x_s = k_u y, \tag{14.158}$$

$$x_u - y < x_T, \tag{14.159}$$

where $y = y(t)$ is a given function. In free-fall state $(x_u - y \geq x_T)$ we have

$$m_u \ddot{x}_u + c_s (\dot{x}_u - \dot{x}_s) + k_s (x_u - x_s) + (m_u + m_s) g = 0, \tag{14.160}$$

$$x_u - y \geq x_T. \tag{14.161}$$

The equations of motion are better in nondimensionalized form to expand the result of the study on all vehicles. The equation of motion of the sprung mass will be

$$X_s'' + 2\xi\left(X_s' - X_u'\right) + \left(X_s - X_u\right) = 0, \tag{14.162}$$

while the equation of motion of the unsprung mass will be different, considering tire–road separation:

$$\begin{aligned} X_u'' + 2\varepsilon\xi\left(X_u' - X_s'\right) + \varepsilon\left(X_u - X_s\right) + \alpha^2 X_u = \alpha^2 Y, \\ X_u - Y \geq 1, \end{aligned} \tag{14.163}$$

$$\begin{aligned} X_u'' + 2\varepsilon\xi\left(X_u' - X_s'\right) + \varepsilon\left(X_u - X_s\right) + \alpha^2 = 0 \\ X_u - Y < 1 \end{aligned} \tag{14.164}$$

The following variables and parameters have been used to make the equations dimensionless:

$$\begin{aligned} &\alpha = \frac{\omega_u}{\omega_s}, \qquad \xi = \frac{c_s}{2m_s\omega_s}, \qquad \varepsilon = \frac{m_s}{m_u}, \\ &\omega_s = \sqrt{\frac{k_s}{m_s}}, \qquad \omega_u = \sqrt{\frac{k_u}{m_u}}, \\ &x_0 = x_T = \frac{\left(m_u + m_s\right)g}{k_u}, \qquad t_0 = \frac{1}{\omega_s}, \\ &Y = \frac{y}{x_0}, \qquad X_u = \frac{x_u}{x_0}, \qquad X_s = \frac{x_s}{x_0}, \qquad \tau = \frac{t}{t_0}, \\ &X_T = \frac{x_T}{x_0} = 1, \qquad X_R = \frac{x_R}{x_0} = n. \end{aligned} \tag{14.165}$$

To determine the boundary of tire–road separation in the frequency response domain, we define a relative displacement variable $x_t$ between the unsprung mass and the road:

$$x_t = x_u - y_0 \cos r\tau. \tag{14.166}$$

As long as the unsprung mass and road relative displacement are less than the static deflection of the tire, the tire will be in contact with the road. The dimensionless frequency response of $x_t$ will be $X_t$.

$$X_t^2 = \frac{4\xi^2 r^2 (\varepsilon + 1)^2 + \left(r^2 - 1\right)^2 + \varepsilon^2 - 2\varepsilon\left(r^2 - 1\right)}{Z_1^2 + Z_2^2} r^4 Y_0^2, \tag{14.167}$$

$$Z_1 = r^2\left(r^2\alpha^2 - 1\right) + \left(1 - (1 + \varepsilon) r^2\alpha^2\right), \tag{14.168}$$

$$Z_2 = 2r\xi\left(1 - (1 + \varepsilon) r^2\alpha^2\right). \tag{14.169}$$

The relative frequency response of $X_t/Y_0$ will show the zone of in-contact, $X_t/Y_0 < 1$, and tire–road separation, $X_t/Y_0 > 1$.

To detect tire–road separation in the time domain, we can define a separation function $S$ that becomes one when the tire separates from the road and remains zero when the tire is in contact with the road:

$$S = H\left(X_u - Y - 1\right) = \begin{cases} 1 & X_u - Y \geq 1, \\ 0 & X_u - Y < 1. \end{cases} \tag{14.170}$$

## 14.6 Key Symbols

| | |
|---|---|
| $a, \ddot{x}$ | acceleration |
| $[a], [A]$ | coefficient matrix |
| $[A] = [m]^{-1}[k]$ | coefficient matrix of characteristic equation |
| $c$ | damping |
| $c_e$ | equivalent damping |
| $c_{ij}$ | element of row $i$ and column $j$ of $[c]$ |
| $c_s$ | main suspension damper |
| $[c]$ | damping matrix |
| $D$ | dissipation function, denominator |
| $e$ | eccentricity arm |
| $E$ | mechanical energy |
| $f$, $\mathbf{F}$ | harmonic force |
| $f = \frac{1}{T}$ | cyclic frequency [Hz] |
| $f_c$ | damper force |
| $f_k$ | spring force |
| $F$ | amplitude of harmonic force $f$ |
| $F_r$, $Q_r$ | generalized force |
| $g$ | gravitational acceleration |
| $g(r)$ | equation of frequency ratio |
| $H$ | Heaviside function |
| $I$ | mass moment of inertia |
| $\mathbf{I}$ | identity matrix |
| $k$ | stiffness |
| $k_{eq}$ | equivalent stiffness |
| $k_{ij}$ | element of row $i$ and column $j$ of $[k]$ |
| $k_R$ | antiroll bar torsional stiffness |
| $k_s$ | main suspension spring stiffness |
| $k_u$ | tire stiffness |
| $[k]$ | stiffness matrix |
| $[\underline{k}]$ | symmetric stiffness matrix |
| $K$ | kinetic energy |
| $l$ | length |
| $l$ | wheelbase |
| $\mathcal{L}$ | Lagrangean |
| $m$ | mass |
| $m_e$ | eccentric mass |
| $m_{ij}$ | element of row $i$ and column $j$ of $[m]$ |
| $m_s$ | sprung mass |
| $m_u$ | unsprung mass |
| $[m]$ | mass matrix |
| $n$ | number of $DOF$ |
| $N$ | numerator |
| $\mathbf{p}$ | momentum |
| $P$ | potential energy |
| $r = \omega/\omega_s$ | frequency ratio |
| $r_i$, $i \in N$ | nodal frequency ratio |
| $r_n = \omega_n/\omega_s$ | natural frequency ratio |
| $t$ | time |
| $T$ | period |
| $u_{ij}$ | $j$th element of the $i$th mode shape |
| $\mathbf{u}$ | mode shape, eigenvector |

| | |
|---|---|
| $\mathbf{u}_i$ | $i$th eigenvector |
| $v, \mathbf{v}, \dot{x}, \dot{\mathbf{x}}$ | velocity |
| $W$ | work |
| $x, \mathbf{x}$ | absolute displacement |
| $x_s$ | sprung mass displacement |
| $x_u$ | unsprung mass displacement |
| $X$ | steady-state amplitude of $x$ |
| $X_s$ | steady-state amplitude of $x_s$ |
| $X_u$ | steady-state amplitude of $x_u$ |
| $y$ | base excitation displacement |
| $Y$ | steady-state amplitude of $y$ |
| $z$ | relative displacement |
| $Z$ | steady-state amplitude of $z$ |
| $Z_i$ | short notation parameter |

Greek

| | |
|---|---|
| $\alpha = \omega_s/\omega_u$ | frequency mass ratio |
| $\varepsilon = m_s/m_u$ | sprung mass ratio |
| $\xi$ | damping ratio |
| $\tau$ | dimensionless time |
| $\omega = 2\pi f$ | angular frequency [rad/s] |
| $\omega_i$ | $i$th natural frequency |
| $\omega_s = \sqrt{k_s/m_s}$ | sprung mass frequency |
| $\omega_u = \sqrt{k_u/m_u}$ | unsprung mass frequency |
| $\omega_n$ | natural frequency |
| $\delta$ | deflection |
| $\lambda$ | eigenvalue |
| $\lambda_i$ | $i$th eigenvalue |

Subscript

| | |
|---|---|
| $f$ | front |
| $i \in N$ | node number |
| $n$ | natural |
| $r$ | rear |
| $R$ | physical limit of wheel travel |
| $s$ | sprung |
| $T$ | tire |
| $u$ | unsprung |

Symbols

| | |
|---|---|
| $DOF$ | degree of freedom |
| $FBD$ | free body diagram |
| $ODF$ | one degree of freedom |
| Re | real part |
| $RMS$ | root mean square |
| det | determinant |
| Hz | Hertz |
| [ ] | matrix |
| $\int$ | integral |
| $\| \|$ | absolute value |
| $d$ | differential |
| $\partial$ | partial derivative |
| $\mathcal{L}$ | Lagrangean |

## References

1. M. Akraminia, M. Tatari, M. Fard, R.N. Jazar, Designing active vehicle suspension system using critic-based control strategy. Nonlinear Eng. **4**(3), 141–154 (2015)
2. F. Alam, A. Faheem, R.N. Jazar, L.V. Smith, A study of vehicle ride performance using a quarter car model and half car model, in *ASME 2010 International Mechanical Engineering Congress and Exposition*, November 12–18, Vancouver, British Columbia, Canada (2010)
3. R.S. Barbosa, Vehicle dynamic safety in measured rough pavement. J. Transport. Eng. **137**(5), 305–310 (2010)
4. A. Best, Vehicle ride-stages in comprehension. Phys. Technol. **15**(4), 205 (2002)
5. K. Captain, A. Boghani, D. Wormley, Analytical tire models for dynamic vehicle simulation. Vehic. Syst. Dyn. **8**(1), 1–32 (1979)
6. J. Christopherson, R.N. Jazar, Optimization of classical hydraulic engine mounts based on RMS method. J. Shock Vibrat. **12**(2), 119–147 (2005)
7. J. Christopherson, R.N. Jazar, Dynamic behavior comparison of passive hydraulic engine mounts, part 2: finite element analysis. J. Sound Vibrat. **290**, 1071–1090 (2006)
8. L. Dai, R.N. Jazar, *Nonlinear Approaches in Engineering Applications* (Springer, New York, 2012)
9. J.P. Den Hartog, *Mechanical Vibrations* (Courier Corporation, North Chelmsford, 1985)
10. S. Deshpande, S. Mehta, R.N. Jazar, Jump avoidance conditions for piecewise linear vibration isolator, in *International Design Engineering Technical Conferences and Computers and Information in Engineering Conference, DETC2005-85396* (2008), pp. 1955–1962
11. S. Deshpande, S. Mehta, R.N. Jazar, Sensitivity of jump avoidance condition of piecewise linear vibration isolator to dynamical parameters, in *ASME 2005 International Mechanical Engineering Congress and Exposition, November 5–11, Orlando, Florida, USA, IMECE2005-81853* (2008), pp. 1103–1113
12. K. Dobaj, Influence of car wheel suspension parameters on improvement of active safety and ride comfort. Czasopismo Techniczne **2015**(Mechanika Zeszyt 3-M)(21) 2015, 13–22 (2016)
13. E. Esmailzadeh, R.N. Jazar, Periodic solution of a Mathieu-Duffing type equation. Int. J. Non-Linear Mech. **32**(5), 905–912 (1997)
14. E. Esmailzadeh, B. Mehri, R.N. Jazar, Periodic solution of a second order, autonomous, nonlinear system. Nonlinear Dyn. **10**, 307–316 (1996)
15. M. Guiggiani, *The Science of Vehicle Dynamics* (Springer Netherlands, Pisa, 2014)
16. R.N. Jazar, Stability chart of parametric vibrating systems, using energy method. Int. J. Nonlinear Mech. **39**(8), 1319–1331 (2004)
17. R.N. Jazar, *Advanced Dynamics: Rigid Body, Multibody, and Aerospace Applications*(Wiley, New York, 2011)
18. R.N. Jazar, *Advanced Vibrations: A Modern Approach* (Springer, New York, 2013)
19. R.N. Jazar, *Vehicle Dynamics: Theory and Application* (Springer, New York, 2017)
20. R.N. Jazar, *Advanced Vehicle Dynamics* (Springer, New York, 2019)
21. R.N. Jazar, *Approximation Methods in Science and Engineering* (Springer, New York, 2020)
22. R.N. Jazar, *Perturbation Methods in Science and Engineering* (Springer, New York, 2021)
23. R.N. Jazar, *Advanced Vibrations: Theory and Application* (Springer, New York, 2023)
24. R.N. Jazar, M. Kazemi, S. Borhani, *Mechanical Vibrations*, (in Persian) (Ettehad Publications, Tehran, 1992)
25. R.N. Jazar, E. Esmailzadeh, B. Mehri, Vibration of road vehicles with non linear suspensions. Int. J. Eng. **10**(4), 209–218 (1997)
26. R.N. Jazar, A. Narimani, F. Golnaraghi, D.A. Swanson, Practical frequency and time optimal design of passive linear vibration isolation mounts. Vehic. Syst. Dyn. **39**(6), 437–466 (2003)
27. R.N. Jazar, R. Houim, A. Narimani, M.F. Golnaraghi, Frequency response and jump avoidance in a nonlinear passive engine mount. J. Vibrat. Control **12**(11), 1205–1237 (2006)
28. R.N. Jazar, M. Mahinfalah, S. Deshpande, Design of a piecewise linear vibration isolator for jump avoidance. IMechE Part K J. Multi-Body Dyn. **221**(K3),441–450 (2007)
29. R.N. Jazar, M. Mahinfalah, N. Mahmoudian, M. Rastgaar, Energy-rate method and stability chart of parametric vibrating systems. J. Braz. Soc. Mech. Sci. Eng. **30**(3), 182–188 (2008)
30. A. Khazaie, N. Hussaini, H. Marzbani, R.N. Jazar, Quarter car suspension model with provision for loss of contact with the road, in *Nonlinear Approaches in Engineering Applications* (Springer, New York, 2018), pp. 167–208
31. C. Kim, P.I. Ro, Effect of the suspension structure on equivalent suspension parameters. HJPotIoME, Part D J. Autom. Eng. **213**(5), 457–470 (1999)
32. C. Kim, P.I. Ro, Reduced-order modelling and parameter estimation for a quarter-car suspension system. PJPotIoME, Part D J. Autom. Eng. **214**(8), 851–864 (2000)
33. B. Liebowitz, Dynamics of vehicle suspensions. Trans. Soc. Autom. Eng. **11**, PART II, 186–222 (1916)
34. G. Lindgren, Jumps and bumps on random roads. J. Sound Vibrat. **78**(3), 383–395 (1981)
35. O. Lukoðevièienë, E. Sokolovskij, Movement of a vehicle after loosing the touch of the wheels with road surface (Transporto priemonës judějimas, ratams atitrûkus nuo kelio pavirðiaus). Transp. Eng. (Transportas) **16**(1), 29 (2001)
36. O. Lukoševičiene, M. Bogdevičius, S. Nagurnas, R. Pečeliūnas, Comparative research of a motor car motion in the case of the loss of contact with the road surface. Transport **19**(1), 20–23 (2004)
37. O. Lukoševičiene, E. Sokolovskij, Movement of the vehicle being braked when some wheels are incapable of braking or have lost touch with the road surface. Transport **21**(1), 8–11 (2006)
38. H. Marzbani, R.N. Jazar, A. Khazaei, Smart passive vibration isolation: requirements and unsolved problems, J. Appl. Nonlinear Dyn. **1**(4), 341–386 (2012). https://doi.org/10.5890/JAND.2012.09.002
39. H. Marzbani, R.N. Jazar, M. Fard, Hydraulic engine mounts: a survey. J. Vibrat. Control **19**(16), 1439–1463 (2013)
40. H. Marzbani, R.N. Jazar, Smart flat ride tuning, in *Nonlinear Approaches in Engineering Applications*, ed. by R.N. Jazar, L. Dai, vol. 2 (Springer, New York, 2014)
41. W.F. Milliken, D.L. Milliken, M. Olley, *Chassis Design: Principles and Analysis* (Society of Automotive Engineers Warrendale, Warrendale, 2002)
42. A. Narimani, F. Golnaraghi, R.N. Jazar, Frequency response of a piecewise linear vibration isolator. J. Vibrat. Control **10**(12), 1775–1794 (2004)

43. A. Narimani, F. Golnaraghi, R.N. Jazar, Sensitivity analysis of the frequency response of a piecewise linear system in a frequency Island. J. Vibrat. Control **10**(2), 175–198 (2004)
44. D.Q. Nguyen, R.N. Jazar, Numerical simulation and experimental validation of the tire-road separation in quarter-car model, in *ICAME 2022: International Conference on Automotive and Mechanical Engineering*, Sydney, Australia, December 02–03 (2022)
45. Q.D. Nguyen, S. Milani, H. Marzbania, R.N. Jazar, Vibration analysis of the bicycle-car model considering tire-road separation. SAE Int. J. Commer. Vehic. **15**(02-15-02-0004), 111–124 (2021)
46. Q.D. Nguyen, S. Milani, H. Marzbani, R.N. Jazar, Tire-road separation time reduction by an adaptive proportional-integral-derivative controller utilizing particle swarm optimization algorithm. SAE Int. J. Commer. Vehic. **15**(02-14-04-0033), 37–50 (2021)
47. D.Q. Nguyen, S. Milani, H. Marzbani, R.N. Jazar, Tire-road separation time reduction by an adaptive PID controller utilizing particle swarm optimization algorithm. SAE Int. J. Commer. Vehic. **14**(4), 1–14 (2022)
48. D.Q. Nguyen, S. Milani, H. Marzbani, R.N. Jazar, Vehicle vibration analysis of the quarter-car model considering tire-road separation, in *Nonlinear Approaches in Engineering Application: Design Engineering Problems* (Springer, New York, 2022), pp. 429–465
49. K. Reza-Kashyzadeh, M.J. Ostad-Ahmad-Ghorabi, A. Arghavan, Investigating the effect of road roughness on automotive component. Eng. Fail. Analy. **41**, 96–107 (2014)
50. M. Roseau, *Vibrations in Mechanical Systems* (Springer, Berlin, 1987)
51. J.C. Snowdon, *Vibration and Shock in Damped Mechanical Systems* (Wiley, New York, 1968)
52. F. Svahn, J. Jerrelind, H. Dankowicz, Suppression of bumpstop instabilities in a quarter-car model, in *Non-smooth Problems in Vehicle Systems Dynamics* (Springer, New York, 2009), pp. 137–147
53. P.G. Thomsen, *Discontinuities in ODEs: Systems with Change of State* (Springer, Berlin, 2010)
54. F.S. Tse, I.E. Morse, R.T. Hinkle, *Mechanical Vibrations Theory and Applications* (Allyn and Bacon Inc, Boston, 1978)
55. J.Y. Wong, *Theory of Ground Vehicles* (Wiley, Hoboken, 2008)
56. D. Young, P. Maher, An insight into linear quarter car model accuracy. Vehic. Syst. Dyn. **49**(3), 463–480 (2011)

# Frequency Response Curves

A

There are four types of one $DOF$ harmonically excited systems as shown in Fig. 7.1:

1 Base excitation
2 Eccentric excitation
3 Eccentric base excitation
4 Forced excitation

The frequency responses of the four systems can be summarized, labeled, and shown as follows (Figs. A 1, A.2, A.3, A.4, A.5, A.6, A.7, and A.8):

$$S_0 = \frac{X_F}{F/k} \tag{A.1}$$

$$= \frac{1}{\sqrt{\left(1 - r^2\right)^2 + (2\xi r)^2}} \tag{A.2}$$

$$S_1 = \frac{\dot{X}_F}{F/\sqrt{km}} \tag{A.3}$$

$$= \frac{r}{\sqrt{\left(1 - r^2\right)^2 + (2\xi r)^2}} \tag{A.4}$$

$$S_2 = \frac{\ddot{X}_F}{F/m} = \frac{Z_B}{Y} = \frac{X_E}{e\varepsilon_E} = \frac{Z_R}{e\varepsilon_R} \tag{A.5}$$

$$= \frac{r^2}{\sqrt{\left(1 - r^2\right)^2 + (2\xi r)^2}} \tag{A.6}$$

$$S_3 = \frac{\dot{Z}_B}{\omega_n Y} = \frac{\dot{X}_E}{e\varepsilon_E \omega_n} = \frac{\dot{Z}_R}{e\varepsilon_R \omega_n} \tag{A.7}$$

$$= \frac{r^3}{\sqrt{\left(1 - r^2\right)^2 + (2\xi r)^2}} \tag{A.8}$$

$$S_4 = \frac{\ddot{Z}_B}{\omega_n^2 Y} = \frac{\ddot{X}_E}{e\varepsilon_E \omega_n^2} = \frac{\ddot{Z}_R}{e\varepsilon_R \omega_n^2} \tag{A.9}$$

$$= \frac{r^4}{\sqrt{\left(1 - r^2\right)^2 + (2\xi r)^2}} \tag{A.10}$$


R. N. Jazar, H. Marzbani, *Vehicle Vibrations*, https://doi.org/10.1007/978-3-031-43486-0

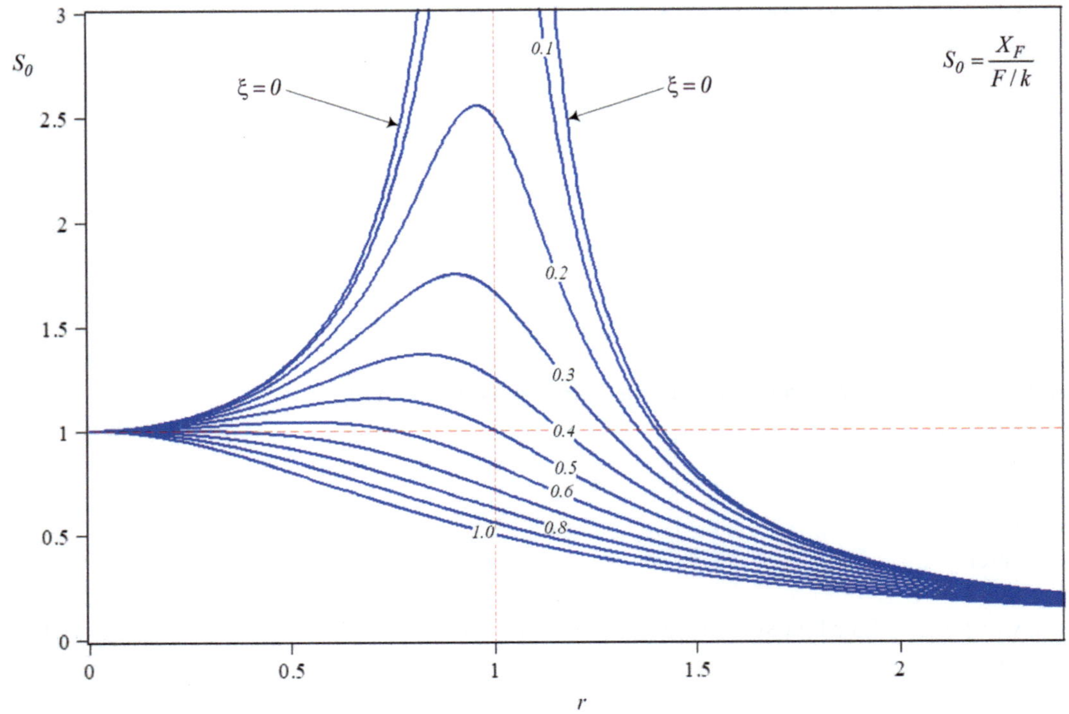


**Fig. A.1** Frequency response for $S_0$

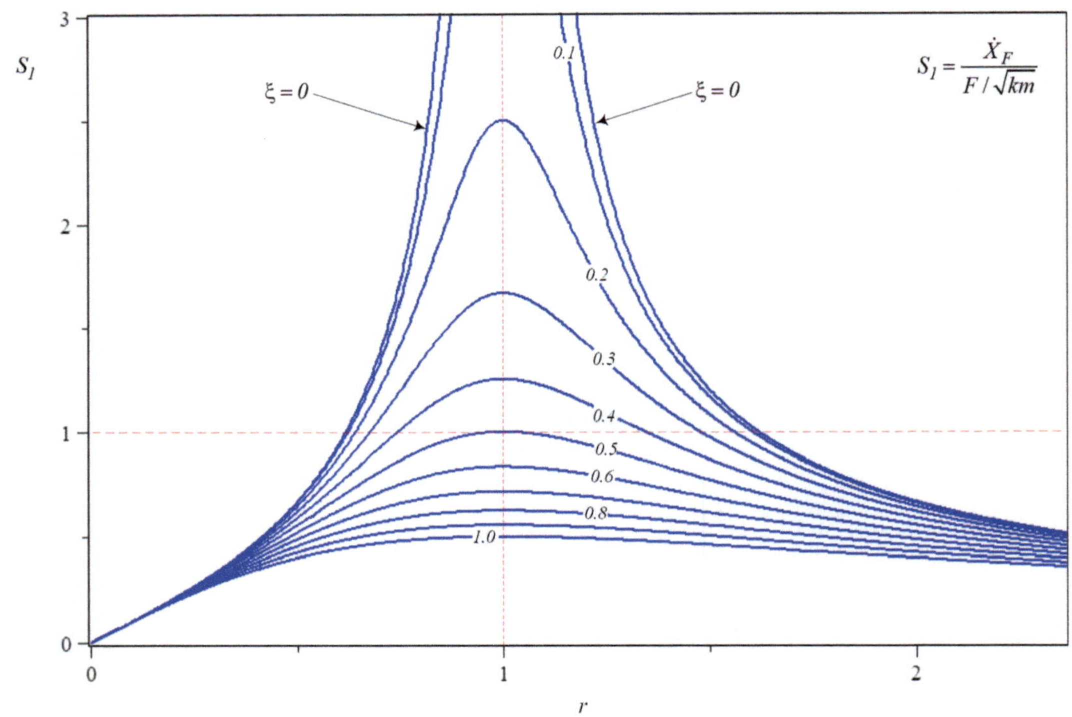


**Fig. A.2** Frequency response for $S_1$

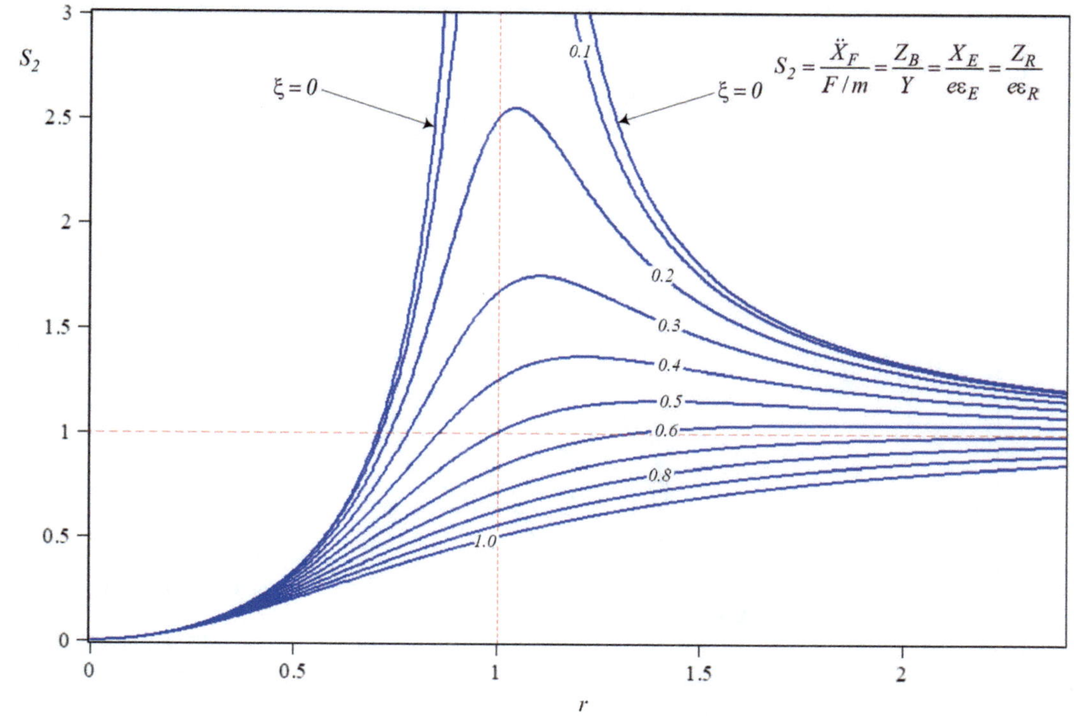


**Fig. A.3** Frequency response for $S_2$

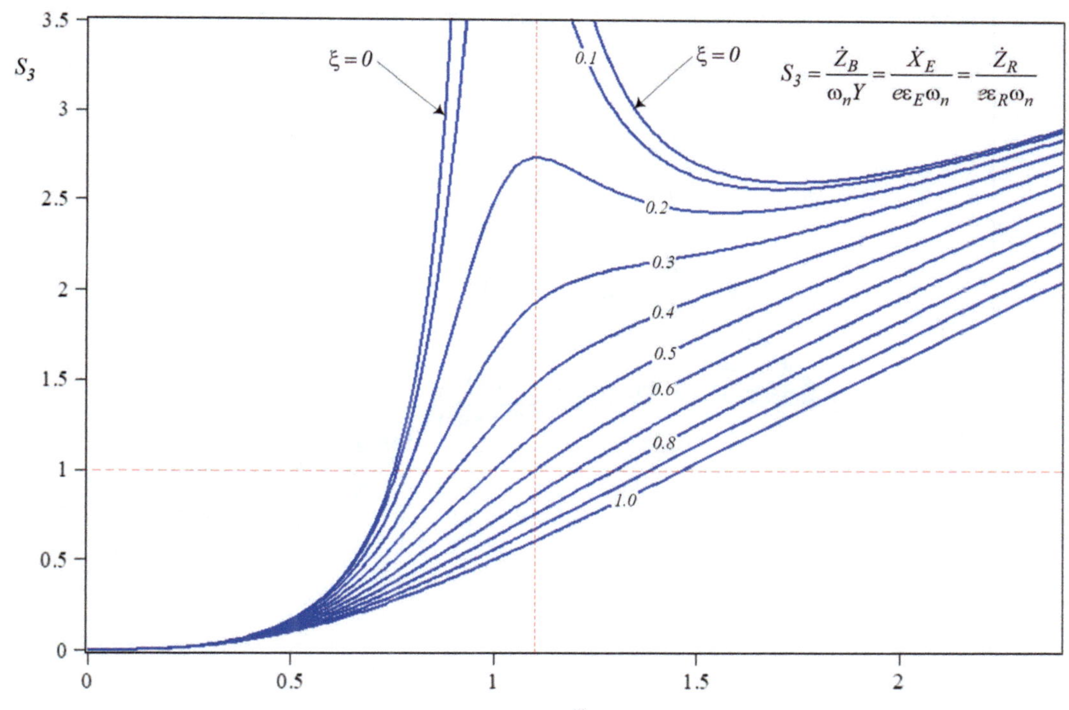


**Fig. A.4** Frequency response for $S_3$

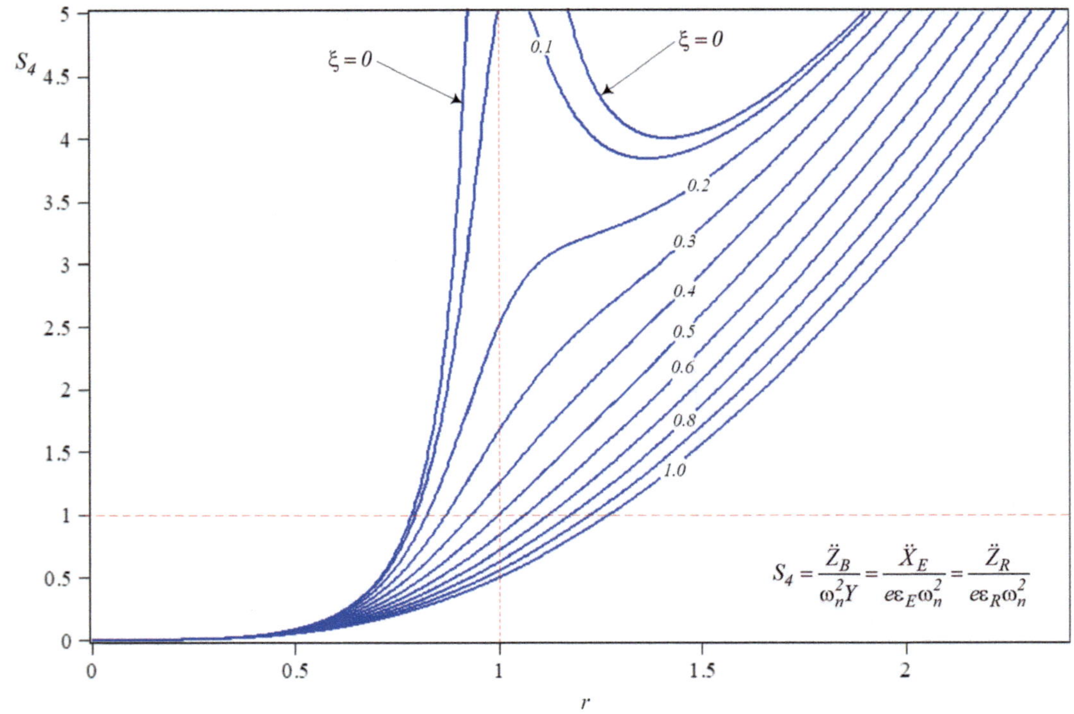


**Fig. A.5** Frequency response for $S_4$

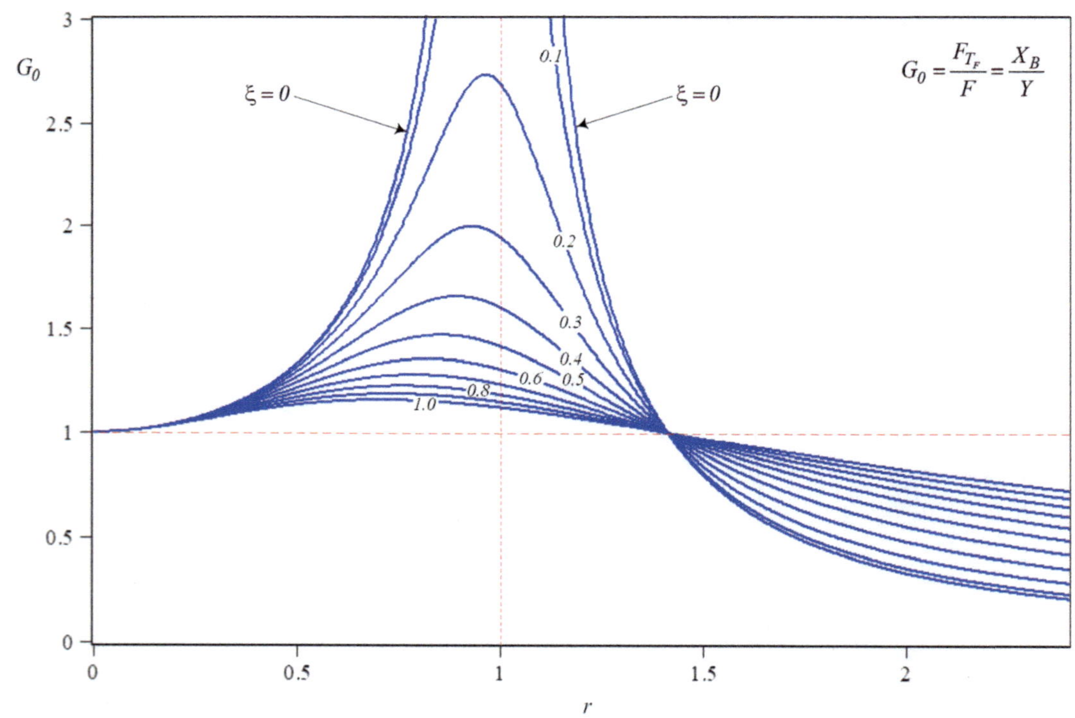


**Fig. A.6** Frequency response for $G_0$

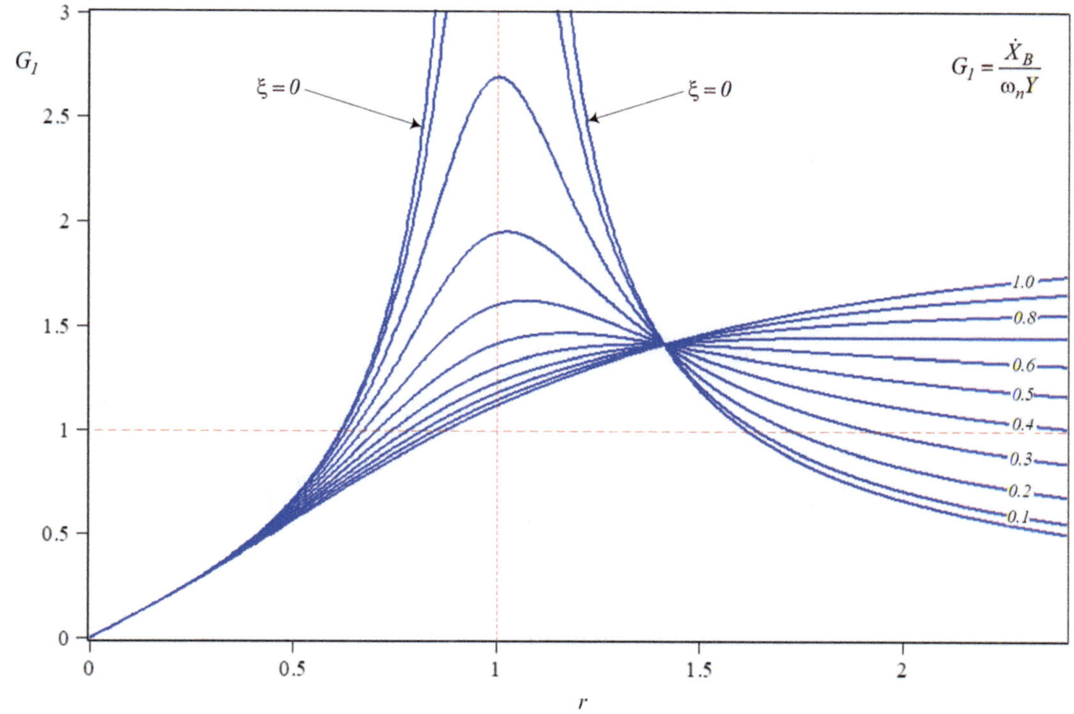


**Fig. A.7** Frequency response for $G_1$

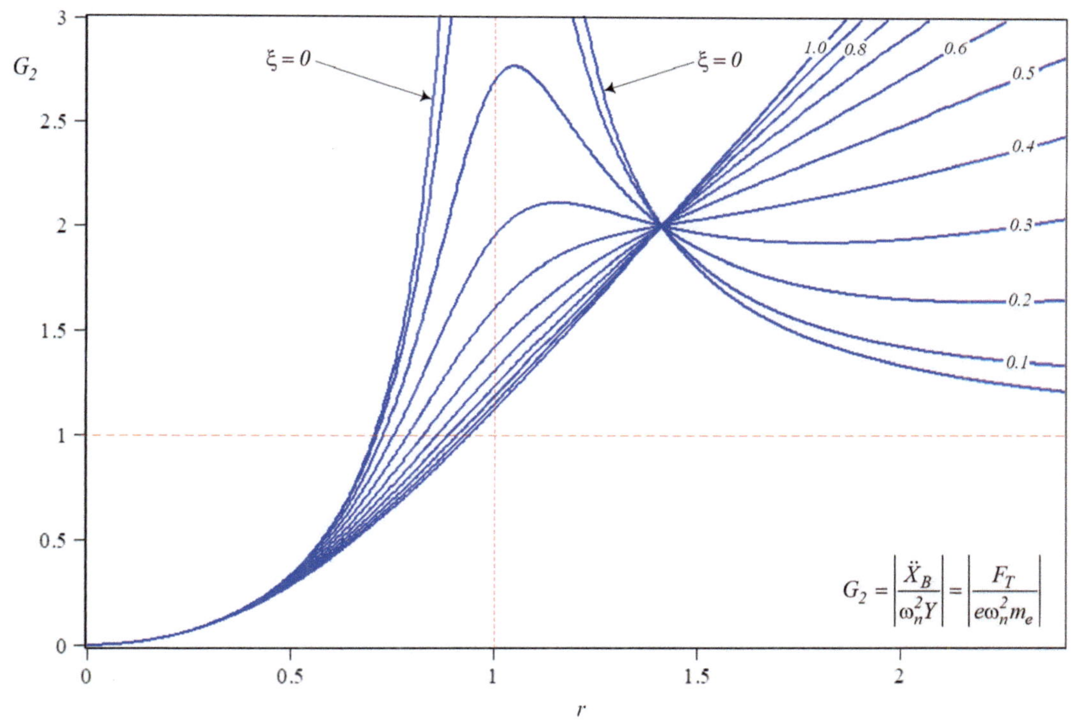


**Fig. A.8** Frequency response for $G_2$

$$G_0 = \frac{F_{T_F}}{F} = \frac{X_B}{Y} \tag{A.11}$$

$$= \frac{\sqrt{1 + (2\xi r)^2}}{\sqrt{\left(1 - r^2\right)^2 + (2\xi r)^2}} \tag{A.12}$$

$$G_1 = \frac{\dot{X}_B}{\omega_n Y} \tag{A.13}$$

$$= \frac{r\sqrt{1 + (2\xi r)^2}}{\sqrt{\left(1 - r^2\right)^2 + (2\xi r)^2}} \tag{A.14}$$

$$G_2 = \frac{\ddot{X}_B}{\omega_n^2 Y} = \frac{F_{T_B}}{kY} = \frac{F_{T_E}}{e\omega_n^2 m_e} = \frac{F_{T_R}}{e\omega_n^2 m_e}\left(1 + \frac{m_b}{m}\right) \tag{A.15}$$

$$= \frac{r^2\sqrt{1 + (2\xi r)^2}}{\sqrt{\left(1 - r^2\right)^2 + (2\xi r)^2}} \tag{A.16}$$

# Matrix Calculus

# B

An $m \times n$ matrix $\mathbf{A}$ consists of $mn$ numbers arranged in $m$ rows and $n$ columns. The element in row $i$ and column $j$ of the matrix $\mathbf{A}$ is denoted by $a_{ij}$.

$$\mathbf{A} = \begin{bmatrix} a_{11} & a_{12} & \cdots & a_{1n} \\ a_{21} & a_{22} & \cdots & a_{2n} \\ \cdots & \cdots & \cdots & \cdots \\ a_{m1} & a_{m2} & \cdots & a_{mn} \end{bmatrix} = \begin{bmatrix} a_{ij} \end{bmatrix} \tag{B.1}$$

An $m \times 1$ matrix is a column vector $\mathbf{v}$ of dimension $m$, and a $1 \times n$ matrix is a row vector $\mathbf{w}$ of dimension $n$. An $m \times n$ matrix is called a square matrix if $m = n$.

$$\mathbf{v} = \begin{bmatrix} v_{11} \\ v_{21} \\ \cdots \\ v_{m1} \end{bmatrix} \tag{B.2}$$

$$\mathbf{w} = \begin{bmatrix} w_{m1} & w_{m2} & \cdots & w_{mn} \end{bmatrix} \tag{B.3}$$

Two matrices $\mathbf{A} = \begin{bmatrix} a_{ij} \end{bmatrix}$ and $\mathbf{B} = \begin{bmatrix} b_{ij} \end{bmatrix}$ are equal only if $a_{ij} = b_{ij}$.

## Diagonal Matrix

A diagonal matrix is a square matrix $\mathbf{\Lambda} = \begin{bmatrix} \lambda_{ij} \end{bmatrix}$ such that $\lambda_{ij} = 0$ for $i \neq j$. The main diagonal of a square matrix is elements from top left to bottom right.

$$\mathbf{\Lambda} = \begin{bmatrix} \lambda_{11} & 0 & \cdots & 0 \\ 0 & \lambda_{22} & \cdots & 0 \\ \cdots & \cdots & \cdots & \cdots \\ 0 & 0 & \cdots & \lambda_{nn} \end{bmatrix} \tag{B.4}$$

A diagonal matrix may also be abbreviated as

$$\mathbf{\Lambda} = \{\lambda_{11}, \lambda_{22}, \lambda_{22}, \cdots, \lambda_{nn}\} \tag{B.5}$$

A scalar matrix $\mathbf{S} = \begin{bmatrix} s_{ij} \end{bmatrix}$ is a diagonal matrix whose diagonal elements are equal $s_{ii} = s$.

$$\mathbf{S} = \begin{bmatrix} s & 0 & \cdots & 0 \\ 0 & s & \cdots & 0 \\ \cdots & \cdots & \cdots & \cdots \\ 0 & 0 & \cdots & s \end{bmatrix} \tag{B.6}$$



R. N. Jazar, H. Marzbani, *Vehicle Vibrations*, https://doi.org/10.1007/978-3-031-43486-0

If all elements of the diagonal of **S** are unity, the matrix is the identity matrix, or unit matrix, and shown by **I**.

$$\mathbf{I} = \begin{bmatrix} 1 & 0 & \cdots & 0 \\ 0 & 1 & \cdots & 0 \\ \cdots & \cdots & \cdots & \cdots \\ 0 & 0 & \cdots & 1 \end{bmatrix} \tag{B.7}$$

A null matrix is a matrix of any shape in which every entry is zero.

$$\mathbf{0} = \begin{bmatrix} 0 & 0 & \cdots & 0 \\ 0 & 0 & \cdots & 0 \\ \cdots & \cdots & \cdots & \cdots \\ 0 & 0 & \cdots & 0 \end{bmatrix} \tag{B.8}$$

## Trace

Trace of a square matrix $\mathbf{A} = \left[a_{ij}\right]$ is the sum of its diagonal elements.

$$\operatorname{tr} \mathbf{A} = a_{11} + a_{22} + \cdots + a_{nn} \tag{B.9}$$

$$\operatorname{tr} [\mathbf{AB}] = \operatorname{tr} [\mathbf{BA}] \tag{B.10}$$

## Matrix Arithmetic

$$k\mathbf{A} = k\left[a_{ij}\right] = \left[ka_{ij}\right] \tag{B.11}$$

$$\mathbf{A} + \mathbf{B} = \mathbf{B} + \mathbf{A} = \mathbf{C} \tag{B.12}$$

$$c_{ij} = a_{ij} + b_{ij} \tag{B.13}$$

$$\mathbf{AB} = \mathbf{C} \qquad c_{ij} = \sum_m a_{im} b_{mj} \tag{B.14}$$

$$\mathbf{AB} \neq \mathbf{BA} \tag{B.15}$$

$$[\mathbf{AB}]\,\mathbf{C} = \mathbf{A}\,[\mathbf{BC}] = \mathbf{D} \tag{B.16}$$

$$d_{ij} = \sum_{m,n} (a_{im} b_{mn})\, c_{jn} = \sum_{m,n} a_{im} \left(b_{mn} c_{jn}\right) \tag{B.17}$$

$$[\mathbf{A} + \mathbf{B}]\,\mathbf{C} = \mathbf{AC} + \mathbf{BC} = \mathbf{D} \tag{B.18}$$

$$d_{ij} = \sum_m (a_{im} + b_{im})\, c_{nj} = \sum_m \left(a_{im} c_{nj} + b_{im} c_{nj}\right) \tag{B.19}$$

## Determinants

The determinant of a 2 × 2 matrix **A**, with elements $a_{ij}$ comprising real or complex numbers, or functions,

$$\mathbf{A} = \begin{bmatrix} a_{11} & a_{12} \\ a_{21} & a_{22} \end{bmatrix} = \left[a_{ij}\right] \tag{B.20}$$

denoted by $|\mathbf{A}|$ or by $\det \mathbf{A}$, is defined by

$$\det \mathbf{A} = |\mathbf{A}| = a_{11}a_{22} - a_{12}a_{21} \tag{B.21}$$

The determinant of a 3 × 3 matrix **A**,

$$\mathbf{A} = \begin{bmatrix} a_{11} & a_{12} & a_{13} \\ a_{21} & a_{22} & a_{23} \\ a_{31} & a_{32} & a_{33} \end{bmatrix} = \left[a_{ij}\right] \tag{B.22}$$

with elements $a_{ij}$ comprising real or complex numbers, or functions, denoted by $|\mathbf{A}|$ or by $\det \mathbf{A}$, is defined by

$$\begin{aligned} \det \mathbf{A} = {} & a_{11}a_{22}a_{33} + a_{12}a_{23}a_{31} + a_{21}a_{32}a_{13} \\ & - a_{13}a_{22}a_{31} - a_{23}a_{32}a_{11} - a_{12}a_{21}a_{33} \end{aligned} \tag{B.23}$$

To calculate the determinant of an $n \times n$ matrix **A**,

$$\mathbf{A} = \begin{bmatrix} a_{11} & a_{12} & \cdots & a_{1n} \\ a_{21} & a_{22} & \cdots & a_{2n} \\ \cdots & \cdots & \cdots & \cdots \\ a_{n1} & a_{n2} & \cdots & a_{nn} \end{bmatrix} = \left[a_{ij}\right] \tag{B.24}$$

we need first to define the minor $M_{ij}$ associated with the element $a_{ij}$ as determinant the order $(n-1)$ derived from by deletion of its $i$th row and $j$th column, and second, to define the cofactor $C_{ij}$ associated with the element $a_{ij}$ as

$$C_{ij} = (-1)^{i+j}\, M_{ij} \tag{B.25}$$

The determinant of **A** may be based on the expansion of elements of a row or of a column of **A**. The expansion of $\det \mathbf{A}$ by elements of the $i$th row is

$$\det \mathbf{A} = \sum_{j=1}^{n} a_{ij}C_{ij} \qquad i = 1, 2, 3, \cdots, n \tag{B.26}$$

and the expansion of $\det \mathbf{A}$ by the elements of the $j$th column is

$$\det \mathbf{A} = \sum_{i=1}^{n} a_{ij}C_{ij} \qquad j = 1, 2, 3, \cdots, n \tag{B.27}$$

The expansion method to determine the determinant is called the Laplace expansion method.

As an example, let us calculate $\det \mathbf{A}$, where

$$\mathbf{A} = \begin{bmatrix} -2 & 1 & 3 \\ 1 & 2 & -1 \\ 4 & -1 & -2 \end{bmatrix} \tag{B.28}$$

Expanding $|\mathbf{A}|$ by the elements of its second row gives

$$|\mathbf{A}| = C_{21} + 2C_{22} - C_{23} \tag{B.29}$$

$$C_{21} = (-1)^{2+1} \begin{vmatrix} 1 & 3 \\ -1 & -2 \end{vmatrix} = -1 \tag{B.30}$$

$$C_{22} = (-1)^{2+2} \begin{vmatrix} -2 & 3 \\ 4 & -2 \end{vmatrix} = -8 \tag{B.31}$$

$$C_{22} = (-1)^{2+3} \begin{vmatrix} -2 & 1 \\ 4 & -1 \end{vmatrix} = 2 \tag{B.32}$$

Therefore,

$$|\mathbf{A}| = -1 + 2\,(-8) - 2 = -19 \tag{B.33}$$

Alternatively, expanding $|\mathbf{A}|$ by the elements of its third column gives

$$|\mathbf{A}| = 3C_{13} - C_{23} - 2C_{33} \tag{B.34}$$

$$C_{21} = (-1)^{1+3} \begin{vmatrix} 1 & 2 \\ 4 & -1 \end{vmatrix} = -9 \tag{B.35}$$

$$C_{22} = (-1)^{2+3} \begin{vmatrix} -2 & 1 \\ 4 & -1 \end{vmatrix} = 2 \tag{B.36}$$

$$C_{22} = (-1)^{3+3} \begin{vmatrix} -2 & 1 \\ 1 & 2 \end{vmatrix} = -5 \tag{B.37}$$

Therefore,

$$|\mathbf{A}| = 3\,(-9) - 2 - 2\,(-5) = -19 \tag{B.38}$$

If $\mathbf{A} = [a_{ij}]$ and $\mathbf{B} = [b_{ij}]$ are two $n \times n$ matrices, then the following results are true for determinants of $\mathbf{A}$ and $\mathbf{B}$:

1. If any two adjacent rows (or columns) of $|\mathbf{A}|$ are interchanged, the sign of the resulting determinant is changed.
2. If any two rows (or columns) of $|\mathbf{A}|$ are identical, then $|\mathbf{A}| = 0$.
3. The value of a determinant is not changed if any multiple of a row (or column) is added to any other row (or column) of $|\mathbf{A}|$.
4. $|c\mathbf{A}| = c^n\,|\mathbf{A}|$, $c \in \mathbb{N}$.
5. $|\mathbf{A}^T| = |\mathbf{A}.|$
6. $|\mathbf{A}|\,|\mathbf{B}| = |\mathbf{AB}|$.
7. $|\mathbf{A}^{-1}| = 1/\,|\mathbf{A}.|$

## Transpose, Symmetric, and Skew Matrices

The matrix $\mathbf{A}^T = [a_{ji}]$ obtained from $\mathbf{A} = [a_{ij}]$ by changing rows to columns is the transpose of $\mathbf{A}$.

$$\mathbf{A} = [a_{ij}] \qquad \mathbf{A}^T = [a_{ji}] \tag{B.39}$$

$$[\mathbf{A}^T]^T = \mathbf{A} \tag{B.40}$$

$$[\mathbf{A} + \mathbf{B} + \cdots + \mathbf{C}]^T = \mathbf{A}^T + \mathbf{B}^T + \cdots + \mathbf{C}^T \tag{B.41}$$

$$[\mathbf{AB} \cdots \mathbf{C}]^T = \mathbf{C}^T \cdots \mathbf{B}^T\,\mathbf{A}^T \tag{B.42}$$

$$[\mathbf{AB}]^T = \left[\sum_n a_{in} b_{nj}\right]^T = \left[\sum_n b_{ni} a_{jn}\right] = \mathbf{B}^T\,\mathbf{A}^T \tag{B.43}$$

A matrix $\mathbf{A}$ such that $\mathbf{A}^T = \mathbf{A}$ is a symmetric matrix. A matrix $\mathbf{A}$ such that $\mathbf{A}^T = -\mathbf{A}$ is skew, skew-symmetric, or alternating matrix.

## Inverse

The inverse of a square matrix $\mathbf{A}$ is indicated by $\mathbf{A}^{-1}$

$$\mathbf{A}^{-1} = \frac{\operatorname{adj}\mathbf{A}}{|\mathbf{A}|} \qquad \mathbf{A}^{-1}\,\mathbf{A} = \mathbf{A}\,\mathbf{A}^{-1} = \mathbf{I} \tag{B.44}$$

where adj $\mathbf{A}$ is the adjoint of the matrix $\mathbf{A}$, and $C_{ij}$ are cofactors associated with element $a_{ij}$ of $\mathbf{A}$.

$$\operatorname{adj}\mathbf{A} = \left[C_{ij}\right]^T \tag{B.45}$$

$$[\mathbf{AB}\cdots\mathbf{C}]^{-1} = \mathbf{C}^{-1}\cdots\mathbf{B}^{-1}\,\mathbf{A}^{-1} \tag{B.46}$$

$$\left[\mathbf{A}^{-1}\right]^{-1} = \mathbf{A} \tag{B.47}$$

$$\left[\mathbf{A}^{-1}\right]^T = \left[\mathbf{A}^T\right]^{-1} \tag{B.48}$$

$$[k\mathbf{A}]^{-1} = k^{-1}\mathbf{A}^{-1} \tag{B.49}$$

Non-square matrices of order $m \times n$, $m \neq n$ do not have inverse. However, if $\mathbf{A}$ is $m \times n$ and the rank of $\mathbf{A}$ is equal to $n$, then $\mathbf{A}$ has a left inverse. An $n \times m$ matrix $\mathbf{B}$ such that $\mathbf{BA} = \mathbf{I}$. If $\mathbf{A}$ has rank $m$, then it has a left inverse, which is an $n \times m$ matrix $\mathbf{B}$ such that $\mathbf{AB} = \mathbf{I}$.

Sherman-Morrison-Woodbury formula.

$$\left[\mathbf{A} + \mathbf{v}^T\,\mathbf{w}\right]^{-1} = \mathbf{A}^{-1} - \mathbf{A}^{-1}\mathbf{v}\left[\mathbf{I} + \mathbf{v}^T\mathbf{A}^{-1}\mathbf{w}\right]^{-1}\mathbf{v}^T\mathbf{A}^{-1} \tag{B.50}$$

## Differentiation and Integration

If $\mathbf{A}(t) = \left[a_{ij}(t)\right]$ and $\mathbf{B}(t) = \left[b_{ij}(t)\right]$ are two $n \times n$ matrices, then

$$\frac{d}{dt}\mathbf{A}(t) = \left[\frac{d}{dt}a_{ij}(t)\right] \tag{B.51}$$

$$\frac{d}{dt}\left[a\mathbf{A}(t) \pm b\mathbf{B}(t)\right] = a\frac{d}{dt}\mathbf{A}(t) \pm b\frac{d}{dt}\mathbf{B}(t) \tag{B.52}$$

$$\frac{d}{dt}\left[\mathbf{A}(t)\,\mathbf{B}(t)\right] = \left[\frac{d}{dt}\mathbf{A}(t)\right]\mathbf{B}(t) + \mathbf{A}(t)\left[\frac{d}{dt}\mathbf{B}(t)\right] \tag{B.53}$$

$$\frac{d}{dt}\left[\mathbf{A}(t)\,\mathbf{B}(t)\right]^T = \left[\frac{d}{dt}\mathbf{B}(t)\right]^T\mathbf{A}^T(t) + \mathbf{B}^T(t)\left[\frac{d}{dt}\mathbf{A}(t)\right]^T \tag{B.54}$$

$$\int_{t_0}^{t}\mathbf{A}(s)\;ds = \left[\int_{t_0}^{t}a_{ij}(s)\;ds\right] \tag{B.55}$$

$$\int_{t_0}^{t}a\mathbf{A}(s) \pm b\mathbf{B}(s)\;ds = a\int_{t_0}^{t}\mathbf{A}(s)\;ds \pm b\int_{t_0}^{t}\mathbf{B}(s)\;ds \tag{B.56}$$

If $|\mathbf{A}| \neq 0$, then

$$\frac{d}{dt}\mathbf{A}^{-1}(t) = -\mathbf{A}^{-1}(t)\left[\frac{d}{dt}\mathbf{A}(t)\right]\mathbf{A}^{-1}(t) \tag{B.57}$$

$$\frac{d}{dt}\mathbf{A}^T(t) = -\mathbf{A}^T(t)\left[\frac{d}{dt}\mathbf{A}(t)\right]\mathbf{A}^T(t) \tag{B.58}$$

If $\mathbf{A}(x) = \left[a_{ji}(x)\right]$, then

$$\frac{d\,|\mathbf{A}|}{dx} = \sum_{i,j=1}^{n} \frac{da_{ij}}{dx} C_{ij} \tag{B.59}$$

where $C_{ij}$ are cofactors associated with element $a_{ij}$ of $\mathbf{A}$.

## The Matrix Exponential

If $\mathbf{A}$ is a square matrix and $z$ is any complex variable, then the matrix exponential $\exp(\mathbf{A}z)$ is defined as

$$e^{\mathbf{A}z} = \mathbf{I} + \mathbf{A}z + \frac{\mathbf{A}^2 z^2}{2!} + \cdots + \frac{\mathbf{A}^n z^n}{n!} = \sum_{k=0}^{\infty} \frac{\mathbf{A}^k z^k}{k!} \tag{B.60}$$

$$e^0 = \mathbf{I} \qquad e^{\mathbf{I}z} = \mathbf{I}\, e^z \qquad e^{\mathbf{A}(z_1+z_2)} = e^{\mathbf{A}z_1} e^{\mathbf{A}z_2} \tag{B.61}$$

$$e^{-\mathbf{A}z} = \left[e^{\mathbf{A}z}\right]^{-1} \qquad e^{\mathbf{A}z} e^{\mathbf{B}z} = e^{(\mathbf{A}+\mathbf{B})z} \tag{B.62}$$

$$\frac{d^k}{dz^k} e^{\mathbf{A}z} = \mathbf{A}^k\, e^{\mathbf{A}z} = e^{\mathbf{A}z}\, \mathbf{A}^k \tag{B.63}$$

# Ordinary Differential Equations

# C

The mathematical formulation of problems in engineering and science usually leads to differential equations.

**Definition 201** A **differential equation** is an equation containing one or more derivatives of the function under consideration. The equation

$$\frac{dx}{dt} = f(x, t) \tag{C.1}$$

is a differential equation of the function $x$, as a function of $t$.

**Definition 202** The **order** of a differential equation is the order of the highest derivative therein. The general differential equation

$$F\left(t, x(t), \frac{dx(t)}{dt}, \frac{d^2x(t)}{dt^2}, ..., \frac{d^nx(t)}{dt^n}\right) = 0 \tag{C.2}$$

is an nth-order differential equation on the unknown $x(t)$.

**Definition 203** A **first-order ordinary differential equation** is an equation that involves at most the first derivative of an unknown function, $x$. If $x$ is a function of $t$, then the first-order ordinary differential equation is shown as

$$\frac{dx}{dt} = f(x, t) \tag{C.3}$$

where $f(x, t)$ is a given function of the variables $x$ ana $t$ which is defined in some interval $a < t < b$.

**Definition 204** An **explicit** solution of the first-order ordinary differential equation

$$\frac{dx}{dt} = f(x, t) \tag{C.4}$$

in some interval $a < t < b$, is any function $x = x(t)$ that identically satisfied the differential equation (C.4), i.e., its substitution into the equation reduces the differential equation to an identity everywhere within that interval.

**Definition 205** Conditions specified at a single point are called **initial conditions** if the solution of differential equation is subjected to be satisfied. The differential equation together with those initial conditions is an **initial-value problem**. Conditions at two points are called **boundary conditions** and the differential equation together with the boundary conditions is an **boundary-value problem**.

**Definition 206** A first-order ordinary differential equation of the form

$$\frac{dx}{dt} = f(x) \tag{C.5}$$

where $f(x)$ is a function of $x$ alone, and not $t$, is called **autonomous**. Therefore, a first-order ordinary differential equation of the form



R. N. Jazar, H. Marzbani, *Vehicle Vibrations*, https://doi.org/10.1007/978-3-031-43486-0

$$\frac{dx}{dt} = f(x, t) \tag{C.6}$$

where $f(x, t)$ is a function of $x$ and $t$, is called **nonautonomous**.

**Definition 207** If the function $x(t) = cte$ is a solution of the differential equation

$$\frac{dx}{dt} = f(x, t) \tag{C.7}$$

then the solution is called an **equilibrium solution** of (C.7).

**Definition 208** The equilibrium solution $x(t) = cte$ of the differential equation

$$\frac{dx}{dt} = f(x, t) \tag{C.8}$$

is called **stable** if all solutions of the differential equation that start near this equilibrium solution remain near this equilibrium solution as $t \to \infty$. An equilibrium solution is called **unstable** if all solutions of the differential equation that start near this equilibrium solution move from this equilibrium solution as $t \to \infty$.

**Theorem 209** *If $f(x, t)$ and $\partial f/\partial x$ are defined and continuous in a finite rectangular region of $x - t$ plane including the point $(x_0, t_0)$, then the differential equation*

$$\frac{dx}{dt} = f(x, t) \tag{C.9}$$

*has a unique solution $x(t)$, passing through the point*

$$x(t_0) = x_0 \tag{C.10}$$

*This solution is valid for all $x$, and $t$ which the solution remains inside a rectangle.*

**Definition 210** A particular of the differential equation

$$\frac{dx}{dt} = f(x, t) \tag{C.11}$$

which we cannot find from the family of regular solutions by selecting a finite value for the arbitrary constant $c$ of the family of solution, is called a **singular solution**.

**Definition 211** The set of differential equations

$$\frac{dx}{dt} = P(x, y, t) \qquad \frac{dy}{dt} = Q(x, y, t) \tag{C.12}$$

is called a **system of first-order differential equations**, where $P(x, y, t)$ and $Q(x, y, t)$ are given functions of $x, y, t$. When $P$ and $Q$ do not contain $t$ explicitly, the system (C.12) is called **autonomous**, otherwise it is called **nonautonomous**.

**Definition 212** The system of autonomous differential equations

$$\frac{dx}{dt} = P(x, y, t) \qquad \frac{dy}{dt} = Q(x, y, t) \tag{C.13}$$

has an **equilibrium point** at $(x_0, y_0)$, if

$$P(x_0, y_0) = 0 \qquad Q(x_0, y_0) = 0 \tag{C.14}$$

# Trigonometric Formula

# D

## Numeric Values and Functions

$$\pi \simeq 3.14159265359\cdots \simeq \frac{355}{113} \simeq \frac{103993}{33102} \tag{D.1}$$

$$1\,\text{rad} \simeq 57.296\,\text{deg} \simeq 57^\circ\ 17'\ 44.8'' \tag{D.2}$$

## Trigonometric Functions

$\theta$ is an acute angle of a right-angled triangle, and the hypotenuse $h$ is the long side that connects the two acute angles. The side $b$ adjacent to $\theta$ is the side of the triangle that connects $\theta$ to the right angle. The third side $a$ is opposite to $\theta$.

$$h = hypotenuse \qquad a = opposite \qquad b = adjacent \tag{D.3}$$

$$\sin\theta = \frac{a}{h} \qquad \cos\theta = \frac{b}{h} \qquad \tan\theta = \frac{a}{b} = \frac{\sin\theta}{\cos\theta} \tag{D.4}$$

$$\csc\theta = \frac{1}{\sin\theta} = \frac{h}{a} \qquad \sec\theta = \frac{1}{\cos\theta} = \frac{h}{b} \qquad \cot\theta = \frac{b}{a} \tag{D.5}$$

## Definitions in Terms of Exponentials

$$\cos z = \frac{e^{iz} + e^{-iz}}{2} = \cosh(iz) \tag{D.6}$$

$$\sin z = \frac{e^{iz} - e^{-iz}}{2i} = -i\sinh(iz) \tag{D.7}$$

$$\tan z = \frac{e^{iz} - e^{-iz}}{i\left(e^{iz} + e^{-iz}\right)} = \frac{1}{i}\tanh z \tag{D.8}$$

$$e^{iz} = \cos z + i\sin z \tag{D.9}$$

$$e^{-iz} = \cos z - i\sin z \tag{D.10}$$

$$(\cos\theta + i\sin\theta)^n = \cos n\theta + i\sin n\theta \tag{D.11}$$

## Angle Sum and Difference

$$\sin(\alpha \pm \beta) = \sin\alpha\cos\beta \pm \cos\alpha\sin\beta \tag{D.12}$$

$$\cos(\alpha \pm \beta) = \cos\alpha\cos\beta \mp \sin\alpha\sin\beta \tag{D.13}$$


R. N. Jazar, H. Marzbani, *Vehicle Vibrations*, https://doi.org/10.1007/978-3-031-43486-0

$$\sinh(\alpha \pm \beta) = \sinh\alpha\cosh\beta \pm \cosh\alpha\sinh\beta \tag{D.14}$$

$$\cosh(\alpha \pm \beta) = \cosh\alpha\cosh\beta \mp \sinh\alpha\sinh\beta \tag{D.15}$$

$$\tan(\alpha \pm \beta) = \frac{\tan\alpha \pm \tan\beta}{1 \mp \tan\alpha\tan\beta} \tag{D.16}$$

$$\cot(\alpha \pm \beta) = \frac{\cot\alpha\cot\beta \mp 1}{\cot\beta \pm \cot\alpha} \tag{D.17}$$

$$\tanh(\alpha \pm \beta) = \frac{\tanh\alpha \pm \tanh\beta}{1 \mp \tanh\alpha\tanh\beta} \tag{D.18}$$

## Periodicity

$$\sin(\alpha + 2n\pi) = \sin\alpha \tag{D.19}$$

$$\cos(\alpha + 2n\pi) = \cos\alpha \tag{D.20}$$

$$\tan(\alpha + 2n\pi) = \tan\alpha \tag{D.21}$$

## Symmetry

$$\sin(-\alpha) = -\sin\alpha \tag{D.22}$$

$$\cos(-\alpha) = \cos\alpha \tag{D.23}$$

$$\tan(-\alpha) = -\tan\alpha \tag{D.24}$$

## Displacement

$$\sin\left(\frac{\pi}{2} - \alpha\right) = \cos\alpha \tag{D.25}$$

$$\cos\left(\frac{\pi}{2} - \alpha\right) = \sin\alpha \tag{D.26}$$

$$\tan\left(\frac{\pi}{2} - \alpha\right) = \cot\alpha \tag{D.27}$$

$$\sec\left(\frac{\pi}{2} - \alpha\right) = \csc\alpha \tag{D.28}$$

$$\csc\left(\frac{\pi}{2} - \alpha\right) = \sec\alpha \tag{D.29}$$

$$\cot\left(\frac{\pi}{2} - \alpha\right) = \tan\alpha \tag{D.30}$$

## Multiple Angles

$$\sin(2\alpha) = 2\sin\alpha\cos\alpha = \frac{2\tan\alpha}{1 + \tan^2\alpha} \tag{D.31}$$

$$\cos(2\alpha) = 2\cos^2\alpha - 1 = 1 - 2\sin^2\alpha = \cos^2\alpha - \sin^2\alpha \tag{D.32}$$

$$\tan(2\alpha) = \frac{2\tan\alpha}{1 - \tan^2\alpha} \tag{D.33}$$

$$\cot(2\alpha) = \frac{\cot^2\alpha - 1}{2\cot\alpha} \tag{D.34}$$

$$\sin(3\alpha) = -4\sin^3\alpha + 3\sin\alpha \tag{D.35}$$

$$\cos(3\alpha) = 4\cos^3\alpha - 3\cos\alpha \tag{D.36}$$

$$\tan(3\alpha) = \frac{-\tan^3\alpha + 3\tan\alpha}{-3\tan^2\alpha + 1} \tag{D.37}$$

$$\sin(4\alpha) = -8\sin^3\alpha\cos\alpha + 4\sin\alpha\cos\alpha \tag{D.38}$$

$$\cos(4\alpha) = 8\cos^4\alpha - 8\cos^2\alpha + 1 \tag{D.39}$$

$$\tan(4\alpha) = \frac{-4\tan^3\alpha + 4\tan\alpha}{\tan^4\alpha - 6\tan^2\alpha + 1} \tag{D.40}$$

$$\sin(5\alpha) = 16\sin^5\alpha - 20\sin^3\alpha + 5\sin\alpha \tag{D.41}$$

$$\cos(5\alpha) = 16\cos^5\alpha - 20\cos^3\alpha + 5\cos\alpha \tag{D.42}$$

$$\sin(n\alpha) = 2\sin((n-1)\alpha)\cos\alpha - \sin((n-2)\alpha) \tag{D.43}$$

$$\cos(n\alpha) = 2\cos((n-1)\alpha)\cos\alpha - \cos((n-2)\alpha) \tag{D.44}$$

$$\tan(n\alpha) = \frac{\tan((n-1)\alpha) + \tan\alpha}{1 - \tan((n-1)\alpha)\tan\alpha} \tag{D.45}$$

## Half Angle

$$\cos\frac{\alpha}{2} = \pm\sqrt{\frac{1+\cos\alpha}{2}} \tag{D.46}$$

$$\sin\frac{\alpha}{2} = \pm\sqrt{\frac{1-\cos\alpha}{2}} \tag{D.47}$$

$$\tan\frac{\alpha}{2} = \frac{1-\cos\alpha}{\sin\alpha} = \frac{\sin\alpha}{1+\cos\alpha} = \pm\sqrt{\frac{1-\cos\alpha}{1+\cos\alpha}} \tag{D.48}$$

$$\sin\alpha = \frac{2\tan\frac{\alpha}{2}}{1+\tan^2\frac{\alpha}{2}} \tag{D.49}$$

$$\cos\alpha = \frac{1-\tan^2\frac{\alpha}{2}}{1+\tan^2\frac{\alpha}{2}} \tag{D.50}$$

## Powers of Functions

$$\cos^2\alpha + \sin^2\alpha = 1 \tag{D.51}$$

$$\cosh^2\alpha - \sinh^2\alpha = 1 \tag{D.52}$$

$$\sin^2\alpha = \frac{1}{2}(1 - \cos(2\alpha))$$

$$\sin\alpha\cos\alpha = \frac{1}{2}\sin(2\alpha) \tag{D.53}$$

$$\cos^2\alpha = \frac{1}{2}(1 + \cos(2\alpha)) \tag{D.54}$$

$$\sin^3\alpha = \frac{1}{4}(3\sin(\alpha) - \sin(3\alpha)) \tag{D.55}$$

$$\sin^2\alpha\cos\alpha = \frac{1}{4}(\cos\alpha - 3\cos(3\alpha)) \tag{D.56}$$

$$\sin\alpha\cos^2\alpha = \frac{1}{4}(\sin\alpha + \sin(3\alpha)) \tag{D.57}$$

$$\cos^3\alpha = \frac{1}{4}(\cos(3\alpha) + 3\cos\alpha)) \tag{D.58}$$

$$\sin^4\alpha = \frac{1}{8}(3 - 4\cos(2\alpha) + \cos(4\alpha)) \tag{D.59}$$

$$\sin^3\alpha\cos\alpha = \frac{1}{8}(2\sin(2\alpha) - \sin(4\alpha)) \tag{D.60}$$

$$\sin^2\alpha\cos^2\alpha = \frac{1}{8}(1 - \cos(4\alpha)) \tag{D.61}$$

$$\sin\alpha\cos^3\alpha = \frac{1}{8}(2\sin(2\alpha) + \sin(4\alpha)) \tag{D.62}$$

$$\cos^4\alpha = \frac{1}{8}(3 + 4\cos(2\alpha) + \cos(4\alpha)) \tag{D.63}$$

$$\sin^5\alpha = \frac{1}{16}(10\sin\alpha - 5\sin(3\alpha) + \sin(5\alpha)) \tag{D.64}$$

$$\sin^4\alpha\cos\alpha = \frac{1}{16}(2\cos\alpha - 3\cos(3\alpha) + \cos(5\alpha)) \tag{D.65}$$

$$\sin^3\alpha\cos^2\alpha = \frac{1}{16}(2\sin\alpha + \sin(3\alpha) - \sin(5\alpha)) \tag{D.66}$$

$$\sin^2\alpha\cos^3\alpha = \frac{1}{16}(2\cos\alpha - 3\cos(3\alpha) - 5\cos(5\alpha)) \tag{D.67}$$

$$\sin\alpha\cos^4\alpha = \frac{1}{16}(2\sin\alpha + 3\sin(3\alpha) + \sin(5\alpha)) \tag{D.68}$$

$$\cos^5\alpha = \frac{1}{16}(10\cos\alpha + 5\cos(3\alpha) + \cos(5\alpha)) \tag{D.69}$$

$$\tan^2\alpha = \frac{1 - \cos(2\alpha)}{1 + \cos(2\alpha)} \tag{D.70}$$

## Products of $\sin$ and $\cos$

$$\cos\alpha\cos\beta = \frac{1}{2}\cos(\alpha - \beta) + \frac{1}{2}\cos(\alpha + \beta) \tag{D.71}$$

$$\cos n\theta\cos\theta = \frac{1}{2}\cos(n + 1)\theta + \frac{1}{2}\cos(n - 1)\theta \tag{D.72}$$

$$\cos m\theta\cos n\theta = \frac{1}{2}\cos(m + n)\theta + \frac{1}{2}\cos(m - n)\theta \tag{D.73}$$

$$\sin\alpha\sin\beta = \frac{1}{2}\cos(\alpha - \beta) - \frac{1}{2}\cos(\alpha + \beta) \tag{D.74}$$

$$\sin\alpha\cos\beta = \frac{1}{2}\sin(\alpha - \beta) + \frac{1}{2}\sin(\alpha + \beta) \tag{D.75}$$

$$\cos\alpha\sin\beta = \frac{1}{2}\sin(\alpha + \beta) - \frac{1}{2}\sin(\alpha - \beta) \tag{D.76}$$

$$\sin(\alpha + \beta)\sin(\alpha - \beta) = \cos^2\beta - \cos^2\alpha = \sin^2\alpha - \sin^2\beta \tag{D.77}$$

$$\cos(\alpha+\beta)\cos(\alpha-\beta)=\cos^2\beta+\sin^2\alpha \tag{D.78}$$

## Sum of Functions

$$\sin\alpha\pm\sin\beta=2\sin\frac{\alpha\pm\beta}{2}\cos\frac{\alpha\pm\beta}{2} \tag{D.79}$$

$$\cos\alpha+\cos\beta=2\cos\frac{\alpha+\beta}{2}\cos\frac{\alpha-\beta}{2} \tag{D.80}$$

$$\cos\alpha-\cos\beta=-2\sin\frac{\alpha+\beta}{2}\sin\frac{\alpha-\beta}{2} \tag{D.81}$$

$$\sinh\alpha\pm\sinh\beta=2\sinh\frac{\alpha\pm\beta}{2}\cosh\frac{\alpha\pm\beta}{2} \tag{D.82}$$

$$\cosh\alpha+\cosh\beta=2\cosh\frac{\alpha+\beta}{2}\cosh\frac{\alpha-\beta}{2} \tag{D.83}$$

$$\cosh\alpha-\cosh\beta=-2\sinh\frac{\alpha+\beta}{2}\sinh\frac{\alpha-\beta}{2} \tag{D.84}$$

$$\tan\alpha\pm\tan\beta=\frac{\sin(\alpha\pm\beta)}{\cos\alpha\cos\beta} \tag{D.85}$$

$$\cot\alpha\pm\cot\beta=\frac{\sin(\beta\pm\alpha)}{\sin\alpha\sin\beta} \tag{D.86}$$

$$\frac{\sin\alpha+\sin\beta}{\sin\alpha-\sin\beta}=\frac{\tan\frac{\alpha+\beta}{2}}{\tan\frac{\alpha-+\beta}{2}} \tag{D.87}$$

$$\frac{\sin\alpha+\sin\beta}{\cos\alpha-\cos\beta}=\cot\frac{-\alpha+\beta}{2} \tag{D.88}$$

$$\frac{\sin\alpha+\sin\beta}{\cos\alpha+\cos\beta}=\tan\frac{\alpha+\beta}{2} \tag{D.89}$$

$$\frac{\sin\alpha-\sin\beta}{\cos\alpha+\cos\beta}=\tan\frac{\alpha-\beta}{2} \tag{D.90}$$

## Trigonometric Relations

$$\sin^2\alpha-\sin^2\beta=\sin(\alpha+\beta)\sin(\alpha-\beta) \tag{D.91}$$

$$=\cos^2\beta-\cos^2\alpha \tag{D.92}$$

$$\cos^2\alpha-\cos^2\beta=-\sin(\alpha+\beta)\sin(\alpha-\beta) \tag{D.93}$$

$$\cos^2\alpha-\sin^2\beta=\cos(\alpha+\beta)\cos(\alpha-\beta) \tag{D.94}$$

$$=\cos^2\beta-\sin^2\alpha \tag{D.95}$$

## Inverse of Trigonometric Functions

$$\sin x = y \qquad \cos x = y \qquad \tan x = y \tag{D.96}$$

$$\arcsin y = (-1)^k x + k\pi \tag{D.97}$$

$$\arccos y = \pm x + 2k\pi \tag{D.98}$$

$$\arcsin y = x + k\pi \tag{D.99}$$

$$k = 0, 1, 2, 3, \cdots$$

$$\begin{array}{ll} -1 \le y \le 1 & -\frac{\pi}{2} \le \arcsin y \le \frac{\pi}{2} \\ -1 \le y \le 1 & 0 \le \arccos y \le \pi \\ -\infty \le y \le \infty & -\frac{\pi}{2} \le \arctan y \le \frac{\pi}{2} \\ -\infty \le y \le \infty & 0 \le \operatorname{arccot} y \le \pi \end{array} \tag{D.100}$$

$$\begin{array}{ll} 1 \le y & 0 \le \operatorname{arccsc} y \le \frac{\pi}{2} \\ y \le -1 & -\frac{\pi}{2} \le \operatorname{arccsc} y \le 0 \\ 1 \le y & 0 \le \operatorname{arcsec} y \le \frac{\pi}{2} \\ y \le -1 & \frac{\pi}{2} \le \operatorname{arcsec} y \le \pi \end{array} \tag{D.101}$$

## Derivatives of Trigonometric Functions

$$\frac{d}{d\theta} \sin\theta = \cos\theta \tag{D.102}$$

$$\frac{d}{d\theta} \cos\theta = -\sin\theta \tag{D.103}$$

$$\frac{d}{d\theta} \tan\theta = 1 + \tan^2\theta = \sec^2\theta \tag{D.104}$$

$$\frac{d}{d\theta} \csc\theta = -\csc\theta \cot\theta \tag{D.105}$$

$$\frac{d}{d\theta} \sec\theta = \sec\theta \tan\theta \tag{D.106}$$

$$\frac{d}{d\theta} \cot\theta = -\left(1 + \cot^2\theta\right) = -\csc^2\theta \tag{D.107}$$

## Integrals of Trigonometric Functions

$$\int \sin\theta \, d\theta = -\cos\theta + C \tag{D.108}$$

$$\int \cos\theta \, d\theta = \sin\theta + C \tag{D.109}$$

$$\int \tan\theta \, d\theta = -\ln|\cos\theta| + C \tag{D.110}$$

$$\int \csc\theta \, d\theta = \frac{1}{2}\ln\left(-\frac{\cos\theta - 1}{\cos\theta + 1}\right) + C$$
$$= -\ln|\csc\theta + \cot\theta| + C \tag{D.111}$$

$$\int \sec\theta \, d\theta = \frac{1}{2}\ln\left(\frac{\sin\theta + 1}{\sin\theta - 1}\right) + C$$
$$= \ln|\sec\theta + \tan\theta| + C \tag{D.112}$$

$$\int \cot\theta \, d\theta = \frac{1}{2}\ln(2 - 2\cos 2\theta) + C = \ln|\sin\theta| + C \tag{D.113}$$

$$\int \cos^2 x \, dx = \frac{1}{2}x - \frac{1}{4}\sin 2x \tag{D.114}$$

$$\int \sin^2 x \, dx = \frac{1}{2}x + \frac{1}{4}\sin 2x \tag{D.115}$$

$$\int \cos mx \cos nx \, dx = \frac{\sin(m-n)x}{2(m-n)} + \frac{\sin(m+n)x}{2(m+n)} \qquad m^2 \neq n^2 \tag{D.116}$$

$$\int \sin mx \sin nx \, dx = \frac{\sin(m-n)x}{2(m-n)} - \frac{\sin(m+n)x}{2(m+n)} \qquad m^2 \neq n^2 \tag{D.117}$$

$$\int \sin mx \cos nx \, dx = -\frac{\cos(m-n)x}{2(m-n)} - \frac{\cos(m+n)x}{2(m+n)} \qquad m^2 \neq n^2 \tag{D.118}$$

$$\int_{-\pi}^{\pi} \cos mx \cos nx \, dx = \pi\delta_{mn} \qquad m^2 \neq n^2 \qquad m, n \in \mathbb{N} \tag{D.119}$$

$$\int_{-\pi}^{\pi} \sin mx \sin nx \, dx = \pi\delta_{mn} \qquad m^2 \neq n^2 \qquad m, n \in \mathbb{N} \tag{D.120}$$

$$\int_{-\pi}^{\pi} \sin mx \cos nx \, dx = 0 \qquad m^2 \neq n^2 \qquad m, n \in \mathbb{N} \tag{D.121}$$

$$\int x\cos x \, dx = \cos x + x\sin x \tag{D.122}$$

$$\int x\sin x \, dx = \sin x - x\cos x \tag{D.123}$$

$$\int x^2\cos x \, dx = 2x\cos x + \left(x^2 - 2\right)\sin x \tag{D.124}$$

$$\int x^2\sin x \, dx = 2x\sin x - \left(x^2 - 2\right)\cos x \tag{D.125}$$

$$\int \cos mx \cos nx \, dx = \frac{\sin(m-n)x}{2(m-n)} + \frac{\sin(m+n)x}{2(m+n)} \qquad m^2 \neq n^2 \tag{D.126}$$

$$\int \sin mx \sin nx \, dx = \frac{\sin(m-n)x}{2(m-n)} - \frac{\sin(m+n)x}{2(m+n)} \qquad m^2 \neq n^2 \tag{D.127}$$

$$\int \sin mx \cos nx \, dx = -\frac{\cos(m-n)x}{2(m-n)} - \frac{\cos(m+n)x}{2(m+n)} \qquad m^2 \neq n^2 \tag{D.128}$$

$$\int_{-\pi}^{\pi} \cos mx \cos nx \, dx = \pi\delta_{mn} \qquad m^2 \neq n^2 \qquad m, n \in \mathbb{N} \tag{D.129}$$

$$\int_{-\pi}^{\pi} \sin mx \sin nx \, dx = \pi \delta_{mn} \qquad m^2 \neq n^2 \qquad m, n \in \mathbb{N} \tag{D.130}$$

$$\int_{-\pi}^{\pi} \sin mx \cos nx \, dx = 0 \qquad m^2 \neq n^2 \qquad m, n \in \mathbb{N} \tag{D.131}$$

$$\int x \cos x \, dx = \cos x + x \sin x \tag{D.132}$$

$$\int x \sin x \, dx = \sin x - x \cos x \tag{D.133}$$

$$\int x^2 \cos x \, dx = 2x \cos x + \left(x^2 - 2\right) \sin x \tag{D.134}$$

$$\int x^2 \sin x \, dx = 2x \sin x - \left(x^2 - 2\right) \cos x \tag{D.135}$$

# Algebraic Formula

# E

$$(a \pm b)^2 = a^2 \pm 2ab + b^2 \tag{E.1}$$

$$(a \pm b)^3 = a^3 \pm 3a^2b + 3ab^2 \pm b^2 \tag{E.2}$$

$$(a \pm b)^4 = a^4 \pm 4a^3b + 6a^2b^2 \pm 4ab^3 + b^4 \tag{E.3}$$

$$(a \pm b)^n = \sum_{k=0}^{n} \binom{n}{k} a^k b^{n-k} \qquad \binom{n}{k} = \frac{n!}{k!\,(n-k)!} \tag{E.4}$$

$$(a + b + c)^2 = a^2 + b^2 + c^2 + 2\,(ab + bc + ca) \tag{E.5}$$

$$(a + b + c)^3 = a^3 + b^3 + c^3 + 3a^2\,(b + c) + 3b^2\,(c + a) + 3c^2\,(a + c) + 6abc \tag{E.6}$$

$$a^2 - b^2 = (a - b)\,(a + b) \tag{E.7}$$

$$a^2 + b^2 = (a - ib)\,(a + ib) \tag{E.8}$$

$$a^3 - b^3 = (a - b)\left(a^2 + ab + b^2\right) \tag{E.9}$$

$$a^2 + b^2 = (a + b)\left(a^2 - ab + b^2\right) \tag{E.10}$$



R. N. Jazar, H. Marzbani, *Vehicle Vibrations*, https://doi.org/10.1007/978-3-031-43486-0

# Unit Conversions

# F

## General Conversion Formulas

$$\begin{aligned}
\mathrm{N}^a\,\mathrm{m}^b\,\mathrm{s}^c &\approx 4.448^a \times 0.3048^b \times \mathrm{lb}^a\,\mathrm{ft}^b\,\mathrm{s}^c \\
&\approx 4.448^a \times 0.0254^b \times \mathrm{lb}^a\,\mathrm{in}^b\,\mathrm{s}^c \\
\mathrm{lb}^a\,\mathrm{ft}^b\,\mathrm{s}^c &\approx 0.2248^a \times 3.2808^b \times \mathrm{N}^a\,\mathrm{m}^b\,\mathrm{s}^c \\
\mathrm{lb}^a\,\mathrm{in}^b\,\mathrm{s}^c &\approx 0.2248^a \times 39.37^b \times \mathrm{N}^a\,\mathrm{m}^b\,\mathrm{s}^c
\end{aligned}$$

## Conversion Factors

### Acceleration

$$1\,\mathrm{ft/s^2} \approx 0.3048\,\mathrm{m/s^2} \quad 1\,\mathrm{m/s^2} \approx 3.2808\,\mathrm{ft/s^2}$$

### Angle

$$1\,\mathrm{deg} \approx 0.01745\,\mathrm{rad} \quad 1\,\mathrm{rad} \approx 57.307\,\mathrm{deg}$$

### Area

$$\begin{array}{ll}
1\,\mathrm{in^2} \approx 6.4516\,\mathrm{cm^2} & 1\,\mathrm{cm^2} \approx 0.155\,\mathrm{in^2} \\
1\,\mathrm{ft^2} \approx 0.09290304\,\mathrm{m^2} & 1\,\mathrm{m^2} \approx 10.764\,\mathrm{ft^2} \\
1\,\mathrm{acre} \approx 4046.86\,\mathrm{m^2} & 1\,\mathrm{m^2} \approx 2.471 \times 10^{-4}\,\mathrm{acre} \\
1\,\mathrm{acre} \approx 0.4047\,\mathrm{hectare} & 1\,\mathrm{hectare} \approx 2.471\,\mathrm{acre}
\end{array}$$

### Damping

$$\begin{array}{ll}
1\,\mathrm{N\,s/m} \approx 6.85218 \times 10^{-2}\,\mathrm{lb\,s/ft} & 1\,\mathrm{lb\,s/ft} \approx 14.594\,\mathrm{N\,s/m} \\
1\,\mathrm{N\,s/m} \approx 5.71015 \times 10^{-3}\,\mathrm{lb\,s/in} & 1\,\mathrm{lb\,s/in} \approx 175.13\,\mathrm{N\,s/m}
\end{array}$$



R. N. Jazar, H. Marzbani, *Vehicle Vibrations*, https://doi.org/10.1007/978-3-031-43486-0

## Energy and Heat

$1\,\text{Btu} \approx 1055.056\,\text{J}$ $1\,\text{J} \approx 9.4782 \times 10^{-4}\,\text{Btu}$
$1\,\text{cal} \approx 4.1868\,\text{J}$ $1\,\text{J} \approx 0.23885\,\text{cal}$
$1\,\text{kW h} \approx 3600\,\text{kJ}$ $1\,\text{MJ} \approx 0.27778\,\text{kW h}$
$1\,\text{ft lbf} \approx 1.355818\,\text{J}$ $1\,\text{J} \approx 0.737562\,\text{ft lbf}$

## Force

$1\,\text{lb} \approx 4.448222\,\text{N}$ $1\,\text{N} \approx 0.22481\,\text{lb}$

## Fuel Consumption

$1\,\text{l}/100\,\text{km} \approx 235.214583\,\text{mi/gal}$
$1\,\text{mi/gal} \approx 235.214583\,\text{l}/100\,\text{km}$
$1\,\text{l}/100\,\text{km} = 100\,\text{km/l}$ $1\,\text{km/l} = 100\,\text{l}/100\,\text{km}$
$1\,\text{mi/gal} \approx 0.425144\,\text{km/l}$ $1\,\text{km/l} \approx 2.352146\,\text{mi/gal}$

## Length

$1\,\text{in} \approx 25.4\,\text{mm}$ $1\,\text{cm} \approx 0.3937\,\text{in}$
$1\,\text{ft} \approx 30.48\,\text{cm}$ $1\,\text{m} \approx 3.28084\,\text{ft}$
$1\,\text{mi} \approx 1.609347\,\text{km}$ $1\,\text{km} \approx 0.62137\,\text{mi}$

## Mass

$1\,\text{lb} \approx 0.45359\,\text{kg}$ $1\,\text{kg} \approx 2.204623\,\text{lb}$
$1\,\text{slug} \approx 14.5939\,\text{kg}$ $1\,\text{kg} \approx 0.068522\,\text{slug}$
$1\,\text{slug} \approx 32.174\,\text{lb}$ $1\,\text{lb} \approx 0.03.1081\,\text{slug}$

## Moment and Torque

$1\,\text{lb ft} \approx 1.35582\,\text{N m}$ $1\,\text{N m} \approx 0.73746\,\text{lb ft}$
$1\,\text{lb in} \approx 8.85075\,\text{N m}$ $1\,\text{N m} \approx 0.11298\,\text{lb in}$

## Mass Moment

$1\,\text{lb ft}^2 \approx 0.04214\,\text{kg m}^2$ $1\,\text{kg m}^2 \approx 23.73\,\text{lb ft}^2$

## Power

| | |
|---|---|
| 1 Btu/h ≈ 0.2930711 W | 1 W ≈ 3.4121 Btu/h |
| 1 hp ≈ 745.6999 W | 1 kW ≈ 1.341 hp |
| 1 hp ≈ 550 lb ft/s | 1 lb ft/s ≈ 1.8182 × $10^{-3}$ hp |
| 1 lb ft/h ≈ 3.76616 × $10^{-4}$ W | 1 W ≈ 2655.2 lb ft/h |
| 1 lb ft/min ≈ 2.2597 × $10^{-2}$ W | 1 W ≈ 44.254 lb ft/min |

## Pressure and Stress

| | |
|---|---|
| 1 lb/in$^2$ ≈ 6894.757 Pa | 1 MPa ≈ 145.04 lb/in$^2$ |
| 1 lb/ft$^2$ ≈ 47.88 Pa | 1 Pa ≈ 2.0886 × $10^{-2}$ lb/ft$^2$ |
| 1 Pa ≈ 0.00001 atm | 1 atm ≈ 101325 Pa |

## Stiffness

| | |
|---|---|
| 1 N/m ≈ 6.85218 × $10^{-2}$ lb/ft | 1 lb/ft ≈ 14.594 N/m |
| 1 N/m ≈ 5.71015 × $10^{-3}$ lb/in | 1 lb/in ≈ 175.13 N/m |

## Temperature

$$°C = (°F - 32)/1.8$$
$$°F = 1.8\,°C + 32$$

## Velocity

| | |
|---|---|
| 1 mi/h ≈ 1.60934 km/h | 1 km/h ≈ 0.62137 mi/h |
| 1 mi/h ≈ 0.44704 m/s | 1 m/s ≈ 2.2369 mi/h |
| 1 ft/s ≈ 0.3048 m/s | 1 m/s ≈ 3.2808 ft/s |
| 1 ft/ min ≈ 5.08 × $10^{-3}$ m/s | 1 m/s ≈ 196.85 ft/min |

## Volume

| | |
|---|---|
| 1 in$^3$ ≈ 16.39 cm$^3$ | 1 cm$^3$ ≈ 0.0061013 in$^3$ |
| 1 ft$^3$ ≈ 0.02831685 m$^3$ | 1 m$^3$ ≈ 35.315 ft$^3$ |
| 1 gal ≈ 3.785 l | 1 l ≈ 0.2642 gal |
| 1 gal ≈ 3785.41 cm$^3$ | 1 l ≈ 1000 cm$^3$ |

# Index




R. N. Jazar, H. Marzbani, *Vehicle Vibrations*, https://doi.org/10.1007/978-3-031-43486-0

**W**

FSC
www.fsc.org
MIX
Papier aus verantwortungsvollen Quellen
Paper from responsible sources
FSC® C105338

If you have any concerns about our products,
you can contact us on
**ProductSafety@springernature.com**

In case Publisher is established outside the EU,
the EU authorized representative is:
**Springer Nature Customer Service Center GmbH**
**Europaplatz 3, 69115 Heidelberg, Germany**

Printed by Libri Plureos GmbH
in Hamburg, Germany